Methods of Mathematics Applied to Calculus, Probability, and Statistics

Methods of Mathematics Applied to Calculus, Probability, and Statistics

RICHARD W. HAMMING

Naval Postgraduate School
Monterey, California

Dover Publications, Inc.
Mineola, New York

Bibliographical Note

This Dover edition, first published in 2004, is an unabridged republication of
the edition published by Prentice-Hall, Inc., Englewood Cliffs, New Jersey, 1985.

Library of Congress Cataloging-in-Publication Data

Hamming, R. W. (Richard Wesley), 1915–
    Methods of mathematics applied to calculus, probability, and statistics /
Richard W. Hamming.
        p. cm.
    Originally published: Englewood Cliffs, N.J. : Prentice-Hall, c1985.
    Includes index.
    ISBN-13: 978-0-486-43945-7 (pbk.)
    ISBN-10: 0-486-43945-3 (pbk.)
        1. Mathematics I. Title.

QA37.3H35 2004
510—dc22

                                                              2004053443

Manufactured in the United States by LSC Communications
4500057135
www.doverpublications.com

Contents

II   THE CALCULUS OF ALGEBRAIC FUNCTIONS

7   Derivatives in Geometry   169

8   Geometric Applications   204

9   Nongeometric Applications   243

10 Functions of Several Variables 282

11 Integration 318

12 Discrete Probability 366

IV   MISCELLANEOUS TOPICS

20   Infinite Series   637

21   Applications of Infinite Series   659

22*   Fourier Series   686

Contents

Preface

Why another calculus book? One reason is that this is not the conventional calculus book. I agree with de Finetti (1906–), who says in his *Theory of Probability:*

> To add one more [book] would certainly be a presumptuous undertaking if I thought in terms of doing something better, and a useless undertaking if I were to content myself with producing something similar to the "standard" type.

My second reason comes from a remark L. E. Dickson (1874–1954) once made in class with respect to his *History of the Theory of Numbers:*

> Every scientist owes a labor of love to his field.

This book is my labor of love.

The title, when the individual words are examined, explains the differences between this book and the current standard calculus texts.

1. Methods

By methods I mean those methods that are widely used in mathematics, science, and the applications of mathematics. Methods are important. It is by applying the methods that we obtain the results of mathematics. We are rapidly approaching an infinite amount of knowledge in the form of results, both in what is known and in what is useful in the applications of mathematics to other fields. Thus it is increasingly futile to teach mathematics by trying to cover the needed material; instead we must teach the methods and how to re-create the material as it is needed. We must abandon the methods of retrieval for those of regeneration. This may be an unscholarly attitude, but there seems little hope for the future if we persist in going down the path of the retrieval of knowledge.

2. of Mathematics

This means that the emphasis is on those methods that are peculiar to mathematics, that is, its abstractions and its methods of reasoning. In my opinion these should no longer be neglected. Some of the methods used in science generally are also specifically mentioned. Finally, many of the philosophical questions that naturally occur to the student while learning mathematics will be examined rather than cleverly evaded or simply glossed over.

3. Applied to

The book shows how the methods of doing mathematics are applied to both mathematics itself and to various uses of mathematics in other fields. Mathematicians tend to view mathematics as an art done for art's sake, but it is well known that the majority of the students in the typical calculus class expect to use mathematics rather than merely admire it. The neglect of either aspect, the innate beauty or the richness of applications, is foolish.

4. Calculus

The calculus is probably the most useful single branch of mathematics. During the many years of using mathematics daily in industry I found that the ability to do simple calculus, easily and reliably, was the most valuable part of all the mathematics I ever learned (of course, I also used more advanced mathematics whenever it was appropriate). My views, therefore, were shaped by the almost daily use of the calculus rather than by occasionally teaching the conventional courses. As a result this book is different from the standard, thick, exhausting text that is crammed with specific results and is short on the understanding of mathematics. The current form of calculus texts has apparently been popular since at least the first appearance of Granville's *Calculus* in 1904.

Understanding the methods of the calculus is vital to the creative use of mathematics in many areas even today. Without this mastery the average scientist or engineer, or any other user of mathematics, will be perpetually stunted in development, and will at best be able to follow only what the textbooks say; with mastery, new things can be done, even in old, well-established fields. Progress involves, among other things, the constant revision of the elements of a field, as well as the creation of significant new results.

5. Probability

The calculus arose from problems in mechanics, and for all of the nineteenth century and more, mechanical and electrical applications have dominated its use.

These problems involve almost no probability. At present, probability plays a central role in many fields, from quantum mechanics to information theory, and even older fields use probability now that the presence of "noise" is officially admitted. The newer aspects of many fields start with the admission of uncertainty.

Although many great mathematicians contributed to the early development of probability theory, it was not generally considered an integral part of mathematics until the publication of Feller's (1906–1970) book (1950) [F] (references are found at the end of Chapter 1). Since then probability has played an increasing role in mathematics.

6. and Statistics

Statistics has had an even slower acceptance in mathematics than probability. Yet statistics is central to much of our lives. We are deluged by statistics from surveys, polls, advertisements, government publications, and even data from laboratory experiments where we have gone beyond the deterministic world view to the acceptance of randomness as being fundamental.

Furthermore, it is probable that a large fraction of the students enrolled in the calculus course are there because the course is needed for statistics. Statistics without the calculus can only be of the cookbook variety. If the student is ever to master and use the simple concept of the distribution of a statistic (the values of a statistic that can be expected from repeated trials of the same experiment), then it is vital that the calculus be mastered. Continuous distributions are basic to the theory of probability and statistics, and the calculus is necessary to handle them with any ease.

It is understandable that many mathematicians do not like statistics, but it should be taught early so that the concepts are absorbed by the student's flexible, adaptable mind before it is too late. I believe that the student's needs require that some parts of mathematical statistics be included in the mathematics curriculum.

This book is not a course in probability or statistics; only the more mathematical parts are discussed at all. The much more difficult part of statistics, often called *data analysis,* is left entirely to the professional statisticians to teach. But using some of the functions and ideas that arise in statistics to illustrate various principles of the calculus and mathematics generally seems a sensible thing to do, rather than using arbitrarily made up functions and artificial problems.

This book emphasizes discrete mathematics (mathematics associated mainly with the integers) much more than does the usual calculus text. Increasingly these days the application of mathematics to the real world involves discrete mathematics. As most mathematicians who have examined the question at all closely know, discrete mathematics often involves the use of continuous mathematics—the nature of the discrete is often most clearly revealed through the continuous models of both calculus and probability. Without continuous mathematics, the study of discrete mathematics soon becomes trivial and very limited. Hence such topics as difference equations, generating functions, and numerical methods are scattered throughout the book as well as in special sections and chapters. In particular, the idea is rejected that for computer scientists a course in discrete mathematics without the calculus is adequate. The two topics, discrete and continuous mathematics, are both ill served by being rigidly separated.

All the material that is in the current standard calculus course cannot be covered together with all these new things; the question is, can an equivalent amount be done? It is going to be hard to find professors of mathematics with the needed backgrounds and interests. Furthermore, they will have to feel that the classical course is now too much out of date to be worth trying to save. The methods of mathematics are the main topic of the course, not a long list of finished mathematical results with such highly polished proofs that the poor student can only marvel at the results, and have no hope of understanding how mathematics is actually created by practicing mathematicians.

The question remains, can all this be done within a course that is somewhat equivalent to the standard calculus sequence? Many of the chapters in Part IV may be omitted depending on the interests of the students, the needs of the curriculum, and the desires of the professor. Will the attempt to teach the essence of mathematics, extension, generalization, and abstraction, take more time from the course than it will later save? Will the attempt to teach so many different ideas in the same course be too much? On the other hand, what else is there to try in this age which is dominated by both probability and statistics?

ACKNOWLEDGMENTS

It is customary to express one's indebtedness to those who have helped in the writing of a book. Certainly, to my colleagues and the management at Bell Telephone Laboratories during the 30 years I spent there I owe most of my attitudes toward mathematics and its uses. The past seven years at the Naval Postgraduate School have been an opportunity to ponder and wonder about the problems of teaching what is known and useful—especially what will be useful in the future!

Specific debts are to Professor Roger Pinkham of Stevens Institute, Professor John W. Tukey of Princeton University, and to Dean Max Woods of the Naval Postgraduate School for both inspiration and knowledge. Finally, without the encouragement of Karl Karlstrom of Prentice-Hall, this book would not have even been started. My thanks go to all the above and to the many people who have contributed so much to my education and supplied numerous comments (as did Allan Vasenius) on various drafts of the book.

R. W. Hamming

Methods of Mathematics Applied to Calculus, Probability, and Statistics

PART I

ALGEBRA AND ANALYTIC GEOMETRY

1

Prologue

1.1 THE IMPORTANCE OF MATHEMATICS

You live in an age that is dominated by science and engineering. Whether you like it or not, they have significant effects on your life. And it seems probable that in the near future their effects will become even greater than they are now. Thus if you wish to be effective in this world and to achieve the things you want, it is necessary to understand both science and engineering (and these require mathematics).

Long ago Pythagoras (died about 492 B.C.) said:

Number is the measure of all things.

(I am told that the strict translation is simply "everything is number.") Galileo (1564–1642) similarly said:

Mathematics is the language of science.

Mathematics clearly plays a fundamental role in the older sciences, such as astronomy, physics, and chemistry, and is of increasing importance in the other "hard" sciences. But mathematics is also rapidly invading all the biological sciences, especially such fields as genetics and molecular biology. Even in the humanities we find that questions of authorship and style are being decided by applying statistical tests to the written material. For a long time business administration has been using more and more mathematics as people have tried to understand both the workings of the large, complex organizations they have to manage and the competition between such organizations. Mathematics is similarly needed in modern government administration. The social sciences are also heavily dependent on the statistical approach to many of their problems. Indeed, one may say

> that Science is a habit of the mind as well as a way of life, and that mathematics is an aspect of culture as well as a collection of algorithms. (C. B. Boyer, 1906–1976)

It appears, therefore, that mathematics, in one form or another, will invade most fields of knowledge as we try to make them more reliable. There is an inevitability of this happening. But it has long been observed that the mathematics that is not learned in school is very seldom learned later, no matter how valuable it would be to the learner. Any unwillingness to learn mathematics today can greatly restrict your possibilities tomorrow.

This is not an assertion that all of mathematics will be useful; all that can be done is to look at both the past and the present, and then make educated guesses as to future needs for mathematics. This book covers what is believed to be the most useful applications (on the average). The three fields, calculus, probability, and statistics are all in constant use. Mathematicians in the past have tended to avoid the latter two, but probability and statistics are now so obviously necessary tools for understanding many diverse things that we must not ignore them even for the average student.

Calculus is the mathematics of change. The mathematics you have learned up to this point has served mainly to describe *static* (unchanging) situations; the calculus handles *dynamic* (changing) situations. Change is characteristic of the world. As Heraclitus (sixth to fifth century B.C.) said,

> You cannot step in the same river twice.

and

> Everything is in a state of becoming and flux.

Probability is the mathematics of uncertainty. Not only do we constantly face situations in which there is neither adequate data nor an adequate theory, but many modern theories have uncertainty built into their foundations. Thus learning to think in terms of probability is essential.

Statistics is the reverse of probability (glibly speaking). In probability you go from the model of the situation to what you expect to see; in statistics you have the observations and you wish to estimate features of the underlying model. There is, of course, much more to statistics than this.

This book is *not* mainly about the results obtained in mathematics; rather it is concerned with mathematics itself. There is simply too much mathematics in current use, let alone what will be in use in the near future, to try to cover all the applications of mathematics. Instead of concentrating on the *results,* we will concentrate more on the *methods* from which the results follow. Thus this book is fundamentally different from the other books on mathematics you have studied. Most mathematics books are filled with finished theorems and polished proofs, and to a surprising extent they ignore the methods used to create mathematics. It is as if you were merely walked through a picture gallery and never told how to mix paints, how to compose pictures, or all the other "tricks of the trade." Of course it would be simpler if I could tell you all the things you need to know about mathematics, but this approach seems to be hopeless for the coming years. Showing you the methods for doing mathematics covers a wider range of applications, but it does leave more creativity to you when you need some specific result. You will be left more able to do mathematics, to create mathematics as you need it, but less able to recall some specific result you happen to need. In short, in the face of almost infinite useful knowledge, we have adopted the strategy of "information regeneration rather than information retrieval." This means, most importantly, you should be able to generate the result you need even if no one has ever done it before you—you will not be dependent on the past to have done everything you will ever need in mathematics.

I have also chosen to raise many questions about the relationship between the mathematical models developed in the text and the physical world in which you live. I have not attempted to supply you with all the answers; rather it is up to you to think about them and come to your own opinions of how far you can trust the results of applying mathematics in the real world. If you are to go very far in your chosen field, it is doubtful if you can long avoid some new applications of mathematics; yet, as you will see, not all applications give sensible answers!

The assumptions and definitions of mathematics and science come from our intuition, which is based ultimately on experience. They then get shaped by further experience in using them and are occasionally revised. They are not fixed for all eternity. In many applications it is essential that you be able to trace the effects of various assumptions and definitions on any conclusions you draw—perhaps the particular mathematics you used was inappropriate for your case! New applications of mathematics will, from time to time, require new assumptions and altered definitions, and it is the intent of the text to prepare you to make them when needed, but naturally we cannot tell what they will be.

It is not claimed that one course will make you a great mathematician able to create all of mathematics for yourself; all that can be done is to start you down the path of learning to create mathematics. In a very real sense, all we can do is coach you; you must have both some talent and the willingness to practice what is being taught. If you expect to continue learning all your life, you will be teaching yourself much of the time. You must learn to learn, especially the difficult topic of mathematics.

1.2 THE UNIQUENESS OF MATHEMATICS

At first glance much of mathematics seems arbitrary, but, at least for most useful mathematics, this is not so. To study the essential uniqueness of mathematics, I asked many expert scientists and engineers the following question: "If we ever find ourselves in two-way communication with a distant world, will they have essentially the same mathematics as we do?" The answers were all a definite "Yes." They generally reasoned along the following lines:

1. The physical phenomena we see in space resemble those we see on Earth.
2. From this we infer that the same laws of physics apply everywhere.
3. Since mathematics is the language of science, it too must be essentially the same.

Their arguments were much more detailed and complete than this, but at present we are in no position to explain the technical details they used, nor the breadth of their arguments.

The meaning of the word "essentially" needs some explanation. In Euclidean geometry (approximately 300 B.C.), for example, the Greeks apparently chose to ignore *orientation*. They said that a left-handed and a right-handed triangle could be congruent (Figure 1.2-1). To carry out the proof, they allowed the flopping over in three dimensions of a triangle, even though a triangle is a two-dimensional figure. As a result of this choice, when they came to three-dimensional geometry they could not get the important theorem that in three dimensions there are only two orientations, the left-handed and the right-handed spirals. The distant intelligent beings we are imagining might have chosen to include orientation in their geometry. But it is claimed that this is not an *essential* difference. Again, Ptolemy (second century A.D.) used the chords of the double angle where we now use sines (Figure 1.2-2), but this is hardly an essential difference; it is merely a notational difference, although an important one in practice.

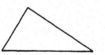

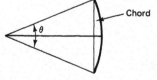

Figure 1.2-1   Congruent triangles

Figure 1.2-2   Ptolemy's chord = 2 sin $\theta/2$

Although the major users of mathematics are almost all in agreement that mathematics is essentially unique, we need to consider the idea often expressed by pure mathematicians that

Mathematics is the free creation of the human mind.

In a sense the users of mathematics are saying that mathematics, like science itself, is *discovered* rather than *created*. The pure mathematicians are saying that mathematics resembles the other arts in the sense that creation is a personal thing. In science creativity is comparatively impersonal; if one scientist does not discover something then another one will. But if Shakespeare had not written the play *King Lear*, we would probably have no closely similar play.

There may be less conflict between these two extreme opinions than appears; perhaps they are talking about different things. The users are talking about the mathematics they have found to be useful, while the pure mathematician may be talking about the mathematics now being created.

Another explanation of the difference is that those who daily work closely with the real world tend to believe that our senses, while occasionally deceived, report fairly accurately what is out there. On the other hand, those who work more with their imaginations tend to believe that our senses are rather unreliable, and the world out there could be very different from what we think it is. All are agreed that we cannot know with absolute certainty.

The mathematical results in this book are user oriented; they are the kinds that have been found to be useful in helping us to understand the universe in which we live. Textbooks in the past, especially at the calculus level, have concentrated on physical science applications. This was appropriate both because they are the historical source of much of the mathematics and because they are usually easier to understand. But the applications of mathematics of interest to the typical student are of far wider range than this. We will do a little in this direction of selecting suitable applications, but we are forced to minimize them lest they get in the way of the essential part of the book—how *you do mathematics*. The applications are often illustrated in the case histories.

However, we do not neglect the beauty of mathematics; it often makes the subject matter much more attractive and hence easier to master. The student should often find beauty in this book and gradually learn to do mathematics in attractive ways. As Edna St. Vincent Millay (1892–1950) wrote,

Euclid alone has looked on beauty bare.

1.3 THE UNREASONABLE EFFECTIVENESS OF MATHEMATICS

There is a universality about mathematics; what was created to explain one phenomenon is very often later found to be useful in explaining other, apparently unrelated, phenomena. Theories that were developed to explain some poorly measured effects are often found to fit later, much more accurate measurements. Furthermore, from measurements over a limited range the theory is often found to fit a far wider range. Finally, and perhaps most unreasonably, quite regularly from the mathematics alone new pehnomena, previously unknown and unsuspected, are successfully predicted. This universality of mathematics could, of course, be a reflection of the way the human mind works and not of the external world, but most people believe it reflects reality.

This remarkable effectiveness is difficult to explain unless one makes some kind of assumption, such as that there exists, in some sense, both a physical and a logical universe to be discovered, and that these two universes are somehow intimately related. Mathematics is then seen to be a description of the logical structure of the universe.

A second explanation is that the useful mathematics, being based on long experience, follows the scientific approach. Only after being well established were the postulates (or axioms) carefully abstracted. Currently, mathematicians make no distinction between axiom and postulate; even Archimedes (287?–212 B.C.) did not bother to distinguish between them. Mathematics based on arbitrary postulates, or possibly picked either for their elegance or else for the elegance of what follows from them, would seem to have little chance of being successfully used in explaining the universe.

The postulational approach is widely used in mathematics and is very valuable when used properly. There is an innate elegance about the postulational approach. But it has been well said, and repeated by several well-known mathematicians, that

> A book which starts off with axioms should be preceded by another volume explaining how and why these axioms have been chosen, and with what end in view.

We will carefully observe this rule and examine closely any assumptions we lay down.

Yet it must be acknowledged that an eminent physicist, the Nobel laureate P. A. M. Dirac (1902–), has said

> I want to emphasize the necessity for a sound mathematical basis for any fundamental physical theory. Any philosophical ideas that one may have play only a subordinate role. Unless such ideas have a mathematical basis they will be ineffective.

Apparently, the elegance of mathematics should not be dismissed lightly.

When a theory is sufficiently general to cover many fields of application, it acquires some "truth" from each of them. Thus there is a positive value for generalization in mathematics that may not be apparent to the beginner. This is one of the many reasons this book emphasizes the processes of extension, generalization, and abstraction. They often bring increased confidence in the results of a specific application, as well as new viewpoints. Notice that we are mainly interested in the *processes,* and we are not interested in presenting mathematics in its most abstract form. On the contrary, we will often begin with concrete forms and then exhibit the *process* of abstraction.

It is necessary to note that science and mathematics do not explain everything. In more than 2000 years they have added little to our understanding of such things as

Truth, Beauty, and Justice. There may be definite limits to the applicability of the scientific method.

1.4 MATHEMATICS AS A LANGUAGE

As Galileo said,

> Mathematics is the language of science.

Not only must you learn to think in the language of mathematics, you also need to read it. To do this, you must learn the alphabet, the vocabulary, and the grammar. In this language there are no syllables, and the individual characters tend to be the words, while the equations tend to be the sentences. Much of mathematics consists of rewriting a sentence in another logically equivalent form.

This book is filled with strange symbols that you need to recognize easily. For example, you have already met in school the Greek lowercase letter π (pi). Because the individual letters often play the role of words, we need a rich alphabet of symbols in mathematics, and for this reason mathematicians have been driven to the use of the Greek alphabet, both upper- and lowercase. It would be foolish to try to avoid the Greek alphabet because it is in constant use in mathematics and its applications. Thus you should learn to recognize these symbols, and for this purpose they are given in Appendix C. Learn to recognize them rapidly so that when they occur in the middle of some difficult passage you need not be distracted by, "What symbol is that?" There are also other strange symbols such as the elongated S ($\int$), which is an operation called "integration." Thus the language has both processes (verbs) and things (nouns).

It is fairly obvious that you need to both read and write mathematics. But it was a long time before I learned to listen to what a formula had to tell me. Formulas should be studied with this attitude in mind: what is it trying to say? Why is the formula the way it is? This attitude is much like that of the experimentalist who listens to what the physical (material) world says; the mathematician is similarly trying to learn what the world of abstract symbols has to say.

It is a common observation that in translating from one natural language, say Russian or Chinese to another, say English, the precise meaning cannot always be preserved. Partly this is because there may be no appropriate word, and partly because each language, to some extent, imposes its own patterns of thought. Mathematics, being very different from the natural languages, has its corresponding patterns of thought. Learning these patterns of thought is much more important than any particular result in the text. They are learned by the constant use of the language and cannot be easily taught in any other fashion. The mastery of the mathematical way of thinking is one of the main goals of this book.

1.5 WHAT IS MATHEMATICS?

In a conversation a friend of mine once said:

> Mathematics is merely clear thinking.

There is a great deal of truth in his remark. Clear thinking can be done without the use of formal mathematical symbols, but in his view the clear thinker is doing mathematics. The use of special symbols is merely an economy in thinking and expressing these thoughts (a shorthand if you prefer). Algebra is not a random collection of arbitrary rules; rather it is based on a few simple rules that came from long experience in handling numbers plus drawing logical conclusions (see Section 2.7 for two examples). You probably remember many of the deductions from this simple base as rules for handling algebraic expressions with ease and rapidity. Theorems are further economies because they record more complex patterns of thinking that once shown to be valid need not be repeated everytime they are needed. As George Temple (1901–) wrote,

> there is some element of truth in maintaining that mathematics is not so much a subject as a way of studying any subject, not so much a science as a way of life.

When you began studying algebra the teacher probably said that the mysterious symbol x was some number that was as yet undetermined. Perhaps it was called "a general number" or some similar thing. But as you got further into the course, you began to see that often the x was never going to be a number. For example, in the equation

$$(x + 1)^2 = x^2 + 2x + 1$$

you really have an equality for any x (meaning for all x), and further thinking and watching what happens when you use algebra suggest that x is merely a symbol with certain rules of combination and need not always be thought of as a number. Thus the x is sometimes thought of as an abstract symbol having no particular meaning.

As you go through this book you will find more and more that mere numbers are left behind. Occasionally, specific numbers will be used to find a specific answer or to illustrate a point, but often the result will be left in a general form. For example, the area of a triangle is usually given by the general formula

$$A = \frac{(\text{base})(\text{altitude})}{2}$$

without any specific numbers. At this stage in your education, substituting specific numbers into a general formula is one of the things we suppose you can do and is hardly to be considered an essential part of your further education. Thus, computing with numbers will have surprisingly little to do with the material in this book. As you will gradually see, calculus systematically evades a great deal of numerical calculation.

If mathematics is not just numbers, and x is not just a general number, then what is mathematics? There is no agreed upon definition of mathematics, but there is widespread agreement that the essence of doing mathematics is:

1. Extension
2. Generalization
3. Abstraction

These are all somewhat the same thing. They are important because faced with almost an infinity of details you cannot afford to deal constantly with the specific; you must learn to embrace more and more detail under the cover of generality.

We will not generalize for the sake of generalization, nor abstract for the fun of it. As Mark Kac (1914–) has written,

> For unrestricted abstraction tends also to divert attention from whole areas of application whose very discovery depends on features that the abstract point of view rules out as being accidental.

We will deliberately give you practice in these arts by giving both examples and exercises in the form: "Extend the above results," or "Generalize the above," or "Give a reasonable abstraction of the above." The difficulty facing you is that there is no unique, correct answer in most cases. It is a matter of taste, depending on the circumstances, your own personal traits and needs, and the particular age you live in. At first you will have to depend on the taste of this book and the professor. Gradually, you will develop your own taste, and along the way you may occasionally recognize that your taste may be the best one! It is the same as in an art course. During the course you must pay reasonable attention to the taste of the professor, but you should not neglect your own. The art is not simple, as the following remark shows:

> As opposed to abstraction the art of doing mathematics consists in finding special cases which contain all the germs of generality. David Hilbert (1862–1943).

In mathematics, at first you are presented with specific problems. Gradually, you learn that there are many things you could do that have not been done. The taste to work on the right problem at the right time and in the right way is the secret of doing significant things. It is a matter of taste as to when and to what you apply the techniques you have mastered. A central problem in teaching mathematics is to communicate a reasonable sense of taste—meaning often when to, or not to, generalize, abstract, or extend something you have just done. Taste is the main difference between first- and second-rate mathematicians. Of necessity the exercises in this book must be fairly easy and of no great importance. When doing significant things, it often takes days, weeks, months, and at times even years to get the result, but such time is not available in a course like this. Thus you must develop your abilities by practicing on simple, often trivial, exercises. Occasionally, short *case studies* are included to illustrate how a number of different ideas can be combined into a larger piece of mathematics.

1.6 MATHEMATICAL RIGOR

When you yourself are responsible for some new application of mathematics in your chosen field, then your reputation, possibly millions of dollars and long delays in the work, and possibly even human lives, may depend on the results you predict. It is then that the *need* for mathematical rigor will become painfully obvious to you. Before this time, mathematical rigor will often seem to be needless pedantry (a pedant is one who unduly explains minutiae). Thus rigor requires some practice before the need first arises. It is not something that you can learn suddenly, but it requires the gradual development of your abilities.

Mathematical rigor is the clarification of the reasoning used in mathematics. Usually, mathematics first arises in some particular situation, and as the demand for rigor becomes apparent more careful definitions of what is being reasoned about are required, and a closer examination of the numerous "hidden assumptions" is made. The student should be aware that often the definitions are gradually changed so some desired results can be obtained, but *sometimes* the earlier definitions are retained and different results are found to follow once a closer examination is made of all the details involved.

Over the years there has been a gradually rising standard of rigor; proofs that satisfied the best mathematicians of one generation have been found to be inadequate by the next generation. Rigor is not a yes–no property of a proof, much as you might like it to be; rather it is a vague standard of careful treatment that is currently acceptable to a particular group. [K–2] and [L] (references are at the end of this chapter).

Ideally, when teaching a topic the degree of rigor should follow the student's perceived need for it; but as just noted it is not something that you can suddenly master. It is necessary to require a gradually rising level of rigor so that when faced with a real need for it you are not left helpless. As a result, this book does not contain a uniform level of rigor, but rather a gradually rising level. Logically, this is indefensible, but psychologically there is little else that can be done.

The truth is that we do not always know what we are talking about. For example, the statement

All statements are false.

appears to contradict itself. First, it is surely a statement; therefore, accordingly, it must a be false statement. But if it is false, then the statement is true. There is a built-in contradiction in the grammatically correct statement.

Another example is the rule

Every rule has an exception.

If this rule is always true, then there are rules without an exception, and therefore this rule is false. Again, a contradiction!

Troubles like this can be made to arise whenever what is being said includes itself—a self-referral situation. Expressions like "the set of all sets" have this self-referral property and hence are dubious expressions; *you* are by no means sure that it does not contain a contradiction within itself.

The function of rigor is mainly critical and is seldom constructive. Rigor is the hygiene of mathematics, which is needed to protect us against careless thinking.

1.7 ADVICE TO YOU

The preceding discussion indicates that this book is unlike almost all other mathematics books, both in content and in aims. We intend to teach the doing of mathematics (of course at an elementary level). The applications of these methods produce the results of mathematics (which is usually only what is taught); the results are the worked out examples in the book and in the exercises to be solved. There is also a deliberate policy to force you to think abstractly. Not only is mathematics by its very nature abstract, but it is only through abstraction that any reasonable amount of the useful mathematics can be covered. There is simply too much known to continue the older approach of giving detailed results.

Consequently, you will find that these unconventional goals cause you a good deal of confusion. Yes, many of the results are important, but more important are the underlying methods. Things are scattered and repeated throughout the text; they are not organized in the conventional way, since the conventional goals are not the ones we have in mind. It is easy to measure your mastery of the results via a conventional examination; it is less easy to measure your mastery of doing mathematics, of creating new (to you) results, and of your ability to surmount the almost infinite details to see the general situation. In the long run the methods are the important parts of the course. Again, the specific results that are obtained have been carefully selected to be highly useful, but they are not the heart of the book.

It is not sufficient to know the theory; you should be able to apply it. There are two extremes, both of which are bad: (1) you know what to do but are unable to do it, and (2) you know how to do things but not what to do. From my more than 30 years of experience using mathematics in large scientific–engineering organizations, I believe that if you cannot handle both equally well then it is better to know what to do. You often, but not always, have some time to learn how to do it, or else you can find someone to help you. But if you do not know what to do, then the things that you do will often cause more harm than good. On the other hand, if you are not a master of details, you can seldom create what is needed.

It is currently believed that the human mind progresses from simple to more difficult ideas by "chunking" isolated small results into larger units of thought. Unfortuntely, these elementary results must be thoroughly learned *before* the chunking can occur. If this theory is true, then a lot of this course consists in first mastering many small details and then gradually chunking them into larger units. Still later, when these chunks are mastered, they are combined into even larger chunks. You should be aware of this process and do what you can to facilitate the chunking. In this way you gradually gain a mastery over the sea of details, and thus come to the mastery of the whole as a whole. But you must begin by *mastering* the small chunks thoroughly. It is only then that you can go to the larger chunks.

Because mathematics is abstract, it covers many special cases and, therefore, while studying large, complex systems, mathematics often enables you to cut through the sea of details to get at the essential nature of the problem. It is necessary that you learn both to be comfortable with abstractions and to see the details that are embraced by the sweeping statements of the abstractions. A careful examination of the great contributions to science shows that usually they require both the abstract point of view, so you can see the whole, and mastery of the details, so you can create a relevant theory.

Finally, I refuse to "write down" to you. It would be easy to replace all the long words and sentences with shorter ones, to elide or evade all difficult ideas, and to use only simple examples and simple problems. To do so would not contribute to your general education. At a minimum an educated person should be able to read reasonably sophisticated writing, and this cannot be left solely to the English department. Similarly, I could avoid all references, however slight, that fall outside the strictest bounds of the subject, and again leave you stunted in your intellectual growth. I believe that *every* course has some responsibility to contribute to your general education lest it slip between the individual courses, and from all the specific courses you take you still emerge uneducated. Thus no matter how you look at it, this book is not going to be easy to read and digest. I am ambitious for you, but I can only coach you; you must do the actual work of learning the many things that are being presented.

1.8 REMARKS ON LEARNING THE COURSE

Universities are organized into schools, the schools by departments, and the departments often have subdepartments. The subdepartments further divide up knowledge into separate courses. All this (and more) breaks up the unity of knowledge into disconnected pieces, and because everyone, when they were in school, was exposed to this arrangement of knowledge, they tend to believe it and repeat these somewhat artificial divisions in their teaching. However, the applications of knowledge, especially mathematics, reveal the unity of all knowledge. In a new situation almost anything and everything you ever learned might be applicable, and the artificial divisions seem to vanish. One of the main purposes of this book is to show some of the unity of the traditionally separated fields of mathematics and its applications.

In a course the author is forced to present the material in a linear sequence (in time). One topic must follow another in spite of the fact that each may illuminate the other. As a result, if unity is to be emphasized, the author must provide a large number of cross references. It is necessary, therefore, to refer many times to things which are physically remote in the book. To facilitate this matter of referencing things, the book is divided into 26 chapters with each chapter further subdivided into around 10 sections. Within a section the equations, figures, tables, and examples are each individually further labeled for easy reference. The notation

Equation (2.7-5)

means Chapter 2, Section 7, Equation 5. We use a similar notation for the other kinds of objects. This produces a flavor of excessive numbering, but it is very convenient to refer the reader to exactly what is meant or at least to the immediate region of the book where it can be found. References to books are indicated by [ ] with the letter inside generally the first letter of the author's name. The references are at the end of this chapter.

Second, elementary mathematical courses are often taught in a form that allows, or even encourages, learning the methods for solving particular classes of problems. The calculus, probability, and statistics can be applied so broadly that it is hopeless for you to try to learn all the "tricks" for solving specific problems. You must learn to create the details of how to solve the problem at hand; you must learn to do mathematics, to create the small tools that you need as you go along. To do this, you must learn to read the language of mathematics, to understand the meaning of the words and symbols, and to listen to the formulas.

Third, at this level of mathematics it is a common remark by students that they learn each course in the following course when they use the mathematics constantly. It is through repetitive use of the ideas that you finally master and make your own the major ideas as well as the skills of doing mathematics. Therefore, this text is organized in the "spiral" for learning (Figure 1.8-1). A topic such as probability is returned to again and again, each time higher up on the spiral (and perhaps including some new material). The first time around you may not be completely sure of what is going on, but on the repeated returns to the topic it should gradually become clear. This is

The
same topic ⟶

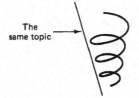

Figure 1.8-1 The spiral of learning

necessary when the ideas are not simple but require a depth of understanding, and this includes the three main topics of the course: calculus, probability, and statistics.

Fourth, unfortunately, besides the theory there are a lot of small technical details that must be learned so well that you can recall them almost instantaneously, such as the trigonometric identities that you once knew and forgot as soon as possible. One way of learning this kind of small detail for immediate recall when needed is to put one part of the identity on one side of a 3 by 5 inch card and the other part on the other side. Using these flash cards you can, in odd moments of your daily life, learn the mechanical parts of the course (small flash cards are often available locally in book stores). If you must stop and look up such details in the middle of a large problem, or even pause to try to recall them, then you will not be able to create the mathematics you will need to solve the problem at hand.

Flash cards may be used to learn many of the formulas and definitions that arise in this course and in later years. It is well known that *for this kind of low-level material* many short learning sessions are much more efficient than a few long, intense ones; but this is not necessarily true for larger ideas. No matter how obvious it is, most students will not use such trivial devices as flash cards; it seems to be beneath their dignity. They suffer accordingly.

You are advised to read this book with paper and pen handy so you can frequently check your understanding of what you are reading. This advice is repeatedly given to mathematics students, and it is repeatedly ignored by them. Remember, the book is only the coach; *you* must learn to do the mathematics. As in so many other situations, you only learn by doing it yourself; you must *involve* yourself in mathematics if you are to master it as a way of thinking about the universe.

Whenever possible, the exercises are arranged as follows:

1. Check that you have the general ideas.
2. Practice on the more mechanical parts.
3. Doing some mathematics using the ideas.
4. Extensions of these ideas (which, if hard, will be starred).

You cannot learn to do mathematics without effort. "There is no royal road to geometry" was said in ancient Greek times, (Menaechmus to Alexander, around 300 B.C.). Because we expect you to make progress, the later parts of the book are deliberately arranged to require greater effort on your part. Often it is not the particular result that is important; it is the development of your abilities that matters.

Measurements on outstanding mathematical and scientific students show that the successful students regularly go over their successful problems and examine why they succeeded. Thus you are *strongly* advised to ask yourself at the end of a problem, "Why did it work?", "What other problems can be solved in this way?", and, "How else could it be solved?" The author believes that the old saying, "You should learn from your mistakes," is less important than, "You should learn from successes." After all, there are so many ways of being wrong and so few of being right that it is much more economical to study successes. The constant recommendation of successful scientists is to

Go to the masters, not the commentators.

It is the master who, by definition, has the right style, and often the commentators give the results without the essence—style! A common error of students is to do too many problems and too little study of why things went right or wrong. Chunking must be accomplished at each level if progress to higher levels of thought is to be made, and reflection on the details is a good way of accomplishing this goal.

Fifth, although textbooks (and professors) like to make definite statements indicating that they know what they are talking about, there is in fact a great deal of uncertainty and ambiguity in the world. Contrary to normal practice, we will not evade this question but rather explore (overexplore?) it. It is hard not to be told definitely that something is or is not so, but great progress is often made when what was long believed to be true is now seen to be perhaps not the whole truth. Thus the text often uses words like "It is currently believed that . . ." and "Mathematicians believe that . . . ," rather than the conventional dogmatic statements that you usually see. They are there to cause you to think about the uncertainess and even the arbitrariness of much of our current conventions and definitions, to ponder about your acceptance of them. Again, this book is concerned with your total education, not just a small part of your mathematical training, and this makes it hard on you as you go along. It is not easy to become an educated person.

REFERENCES

The following references are for general reading. [N] gives, in four volumes, selected writings of great mathematicians and scientists, translated when necessary into English. Both [B–2] and [K–1] are histories of mathematics. [B–1] refers especially to the history of the calculus, and [T] to the history of probability. [P] is the classic on how to solve problems and has not been improved on in spite of recent computer-based books.

[D] gives a survey of what it is to be a mathematician. [L] is a powerful dialogue on the actual nature of mathematics as opposed to the official versions that most people have in their minds, and [K–2] is devoted to the subtitle, "the loss of certainty" in mathematics since the early Greek times, when mathematics was regarded as certain knowledge.

Any reasonable library can supply many other books on mathematics, generally, and calculus, probability, and statistics specifically. [C] is a classic on the calculus and is well worth consulting, and [A] is an elegant calculus book at a slightly more advanced level; both are in two volumes. [F] is the classic in probability, especially the first volume. It is hoped that this course arouses your interest enough for you to look beyond it for further information of interest to you.

[A]   Apostol, T. M. *Calculus,* 2 vols., Blaisdell Publishing Co., Waltham, Mass., 1962.

[B-1]   Boyer, C. B. *The History of the Calculus,* Dover Publishing Co., New York, 1949.

[B-2]   Boyer, C. B. *A History of Mathematics,* John Wiley & Sons, Inc., New York, 1968.

[C]   Courant, R. *Differential and Integral Calculus*, 2 vols., revised, Nordeman Publishing Company, Inc., New York, 1937.

[D]   Davis, P., and R. Hersh. *The Mathematical Experience*, Birkhauser, Boston, 1980.

[F]   Feller, W. *An Introduction to Probability Theory and Its Applications*, 3rd. ed. vol. 1, John Wiley & Sons, Inc., New York, 1957.

[G-R] Gradshteyn, I., and I. Ryzhik. *Table of Integrals, Series, and Products*, 2nd ed., Academic Press, New York, 1980.

[K-1] Kline, Morris. *Mathematical Thought from Ancient to Modern Times*, Oxford University Press, New York, 1972.

[K-2] Kline, Morris. *Mathematics: The Loss of Certainty*, Oxford University Press, New York, 1980.

[L]   Lakatos, Imre. *Proofs and Refutations*, Cambridge University Press, New York, 1976.

[N]   Newman, James. *The World of Mathematics*, Simon and Schuster, New York, 1956.

[P]   Polya, George. *How to Solve It*, Princeton University Press, Princeton, N. J., 1945.

[T]   Todhunter, I. *A History of the Mathematical Theory of Probability*, Chelsea House Publishers, New York, 1949.

2

The Integers

2.1 THE INTEGERS

There is a famous saying due to Kronecker (1823–1891):

> God made the integers, man did the rest.

So fundamental are the integers that whenever you see a discussion of communication with an alien intelligence it is always assumed that they too have the integers, at least the positive ones. Counting is so basic that a high level of technology seems impossible without the integers. Counting is usually taught as 1, 2, 3, . . . , but experience is beginning to favor 0, 1, 2, . . . as the preferable way. You will see numerous examples of this change as we go along.

The most startling aspect of the positive integers is that there is no last integer. It takes time to grasp the implications of this fact. Young children often fail to appreciate the logic that there can be no largest integer, because, if there were, then adding 1 to it would make a still larger integer. A little later in life they are often fascinated by this observation.

It is often said that the integers *approach infinity*, which is a simple way of describing the *process* of becoming arbitrarily large, that no matter how large an integer you mention, I can always name a larger one. By viewing it as a *process*, you can avoid the question of the *set* of all integers and escape such questions are "Is there a largest integer and if so what is its name?"

There seems to be a significant psychological (and logical) difference between the simple process of admitting that there is always a next integer, and the assumption that there exists a *set* consisting of all the integers. The first is much closer to experience, and the latter raises the question of how anyone could form the entire set of all positive integers.

Between these two concepts, of process and set, lies the concept of the *class* of all positive integers. For a class you are only required to decide if a given object is or is not in the class; you are not asked to form the entire set. For example, the class of even integers seems to be a reasonable idea. The concept of class seems to be closer to most people's intuitions than is the concept of an infinite set.

The fact that there is no largest integer produces some rather surprising observations. Long ago, Galileo (1564–1642) observed that there are as many integers as there are squares of integers, since we clearly have a term-by-term correspondence between the following two lines;

$$1 \quad\quad 2 \quad\quad 3 \quad\quad 4 \quad\quad 5 \quad\quad \ldots$$

$$1 \quad\quad 4 \quad\quad 9 \quad\quad 16 \quad\quad 25 \quad\quad \ldots$$

The three dots imply that the sequence is indefinitely extended. There is a one-to-one correspondence between the entries in the two lines of numbers. Using this method of comparing, neither row can have more numbers than the other row; both have the same number of numbers.

The power of a simple contradiction is very great in mathematics. For example, using only the tool of contradiction, Euclid (flourished about 300 B.C.) proved that there is no largest prime number (a number with no factors other than itself and 1). He reasoned as follows. Suppose there were a largest prime number. Then let all the prime numbers be labeled in increasing order as $p_1 = 2, p_2 = 3, p_3 = 5, \ldots$, on up to p_n, where p_n is the assumed largest prime number. Now consider the number

$$N = (p_1 p_2 \cdots p_n) + 1$$

(where the three dots represent the missing prime numbers). Since no integer (other than 1) can divide each of two consecutive integers, the number N cannot be divided by p_1, or p_2, $\ldots$, or p_n. Therefore, N is either a prime number larger than p_n or else is composed of factors all of which are larger than p_n. In either case we have shown that there exists a larger prime number, a contradiction. Thus Euclid concluded that there is no largest prime number; there are an infinite number of prime numbers.

Much of mathematics is involved with the infinite. D. H. Lehmer (1905–) says somewhere that if a result is to be called a theorem in mathematics then it must involve an infinite number of cases. For example, the statement that

$$1 + 2 + 3 + 4 = 10$$

cannot be called a theorem. But the statement that

$$1 + 2 + \cdots + n = \frac{n(n + 1)}{2}$$

is true for all integers n is a theorem. Thus, somehow, in mathematics we must learn to cope with an infinite number of cases. We have already seen that there is no largest prime number. As another example, consider the process of getting more and more accurate approximations to the number pi, $\pi = 3.14159265358979 \ldots$. . . . We can focus on the process of getting more and more digits, but we are soon compelled to think about the existence of the number itself, which is represented by *all* its digits (although there is no last digit).

No one has any direct experience with the actual infinite. Indeed, it is currently believed that the whole universe is finite, and that even space is finite. But whether the universe is finite or not, within the framework of current astronomy and physics, there is a sphere, drawn with you as the center, of say 100-light-year radius (the distance light travels in 100 years) such that anything now outside this sphere cannot affect your life for the next 100 years. Thus the physical universe you can hope to experience is bounded and finite.

But the concept of the infinite is central to much of mathematics. Allowing no integers beyond some point would be very awkward in practice. Infinity may be only an idealization of our experience, but it is a very useful idealization. Many things are much easier to think about and handle in our minds when we allow the concept of infinity. The basic idea comes from the experience of "There is always a next integer." But how we get to an abstract idea needs to be watched so that we understand both its usefulness and the dangers of believing too much in the results later obtained. Often it is sufficient to admit only the process, but occasionally we have to deal with the infinite set itself.

The concept of the number zero is especially difficult at first. The idea that something (the symbol 0) can stand for nothing (no things) seems a bit paradoxical to the untrained mind. It also takes experience to avoid the logical troubles with the two concepts, nothing and the symbol for nothing. Often the symbol $\emptyset$ (resembling ϕ, the Greek lowercase phi) is used to denote the empty set just to avoid this confusion. The symbol 0 is something and must be treated accordingly. We must be careful to distinguish between the symbol and its meaning.

When you include the negative integers, then there is no first (least) integer. This infinite regression is also very difficult to grasp. Many of the ancient myths that discuss the creation of the world have the world floating in a sea, resting on the back of an elephant, or on a tortoise that swims in a sea, or some such explanation for what holds the world up. The fact that the myths have come down to us in this form indicates that neither the audience nor the speakers themselves often asked, "And on what does that rest?" It takes some sophistication to recognize the infinite regression and to appreciate that there is neither a first nor a last integer.

2.2 ON PROVING THEOREMS

Many results in mathematics involve a potentially infinite number of cases. As examples, consider the following equations, which, because each depends on an arbitrary integer n, are true for all positive integers n.

$$S_1(n) = 1 + 2 + 3 + \cdots + n = \frac{n(n + 1)}{2}$$

$$O_1(n) = 1 + 3 + 5 + \cdots + (2n - 1) = n^2$$

$$S_2(n) = 1^2 + 2^2 + 3^2 + \cdots + n^2 = \frac{n(n + 1)(2n + 1)}{6}$$

$$s_n = \frac{1}{2} + \cos x + \cos 2x + \cdots + \cos nx = \frac{\sin [(2n + 1)/2]x}{2 \sin (x/2)}$$

The three dots in each equation are called an *ellipsis* (ellipsis means leaving out, omitting), meaning that we have not written the obvious terms. When n is very small, we have sometimes written too many terms. In particular, in the first equation if $n = 2$, then we have only

$$S_1(2) = 1 + 2 = \frac{2(2 + 1)}{2} = 3$$

The four expressions on the left sides of the equations are functions whose values are computed only for positive integers; they are functions over the (class of) positive integers. However, many times it is very convenient to evaluate them for the integer 0, often meaning no terms in the sums. Usually the value is then 0, the sum of no terms. However, for the fourth equation the sum is $\frac{1}{2}$ when $n = 0$. Furthermore, the fourth statement depends on an arbitrary x. It is a very strong statement; for all positive integers n, and for all real x, the left-hand side equals the right-hand side of the equation. Occasionally, we will want to extend a function to the negative integers, but the above examples are not such cases.

How could we ever hope to prove that such formulas are true? We cannot do this by testing them on any finite number of cases as the following examples show.

Example 2.2-1

A formula for primes. Consider the following well-known formula:

$$f(n) = n^2 - n + 41$$

Suppose it is claimed that for any positive (or possibly 0) integer this formula gives a prime number (a number having no factors other than 1 and itself). Well, we start with $n = 0$. We get 41, which is a prime. Next we try $n = 1$, and again get the number 41. Next we make a short table

$n = 2$	43
$n = 3$	47
$n = 4$	53
$n = 5$	61
$n = 6$	71
. . .	. . .
$n = 40$	1601

and by some careful checking we decide that all the entires in the table are indeed primes.

We have verified a formula for 41 consecutive cases; are these cases enough to prove that the formula always gives primes? No, because the very next case, $n = 41$, clearly gives a *composite number* (a composite number is one that has factors other than 1 and itself): $41 \times 41 = 41^2$, which is not a prime. Thus testing any finite number of cases is not enough to prove that a theorem *in mathematics* is true for all integers. On the other hand, in *science* we must always stop after a finite number of experiments and simply assert the universal law. In both mathematics and science, one counterexample can be enough to disprove a conjecture.

Example 2.2-2

An obvious example of the same thing is the function

$$f(n) = n(n - 1)(n - 2) \cdots (n - k)$$

This expression in the variable n vanishes for $n = 0, 1, 2, \ldots, k$ and for no other cases. Thus $k + 1$ verifications that a function is zero do not prove that it is always zero. This function could be added to any other function $g(n)$ to get the sum $f(n) + g(n)$, and you would get the same values for both $g(n)$ and $f(n) + g(n)$ for the first $k + 1$ values, and then the two would differ.

Example 2.2-3

The purpose of this example is to show you the triviality of making up such examples as the above. Suppose, to be specific, you want the first million cases to be true, but no others. We have, for the million cases $n = 0, 1, \ldots, 999,999$, the following obvious theorem:

> All the nonnegative integers require less than 7 decimal digits to represent them (in decimal form).

Even a child can see the falsity of the theorem and generalize to the futility of using any finite number of true cases to prove such a theorem.

Many methods for proving statements involve a potentially infinite number of cases. The best known method is called *mathematical induction,* which we take up in the next section.

EXERCISES 2.2

1. How far does the formula $f(n) = n^2 - 79n + 1601$ generate primes? *Ans.:* It fails at $n = 80$ since $f(80) = 41^2$.
2. Prove that the square of an odd integer is an odd integer. *Hint:* Write the odd number as $2k + 1$.
3. Create your own example of a large number of true cases but that are not universally true.
4. Extend the formula $n^2 - n + 41$ to negative integers.
5. Is it true that every positive integer is either a prime or a composite number?
6. Do *you* think that -7 should be called a prime? Why?

2.3 *MATHEMATICAL INDUCTION*

How can we hope to prove that some statement is true for all positive integers (and possibly 0, too)? Certainly not by testing a finite number of cases. But suppose we can show that if we take any one arbitrary case as being true then the next case must also be true. This is like a string of dominos arranged so that each in falling down knocks over the next. Then it would only remain to show that the first case was true (tipped over). Can we do this sort of thing in mathematics? As an illustration that we can, consider the following example.

Example 2.3-1

Sum of the first n integers. The first formula in the last section states that

$$S_1(n) = 1 + 2 + 3 + \cdots + n = \frac{n(n + 1)}{2} \qquad (2.3\text{-}1)$$

We first, prudently, check the first few cases.

$$n = 1, \qquad 1 = \frac{1(2)}{2} = 1$$

$$n = 2, \qquad 1 + 2 = 3 = \frac{2(3)}{2}$$

$$n = 3, \qquad 1 + 2 + 3 = 6 = \frac{3(4)}{2}$$

Now we use the falling domino plan. We *assume* that the formula is true for the mth case; that is, the following equation is true:

$$1 + 2 + 3 + \cdots + m = \frac{m(m + 1)}{2}$$

To advance the left-hand side to the next case, we must (for this problem) add $m + 1$ to *both* sides (thus keeping the equality):

$$[1 + 2 + 3 + \cdots + m] + (m + 1) = \frac{m(m + 1)}{2} + (m + 1)$$

The left-hand side is now the left-hand side of (2.3-1) for the next case $n = (m + 1)$. How about the right-hand side? We can factor out the $m + 1$ from each term on the right-hand side and get

$$(m + 1)\left[\left(\frac{m}{2}\right) + 1\right] = \frac{(m + 1)(m + 2)}{2}$$

This can now be seen to be the right-hand side of the original equation (2.3-1) with n replaced by $m + 1$. Thus from the statement for m we have *derived* the statement for $m + 1$; we have successfully done the domino trick. Since we have already checked the formula for the first case, $n = 1$, we conclude that the formula is true for all positive integers n.

This method of *mathematical induction* is very important and has many variations. One variation is to introduce a slight change that sometimes makes the algebra significantly easier (because the target formula is simpler); we assume the truth of the $(m - 1)$th case and derive from it the truth of the mth case (instead of going from m to $m + 1$). This is one of the many "tricks of the trade" that the experts all know but that are seldom explicitly stated.

Example 2.3-2

Sum of the squares of the integers. We will prove the third equation in the previous section,

$$S_2(n) = 1^2 + 2^2 + 3^2 + \cdots + n^2 = \frac{n(n + 1)(2n + 1)}{6} \tag{2.3-2}$$

First we must verify it for some particular case, say $n = 1$.

$$S_2(1) = 1^2 = 1 = \frac{1(2)(3)}{6}$$

With experience we might have noted that the trivial case $n = 0$ is a simpler basis for the induction. To advance from the $(m - 1)$th case to the mth case, we need, for this problem, to add the quantity m^2 to both sides of the $(m - 1)$th case (put $n = m - 1$ in Equation (2.3-2) and then add m^2 to *both* sides). We get

$$[1^2 + 2^2 + \cdots + (m - 1)^2] + m^2 = \frac{(m - 1)m(2m - 1)}{6} + m^2$$

which is the proper left-hand side of (2.3-2) for the mth case. To get the proper form on the right-hand side, we multiply out the parentheses on the right-hand side and rearrange into the appropriate form

$$\frac{2m^3 - 3m^2 + m}{6} + m^2 = \frac{(2m^3 - 3m^2 + m) + 6m^2}{6}$$

$$= \frac{2m^3 + 3m^2 + m}{6}$$

$$= \frac{m(2m^2 + 3m + 1)}{6}$$

$$= \frac{m(m + 1)(2m + 1)}{6}$$

which is the required form ($n = m$) for the right-hand side of (2.3-2).

Sometimes when we apply mathematical induction we need more than one starting case. We can also abbreviate the steps written down in the proof by eliding (omitting) the formality "assume it is true for m."

Example 2.3-3

Cos nx is a polynomial in cos x. Suppose we wish to prove that

$$\cos nx = P_n(\cos x)$$

where $P_n(\cos x)$ is a polynomial (many terms, "poly" means many) of degree n in the variable cos x. Notice that we are being rather abstract. We are not going to produce the explicit polynomial; we are only going to show that one exists. This is part of the plan to force you to learn to handle abstraction; we are deliberately withholding the concrete expression.

　　　　To carry out the proof, we need a particular trigonometric identity that is easy to derive from the basic addition formulas (16.1-7) and (16.1-6)

$$\cos (n + 1)x = \cos nx \cos x - \sin nx \sin x$$

$$\cos (n - 1)x = \cos nx \cos x + \sin nx \sin x$$

Adding these two identities causes the sine terms to cancel, and then transposing one term gives us

$$\cos (n + 1)x = 2 \cos x \cos nx - \cos (n - 1)x \qquad (2.3\text{-}3)$$

Suppose, now, that *both* cos nx and cos $(n - 1)x$ were polynomials in cos x of degrees n and $n - 1$, respectively. (Notice that we have not bothered to introduce the "dummy letter" m but assume that the process is sufficiently familiar that you can follow the argument in the abbreviated form.) The reasoning goes as follows: since cos nx is a polynomial of degree n in cos x and this is, by the identity (2.3-3), to be multiplied by $2 \cos x$, it follows that the product is now of degree $n + 1$ in cos x. The second term, cos $(n - 1)x$, adds a polynomial of degree $n - 1$ in cos x and cannot affect the *degree* of the sum. Thus from our identity we see that cos $(n + 1)x$ is a polynomial of degree $n + 1$ in cos x. But to get to the case $n + 1$ we need *both* the case n and the case $n - 1$. Therefore, this induction proof needs two consecutive starting cases. For $n = 1$ and $n = 2$, we have [again used (16.1-7)]

$$\cos x = \cos x \qquad \text{(has degree 1)}$$

$$\cos 2x = \cos x \cos x - \sin x \sin x$$

$$= \cos^2 x - \sin^2 x = \cos^2 x - (1 - \cos^2 x)$$

$$= 2 \cos^2 x - 1 \qquad \text{(has degree 2)}$$

We might have used the simpler cases $n = 0$ and $n = 1$ as a basis for the induction since

$$\cos 0 = 1 \qquad \text{(has degree 0)}$$

Occasionally, in going to the next case it is necessary to assume the truth of all the preceding cases, but this does not change the argument very much.

There is a more sophisticated argument for the basis for the mathematical induction proof that some people find more convincing than the domino argument. This argument goes as follows. Suppose the statement is not true for every positive (or every nonnegative) integer. If so, then there are some false cases. Now consider the set for which the statement is false. *If* this is a nonempty set, then it would have a least integer, call this integer m. Now consider the preceding case, which is $m - 1$. This $(m - 1)$th case must be true by definition, and we know that there is such a case because as a basis for the induction we showed that there was at least one true case. We now apply the step forward, starting from this true case $m - 1$, and conclude that the next case, case m, must be true. But we assumed that it was false! A contradiction. So we go back up the argument until we find the *if* and decide that what immediately follows the "if" must be false. Thus there are no false cases and the formula is true for all integers above the integer used as the basis for induction. Logical consistency is a foundation stone of both mathematics and science.

The technical difficulty the beginning student has with mathematical induction is in the matter of how to go from one step, say the $(m - 1)$th case, to the next, the mth. When applying mathematical induction to sums, as we did in the above cases, you simply add the next term of the series to both sides of the equation (you must preserve the equality you are assuming for the induction step). As we will later see, in other cases you must do the appropriate thing to get to the next case.

The theoretical difficulty the student has with mathematical induction arises from the reluctance to ask seriously, "How could I prove a formula for an infinite number of cases when I know that testing a finite number of cases is not enough?" Once you really face this question, you will then see how mathematical induction gives an answer; you will understand the ideas behind mathematical induction. It is only when you grasp the problem clearly that the method becomes clear.

Example 2.3-4

Number of subsets in a set. Given a set of n elements (items), how many subsets can be formed? It is customary to include the two *improper subsets* consisting of the null set (no elements) and the whole set (all the elements). The reason for this will become apparent from the following proof.

We assert that a set of n elements has 2^n subsets. This is certainly true for the case $n = 1$, since there we have only the two improper subsets. Suppose it is true for the case of m elements. Then add one element to make the set with $m + 1$ elements. The subsets that can be formed are all those without the new element, which total, by hypothesis, 2^m, and those with the new element added to each old subset, again 2^m. No two of these sets are the same, so for the case $m + 1$ there are $2(2^m) = 2^{m+1}$ subsets, and the proof is done.

Consider how awkward the proof would be if the improper subsets were excluded. There would be $2^m - 2$ subsets for m elements. Adding the new element would produce another $2^m - 2$ subsets. We would then have to argue that the new element itself was a subset, as well as that the original whole set is now a subset. This would give the proof, since

$$2(2^m - 2) + 1 + 1 = 2^{m+1} - 2$$

as required for the induction step.

This shows how the details of mathematical proofs can sometimes shape the very definitions to be made; definitions do not arise from the vacuum of pure thought, but from the real world of experience plus the details of working with the definitions. This example again shows the importance of thinking of the case $n = 0$ (counting 0, 1, 2, . . .).

As the following example shows, sometimes it is easier to think about the downward induction.

Example 2.3-5

Sum of the angles of a convex polygon. Prove that the sum of the interior angles of a convex polygon having n sides is $180° (n - 2)$.

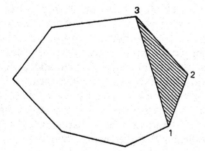

Figure 2.3-1   Convex polygon

Given a convex polygon we pick three consecutive vertices along the edge and draw a line from the first to the third. See Figure 2.3-1. This forms a triangle. If we remove the triangle (of course we leave the edge between the first and third points), we decrease the number of sides of the polygon by 1 and the sum of the interior angles by 180°. We can continue this as long as we have at least four sides. When we get down to the case of three vertices, we have a triangle and know that the sum of the interior angles is 180°. Thus we have a proof in the downward form; if the formula is not true for some n, then an earlier case $(n - 1)$ is not true, and when we get down to the first case which we know to be true we decide (since the steps are reversible) that all the other cases must also be true.

EXERCISES 2.3

1. Find the sum of the first 100 positive integers. (It is said that Gauss at the age of 10 solved this in his head, which started him on the path to being a mathematician.)

2. Find the sum of the squares of the first 100 positive integers.

3. For the series $1 + 2^3 + 3^3 + \cdots + (m - 1)^3$, what term would you add in an inductive proof?

4. For the series $1 + 4 + 7 + \cdots + 3m - 2$, what term would you add in an induction proof?

5. For the series $1 + \frac{1}{2} + \frac{1}{3} + \cdots + 1/(m - 1)$, what term would you add in an induction proof?

6. Prove that $1 + 3 + 5 + \cdots + (2n - 1) = n^2$.

7. Show that

$$\frac{1}{(1)(2)} + \frac{1}{(2)(3)} + \cdots + \frac{1}{(n)(n + 1)} = 1 - \frac{1}{n + 1}$$

   Hint: $1/k(k + 1) = 1/k - 1/(k + 1)$.

8. Prove that $\sin (nx)/\sin x$ is a polynomial of degree $n - 1$ in $\cos x$. *Hint:* Expand $\sin (m + 1)x$.

9. Guess the coefficient of the leading term of the polynomial representation of $\cos nx$. Prove it by induction.

10. Show that 6 divides $n^3 - n$ for all $n > 1$.

11. By mathematical induction, prove that $2^n > n^2$ for all $n > 4$.

12. Suppose that all you can prove is that from the case m the case $m + 3$ follows. How would you make the proper induction argument?

13. Suppose you want to prove an assertion for all integers positive, zero, and negative. Give a suitable pattern of logic.

14. Consider the following argument. *Theorem:* All positive integers are interesting. *Proof:* Certainly the integer 1 is an interesting number. If all the integers are not interesting, then there is a least not interesting integer. But this is an interesting property! A contradiction.

15. Marbles in a bag example: The theorem states that in any bag of black and white marbles all the marbles are the' same color. The proof by induction starts with the case of a bag with one marble. Certainly all the marbles in the bag are of the same color. Next we assume the theorem is true for any bag of m marbles. Take a bag with $m + 1$ marbles and remove one. By the induction hypothesis, they are all of the same color. Replace it and remove another marble. Again they are all of the same color. Thus all the marbles in the bag are of the same color. *Hint:* In all such cases of an obvious error, it is wise to examine closely the first case you believe to be false.

16. Extend Example 2.3-5 to nonconvex polygons. What restrictions do you want to observe?

17.* Two different arguments are presented for the logic used in connection with mathematical induction. Write an essay comparing them in detail. Which do you prefer and *why?*

18. Prove $O_2(n) = 1^2 + 3^2 + 5^2 + \cdots + (2n - 1)^2 = n(4n^2 - 1)/3$.

2.4 THE BINOMIAL THEOREM

A very important application of mathematical induction is the proof of the binomial (two term; "bi" means two) theorem:

$$(a + b)^n = a^n + na^{n-1}b + \frac{n(n-1)}{2}a^{n-2}b^2 + \cdots + b^n \qquad (2.4\text{-}1)$$

Again we begin with a short table to get a feel for the results. This habit of doing a number of special cases first was very characteristic of Leonhard Euler (1707–1783), one of the greatest mathematicians who ever lived. By this time we have learned the value of starting with the zeroth case. Notice that for this induction each line is obtained from the preceding line by multiplying by $(a + b)$.

$$(a + b)^0 = 1$$
$$(a + b)^1 = a + b$$
$$(a + b)^2 = a^2 + 2ab + b^2$$
$$(a + b)^3 = a^3 + 2a^2b + \quad ab^2 +$$
$$\qquad\qquad\quad + \quad a^2b + 2ab^2 + b^3$$
$$\qquad = a^3 + 3a^2b + 3ab^2 + b^3$$

We need a notation for describing the coefficients. For the nth power, let us write the kth coefficient along the nth line as the *binomial coefficient*

$$C(n, k)$$

Read this as "Binomial coefficient of order n and index k," or more simply, "C of n and k." The first integer n names the row, and the second integer k names the position along the row. The values of k run from 0 through n. For any other integer k, it is conventional to set the binomial coefficient equal to zero. A currently popular notation for the binomial coefficients is

$$\binom{n}{k}$$

but this is hard to type, takes extra space whenever it occurs in a running text of words, and is difficult to handle using a computer terminal, so we use the older, more convenient, $C(n, k)$ notation here.

The binomial coefficients are the numerical coefficients in the above equations, and they may be written neatly in the form of the *Pascal* (1623–1662) *triangle* as follows:

$$
\begin{array}{ccccccccccccccccc}
 & & & & & & & & 1 & & & & & & & & \\
 & & & & & & & 1 & & 1 & & & & & & & \\
 & & & & & & 1 & & 2 & & 1 & & & & & & \\
 & & & & & 1 & & 3 & & 3 & & 1 & & & & & \\
 & & & & 1 & & 4 & & 6 & & 4 & & 1 & & & & \\
 & & & 1 & & 5 & & 10 & & 10 & & 5 & & 1 & & & \\
 & & 1 & & 6 & & 15 & & 20 & & 15 & & 6 & & 1 & & \\
 & 1 & & 7 & & 21 & & 35 & & 35 & & 21 & & 7 & & 1 & \\
1 & & 8 & & 28 & & 56 & & 70 & & 56 & & 28 & & 8 & & 1 \\
\end{array}
$$

(2.4-2)

$$\bullet \qquad \bullet \qquad \bullet$$

This triangle also appears in Omar Khayyam's (1048–1131) *Algebra,* as well as in Chu Shih-Chien's *Precious Mirror* (1303).

As is easily seen, each (upper) edge number is a 1, and each inner number is the sum of the two numbers immediately above it. To show that this last statement is true in general (meaning for all cases), it suffices to note the way a term in one line arises from those in the line above when the corresponding equation is multiplied by $(a + b)$. Looking at the coefficient in the triangle corresponding to the kth term in the nth row (beyond the $n = 1$ row),

$$a^{n-k}b^k$$

we see that one part of the term comes from the $C(n - 1, k - 1)$ when we multiply it by the a (the previous row and one lower power in a), and the other part comes from the $C(n - 1, k)$ when we multiply it by the b (the previous row and the same power of a). Thus we have the fundamental relation

$$C(n, k) = C(n - 1, k - 1) + C(n - 1, k) \tag{2.4-3}$$

Each number is the sum of the two numbers in the row above. This equation, together with the boundary 1's, *implicitly* defines the binomial coefficients. It is easily seen that these are the mathematical induction conditions that make the binomial theorem true for all positive integer components.

In more mathematical notation, this identity is

$$(a + b)^n = (a + b)^{n-1}(a + b)$$
$$= a(a + b)^{n-1} + b(a + b)^{n-1}$$

and we select the corresponding term, $a^k b^{n-k}$, from both sides.

To get an *explicit* representation for the $C(n, k)$ we need to introduce the factorial notation for the product of the first n integers:

$$1(2)(3)(4) \cdots (n) = n!$$

It is easy to see that $1! = 1$ and that

$$n! = n[(n - 1)!] \tag{2.4-4}$$

for all n greater than 1. This is a *recursive* definition; each factorial (after the first) is defined in terms of an earlier factorial.

It is convenient to *extend* the definition of the factorial function in a consistent way to the case of 0!. We simply ask what 0! would have to be to satisfy the Equation (2.4-4). Put $n = 1$ and we get

$$1! = 1(0!)$$

from which we conclude that

$$0! = 1$$

is a reasonable extension of the definition of a factorial.

We now use mathematical induction to prove that the value of $C(n, k)$ is given by

$$C(n, k) = \frac{n!}{k! \, (n - k)!} \qquad (2.4\text{-}5)$$

It is certainly true for the case $n = 1$, where the two cases to check are $k = 0$ and $k = 1$.

$$C(1, 0) = \frac{1!}{0! \, 1!} = 1$$

$$C(1, 1) = \frac{1!}{1! \, 0!} = 1$$

and this serves as the basis for the induction reasoning. Assuming that the factorial representation (2.4-5) of the binomial coefficient is true for the case $m - 1$ (for all k), we compute $C(m, k)$ from the earlier identity (2.4-3) between the binomial coefficients. The right-hand side of (2.4-3) is

$$C(m - 1, k - 1) + C(m - 1, k) = \frac{(m - 1)!}{(k - 1)! \, (m - k)!} + \frac{(m - 1)!}{k! \, (m - 1 - k)!}$$

We factor out the first term on the right-hand side (notice that we have an extra factor k in the denominator of the second term and lack a factor $m - k$ in the denominator, which must therefore also be supplied in the numerator). Hence the right-hand side is

$$= \frac{(m - 1)!}{(k - 1)! \, (m - k)!} \left[1 + \frac{m - k}{k} \right]$$

We now rearrange the expression in brackets.

$$= \frac{(m - 1)!}{(k - 1)! \, (m - k)!} \left[\frac{k + m - k}{k} \right]$$

$$= \frac{(m - 1)!}{(k - 1)! \, (m - k)!} \left[\frac{m}{k} \right]$$

$$= \frac{m!}{k! \, (m - k)!} = C(m, k)$$

The last line arises when we move the trailing factors of m and k forward to their corresponding factorial terms. Thus we have completed the induction step for Equation (2.4-5). Since we verified the induction for the case $k = 1$, we seem to have a complete proof. But let us check all the logic carefully. We have the basis for the induction, and we moved from the $(m - 1)$th case to the mth case for $k = 1, 2, \ldots, m - 1$. We missed the two end values, $k = 0$ and $k = m$. Examining the proof, we see from (2.4-5) that both end values $C(m, 0)$ and $C(m, m)$ are 1. Thus we have completed the proof to cover all $m + 1$ terms for the new line of binomial coefficients.

It is worth noting at this point that

$$C(n, k) = \frac{n!}{k! \, (n - k)!} = \frac{n!}{(n - k)! \, k!} = C(n, n - k) \qquad (2.4\text{-}6)$$

which simply says that the binomial coefficients of order n are symmetric in k about the middle value (or the two middle values); replacing k by $n - k$ does not change the numerical value of the binomial coefficient of order n. Note that in spite of the indicated division the binomial coefficients are integers.

Example 2.4-1

$$C(17, 12) = C(17, 5) \qquad C(17, 8) = C(17, 9)$$

For a given order n, the binomial coefficients along a row can be found directly rather than by laboriously filling in all the Pascal triangle down to that level. There is a simple recursive rule, which is clear from the following. In each case the brackets [ ] are the preceding coefficient in the row:

$$1, \quad [1]\frac{n}{1}, \quad \left[\frac{n}{1}\right]\frac{n - 1}{2}, \quad \left[\frac{n(n - 1)}{2!}\right]\frac{(n - 2)}{3}, \quad \left[\frac{n(n - 1)(n - 2)}{3!}\right]\frac{(n - 3)}{4}, \ldots$$

In practice you write down a 1 for the coefficient of the first term, multiply this by n and divide by 1 to get the next coefficient, multiply this by $n - 1$ and divide by 2 to get the next coefficient, multiply this by $n - 2$ and divide by 3 to get the next, and so on, and you end up with 1 for the final coefficient. Actually, since you know that the coefficients are symmetric about the middle coefficient(s), you need only compute one-half of them *if* you are confident; computing them all gives a check by using the symmetry.

Rule.   In passing from one term in the row to the next term in the same row, you simply multiply by a number one smaller than the one you multiplied by in the previous step and divide by a number one greater than you divided by in the previous step. In symbols the relationship we are using is

$$[C(n, k - 1)]\frac{n - (k - 1)}{k} = C(n, k)$$

If in the binomial expansion (2.4-1) we replace b by $-b$ (the expansion is true for any a and b), then since

$$(-b)^2 = b^2, \quad (-b)^3 = -b^3, \quad (-b)^4 = b^4, \quad \text{etc.}$$

we get the same terms, but now with alternating signs:

$$(a - b)^n = a^n - na^{n-1}b + \frac{n(n-1)}{2}a^{n-2}b^2 - \cdots$$

The alternation in sign is very common in mathematics and is worth a moment's extra attention. The quantity

$$(-1)^n$$

has the following useful properties:

$$(-1)^{n+1} = (-1)^{n-1}$$

$$(-1)^{2n+k} = (-1)^k$$

$$\frac{1}{(-1)^n} = (-1)^n$$

Example 2.4-2

$$(-1)^7 = (-1)^{2(3)+1} = (-1) = -1$$

$$(-1)^8 = 1$$

$$(-1)^9 = -(-1)^8 = -1$$

We now return to further examples of mathematical induction.

Example 2.4-3

Fibonacci number representation. The famous *Fibonacci* (also known as Leonardo of Pisa, 1170–1250) *numbers* f_k are defined by the rules

$$f_0 = 0$$

$$f_1 = 1$$

$$f_n = f_{n-1} + f_{n-2} \quad \text{(for } n > 1\text{)}$$

Let us compute a number of early cases.

$$f_2 = f_1 + f_0 = 1 + 0 = 1$$

$$f_3 = f_2 + f_1 = 1 + 1 = 2$$

$$f_4 = f_3 + f_2 = 2 + 1 = 3$$

$$f_5 = f_4 + f_3 = 3 + 3 = 5$$

$$f_6 = f_5 + f_4 = 5 + 3 = 8$$

Suppose you are now told that the numbers can be represented by the form (which we will derive in example 24.8-1)

$$f_n = \frac{1}{\sqrt{5}}\left[\left(\frac{1 + \sqrt{5}}{2}\right)^n - \left(\frac{1 - \sqrt{5}}{2}\right)^n\right]$$

The presence of a $\sqrt{5}$ looks like the formula cannot lead to integers, but from the definition the Fibonacci numbers are clearly integers! A closer look shows that the first of the two binomial expansions of order n has all positive terms, while the second binomial expansion has alternating signs. These are being subtracted, and therefore only terms with the square root will persist inside the brackets [remember that $(\sqrt{5})^2 = 5$]. The divisor outside will exactly take care of them! Thus you see that it is possible that this formula leads to integers. But we still face the problem of proving a formula for an infinite number of cases. As a basis for the conventional mathematical induction, we check that the cases $n = 0$ and $n = 1$ are correct. In the process we see how the square roots exactly cancel, confirming our earlier reasoning. If you not do see this, then do the next several cases until you see why $\sqrt{5}$ cancels out.

We therefore assume that the formula is correct for all cases up to n, and form the right-hand side of the defining equation for the f_{n+1}.

$$f_{n+1} = \frac{1}{\sqrt{5}}\left[\left(\frac{1 + \sqrt{5}}{2}\right)^n - \left(\frac{1 - \sqrt{5}}{2}\right)^n\right] + \frac{1}{\sqrt{5}}\left[\left(\frac{1 + \sqrt{5}}{2}\right)^{n-1} - \left(\frac{1 - \sqrt{5}}{2}\right)^{n-1}\right]$$

To expand these seems like too much work. However, it might possibly work out that the terms that contain the $+\sqrt{5}$ sign will separately combine to produce the new term that contains the $+\sqrt{5}$ sign. And, therefore, the terms that contain the $-\sqrt{5}$ sign would likewise combine properly. It is worth a try! We have

$$f_n + f_{n-1} = \frac{1}{\sqrt{5}}\left[\left(\frac{1 + \sqrt{5}}{2}\right)^n + \left(\frac{1 + \sqrt{5}}{2}\right)^{n-1}\right] + \text{terms with minus signs}$$

We next factor out the lowest power of the two terms that contain the $+\sqrt{5}$ to get for the $f_n + f_{n-1}$ terms; and writing *only* the terms with the positive $\sqrt{5}$, we have

$$\left(\frac{1 + \sqrt{5}}{2}\right)^{n-1}\left[\frac{1 + \sqrt{5}}{2} + 1\right]$$

But this is

$$\left(\frac{1 + \sqrt{5}}{2}\right)^{n-1}\left[\frac{3 + \sqrt{5}}{2}\right]$$

If this is to be the proper term for the f_{n+1} expression, then we must have the bracketed expression in this equation:

$$\frac{3 + \sqrt{5}}{2} = \left(\frac{1 + \sqrt{5}}{2}\right)^2$$

Therefore, let us multiply out the right-hand side to check this condition. We get

$$\frac{1 + 2\sqrt{5} + 5}{4} = \frac{6 + 2\sqrt{5}}{4} = \frac{3 + \sqrt{5}}{2}$$

as it should! The corresponding terms with the $-\sqrt{5}$ behave similarly. Thus we have completed the induction step (as well as verified the induction for two special cases) and believe that the representation of the Fibonacci numbers in this form is correct.

From the recurrence relation defining the numbers, we get $f_0 = 0$, $f_1 = 1$, $f_2 = 1$, $f_3 = 2$, $f_4 = 3$, $f_5 = 5$, $f_6 = 8$, ... Evidently, it is much easier to compute the first few Fibonacci numbers from the recurrence relation than from the formula we just proved. The formula is useful, however, when estimating the values of the Fibonacci numbers for large n. You have only to compute numerically

$$\frac{1 + \sqrt{5}}{2} = 1.618033989 \ldots$$

If you raise this number to the appropriate power n and finally divide by $\sqrt{5}$, you get an estimate of the size of the nth number, because the term with the minus sign

$$\frac{1 - \sqrt{5}}{2} = -0.618033989 \ldots$$

is less than 1 in size, and for high powers it is very small (see Section 7.2). If you round to the nearest integer, the first term *alone* gives f_n.

The number $(1 + \sqrt{5})/2$ is called the *golden ratio* and is believed (by some people) to be the ratio of the sides of the most esthetically pleasing rectangle. (This can be seen in some packaging!) It also has the property that if you remove the largest square that you can from the rectangle then the remaining rectangle has the same ratio of its two sides (Figure 2.4-1).

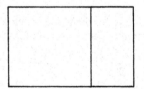

Figure 2.4-1   Golden rectangle

EXERCISES 2.4

1. From the Pascal triangle, expand $(a + b)^5$.
2. From the Pascal triangle, expand $(a - b)^5$.
3. Write out the next line of the Pascal triangle.
4. Expand $(1 + x)^{10}$.
5. Expand $(a - b)^7$.
6. Expand $(p + q)^n$.
7. Expand $(p - q)^n$.

8. Expand and evaluate $(1 + 1)^5$.
9. Expand and evaluate $(1 - 1)^5$.
10. Add the expansions in Exercises 8 and 9 and divide by 2 to get the sum of $C(5, 0) + C(5, 2) + C(5, 4)$.
11. Add the expansions in Exercises 6 and 7 and divide by 2 to get the sum of the terms of even index terms $C(n, 0)p^n + C(n, 2)p^{n-2}q^2 + \cdots$.
12. In Exercise 11 set $p = q = \frac{1}{2}$ and find the sum of the even index binomial coefficients of order n.
13. Expand $(1 + x)^n$ and then set $x = 1$ to get the sum of all the binomial coefficients of order n.
14. As in Exercise 13, but replace x by -1. Then add the two equations and divide by 2 to get the sum of the even index coefficients

$$C(n, 0) + C(n, 2) + C(n, 4) + \cdots = 2^{n-1}$$

By subtraction find the sum of the odd index terms.
15. Show that $1(1!) + 2(2!) + 3(3!) + \cdots + n(n!) = (n + 1)! - 1$.
16. Show that the edge values of the Pascal triangle satisfy the recurrence relation for the binomial coefficients. *Hint:* Outside the printed triangle the values are all zero.
17. Show that

$$\frac{1 + \sqrt{5}}{2} = \frac{-1}{(1 - \sqrt{5})/2}$$

18. Show that $(2k + 1)!/(2^k k!) = (2k + 1)(2k - 1)(2k - 3) \cdots (1)$.
19. Show that $(2k)!/(k! \, k!) = 2^k[2 - 1/k][2 - 1/(k - 1)][2 - 1/(k - 2)] \cdots [2 - 1]$.
20. Prove the last statement of this section.
21. Why would you not want to extend the definition of the factorial function to negative integers?

2.5 MATHEMATICAL INDUCTION USING UNDETERMINED COEFFICIENTS

The glaring weakness of mathematical induction is that you must know the answer before you start. All you do is verify the truth; you do not find the formula. Sometimes you do know the answer before you start, but usually you are not so lucky. Often much of your problem is to find the formula. Having found a result by some method, mathematicians often then publish only the polished inductive proof, which completely conceals the method they used originally to find the formula.

How can you find the right-hand side of a formula for starting the mathematical induction? Often guessing from the results of a few cases will work, but sometimes it will not. The following *method of undetermined coefficients* requires only that you know the general form of the answer, not all the details of it.

The idea is as follows:

1. Assume the general form of the answer.
2. Impose the condition that the induction be true for $n = 0$ (or any other convenient value).
3. Impose the induction condition that the step from $m - 1$ to m be true.
4. From steps 2 and 3, determine the unknown coefficients.

If we can do all these steps, then we are assured, *without carrying it out*, that the steps of the induction proof will be true; we have already in steps 2 and 3 imposed the conditions that they will!

If we cannot carry out all the steps, we have assumed the wrong form, but the failure will probably suggest modifications to the assumed form, and we can try again (and again).

Extension 2.5-1

Sum of the third powers of the integers. When computing the sums $S_k(n)$ of the kth powers of the successive integers for the cases $k = 1$ and $k = 2$, we found that the right-hand sides were polynomials of degrees 2 and 3, respectively. The obvious conjecture is that the sum of the kth powers must be a polynomial of degree $k + 1$. You can make an additional check on the hypothesis by dropping back to the case $k = 0$. The positive integers to the zeroth power are always 1, so the sum should be exactly n, a polynomial of degree 1. This gives you further evidence that the conjecture is likely to be true.

Therefore, for step 1 we write out the sum of the third powers of the integers as a polynomial of degree 4 with five *undetermined coefficients*, a, b, c, d, and e.

$$1^3 + 2^3 + 3^3 + \cdots + n^3 = an^4 + bn^3 + cn^2 + dn + e \qquad (2.5\text{-}1)$$

For step 2 we use as a basis for the induction the case of $n = 0$, no terms, and this forces us to conclude that

$$e = 0$$

For step 3 we apply the induction step (going from $m - 1$ to m) to this polynomial. To do this we simply add m^3 to both the left- and right-hand sides. As usual, we then need to transform the right-hand side to the proper form. The right-hand side of (2.5-1) is

$$[a(m - 1)^4 + b(m - 1)^3 + c(m - 1)^2 + d(m - 1)] + m^3 \qquad (2.5\text{-}2)$$

and this must be, if the induction is to work, exactly equal to

$$am^4 + bm^3 + cm^2 + dm \qquad (2.5\text{-}3)$$

For step 3 we equate these two expressions (2.5-2) and (2.5-3).

$$a(m - 1)^4 + b(m - 1)^3 + c(m - 1)^2 + d(m - 1) + m^3$$
$$= am^4 + bm^3 + cm^2 + dm$$

We now rearrange the left-hand side of this equation. Apply the binomial expansion to each term, and organize the results by the powers of m to match the form of the right-hand side. The equation is

$$m^4[\quad a]$$
$$+\; m^3[-4a +\quad b +\quad 1]$$
$$+\; m^2[\quad 6a -\quad 3b +\quad c] \tag{2.5-4}$$
$$+\; m\;[-4a +\; 3b -\; 2c +\; d]$$
$$+\quad[\quad a -\quad b +\quad c -\; d] = am^4 + bm^3 + cm^2 + dm$$

We see the binomial coefficients in the columns. The extra 1 in the m^3 term arises from the added m^3 to each side to make the induction step.

Notice that each term on the right-hand side of this equation will exactly cancel a corresponding term on the left. We have, therefore, a polynomial of degree 3, which is to be zero for all positive integers m.

$$m^3[-4a +\; 1]$$
$$+\; m^2[\quad 6a -\quad 3b]$$
$$+\; m\;[-4a +\; 3b -\; 2c] \tag{2.5-5}$$
$$+\quad[\quad a -\quad b +\quad c -\; d] = 0$$

Can any polynomial be zero for *all* positive integers unless all its coefficients are zero? What is the equation except the assertion that the sum of all its positive terms will exactly cancel the sum of the negative terms? More tersely, the sum of all the terms must cancel out completely. This being so, we see immediately that, if we pick m sufficiently large, then the first term cannot possibly be canceled by all the other terms in the equation *unless* the coefficient $-4a + 1$ is zero. We apply the same argument to the next lower power, then the next lower, and so on, and thus we have the following important rule:

If a polynomial is identically zero for all positive integers, then each coefficient must be zero separately.

We apply this to the long equation (2.5-5) in the variable m (partially simplified by the canceling of like terms). Equate the coefficients of each power of m to zero separately and get the set of equations

$$-4a + 1 = 0$$
$$6a - 3b = 0$$
$$-4a + 3b - 2c = 0 \tag{2.5-6}$$
$$a - b + c - d = 0$$

To complete step 4, we solve these equations in turn, top down. Notice that the first equation involves only a. The second uses this value of a to find the value of b. The third equation involves only the new value c and uses the two values a and b just found. The system of equations (2.5-6) is *triangular* (a common case) and is therefore easily solved. The undetermined coefficients are

$$a = \frac{1}{4} \qquad b = 2a = \frac{1}{2}$$

$$c = -2a + \frac{3b}{2} = \frac{-1}{2} + \frac{3}{4} = \frac{1}{4}$$

$$d = a - b + c = \frac{1}{4} - \frac{1}{2} + \frac{1}{4} = 0$$

Thus the polynomial we started with, and wrote with undetermined coefficients, now has its coefficients determined. The formula for the powers of the cubes of the consecutive integers is now known to be

$$\frac{n^4 + 2n^3 + n^2}{4} = \left[\frac{n(n + 1)}{2}\right]^2 \tag{2.5-7}$$

Without directly demonstrating it, we *know* that the mathematical induction process will apply to this equation, and therefore it is a correct expression.

We have shown that the sum of the cubes of the first n positive integers is exactly the square of the sum of the first powers of the first n positive integers. This is a remarkable feature that apparently is just a chance relationship, as there seems to be no comparable feature among the other powers of the integers. But it is true that $S_5 + S_7 = 2(S_3)^2$.

Let us review this method of undetermined coefficients.

1. We assumed the general polynomial of degree $k + 1$ for the sum of the kth powers of the integers (2.5-1).
2. We set up a basis for the induction (this forced $e = 0$).
3. Then we applied the induction step to the assumed general form and rearranged the terms as powers of m (2.5-5). This polynomial was identically zero for all values m. Using an important result, we concluded that we could equate the coefficients to zero separately.
4. When we carry out the steps, we are assured that the formula will obey the induction step. We are sure that the formula is correct provided we accept the pattern of mathematical induction as being valid reasoning.

Generalization 2.5-2

Sum of the kth powers of the integers. For practice in handling general expressions, let us carry out the first few steps for the *general* case of the kth powers. The long spread-out equation (2.5-4) we first found will now look like (we see the binomial coefficients in the appropriate columns)

$$m^{k+1}\{a\}$$

$$+ m^k\{-[k + 1]a + b + 1\}$$

$$+ m^{k-1}\left\{\left[\frac{(k + 1)k}{2}\right]a - kb + c\right\}$$

$$+ m^{k-2}\left\{-\left[\frac{(k + 1)k(k - 1)}{6}\right]a + \left[\frac{k(k - 1)}{2}\right]b - (k - 1)c + d\right\}$$

$$\cdots \qquad\qquad\qquad\qquad [2.5\text{-}8]$$
$$= am^{k+1} + bm^k + cm^{k-1} + dm^{k-2} + \cdots$$

Next, cancel each term on the right with the corresponding term on the left, and then equate each coefficient to zero separately. As before, from the basis for the induction with $n = 0$ we get that the constant term is zero. The individual equations are in a *triangular form*, which is easy to solve one equation at a time; compare with (2.5-5) and (2.5-6).

$$-(k + 1)a + 1 = 0$$

$$\left[\frac{(k+1)k}{2}\right]a - kb = 0 \qquad (2.5\text{-}9)$$

$$-\left[\frac{(k+1)k(k-1)}{6}\right]a + \left[\frac{k(k-1)}{2}\right]b - (k-1)c = 0$$

$$\cdots$$

The first equation of (2.5-9) gives (for all $k \ge 0$)

$$a = \frac{1}{k+1}$$

The second equation gives (for all $k \ge 1$)

$$b = \frac{1}{2}$$

After some algebra, the third equation gives (for all $k \ge 2$)

$$c = \frac{k}{12}$$

Still further algebra gives $d = 0$ ($k \ge 3$). Evidently, we could go as far as we pleased (and our patience held out). As a result we have the general formula

$$1^k + 2^k + 3^k + \cdots + n^k = \frac{1}{k+1}n^{k+1} + \frac{1}{2}n^k + \frac{k}{12}n^{k-1} + \cdots$$

for the summation of the kth powers of the consecutive integers.

We can easily check this result against our earlier special cases of $k = 0, 1, 2,$ and 3. We have the following table:

k	$a = \dfrac{1}{k+1}$	$b = \dfrac{1}{2}$	$c = \dfrac{k}{12}$	$d = 0$	Polynomial
1	$\dfrac{1}{2}$	$\dfrac{1}{2}$			$\dfrac{n(n+1)}{2}$
2	$\dfrac{1}{3}$	$\dfrac{1}{2}$	$\dfrac{1}{6}$		$\dfrac{n(n+1)(2n+1)}{6}$
3	$\dfrac{1}{4}$	$\dfrac{1}{2}$	$\dfrac{1}{4}$	0	$\left[\dfrac{n(n+1)}{2}\right]^2$

One purpose for finding part of the general solution was to give you practice in the art of working with abstract symbols. The approach we used to find the sum of the powers of the consecutive integers is the same as that used by Jacques (James) Bernoulli (1654–1705). The results of the general case checked the several special cases. If there had been a disagreement, then either the special case would have been wrong or else there would have been an error in the general case. Thus, even if you are only interested in some special case, working out the details of the general case gives increased confidence in the particular result. You are then more willing to risk your reputation on the result.

We see that the method of undetermined coefficients depends on guessing the proper form. If you miss guessing the right form, this error is revealed to you by the subsequent work; you will not be able to complete all the algebra properly. But often the form of the failure will suggest a new guess at the general form of the solution.

This method of *undetermined coefficients* is a powerful method for finding answers to problems when all you have is a general idea of the form of the answer. If you guess right, when you finish you will have a proof by induction. We will use this method of undetermined coefficients many times in various ways to demonstrate its versatility. We applied the method of undetermined coefficients to mathematical induction, and therefore we applied the conditions that the steps of mathematical induction be true; for other applications of the method we would apply other, appropriate conditions. The basic idea is that we assume the form we think we want and then apply the conditions that define the object we are searching for. When the method works, the undetermined coefficients are determined by the applied conditions and we have our answer.

EXERCISES 2.5

1. Compute the sum of the first 100 integers cubed.
2. Sum $101 + 102 + 103 + \cdots + 200$.
3. Sum $101^2 + 102^2 + 103^2 + \cdots + 200^2$.
4.* In the induction we did for finding the sum of the third powers of the integers in this section, we went from $m - 1$ to m. Discuss and carry out the induction when you go from m to $m + 1$. In this case, which one do you prefer?
5. Find one more term in the fourth power summation of the integers, and then find the real factors of the resulting polynomial for the sum of the fourth powers.
6.* Repeat Exercise 5 but for the generalization to the kth powers.
7. Show that $S_k(2n) - 2^k S_k(n)$ equals the sum of the kth powers of the odd integers

$$1^k + 3^k + 5^k + \cdots + (2n - 1)^k$$

8. Adapt the idea in Exercise 7 to find the first three terms of the sum of the kth powers of the integers with alternating signs, $1^k - 2^k + 3^k - \cdots - (2n - 1)^k$.
9.* Prove that $S_5 + S_7 = 2(S_3)^2$.

2.6 THE ELLIPSIS METHOD

The ellipsis method is another way of handling the problem of proving an infinite number of cases. In this method we write the equation in the general form with the conventional three dots, "$\cdots$."

Example 2.6-1

Sum of a geometric progression. Consider the sum of a *geometric progression* where the first term is a, and each term after the first is found by multiplying the preceding term by the fixed *rate* r. The first n individual terms are, in sequence,

$$a, ar, ar^2, ar^3, \ldots, ar^{n-1}$$

The sum of the first n terms is

$$S_n = a + ar + ar^2 + ar^3 + \cdots + ar^{n-1}$$

A bit of study suggests that multiplying this equation by r will produce another equation with many of the same terms:

$$rS_n = \quad ar + ar^2 + ar^3 + ar^4 + \cdots + ar^n$$

Subtracting the second of these equations from the first produces a great deal of cancellation of terms:

$$(1 - r)S_n = a - ar^n = a(1 - r^n)$$

Provided $r \neq 1$, we can divide both sides by $1 - r$ to solve for S_n:

$$S_n = \frac{a(1 - r^n)}{1 - r} \tag{2.6-1}$$

If $r = 1$, then this formula has a division by 0 and is invalid. In this case we note that the sum of the terms of the geometric progression is

$$S_n = a + a + a + \cdots + a = na$$

When r is greater than 1 in size, the conventional form for writing the sum is (it is obtained by multiplying both numerator and denominator by -1)

$$S_n = \frac{a(r^n - 1)}{r - 1} \tag{2.6-2}$$

We have a formula, (2.6-1) or (2.6-2), for the sum of a geometric progression for all n; by using the ellipsis notation all the equations have been solved in parallel, all at the same time.

An especially useful case of this result occurs when $a = 1$ and r is replaced by x.

$$1 + x + x^2 + \cdots + x^{n-1} = \frac{1 - x^n}{1 - x} = \frac{x^n - 1}{x - 1}$$

This shows for $n = $ positive integer that $1 - x^n$ is always divisible by $1 - x$. Now that we know the formula we could, of course, give an inductive proof; but this is not necessary since our derivation proved the truth of the formula.

Example 2.6-2

Sum of an arithmetic progression. Another commonly occurring progression is the *arithmetic progression* where each term is found from the preceding term by adding a constant amount labeled d (which might be a negative number). The first n terms are, in sequence,

$$a, a + d, a + 2d, \ldots, a + (n - 1)d$$

The sum of the terms is

$$T_n = a + (a + d) + (a + 2d) + (a + 3d) + \cdots + [a + (n - 1)d]$$

We can write this sum as (arrange the terms by the parameters a and d)

$$T_n = na + d[1 + 2 + 3 + \cdots + (n - 1)]$$

Now using our earlier sum for the consecutive integers, we get for the sum of an arithmetic progression

$$T_n = na + \frac{d(n - 1)n}{2} = \frac{n}{2}[2a + (n - 1)d]$$

$$= \frac{n[a + \{a + (n - 1)d\}]}{2} = \frac{n(a + l)}{2} \tag{2.6-3}$$

where l is the last term. This formula shows that the sum T_n of an arithmetic progression is the average of the first and last terms multiplied by the number of terms n.

An alternative proof of the formula for the sum of an arithmetic progression [and the proof that the young Gauss (1777–1855) probably used] is to write out the progression twice, once forward and once in reverse order (where again l is the last term):

$$T_n = a + (a + d) + (a + 2d) + \cdots + [a + (n - 1)d]$$

$$T_n = l + (l - d) + (l - 2d) + \cdots + [l - (n - 1)d]$$

Then add the two equations:

$$2T_n = (a + l) + (a + l) + \cdots + (a + l)$$

There are exactly n of the terms $a + l$, so the sum must be

$$T_n = \frac{n(a + l)}{2}$$

The first proof of the general arithmetic progression was based on the very simple special case of $a = 0$, $d = 1$. This is a useful method of proof—examine the simplest case you think will include the main difficulty, and with luck its proof will be easy as well as serve as a base for proving the general case. (See Hilbert's statement quoted in Section 1.5.) The second method of proof was based on a sudden insight into a trick that would work in this case; in a sense, once this derivation is seen, it shows "why" the answer is what it is.

Example 2.6-3

Sum of cosines. We can also use the ellipsis method on the trigonometric identity of Section 2.2.

$$s_n = \frac{1}{2} + \cos x + \cos 2x + \cos 3x + \cdots + \cos nx$$

We need to know a trick (which must have been hard to discover the first time, but once seen is easy to use again in similar situations). It is based on the trigonometric identity

$$\sin a \cos b = \frac{1}{2}[\sin (a + b) + \sin (a - b)]$$

which can be checked by expanding the basic addition formulas (16.1-8) and (16.1-9) on the right-hand side of the equation. Multiply the given trigonometric sum by $\sin (x/2)$, and apply the above identity to each product:

$$\left(\sin \frac{x}{2}\right)s_n = \frac{1}{2}\left[\sin \frac{x}{2}\right.$$
$$+ \sin \frac{3x}{2} - \sin \frac{x}{2}$$
$$+ \sin \frac{5x}{2} - \sin \frac{3x}{2}$$
$$\cdots$$
$$\left. + \sin \left(n + \frac{1}{2}\right)x - \sin \left(n - \frac{1}{2}\right)x\right]$$

The right-hand side collapses due to cancellations (as did the geometric progression), and this time we have only one term:

$$\frac{1}{2} \sin \left\{\left(n + \frac{1}{2}\right)x\right\}$$

Therefore, the original sum is (solve for s_n)

$$s_n = \frac{\sin (n + 1/2)x}{2 \sin (x/2)}$$

Again using the ellipsis method, we have found the solution for all the equations in parallel; we know the sum for any n and all x for which $\sin (x/2) \neq 0$. (The symbol $\neq$ means "not equal to.")

There are many other methods for coping with an infinite number of cases but we will not take them up here. Some other methods will appear as we go along in the text.

EXERCISES 2.6

1. Find the sum of $1 + 2 + 4 + 8 + \cdots + 2^n$.
2. Sum $3 + 6 + 12 + \cdots$ for n terms.
3. Sum $1 + \frac{1}{2} + \frac{1}{4} + \frac{1}{8} + \cdots + \frac{1}{2^n}$.
4. Find the sum of $1 + 4 + 7 + \cdots + 301$.
5. Find the sum of $100 + 96 + 92 + \cdots + 0$.
6. It is said that the inventor of the game of chess, when asked to name his reward, asked for one grain of wheat for the first square, 2 for the second, 4 for the third, and so on. What was the sum he asked for? Using the approximation that 2^{10} is approximately 10^3, estimate the sum.
7. When $r = 2$, compare the last term with the sum of the geometric progression. (The last term is the amount that happens in the last doubling period of the geometric progression.)
8. If the rate r is greater than 1 (and the starting value is positive), compute the value of n for which the next term exceeds a given number, say C, no matter how large C may be.
9. It is said that currently knowledge is doubling every 17 years. If this rate continued what would be the increase in 340 years?
10. In Example 2.6-3 analyze the case when $\sin (x/2) = 0$.

2.7 REVIEW AND FALLACIES IN ALGEBRA

Algebra arose from arithmetic, and the letters used in algebra were originally assumed to be numbers. Hence the rules that evolved for algebra were those used in arithmetic. It was comparatively late that these rules were formalized into postulates that summarize the practice in arithmetic.

Let the letters a, b, and c be any quantities in the algebra, then we have the following:

1. *Associative rules* (meaning you can associate the quantities in any way):
$$a + (b + c) = (a + b) + c, \qquad a(bc) = (ab)c$$

2. *Commutative rules* (meaning you can interchange the order of the letters):
$$a + b = b + a \qquad ab = ba$$

3. There exist unique identity elements 0 and 1 together with inverses to each letter (meaning subtraction, and division by other than 0, can always be done):
$$a + (-a) = 0, \qquad a\left(\frac{1}{a}\right) = 1 \qquad (a \neq 0)$$

with the corresponding properties
$$a + 0 = a, \qquad a(0) = 0, \qquad a(1) = a$$

4. Distributive rule (connects addition and multiplication):

$$a(b + c) = ab + ac$$

From these simple rules all the rules of algebra follow, provided you use the standard forms of logic, including equals plus equals are equal. For example, the rule of transposition is easily deduced.

Example 2.7-1

Let

$$a + b = 0$$

Then use (the identity)

$$-b = -b$$

Add both equations together and apply the associative rule:

$$a + (b - b) = 0 - b$$

or
$$a = -b$$

and we have the rule for transposing a term to the other side of an equation.

Example 2.7-2

We can similarly derive the rule that a negative times a negative is a positive. We begin with

$$a(0) = a(b - b) = a(b) + a(-b) = 0$$

Hence

$$a(-b) = -ab$$

We next begin with

$$-a(0) = -a(b - b) = (-a)b + (-a)(-b) = 0$$

from which it follows that

$$(-a)(-b) = -(-ab) = ab$$

Our desire for general rules without exceptions forces these results.

We repeat, these rules were abstracted from the rules of arithmetic that developed over centuries. They represent what you have learned over many years, but being in a compact form they summarize a lot of individual rules you learned when you studied arithmetic. It is because the symbols in algebra were at first seen as merely generalized numbers that the rules for algebra are the same as those for arithmetic.

We have reviewed only the parts of algebra that have reasoning patterns which are needed later, and have neglected many other parts with which we assume you are familiar. However, we need to warn the reader against a number of standard errors that arise when doing algebra without thinking.

The simplest fallacy is supposing that

$$x^{1/2} = \sqrt{x}$$

for all x. There is a sign missing. We see this error in the equation

$$\sqrt{x^2} = x$$

which is not true when x is negative. What is true is that

$$\sqrt{x^2} = \pm x$$

Example 2.7-3

A better disguise of picking the wrong sign of the radical is the following "proof." Surely

$$-15 = -15$$

Rewrite this as

$$25 - 40 = 9 - 24$$
$$5^2 - 8(5) = 3^2 - 8(3)$$
$$5^2 - 8(5) + 16 = 3^2 - 8(3) + 16$$
$$(5 - 4)^2 = (3 - 4)^2$$
$$(5 - 4) = (3 - 4)$$
$$5 = 3$$

which is clearly false!

Another class of fallacies arises from the division by zero. There are many such examples, and perhaps the simplest is the following.

Example 2.7-4

Let

$$x = 1$$

Then

$$x - x^2 = 1 - x^2$$
$$x(1 - x) = (1 - x)(1 + x)$$
$$x = x + 1$$
$$0 = 1$$

Example 2.7-5

For a more subtle example, consider the equation

$$\frac{1}{x(x - a)} - \frac{1}{x} = \frac{a}{(x - a)}$$

Cleared of fractions,

$$1 - (x - a) = ax$$
$$1 - x + a = ax$$
$$1 + a = (1 + a)x$$
$$x = 1$$

But when $a = 1$ the original equation is not defined for $x = 1$; hence $x = 1$ cannot be a solution. Also, what happens when $a = -1$? When $a = -1$, the equation is an identity and any x other than 0 and -1 is a solution!

Clearly, we need to be careful, when we do a long problem, to see that we have not included one or more of these fallacies. Whenever the result is going to be used, it should receive a careful inspection that searches for these kinds of blunders.

2.8 SUMMARY

This chapter looked at the integers and examined how to prove a result for an infinite number of values. The classic tool for doing this is mathematical induction, which has a number of simple variations. The ellipsis method is widely used as an alternative method of proof. Occasionally, a simple indirect proof by contradiction can be used (for example the proof in Section 2.1 that there is no largest prime number).

Thus we came to grips almost immediately with a fundamental problem of mathematics, coping with the infinite. One important tool introduced is the device of undetermined coefficients: if you know the form, but not the details, then you can impose the conditions you want to apply to the form and determine the details. In one particular case of using the form of a mathematical induction proof we imposed both the initial conditions and the induction step, and these conditions determined the particular coefficients. Along the way we did a number of examples (including geometric and arithmetic progressions) whose results are useful both in using mathematics generally and in later parts of the book. The examples are useful, but the methods are more important.

The central idea behind all of them is *recursion*; just as the integers are defined recursively, so too must the proof somehow be recursive in order to exhaust all the cases.

We have not tried to develop the material in an orderly fashion; rather we have used various properties of numbers that you have learned earlier. We concentrated on those that involve the concept of "infinity" and some of its consequences, so that as we go forward things will not be "slipped by" you without your realizing what is implied.

We also included a number of fallacies in algebra to warn you against idiotic results that may arise when you are careless.

Fractions—Rational Numbers

3.1 RATIONAL NUMBERS

In school, after studying the integers, you turned to the fractions, which were numbers of the form

$$\frac{p}{q}$$

where p and q are integers, and $q \neq 0$.

Unfortunately, the word "fraction" has been generalized to include other numerators and denominators than integers. The reason for this is that the rules for handling fractions turn out to include the more general numbers like

$$\frac{\sqrt{5}}{\sqrt{7}}$$

which are often called *surds*. The name *rational numbers* (ratio numbers) was introduced to describe the case of one integer divided by another.

A number of conventions are used when handling rational numbers. A *proper rational number* (*fraction*) is one whose numerator is less in size than is the denominator. For example,

$$\frac{5}{7}$$

is a *proper* rational number, while

$$\frac{7}{5}$$

is an improper rational number. A rational number whose denominator is 1, or whose numerator is 0, is equivalent to an integer. Similarly, rational numbers like

$$\frac{6}{3}$$

are equivalent to integers. The integers are, then, special cases of rational numbers. Expressions like

$$\frac{5}{1}, \frac{0}{7}, \quad \text{and} \quad \frac{6}{3}$$

are integers as well as rational numbers.

There is a natural association of numbers with points on a line (Figure 3.1-1). We pick an endless straight line, mark an arbitrary point as 0, and then place the positive integers 1, 2, 3, . . . at equal spaces toward the right. The negative integers are placed correspondingly to the left. The fractions then find their appropriate places; for example $\frac{3}{2}$ is midway between 1 and 2. The line is unending in either direction.

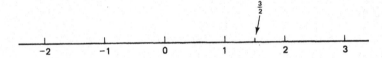

Figure 3.1-1 Numbers arranged on a line

A basic trouble with rational numbers soon arises. For example,

$$\frac{6}{8} = \frac{3}{4}$$

shows two different representations of the same number. In general

$$\frac{kp}{kq} = \frac{p}{q}$$

for any integer $k \neq 0$. Thus the same rational number (the same point on the line) has infinitely many different representations. Note that k may be either positive or negative. It is natural to regard the representation with the smallest positive denominator as the *lowest, canonical form* (canonical means reduced to the simplest form).

We must, therefore, investigate how to reduce a given rational number to its lowest form. The answer is given by applying *Euclid's algorithm* to the numerator and denominator to find the greatest common divisor k (see the next section). Once we have the k, we can divide it out of both the numerator and denominator and obtain the lowest form. The word *algorithm* means an explicitly described process that is certain to end in a finite number of steps.

3.2 EUCLID'S ALGORITHM

Euclid's algorithm finds the greatest common divisor (the largest factor common to both) of two integers, say p and q. This is often abbreviated as GCD (greatest common divisor). If we suppose that p is larger than q (and both are positive for convenience), the plan is to divide p by q to get a quotient and a remainder. We will then divide the q by this remainder, getting another quotient and remainder, and repeat this process an indefinite number of times.

We have a problem of notation. Let us relabel the first number p as r_0 and the second number q as r_1, with r_0 larger than r_1. We divide r_1 into r_0 to get the quotient q_1 and a remainder r_2; that is,

$$\frac{r_0}{r_1} = q_1 + \frac{r_2}{r_1}$$

or, as mathematicians prefer to write the equation,

$$r_0 = q_1 r_1 + r_2$$

where, of course, the remainder r_2 is less than the divisor r_1. Note that any number that divides both r_0 and r_1 must also divide r_2 (since a fraction cannot be equal to an integer, and q_1 is an integer). Since we have reduced the size of one of the numbers, we next divide r_1 by r_2 to get a new quotient q_2 and a new remainder r_3.

$$r_1 = q_2 r_2 + r_3$$

Again, any number dividing r_1 and r_2 must also divide r_3. We can repeat this process until we get a remainder of 0. This *must* happen, because at each step we are reducing the size of the (integer) remainder and therefore we must ultimately reach 0.

We write out all the equations in a single block so we can inspect them. Let r_{n+1} be the zero remainder. We have

$$r_0 = q_1 r_1 + r_2$$
$$r_1 = q_2 r_2 + r_3$$
$$\cdots \tag{3.2-1}$$
$$r_{n-2} = q_{n-1} r_{n-1} + r_n$$
$$r_{n-1} = q_n r_n$$

Any number that divides both r_0 and r_1 must finally divide r_n. But r_n might have other factors, and hence we must investigate this.

We see from the bottom line that r_n divides r_{n-1}. Looking at the next equation up the array, we see that since r_n divides both itself and r_{n-1} it must also divide r_{n-2}. And so on up to the statement that r_n divides r_1 and therefore from the top line must divide r_0. Thus r_n divides both r_0 and r_1, and r_n is the greatest common divisor (GCD) of both of the original numbers r_0 and r_1, which is what the algorithm is designed to produce.

Example 3.2-1

It is often wise to try out a general formula on a few special cases. Take the integers 24 and 18. The set of equations (3.2-1) becomes

$$24 = 1(18) + 6$$

$$18 = 3(6) + 0$$

and we see that 6 is the greatest common divisor of 24 and 18. Next, suppose the starting numbers are *relatively prime* (have no common factor), for example 32 and 27. We have, corresponding to (3.2-1),

$$32 = 1(27) + 5$$

$$27 = 5(5) + 2$$

$$5 = 2(2) + 1$$

$$2 = 2(1) + 0$$

The last line is trivial and is not worth the effort. Once a remainder of 1 occurs you know there is no common factor except the trivial factor 1.

This algorithm is included for a number of reasons: to solve the problem of reducing a rational number to its lowest form; to illustrate the concept of an algorithm, which is a central idea in mathematics and computing; and for several other purposes, which will emerge later (Abstraction 4.6-1; multiple zeros, Section 9.8; and partial fractions, Sections 17.2 and 17.3). An algorithm is a definite process that is known to terminate in a finite number of steps. Especially in the field of computer programs, the algorithm must be exactly described for all the possibilities that can arise and must be shown to terminate in a finite number of steps.

EXERCISES 3.2

1. Find the GCD of 12 and 28.
2. Find the GCD of 22 and 39.
3. Find the GCD of 287 and 343.
4. Find the GCD of 256 and 243.
5. Was it necessary to assume that the first number r_0 was larger than the second, r_1? What would happen if it were not?
6. How would *you* handle negative numbers in Euclid's algorithm?

3.3 THE RATIONAL NUMBER SYSTEM

When we consider the rational numbers, we find that there are a couple of surprises in store. First, unlike the integers, there is no *next number* along the line on which they are conventionally placed, because between any two distinct rational numbers we can place as many more numbers, equally spaced or otherwise, as we please. We have only to take the difference between the two given numbers, break it into N equally spaced intervals each of length $1/N$ times the difference, and then repeatedly add this length onto the starting number. These $N - 1$ numbers will fall between the given two rational numbers. Generally, these *interpolated* numbers will be fractions, but some of them could be integers.

Example 3.3-1

Interpolation of n numbers between two given numbers. Suppose $\frac{1}{3}$ and $\frac{1}{2}$ are the given numbers, and we want to insert 10 numbers between them. We therefore need 11 intervals, $N = 11$. The arithmetic is

$$\text{difference} = \frac{1}{2} - \frac{1}{3} = \frac{3}{6} - \frac{2}{6} = \frac{1}{6}$$

$$\text{step size} = \frac{1/6}{11} = \frac{1}{66}$$

The 10 desired numbers are, therefore,

$$\frac{1}{3} + \frac{1}{66}, \frac{1}{3} + 2\left(\frac{1}{66}\right), \frac{1}{3} + 3\left(\frac{1}{66}\right), \ldots, \frac{1}{3} + 10\left(\frac{1}{66}\right)$$

The next number,

$$\frac{1}{3} + \frac{11}{66} = \frac{1}{3} + \frac{1}{6} = \frac{1}{2}$$

is the other end number. This is a convenient check on the arithmetic.

We see that the rational numbers are *everywhere dense* along the line, meaning just what we proved: no matter how finely spaced you think the numbers are, there are as many more numbers in any given interval as you please—and even more than that! Furthermore, in some sense, the numbers are *uniformly dense* on the line; no interval of the line has more rational numbers than any other interval of the same length or any other length. Figure 3.3-1 shows that in some sense each point on the short segment corresponds uniquely to a point on the long line segment.

The rational numbers have the further property that any additions, subtractions,

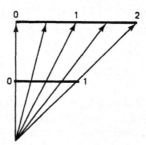

Figure 3.3-1 The interval [0, 1] mapped onto [0, 2]

multiplications, or divisions (other than by 0) can be done and the answer is another rational number. This property of the rational numbers is called *closed with respect to the operations mentioned*. As far as these four operations are concerned there are no gaps in the rational number system; no need for further numbers can arise within the system.

All decimals that have only a finite number of nonzero digits (with all 0's past some point if you wish) are clearly rational numbers whose denominators are suitable powers of 10. For example,

$$3.1416 = \frac{31,416}{10,000}$$

$$0.125 = \frac{125}{1000} = \frac{1}{8}$$

Similarly, the integers are closed with respect to the three operations of addition, subtraction, and multiplication. It is the operation of division that forces us to consider the rational numbers.

EXERCISES 3.3

1. What would happen in Example (3.3-1) if the ends of the interval were reversed?
2. Insert 7 numbers between $\frac{1}{4}$ and $\frac{1}{3}$.
3. Describe all numbers having a terminating decimal number representation.
4. Describe all numbers having a terminating binary (base 2 instead of 10) representation.
5. Write the formulas for interpolating N numbers between a and b.

3.4 IRRATIONAL NUMBERS

We return to Pythagoras again. It is often claimed that he (or one of the Pythagorean sect) first discovered that the square root of 2 is not a fraction. The proof probably went somewhat as follows and arose from the problem of the length of the diagonal of a square of unit size (Figure 3.4-1).

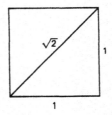

Figure 3.4-1 The diagonal of a unit square

Example 3.4-1

The $\sqrt{2}$ is an irrational number. We want to prove that $\sqrt{2}$ is not a fraction p/q. Most impossibility problems begin by assuming that the opposite is true and then produce a contradiction, thus showing that the assumption is false and hence that the representation is impossible. Therefore, we *suppose*, contrary to what we wish to prove, that the $\sqrt{2}$ is a rational number; that is,

$$\sqrt{2} = \frac{p}{q} \tag{3.4-1}$$

and the fraction p/q is in its lowest (canonical) form (p and q have no common factors). To get rid of the new operation (the square root), we square both sides of (3.4-1) (what else could we do?). Thus we get

$$2 = \frac{p^2}{q^2}$$

Cleared of fractions,

$$2q^2 = p^2 \tag{3.4-2}$$

The left side is clearly an even integer.

Every odd integer is of the form $2k + 1$ (for some integer k), and the square of the odd integer $2k + 1$ is

$$(2k + 1)^2 = 4k^2 + 4k + 1$$

which is again an odd integer. It follows that p could not be an odd number; it must be an even integer. Therefore, let

$$p = 2m$$

and put this into Equation (3.4-2):

$$2q^2 = 4m^2$$

Dividing out the 2, we have

$$q^2 = 2m^2$$

A similar argument shows that q is also even. We started with the fraction in lowest form and have found a contradiction, since 2 divides both p and q. Mathematics generally insists on no contradictions, so we look back to find the *suppose* and decide that what immediately follows must be false.

Since we have just shown that $\sqrt{2}$ is not a ratio number, it is therefore called an *irrational number* (not a ratio number).

Generalization 3.4-2

Irrational numbers in general. A far more direct, and more general, proof of the irrationality of various roots is based on the simple observation that when you take a fraction (that is not an integer) and square it you will not get an integer, because the denominator cannot completely cancel out. Therefore, you can see that for all square roots, indeed for all cube roots, fourth roots, and so on, if the root is not an integer, then it cannot be a fraction.

The irrationality of $\sqrt{2}$ certainly came as a surprise to the ancient Greeks. The rational numbers are everywhere dense along the line, yet $\sqrt{2}$ is not one of them! In *The Elements*, Euclid effectively decided that $\sqrt{2}$ was a *magnitude* but not a number. Eudoxus had already (around 360 B.C.) created an elaborate theory to handle magnitudes. We, on the other hand, have decided that there should be a *number* that measures the diagonal of a unit square, that is, $\sqrt{2}$. See Figure 3.4-1. Similarly, for other such lengths that are not represented by rational numbers, we feel that there should be a number that measures the length. To get a deeper look at this apparent paradox that the everywhere dense rational numbers do not exhaust the points on the line, we look at the decimal representations of irrational and rational numbers in the next two sections.

EXERCISES 3.4

1. Define an irrational number.
2. Prove that $\sqrt{3} \neq p/q$ using the method of Example 3.4-1.
3. Same for $\sqrt[3]{2}$.
4. Prove that $\sqrt{5}$ is irrational.
5. Why does the proof fail for $\sqrt{4}$? At what step does this occur?
6. Prove that, if the polynomial with integer coefficients

$$x^n + a_{n-1}x^{n-1} + \cdots + a_1 x + a_0 = 0$$

   has a root that is an integer, then the root divides a_0.
7.* Prove that, if the polynomial with integer coefficients

$$a_n x^n + a_{n-1}x^{n-1} + \cdots + a_0 = 0$$

   has a rational root p/q, then p divides a_0 and q divides a_n.

3.5 ON FINDING IRRATIONAL NUMBERS

We now investigate these irrational numbers, in particular, how to find their decimal representations.

Example 3.5-1

An algorithm for finding some irrational numbers. How might we find the decimal representation of $\sqrt{2}$? The only property of the number we know is that its square is the number 2. Let us use this. We make a first guess that the square root is 1. We observe that the square of 1 is 1 and that this is too small. This suggests that we try the number 2 as the next guess, and we find that 2 squared is 4, which is too large. We suspect that a better guess would be around 1.5. The square of 1.5 is 2.25, which is too large by a small amount. So we try 1.4 whose square is 1.96, a bit too small.

We see that there exists an organized method of search for better and better approximations to $\sqrt{2}$. At each stage of the process we try a reasonable guess between a number known to be too large and one known to be too small. By this method we can come as close as we please to the square root of any positive number.

We continue the search method to be sure you understand it. Since 1.96 is much closer to 2 than is 2.25, we first try 1.41.

Guess	Square	Comment
1.41	1.9881	Too small
1.42	2.0164	Too large
1.415	2.002225	Too large
1.414	1.999396	Too small
. . .	. . .	. . .

We see that we have a recursive *process* that can improve any approximation we have at hand; we merely *interpolate* a good guess between the current known upper and lower bounds and thus get an improved bound on one side or the other. The process will never end for the square root of 2, since we know that the number we are approximating is an irrational number, and every terminating decimal is clearly a rational number (since it is an integer divided by some power of 10). We are not, at present, interested in an efficient search method, although we will be later (Section 9.7). Our interest is in the *concept* of *recursively* finding better and better approximations to a number when we are given a property that suitably defines the number. This is an algorithm, *provided* we agree to stop at some preassigned level of accuracy.

Example 3.5-2

Suppose we want to find the numerical value of the positive zero of the equation

$$x^2 - x - 1 = 0$$

Consider the left-hand side as a function of x; that is, write

$$f(x) = x^2 - x - 1$$

For $x = 1$ we get the value -1; for $x = 2$ we get the value $+1$; hence the value of x we want (which makes the left-hand side zero) lies between 1 and 2. Since the two polynomial values are equal in size and opposite in sign we naturally guess that the true value lies near 1.5. We get from the left-hand side of the equation the corresponding value -0.25, so we need a slightly larger value. Let us try 1.6. By now we need to tabulate things:

x value	Polynomial value
1.6	−0.04
1.7	0.19
1.62	0.0044
1.61	−0.0179
1.618	−0.000076
1.6181	0.00014761

And so on. This process can find as many decimal places as you desire, provided you are willing to do all the arithmetic. The actual number is the golden ratio of Section 2.4:

$$\frac{1 + \sqrt{5}}{2} = 1.618033989 \ldots$$

as can be seen from the formula for the roots of a quadratic (if you remember it, and if you do not remember the formula, then see Example 6.3-2).

Generalization 3.5-3

Finding zeros of simple (smooth) equations. This method of finding the decimal expansion of a number from its defining equation by using successive approximations, in the above cases $x^2 - 2 = 0$ and $x^2 - x - 1 = 0$, can be applied to many (but not all) numbers. We simply use the defining equation, and by noting the sign of the value of the corresponding function we generate a new estimate of the number. By sufficient work we can reduce the size of the interval in which a sign change occurs to be as small as we please. Thus we have a bound on the accuracy of the result, one number is smaller and one number is larger than the sought for root of the equation, and they are as close together as you wish (and are willing to compute).

The central problem in finding the representation of a number is to find an equation of which it is a solution. This is a matter of inspiration in the general case, but often it is fairly easy to do. Once we have its defining equation, we are on the path to its representation (in any number base we wish).

EXERCISES 3.5

Using three-decimal arithmetic, do the following problems.

1. Find $\sqrt{3}$.
2. Find $\sqrt{5}$.
3. Find $\sqrt[3]{2}$.
4. Find $\sqrt[3]{5}$.
5. Find the real zero of $9x^3 + 9x^2 + 3x - 1 = 0$
6. Find the real root of $x^3 + 7x^2 + x - 4 = 0.4$.
7. Find the positive zero of Newton's cubic $x^3 - 2x - 5$. *Ans.:* 2.0945 . . . .
8. Find the two positive roots of $x^3 - 3x + 1 = 0$.
9. When using a computing machine, it is customary to bisect the interval at each stage rather than try to interpolate a next guess. Show that for every ten iterations of this bisection method you gain at least three decimal digits in accuracy.
10.* Describe the bisection method in some detail, including the starting assumptions. Note that you must allow for hitting the zero in the middle of the process. Apply to $\sqrt{2}$ to three decimal places.

3.6 DECIMAL REPRESENTATION OF A RATIONAL NUMBER

To find the decimal form of a rational number, we merely divide the numerator by the denominator.

Example 3.6-1

When the number $\frac{1}{3}$ is divided out, we get

$$\frac{1}{3} = 0.33333 \ldots = 0.\overline{3}$$

where the ellipsis dots mean an unending string of 3's, and the bar over the 3 means repeat endlessly. This decimal representation is periodic with a period of one decimal digit, 3. If we try $\frac{1}{8}$, we get

$$\frac{1}{8} = 0.125 = 0.125000 \ldots = 0.125\overline{0}$$

By filling in the endless string of 0's, we write a comparable form to the endless string of 3's. Again we have a period of one.

Next consider a rational number like $\frac{1}{7}$. Here we need to explicitly divide it out:

$$
\begin{array}{r}
0.142857 \\
7{\overline{)1.000000}} \\
\underline{7} \\
30 \\
\underline{28} \\
20 \\
\underline{14} \\
60 \\
\underline{56} \\
40 \\
\underline{35} \\
50 \\
\underline{49} \\
1
\end{array}
$$

and we see that the remainder 1 will start the same period again. Thus

$$\frac{1}{7} = 0.142857 \quad 142857 \ldots = 0.\overline{142857}$$

where this time the ellipsis means the endless repetition of the block of six digits, 142857.

A little study will convince you that once you have passed beyond all the digits of the numerator and begin the "bringing down of the zeros" then the periodicity is inevitable. Each remainder is less than the divisor, and therefore in not more steps than the size of the divisor the same remainder must occur again. (For the denominator q

there cannot be q consecutive remainders without a duplicate or else a 0. A zero remainder, of course, produces all zeros after this point and has a period of one.) Thus we see that *rational numbers must have periodic decimal expansions,* once we get past a finite number of leading digits, which may be almost anything depending on the numerator. But the numerator being a finite number can affect only a finite number of places, and then the "bringing down of the zeros" begins.

Example 3.6-2

As a very simple example of the effect of nonzero digits in the numerator, consider the fraction

$$\frac{23}{30}$$

Divide this out to get

$$
\begin{array}{r}
0.7666\ldots \\
30\overline{)23.000}\ldots \\
21\ 0 \\
\hline
2\ 00 \\
1\ 80 \\
\hline
\ldots
\end{array}
$$

Hence

$$\frac{23}{30} = 0.7\overline{6}$$

We next ask the *converse:* Is a periodic decimal number always a rational fraction? To answer this question, we need to look closer at the meaning of a decimal number. The number 0.333 . . . is defined to be the expression

$$0.3333\ldots = \frac{3}{10} + \frac{3}{10^2} + \frac{3}{10^3} + \frac{3}{10^4} + \cdots \qquad (3.6\text{-}1)$$

For a finite number of terms, this resembles the geometric progression

$$a + ar + ar^2 + \cdots + ar^{n-1}$$

of Example 2.6-1 with $a = \frac{3}{10}$ and the rate $r = \frac{1}{10}$. If we take only the first n digits of the decimal expansion, then the sum is, according to the formula for a geometric progression (2.6-1),

$$
\begin{aligned}
S_n &= \frac{a(1 - r^n)}{1 - r} \\
&= \frac{3/10}{1 - 1/10}\left[1 - \left(\frac{1}{10}\right)^n\right] \\
&= \frac{1}{3}\left[1 - \left(\frac{1}{10}\right)^n\right] = \frac{1}{3} - \frac{1}{3}\left(\frac{1}{10}\right)^n
\end{aligned}
$$

But the term $(\frac{1}{10})^n$ approaches zero as n gets large. Hence as n gets larger and larger the numerical value of the number represented by the first n digits gets closer and closer to the first term, which is

$$\frac{1}{3}$$

If you believe that there is a difference between the endless string of 3's and the fraction $\frac{1}{3}$, then when you state this (nonzero) size, no matter how small a number you choose, it is clear that I can tell you how far out to go along the endless string of 3's so that you will then get and remain closer (if you go still farther) than you required. Therefore, there is a contradiction with your assumption that you could name a difference. The difference between $\frac{1}{3}$ and the n digit terminated decimal number becomes arbitrarily small as n becomes large. Thus one is inclined to say that there is no difference between the number represented by the infinite (endless) decimal expansion $0.3333\ldots$ and the number $\frac{1}{3}$.

We can also go back to the fundamental "trick" for summing a geometric progression and use it directly to shorten the argument. Let

$$x = 0.333333\ldots$$

then

$$10x = 3.33333\ldots$$

Subtract the upper from the lower expression (it is not exactly clear how you can do this for the infinitely long string of digits but it seems "reasonable" that the difference is 3):

$$10x - x = 3$$

We now solve for x

$$9x = 3$$

$$x = \frac{1}{3}$$

Example 3.6-3

We may apply the same method to any repeating decimal. For example, let

$$x = 4.6373737\ldots = 4.6\overline{37}$$

The period for repetition (37) is two digits, so we write

$$100x = 463.737\ldots$$

and subtract from this the original value

$$(100 - 1)x = 463.7\overline{37} - 4.6\overline{37} = 459.1$$

from which we get

$$x = \frac{459.1}{99} = \frac{4591}{990}$$

Direct division will produce the original string of digits.

Example 3.6-4

If we apply this same reasoning to the decimal expansion

$$x = 0.9999\ldots = 0.\overline{9}$$

and if we use the geometric progression argument, then we have $a = \frac{9}{10}$ and $r = \frac{1}{10}$. The formula for the sum of the finite geometric progression involving only the first n digits is

$$\frac{9/10}{1 - 1/10}\left[1 - \left(\frac{1}{10}\right)^n\right] = \frac{9/10}{9/10}\left[1 - \left(\frac{1}{10}\right)^n\right]$$

$$= 1 - \frac{1}{(10)^n}$$

Again, as n becomes larger and larger, the r^n term rapidly approaches zero. Thus a string of endless 9's appears to be the number 1. After all, there can be no numerical difference between the two represented numbers, $0.9999\ldots$ and 1, no matter how small you think the difference is. Hence we conclude that they are just different representations of the same number.

If we use the short version, we get

$$10x = 9.99\ldots$$

and on subtracting the original expression we get

$$9x = 9$$

$$x = 1$$

as before.

Generalization 3.6-5

If the preceding example seems strange, it merely means that some numbers may have two different appearing decimal representations. A number like

$$2.3459999\ldots$$

where the three dots mean all 9's, may also be written as

$$2.3460000\ldots$$

Calling the two different representations the same number is convenient because there is no difference in their numerical values. If you think that there is, then (for yet one more time) it is easy to show that, no matter how small you think the difference is, the sequence of digits can be pushed until the difference is less than you claimed. Hence your "*if*" must be wrong.

To summarize where we are, we have found that the rational numbers have periodic decimal expansions (and by extension this is true for any other integer base, including base 2). Conversely, periodic decimal expansions (or for any other integer base) correspond to rational numbers. The irrational numbers, therefore, cannot have periodic decimal expansions. Conversely, nonperiodic decimal expansions correspond to irrational numbers.

This is an explanation of how there can be numbers other than the everywhere dense rationals—all those numbers whose decimal expansions do not end with a periodic structure are not rational numbers; they are by definition the irrational numbers. Periodicity in the decimal expansion of a number is a very great restriction; once past the nonperiodic part, *every* digit is determined by the periodicity. For other numbers the digits are not so restricted, so apparently the nonperiodic expansions are, in some sense, much more numerous. But recall Galileo's observation that in a certain sense the integers and their squares are equally numerous (Section 2.1). We are not in a position to prove that the irrational numbers are very much more numerous than the rationals; we are reduced to stating the fact and indicating loosely why this might be so.

Along the way we have seen enough special cases of a geometric progression with the rate r less than 1 in size to realize that it is reasonable to say that the infinite series of terms has the sum

$$S = \frac{a}{1 - r}$$

A more careful proof of this will be given in Section 7.2.

EXERCISES 3.6

1. Find the decimal representation of $\frac{3}{11}$.
2. Find the decimal representation of $\frac{473}{6}$ and the period.
3. Find the fraction equivalent to $0.121212\ldots$.
4. Find the fraction equivalent to $3.145145\ldots$.
5. Find the fraction equivalent to $1.51515\ldots$.
6. Find the fraction equivalent to $0.0011111\ldots$.
7. Show that adding the corresponding decimal fractions agrees with the equation $\frac{1}{3} + \frac{1}{9} = \frac{4}{9}$. How about $\frac{1}{3} + \frac{1}{3} = \frac{2}{3}$?
8. Write out the details for finding the binary representation of a decimal number less than 1.
9. Show that $\frac{1}{10}$ in binary notation has an endless periodic representation, and hence that $\frac{1}{10} + \frac{1}{10} + \cdots + \frac{1}{10}$ (ten times) *when rounded off* to a fixed number of binary places will not total up to 1.
10. An ideal bouncing ball rebounds each time to a height k (less than 1) times the height it fell from on that bounce. If the initial height is h, find the total distance the ball travels.

3.7 INEQUALITIES

When dealing with integers, there is a "next number," but when dealing with the rational numbers, which are everywhere dense, there is not a "next number" (since, for example, the average of two numbers lies between them). The irrational numbers have the same property that there is no "next number." In place of "next" we introduce the *ordering* of the numbers by the use of the inequality. For the moment, think of a and b as rational numbers.

$$a < b$$

means a is *less than* b, that a lies *to the left* of b on the line that we conventionally use to represent the values of numbers. For example,

$$7 < 8$$

and for negative numbers

$$-8 < -7$$

We also write these as

$$8 > 7 \quad \text{and} \quad -7 > -8$$

(8 is *greater than* 7). In both cases the pointed end of the symbol points to the *smaller* number. It is often convenient to say "greater than or equal to" which in mathematical notation takes the form

$$8 \geq 7$$

In general symbols we have both

$$a \leq b \quad \text{and} \quad b \geq a$$

as the same inequality.

We can add inequalities *in the same sense*. For example, if

$$a \geq b \quad \text{and} \quad c \geq d$$

then

$$a + c \geq b + d$$

A slightly more mathematical proof of this can be given. The inequality

$$a \geq b$$

means

$$a - b \geq 0$$

Similarly,

$$c - d \geq 0$$

The sum of two positive numbers is positive, so we have

$$a + c - b - d \geq 0$$

and then we have the result when we transpose the two negative terms.

We *cannot* subtract inequalities, as you can see from making up simple examples. Of course, the *same number* can be subtracted from both sides of an inequality.

We can multiply both sides of an inequality by the same *positive* number and retain the inequality. For example,

$$6 < 7$$

becomes, when multiplied by 3,

$$18 < 21$$

But if we multiply by a *negative* number, then the sense of the inequality is reversed. In the same example, multiplying by -3 produces

$$-18 > -21$$

In mathematical symbols, if $a \geq b$ and $c \geq 0$, then, using the fact that the product of two positive numbers is positive, we have

$$(a - b)c \geq 0$$

$$ac - bc \geq 0$$

from which it follows that

$$ac \geq bc$$

Multiplication by a negative number is easily seen from the above to reverse the inequality.

We can also take the positive square roots of quantities greater than 0 and maintain inequalities. To prove this we need to show that if

$$0 < a < b$$

then

$$0 < \sqrt{a} < \sqrt{b}$$

One method of proof is to assume that the opposite is true, that

$$\sqrt{b} \leq \sqrt{a}$$

Then multiplying each term by the positive number $\sqrt{b}$ will give

$$0 < \sqrt{b}\sqrt{b} \leq \sqrt{b}\sqrt{a} \leq \sqrt{a}\sqrt{a}$$

where we have used the assumption that $\sqrt{b} \leq \sqrt{a}$ in the last step. From this we have

$$0 < b \leq a$$

which is an immediate contradiction.

Example 3.7-1

An important inequality. We will use these ideas to prove the important inequality that if $a > 0$ then

$$(1 + a)^n \geq 1 + na$$

when n is an integer and $n \geq 1$. The proof is by induction. For the case $n = 1$ it is clearly true (indeed the case $n = 0$ would also serve). We therefore assume it is true for the case $m - 1$; that is, we assume

$$(1 + a)^{m-1} \geq 1 + (m - 1)a$$

with $m \geq 2$. To get to the next case we must, for this induction, *multiply* both sides by the positive number $1 + a$. We get

$$(1 + a)^m \geq [1 + (m - 1)a] + [a + (m - 1)a^2]$$

If we drop the nonnegative number $(m - 1)a^2$ from the right-hand side, we can only strengthen the inequality. The result is

$$(1 + a)^m \geq 1 + ma - a + a = 1 + ma$$

as required for the induction process.

We could also have proved the inequality by using the binomial theorem expansion

$$(1 + a)^n = 1 + na + \frac{n(n - 1)}{2}a^2 + \cdots$$

and dropping all the positive terms after the first two terms. Remember that $a > 0$.

This shows that often the same result can be found in different ways. Since we are concerned with *methods*, we shall often find the same result by several different methods. One advantage of alternative proofs is that when we "listen to the mathematics" involved in the proofs we can then learn different things about the same result. Furthermore, when we come to extend, or generalize, a formula, one derivation may be more appropriate than another.

EXERCISES 3.7

Hint: Try specific numbers first before discussing the general case. Be sure to try positive, zero, and negative numbers.

1. Under what conditions can inequalities be multiplied term by term?
2. Under what conditions can inequalities be divided term by term?
3. Show that $1 < \sqrt{2} < \sqrt{3} < 2$.
4. Prove that $2xy \leq x^2 + y^2$.
5.* Schwartz Inequality: In Exercise 4 set $x_i = a_i/\{a_1^2 + a_2^2 + \cdots + a_n^2\}^{1/2}$ and $y_i = b_i/\{b_1^2 + b_2^2 + \cdots + b_n^2\}^{1/2}$. Apply the result of Exercise 4. Then add for all i to get $x_1y_1 + x_2y_2 + \cdots + x_ny_n \leq 1$. Finally, rewrite in terms of a_i and b_i to get

$$a_1b_1 + \cdots + a_nb_n \leq [a_1^2 + \cdots + a_n^2]^{1/2}[b_1^2 + \cdots + b_n^2]^{1/2}$$

3.8 EXPONENTS—AN APPLICATION OF RATIONAL NUMBERS

We have already used some simple properties of exponents with which you are familiar. We are reviewing the topic in more detail to provide a pattern of abstraction and generalization that we will often refer to. Thus it is the methods (as well as the results) that require your attention.

Example 3.8-1

The laws of exponents are a good example of simply listening to what a formula says. Descartes (1596–1650) popularized the compact notation for writing a sequence of copies of the same symbol (notice we are not asserting that a is a number):

$$aaaaa \ldots a = a^n \tag{3.8-1}$$

where there are n symbols a on the left. Just reading what the following symbols mean (m copies of the symbol a followed by n copies), it is clear that

$$a^m a^n = a^{m+n} \tag{3.8-2}$$

Similarly,

$$(a^m)^n = (a^m)(a^m)(a^m) \ldots (a^m) = a^{mn} \tag{3.8-3}$$

We are clearly using the ellipsis method of proof. The more formal method of mathematical induction can be used if you wish, but it hardly makes things more convincing.

Extension 3.8-2

To *extend* the concept of exponents to the integer 0, we simply put $n = 0$ in Equation (3.8-2) to get

$$a^m a^0 = a^m$$

from which clearly (assuming, of course, that $a \neq 0$)

$$a^0 = 1 \tag{3.8-4}$$

(where the symbol "1" is the identity if you think that a is a general symbol, and is the number 1 if you think of a as a number). We get a consistent result from (3.8-3) when $m = 0$. Thus this seems to be a reasonable extension of the meaning of an exponent.

The next *extension* of the idea is to negative exponents. We do this by setting $n = -m$ in Equation (3.8-2):

$$a^m a^{-m} = a^0 = 1 \qquad (a \neq 0)$$

from which it follows that (dividing by a^m)

$$a^{-m} = \frac{1}{a^m} \tag{3.8-5}$$

We next *extend* the idea of exponents to the rational numbers. This was first done by Nicole Oresme (1323–1382). What could

$$a^{1/2}$$

possibly mean? Again using Equation (3.8-2) with $m = n = \frac{1}{2}$, we have

$$a^{1/2}a^{1/2} = a^1 = a$$

Alternately, from Equation (3.8-3),

$$(a^{1/2})^2 = a^1 = a$$

These equations both ask, "What symbol multiplied by itself gives a?" The answer (when adjacency of the symbols means multiplication of numbers) is, of course, the square root of a; that is,

$$a^{1/2} = \text{square root of } a = \pm\sqrt{a} \qquad (3.8\text{-}6)$$

(always supposing that the root exists). It follows from a similar argument (it is supposed that you can now supply the details for this simple extension) that

$$a^{1/q} = q\text{th root of } a \qquad (3.8\text{-}7)$$

Furthermore, since

$$(a^{1/q})^{pq} = (a^{q/q})^p = (a)^p = a^p$$

it follows that

$$a^{p/q} = (q\text{th root of } a)^p$$

It takes only a little thinking of what you mean by the qth root of a product to convince yourself that, for positive numbers,

the qth root of a product is the product of the qth roots

$$(abc \ldots)^{1/q} = a^{1/q}b^{1/q}c^{1/q} \ldots$$

We have only to multiply the right-hand side by itself q times and regroup the letters (assuming that the symbols commute with one another) to see the truth of this statement.

For negative numbers, you must do some careful thinking first before using the laws of exponents freely. Odd-order roots of a negative number are negative, while even-order roots are imaginary (see Section 4.8).

It was necessary to convince ourselves that the equality of the symbols

$$p\frac{1}{q} = \frac{1}{q}p = \frac{p}{q}$$

when viewed as rational numbers is also true when they are used as exponents. The formal writing of the symbol is not enough; you must verify that the extension is valid in all the properties you use.

EXERCISES 3.8

1. Evaluate (a) x^7x^6; (b) x^7/x^6; (c) $(x^7)^6$.
2. Evaluate (a) $x^{3/2}x^{1/2}$; (b) $x^{-3}x^{-4}$; (c) $x^{-(2/3)}x^{-(1/2)}$.
3. Show that $\sqrt{2}\sqrt{3}\sqrt{6} = 6$.

4. Show that $(\sqrt{2})^5 = 4\sqrt{2}$.

5. Show that $(\sqrt{2}/2) = 1/\sqrt{2}$.

6. Show that $a/\sqrt{a} = \sqrt{a}$.

7. Show that $(\sqrt{2}/2)(\sqrt{3}/2)(\sqrt{2}/2) = \sqrt{3}/4$.

8. Show that $(\sqrt{1/2})(\sqrt{2/2})(\sqrt{3/2}) \ldots (\sqrt{n/2}) = \sqrt{(n!)}/2^{n/2}$.

9. Show that $[(a^b)^c]^d = [(a^d)^c]^b$

3.9 SUMMARY AND FURTHER REMARKS

This chapter examined a few of the characteristics of rational numbers, in particular their everywhere dense properties. We showed that rational numbers are equivalent to periodic decimal expansions, and, conversely, any decimal expansions that are ultimately periodic correspond to rational numbers. Next we saw that there are other numbers, the much more numerous irrationals (having nonrepeating expansions), which fall between the everywhere dense rationals, and that correspondingly their decimal expansions are not periodic.

Inequalities were examined briefly and their general properties established. Inequalities are very useful, especially in more advanced mathematics where exact equality is often not possible to establish. Often we must settle for a bound.

We also examined a simple case of "listening to what a formula says" when we repeated the familiar derivations of the laws of exponents. The extension of the concept of an exponent to the rationals was an example of the typical mathematical pattern of extending and generalizing specific results. Along the way we got further practice in using both mathematical induction and the ellipsis method.

When we *extend* a concept, we try to do it so that we need to learn as little new as possible; we *try* to keep consistency. When we *abstract*, we try to reduce the amount of independent material to again aid the memory. The results, when needed, are easily deduced *provided* you understand the processes of extension and abstraction.

4

Real Numbers, Functions,
and Philosophy

We have an instinctive feeling that time flows "continuously," although we may not be clear exactly what we mean by the word "continuous." Similarly, but slightly less strongly, we feel that when we draw a line without lifting the pen or pencil we make a continuous, uniform line (in our imagination, of course, since the real world is apparently made of atoms and molecules and hence has a basic discreteness). Thus the mental line along which we put the numbers is continuous, although the points (numbers) themselves are discrete. Pythagoras found (Section 3.4) that there were gaps in the "everywhere dense" rationals, and we decided that the irrational roots of number should also be included in the number system. When we add to the class of rationals all numbers that are roots of polynomials with integer coefficients, we get the *algebraic numbers*. An algebraic number is thus defined to be a solution of the equation

$$c_0 + c_1 x + c_2 x^2 + \cdots + c_n x^n = 0 \qquad (4.1\text{-}1)$$

where all the c_i are integers (or fractions if you wish). The algebraic numbers include the integers and rationals since they are solutions of linear equations with integer coefficients.

Sums, differences, products, and quotients of algebraic numbers are still algebraic numbers, although we do not prove it here. Do the algebraic numbers fill out the line (the number system)? No! It is not easy to prove, but it can be shown that the number $\pi = 3.14159265358979\ldots$, for example, is not an algebraic number; π is not the solution of *any* polynomial with rational (or even algebraic) coefficients

(Figure 4.1-1). Such numbers are called *transcendental numbers*. There are many, many transcendental numbers *provided* we decide that:

1. Every decimal representation corresponds to a number.
2. Conversely, every number must have a decimal representation. As noted in Section 3.6, some numbers may have more than one representation.

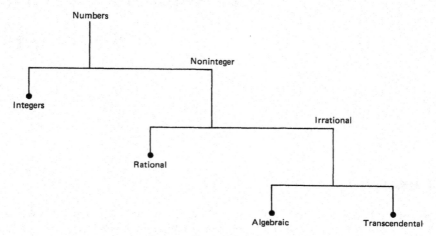

Figure 4.1-1   Types of numbers

The decision that a decimal representation corresponds to a number is in close agreement with what most people feel about what a number is. Never mind any elaborate definitions that may be made; such definitions come ultimately from the common feeling that numbers have decimal (or other number base) representations, and, conversely, decimal representations correspond to numbers. This approach leads to difficulties whose resolution we cannot go into here; that is, how we can be sure that these decimal representations obey the common rules of arithmetic? The average person believes they do, and we will leave it at that.

A polynomial is a *linear combination* of the powers of x with coefficients selected from some class. At the moment this class is the integers; later it may be the class of algebraic numbers or even the class of all real numbers. This is a special case of the general concept of a *linear combination* and is fundamental in mathematics. This concept is worth abstracting since it will occur many times in this book. First, we are given a collection of entities, call them G_i (these are the powers of x for a polynomial). Then a linear combination of the G_i can be written as

$$c_0 G_0 + c_1 G_1 + \cdots + c_n G_n$$

where the c_i are selected from some other class of entities (they were the integers in the case from which we are abstracting).

The difference between algebraic and transcendental numbers can be described neatly in terms of polynomials with integer coefficients. Because an algebraic number is a solution of a polynomial of some degree n with integer coefficients, it follows that there is a linear combination (using integer coefficients) of the powers of an algebraic number such that the sum is identically zero. Therefore, all high powers of a particular algebraic number can be reduced, using its particular defining polynomial, to a sum of powers less than n. On the other hand, no finite linear combination of the powers of π using integer coefficients can be exactly zero. In mathematical symbols,

$$c_0 + c_1 \pi + c_2 \pi^2 + \cdots + c_n \pi^n \neq 0$$

for any integer n and integers c_i ($i = 0, 1, \ldots, n$), unless all c_i are zero. See Section 4.7 for more details.

4.2 PHILOSOPHY

The idea that a continuous line is composed of points (which have no size) seems to be a contradiction. How can a line segment that has some length be composed of points that have no length?

As Zeno (336?–264? B.C.) dramatically put it in one of his famous paradoxes,

If at every instant the flying arrow is at some point, how is motion possible?

He created his paradoxes, apparently, not as denials of what we think we see in the real world, but rather to dramatize the *logical* difficulties that arise between the discrete and continuous views we have of the world.

A similar situation involving logical difficulties arises in the field of quantum mechanics in modern physics. In the famous two-slit experiment (Figure 4.2-1), the

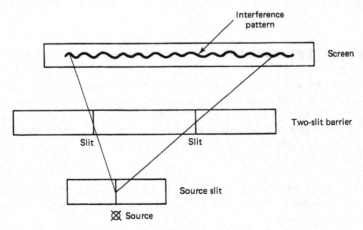

Figure 4.2-1   Two-slit experiment

photon of light in passing through the two open slits seems to "know" how far the slits are apart because the resulting pattern on the photographic plate shows interference fringes whose detailed shape depends on the separation of the two slits. If one slit is closed, the photon goes directly through the other open slit and strikes the screen in the pattern of a small slit. Thus the photon exhibits "wave-like" properties, which produce the interference pattern. On the other hand, when the photon hits the photographic plate, apparently it hits as a particle, because when the photographic plate is processed we see that the corresponding molecule of silver is developed. Thus the photon exhibits "particlelike" properties. The explanation of this apparent paradox has not been clearly described in the more than 50 years since the start of modern quantum mechanics in 1925. The apparent contradiction between the two views is often resolved by what is called the *complementarity principle*, meaning that the two descriptions are to be regarded as complementary rather than competing descriptions.

So, too, we have the two descriptions: the continuous line has both length and is composed of points that do not have length. Mathematicians have found ways of defining things to evade the apparent paradox. This does not mean that the paradox has been solved, only that they have found ways of logically consistently evading the problem. Thus in mathematics you also face a complementarity principle: the continuous line is composed of discrete points.

As a result you are advised to meditate on the matter and to make up your own mind as to what you think about Zeno's paradox of the flying arrows and the line being composed of points. As the student is usually told in a quantum mechanics course, "You will get used to it."

By the identification (mapping) of the numbers with the points on a straight line, we establish an *isomorphism* (iso- means same; morphism means shape, form) between numbers and points. We want some features of one to be matched by similar features in the other; for example, each point on the line corresponds to a number and each number to a point on the line. The feature "greater than" corresponds to lying to the right.

4.3 THE IDEA OF A FUNCTION

In principle, the idea of a function is very simple. For each value of the independent variable (usually x) in some range of values (usually over the entire real line or some part of it, although, as we have already seen, sometimes only over a subset, say the integers), there is a corresponding value of the function (usually the value of y). In mathematical notation this is

$$y = f(x)$$

In other words, for a *numerical function* $f(x)$ there are pairs of numbers (x, y) that give the correspondence between x and y. For each x in some admissible range, there is a single, unique value y.

This is a very large abstraction of your experience with particular functions. Consider the rich variety of functions you have already met in your life. If they are to be handled in any systematic fashion, together with all those you are going to meet in the future, some general concept and notation for a function are necessary.

Example 4.3-1

A number of familiar examples of functions are:

$$y = x^2 + 3, \qquad y = \cos x, \qquad y = \frac{x^2 - 1}{x^2 + 1}$$

Their usual graphs are sketched in Figure 4.3-1.

If we pick a particular function, say

$$y(x) = x^2 + 3$$

then you need to understand that the notation

$$y(x + 1) = (x + 1)^2 + 3$$

means "in place of the original x you now use $x + 1$." Again,

$$y(x^3) = (x^3)^2 + 3 = x^6 + 3$$

means in place of the original x you now use x^3. Similarly,

$$y(a) = a^2 + 3, \qquad y(a + 1) = (a + 1)^2 + 3$$

The difficulty with the abstract idea of a *general function* is the same difficulty you met in beginning algebra. You were told that the symbol x was a general number. It took a long time for you to get used to the idea of a general number. We are in the same position now. You have had much experience with particular functions, even functions that were general in the sense of being a polynomial of degree n with both the degree and the coefficients of the function unspecified. But when we say no more than that y is a function of x, then we are dealing with a general function, and you need to make the same mental adjustment to this abstraction as you did for the general number. Many times we will finally get down to a specific function, but there will be times when we never specify much more than that y is a function of x. We have already deliberately used general polynomials on a number of occasions just to give you practice in handling generality. The generality gives vagueness, but it also gives power because the abstract functional dependence covers many special cases.

We will not be concerned with the most general possible functions; rather we deal with a limited class called *piecewise continuous*. This means that each function is composed of pieces, and each piece is continuous, where "continuous" at present means just what you think; when you draw a curve without lifting the pen or pencil, the curve is continuous (see Figure 4.3-2). "Piecewise" means that there may be discontinuities (jumps) between the pieces. We further restrict the definition to permit only a finite number of pieces for a finite range of x.

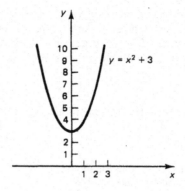

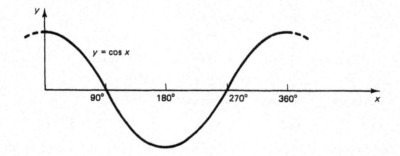

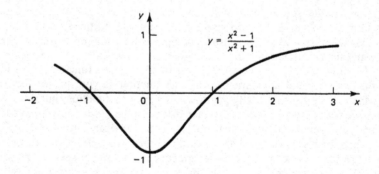

Figure 4.3-1   Some graphs of functions

      Furthermore, we assume that each continuous piece is either nondecreasing or else is nonincreasing, usually called monotone increasing or monotone decreasing. See Figure 4.3-3 for an example. Occasionally, the words "strictly monotone increasing" are used to mean that there are no horizontal parts to the corresponding graph, and similarly for strictly monotone decreasing.

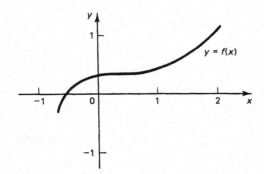

Figure 4.3-2   A continuous function

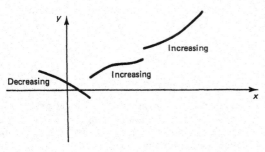

Figure 4.3-3   Piecewise monotone

Example 4.3-2

Consider the curve $y = \sin x$ for $0 < x < 360°$. Figure 4.3-4 shows the graph. It can be broken into three parts. In the first one-fourth of the range, the values of the function form a monotone increasing function. In the middle half of the interval, the function sin x is monotone decreasing. Finally, in the last fourth the function is again monotone increasing.

Functions of a discrete variable together with functions that are piecewise monotone and continuous (Figure 4.3-5) can describe all the situations you are apt to meet when you apply mathematical models to real-world situations. More advanced mathematics is necessary to cope with the few situations that might arise when more

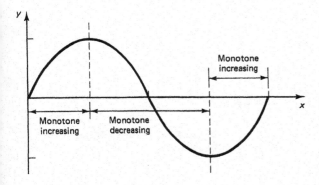

Figure 4.3-4 $y = \sin x$

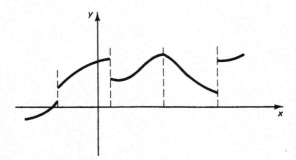

Figure 4.3-5 Piecewise monotone function

complicated functions are studied. At this point, the mathematical technicalities to handle them do not seem worth the effort.

Example 4.3-3

The function

$$y(x) = \{\text{the greatest integer} \leq x\}$$

is easily defined for all positive x and has a infinite number of horizontal pieces (see Figure 4.3-6). It is a monotone increasing function. For a finite range we allow only a finite number of places where there are discontinuities, where you can lift the pen or pencil and begin again. Evidently, we are *excluding* functions like

$$y(x) = \left\{\text{the greatest integer} \leq \frac{1}{x}\right\}$$

because near the origin the function has an arbitrarily large number of jumps. See Figure 4.3-7.

The reason for admitting discontinuous functions is that they occur frequently in practice. Examples are an on-off switch that controls a device, laws that become effective at a given date so that the corresponding social effects are being driven by a discontinuous cause, and many other situations in modern society where there is the

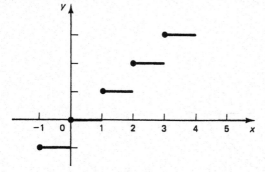

Figure 4.3-6 "Greatest integer in" function

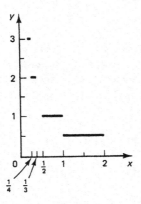

Figure 4.3-7   Greatest integer $\leq 1/x$

sudden change of a cause or of an effect. If we are to model these with our mathematics, we need to allow gracefully for such functions.

Several troubles arise from the requirement that a function be single valued.

1. If you rotate the coordinate axes (Section 6.9*), then you may introduce multiple values for the same value x, and hence you no longer will have a function $y = f(x)$.

2. At a point of discontinuity you are forced to omit at least one end value and hence later must compensate by supplying some technical details to obtain the missing value (Section 7.4).

3. As we will see in Chapter 22*, the sequence of approximating functions that approximate a discontinuous function comes closer and closer to supplying the vertical line bridging the discontinuity (see the figures in Section 22.3* for example).

4. We often need to deal with what would naturally be multiple-valued functions and are therefore forced to make artificial distinctions, which few mathematicians actually do in the "heat of doing mathematics." It is only in the refined proofs that the care to restrict functions to single-valued functions is maintained.

On the other side of the argument, a number of troubles arise when you let functions have multiple values, and it is presently the custom to restrict the definition of a function to those that, for each x in some given range, have a single value.

In these days of computers, a function is often compared to a subroutine that takes one number in and delivers one number out. For example, some typical numerical functions are the square root, the sine, and the logarithm. In these words, the function of a function

$$y = \sin \sqrt{x}$$

means that the x is supplied to the square-root routine, and the square-root routine output is then supplied as input to the sine routine, whose output in turn is y.

EXERCISES 4.3

1. Extend "the greatest integer in" function to negative numbers.
2. Discuss the corresponding function, the smallest integer greater than or equal to x.
3. Discuss how the function $\cos x$ is composed of monotone pieces.
4. Is the function $\sin(1/x)$ an admissible function? *Hint:* Consider the neighborhood of the origin.
5. If $f(x) = (1 - x)/(1 + x)$, find $f(1/x)$.
6. If $g(x) = (x^2 + 2)/(x^2 - 2)$, show that $g(\sqrt{x}) = (x + 2)/(x - 2)$.
7. If $f(t) = a^t$, show that $f(x)f(y) = f(x + y)$.
8. If $y(x) = (1 - x)/(1 + x)$, show that $y((1 - x)/(1 + x)) = x$.

4.4 THE ABSOLUTE VALUE FUNCTION

Often, instead of wanting to know the position of a point on the line, we merely want to know how far it is from the origin. We measure this distance by the *size* of the number, the *absolute value* of the number, meaning

$$|x| = \begin{cases} -x, & x < 0 \\ 0, & x = 0 \\ x, & x > 0 \end{cases}$$

See Figure 4.4-1. Specific examples are

$$|6| = 6$$
$$|-6| = 6$$

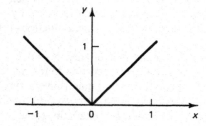

Figure 4.4-1 Absolute value function

Often we want to know the *distance* between two points a and b on a straight line. We have the simple formula for the distance function $d(\ ,\ )$:

$$d(a, b) = |a - b| = |b - a|$$

This seems to be a reasonable measure of the distance along a line since:

1. It is nonnegative, $d(a, b) \geq 0$.
2. It is zero if and only if the two points are the same point, $d(a, a) = 0$.
3. It is symmetric, meaning that the distance from one point to the other is the same as the distance from the second to the first point, $d(a, b) = d(b, a)$.

The absolute value function provides an isomorphism (end of Section 4.2) between "the distance between numbers" and "the distance between points along the real line."

Example 4.4-1

There are a number of simple relationships involving the absolute value function that will develop your skills in handling the idea. The first is the convenient relationship

$$|a| = \sqrt{a^2}$$

since $\sqrt{\ }$ always means the positive square root. From the obvious relation

$$a \leq |a|$$

we can deduce that

$$0 \leq (a - b)^2 = a^2 - 2ab + b^2$$
$$\leq |a|^2 + 2|a||b| + |b|^2$$
$$\leq [|a| + |b|]^2$$

Therefore, we have the right-hand side of the following equation; by similar reasoning we get the left-hand side of

$$|a| - |b| \leq |a - b| \leq |a| + |b|$$

Since b could be any number, positive, or negative, or zero, we may change the sign of the b term if we wish. Thus we have, in general, after removing some minus signs,

$$|a| - |b| \leq |a + b| \leq |a| + |b| \tag{4.4-1}$$

Example 4.4-2

The particular function

$$y(x) = \frac{x}{|x|}$$

For $x > 0$, it has the value $y(x) = 1$. For $x < 0$, it has the value $y(x) = -1$. For $x = 0$, it is technically not defined because of the division by zero that would occur; but conventionally it is given the value $y(0) = 0$. See Figure 4.4-2. This is also called the *signum function* (the sign function) and is written

$$\text{sgn}(x) = \frac{x}{|x|}$$

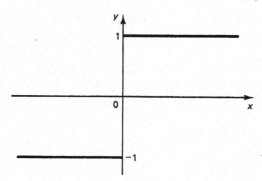

Figure 4.4-2 $y = x/|x|$

EXERCISES 4.4

1. Show that the function $y(x) = |x|/x$ is the same function as that in Example 4.4-2.
2. Discuss the difference between $y(|x|)$ and $|y(x)|$.
3. Discuss the Heaviside function $y(x) = \{x + |x|\}/2$.
4. Plot the function $y(x) = \{|x| - |x - 1| + 1\}/2$.
5.* Devise a function that is always 0 except that from 0 to 1 it rises along a straight line to the height 1, and then from 1 to 2 it goes back down to 0 along another straight line.
6. Is there a difference between x^2 and $|x|^2$?

4.5 ASSUMPTIONS ABOUT CONTINUITY

The concept of *continuity* arises from our intuition about drawing lines without lifting the pen (or pencil) and from our feelings about the continuous flow of time. We feel that at any place on the line, or at any moment in time, there should be a suitable number to mark it. We need to make this vague concept more definite.

We are not in a position to prove rigorously a number of properties about numbers and continuous functions, so we shall assume the following (in effect they define continuity):

1. If a formula (of the kind we will use) involving only continuous functions is true for all rational numbers, then it is also true for the irrational numbers. This is vague at present, but will become clear as we use it.
2. If in a closed interval $a \leq x \leq b$, also written as $[a, b]$, a continuous function changes sign an odd number of times, that is, if

$$f(a)f(b) < 0$$

then there exists a number θ (Greek lowercase theta) such that

$$f(\theta) = 0$$

(meaning there exists at least one number θ where the function is zero).

3. In a closed interval $[a, b]$ a continuous function takes on its maximum; that is, there exists at least one number θ (although it is not the same θ as above) for which $f(x)$ is at least as large as for any other number in the interval. In mathematical notation, there exists a θ such that

$$f(\theta) \geq f(x), \quad \text{for all } x \text{ in the interval } [a, b]$$

These are the three properties we assume about continuous functions and the existence of numbers; they assure us that when we use continuous functions in these ways there is always a number waiting for us. We will therefore not be in the position of Pythagoras who found that among the everywhere dense rationals there was no $\sqrt{2}$. This does not mean that there could not be other processes for which there would be no corresponding numbers, merely that if we use only the above processes we will always find a suitable number waiting for us.

We need one further assumption:

4. For any convergent sequence of numbers, the limit is a number, where by the words "convergent sequence" at the moment we mean something similar to taking more and more decimal digits of a number and having it "converge" to the number, or similar to a geometric progression when the rate r is less than 1 in size and the number of terms is indefinitely increased. See Examples 3.6-2 and 3.6-7, as well as Section 7.2 and Chapter 20.

We need to examine these assumptions to see how well they agree with our intuitive feelings. If they do not agree then we must either change the assumptions or else our feelings as to what continuity is. The first assumption says that since the rational numbers are everywhere dense any further numbers that we put on the line should not have wildly different properties when we use continuous functions. In this way we assert that we can operate with the new numbers without proving that the rules for rationals also apply to the irrationals. A rigorous treatment of the subject would require proofs of these properties, but we are not in a position to prove them in a first course in calculus. We need this principle to extend the laws of exponents (Section 3.8) from the rational exponents to irrational exponents, for example, that there exists a number (given an $x > 0$)

$$x^{\sqrt{2}} \quad \text{such that} \quad (x^{\sqrt{2}})^{\sqrt{2}} = x^2$$

We also need it so we can say that the rule

$$x^a x^b = x^{a+b}$$

is true for all real numbers a and b.

The second assumption merely says that a continuous curve cannot pass from positive to negative values without taking on the value 0 for some number x. This is a reasonable description of our idea of continuity. We are assured by this assumption that there is always a number where a continuous function that changes sign takes on the value 0, that there will be a number waiting for us; we will not be in the position of Pythagoras when he first met $\sqrt{2}$. We used this assumption in Section 3.5 when we found certain irrational numbers.

The third assumption also seems reasonable, that in a closed interval a continuous function takes on an extreme value somewhere in the interval, perhaps at the ends. The assumption assures us that there is a number where the function takes on its extreme value, something that we will often need later.

The last assumption about the existence of a limit that is a number requires a clearer idea of convergence (see Chapter 20) before we can say that we believe it. It does seem, however, to be a plausible assumption.

4.6 POLYNOMIALS AND INTEGERS

The familiar polynomials are the simplest continuous functions, and they play a role very much like the integers. In both cases they are the elementary pieces from which more complex expressions are built. Both the integers and polynomials are closed under the operations of addition, subtraction, and multiplication; the results of these operations lie within their own class. There is a strong similarity between polynomials and the integers. In this book, most of the time, we are concerned with real numbers and real polynomials. The exceptions are specifically mentioned when they occur.

It is the operation of the division of two integers that requires the extension to the rational numbers (to fractions), and, similarly for polynomials, division leads to the class of *rational functions,*

$$\frac{N(x)}{D(x)}$$

where $N(x)$ and $D(x)$ are polynomials. Both the integers and polynomials have the ideas of factorization, primeness, and reduction to lowest form. For example, in the class of polynomials with real coefficients, $x^2 + 1$ is a prime polynomial (has no real factors other than 1 and itself).

Rational functions dominate many parts of mathematics and statistics. The reasons that make the rational numbers prominent in arithmetic are similar to those that cause the rational functions to play a prominent role in much of mathematics. There are, similar to the case of numbers, functions that are not rational functions. For example, the trigonometric functions are transcendental functions, meaning that they are not the solution of any algebraic equation

$$f(x, y) = 0$$

where $f(x, y)$ is a polynomial in the two variables x and y. See end of Section 4.7.

Abstraction 4.6-1.

Euclid's algorithm. When we see the preceding similarities, it is natural to ask, "How far can we carry the similarities between the integers and the polynomials?" We have already compared a great many properties. What else could we compare? Euclid's algorithm comes to mind as a method of finding the GCD (greatest common divisor) of two integers. Does it apply to polynomials? We have only to reexamine the proof in Section 3.2 and identify the r_i and the q_i in the section with the corresponding polynomials. Thus, as we reread the proof, each time we see an r_i or a q_i we have to think the polynomials $r_i(x)$ and $q_i(x)$.

Following along the proof, we see that dividing one polynomial by another leaves a remainder of *degree* less than that of the divisor. Evidently, the integer being forced to zero in the original proof corresponds to the degree of the polynomial being forced to zero (a constant) in the new proof. An examination of the proof shows that all the divisibility arguments are the same in both cases.

We conclude that, with a simple change in understanding, Euclid's algorithm applies to polynomials just as it did for the integers. Evidently, Euclid's algorithm is not concerned with just integers; it applies to any entities that have certain properties. We will not, at this point, give the least properties that suffice for the algorithm to work, but the above shows that methods developed for some things (integers) can often apply, sometimes with slight changes, to other things (polynomials).

EXERCISES 4.6

1. Define a polynomial.
2. If $P(x) = x^3 + x + 1$ and $Q(x) = x^2 - 1$, find $P(x) + Q(x)$, $P(x) - Q(x)$, and $P(x)Q(x)$.
3. Find the GCD of $x^3 + 2x^2 + 2x + 1$ and $x^2 - 1$. *Ans.:* $x + 1$.
4. Find the GCD of $x^4 + 1$ and $x^2 - \sqrt{2}x + 1$. Thus you can factor $x^4 + 1$ in terms of real polynomials.
5.* Generalize Exercise 4. *Ans.:* You can always factor $x^k + 1$ for $k > 2$ in terms of real polynomials. *Hint:* The odd powers always have a factor of $x + 1$, and in the even exponent cases, writing x to some suitable even power as t, you can get either an odd power for the t or else an exponent of 4.
6.* Fill in all the details of Euclid's algorithm for polynomials.

4.7 LINEAR INDEPENDENCE

In Section 2.5 we proved that a polynomial in x (a linear combination of a finite number of the powers of x) cannot be identically zero for all integer values of x unless all its coefficients are zero. This is a very useful result and is worth looking at more closely to see how much the requirements can be weakened. When a linear combination of functions (Section 4.1) is identically zero (for all values of the variable), we say that the *functions are linearly dependent*. In the abstract notation of Section 4.1, if the c_i (not all zero) and the $G_i(x)$ are given, then if

$$c_0 G_0(x) + c_1 G_1(x) + \cdots + c_n G_n(x) = 0$$

for all x in some set (possibly an interval), then the $G_i(x)$ are *linearly dependent;* there is a linear relationship between the $G_i(x)$. If there is no set of coefficients c_i, not all zero, such that the linear combination is identically zero, then the functions are said to be *linearly independent* (over that set of points).

In the case of polynomials, the G_i are the powers of x, and c_i are the constant coefficients

$$c_0 + c_1 x + c_2 x^2 + \cdots + c_n x^n$$

Linear independence is a powerful tool. For example (as we will soon prove), from the statement that a polynomial of degree n is zero for at least $n + 1$ values, you can set each coefficient equal to zero separately and get $n + 1$ separate equations.

We turn to the proof of this important statement. Most students remember something about a polynomial of degree n having exactly n zeros. This suggests that if the polynomial were zero (vanished) at $n + 1$ points it would have to be identically zero; every coefficient would have to be zero. This kind of proof would be based on the fundamental theorem of algebra that a polynomial of degree n has exactly n zeros, real or complex. This theorem is difficult to prove, so we will base our examination of the situation on a simpler result, the factor theorem.

The factor theorem states that if for a value $x = a$ the polynomial $P_n(x)$ is zero, that is, if

$$P_n(a) = 0$$

then the polynomial is divisible by the factor $x - a$.

The proof is based on the observation that (see Section 2.6 under geometric progressions)

$$x^k - a^k = (x - a)(x^{k-1} + x^{k-2}a + x^{k-3}a^2 + \cdots + a^{k-1})$$

This can be seen directly by multiplying out the right-hand side and noting that all the terms, except the first and last, cancel. This is a typical ellipsis proof. If a is a zero of $P_n(x)$, meaning that $P_n(a) = 0$, then we can write

$$P_n(x) = P_n(x) - P_n(a)$$

since the second term on the right is zero by assumption. If we now write out the general polynomial $P(x)$ in some detail, we have

$$P_n(x) = c_n x^n + c_{n-1} x^{n-1} + \cdots + c_1 x + c_0$$

then

$$P_n(x) - P_n(a) = c_n(x^n - a^n) + \cdots + c_1(x - a)$$

and each term is divisible by $x - a$. Hence

$$P_n(x) = (x - a)P_{n-1}(x)$$

where $P_{n-1}(x)$ is some polynomial of degree $n - 1$. Thus, if a polynomial vanishes for some value $x = a$, then it has the factor $x - a$.

A variant of this result is called the *remainder theorem,* which states that, if you divide a polynomial $P(x)$ by the factor $x - a$, then the remainder is $P(a)$. This is easily seen from the division when written in the form

$$P_n(x) = Q(x)(x - a) + r$$

where $Q(x)$ is the quotient and r is the remainder (which is a constant). If you set $x = a$ in this equation, you get the result

$$P(a) = r$$

From the factor theorem we see that, corresponding to *each* distinct number x_k for which the polynomial vanishes, $P(x_k) = 0$, there is a factor

$$x - x_k$$

of the polynomial $P_n(x)$. Since each (distinct) number x_k cannot make any one of the already found factors zero, and since the product of them with the remaining polynomial must vanish at any other zero, it follows that the new polynomial has all the other zeros that have not already been used up to this point. We can therefore apply the factor theorem to the result. Now a polynomial of degree n cannot possibly have more than n factors of degree 1. Therefore, if there were n distinct values a_k for which the original polynomial $P_n(x)$ vanished, it would have the form

$$P_n(x) = c_n(x - x_1)(x - x_2) \cdot \cdot \cdot (x - x_n)$$

If there are $n + 1$ distinct zeros, then $c_n = 0$ and the polynomial is identically zero. We have now proved that:

If a polynomial of degree n *has* n $+$ 1 *distinct zeros, then it is identically zero.*

Since a polynomial is a linear combination of the powers of x, another way of stating the above result is that the $n + 1$ powers of x staring from $x^0 = 1$ and going up to (and including) x^n are *linearly independent* over any set of $n + 1$ points. No linear combination of them with numerical coefficients can be zero at $n + 1$ points unless all the coefficients are zero. We have reduced the condition of Section 2.5 that the polynomial be zero for all sufficiently large integers to the much simpler condition that the polynomial vanish at $n + 1$ points. This has many important consequences.

Example 4.7-1

Unique representation. One example of the use of this result tells us that any $n + 1$ distinct samples of a polynomial of degree n suffice to determine the polynomial; we are assured that we will always be able to solve for the coefficients when we use undetermined coefficients.

To prove this statement, two things must be shown: first, that there is at least one polynomial through the given data, and, second, that there is not more than one such polynomial. To prove the first part, we use mathematical induction. As a basis for the induction, we observe that given $n + 1$ samples

$$(x_i, y_i) \qquad (i = 0, 1, \ldots, n)$$

then for $n = 0$ we have

$$y = y_0$$

as the required polynomial. When $n = 1$ there are only two points, and the first-degree equation

$$y = y_0 + \frac{(y_1 - y_0)(x - x_0)}{x_1 - x_0}$$

is the solution. This can be seen by substituting x_0 and x_1 and evaluating the right-hand side. We see the beginning of a recursive proof.

In the mathematical induction pattern, we now assume that there is a polynomial of degree $n - 1$ through any n points, and then start with the $n + 1$ points and a polynomial of degree n. Let this polynomial be

$$P_n(x)$$

We then write this as

$$P_n(x) = y_n + (x - x_n)Q_{n-1}(x)$$

where $Q_{n-1}(x)$ is to be determined. We see that this expression is satisfied at the value x_n. For all n other values, we compute the value of

$$Q_{n-1}(x_i) = \frac{P_n(x_i) - y_n}{x_i - x_n}, \qquad \text{for } i = 0, 1, \ldots, n - 1$$

By the induction hypothesis, this polynomial $Q_{n-1}(x)$ can be found and the induction is proved (since we have the basis for the induction for both $n = 0$ and $n = 1$).

For the second part of the proof, if there were two or more such polynomials, say $P(x)$ and $R(x)$, that satisfied the given data of $n + 1$ points, then the difference $P(x) - R(x)$ would be zero at the $n + 1$ points; hence it would be identically zero. Thus the result is proved.

Thus any $n + 1$ samples of a polynomial of degree n are equivalent to the polynomial. This means that we may use the samples of the polynomials as if they were the polynomial with no loss of information. The result says, in different words, that if two polynomials are equal to each other for enough values, one more than the highest degree term occurring, then they are equal for all values. We can use the samples of the function to determine the polynomial.

Example 4.7-2

Given the samples $(-1, 0)$, $(0, 2)$, $(1, 2)$, and $(2, 4)$, construct the polynomial having these values (Figure 4.7-1). We could proceed as in the above proof, which involves a lot of special memory on your part, or else we could use the method of undetermined coefficients, which is a very general method. Given four samples, we need a polynomial with four undetermined coefficients, that is,

$$P(x) = ax^3 + b^2 + cx + d$$

We next impose the four conditions that the data satisfy the equations.

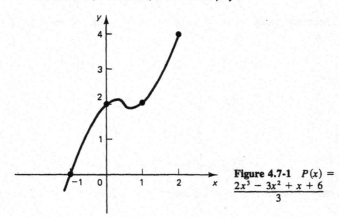

Figure 4.7-1 $P(x) = \dfrac{2x^3 - 3x^2 + x + 6}{3}$

Point	Condition
$(-1, 0)$:	$a(-1)^3 + b(-1)^2 + c(-1) + d = 0$
$(0, 2)$:	$a(0)^3 + b(0)^2 + c(0) + d = 2$
$(1, 2)$:	$a(1)^3 + b(1)^2 + c(1) + d = 2$
$(2, 4)$:	$a(2)^3 + b(2)^2 + c(2) + d = 4$

From the second equation, it is immediately clear that $d = 2$. Using this, rewriting the other three equations, and dividing out the factor 2 in the bottom equation, we get

$$-a + b - c = -2$$
$$a + b + c = 0$$
$$4a + 2b + c = 1$$

Add the top two of these three equations to get immediately $b = -1$. Take the top and third equation with this value of b substituted in, and we have

$$a + c = 1$$
$$4a + c = 3$$

Subtract the top from the second of these, and we have

$$3a = 2 \quad \text{or} \quad a = \frac{2}{3}$$

This gives, from the upper of the two equations,

$$c = \frac{1}{3}$$

and the coefficients are now known:

$$P(x) = \frac{1}{3}(2x^3 - 3x^2 + x + 6)$$

We can check this easily by substituting the sample values to see if we get the right values; we get

$$P(-1) = \frac{1}{3}(-2 - 3 - 1 + 6) = 0$$

$$P(0) = \frac{1}{3}(6) = 2$$

$$P(1) = \frac{1}{3}(2 - 3 + 1 + 6) = 2$$

$$P(2) = \frac{1}{3}(16 - 12 + 2 + 6) = 4$$

Thus we have found the polynomial and verified the fact that it satisfies the given sample values.

Linear independence is a negative property; it asserts the nonexistence of a set of coefficients. When a set of nonzero coefficients exists such that the sum of some set of functions is zero for the corresponding interval (or set of points), then the functions are *linearly dependent* (over some interval or set of discrete points). Linear dependence asserts that there is a linear relationship between the functions, and therefore they are not suitable for the representation of the general function of some class because there are one or more hidden linear relationships among the representing functions. Linearly independent functions, on the other hand, are often a suitable basis for representing things. Linear independence is an idea that we will often use and is a central concept in mathematics.

We can now prove that the function $y = \sin x$ cannot be the solution of an algebraic equation (a polynomial in two variables x and y)

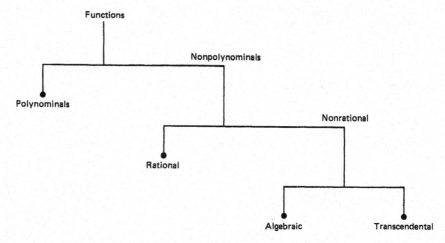

Figure 4.7-2 Types of functions

$$f(x, y) = 0$$

since for a fixed y less than 1 in size there are infinitely many solutions x of $y = \sin x$ ($x_k = x_0 + 360°k$ for all integers k and x_0 is any particular solution), while a polynomial can have (for a fixed y) only a finite number of solutions (zeros). Thus the trigonometric function

$$y = \sin x$$

must be a *transcendental function* (it cannot satisfy an algebraic equation in two variables). A similar proof applies to any other periodic function (except a constant). Figure 4.7-2 is to be compared with Figure 4.1-1.

EXERCISES 4.7

1. Define linear independence.
2. Given that $x = 2$ and $x = 3$ are zeros of $P(x) = x^3 - 9x^2 + 26x - 24$, find the factors of $P(x)$.
3. Given that $x = 1$ and $x = -1$ are factors of $P(x) = x^4 - 5x^3 + 5x^2 + 5x - 6$, find the factored form of $P(x)$.
4. Find the polynomial through the data $(-1, 2)$, $(0, 1)$, $(1, 5)$.
5. Find the polynomial through $(-2, -8)$, $(-1, -1)$, $(0, 0)$, $(1, 1)$, and $(2, 8)$.
6. Show that $y = \cos x$ is a transcendental function.
7. Show that in general any periodic function (other than $y = c$) is a transcendental function.

4.8 COMPLEX NUMBERS

Mathematicians, scientists, and engineers are all agreed that it is very unfortunate that $\sqrt{-1} = i$ is called an *imaginary number*, and that numbers like $2 + 3i$ are called *complex numbers*. Had they been called *numbers of the second kind*, then probably there would be much less resistance on the part of the beginner to accepting them as numbers. Experience has shown that these numbers are as "real" as any other numbers. In particular, in many fields of knowledge, complex numbers explain and unite many different things that at first seem to be unrelated.

Complex numbers first arose in connection with the solution of the quadratic equation

$$ax^2 + bx + c = 0 \qquad (4.8\text{-}1)$$

using the formula for the roots (derived in Example 6.3-2 if you have forgotten it)

$$x = \frac{-b \pm \sqrt{b^2 - 4ac}}{2a} \qquad (4.8\text{-}2)$$

If the *discriminant* $b^2 - 4ac > 0$, then the quadratic equation has two real and distinct roots: If $b^2 - 4ac = 0$, then both roots are equal to $-b/2a$. If $b^2 -$

$4ac < 0$, then the formula for the roots can be written as

$$x = \frac{-b \pm i\sqrt{4ac - b^2}}{2a} \tag{4.8-3}$$

where $i = \sqrt{-1}$. The roots are then complex numbers, each root being the *complex conjugate* of the other ("conjugate" means that the sign of the imaginary part is changed).

As a general rule, you handle complex numbers just as you would handle rational numbers, when you also had a $\sqrt{2}$. In such a system, we let a and b be rational numbers and *adjoin* the irrational number $\sqrt{2}$. Suppose

$$a + b\sqrt{2} \quad \text{and} \quad c + d\sqrt{2}$$

are two such numbers with rational coefficients a, b, c, and d. The two terms of each number cannot be combined because, as we proved in Section 3.4, a fraction cannot be equal to $\sqrt{2}$. If we add the two numbers, we get

$$(a + c) + (b + d)\sqrt{2}$$

as the sum, and similarly for the difference. When we multiply the two numbers, we get

$$(a + b\sqrt{2})(c + d\sqrt{2}) = ac + ad\sqrt{2} + bc\sqrt{2} + bd(2)$$
$$= (ac + 2bd) + (ad + bc)\sqrt{2}$$

We merely replace $(\sqrt{2})^2 = 2$ whenever it arises. In a similar fashion, all numbers involving higher powers of $\sqrt{2}$ can be reduced to a rational number plus a rational number multiplied by $\sqrt{2}$.

For division the matter is slightly more complicated. Given the quotient

$$\frac{a + b\sqrt{2}}{c + d\sqrt{2}}$$

we multiply both numerator and denominator by $c - d\sqrt{2}$. We get

$$\frac{a + b\sqrt{2}}{c + d\sqrt{2}} = \left(\frac{a + b\sqrt{2}}{c + d\sqrt{2}}\right)\left(\frac{c - d\sqrt{2}}{c - d\sqrt{2}}\right)$$

$$= \frac{ac - ad\sqrt{2} + bc\sqrt{2} - 2bd}{c^2 - 2d^2}$$

$$= \frac{ac - 2bd}{c^2 - 2d^2} + \frac{bc - ad}{c^2 - 2d^2}\sqrt{2}$$

Thus we have reduced the quotient to the form of a rational number plus a rational number multiplied by $\sqrt{2}$.

Complex numbers are handled *exactly the same way* except that $i^2 = -1$. We do not want exceptions in our methods when we can avoid them. Given

$$a + ib \quad \text{and} \quad c + id$$

their sum is

$$(a + c) + i(b + d)$$

Their product is

$$(a + ib)(c + id) = ac + iad + ibc + i^2bd$$

$$= (ac - bd) + i(ad + bc)$$

(4.8-4)

For the quotient we use the conjugate of the denominator $(c - id)$ as the multiplicative factor for both the numerator and denominator:

$$\frac{a + ib}{c + id} = \left(\frac{a + ib}{c + id}\right)\left(\frac{c - id}{c - id}\right)$$

$$= \frac{ac + bd}{c^2 + d^2} + i\frac{bc - ad}{c^2 + d^2}$$

(4.8-5)

One rule that complex numbers do *not* obey is the old rule for real numbers:

$$\sqrt{(ab)} = \sqrt{a}\sqrt{b}$$

(4.8-6)

For complex numbers we would have, if they did,

$$1 = \sqrt{1} = \sqrt{(-1)(-1)} = \sqrt{-1}\sqrt{-1} = ii = -1$$

(4.8-7)

which is obviously false.

We see that these new numbers satisfy most, but not all, of the expected rules of algebra. For example, the quadratic equation (4.8-1)

$$y = ax^2 + bx + c = 0$$

can be factored into the two factors

$$y = a(x - x_1)(x - x_2) = 0$$

When multiplied out, this gives

$$a[x^2 - (x_1 + x_2)x + x_1x_2] = 0$$

and a comparison with the original equation (4.8-1) shows that the sum of the roots is $-b/a$, while the product of the roots from (4.8-2) is c/a. In the complex form we easily see that the sum of the two conjugate roots is indeed $-b/a$. It remains to check the product of the two roots (4.8-2):

$$\frac{(-b + i\sqrt{4ac - b^2})}{2a}\frac{(-b - i\sqrt{4ac - b^2})}{2a}$$

$$= \frac{b^2 + ib\sqrt{4ac - b^2} - ib\sqrt{4ac - b^2} - i^2(4ac - b^2)}{4a^2}$$

$$= \frac{b^2 - (-1)(4ac - b^2)}{4a^2} = \frac{4ac}{4a^2} = \frac{c}{a}$$

as it should. Thus the formal replacement of $i^2 = -1$ is all that is new.

Using the concept of linear independence, we can say that the real and imaginary parts of a complex number are linearly independent since if

$$a + ib = 0$$

it follows that (since a real number a cannot be equal to an imaginary number b) both $a = 0$ and $b = 0$. Similarly, if two complex numbers are equal

$$a + ib = c + id$$

Then

$$a = c \quad \text{and} \quad b = d$$

As we noticed before, when we extend a class of numbers to a larger class we often lose one or more properties. For example, in going to complex numbers, we lost (4.8-6)

$$\sqrt{ab} = \sqrt{a}\sqrt{b}$$

In going from the real numbers to the complex numbers, we also lose the order relationship. Real numbers have the property that, given any two numbers a and b, we have one of the three possibilities

$$a < b, \quad a = b, \quad \text{or} \quad a > b$$

We further have (Section 3.3) that, if $c > 0$ and $a > b$, then

$$ac > bc$$

The order relationship is maintained if we multiply by a positive number.

For complex numbers there is no such ordering. To prove this, we argue as follows. Suppose there were an order relation among the complex numbers; then consider the quantity i. It is surely not equal to zero. If we first assume that

$$i > 0$$

then multiplying by $i > 0$ we would have

$$i^2 = -1 > 0$$

which is false. If we next assume that

$$i < 0$$

then multiplying by i would reverse the inequality, and we would again have

$$(-i)^2 = i^2 = -1 > 0$$

Thus the complex numbers cannot be ordered.

In place of ordering, we still have the concept of size, called the *absolute value* or the *modulus* (modulus refers to distance, Roman origin) of the number. Given $a + ib$, we define the absolute value as

$$|a + ib| = \sqrt{(a^2 + b^2)} = |a - ib| \qquad (4.8\text{-}8)$$

Thus the number and its conjugate have the same "size" (modulus). We often use the convenient form

$$|a + ib|^2 = (a + ib)(a - ib) \qquad (4.8\text{-}9)$$

In words, the square of the modulus of a complex number is the product of the number and its conjugate.

We see that the complex numbers are really not so strange as they first seem. You handle them much as you handle a quadratic irrationality like the $\sqrt{2}$ when dealing with rational numbers. The only two things to watch for are that the square root of a product is not necessarily the product of the square roots, and that the numbers cannot be strictly ordered; but the modulus, or absolute value, plays a very similar role in their arithmetic and algebra.

EXERCISES 4.8

Evaluate the following:

1. $(3 + 4i) + (4 - 3i)$.
2. $(3 + 4i)(4 - 3i)$.
3. $(3 + 4i)(3 - 4i)$.
4. $(2 + 3i)/(4 - 3i)$.
5. $(6 + 3i)/(2 + i)$.
6. $(a + ib)^2$.
7. $(a + ib)^3$.
8. $(a + ib)^4(a - ib)^4$.
9. Show that $(a + ib)^2(a - ib)^2 = |a + ib|^4$.
10. If $a + b\sqrt{2} = c + d\sqrt{2}$ for rational a, b, c, d, then show that $a = c$ and $b = d$.
11. Show that $(1/2 + i\sqrt{3}/2)^3 = -1$.
12. Show that $(\sqrt{3}/2 + i/2)^3 = i$.
13.* Find the square root of $a + ib$. *Hint:* Set it equal to $c + id$, square both sides, equate coefficients, eliminate one letter, solve the resulting quadratic in the square of the variable, and take the positive root so that the variable itself will be real.
14.* Show how to reduce any high power of a complex number to the form $u + iv$.
15.* Show that any polynomial in $a + ib$ can be written as $u + iv$ where u and v are real.
16.* Show that $Q(x) = [x - (a + ib)][x - (a - ib)]$ is a real quadratic.
17.* Generalized remainder theorem. Show that for complex numbers x, $P(x) = Q(x)q(x) + r_1x + r_0$, where $Q(x)$ is a real quadratic, $q(x)$ is the quotient, and $r_1x + r_0$ is the remainder.
18.* Show that to evaluate a polynomial for a complex number $P(a + ib)$ you can (1) form the real quadratic as in Exercise 16, (2) divide the polynomial by the quadratic as in Exercise 17, and then (3) the remainder at $x = a + ib$ is the value of the polynomial.

4.9 *MORE PHILOSOPHY*

Mathematics often gives the appearance of certainty. Furthermore, people wish to believe that somewhere there is certainty in this world of constant change—from moment to moment even science changes its rules and pronouncements—and mathematics seems to offer a haven from constant change. But this certainty is an illusion! See references [D], [K–2], and [L].

Consider the sentence (compare with Section 1.6)

This statement is false.

What are we to make of it? It seems to contradict itself! If it is true, then it asserts that it is false, and if the sentence is false, then it is a true statement. Even a simple grammatical sentence can contradict itself. How about mathematics? In mathematics we have occasionally managed to show that, if one set of assumptions is not self-contradictory, then another set is not, but we have not (as yet) shown that simple arithmetic is consistent, that from the same problem, and using only legitimate arithmetic, we cannot get two different answers. We also know that if a set of assumptions contains an inconsistency then any result can be obtained!

We do know that:

The
postulates
of mathematics
were not on
the stone tablets
that Moses
brought down
from Mount Sinai

The postulates of mathematics were made by humans and are subject to human frailty.

Mathematicians also know that at the logical foundations of mathematics there are many paradoxes, meaning that we can get, by reasoning in accepted ways, results that we do not like. Although mathematics is heavily concerned with symbols, we know that we cannot even define the concept of a symbol; we can only give examples of what we mean. Furthermore, from the Gödel theorems we believe that there are strings of symbols whose truth or falsity we cannot determine within any reasonably rich system of symbols ("cannot determine" in the same sense that $\sqrt{2}$ cannot be a rational number).

Add to all this the simple fact that we use mathematics to "model" the real world of our senses. We begin with experience, model it in mathematical symbols, apply more or less standard methods of mathematics, and then try to reinterpret the results back to the real world of experience. The gaps between the external world of our senses and the internal world of our symbols are very large indeed, and we have (as yet) no idea of how to make a secure connection in either direction. See Figure 4.9-1.

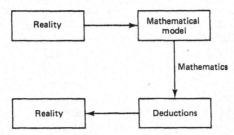

Figure 4.9-1 Use of mathematics

But when all this is said and accepted, we also know from experience (as given briefly in Section 1.3) the effectiveness of mathematics. We do risk human lives and large sums of money on the results we obtain from mathematics. It is useless to say that since we cannot be sure then we should take no action. To take no action is as definite an action as any other action! Passive resistance in the hands of Gandhi was a very aggressive weapon indeed. The refusal to take actions in the presence of uncertainty can be very dangerous and harmful to you and/or others.

What can the reasonable person do? You can neither sensibly believe everything that comes from mathematics nor believe nothing. You are left to think about the assumptions in the form of postulates and about the logic you use along the way, and then decide when it is prudent to believe and when it is prudent not to believe. Mathematics, much as you may wish that it would, does not give certainty; it is by no means what it appears on the surface. There is a strong tendency for most people to gloss over these awkward aspects of mathematics and to go forward without questioning. But for those who expect to use mathematics to shape actions in the world, serious attention to these difficulties is necessary.

We have tried to clearly label assumptions and the forms of logic we have used, and we will from time to time highlight results that are often ignored. It seems necessary to actively combat both the illusion of, as well as the desire for, infallible mathematics.

4.10 SUMMARY

In this chapter we have carefully introduced the geometric representation of numbers (although we have used it implicitly before). We have not slighted the difference between our ideas of a uniform continuous line and the discrete nature of the number system and have left the matter for your consideration. If you are to develop new applications of mathematics, it may well depend on your understanding how numbers do or do not represent the things of interest in your chosen field. You may have to convince both yourself and others that the classical model of mathematics is, or is not, relevant.

The next concept, the idea of a function, is not easy to master, but it is essential that we be able to deal with broad classes of functions and not get caught endlessly with learning about each specific function. The failure to grasp this central idea will greatly hinder you later. We applied the concept of a function to a number of special cases to illustrate it, in particular, the useful absolute value function $y = |x|$.

We then took up the concept of continuity, which is simply an abstraction of your experiences with time and a continuous line. We could not use much rigor and were reduced to "hand waving" and saying, "You see that . . . ." But if the rigorous approach did not agree with our intuition, then it would (probably) be altered.

We also took up the topic of complex numbers, which needlessly frightens the beginner (mainly because of the names "imaginary" and "complex"). We used the analogy with rational numbers and $\sqrt{2}$. Indeed, if you know how to handle these, then complex numbers are exactly the same except that you use $i^2 = -1$. Complex numbers are important for many reasons. One reason is that the two parts, the real and the imaginary, are linearly independent so that a single complex equation is equivalent to two real equations, a great gain in many situations. But more important, as you will see in Section 21.5 and Chapters 22* and 24, complex numbers provide a unity for what on the surface seems to be diversity.

Finally, we again remind the reader that mathematics is not the certain thing that many people believe, or at least wish were true. It is essential that you learn to both believe and disbelieve at the same time; I have long observed that great scientists can tolerate ambiguity, and you need practice in this important trait.

5

Analytic Geometry

5.1 CARTESIAN COORDINATES

We have just seen in Chapter 4 that there is an *isomorphism* (iso-, same; morphism, form) between the real number system and the *geometry* of a straight line (a one-to-one correspondence between parts of each with some properties also corresponding). For example, the absolute value function measures both the distance between points and the distance between numbers.

We now turn to a review of the *geometry* of the plane. It is natural to use a pair of mutually perpendicular lines as the basis for locating points. These lines are called the *coordinate* (co-ordinate) axes and are typically labeled x for the horizontal and y for the vertical line. As shown in Figure 5.1-1 the values on the coordinate axes of a point in the plane are the numbers where the perpendicular *projections* of the point fall on the axes. Thus, to each point in the plane there is a unique pair of numbers, and to each pair of numbers there is a unique point. In the jargon of mathematics, there is a *one-to-one correspondence* (isomorphism) between the points of the plane and pairs of numbers. The number pairs are conventionally written

$$(x, y)$$

The names *abscissa* and *ordinate* are given to the x and y coordinates. (Note that in the parentheses and in the dictionary x precedes y and abscissa precedes ordinate.)

Coordinate systems with nonperpendicular axes are occasionally convenient, but the advantages of the standard perpendicular ones are so great that they are almost always used. Once in awhile curves rather than straight lines are used as the coordinate

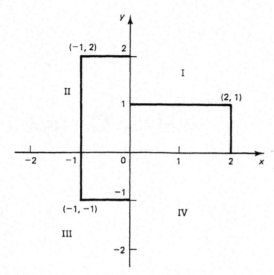

Figure 5.1-1 Coordinate system

axes. Unless otherwise stated, the coordinate systems used in this book will always be the mutually perpendicular straight lines.

In geometry we think of the plane as being *isotropic* (iso-, same; tropic, turning, changing), the same in all directions and all positions. Thus a translation of the coordinate system or a rotation of the axes does not change the *geometric properties* of things, but clearly does change the numerical values of the coordinates of a particular point. Felix Klein (1849–1925) in 1892 suggested that the various geometries be classified according to the set of transformations that left the corresponding geometrical elements unchanged. For Euclidean geometry, these transformations are (1) translations, (2) rotations, and (3) a reflection in a line. The geometrical elements are points, lines, triangles, circles, areas, and so on. Orientation is not an element of classical Euclidean geometry, since reflections are permitted and they reverse the orientation in the plane (recall Section 1.2).

As your algebra teacher probably said, "You cannot add apples to oranges," and hence the ability to rotate the coordinate system implies that *the same size and kind of units* are used on both axes. If both axes are lengths, then rotation is a possible transformation, but if one axis is measured in dollars and the other is measured in units of time, then a rotation makes no sense. As noted earlier (end of Section 4.3), the definition of a function says that there must be a single value for y for each value of x that is permitted, and this can cause trouble when you rotate the axes of the coordinate system.

The coordinate axes divide the plane into four *quadrants*. Conventionally, they are numbered, I, II, III, and IV in a counterclockwise direction, as shown in the Figure 5.1-1. Thus, in the quadrant II, $x < 0$ and $y > 0$, while in quadrant IV, $x > 0$ and $y < 0$. In quadrant III, both coordinates are negative.

EXERCISES 5.1

1. Plot the points (a) $(3, -4)$; (b) $(4, -3)$; (c) $(0, 0)$; (d)$(0, -3)$.
2. Describe the coordinates of points on the x-axis. On the y-axis.
3. Describe the coordinates of points on the 45° line. On the 135° line.
4. Connect the four points $(1, 0)$, $(0, 1)$, $(-1, 0)$, and $(0, -1)$ in order. What is the shape?
5. Connect the four points $(1, 1)$, $(1, -1)$, $(-1, -1)$, and $(-1, 1)$ in the natural order. What is the shape?

5.2 THE PYTHAGOREAN DISTANCE

In geometry it is not the values of particular coordinates that matter, but only the relationship between coordinates of different points. Therefore, we need a measure of the distance between two points p_1 and p_2. The coordinate axes are perpendicular, so we use the *Pythagorean (Euclidean) distance*, the length of the diagonal of a rectangle, as the measure of this distance. See Figure 5.2-1. We will later see that this is the only distance measure that makes sense in geometry when both rotation and translation are allowed. If the two particular points are $p_1 = (x_1, y_1)$ and $p_2 = (x_2, y_2)$, then the square of the distance between them is given by

$$d^2(p_1, p_2) = (x_2 - x_1)^2 + (y_2 - y_1)^2 \qquad (5.2\text{-}1)$$

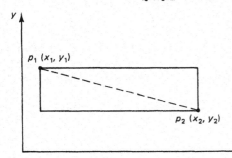

Figure 5.2-1   Pythagorean distance

Example 5.2-1

If the two points are $(3, 4)$ and $(5, -2)$, then the distance squared is

$$d^2(p_1, p_2) = (5 - 3)^2 + (-2 - 4)^2 = 4 + 36 = 40$$

This distance function (5.2-1) effectively commits you to using the same units on each of the coordinate axes; you cannot speak intelligently about the distance between different points on a dollar versus year curve.

This Pythagorean distance function has the properties you would expect a *distance function* to have:

1. $d(p_1, p_2) \geq 0$
2. $d(p_1, p_2) = 0$ if and only if $p_1 = p_2$
3. $d(p_1, p_2) = d(p_2, p_1)$
4. $d(p_1, p_2) + d(p_2, p_3) \geq d(p_1, p_3)$

In words, the first says that the distance between points is never negative; the second says that the distance is zero if and only if the two points are the same point; the third says that the distance from one point to a second point is the same as in the opposite direction; while the fourth says that the distance obeys the *triangle inequality*, that is, the sum of any two sides is greater than, or equal to, the third side of a triangle since a straight line is the shortest distance between two points. Recall that in Section 4.4 the absolute value function had all of these properties (property 4 trivially).

If we look at the definition (5.2-1) of the distance between two points, from the algebraic point of view we see immediately that, since the square root is always taken to be positive, the first three conditions are satisfied. The fourth is harder to prove algebraically.

EXERCISES 5.2

1. Find the distance from the origin to each of the four points in Exercise 5.1-1.
2. Find the distance from $(1, -1)$ to $(-1, 1)$. *Ans.: $d = 2\sqrt{2}$*.
3. Find the distance between each pair of points: $(2, 2)$; $(0, 1)$; $(1, 0)$.
4. Check the triangle inequality for the points $(0, 0)$, $(3, 4)$, and $(6, 8)$. What does this mean geometrically?
5. Show that the three points $(1, 0)$, $(0, 3)$, and $(2, 2)$ form an isosceles triangle (at least two sides equal in length).
6.* Consider the function $D(x, y) = |x_2 - x_1| + |y_2 - y_1|$. Is this a suitable distance function? Why?

5.3 CURVES

Given a function

$$y = f(x)$$

then, for each real number x (of those that are suitable), there is a corresponding y, and the x and y may be regarded as the coordinates of a point. Thus corresponding to a function there is a set of points in the plane. We call the set of points the *locus* (loci is the plural of locus). There is, therefore, a correspondence between equations and single-valued curves in the plane; each locus is a curve consisting of *all* pairs of numbers (points) that satisfy its equation

$$y = f(x)$$

By convention, straight lines are regarded as being curves. Note that there are equations that are not single valued.

Example 5.3-1

The equation $x^2 + y^2 = 1$ is the locus of all points one unit from the origin (Figure 5.3-1).

This is Descartes' (1596–1650) great observation and is the basis of *analytical geometry*. He noted that algebra and geometry correspond, that there is (almost) a complete isomorphism between them. For example, in Figure 5.3-2 there are two curves and they have a point in common. It follows that the corresponding equations have a solution in common. And conversely. This simple, elegant, and fundamental observation by Descartes is why the perpendicular coordinate system is called *Cartesian coordinates*.

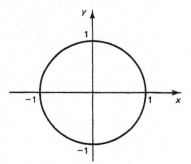

Figure 5.3-1 $x^2 + y^2 = 1$ **Figure 5.3-2**   Intersection of two curves

Example 5.3-2

As a concrete example, consider the two linear equations (linear means that all the terms in the equation are of first degree or lower)

$$y = 3x - 4$$

$$y = -2x + 1$$

Figure 5.3-3 shows the corresponding curves (loci) of these equations; in this case they are straight lines, as we will soon prove. (This is probably the source of the word "linear" that is used so much in mathematics.) In the algebraic approach, we subtract the second equation from the first to eliminate the y term. We get

$$0 = 5x - 5 \quad \text{or} \quad x = 1$$

Put this solution (coordinate) $x = 1$ in *either* equation and you find that $y = -1$. Thus the algebraic solution of the two equations $(1, -1)$ also gives the coordinates of the common point of the two lines $(1, -1)$.

To repeat the idea, Descartes showed that we can take problems in algebra and convert them into problems in geometry, and, conversely, we can take problems in

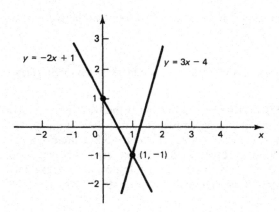

Figure 5.3-3 Intersection of two straight lines

geometry and convert them to algebra. For a given problem, we typically will convert one way and then back again to get the appropriate answer. When we start with the algebra, the values of the coordinates have significance, but when we start with the geometry, the values of the algebraic solution are not significant in themselves, since in geometry there is no meaning to the absolute position; only relative positions matter. For geometry problems we can pick the location of the coordinate system to make the algebra easy to carry out.

In the Cartesian approach to geometry we use the *analytical* methods of algebra, rather than the *synthetic* methods that are taught in the conventional first course in geometry. That is why it is called "analytic geometry." In principle, either method (analytic or synthetic) can be used, but in practice each has its appropriate domain of application. The analytical approach to geometry in the long run has great power, and we will examine it carefully. Correspondingly, "geometric algebra" is highly useful; we can form a geometric picture in our minds of the algebraic problem.

EXERCISES 5.3

1. What does it mean mathematically when you say that a point is on a line?
2. What does it mean when you say that a point is on a pair of lines?
3. Find a point on the line $2x + 3y = -4$. *Hint:* Pick any convenient x value, say $x = 1$, and solve for the y value.
4. Find two points on the line of Exercise 3 and plot the line.
5. Plot the following lines (by finding a pair of points on each): $x + y = 1$, $x - y = 2$. Find their common algebraic solution and check that it lies on the intersection of the lines you have drawn.
6. As in Exercise 5, but use the lines $3x - 4y = 7$, $4x + 3y = 0$.
7. Describe the line $x = 0$. The line $y = 0$.
8. Describe the line $y = x$.
9. Show that the line $Ax + By = 0$ goes through the origin.

5.4 *LINEAR EQUATIONS—STRAIGHT LINES*

The two linear equations in Figure 5.3-3 appear in the picture as straight lines. Since we cannot afford to study each specific line, we immediately turn to the *general case*. We now claim that any (general) linear equation in x and y

$$Ax + By + C = 0 \qquad (5.4\text{-}1)$$

(where A, B, and C are arbitrary constants often called *parameters*) will have a locus of points that is a straight line. This is saying that any pair of numbers that satisfies the equation corresponds to a point that lies on the line, and, conversely, any point on the line has coordinates that satisfy the equation. If this generalization bothers you, try using $A = 3$, $B = 5$, and $C = 7$ (prime numbers are usually best since they are easily followed through most arguments). General cases can be mastered by trying various specific cases until *you* "see" the general case.

The form (5.4-1) appears to have three arbitrary parameters, A, B, and C; but since we can divide the whole equation by any nonzero coefficient without changing the solutions of the equation, we really have only two parameters in the equation. The main reason for using the above form is that at the start of a problem we usually do not know which coefficient is not zero.

To plot the points, it is convenient to solve (5.4-1) for y (provided $B \neq 0$). We get

$$y = \left(\frac{-A}{B}\right)x + \left(\frac{-C}{B}\right) \qquad (5.4\text{-}2)$$

If we try various values of x, we get corresponding values of y, and each pair of numbers can be plotted as a point.

Example 5.4-1

As a concrete example, consider the equation corresponding to (5.4-1):

$$4x - 2y + 6 = 0$$

Solve this equation for y [compare with Equation (5.4-2)]:

$$y = 2x + 3$$

We now pick $x = 0$ and find $y = 3$. See Figure 5.4-1. The point $(0, 3)$ is on the line. Another value might be $x = -2$. Then $y = -1$, and the point is $(-2, 1)$ and is also on the line. To check the idea that the locus of points forms a straight line, we compute a few other points, say $x = 1$, which gives $y = 5$; $x = -1$, $y = 1$; and $x = 2$ gives $y = 7$. Yes, these points all *appear* to fall on a straight line.

Example 5.4-2

In Equation (5.4-2), set $x = 0$; then $y = (-C/B)$. This gives the point $(0, -C/B)$. Since two points determine a straight line, any other value of x, say $x = 1$, will give the second point, $(1, -(A + C)/B)$. We can then try any further points we wish, say $x = 2$, from which we get the point $(2, -(2A + C)/B)$, and we will see that the points *appear* to lie on the line.

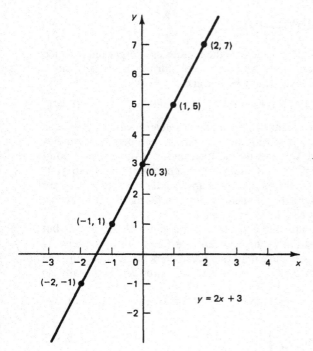

Figure 5.4-1 $y = 2x + 3$

But it is a characteristic of mathematicians that they have learned not to believe the obvious. Are we sure that *all* pairs of numbers satisfying the equation will lie on a straight line, and conversely? The picture looks nice and convincing, but some algebra would be more so. How could we convince ourselves of this statement? Never mind that we said that a linear equation corresponds to a straight line. Are *you* completely sure? What do you know about a straight line? The one thing you surely remember is that a straight line is the shortest distance between two points. How can you use this piece of information (which amounts to the basic definition of a straight line)? A little thought about the very few relevant tools you have available suggests that the triangle inequality is the clue; for a straight line it must be an equality. Therefore, we will pick three distinct points, any three points, on the line. Let them be p_1, p_2, and p_3 with p_2 between the other two. "Between" means, in this case, that the x coordinates of the three points have the "betweenness" property. Of course, we *expect* that the y coordinates will have the same property.

Given the general linear equation of the form (5.4-1),

$$Ax + By + C = 0$$

the assumption that these three points p_1, p_2, p_3 each lie on the line means that we have the following three equations (the coordinates of each point must separately satisfy the equation if it lies on the line):

$$Ax_1 + By_1 + C = 0$$
$$Ax_2 + By_2 + C = 0 \qquad (5.4\text{-}3)$$
$$Ax_3 + By_3 + C = 0$$

where the point p_i has the coordinates (x_i, y_i). We want to show that

$$d(p_1, p_2) + d(p_2, p_3) = d(p_1, p_3)$$

(Remember that p_2 is between p_1 and p_3.) Rather than write out the distance function (5.2-1) for each pair of points, we write a typical distance from p_i to p_j.

$$d(p_i, p_j) = \sqrt{(x_i - x_j)^2 + (y_i - y_j)^2} \qquad (5.4\text{-}4)$$

If $B \neq 0$ (we will take care of the case $B = 0$ in the next paragraph), then subtract the jth equation of (5.4-3) from the ith equation (note that C cancels out), and divide by B to get

$$y_i - y_j = -\left(\frac{A}{B}\right)(x_i - x_j)$$

We have, therefore, when we put this into the distance function (5.4-4),

$$d(p_i, p_j) = \sqrt{(x_i - x_j)^2 + (y_i - y_j)^2}$$
$$= \sqrt{[1 + (A/B)^2][x_i - x_j]^2}$$
$$= \sqrt{[1 + (A/B)^2]}\, |x_i - x_j|$$

The distance along the line is proportional to the distance between the x coordinates of the points. It is the same constant of proportionality for each pair of points. The distance along the x-axis obeys the triangle inequality trivially; the distance from p_1 to p_2 plus the distance from p_2 to p_3 along a straight line equals the distance from p_1 to p_3 when p_2 is between the other two points. From this it follows that the result is proved, that a linear equation corresponds to a straight line.

The special case of $B = 0$ (which was earlier set aside) means that the equation is

$$Ax + C = 0$$

The only admissible x value is $-C/A$ (assuming that $A \neq 0$). The line is a vertical line parallel to the y-axis, since the points $(-C/A, y)$, for any y, all satisfy the equation. The term in x in the distance function drops out, and we have the same argument along the y-axis as we had along the x-axis. Notice that we are in trouble when we have a vertical line; we do not have a function corresponding to the line since functions must be single valued.

The (degenerate) case that both A and B are zero forces $C = 0$, and we do not have an equation at all. In this degenerate case we have the whole plane; there are no constraints on the coordinates!

A linear equation is a straight line, but does every straight line correspond to a linear equation? Suppose we did have a straight line in the plane. Pick any two distinct points, say p_1 and p_2, on the line (with $x_1 \neq x_2$). We can use these two points to find the unique linear equation that passes through them. To actually find the equation of the line, we again use the method of undetermined coefficients. Assume the general form of the equation (5.4-1),

$$Ax + By + C = 0$$

The coordinates of the two distinct points that lie on this line generate the two equations

$$Ax_1 + By_1 + C = 0$$
$$Ax_2 + By_2 + C = 0 \tag{5.4-5}$$

To eliminate A from these two equations, we multiply the top equation by $-x_2$, the second equation by x_1, and add the two equations. We get

$$B(x_1y_2 - x_2y_1) = -C(x_1 - x_2) \tag{5.4-6}$$

If the important quantity, called the *discriminant*,

$$x_1y_2 - x_2y_1 = 0 \tag{5.4-7}$$

then clearly from (5.4-6) $C = 0$ (unless $x_1 = x_2$, which is a vertical line). The case $C = 0$ means that the line has the point $(0, 0)$ on it. This in turn means that the line goes through the origin of the coordinate system. To complete this special case, assume that $x_1 \neq 0$. Then the equation involving this point becomes

$$A = -B\frac{y_1}{x_1}$$

and putting this in the general form (remember that for this case $C = 0$) we have, finally,

$$y = \frac{y_1}{x_1}x$$

as the equation through the two points.

If we look at this result, we see how obvious it is. Certainly, the point $(0, 0)$ satisfies the equation (we are now being careless with the words "point" and "coordinates" since we have identified a correspondence between them), and almost as clearly the point (x_1, y_1) satisfies the equation. We could practically write the equation directly from the given conditions.

We now return to the main problem. If the discriminant in (5.4-7) is not zero, that is, if

$$x_1y_2 - x_2y_1 \neq 0 \tag{5.4-8}$$

then we solve for the unknown quantity B:

$$B = \frac{-C(x_1 - x_2)}{x_1y_2 - x_2y_1} \tag{5.4-9}$$

If we use multipliers on the two original equations (5.4-5) of y_2 and $-y_1$ and add, we eliminate B. We get, by similar arguments,

$$A = \frac{-C(y_2 - y_1)}{x_1 y_2 - x_2 y_1}$$

Note that the denominator is the same as before (5.4-9). Also note that A and B are both proportional to C, and therefore if $C \neq 0$ it can be divided out of each of the three terms of the equation. The equation becomes (when we multiply by the discriminant)

$$-(y_2 - y_1)x - (x_1 - x_2)y + (x_1 y_2 - x_2 y_1) = 0 \qquad (5.4\text{-}10)$$

It is easy to check the algebra by noting that the two points p_1 and p_2 both lie on this line. To do this, merely set $x = x_1$ and $y = y_1$ and get from (5.4-10)

$$-x_1 y_2 + x_1 y_1 - x_1 y_1 + x_2 y_1 + x_1 y_2 - x_2 y_1 = 0$$

and all the terms cancel. And similarly for the second point.

There are a couple of special cases to note in our analysis. If the two x values are the same, that is, $x_1 = x_2$, then the equation is that of a line that is vertical; while if the two y values are the same, then the line is horizontal.

If $C = 0$, then from (5.4-6) the discriminant equals 0 (or else $B = 0$, which leaves a degenerate equation $Ax = 0$), and we have a line through the origin. The condition $C = 0$ means that there is a linear relation between x and y of the form $y = (-A/B)x$.

Now consider any other point on the original straight line. Since we proved that a linear equation is a unique straight line, it follows that this point must also be on the line we calculated through the two points. Thus any point on the original line is on the computed line; the line and the equation are in some sense the same thing!

EXERCISES 5.4

1. Verify that the second point lies on the line (5.4-10) we derived above.

Find the line through the two points:

2. $(2, -6)$ and $(-1, 1)$.
3. $(6, 4)$ and $(0, 0)$.
4. $(-1, -1)$ and $(-3, -5)$.
5. Find the equations of the sides of the triangle having vertices $(0, 0)$, $(1, 3)$, and $(-2, -3)$.
6. Draw a diagram of the analysis of this section and include all special cases. Complete the missing arguments.
7. Using Equation (5.4-1), what is the condition that the line be vertical? Be horizontal? Pass through the origin?
8. In Equation (5.4-1), give the geometric meaning of $A = 0$. Of $B = 0$. Of $C = 0$.

5.5 SLOPE

Geometrically speaking, we feel that a straight line has a constant *slope*, where by the word "slope" we mean the *rise* that occurs in going from one point p_1 to another (different) point p_2 on the line, *divided by* the *horizontal distance* between the two points; that is,

$$\frac{\text{change in } y}{\text{change in } x} = \text{slope}$$

Also, it is said to be "rise over the run." In mathematical notation, this is

$$\frac{y_2 - y_1}{x_2 - x_1} = m \tag{5.5-1}$$

where m is the conventional notation for the slope (see Figure 5.5-1). A horizontal line has zero slope. Notice that if we interchange the two points p_1 and p_2 in the formula (5.5-1) we get the *same* slope, as we should.

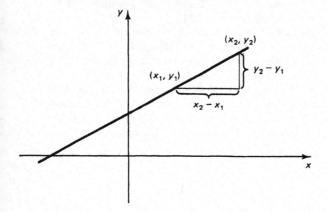

Figure 5.5-1   Slope

Are we sure that the slope of a straight line is independent of the choice of the two points we use? It ought to be, but let us check this algebraically. We have for the general equation of a line

$$Ax + By + C = 0 \tag{5.5-2}$$

The following two equations represent the fact that the two points p_1 and p_2 lie on the line:

$$Ax_1 + By_1 + C = 0$$

$$Ax_2 + By_2 + C = 0$$

To eliminate C, subtract the second equation from the first:

$$A(x_2 - x_1) + B(y_2 - y_1) = 0 \tag{5.5-3}$$

Divide both sides by $B(x_2 - x_1)$, supposing that it is not zero, and rearrange slightly to get

$$\frac{y_2 - y_1}{x_2 - x_1} = \frac{-A}{B} = m \qquad (5.5\text{-}4)$$

Thus the slope m depends only on the coefficients A and B and does not depend on the two particular points p_1 and p_2 chosen to compute the slope. Nor does the slope depend on C. Changing the value of C only raises or lowers the line, but does not change the slope of the line.

The slope of m is related to the angle ϕ (Greek lowercase phi) between the line and the x-axis, as shown in Figure 5.5-2.

$$m = \tan \phi \qquad (5.5\text{-}5)$$

Since the line is not usually thought of as having a direction along itself, it is often convenient to limit the angle to a range of 180°. The choice can be $-90° < \phi \le 90°$, or $0 \le \phi < 180°$, depending on the needs of the situation. It is convenient to make no general restriction on the angle but let the problem dictate which angles to use.

When the line is vertical, then $x_1 = x_2$ and there is technically no slope since we cannot divide by zero in form (5.5-4). See Figure 5.5-3 for the corresponding vertical line. It is convenient to say "the slope is infinite," meaning that there is no finite number to describe the slope.

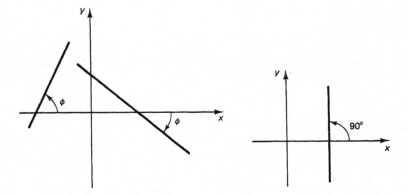

Figure 5.5-2 Angle of line   Figure 5.5-3 Infinite slope

We next look at the angle between two lines having slopes m_1 and m_2 (Figure 5.5-4). The angle between them is

$$\phi = \phi_2 - \phi_1$$

To get things in terms of the slopes, we naturally take the tangent of both sides. The trigonometric identity for the tangent of the difference between two angles (16.1-17) is

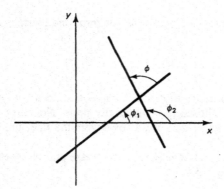

Figure 5.5-4   Angle between two lines

$$\tan \phi = \frac{\tan \phi_2 - \tan \phi_1}{1 + \tan \phi_2 \tan \phi_1}$$

Hence, on substituting the slopes m_i from each line,

$$\tan \phi = \frac{m_2 - m_1}{1 + m_2 m_1} \qquad (5.5\text{-}6)$$

Example 5.5-1

Consider the two lines

$$2x + 3y - 5 = 0$$
$$5x - 2y + 9 = 0$$

From Equation (5.5-4) we know that the slope of the line described by the first equation is (solve for y and the slope is the coefficient of x)

$$m_1 = \frac{-A}{B} = \frac{-2}{3}$$

For the second equation,

$$m_2 = \frac{-A}{B} = \frac{5}{2}$$

Therefore, the tangent of the angle between the two lines is, using (5.5-6),

$$\tan \phi = \frac{5/2 - [-(2/3)]}{1 + (5/2)[-(2/3)]}$$

$$= \frac{19/6}{-(4/6)} = \frac{-19}{4}$$

$$\phi = -78.1113 \ldots °$$

See Figure 5.5-5.

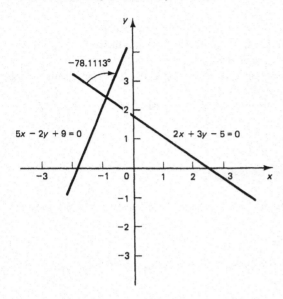

Figure 5.5-5 Angle between two lines

The condition for two lines to be perpendicular is of special interest. We can reason that if $\phi = 90°$ in (5.5-6) then the tangent is infinite, and hence the denominator on the right-hand side must be zero:

$$1 + m_1 m_2 = 0 \qquad (5.5\text{-}7)$$

or

$$m_2 = \frac{-1}{m_1} \qquad (5.5\text{-}8)$$

The slopes of perpendicular lines are the negative reciprocals of each other. When two lines are perpendicular to each other, they are said to be *normal* to each other. Note the symmetry in the formulas.

EXERCISES 5.5

What is the slope of the following lines:

1. $y = 3x - 6$
2. $y = -3x + 6$
3. $x = 2y - 7$
4. $3x + 2y = 9$
5. $px + qy = r$
6. What is the tangent of the angle between the lines $x + y - 1 = 0$ and $2x - y = 0$?
7. What is the tangent of the angle between the lines $3x - 5y + 7 = 0$ and $2x + y = 0$?

8. Show that the lines $5x + 7y = 9$ and $7x - 5y = 0$ are perpendicular.
9. Show that the lines $x + y - a = 0$ and $x - y - b = 0$ are perpendicular.
10. What is the slope of a line perpendicular to the line $x + 3y = 0$?
11. Show by algebra that, if line 1 is perpendicular to line 2, and line 2 is perpendicular to line 3, then line 1 has the same slope as line 3.
12. Show that the perpendicularity condition for two general lines $A_i x + B_i y + C_i = 0$; $(i = 1, 2)$ is $A_1 A_2 + B_1 B_2 = 0$.

5.6 SPECIAL FORMS OF THE STRAIGHT LINE

We have just seen that there is a one-to-one correspondence between straight lines and the general linear equation

$$Ax + By + C = 0 \tag{5.6-1}$$

In a sense this is all we need to know beyond (1) the definition of the slope of a line (5.5-1)

$$m = \frac{y_2 - y_1}{x_2 - x_1} = \tan \phi \tag{5.6-2}$$

and (2) the distance between two points, which is given by (5.2-1):

$$d(p_1, p_2) = \sqrt{(x_1 - x_2)^2 + (y_1 - y_2)^2} \tag{5.6-3}$$

However, straight lines occur so often that it is worth a little time to learn a few special cases.

First, there is the *two-point form* of the line. Two distinct points determine the slope m of the line by Equation (5.6-2). Next, by inspection we see that the equation

$$y - y_1 = \left[\frac{y_2 - y_1}{x_2 - x_1} \right] (x - x_1) \tag{5.6-4}$$

passes through the point (x_1, y_1) and has the right slope. It clearly also passes through the point (x_2, y_2). Of course, the points p_1 and p_2 may be interchanged if desired. In this form (5.6-4) the slope must be a finite number; the line cannot be vertical (see Figure 5.6-1). However, if we write Equation (5.6-4) in the more symmetric form by multiplying by the denominator $x_2 - x_1$, we get

$$(x_2 - x_1)(y - y_1) = (y_2 - y_1)(x - x_1) \tag{5.6-5}$$

and this can represent a vertical line (which occurs when $x_1 = x_2$). In this special case the left-hand side is zero (we assumed that the two points were distinct, so it follows that $y_2 \neq y_1$); therefore, we have

$$x - x_1 = 0$$

as the equation of the vertical line.

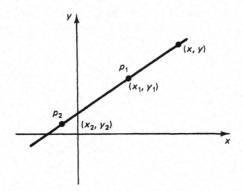

Figure 5.6-1 Two-point form

Second, there is the *point-slope* formula

$$y - y_1 = m(x - x_1) \qquad (5.6\text{-}6)$$

which is the same as (5.6-4) with the slope replaced by the letter m (see Figure 5.6-2). Again, the line cannot be vertical in this form.

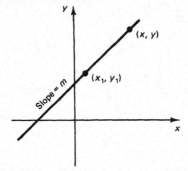

Figure 5.6-2 Point-slope form

Third, there is the *slope-intercept* form

$$y = mx + b \qquad (5.6\text{-}7)$$

where b is the *intercept* on the y-axis (meaning the value of y when $x = 0$, where the line intercepts the y-axis). It comes from the point-slope formula (5.6-6) by transposing the y_1 term to the right and combining it with the other term so that b is defined as

$$b = y_1 + mx_1$$

An alternative approach is to set $y_1 = 0$ and $mx_1 = b$. Again, this line cannot be vertical (see Figure 5.6-3).

Fourth, there is the *intercept* form of the line

$$\frac{x}{a} + \frac{y}{b} = 1 \qquad (5.6\text{-}8)$$

where a is the value of the x intercept and b is the value of the y intercept. The intercept form does not allow for lines through the origin, nor vertical, nor horizontal lines (see Figure 5.6-4).

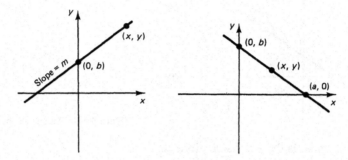

Figure 5.6-3 Slope-intercept form   **Figure 5.6-4** Intercept form

Example 5.6-1

These examples show how to go from geometric properties to equations. We can illustrate each type easily.

Given the two points $p_1 = (2, 3)$ and $p_2 = (4, -2)$ (see Figure 5.6-5), we get from the two-point form (5.6-4)

$$y - 3 = \left[\frac{-2 - 3}{4 - 2}\right](x - 2)$$

which is

$$y - 3 = \frac{-5}{2}(x - 2)$$

Given the point $(3, 5)$ and the slope $m = 4$, the point-slope formula (5.6-6) gives

$$y - 5 = 4(x - 3)$$

Given the slope $m = 4$ and the y intercept -3, we have, using the slope-intercept equation (5.6-7),

$$y = 4x - 3$$

Finally, given the x intercept 3 and the y intercept -2, we have the line, using (5.6-8),

$$\frac{x}{3} + \frac{y}{-2} = 1$$

Multiplying through by 6, we get

$$2x - 3y = 6$$

These are some of the ways you can go from geometric properties to equations.

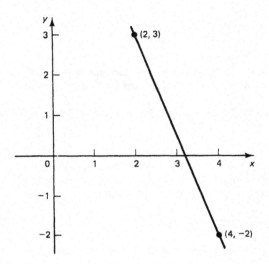

Figure 5.6-5 Two-point line

As a general rule, two conditions are sufficient to determine a line. A condition can be (1) a slope or (2) passing through a point. Note that an intercept is simply a point with one of its coordinates equal to zero. The special forms are merely convenient ways of writing the equation initially or, conversely, having written the equation in that form, of finding the geometric properties of the line. Note that when the special equations are used there can be implied restrictions on the solutions you can find.

Example 5.6-2

These examples show how to go from equations to geometric properties. Given the general line

$$Ax + By + C = 0$$

you can assign any value x (unless $B = 0$), compute a corresponding y, and thus find a point on the line. Suppose $B \neq 0$; you can write the above equation in the form

$$y = -\left(\frac{A}{B}\right)x - \frac{C}{B}$$

and the above computation is easy. If two points are wanted, then choose a second x value. For plotting purposes it is wise to have the points far apart.

For the slope, you clearly have (from the slope-intercept form)

$$m = \frac{-A}{B}$$

and the corresponding y intercept ($x = 0$) is $-C/B$.

The two intercepts can be found from the general equation by dividing by $-C$ (which must not be zero in this case). You then write the equation in the proper intercept form:

$$\frac{x}{-C/A} + \frac{y}{-C/B} = 1$$

Thus the x intercept is $-C/A$ and the y intercept is $-C/B$.

An alternative way is to first put $x = 0$, thus finding the y intercept, $b = -C/B$, followed by putting $y = 0$ to get the x intercept, $a = -C/A$, and finally putting these into the intercept form.

In these ways you can easily pass from the equation to the geometric properties of the line.

Example 5.6-3

Midpoint of a line segment. Find the midpoint of the line segment running from (2, 4) to (−4, 6). A little thought (and a look at Figure 5.6-6) suggests that the coordinates of the midpoint must be the average of the coordinates of the end points:

$$x = \frac{2 + -4}{2} \quad \text{and} \quad y = \frac{4 + 6}{2}$$

This immediately suggests the general rule: the midpoint of the line segment from (a, b) to (c, d) has the coordinates

$$x = \frac{a + c}{2} \qquad y = \frac{b + d}{2}$$

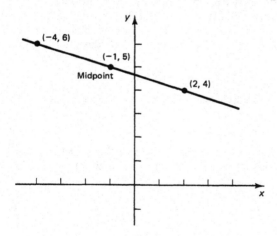

Figure 5.6-6 Midpoint

EXERCISES 5.6

Find the equation through the following points:

1. (2, −3) and (3, −4)
2. (2, 2) and (3, 3)
3. (2, 1) and (5, 1)
4. (−1, 1) and (1, −1)

Find the line through the following:

5. (0, 0) with slope -3.

6. (1, 1) with slope -1.

7. (4, -2) with slope -1.

8. Find the line with intercepts $-a$ and $-a$. What is the slope?

9. Given the line $3x - 4y = 19$, what is the slope? The x intercept?

10. Given the line $x + y + 1 = 0$, what is the slope and the y intercept?

11. Find the intercepts of the line $2x + 3y + 4 = 0$.

12. Find the intercepts of the line $7x + 9y + 5 = 0$.

13. Find the midpoint of the line segment between $(-3, 3)$ and $(3, -1)$.

14. Find the midpoint of the line segment between the points (p, q) and (r, s).

15. Find the formula for the $\frac{1}{3}$ and $\frac{2}{3}$ points of a line segment (trisection).

16.* Generalize Exercise 15.

5.7 ON PROVING GEOMETRIC THEOREMS IN ANALYTIC GEOMETRY

In principle, you can use the analytic geometry approach and prove various theorems in (synthetic) Euclidean geometry, but in fact we use the two methods for different purposes most of the time. There is some overlap, as the following example shows.

Example 5.7-1

Prove that the three perpendicular bisectors of the sides of a triangle meet in one point. Since it is a problem in geometry, the absolute values of the coordinates do not matter; so we can pick the coordinate system to make the algebra as easy for ourselves as possible. This observation suggests we pick the origin of the coordinate system at one of the vertices of the triangle, say A. Next we put the x-axis along one side to the point B, which will now have, say, the coordinates $(b, 0)$. The third vertex, C, will now have coordinates (c, d), which are arbitrary. (See Figure 5.7-1 for one specific arrangement of the points.)

We just saw (Example 5.6-3) that the midpoint of a line segment has coordinates that are the average of the coordinates of the two end points. We now take the sides one at a time and construct the perpendicular bisectors: lines through the midpoints and with the negative reciprocals (5.5-8) of the slopes of the corresponding sides. For the side AB we have the midpoint as $(b/2, 0)$ and the slope $m = 0$. Hence the perpendicular bisector has the equation

$$x = \frac{b}{2}$$

This is a vertical line through the midpoint, and we see that it is correct.

The second side, AC, has the midpoint $(c/2, d/2)$, and the slope of the side is $m = d/c$. The perpendicular bisector is, therefore, from the point-slope formula (5.6-6),

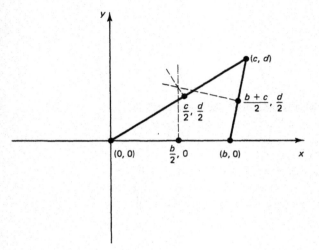

Figure 5.7-1   Perpendicular bisectors

$$y - \frac{d}{2} = \left(\frac{-c}{d}\right)\left(x - \frac{c}{2}\right)$$

This line intersects the earlier perpendicular bisector $x = b/2$ at the y value (put $x = b/2$):

$$y = \frac{d}{2} + \left(-\frac{c}{d}\right)\left(\frac{b}{2} - \frac{c}{2}\right)$$

When we simplify this, we get

$$y = \frac{d^2 + c^2 - bc}{2d} \qquad (5.7\text{-}1)$$

Thus the intersection of the first two perpendicular bisectors is at the point $x = b/2$, and y is equal to the above value (5.7-1).

For the third side, BC, we have the midpoint, $((b + c)/2, d/2)$, and the slope of the side is $m = d/(c - b)$. Hence the perpendicular bisector of this side is given by the equation

$$y - \frac{d}{2} = \left(\frac{b - c}{d}\right)\left(x - \frac{b + c}{2}\right)$$

This line intersects the first perpendicular bisector at $x = b/2$, so the corresponding y value is given by

$$y = \frac{d}{2} + \left(\frac{b - c}{d}\right)\left(\frac{b}{2} - \frac{b}{2} - \frac{c}{2}\right)$$

$$= \frac{d}{2} + \left(\frac{b - c}{d}\right)\left(-\frac{c}{2}\right)$$

$$= \frac{d^2 + c^2 - bc}{2d}$$

as before (5.7-1). Hence the three perpendicular bisectors have the same common point, and we have proved the theorem.

EXERCISE 5.7

1. Prove that the diagonals of a parallelogram bisect each other. *Hint:* Find their midpoints.

5.8* THE NORMAL FORM OF THE STRAIGHT LINE

The *normal form* of the equation is very useful when you are interested in the distance of a point from a given line. From Figure 5.8-1 we see that the distance p from the origin to the line in a perpendicular direction (the shortest distance from the origin to the line) meets the x-axis with an angle θ (theta). What are the intercepts of the line in terms of the distance p? The x intercept a is found from the lower triangle to be

$$a = \frac{p}{\cos \theta}$$

Similarly, from the upper triangle, the y intercept is

$$b = \frac{p}{\sin \theta}$$

Next, using the intercept form of the line (5.6-8),

$$\frac{x}{a} + \frac{y}{b} = 1$$

We have (after some algebra)

$$x \cos \theta + y \sin \theta - p = 0 \qquad\qquad (5.8\text{-}1)$$

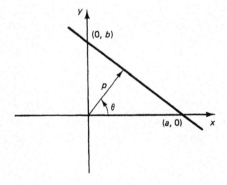

Figure 5.8-1   Normal form

How can we get the general equation of a line (5.5-2) into this form? We note first that

$$\sin^2 \theta + \cos^2 \theta = 1$$

Therefore, if we take the general equation

$$Ax + By + C = 0$$

and divide by

$$[A^2 + B^2]^{1/2}$$

we will have the correct general form, usually called *the normal form*, where all + or all − signs are used:

$$\frac{Ax}{\pm\sqrt{A^2 + B^2}} + \frac{By}{\pm\sqrt{A^2 + B^2}} + \frac{C}{\pm\sqrt{A^2 + B^2}} = 0$$

How do we decide which sign to use? In the form we have taken (5.8-1), the quantity p is positive. Thus the sign of the square root must be taken to make p positive; that is,

$$\frac{C}{\pm\sqrt{A^2 + B^2}} \le 0 \qquad\qquad (5.8\text{-}2)$$

The square root has the opposite sign to that of C: *p is positive is the rule*.

In this form we make the identification by comparing it with (5.8-1):

$$\cos \theta = \frac{A}{\pm\sqrt{A^2 + B^2}}$$

$$\sin \theta = \frac{B}{\pm\sqrt{A^2 + B^2}} \qquad\qquad (5.8\text{-}3)$$

From this we get

$$\tan \theta = \frac{B}{A} \qquad\qquad (5.8\text{-}4)$$

As a check, we note that the slope of the original line is $-A/B$ and that (5.8-4) is the negative reciprocal, as it should be.

Example 5.8-1

Suppose we have the equation

$$3x + 4y + 10 = 0$$

and we want to put this into the normal form. Since $A^2 + B^2 = 3^2 + 4^2 = 5^2$, we divide by ± 5. We need to pick the minus sign to make p positive. Thus we divide through by -5 to get the equation

$$-\left(\frac{3}{5}\right)x - \left(\frac{4}{5}\right)y - 2 = 0$$

as the normal form (5.8-1) of the equation (see Figure 5.8-2). The distance of the origin from the line is 2. From the sign of the coefficients ($\sin \theta < 0$, $\cos \theta < 0$), θ must lie in the third quadrant.

If we hold θ fixed (meaning that the first two coefficients in the normal form are not to be changed) and put into the equation the coordinates of any point $p_1 = (x_1, y_1)$, we will find a new p value for an equation of a line that is parallel to the original given line. It is necessary to distinguish between the subscripted p_1, which is a point, and the distance p of the line from the origin. But p is the perpendicular distance from the line to the origin. The difference between the new p and the old p is exactly how far the point p_1 is from the line (see Figure 5.8-3). The distance of the point from the line can be found directly by simply substituting the coordinates of the point p_1 into the normal form of the line to get

$$d = x_1 \cos \theta + y_1 \sin \theta - p \qquad (5.8\text{-}5)$$

As seen from Figure 5.8-3, this d is the distance of the point from the line. If $d > 0$, then the point is on the far side of the line from the origin, and if $d < 0$, then it is on the near side.

Thus the normal form is useful for finding the distance from a point to a line.

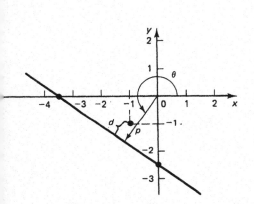

Figure 5.8-2 $3x + 4y + 10 = 0$

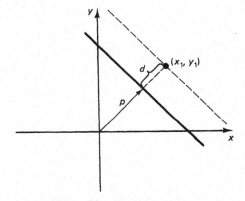

Figure 5.8-3   Distance of a point to a line

Example 5.8-2

Using Example 5.8-1 suppose we find the distance of the point $(-1, -1)$ from the line. We have from (5.8-5)

$$d = -\left(\frac{3}{5}\right)(-1) - \left(\frac{4}{5}\right)(-1) - 2 = \frac{7}{5} - 2 = -\frac{3}{5}$$

The negative distance clearly means that the point is on the side of the line that includes the origin (Figure 5.8-2).

Example 5.8-3

Angle bisectors. How can we find the bisector of the angle between two lines? We know that the bisecting line is a linear equation and that it goes through the common point of the two given lines. What else do we know about the bisector? Some thought and we recall that any point on the bisector is equidistant from the two sides. This suggests that we put both of the given lines into the normal form and then equate the distances, either both positive or one positive and one negative. The common point of the two sides (the vertex) will, of course, lie on this line. The difference in sign determines which of the two bisectors (Figure 5.8-4) we get. This must give the line we want. First, it is linear and so is a straight line, and, second, we know that any linear combination of two lines must pass through the common intersection; so this must be the bisector of the two given lines.

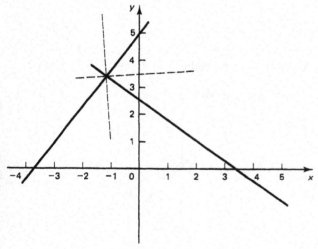

Figure 5.8-4   Bisectors of angles

Example 5.8-4

We illustrate the above with an example. Given the two lines

$$3x + 4y = 10 \quad \text{and} \quad 4x - 3y = -15$$

we put both in the normal form by dividing by the sum of the squares of the coefficients of x and y. Thus we get the equations

$$\left(\frac{3}{5}\right)x + \left(\frac{4}{5}\right)y - 2 = 0 \quad \text{and} \quad -\left(\frac{4}{5}\right)x + \left(\frac{3}{5}\right)y - 3 = 0$$

as the two normal forms. The two solutions are therefore

$$\left(\frac{3}{5}\right)x + \left(\frac{4}{5}\right)y - 2 = \pm\left\{-\left(\frac{4}{5}\right)x + \left(\frac{3}{5}\right)y - 3\right\}$$

We multiply through by 5 to get rid of the fractions:

$$3x + 4y - 10 = \pm\{-4x + 3y - 15\}$$

from which we get the two equations (use the $+$ and the $-$ signs)

$$7x + y + 5 = 0 \quad \text{and} \quad -x + 7y - 25 = 0$$

Beware of the fact that in the example both normalizing factors were of size 5; usually they are different. *Note:* The two bisectors are mutually perpendicular.

EXERCISES 5.8

Put the following equations into normal form:

1. $5x + 12y + 36 = 0$
2. $y = 4x - 5$
3. $x/3 + y/4 = 1$

How far are the following points from the line:

4. $(2, 3)$ and $5x - 12y + 39 = 0$
5. $(1, 1)$ and $3x + 3y = 6$
6. $(1, -1)$ and $x + y = 1$
7. Find the $\tan \theta$ and p for the equation $2x + 2y = 17$.
8.* Develop the general equations for the bisectors of two general lines, $A_1 x + B_1 y + C_1 = 0$ and $A_2 x + B_2 y + C_2 = 0$.
9.* Prove that the bisectors of the angles of a triangle meet in a common point. *Hint:* Pick the coordinates as in Example 5.7-1.
10.* Discuss the normal form of the line when the line passes through the origin.

.9 TRANSLATION OF THE COORDINATE AXES

To eliminate some terms in an equation, it is frequently very convenient to translate the origin of the coordinate axes to some new point in the plane. Conventionally, this new origin is assigned the coordinates (h, k) in the old coordinate system. Figure 5.9-1 shows such a shift. The values of h and k are shown in the picture as being positive, but they may be of either sign or zero. Let the new coordinate system use coordinates (x', y'). Then it is easy to see that

$$x = x' + h$$
$$y = y' + k$$

(5.9-1)

If we solve for the new coordinates in terms of the old, we have

$$x' = x - h$$
$$y' = y - k$$

(5.9-2)

Note the structure of Equations (5.9-1) and 5.9-2).

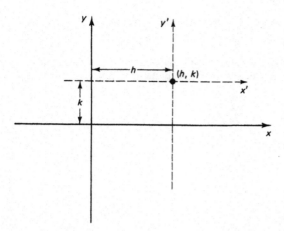

Figure 5.9-1 Translation of coordinates

Example 5.9-1

Suppose we have the line

$$y = 3x - 7$$

and wish to shift the origin of the coordinate system to the point $(1, -2)$. We have, using (5.9-1),

$$y' - 2 = 3(x' + 1) - 7$$

or
$$y' = 3x' - 2$$

Notice that, of course, the slope $m = 3$ stays the same.

EXERCISES 5.9

1. Translate the equations $x - y - 2 = 0$ and $x + y - 6 = 0$ to the coordinate system with the origin at $(4, 2)$.
2. Translate the general equation $Ax + By + C = 0$ to the point (h, k).
3. Translate the origin of the coordinate system to the common point of the two lines $3x + 4y = 5$ and $x - y = -5$.
4. Develop the theory for translating the origin of the coordinate system to the common point of two arbitrary lines, $A_1x + B_1y + C_1 = 0$ and $A_2x + B_2y + C_2 = 0$.
5. Discuss the fact that the translation $x = x' + h$ followed by the translation $x'' = x' - h$ leaves the equation unchanged.
6. Discuss the result of one translation followed by another translation.

5.10* THE AREA OF A TRIANGLE

Case Study 5.10-1

Area of a triangle. A common problem in geometry is to find the area of a triangle. Usually the triangle is defined by the three vertices, although the equations of the three sides might be given instead. If the triangle is in the form of three equations, solving the equations two at a time to find the vertices (the common point of the two lines) reduces the problem to the first form, given three points.

The problem looks new, so let us start by doing the simplest case we can imagine, a standard method of approaching problems in mathematics. (See Section 1.5, Hilbert.) Put the first point at the origin, the second point at $(x_2, 0)$ on the x-axis, and the third point at (x_3, y_3). (We do not pick the coordinates a, b, and c as before because we must end up with three general points with general coordinates.) We look at the picture, Figure 5.10-1. The area A is the base, x_2, times the altitude, y_3, times $\frac{1}{2}$; that is

$$A = \frac{x_2 y_3}{2} \qquad (5.10\text{-}1)$$

and we have the answer.

Next, we approach nearer to the general case by moving the point $(x_2, 0)$ off the x-axis to the general point (x_2, y_2) (see Figure 5.10-2). In this figure the area of the shaded triangle (which is what we want) is the difference between the areas of the two triangles, both with the same base a on the x-axis. Thus the area is now

$$A = \frac{a y_3}{2} - \frac{a y_2}{2} = \frac{a(y_3 - y_2)}{2} \qquad (5.10\text{-}2)$$

where a is the x intercept of the line passing through the third and second points.

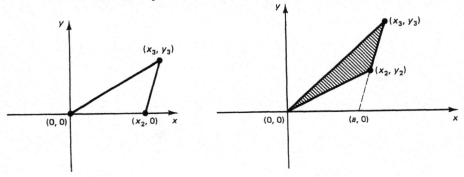

Figure 5.10-1   **Figure 5.10-2**

To find this intercept a, we first find the line through the points p_3 and p_2. We begin by finding the slope (5.5-1)

$$m = \frac{y_3 - y_2}{x_3 - x_2} \qquad (5.10\text{-}3)$$

Then we use the point-slope formula (5.6-6)

$$y - y_2 = m(x - x_2)$$

Where does this line cross the x-axis? At $y = 0$. Therefore, the solution x is our a value (meaning set $y = 0$ and $x = a$). We have

$$-y_2 = m(a - x_2)$$

Now divide both sides by the slope m and then eliminate m using the slope formula above (5.10-3) to get

$$a - x_2 = -y_2 \frac{x_3 - x_2}{y_3 - y_2}$$

Solve for a by transposing x_2 and getting a common denominator:

$$a = \frac{-y_2 x_3 + y_2 x_2 + x_2 y_3 - x_2 y_2}{y_3 - y_2}$$

$$= \frac{x_2 y_3 - x_3 y_2}{y_3 - y_2}$$

Now put this value of a into the earlier formula (5.10-2) for the area. We get

$$A = \frac{x_2 y_3 - x_3 y_2}{2} \tag{5.10-4}$$

The final step in deriving the general formula for the area is to translate the axis to the point $(-x_1, -y_1)$ using (5.8-1) so that in the new coordinate system the point that was at the origin will now have coordinates (x_1, y_1). Dropping, for the moment, the $\frac{1}{2}$ factor, we have

$$x_2 y_3 - x_3 y_2 = (x_2' - x_1)(y_3' - y_1) - (x_3' - x_1)(y_2' - y_1)$$

Expand the right-hand side and drop the primes for convenience:

$$= x_2 y_3 - x_2 y_1 - x_1 y_3 + x_1 y_1 - x_3 y_2 + x_3 y_1 + x_1 y_2 - x_1 y_1$$

The $x_1 y_1$ terms cancel, and rearranging the terms as well as including the $\frac{1}{2}$ factor we get, finally, for the area of the general triangle

$$A = \frac{1}{2}[(x_2 y_3 - x_3 y_2) + (x_3 y_1 - x_1 y_3) + (x_1 y_2 - x_2 y_1)] \tag{5.10-5}$$

The symmetry of this formula is highly appealing. For those who know determinants, this can be written in the form of a determinant

$$A = \frac{1}{2} \begin{vmatrix} 1 & x_1 & y_1 \\ 1 & x_2 & y_2 \\ 1 & x_3 & y_3 \end{vmatrix} \tag{5.10-6}$$

The area will turn out to be positive if the points are taken in a counterclockwise direction and negative if in the clockwise direction. We will later see why we might want to admit negative areas.

The formula passes some tests of reasonableness.

1. If any two points are the same, then the area is 0.
2. As noted, there is a high degree of symmetry in the formula. Thus, if the names of the points are rotated, 1 to 2, 2 to 3, and 3 to 1, the area (and formula) stays the same. And similarly for a rotation of the point labels in the reverse direction.

3. We can verify the formula in a special case to see if the front coefficient is correct. The case $(0, 0)$, $(1, 0)$, and $(0, 1)$ checks this. We get

$$A = \frac{1}{2}[1 - 0 + 0 - 0 + 0 - 0] = \frac{1}{2}$$

as it should.

4. We can also try a second special case, $x_1 = 0$ and $y_1 = 0$. The result agrees with our preliminary formula (5.10-4):

$$A = \frac{1}{2}[x_2y_3 - x_3y_2]$$

Rule: Whenever you derive a general formula, it is wise to subject it to a number of specific tests as well as general ones, such as having the necessary *symmetry and invariance*. While no number of specific tests will *prove* the formula correct for all cases, usually you can find a set that will greatly increase your confidence in the result. Of course, checking all the reasoning and all the algebra is one test; but, unfortunately, if you have made an error in the derivation, you are apt to make the same one again!

EXERCISES 5.10

Find the area of the triangle having the following vertices:

1. $(1, 0)$, $(0, 1)$, and $(1, 1)$
2. $(1, 2)$, $(2, 3)$, and $(-3, -2)$
3. (a, a), (b, b), and $(c, -c)$
4. (a, b), (b, c), and (c, a)
5. Using (5.10-5), translate the x coordinate an amount h. Note the invariance of the area. Do the same for the y coordinate.

5.11* A PROBLEM IN COMPUTER GRAPHICS

A problem that occurs constantly in computer graphics is the "hidden point problem." You are often given the coordinates of the vertices of the various figures whose sides are straight lines, and are asked to show what is seen and what is hidden from some viewpoint. For convenience we first translate the viewpoint to the origin.

Case Study 5.11-1

Hidden point. To decide if a point is in front of an object or behind it (Figure 5.11-1), we need first to find the equation of the line through the given pair of points that define that edge, imagine the equation in normal form, and substitute the coordinates of the point into the line. We do not actually need to find the distances, only the sign. Therefore, there is no need to find the squares and the square root; it is only necessary to get the sign

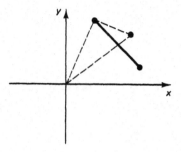

Figure 5.11-1 Hidden point

of the constant term correct, substitute the coordinates of the point into the equation, and note the resulting sign.

To decide if some point is to the right or left of another point, we can merely use the area formula. With the viewpoint put at $(0, 0)$, we have the simple formula

$$2A = x_2 y_3 - x_3 y_2$$

Again, it is only the sign that we want, not the actual angles involved. We must test, of course, that the hidden point is between the ends of the line segment.

These two observations, based on a slight knowledge of analytic geometry, together with the belief that one should not compute more than is necessary, is all that is needed to solve this common computer graphics problem. Neither the actual distances nor the actual angles are needed in many cases, only whether or not some point is visible. In this geometry, it is not the distances that matter, only in front or behind, nor the angles, only to the right or to the left. Both are easily solved using the elements of analytic geometry and greatly decrease the computer time used to solve the hidden point problem.

EXERCISES 5.11

1. Given the line segment from $(1, 1)$ to $(3, -4)$, give the details for finding if the point (x, y) is hidden by the segment or not.

2.* Given the four vertices of a quadrilateral, p_1, p_2, p_3, and p_4 in a counterclockwise order, and a general point (x, y), indicate the main steps you would use to decide if the point (x, y) is inside or outside the quadrilateral. (Beware of the origin's being inside the quadrilateral.)

5.12 THE COMPLEX PLANE

A complex number of the form $a + ib$ ($i = \sqrt{-1}$) has two real numbers, a and b, and this suggests plotting a complex number on a plane similar to the Cartesian plane *except* that we use the axes x and iy.

Example 5.12-1

Plot $2 + 3i$, $-2 - 3i$. See Figure 5.12-1.

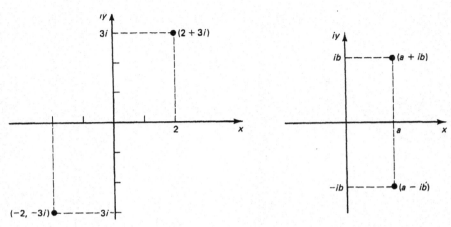

Figure 5.12-1 Complex plane   **Figure 5.12-2** Conjugate points

Example 5.12-2

Plot the complex conjugate points $a + ib$ and $a - ib$. See Figure 5.12-2.

The *absolute value* squared of a complex number is the product of the number and its conjugate (4.8-8):

$$|a + ib|^2 = (a + ib)(a - ib) = a^2 + b^2$$

Thus the absolute value plays a role in the complex *Argand* (1768–1822) *plane* similar to the Pythagorean distance in the Euclidean plane. The absolute value of a complex number is the *modulus* of the number, (4.8-7), and plays the role of the distance from the origin.

The representation of complex numbers in the complex plane gives a natural picture of the numbers, and hence tends to give the student a feeling that complex numbers are less strange than when only the algebraic $i = \sqrt{-1}$ is used. Together with the distance function, which is the absolute value, the Argand plane seems to provide some reasonableness to complex numbers.

For example, when two complex numbers are added, say $a + ib$ and $c + id$, their sum

$$(a + ib) + (c + id) = (a + c) + i(b + d)$$

is represented by the diagonal of the parallelogram with the two sides $a + ib$ and $c + id$, as shown in Figure 5.12-3.

From Figure 5.12-4 we see that

$$x = r \cos \theta \quad \text{and} \quad iy = r \sin \theta$$

The modulus of $x + iy$ is therefore

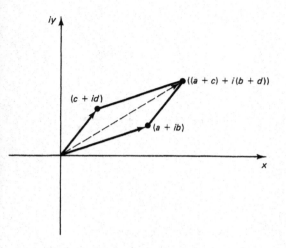

Figure 5.12-3 Additional of complex
numbers

Figure 5.12-4

$$[x^2 + y^2]^{1/2} = [r^2(\cos^2 \theta + \sin^2 \theta)]^{1/2} = r$$

Some of the power of complex numbers arises from the simple observation that a single equation in complex numbers is equivalent to two real equations, since a complex number $a + ib = 0$ requires both $a = 0$ and $b = 0$. There are also many interesting relationships that are best understood when written in the form of complex numbers, and we will see many of them in this book. Trigonometry is especially given some unity when complex numbers are used; it is no longer a miscellaneous collection of random formulas (see Section 21.5).

The Euclidean (Cartesian) plane and the complex (Argand) plane are different. In the Euclidean plane a pair of real numbers (x, y) are used to name a point; in the complex plane a *single* complex number $x + iy$ defines the point.

EXERCISES 5.12

1. Plot the points $3 + 3i$ and $3 - i$.
2. Find the modulus of $3 + 4i$.
3. Show that $|1/2 + i\sqrt{3}/2| = |\sqrt{3}/2 + i/2| = 1$.
4. For real a, show that $|(ai - 1)/(ai + 1)| = 1$.
5. Find the modulus of $1/(1 + i)$.
6. Show that $|\cos x + i \sin x| = 1$.
7. Give the modulus of $i, -i, 1, -1$.
8. Prove that the modulus of a complex number and its conjugate are the same.

9. Prove that the $|x|^2 = |x^2|$.
10. Prove $|x||y| = |xy|$, where x and y are complex numbers.
11. Generalize Exercise 3 and contrast with Exercise 8.

.13 SUMMARY

You have begun the topic of analytic geometry. The basic idea of analytic geometry is that there is an exact correspondence between number pairs (x, y) and points in the plane. Therefore, there is a correspondence between parts of algebra and geometry. You have also seen that the simplest geometric curves (straight lines) correspond to the simplest algebraic equations (linear equations). You have further learned how to go from geometric properties of straight lines to linear algebraic equations and back again to geometric properties. Finally, you learned how to translate the coordinate system from one origin to another. Along the way you found how to compute various useful things such as the distance between two points, the slope of a line, angles between lines, intersections of lines, the distance of a point from a line, and the area of a triangle. You have also used such things as the method of undetermined coefficients, the solution of simultaneous equations, some elements of trigonometry, and others.

We also continued your introduction to complex numbers beyond the usually brief treatment when you first see the quadratic equation. In beginning algebra, complex numbers are used only to explain where the two real roots disappear when the discriminant $b^2 - 4ac$ passes from positive to negative values. Plotting complex numbers in the Argand diagram provides one means of grasping what role they play in mathematics. The complex plane is one of the major tools in advanced mathematics, both continuous and discrete, as well as in many applications in the real world, such as unifying electric and magnetic phenomena.

6

Curves of Second Degree—Conics

6.1 STRATEGY

In the previous chapter we examined linear equations and found that they correspond to straight lines and, conversely, that straight lines correspond to linear equations. It is natural to turn next to the general quadratic (second degree) equation. However, it is sound strategy to approach the subject cautiously and begin with simple things. The simplest geometric figures, after straight lines, are circles, which we examine first. They will turn out to be particularly simple quadratic equations. Then we will gradually examine more complex quadratic equations until we have reached the general case. This is the way new territory is generally explored; the easier cases are tried first so you can get oriented, and when the easy cases are done you are ready to complete a general survey of quadratic equations (see Section 1.5, Hilbert).

6.2 CIRCLES

Consider a circle centered about the origin. From the definition of a circle, all the points on the circumference are at a fixed distance from the origin. That is, any point with coordinates (x, y) on the circle must satisfy the distance condition (5.2-1) from the origin.

$$x^2 + y^2 = r^2$$

where r is the radius of the circle. For example,

$$x^2 + y^2 = 25$$

is a circle of radius 5 centered about the origin.

There is an ambiguity in the common use of the word "circle." We say, "Draw a circle" and clearly mean draw the *circumference*. But we also speak of the area of a circle, meaning everything inside. To describe the set of points, we can say that the *circumference* is the set of points satisfying

$$x^2 + y^2 = r^2$$

The *circumference plus the inside* is the set of points satisfying

$$x^2 + y^2 \le r^2$$

Finally, the set of points satisfying

$$x^2 + y^2 < r^2$$

is the *inside* of the circle.

If you want the center of the circle to be at some other point than the origin, say at the general point (h, k), then the equation is

$$(x - h)^2 + (y - k)^2 = r^2 \qquad (6.2\text{-}1)$$

which describes all points a distance r from the center (h, k). We can either find the equation directly from the definition of the distance, or else we can use the equation of the circle about the origin and translate the origin to the new center (Section 5.9). Notice that these equations are not single-valued functions.

Example 6.2-1

The circle about the point (3, 4) of radius 5 is (see Figure 6.2-1)

$$(x - 3)^2 + (y - 4)^2 = 25$$

This can also be written in the form (expand the two square terms)

$$x^2 + y^2 - 6x - 8y = 0$$

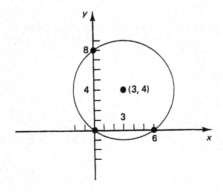

Figure 6.2-1   Circle centered at (3, 4)

Thinking about this particular circle, we see that it should pass through the origin (since the center is exactly 5 units from the origin and 5 is also the radius). To check this, we simply put in the coordinates (0, 0) and then note that they satisfy the equation.

The general circle (6.2-1) of radius r about the point (h, k) can be written in the expanded form

$$x^2 - 2hx + h^2 + y^2 - 2ky + k^2 - r^2 = 0 \tag{6.2-2}$$

which takes the *general form*

$$x^2 + y^2 + Dx + Ey + F = 0 \tag{6.2-3}$$

The reason for the choice of the letters D, E, and F will become apparent later. Of course, this equation could be multiplied through by any nonzero constant, and it would still have the same points as solutions of the equation. Indeed, any of the equations (6.2-1), (6.2-2), or (6.2-3) when multiplied by a nonzero constant represents the same circle. It is conventional to pick the coefficient of $x^2 + y^2$ as 1.

A comparison of the coefficients of the two forms (6.2-3) and (6.2-2), the algebraic and the geometric, shows that they are related by the conditions

$$D = -2h$$

$$E = -2k \tag{6.2-4}$$

$$F = h^2 + k^2 - r^2$$

We now know how to go from the definition of a circle of given center and radius to the equation for a circle. Can we handle other given conditions?

Suppose you are given three points to determine the circle. How will you find the corresponding equation? Easy! Use the method of undetermined coefficients. Because there are only three parameters, we have only to impose the three conditions that the three points lie on the circle when you write it in the general form (6.2-3). Since the general form has three parameters, D, E, and F, you will find that you have three linear equations to solve for the three parameters.

Example 6.2-2

In this first example we pick the points to make the algebra and arithmetic easy. Suppose the three points are (0, 0), (6, 0), and (0, 8). Using the form (6.2-3), these three points will generate the three equations (see Figure 6.2-1)

$$0^2 + 0^2 + 0D + 0E + F = 0$$

$$6^2 + 0^2 + 6D + 0E + F = 0 \tag{6.2-5}$$

$$0^2 + 8^2 + 0D + 8E + F = 0$$

Note that these are linear equations in the unknown coefficients D, E, and F and are therefore comparatively easy to solve.

From the top equation, $F = 0$. From the second equation, $D = -6$. And from the third equation, $E = -8$. Thus we have the equation of the circle through the three points

$$x^2 + y^2 - 6x - 8y = 0 \qquad (6.2\text{-}6)$$

It is easy to check that each of the original points lies on the circle (the coordinates of the points satisfy the equation).

Generally, three given points will not allow such an arithmetically simple solution of the three equations, but the method is the same. The method of undetermined coefficients wil always lead, for circles through three points, to three linear equations in the three unknowns D, E, and F. These three equations can generally be solved for the D, E, and F which define the circle algebraically. The only exception when these three equations cannot be solved for the unknowns is the case when the three points fall on a straight line (see Exercise 6.2-11).

You can now see some of the power of the method of undetermined coefficients as it applies to analytic geometry—assume the general form and apply the conditions to determine the unknown coefficients.

Example 6.2-3

As an example where the algebra is slightly more difficult, but still in terms of nice integers, consider the circle through the three points $(1, 1)$, $(-1, 1)$, and $(2, 3)$ (Figure 6.2-2). Put these coordinates into the general form for a circle, Equation (6.2-3):

$$x^2 + y^2 + Dx + Ey + F = 0$$

and get the three equations corresponding to Equations (6.2-5):

$$1 + 1 + D + E + F = 0$$
$$1 + 1 - D + E + F = 0 \qquad (6.2\text{-}7)$$
$$4 + 9 + 2D + 3E + F = 0$$

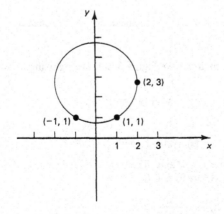

Figure 6.2-2 Circle through three points

Next rewrite these in the conventional form, with the unknowns on the left-hand side and the constants on the right:

$$D + E + F = -2$$
$$-D + E + F = -2 \qquad (6.2\text{-}8)$$
$$2D + 3E + F = -13$$

To use what little symmetry we have, we subtract the second equation from the first, which eliminates the E and F terms. We get the simple equation

$$2D = 0$$

which implies that $D = 0$. We drop one of the two equations we used, say the first equation. The lower two equations of (6.2-8) are now

$$E + F = -2$$
$$3E + F = -13 \qquad (6.2\text{-}9)$$

We subtract the top of these two equations from the lower to eliminate the F term, and we get

$$2E = -11$$

from which it follows that

$$E = \frac{-11}{2}$$

Substitute this value of E into the top (or bottom) equation of the pair (6.2-9) to get

$$F = -2 - E = -2 + \frac{11}{2} = \frac{7}{2}$$

Thus the equation for the circle through the three points is

$$x^2 + y^2 - \left(\frac{11}{2}\right)y + \frac{7}{2} = 0 \qquad (6.2\text{-}10)$$

It is easy to check that Equation (6.2-10) is satisfied by the coordinates of the three points.

Example 6.2-4

Show that the angle in a semicircle (see Figure 6.2-3) is a right angle. We pick the circle as

$$x^2 + y^2 = a^2 \qquad (6.2\text{-}11)$$

and the point on the circle as

$$(x_0, y_0), \quad \text{with } y_0 > 0 \qquad (6.2\text{-}12)$$

with the semicircle above the $y = 0$ axis. The slopes of the lines from this point (6.2-12) to the ends of the diameter $(-a, 0)$ and $(a, 0)$ are

$$m_1 = \frac{y_0}{x_0 + a} \quad \text{and} \quad m_2 = \frac{y_0}{x_0 - a}$$

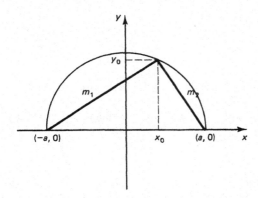

Figure 6.2-3 Angle in a semicircle

The condition for perpendicularity is, from (5.5-7),

$$m_1 m_2 = -1$$

This becomes

$$\frac{y_0}{x_0 + a} \frac{y_0}{x_0 - a} = \frac{y_0^2}{x_0^2 - a^2} = -1$$

We simplify this to

$$y_0^2 = -(x_0^2 - a^2) \quad \text{or} \quad x_0^2 + y_0^2 = a^2$$

which, by (6.2-11), is true since the point lies on the circle. These steps are reversible; therefore, we have shown that the two lines are perpendicular at the point where they intersect the circumference of the semicircle, as required.

These are examples of going from the geometry of the problem to the algebra. We see again the power of the method of undetermined coefficients: you assume the general form and *impose* the conditions to determine the unknown parameters. It is a powerful, general method in many areas of mathematics.

EXERCISES 6.2

Find the circle given:

1. Center $(-1, 1)$ and radius 1
2. Center $(-3, -4)$ and radius 5
3. Center $(1, 1)$ and radius 2

Find the circles through the following points:

4. $(1, 1), (1, -1), (-1, -1)$
5. $(2, 3), (3, 4), (4, 5)$

  6. $(5, 12)$, $(0, 0)$, and radius $6\frac{1}{2}$

  7. $(0, 0)$, $(0, 10)$, $(5, 13)$

  8. $(1, 10)$, $(1, -1)$, and radius 4

  9. $(6, 6)$, $(-6, 6)$, and $r = 2$

  10. Show that, if (a, b) lies on the circle $x^2 + y^2 = r^2$, then the points $(a, -b)$, $(-a, b)$, and $(-a, -b)$ also lie on the circle.

  11.* If the three points you select to fit a circle to happen to lie on a straight line, what happens?

6.3 COMPLETING THE SQUARE

We next examine problems involved in going from the algebra to the geometry. Suppose we have the equation of a circle and want to determine the geometric properties, the center and radius. How can we do this? We compare the expanded general form (6.2-2) about the point (h, k),

$$x^2 + y^2 - 2hx - 2ky + h^2 + k^2 - r^2 = 0 \qquad (6.3\text{-}1)$$

with the general form (6.2-3),

$$x^2 + y^2 + Dx + Ey + F = 0 \qquad (6.3\text{-}2)$$

We see as before, in Equation (6.2-4), that we must have

$$-2h = D, \qquad -2k = E, \qquad h^2 + k^2 - r^2 = F \qquad (6.3\text{-}3)$$

We have only to solve for h, k, and r. Note that it is not a sure thing that we can solve for r, since it might turn out that

$$r^2 = h^2 + k^2 - F = \frac{D^2}{4} + \frac{E^2}{4} - F \qquad (6.3\text{-}4)$$

is a negative number (and you can see that if we persisted then we would have a circle with an imaginary radius!).

Since it is a common problem to find the geometric properties, the h, k, and r in this case, from the algebraic equation, we look at a second approach [who wants to remember the formulas (6.3-3)?]. We try to write the general form, Equation (6.3-2),

$$x^2 + y^2 + Dx + Ey + F = 0$$

in the particular form (6.3-1):

$$(x - h)^2 + (y - k)^2 = r^2$$

This approach requires you to take the x terms,

$$x^2 + Dx$$

and get them into the form

$$(x - h)^2$$

To do this, we *complete the square* by adding to *both* sides of the equation

$$\left(\frac{D}{2}\right)^2$$

(in words, add the square of one half of the x term coefficient to both sides). Now we can write (on the left side)

$$x^2 + Dx + \left(\frac{D}{2}\right)^2 = \left(x + \frac{D}{2}\right)^2$$

Similarly, for the y terms we add $(E/2)^2$ to *both* sides. We get on the left-hand side

$$y^2 + Ey + \left(\frac{E}{2}\right)^2 = \left(y + \frac{E}{2}\right)^2$$

We take the remaining term F of (6.3-2) to the right-hand side, and as a result we have on the right side [see Equation (6.3-4)]

$$\left(\frac{D}{2}\right)^2 + \left(\frac{E}{2}\right)^2 - F = r^2$$

Therefore, the equation of the circle is

$$\left(x + \frac{D}{2}\right)^2 + \left(y + \frac{E}{2}\right)^2 = r^2$$

As before, only if this right-hand side is nonnegative can we have a circle that is real. If it is exactly zero, then the circle is a point at $(-D/2, -E/2)$. If it is positive, then the center of the circle and radius are given by

$$h = \frac{-D}{2}$$

$$k = \frac{-E}{2}$$

$$r^2 = \left(\frac{D}{2}\right)^2 + \left(\frac{E}{2}\right)^2 - F$$

as before. We have derived Equations (6.3-3) by a simple method that is worth learning carefully.

Example 6.3-1

Given the equation

$$x^2 + y^2 + 2x - y - 5 = 0$$

find the center and radius. Transpose the constant term, and complete the square in both x and y (remembering to add the same quantities to both sides to preserve the equality):

$$(x^2 + 2x + 1) + \left(y^2 - y + \frac{1}{4}\right) = 5 + 1 + \frac{1}{4}$$

or

$$(x + 1)^2 + \left(y - \frac{1}{2}\right)^2 = \frac{25}{4} = \left(\frac{5}{2}\right)^2$$

Thus the center is at $(-1, \frac{1}{2})$ and the radius is $\frac{5}{2}$.

Completing the square is a very common process in mathematics and is not confined merely to problems involving coordinate changes.

Example 6.3-2

Roots of a quadratic equation. The *process* of completing the square is exactly what you used when you found the general solution of the quadratic equation in beginning algebra:

$$ax^2 + bx + c = 0 \qquad (6.3\text{-}5)$$

First, you transposed the constant term c to the other side and then divided through by a.

$$x^2 + \left(\frac{b}{a}\right)x = \frac{-c}{a}$$

Next, you completed the square by adding to both sides the square of one-half the x term coefficient:

$$x^2 + \left(\frac{b}{a}\right)x + \left(\frac{b}{2a}\right)^2 = \left[x + \left(\frac{b}{2a}\right)\right]^2 = \left(\frac{b}{2a}\right)^2 - \frac{c}{a}$$

Then you found the common denominator on the right-hand side and took the square root to get, after transposing the $b/2a$ term,

$$x = -\left(\frac{b}{2a}\right) \pm \frac{\sqrt{b^2 - 4ac}}{2a}$$

as the final formula. Conventionally, this is written with a single denominator:

$$x = \frac{-b \pm \sqrt{b^2 - 4ac}}{2a} \qquad (6.3\text{-}6)$$

This is the standard formula for the roots of a quadratic, which we have already used several times. It is important to master both the method and the formula for the roots of a quadratic.

Example 6.3-3

Find the two lines through a given point tangent to a given circle. In such a problem we have the freedom to pick the coordinate system as we please, and it is natural to pick the circle centered about the origin. We pick the scale so that the radius of the circle is a; that is, we let the circle be

$$x^2 + y^2 = a^2 \qquad (6.3\text{-}7)$$

We next pick the equation of the tangent line in a convenient form. Since the circle has rotational symmetry, we are not giving up any generality by picking the point at the place $x = 0, y = b$. Let us take the slope-intercept form (Figure 6.3-1)

$$y = mx + b$$

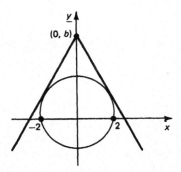

Figure 6.3-1   Tangent lines

What do we mean by the word "tangent"? We mean that the two potential intersections between the circle and the line are actually a single real point. This means that the roots of the quadratic equation, when we have eliminated one variable, say y, will be double. Using (6.3-6), this means that the discriminant $b^2 - 4ac = 0$. To follow this plan, we first eliminate y from (6.3-7) to get the quadratic equation

$$x^2 + (mx + b)^2 = a^2$$

which is

$$(1 + m^2)x^2 + 2mbx + (b^2 - a^2) = 0$$

This quadratic equation has a double root if the discriminant of the quadratic is zero, that is, if

$$(2mb)^2 - 4(1 + m^2)(b^2 - a^2) = 0$$

After some algebra,

$$4m^2b^2 - 4b^2 + 4a^2 - 4m^2b^2 + 4m^2a^2 = 0$$

and canceling the like terms $4m^2b^2$, this comes down to

$$m^2 = \frac{b^2 - a^2}{a^2}$$

Now, taking the square root gives the two slopes (of opposite sign).

$$m = \pm \frac{\sqrt{b^2 - a^2}}{a}$$

We now have both m and b for the tangent lines.

What is the meaning of the situation when b^2 is less than a^2? The answer is that the intercept b is inside the circle and the tangent lines are imaginary!

We apparently have mastered the circle and the art of completing the square. For a circle we now know how to go (1) from the geometry to the equation, and (2) from the equation to the geometry. These are the two fundamental steps in analytic geometry.

When we have completed the square for both x and y, we have the equation in the form

$$(x - h)^2 + (y - k)^2 = r^2$$

If we choose, we can translate the circle to the origin by the transformation of coordinates (see Section 5.9)

$$x' = x - h$$
$$y' = y - k$$

This transformation can be looked at in two ways. We can think of the coordinates being translated from the origin $(0, 0)$ to the center of the circle, or else we can think of the circle being translated to the origin. The first is called *alias*, meaning we have the same circle but under a different name. The second is called *alibi*, meaning not being there (which is the technical use of the word alibi). It does not matter which view you take of the transformation so long as you keep a consistent view. It is when you mix the two (alias and alibi) in the same problem that confusions can arise.

EXERCISES 6.3

1. Complete the square for $x^2 + 6x$.
2. Complete the square for $3x^2 + 9x$.
3. Complete the square for $5x^2 - 3x$.
4. Is the point $(2, 2)$ inside or outside the circle $x^2 + y^2 = 5$.
5. Find the equation of a circle of radius 7 and center $(-2, -3)$.
6. Find the circle of radius 13 about the center $(-5, 12)$. Show that it passes through the origin.
7. Given the equation

   $$x^2 + y^2 + 4x - 6y + F = 0$$

   what is the value of F so that the equation represents a point?
8. Find the circle through the points $(1, 1)$, $(2, 2)$, and $(3, -3)$. Give both center and radius. Check by making a drawing of the geometric problem.
9. The British often write the general quadratic as

   $$Ax^2 + 2Bx + C = 0$$

   Derive the corresponding formula for the roots. Discuss the amount of arithmetic needed in the two cases to compute the roots from the formulas.
10. Discuss the use of the general form for the circle

    $$x^2 + y^2 + 2Dx + 2Ey + F = 0$$

    in place of the one we used.

11. Given two intersecting circles

$$x^2 + y^2 + D_1x + E_1y + F_1 = 0$$

$$x^2 + y^2 + D_2x + E_2y + F_2 = 0$$

subtract the two equations to get the equation of a straight line. Describe geometrically what this line must be. *Ans.:* The line through the two common points of the circle.

12. Given the equations in Exercise 11, multiply one by a constant K and add to the other. Describe what this circle is geometrically. *Ans.:* A circle through the two common points, and when $K = -1$ it is a straight line (as in Exercise 11).

13.* Find the tangents to the circle $x^2 + y^2 = a^2$ going through the point (b, b) $(b > 0)$.

14.* In Example 6.3-3, discuss the case when $a = b$.

6.4 A MORE GENERAL FORM OF THE SECOND-DEGREE EQUATION

For a circle the coefficients of both the square terms are the same, and it is taken conveniently to be 1. More generally, but not completely general as yet (we are still omitting the cross product term xy), we examine the quadratic form

$$Ax^2 + Cy^2 + Dx + Ey + F = 0 \qquad (6.4\text{-}1)$$

We are for the moment assuming that neither A nor C is zero ($AC \neq 0$). The cases when $AC = 0$ will be examined later.

It is natural to complete the square to eliminate the linear terms Dx and Ey. We have, at the first stage of the algebra,

$$A\left(x^2 + \frac{Dx}{A}\right) + C\left(y^2 + \frac{Ey}{C}\right) = -F$$

Notice that we must factor out the leading coefficient of each variable before we start. To complete the squares, we add $A(D/2A)^2$ and $C(E/2C)^2$ to both sides (notice that the leading coefficient must be remembered for each term):

$$A\left[x + \frac{D}{2A}\right]^2 + C\left[y + \frac{E}{2C}\right]^2 = \frac{D^2}{4A} + \frac{E^2}{4C} - F$$

Next, we translate the coordinate system to "clean up" the equations a bit. Set

$$x' = x + \frac{D}{2A}$$

$$y' = y + \frac{E}{2C} \qquad (6.4\text{-}2)$$

$$K = \frac{D^2}{4A} + \frac{E^2}{4C} - F$$

which gives

$$A(x')^2 + C(y')^2 = K$$

We now drop the primes and divide by K (the case $K = 0$ will be taken up later):

$$\frac{Ax^2}{K} + \frac{Cy^2}{K} = 1$$

Divide the numerator and denominator of the first term by A and of the second term by C:

$$\frac{x^2}{K/A} + \frac{y^2}{K/C} = 1$$

To get the expression in a symmetric form, we write the denominators as squares.

$$\left|\frac{K}{A}\right| = a^2 \quad \text{and} \quad \left|\frac{K}{C}\right| = b^2 \tag{6.4-3}$$

We then have the *standard reduced form*

$$\pm\frac{x^2}{a^2} \pm \frac{y^2}{b^2} = 1 \tag{6.4-4}$$

In this form we see the advantage of choosing the quantities as squares of numbers: we have a natural homogeneity in the terms. Had we not done so, then as we moved forward we would have found a great many square roots and would have finally decided to back up and introduce the above notation.

EXERCISES 6.4

Reduce the following to standard form:

1. $x^2 + y^2 + 4x - 6y - 3 = 0$
2. $3x^2 - 4y^2 + 2x - 8y + 4 = 0$
3. $x^2 - 4y^2 - 4x - 8y - 9 = 0$
4. $x^2 + x + y^2 - y = \frac{3}{2}$
5. $2x^2 + 8x - 3y^2 + 12y = 10$
6. Find the quadratic (6.4-1) through the four points $(0, 0)$, $(0, 2)$, $(3, 1)$, $(1, 3)$. *Hint:* Substitute the four points into the general form for the quadratic (6.4-1) (which has essentially four parameters) and solve for the unknown coefficients. One coefficient can be divided out.
7.* Given two general quadratics of the general form 6.4-1, show that any linear combination of them is a quadratic (possibly degenerate) through the common points.

.5 ELLIPSES

First, we consider the standard reduced form (6.4-4) for the case when both signs are positive. For the moment, assume that $a > b > 0$. A plot of selected points on the curve gives a shape like that in Figure 6.5-1. It is called an *ellipse*. We see that the axis crossings (intercepts) are $\pm a$ on the x-axis and $\pm b$ on the y-axis. Since x and $-x$ always lead to the same y, there is symmetry about the y-axis. Similarly, the presence of only even powers of y means that y and $-y$ lead to the same x, and there is symmetry about the x-axis.

If $a = b$, then we clearly have a circle of radius a. The ellipses (for $b < a$) are simply circles shortened in the y direction by the ratio b/a. As you can see by a little thought, an ellipse is the shadow of a circle on a plane when one or the other is tilted with respect to the direction of a parallel beam of light. To sketch the ellipse, we simply draw the box shown in dotted lines in Figure 6.5-1, and sketch the ellipse inside the box. If $b > a$, then the long direction is up the y-axis.

To get the sketch of the original equation, we merely shift the coordinate axes back to the original ones via the translation equations (6.4-2).

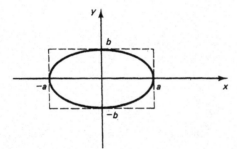

Figure 6.5-1   An ellipse

Example 6.5-1

Consider the equation

$$4x^2 + 9y^2 + 8x - 36y + 4 = 0$$

Complete the square and transpose all other terms to the right-hand side:

$$4(x^2 + 2x + 1) + 9(y^2 - 4y + 4) = -4 + 4 + 36$$

$$4(x + 1)^2 + 9(y - 2)^2 = 36$$

Now divide through by 36:

$$\frac{(x + 1)^2}{9} + \frac{(y - 2)^2}{4} = 1$$

From this you see that the center of the ellipse is $(-1, 2)$, $a = 3$, and $b = 2$. Thus, starting from the equation, you have found the geometrical properties. Figure 6.5-2 shows a sketch of this ellipse.

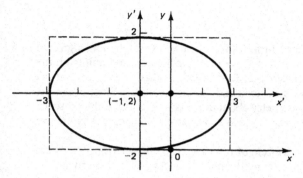

Figure 6.5-2

It is natural to ask, "How can we describe the *shape* of an ellipse?" Clearly, the values of a and b give the fundamental answer, but they depend on the size of the ellipse and we want to describe the *shape*. One way is to simply take the ratio of b to a (where we are assuming, as is natural, that both a and b are positive). Thus one measure is the ratio

$$r = \frac{b}{a}$$

In the above example, $a = 3$ and $b = 2$, so the ratio is $\frac{2}{3}$.

The most commonly used measure is, however, called the *eccentricity*, and is labeled by e:

$$e = \frac{\sqrt{a^2 - b^2}}{a} = \sqrt{1 - r^2}$$

where it is assumed that a is greater than or equal to b (if not, you would naturally interchange the roles of a and b even for the ratio, since the shape does not depend on which direction the larger "diameter" of the ellipse lies). In the above example you have

$$e = \sqrt{1 - \frac{4}{9}} = \frac{\sqrt{5}}{3}$$

When $a = b$, you have a circle whose eccentricity $e = 0$, while the ratio $r = 1$. Table 6.5-1 demonstrates the relationship of r and e.

Note that the eccentricity e measures small departures from circular better than does the ratio r. We have

$$e = \sqrt{1 - r^2} \quad \text{or} \quad e^2 + r^2 = 1$$

as the direct relationship between the two measures.

Notice that in both measures, e and r, the quantity is independent of the size (the units of measurement) of the ellipse. Such quantities are called *dimensionless* since they do not depend on the particular units (centimeters or inches) used to measure the

TABLE 6.5-1

r	e	e	r
1.0	0.000	0.0	1.000
0.9	0.4	0.1	0.995
0.8	0.6	0.2	0.98
0.7	0.7	0.3	0.95
0.6	0.8	0.4	0.92
0.5	0.85	0.5	0.85
0.4	0.92	0.6	0.80
0.3	0.95	0.7	0.70
0.2	0.98	0.8	0.60
0.1	0.995	0.9	0.40
0.0	1.000	1.0	0.000

dimension. Dimensionless quantities play a fundamental role in science. For example, in (6.4-1), by picking a^2 and b^2 we made the form dimensionless when the quantities a, b, x, and y are measured in the same quantities.

EXERCISES 6.5

Sketch the ellipse for the following:

1. $x^2/3^2 + y^2/4^2 = 1$
2. $x^2 + 4y^2 = 16$
3. $x^2/4^2 + y^2 = 1$
4. $x^2 + 4y^2 - 2x + 8y + 5 = 0$
5. $9x^2 + 4y^2 + 36x - 8y + 4 = 0$
6. Find the ellipse through the points $(0, 0)$, $(0, 2)$, $(3, 1)$, and $(-3, 1)$.
7. Discuss the ellipse $x^2/a^2 + y^2/b^2 = k$ as k goes from 1 through 0 to -1.

6.6 HYPERBOLAS

The next case we take up has opposite signs on the two square terms in the reduced form ($AC < 0$). Consider first the case when we have

$$\frac{x^2}{a^2} - \frac{y^2}{b^2} = 1 \tag{6.6-1}$$

The value $y = 0$ gives $x = \pm a$. As x and y both get large, the size of the two terms on the left becomes very large with respect to the number 1 on the right-hand side. The terms approach each other in size, although, of course, their difference must be

1. We are making a small change in the values of x and y when we replace (for large x and y) Equation (6.6-1) by the equation

$$\frac{x^2}{a^2} - \frac{y^2}{b^2} = 0 \qquad (6.6\text{-}2)$$

This equation factors into

$$\left[\frac{x}{a} + \frac{y}{b}\right]\left[\frac{x}{a} - \frac{y}{b}\right] = 0 \qquad (6.6\text{-}3)$$

Since a product is zero only if one (or both) of the factors is zero, we conclude that we have the two separate straight lines, called the *asymptotes* of the curve,

$$y = -\left(\frac{b}{a}\right)x \quad \text{and} \quad y = \left(\frac{b}{a}\right)x$$

The curves of the *hyperbola* approach the asymptotes as x and y become large. In Figure 6.6-1 we see the same box as before in dashed lines with dimensions a and b. The two asymptotes go through the corners as shown. We have the *vertices* of the *hyperbola* at $(\pm a, 0)$ and can sketch the curve fairly easily.

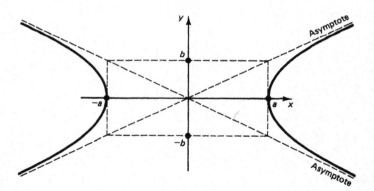

Figure 6.6-1   Hyperbola

If the first term is negative and the second is positive,

$$\frac{-x^2}{a^2} + \frac{y^2}{b^2} = 1$$

then the vertices are clearly $(0, \pm b)$. The asymptotes are the same as before, but the hyperbola goes up the y-axis rather than along the x-axis (Figure 6.6-2).

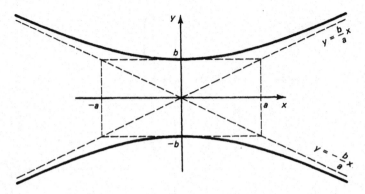

Figure 6.6-2   Hyperbola

Example 6.6-1

To sketch the hyperbola

$$x^2 - y^2 + 2x - 6y + 1 = 0$$

we first complete the square

$$(x + 1)^2 - (y + 3)^2 = 1 - 9 - 1 = -9$$

Next we shift the coordinates and rearrange in conventional form:

$$(x')^2 - (y')^2 = -9$$

Drop the primes, and divide through by the minus 9:

$$\frac{-x^2}{3^2} + \frac{y^2}{3^2} = 1$$

This hyperbola has vertices $(0, \pm 3)$ and therefore goes up the y-axis (see Figure 6.6-3). The sketched box is the same for both ellipses and hyperbolas. The difference is that the ellipse lies wholly inside the box, and the hyperbola lies outside. The diagonals of the box provide the asymptotes for the hyperbola, and then you fix the vertices, whether on the x- or on the y-axis (either by trial and error or by remembering that it is the term with the plus sign that gives the vertices).

In terms of the Argand plane, there is an interesting relationship between the ellipse and the hyperbola. The ellipse is described by the equation

$$\frac{x^2}{a^2} + \frac{y^2}{b^2} = 1$$

If we write this equation as

$$\frac{x^2}{a^2} - \frac{(iy)^2}{b^2} = 1$$

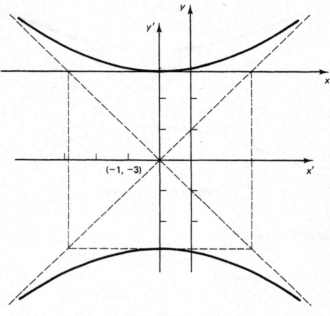

Figure 6.6-3

the result is a hyperbola in the (x, iy) coordinates. Thus what in the Cartesian plane is an ellipse becomes in the Argand plane a hyperbola!

EXERCISES 6.6

1. Plot $x^2 - y^2 + 1 = 0$.
2. Plot $4x^2 - 9y^2 = 36$.
3. Plot $-4x^2 + 9y^2 = 36$.
4. Plot $x^2 - y^2 + 4x - 6y = 3$.
5. Plot $x^2 + 4x - y^2 + 4 = 0$.
6. Discuss the hyperbola $x^2/a^2 - y^2/b^2 = k$ as k goes from 1 through 0 to -1.
7. Discuss hyperbolas in the Argand plane.

6.7 PARABOLAS

In the general quadratic form, Equation (6.4-1), it is possible that one of the square terms is missing. This is expressed mathematically by $AC = 0$. Suppose that it is the

y^2 term that is missing ($C = 0$). We have, therefore, the form

$$Ax^2 + Dx + Ey + F = 0$$

to analyze. First, complete the square on x.

$$A\left[x^2 + \left(\frac{D}{A}\right)x + \left(\frac{D}{2A}\right)^2\right] = \frac{D^2}{4A} - Ey - F$$

Note again the necessity to factor out the A and to remember it when adding the term on both sides. Now consider the choice of the y coordinate translation. The best you can do to get rid of terms is to pick the k in the translation to remove all the constant terms:

$$A\left[x + \frac{D}{2A}\right]^2 = -E\left[y - \left(\frac{D^2}{4AE} - \frac{F}{E}\right)\right]$$

Making the obvious translation of coordinates,

$$x' = x + \frac{D}{2a} \quad \text{and} \quad y' = \frac{y - (D^2 - 4AF)}{4AE}$$

and dropping the primes you have

$$x^2 = -\left(\frac{E}{A}\right)y \quad \text{or} \quad y = \left(\frac{-A}{E}\right)x^2$$

This is called a *parabola* with the *vertex* at $(0, 0)$. It opens along the y-axis, upward if the entire coefficient $-A/E$ in front of the x^2 term is positive, and downward if it is negative. See Figure 6.7-1a and 6.7-1b for the two cases. Since only x^2 terms occur, there is symmetry with respect to the y-axis.

Had you chosen the case of no x^2 term ($A = 0$), then the parabola would have opened left or right (see Figure 6.7-2), and the symmetry would have been with respect to the x-axis.

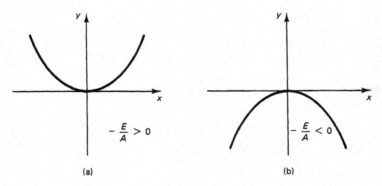

(a)   (b)

Figure 6.7-1   Parabolas

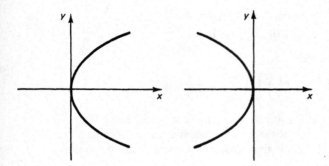

Figure 6.7-2 Parabolas

Example 6.7-1

Given the equation

$$x^2 + 6x - 4y + 1 = 0$$

you first complete the square on the x term and then adjust the y term properly to get

$$(x + 3)^2 = 4y - 1 + 9 = 4(y + 2)$$

Translated to the point $(-3, -2)$, you have the parabola

$$x^2 = 4y \quad \text{or} \quad y = \frac{x^2}{4}$$

as shown in Figure 6.7-3.

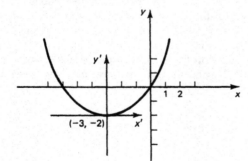

Figure 6.7-3

Example 6.7-2

Given the equation

$$y^2 + x + 4y + 6 = 0$$

completing the square on the y term gives

$$(y + 2)^2 = -x - 2 = -(x + 2)$$

Set $y + 2 = y'$ and $x + 2 = x'$ to get

$$(y')^2 = -x'$$

This clearly requires $x' \leq 0$ and is symmetric with respect to the x-axis (because of the even powers in y). See Figure 6.7-4.

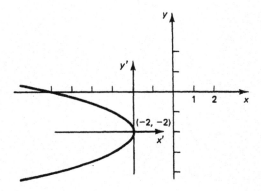

(-2, -2)

Figure 6.7-4

EXERCISES 6.7

1. Plot $x^2 = -y$.
2. Plot $x^2 - 6x + y = 7$.
3. Plot $y^2 + 4y - 6x + 3 = 0$.
4. Plot $y^2 - 6y - x = 4$.
5. Plot $y^2 = 2px$, with $p > 0$. *Hint:* If $x = 2p$, then $y = \pm 2p$.
6. Plot $x^2 = 2py$, with $p < 0$.

MISCELLANEOUS CASES

In our analysis of the general quadratic equation (with the xy term missing), we have now to pick up the missing cases. For example, if both the square terms in the general form (6.4-1) (when reduced by the translation) have negative coefficients, you have

$$-\frac{x^2}{a^2} - \frac{y^2}{b^2} = 1$$

which is impossible for real numbers; it is an imaginary curve. Indeed, we are apt to say that we have "an imaginary ellipse."

In the reduction by completing the square when both square terms are present and we find that the right-hand side is zero, then we have either (1) a sum of two squares equal to zero requiring both to be zero, meaning the curve is only one (real) point, or else (2) a difference of two squares, meaning we have a pair of intersecting lines.

Example 6.8-1

These two cases are realized in the following two specific examples.

(1) $$x^2 + y^2 = 0$$

which has only $(0, 0)$ as a (real) solution. And

(2) $$x^2 - y^2 = 0$$

which has solutions

$$y = \pm x$$

Further degenerate cases can occur, such as

$$x^2 = a^2$$

leading to the two parallel vertical lines

$$x = a \quad \text{and} \quad x = -a$$

The degenerate case

$$x^2 + a^2 = 0$$

is, of course, a pair of imaginary lines. The case

$$x^2 = 0$$

is two coincident lines, $x = 0$ and $x = 0$. Of course, these degenerate cases can come in many disguises, a simple variation being y in place of x.

EXERCISES 6.8

1. Plot $x^2 + 4x + y^2 = -4$.
2. Plot $y^2 - x^2 = 0$.
3. Plot $x^2 = 4$.
4. What is the locus of $x^2 + 1 = 0$?
5. What is the locus of $y^2 = 4$?
6. What is the locus of $x^2 + 4x + 4 = 0$?

6.9* ROTATION OF THE COORDINATE AXES

The complete second-degree form

$$Ax^2 + 2Bxy + Cy^2 + Dx + Ey + F = 0 \qquad (6.9\text{-}1)$$

has, in conventional notation, a cross-product term xy with coefficient $2B$. This term may be removed by a rotation of the coordinate axes. Thus we must look briefly at coordinate axes rotations.

We begin with Figure 6.9-1, connecting the old x–y coordinate system with the new x'–y' coordinate system by a rotation of angle ϕ (Greek lowercase phi). In the original coordinate system we set

$$x^2 + y^2 = r^2$$

$$\theta = \arctan\left(\frac{y}{x}\right) \tag{6.9-2}$$

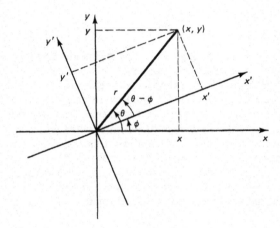

Figure 6.9-1   Rotation

After we rotate an angle ϕ to the new coordinate system, we have the same r, but the angle $\theta - \phi$ as the new angle. Thus we have

$$x' = r\cos(\theta - \phi)$$

$$y' = r\sin(\theta - \phi)$$

We expand the trigonometric functions by the addition formulas (16.1-6) and (16.1-9):

$$x' = r\cos\theta\cos\phi + r\sin\theta\sin\phi$$

$$y' = r\sin\theta\cos\phi - r\cos\theta\sin\phi \tag{6.9-3}$$

We note that in the old coordinate system

$$x = r\cos\theta$$

$$y = r\sin\theta$$

and use these to eliminate the occurrences of r in (6.9-3). Hence

$$x' = x\cos\phi + y\sin\phi$$

$$y' = -x\sin\phi + y\cos\phi \tag{6.9-4}$$

These are the equations of rotation from the old to the new coordinate systems.

However, in use it is often the other way around; given the new coordinates x', y', we use the above substitutions to get the equation into the old coordinates.

We can find the inverse rotation of (6.9-3) (from the new to the old) by simply multiplying the top equation by $\cos \phi$ and the lower by $-\sin \phi$ to eliminate the y term, and by $\sin \phi$ and $\cos \phi$, respectively, to eliminate the x term. Using

$$(\cos \phi)^2 + (\sin \phi)^2 = 1$$

we get the transformations

$$x = x' \cos \phi - y' \sin \phi$$
$$y = x' \sin \phi + y' \cos \phi \qquad (6.9\text{-}5)$$

Observe that (6.9-5) is equivalent to replacing the angle ϕ by $-\phi$ in Equations (6.9-4) for the initial rotation. Thus we know how to transform either way. In particular, given an equation in x, y coordinates, Equations (6.9-5) can be substituted in place of the x and y to get the same geometric features in the new, rotated coordinate system.

It is worth noting that when you want to rotate the axes for an equation you need to have expressions in terms of the old variables, the set (6.9-5); but when you want to transform a point (x, y), you use the other set (6.9-4) and insert the values of the point (x, y) to get the new coordinates (x', y') in the primed variables.

We are now ready to show how the xy term can be eliminated by a suitable rotation. Using the transformation (6.9-5) on the general form, we find that only the quadratic terms are involved in determining the actual angle of rotation. We have, after a lot of algebra,

$$\begin{aligned}
Ax^2 + 2Bxy + Cy^2 = {}& A[(x')^2 \cos^2 \phi - 2x'y' \cos \phi \sin \phi + (y')^2 \sin^2 \phi] \\
& + 2B[(x')^2 \cos \phi \sin \phi + x'y'(\cos^2 \phi - \sin^2 \phi) \\
& - (y')^2 \cos \phi \sin \phi] \\
& + C[(x')^2 \sin^2 \phi + 2x'y' \cos \phi \sin \phi + (y')^2 \cos^2 \phi]
\end{aligned}$$

We want to eliminate the $x'y'$ term, so we must equate its coefficient to zero. Thus we have

$$(-2A + 2C)\cos \phi \sin \phi + 2B(\cos^2 \phi - \sin^2 \phi) = 0$$

which determines the angle of rotation ϕ. Using the following double-angle formulas (16.1-10) and (16.1-11) (obtained from the addition formulas by setting both angles equal to each other),

$$2 \sin \phi \cos \phi = \sin 2\phi$$
$$\cos^2 \phi - \sin^2 \phi = \cos 2\phi$$

we have

$$(-A + C) \sin 2\phi + 2B \cos 2\phi = 0$$

Divide through by $(C - A) \cos 2\phi$ to get

$$\frac{2B}{A-C} = \tan 2\phi \qquad (6.9\text{-}6)$$

This determines the particular angle ϕ of rotation that eliminates the xy term. Having eliminated it, we are now ready to proceed as before to determine what geometric curve corresponds to the given equation.

Rotation of the coordinate axes is clearly a messy problem, which we avoid when possible. Therefore, we will illustrate the coordinate rotation with only one example where the amount of the rotation is already given.

Example 6.9-1

Given the special (equilateral) hyperbola (in a sense this corresponds to the circle among the class of all ellipses)

$$x^2 - y^2 = 2$$

we rotate the coordinates by $-45°$. The rotation (6.9-5) is

$$x = \frac{x' + y'}{\sqrt{2}}$$

$$y = \frac{-x' + y'}{\sqrt{2}}$$

We get

$$\frac{(x')^2 + 2x'y' + (y')^2}{2} - \frac{(x')^2 - 2x'y' + (y')^2}{2} = 2$$

or, since so many terms cancel,

$$x'y' = 1$$

Figures 6.9-2a and b show the curves before and after rotation.

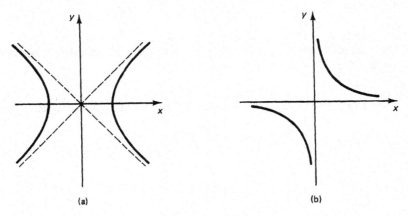

(a)   (b)

Figure 6.9-2

EXERCISES 6.9

1. Reduce the equation $x^2 + 2xy + y^2 = 1$ to standard form.

2. Remove the xy term from $xy = 1$.

3. Show that $B^2 - 4AC = (B')^2 - 4A'C'$ for any rotation.

6.10* *THE GENERAL ANALYSIS*

In these days, when the control of a situation is often done in a computing machine, it is necessary to make a complete analysis of all possible situations. In practice, sooner or later, every possible situation that can occur will occur, and if you have not made adequate preparation for it in the computer program, you will almost surely get nonsense out of the computer. However, it will sometimes require an effort on your part to recognize this! Everyone is familiar with the consequences of incompletely analyzed computer programs, so nothing more need be said of the importance of a complete analysis. We will use the quadratic in two variables to illustrate one way of doing such an analysis.

Given the complete quadratic in x and y,

$$Ax^2 + 2Bxy + Cy^2 + Dx + Ey + F = 0$$

if $B \neq 0$, then we rotate the coordinate axes (Section 6.9) by the angle ϕ defined by (6.9-6),

$$\tan 2\phi = \frac{2B}{A - C}$$

which reduces the quadratic to the form (new letters A, C, D, E, and F in place of the old ones)

$$Ax^2 + Cy^2 + Dx + Ey + F = 0$$

If $A = C$, then $\phi = 45°$. We now have $B = 0$.

We continue with the analysis. At this point there are three cases to be examined, depending on the presence of the square terms.

Case I

Both terms present $(AC \neq 0)$ (Section 6.3). You get after completing the square

$$A\left(x + \frac{D}{2A}\right)^2 + C\left(y + \frac{E}{2C}\right)^2 = K$$

where

$$K = \frac{D^2}{4A} + \frac{E^2}{4C} - F$$

We now have two subcases depending on whether K is or is not 0. If $K \neq 0$, then we can divide through by it and write the equation in the form

$$\pm \frac{x^2}{a^2} \pm \frac{y^2}{b^2} = 1$$

There are three possibilities, depending on the $\pm$ signs.

1. Both plus: ellipse (Section 6.5)
2. Opposite signs: hyperbola (Section 6.6)
3. Both negative: imaginary curve (Section 6.8)

If $K = 0$, then we have the degenerate case of Section 6.8. If both the signs are the same, then it is imaginary, and if they are different then you can factor it into two intersecting lines (Section 6.8).

Case II

One square term only. We have parabolas (Section 6.7) provided both variables x and y are present. If only one variable is present, there are three possible cases having the typical forms:

$$x^2 - a^2 = 0, \quad \text{parallel lines}$$

$$x^2 = 0, \quad \text{double line}$$

$$x^2 + a^2 = 0, \quad \text{imaginary parallel lines}$$

These equations could be in the variable y instead, and remember we are talking about the situation after a rotation to remove any xy term that may be present.

Case III

No square terms. This means that if we translated and rotated back to the original equation we would still have a linear equation, a straight line. Therefore, the original given quadratic was in fact linear in x and y, that there were no original quadratic terms, that $A = B = C = 0$. We must allow, of course, for further degenerate cases including all the coefficients equal to zero.

We now examine our analysis and arrange it in a logical form.

1. If $B \neq 0$, then you rotate the coordinate system to make the cross-product term vanish. Thus you are reduced to the case of $B = 0$ from now on.
2. If $AC \neq 0$ (both square terms are present), you complete the squares in both variables. There are two cases to consider depending on the K term. If $K \neq 0$, then we have (a) the ellipse when both coefficients are positive, (b) the imaginary ellipse when both terms are negative, and (c) the hyperbola when the two terms have opposite signs. If $K = 0$, then if the signs are different the equation factors into two lines (the asymptotes of the hyperbola when the $K \neq 0$); while if the signs are the same, then it is a pair of imaginary lines.

3. If $AC = 0$, then one square term is missing, and you have a parabola when the other variable is still present, and a square of the variable if the other is missing. This square leads to (a) parallel real lines, (b) imaginary parallel lines, or (c) coincident lines depending on the value of the constant term.
4. If both $A = 0$ and $C = 0$ (after the possible rotation), then you did not have a quadratic to begin with.

6.11 SYMMETRY

Symmetry is a fundamental concept in mathematics. At the moment, we are concerned with geometric symmetry. Symmetry of two points with respect to a given line L requires that the line segment joining the two points be bisected perpendicularly by the given line L (Figure 6.11-1). In mathematics, given the points (x_1, y_1) and (x_2, y_2), the midpoint of the line segment (see example 5.6-3) is clearly

$$\left[\frac{x_1 + x_2}{2}, \frac{y_1 + y_2}{2}\right]$$

This point must lie on the line L of symmetry, and the two lines must be perpendicular.

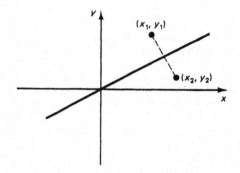

Figure 6.11-1 Line of symmetry passes through origin

The special cases of symmetry with respect to the x-axis require that the y coordinates of symmetric points be equal in size and opposite in sign. If the function depends only on y^2, then both y and $-y$ will (or will not) satisfy the equation (Figure 6.11-2).

Similarly, symmetry with respect to the y-axis requires the formula to have only even powers in x (although, of course, there can be y terms and constant terms) (Figure 6.11-3).

The reduced ellipses and hyperbolas (6.4-4) have both x- and y-axis symmetries. The parabola when reduced has only one of these symmetries.

Next in importance is the case of symmetry with respect to the 45° line. For this case the symmetric points are (x, y) and (y, x); you simply interchange the x and y

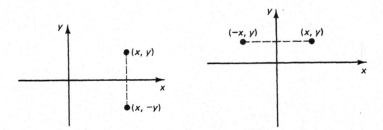

Figure 6.11-2 Symmetry with respect   **Figure 6.11-3** Symmetry with respect
to the x-axis   to the y-axis

coordinates! This is easily seen from Figure 6.11-4. For example, the equation (a quartic)

$$x^2 y^2 = 1$$

has all three types of symmetry. The rotated hyperbola $xy = 1$ has symmetry with respect to the 45° line.

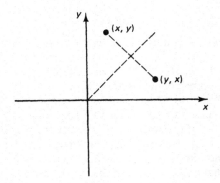

Figure 6.11-4 Symmetry with respect
to the 45° line

Symmetry with respect to the 45° line is easily described by the words, "They are inverse functions of each other." Thus the parabolas

$$x = y^2 \quad \text{and} \quad y = x^2$$

are inverse functions (see Figure 6.11-5). One branch of $x = y^2$ is

$$y = \sqrt{x}$$

and the other branch has the minus sign.

There is also symmetry with respect to a point, typically with respect to the origin. Here if we have the point (x, y), we must also have the corresponding point $(-x, -y)$ (Figure 6.11-6). The ellipses and hyperbolas when reduced have this symmetry. Indeed, if a curve has two of the three symmetries, x-axis, y-axis, and origin symmetry, then it has the third.

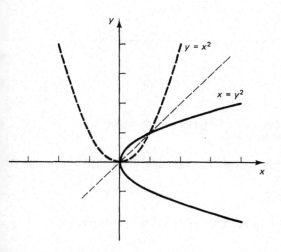

Figure 6.11-5 Inverse functions

Figure 6.11-6 Symmetry with respect
to the origin

There is also rotational symmetry. A circle has perfect rotational symmetry; any rotation must leave the equation unaltered. A square centered about the origin has rotational symmetry of $\pm 90°$, $\pm 180°$, $\pm 270°$, ... Similarly, for any regular polygon about the origin, there are the corresponding symmetries.

How can one remember all these rules? You do not even try! All symmetries come from the invariance of the equation under the appropriate change of variables. Fundamentally, you look for changes of variables that will leave the equation invariant. That is all there is to geometric symmetry! Common invariances are $x \rightarrow -x$, $y \rightarrow -y$, separately or together, and interchange of x and y. You have only to learn to think, "What point goes into what point?" to grasp the corresponding symmetry.

EXERCISES 6.11

1.* Show that a circle centered about the origin has rotational symmetry by making an arbitrary rotation of its equation.

2. Prove by algebra that if two of the symmetries mentioned above occur then the third must occur.

3. Discuss the symmetry for the equation $x^2 = f(y)$. For $y^2 = g(x)$.

4. What symmetry has the equation $f(x, y) = f(y, x)$?

5. What symmetry has the equation $f(x, y) = f(-x, -y)$?

.12 NONGEOMETRIC GRAPHING

The discussion has been, so far, about geometric figures. In geometry we believe that the plane is *isotropic*, the same in every direction. But in practice we often plot nongeometric relationships in rectangular coordinates, say dollars versus years. There is no common measure of such different quantities as dollars and years; either axis can have its units chosen independently of the other. Thus there is no meaning to rotation in such a coordinate system since the rotation formulas would have you adding dollars to years (apples to oranges in the classical example in beginning algebra).

Similarly, there is no meaning to the slope of a line. Instead, we fall back on the *rate of change*, a change of so many dollars per year. The slope itself has no meaning *except* in two cases, zero slope and vertical slope. In such cases the change of the unit size in one variable does not affect the tilt of those lines. Scale transformations

$$x = k_1 x' \qquad y = k_2 y' \qquad\qquad (6.12\text{-}1)$$

are a special case of *affine transformations*, and therefore *scaling geometry* (6.12-1) is a special case of *affine geometry*.

Thus we need to be wary about when we use rotations; they are appropriate *only* when the coordinate axes have the same units.

6.13 SUMMARY OF ANALYTIC GEOMETRY

The basis for analytic geometry is that a pair of numbers corresponds to a unique point in a plane, and conversely. Similarly, equations and curves correspond (almost). In particular, straight lines correspond to linear equations, while quadratic equations correspond to the conics (the ellipses, hyperbolas, parabolas, and degenerate pairs of straight lines). These conics occur in many places: a single small planet about a massive sun moves in an ellipse, as does a satellite about the earth (neglecting perturbing effects of other planets and the sun, and assuming Newtonian mechanics, but not relativity); higher-speed particles from outer space may move in a hyperbolic orbit (similarly a satellite may acquire sufficient energy and then move in a hyperbolic orbit); and telescope mirrors (as well as automobile headlights) have a parabolic shape.

It is necessary to point out that, although you tend to equate in your mind "function" and "graph," there are significant differences. A function, by definition, must be single valued, while many equations have multiple values (for example, a circle).

You should learn to go from the geometric properties to the algebraic properties and back for each kind of object and equation. We have glossed over many of the details and barely discussed the rotation of coordinates because they are not of great interest in this course, although they can be in some situations. We have also ignored many details of the conics, which may be found in texts specially devoted to them.

PART II

THE CALCULUS OF ALGEBRAIC FUNCTIONS

7

Derivatives in Geometry

The calculus is conventionally divided into two parts, the differential and the integral calculus. Historically, the integral calculus arose first, both Euclid and Archimedes having calculated areas and volumes of shapes bounded by other than straight lines and plane surfaces. The differential calculus arose from the problem of finding tangent lines to curves. There are a few early examples of this, but the main development came after the discovery of analytic geometry. So close is the relationship of analytic geometry to the calculus that there is an inevitability to the discovery of the calculus once analytic geometry is generally known (first publicized by Descartes in 1637).

The main development of the calculus came when these two problems, finding areas and finding tangent lines, were recognized as being inverses of each other, which is the fundamental theorem of the calculus. While Isaac Barrow (1630–1677), Newton's teacher, in a sense knew this theorem, it was the systematic application of the methods, rather than the development of special tricks, that marked the real founding of the calculus by Gottfried Wilhelm von Liebniz (1646–1716), published in 1684, and by Isaac Newton (1642–1727), published in 1687.

From a high art the calculus has been converted to a routine, systematic method, hence its name "calculus." Over the subsequent years the presentation of the calculus has been polished until it is now easily mastered by the willing student.

7.2 THE IDEA OF A LIMIT

A central idea in the calculus is the *limit*. Part I has carefully prepared the ground, so it is mainly a matter of reminding the reader of what has already been said with respect to this idea.

We begin with the idea of a *limit of a sequence of numbers*, indexed by the integer n. For example, consider the sequence $\{u_n\}$, where

$$u_n = \frac{1}{n}$$

as n gets larger and larger. Clearly, the sequence approaches the number 0. Indeed, no matter how close to 0 you want me to get (other than a zero distance away), I can find an integer n_0 such that *for all larger* n the values of u_n are at least that close to 0.

The expressions "as large as you please" and "larger and larger" are conveniently replaced by the words "as n approaches infinity." This should *not* be interpreted to mean that the numbers n reach some "number" called "infinity," but rather to describe colorfully what is going on. Furthermore, we are not saying that the values of the sequence $\{1/n\}$ reach 0, but only that the numbers in the sequence ultimately get and stay arbitrarily close to 0. There is a big difference between these two statements.

In words, we say that the sequence u_n (in the above particular case $u_n = 1/n$) approaches a limit, L (which in this case is 0), as n approaches infinity. In symbols we write

$$\lim_{n \to \infty} u_n = L \tag{7.2-1}$$

For the sequence

$$u_n = 1 + \frac{1}{n}$$

the corresponding limit L is clearly 1. The test is that the *difference* between u_n and its limit L approaches 0.

A sequence that approaches a limit is said to *converge* to the limit. A sequence that does not converge is said to *diverge*. For example, the sequence

$$u_n = (-1)^n$$

has values alternately $+1$ and -1. It does not converge because it never *stays* close to a limit; rather it continually oscillates between two numbers $+1$ and -1.

Example 7.2-1

We have already seen (Section 3.6) this idea of convergence several times. When we wrote the decimal digits of the rational number $\frac{1}{3}$ as

$$0.3 \quad = \frac{3}{10} \qquad\qquad\qquad = u_1$$

$$0.33 \quad = \frac{3}{10} + \frac{3}{10^2} \qquad\qquad = u_2$$

$$0.333 \quad = \frac{3}{10} + \frac{3}{10^2} + \frac{3}{10^3} \qquad = u_3$$

$$0.3333 = \frac{3}{10} + \frac{3}{10^2} + \frac{3}{10^3} + \frac{3}{10^4} = u_4$$

we wrote a sequence of approximations u_n. This sequence of approximations u_n approaches (and stays close) to the limit $\frac{1}{3}$. We did not say that any finite number of 3's was equal to $\frac{1}{3}$; we merely said that as we took more and more digits (as the number of digits approached infinity), the sequence of numbers approached the number $\frac{1}{3}$. We said that $\frac{1}{3}$ was the limit of the sequence.

Example 7.2-2

Another illustration (Example 3.6-4) is the sequence $(9/10)^n$. We have for the u_n values

$$0.9 \ = 0.9$$

$$0.9^2 = 0.81$$

$$0.9^3 = 0.729$$

$$0.9^4 = 0.6561$$

$$0.9^5 = 0.59049$$

$$0.9^6 = 0.531441$$

$$0.9^7 = 0.4782969 < \frac{1}{2}$$

To get a better "feel" for what is happening with respect to the *convergence* of the sequence, we look for a bound in the more familiar decimal notation. From the last line of numbers (in this example) we see that

$$0.9^{14} < \left(\frac{1}{2}\right)^2$$

$$0.9^{21} < \left(\frac{1}{2}\right)^3$$

and in particular that

$$(0.9)^{70} < \left(\frac{1}{2}\right)^{10} = \frac{1}{1024} < 10^{-3}$$

Here we see, in the familiar decimal notation, that the sequence $(0.9)^n$ is approaching 0. Increases in the exponent by 70 multiply the bound by a factor smaller than 1/1000.

In Example 3.5-3 we saw how (in principle) we can find a decimal approximation to an algebraic (real) number as accurately as we please. In particular, we constructed a convergent sequence of numbers whose limit was the root of an algebraic equation. As we compute more and more digits of the algebraic number, we approach the root more and more closely.

Sequences that approach zero are called *null sequences*. They play a central role in the theory. The sequence of differences between a convergent sequence u_n and its limit L is a null sequence. When you study a sum of terms, such as a geometric progression, the series is converted into a sequence of partial sums, and *it is the sequence that is tested for convergence*. Thus, when we looked at a decimal number, we regarded it as a sum of the individual digits, each digit divided by its corresponding power of 10, and we then studied the sequence of partial sums.

Example 7.2-3

A similar, and more general, example of the above is the geometric progression when the rate r satisfies the inequality of $|r| < 1$. From Example 2.6-1 we have for the sum of a geometric progression

$$S_n = \frac{a}{1 - r} - \frac{ar^n}{1 - r} \tag{7.2-2}$$

We first want to show that when $|r| < 1$ then r^n approaches 0. (We just saw this was true when $r = 9/10$.) If we can show this, then the last term of the sum will approach 0 (will be a null sequence), and the limit of the sum will be only the first term.

We can prove that the limit of the sequence r^n is 0 for any $|r| < 1$ by using an inequality we derived in Example 3.7-1, that

$$(1 + a)^n \geq 1 + na$$

This a is not to be confused with the a in the geometric progression. To use this inequality, we observe that if $|r| < 1$ then its reciprocal is greater than 1. Hence there is an a such that

$$\left| \frac{1}{r} \right| = 1 + a \qquad (a > 0)$$

or, in a more convenient form for our immediate need,

$$|r| = \frac{1}{1 + a}$$

Therefore, remembering that the $1 + a$ is in the denominator, we have the inequality in the reverse direction:

$$|r|^n = \frac{1}{(1 + a)^n} \leq \frac{1}{1 + na}$$

As n approaches infinity, the denominator approaches infinity, and the right-hand side comes as close as you please to 0. You tell me how close you want $|r|^n$ to be to 0 and I can tell you an n such that for all larger indices it is at least that close. I do not need to tell you the smallest n; I only have to tell you one that "works." We have therefore proved that $|r|^n$ (when $|r| < 1$) is a null sequence. Hence r^n is a null sequence.

Since the r^n approaches 0, it follows that the second term in the sum in (7.2-2) of a geometric progression,

$$\frac{ar^n}{1-r} = \left[\frac{a}{1-r}\right] r^n \qquad (7.2\text{-}3)$$

must also approach 0 (the term in the brackets is a fixed number independent of n). Thus we say that the S_n (the partial sum) of a geometric progression, as we take more and more terms, approaches the limit S (for $|r| < 1$), where

$$S = \frac{a}{1-r} \qquad (7.2\text{-}4)$$

and the S_n are the *partial sums*

$$S_n = a + ar + ar^2 + \cdots + ar^{n-1} = \frac{a(1-r^n)}{1-r}$$

of the series. We do not claim that the sum takes on the value of the limit, only that as you let n approach infinity the values of the partial sum ultimately become and stay arbitrarily close to the limit.

This is a good example of a common phenomenon in science. Often, when progress seems to be blocked, the advance is made by evading the question. The question, "Does the sequence reach its limit?" is evaded; we only claim that the sequence gets and stays arbitrarily close to the limit. It turns out that this is good enough for us; we do not need to know more than that.

We have emphasized that to have a limit requires both (1) that the sequence gets close, and (2) past some point the sequence stays close to the limit. Consider the sequence

$$1, \frac{2}{3}, 1, \frac{3}{4}, 1, \frac{4}{5}, \ldots$$

The limit is 1, and it often takes on that value, and then wanders away, before it finally gets and stays as close as you require. On the other hand, the sequence

$$1, \frac{1}{3}, 1, \frac{1}{4}, 1, \frac{1}{5}, \ldots$$

does not converge. It does not stay close to either 0 or 1 (or any other number).

Abstraction 7.2-4*

The words we have been using to describe a limit are conventionally expressed in the following mathematical notation (don't let the Greek letter ϵ, lowercase epsilon, bother you). Given any $\epsilon > 0$, there exists a number n_0 such that for all integers $n \geq n_0$

$$|u_n - L| < \epsilon$$

When we say "any $\epsilon > 0$" in a sense we imply "for all numbers > 0," no matter how large or small. Notice how neatly the mathematics expresses the precise idea of a limit of a sequence.

7.3 RULES FOR USING LIMITS

Example 7.3-1

In using limits we sometimes need to handle complicated expressions. Thus we need to study combinations of limits. From the simple example, as n approaches infinity,

$$\lim \left\{ a\left(\frac{1}{n}\right) + b\left(1 + \frac{1}{n}\right) \right\} \to b$$

it is easy to see that the limit of a sum is the sum of the limits (provided the individual limits exist). And similarly for differences. Limits of sums and differences are the sums and differences of the limits for any finite number of terms, as you can see, or as you can prove using either ellipsis or mathematical induction.

We next look at products of limits, for example,

$$u_n = \left[2 + \frac{2}{n}\right]\left[2 - \frac{3}{n}\right]$$

Both factors of the product approach 2, and therefore the product approaches 4. To see the details, if you wish to see them, we write the *difference* between the sequence and its limit as a null sequence:

$$u_n - 4 = \left[4 - \frac{2}{n} - \frac{6}{n^2}\right] - 4 = \frac{-2}{n} - \frac{6}{n^2}$$

And as n approaches infinity the right-hand side approaches 0. Therefore,

$$u_n \to 4$$

as its limit.

Finally, we need to look at quotients. Consider the sequence

$$u_n = \frac{1 + (1/n)}{2 - (3/n)}$$

The numerator approaches 1 and the denominator approaches 2, so the quotient approaches $\frac{1}{2}$. If you wish to look at the details, you write the difference as a null sequence

$$
\begin{aligned}
u_n - \frac{1}{2} &= \frac{1}{2}\left[\frac{1 + (1/n)}{1 - (3/2n)} - 1\right] \\
&= \frac{1}{2}\left[\frac{1 + (1/n) - 1 + (3/2n)}{1 - (3/2n)}\right] \\
&= \frac{1}{2}\left[\frac{5}{2n}\right]\frac{1}{1 - (3/2n)} \\
&= \frac{5}{4n}\frac{1}{1 - (3/2n)}
\end{aligned}
$$

and the first term approaches 0 (when $n \to \infty$) as required (and the other term stays away from 0). If you asked me to find a bound on n beyond which the right-hand side was as close to 0 as some number you required, I would reason as follows. Assuming that

$n \geq 2$, the denominator of the second term could be replaced by

$$1 - \frac{3}{4} = \frac{1}{4}$$

because this merely makes the right-hand side larger.

$$\frac{5}{4n} \frac{1}{1 - (3/2n)} < \frac{5}{4n}(4) = \frac{5}{n}$$

Thus I can pick an n such that

$$\frac{5}{n} \leq \text{your number}$$

which is easily solved for n:

$$n \geq \frac{5}{\text{your number}}$$

In summary you see that the limits of sums, differences, products, and quotients of limits have the same value as the corresponding combinations of limits (provided the individual limits exist, and in the case of quotients that the denominator limit is not 0).

EXERCISES 7.3

Find the limit as n approaches infinity of the following (assume that the individual coefficients a, b, c, and d are not zero):

1. $a + b/n + c/n^2 + d/n^3$
2. $[a + b/n]/[c + d/n]$, provided $c \neq 0$
3. $[a + b/n + c/n^3][d + e/n + f/n^2]$
4. $[an + b]/[cn + d]$. *Hint:* First divide both numerator and denominator by n.
5. $[an^3 + 1]/[bn^3]$
6. $[an^2 + 1]/[bn^3 - 1]$
7. Define a limit.

7.4 LIMITS OF FUNCTIONS—MISSING VALUES

We have looked at limits of sequences; now we turn to limits for continuous functions.

Example 7.4-1

Consider the expression

$$y(x) = \frac{x^2 - 4}{x - 2}$$

This is not defined for the value $x = 2$ since division by 0 is normally excluded. However, *for all other values* this is *equivalent* to

$$y(x) = x + 2$$

In the original form we are missing one isolated value at $x = 2$.

Now consider (see Figure 7.4-1) the limit of $y(x)$ as x approaches the value 2, for either larger or smaller values of x. This is simply the value at that point of the equivalent form $x + 2$. Because of the division by 0, we cannot write

$$y(2) = 4$$

for the original form of $y(x)$, but we can write

$$\lim_{x \to 2} y(x) = 4$$

Read this as: the limit as x approaches 2 of $y(x)$ equals 4.

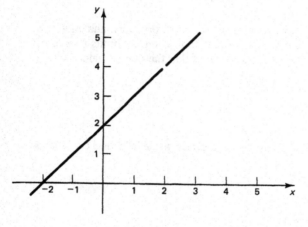

Figure 7.4-1   Missing value at $x = 2$

The necessity for this circumlocution (circum, around; locutio, to speak) is the desire to avoid the division by zero; most mathematicians simply do not like the thought of dividing by zero. For example, if the *unrestricted* division by 0 were allowed, then from the equation

$$(0)(3) = (0)(7)$$

the division by 0 would give

$$3 = 7$$

which is a contradiction! In a sense, the limit *process* can be viewed as a *controlled* division by zero (a potential as contrasted with an actual zero).

Example 7.4-2

Another example of where we are missing an isolated point in the function (because of an indicated division by 0) is

$$f(x) = \frac{x^3 - a^3}{x - a} = x^2 + ax + a^2$$

when $x \neq a$. When we consider the limit as x approaches a from either side, we find, on looking at the right-hand side of the equivalent form, that the limit is

$$a^2 + a^2 + a^2 = 3a^2$$

(merely put $x = a$ in the right-hand side of the above equation).

Example 7.4-3

Consider the trivial example

$$y(x) = \frac{ax}{x} = a$$

except for the value $x = 0$. See Figure 7.4-2. It is natural to take the value at the missing point ($x = 0$) as the corresponding value of the equivalent form, $y = a$. Thus, when there is an indicated division by zero, it is *often* merely a matter of rewriting the expression to eliminate this division to get an equivalent form (equivalent for all values other than the isolated value), and for which the new form gives directly a definite value.

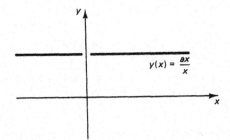

Figure 7.4-2   Missing value at $x = 0$

This situation, where an isolated value is missing, arises when there is an indicated division by 0 and is of common occurrence. The *method of limits* handles the problem of determining this missing value. Usually, the value is easy to find, but on occasions it requires some effort. It is a situation, with some variations, that we will often meet.

Example 7.4-4

Consider the function

$$y = \frac{x}{|x|}$$

There is a division by zero to examine. For $x > 0$ the function is $+1$, and as x approaches zero from the positive side the limit is $+1$; but when approached from the negative side the limit is -1 (see Figure 7.4-3). Thus there is no limit as $x \to 0$.

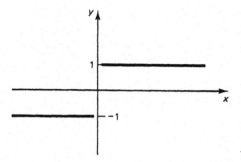

Figure 7.4-3 $y = x/|x|$

It is an exercise in rereading the limits for sequences to see that for functions the limits of sums, products, and quotients are the corresponding sums, products, and quotients of the limits.

Abstraction 7.4-5*

Corresponding to Abstraction 7.2-4*, we can express the idea of a limit as $x \to a$ of a function $f(x)$ as: given any $\epsilon > 0$, there exists a δ (Greek lowercase delta) such that for all x in the interval

$$|x - a| < \delta$$

then

$$|f(x) - L| < \epsilon$$

Notice how this catches the idea in mathematical symbols, which can later be manipulated in mathematical ways; to capture words in the form of symbols is one of the basic processes in doing mathematics.

We can now make the idea of *continuity* a bit clearer. A function is *continuous at a point* $x = a$ if both (1) $f(a)$ exists, and (2) $\lim_{x \to a} f(x) = f(a)$. For continuity at a point $x = a$, the value of the function at the point must exist, and the limit of the function as you approach the point must be this same value.

Continuity in an interval requires continuity at all points in the interval. This does seem to be close to our intuitive feeling of what we mean by continuity. Using it, we can easily prove that, for example, the powers of x are continuous. Let

$$y(x) = x^n$$

Then at any point $a, f(a) = a^n$. When we try part (2), we have

$$|f(x) - f(a)| = |x^n - a^n|$$
$$= |x - a||x^{n-1} + ax^{n-2} + \cdots + a^{n-1}|$$

The second absolute term value is bounded (near $x = a$), and the first term approaches zero, so the function is continuous at the point a. Since a was any point along the real line, the powers of x, and hence polynomials, are continuous for any finite interval.

EXERCISES 7.4

Write the following in their equivalent forms and find the corresponding limits.

1. $(x^2 - 2x + 1)/(x^2 - 3x + 2)$ as $x \to 1$
2. $(x^2 - 1)/(x - 1)$ as $x \to 1$
3. $(x - 1)/(\sqrt{x} - 1)$ as $x \to 1$
4. $(x - 1)/(\sqrt[3]{x} - 1)$ as $x \to 1$
5. $(x - 1)/(\sqrt[n]{x} - 1)$ as $x \to 1$
6. $(x^n - a^n)/(x - a)$ as $x \to a$. *Hint:* See Section 4.7.
7. $(\sqrt{(a + x)} - \sqrt{a})/x$ as $x \to 0$. *Hint:* Multiply numerator and denominator by $(\sqrt{(a + x)} + \sqrt{a})$.
8. Prove that in the interval $1 \le x \le 2$ the function $y = \sqrt{x}$ is continuous.

7.5 THE Δ PROCESS

We first need to introduce a notation. We will mean by the symbols

$$\Delta x$$

a change in x (Δ is the Greek capital delta). The change may be either positive or negative, although we usually picture it as if Δx were positive. It does *not* mean a product of Δ times x; it is an operation on x just as sin x means the operation of taking the sine of x. However, it is more general than is the sine function since it is an *unspecified* amount of change. Sometimes this is written as

$$\Delta x = h$$

Example 7.5-1

Suppose we consider the parabola

$$y(x) = x^2$$

We have for a given change in x, labeled Δx, the corresponding change in y:

$$\Delta y(x) = y(x + \Delta x) - y(x)$$

The ratio of the changes is the slope of the line through the points at x and at $x + \Delta x$:

$$\frac{\Delta y}{\Delta x} = \frac{y(x + \Delta x) - y(x)}{\Delta x}$$

$$= \frac{(x + \Delta x)^2 - x^2}{\Delta x}$$

$$= \frac{x^2 + 2x\Delta x + (\Delta x)^2 - x^2}{\Delta x}$$

$$= [2x + \Delta x]$$

The original expression was not defined for $\Delta x = 0$, but we have rewritten the expression in an equivalent form where this isolated point no longer causes trouble. To find the limit as Δx approaches 0, we evaluate the equivalent form, which does not have the awkward division by 0. The result is

$$2x$$

Figure 7.5-1 shows the parabola $y = x^2$, and the corresponding slopes at selected values of x, such as $x = -1, 0, 1,$ and 2. These limits are the limiting slopes of the secant line of the curve $y = x^2$ at the corresponding points. In the limit the secant line becomes the tangent line (by definition).

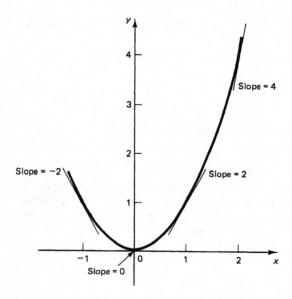

Figure 7.5-1 $y = x^2$

Now consider the general function of x; call it

$$y = y(x)$$

Fix the point $P = (x, y)$ in your mind, and then let x be changed, *incremented*, by the amount Δx; see Figure 7.5-2 for details. The change in x produces a corresponding change in y, called Δy. Thus there are now two points P and Q on the curve of $y(x)$:

$$P = (x, y) \quad \text{and} \quad Q = (x + \Delta x, y + \Delta y)$$

The slope of the secant line through these two points is, of course,

$$\frac{\Delta y}{\Delta x}$$

In words, the slope of the secant line is the change in y, called delta y, divided by the change in x, called delta x.

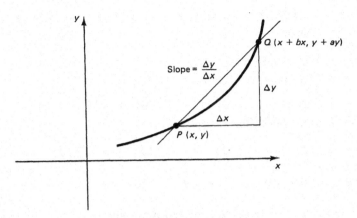

Figure 7.5-2   General slope

We now think of Δx as a variable, and consider the limit of $\Delta y/\Delta x$ as Δx approaches 0. We do not let $\Delta x = 0$ since that would be a division by 0, but rather, as in Section 7.4, we consider the limit of the quotient as Δx approaches 0. We search for the "missing value," the limiting slope of the secant line that is the slope of the tangent line.

Geometrically, we have taken the secant line through two nearby points P and Q and then let the point Q approach P. We consider the *limiting slope* as Δx approaches 0. We see that indeed the limiting slope is the slope of the tangent line to the curve at each point (x, y). In Example 7.5-1 for the parabola $y = x^2$, we found that at any point x the slope is $2x$.

We need a compact symbol to represent this limiting slope. The suggestive notation due to Leibniz is

$$\lim_{\Delta x \to 0} \left(\frac{\Delta y}{\Delta x} \right) = \frac{dy}{dx}$$

Again, the symbol on the right is to be read as a whole, "dee y over dee x," and not as a quotient. The symbol tells you where it came from, that is, the limit of the quotient of the corresponding deltas.

From the original function

$$y(x)$$

we "derived" a new function

$$\frac{dy}{dx}$$

It is called "*the derivative* of y with respect to x."

Since the derivative plays a dominant role in the differential calculus, we need to describe (give an algorithm for) exactly what we are doing. This is the classic *four-step rule.* To compute the derivative of a given function $y(x)$, you do the following four steps:

1. Give x an increment Δx.
2. Compute the corresponding *change* in y:

$$\Delta y = y(x + \Delta x) - y(x)$$

3. Compute the ratio of the changes:

$$\frac{\Delta y}{\Delta x}$$

4. Take the limit as Δx approaches 0.

This last step can be written in the mathematical notation

$$\lim_{\Delta x \to 0} \left(\frac{\Delta y}{\Delta x} \right) = \frac{dy}{dx}$$

The four-step rule uses the limit process to handle the division by zero. Stop here and review this rule until you understand exactly how it expresses *your* ideas of how to find the slope of the tangent line (and hence the local slope of the curve itself).

We turn to some specific examples that are carefully chosen to lead you to more general results. No doubt the first persons did many examples before they saw the general results, and could then avoid the almost endless repetition of the same sort of thing in each special case.

Example 7.5-2

Suppose we try the simple power function

$$y(x) = x^3$$

Following the four-step rule, we find that from step 3 of the rule we have

$$\frac{\Delta y}{\Delta x} = \frac{(x + \Delta x)^3 - x^3}{\Delta x}$$

As always, some algebra is needed *before* we try to take the limit. The purpose of the algebra is to eliminate the appearance of the division by the Δx so that the limit is easily found from the equivalent form. In this case we simply expand the binomial and cancel the x^3 terms:

$$\frac{\Delta y}{\Delta x} = \frac{3x^2 \Delta x + 3x(\Delta x)^2 + (\Delta x)^3}{\Delta x}$$

$$= 3x^2 + 3x\Delta x + (\Delta x)^2$$

In this equivalent form it is easy to take the limit as Δx approaches 0. The result is

$$\frac{dy}{dx} = 3x^2$$

Example 7.5-3

Consider next the function

$$y(x) = x^3 + x$$

and apply the four-step rule. From the third step we get

$$\frac{\Delta y}{\Delta x} = \frac{\{(x + \Delta x)^3 + (x + \Delta x)\} - \{x^3 + x\}}{\Delta x}$$

We have to rearrange this expression before taking the limit. Expand the binomial and cancel the like terms (both the x^3 and the x terms) to get

$$\frac{\Delta y}{\Delta x} = 3x^2 + 3x\Delta x + (\Delta x)^2 + 1$$

In this equivalent form we can take the limit without facing a division by zero. We get

$$\frac{dy}{dx} = 3x^2 + 1$$

Thus the derivative of $y(x) = x^3 + x$ is $dy/dx = 3x^2 + 1$.

Example 7.5-4

Another example is

$$y(x) = \sqrt{x}$$

The first three steps of the rule produce

$$\frac{\Delta y}{\Delta x} = \frac{\sqrt{x + \Delta x} - \sqrt{x}}{\Delta x}$$

Here the rearrangement to remove the division by zero is not so obvious. Some thought (and the first person must have done a lot of thinking) and recalling that

$$a - b = (a - b)\left[\frac{a + b}{a + b}\right] = \frac{a^2 - b^2}{a + b}$$

suggests multiplying *both* numerator and denominator by

$$\sqrt{x + \Delta x} + \sqrt{x}$$

This gives, in the numerator, the difference of the squares of the two square roots:

$$(\sqrt{x + \Delta x})^2 - (\sqrt{x})^2 = (x + \Delta x) - x = \Delta x$$

This Δx will now cancel the one in the denominator. The result is

$$\frac{\Delta y}{\Delta x} = \frac{1}{\sqrt{x + \Delta x} + \sqrt{x}}$$

In this form, as Δx approaches zero, we get the limit

$$\frac{dy}{dx} = \frac{1}{2\sqrt{x}}$$

Example 7.5-5

Because of the importance of the delta process, we look at a final example:

$$y(x) = \sqrt{x^3 + x}$$

The third step of the four-step process produces

$$\frac{\Delta y}{\Delta x} = \frac{\sqrt{(x + \Delta x)^3 + (x + \Delta x)} - \sqrt{x^3 + x}}{\Delta x}$$

Again we multiply both numerator and denominator by the sum of the square roots. This time we get the expression

$$\frac{\Delta y}{\Delta x} = \left[\frac{1}{\sqrt{(x + \Delta x)^3 + (x + \Delta x)} + \sqrt{x^3 + x}} \right][3x^2 + 3x\Delta x + (\Delta x)^2 + 1]$$

In this form we recognize parts of the last two examples. The first [ ] is the immediately preceding, Example 7.5-4, with $x^3 + x$ in place of x, and the second [ ] is exactly Example 7.5-3. The limit is, therefore, the product of the limits (Section 7.3)

$$\frac{dy}{dx} = \frac{1}{2\sqrt{x^3 + x}}[3x^2 + 1]$$

The four-step delta process involves in step 3 a division by Δx and then in step 4 the limiting process as $\Delta x \to 0$. The limiting process requires that you first rearrange the difference quotient $\Delta y/\Delta x$ into an equivalent form so that the limit can be found. This uses the methods of Section 7.4, so we can find the "missing value." We say that the slope of the curve at the given point is the slope of the tangent line at that point.

EXERCISES 7.5

Apply the four-step delta process to the following:

1. $y = x^4$
2. $y = ax^2 + bx + c$
3. $y = 1/x$
4. $y = 1/(x + 1)$
5. $y = \sqrt{x} + 1$
6. $y = 5x^4$
7. $y = \sqrt{ax + b}$
8. $y = 1/(x^2 + 1)$

.6 COMPOSITE FUNCTIONS

The last three examples show why we want to think about the *composition* of functions, such as

$$f(x) = \sin(\sqrt{x})$$

which is *composed* of the function "sine" of the function "square root" of x. Such a function is called a *composite function*. We saw in Example 7.5-5 that the delta process effectively took the derivative of the first function and multiplied this by the derivative of the second function.

To fix notation, suppose we have a pair of functions y and u, such that

$$y(x) = y[u(x)]$$

much like the above example, where we had $y = \sqrt{u}$ and $u = x^3 + x$. It is immediately obvious that (for Δx and Δu both not 0)

$$\frac{\Delta y}{\Delta x} = \left(\frac{\Delta y}{\Delta u}\right)\left(\frac{\Delta u}{\Delta x}\right)$$

and if the separate limits exist, then in the limit we have the fundamental relationship

$$\frac{dy}{dx} = \frac{dy}{du}\frac{du}{dx} \qquad (7.6\text{-}1)$$

This is the important *composite function rule*, also sometimes called the *chain rule*.

Example 7.6-1

In Example 7.5-5, we had

$$y(x) = \sqrt{x^3 + x}$$

We think of $y = \sqrt{u}$ and $u = x^3 + x$. From Example 7.5-4, we found that if

$$y = \sqrt{u}$$

then

$$\frac{dy}{du} = \frac{1}{2\sqrt{u}} = \frac{1}{2\sqrt{x^3 + x}}$$

and from Example 7.5-3 we found that if

$$u(x) = x^3 + x$$

then du/dx is given by

$$\frac{du}{dx} = 3x^2 + 1$$

Therefore, we conclude from Equation (7.6-1) that in this example

$$\frac{dy}{dx} = \frac{dy}{du}\frac{du}{dx} = \frac{1}{2\sqrt{x^3 + x}}[3x^2 + 1]$$

which is what we got before (Example 7.5-5).

Thus *if* we know how to find the derivatives of simple functions, it is easy to pass via the composite function rule (7.6-1) to more complex functions. Indeed, the composite rule for functions of functions of functions, and so on, is clearly

$$\frac{dy}{dx} = \frac{dy}{du}\frac{du}{dv}\frac{dv}{dw} \cdot \cdot \cdot \frac{dz}{dx} \tag{7.6-2}$$

and we can go as deep in the notation as we please. We need to turn, therefore, to finding the derivative of simple functions of some generality.

In the spirit of *generalization* consider, next, the general integral power of x:

$$y(x) = cx^n$$

From step 3 of the four-step rule, we have (factoring out the c)

$$\frac{\Delta y}{\Delta x} = c\frac{(x + \Delta x)^n - x^n}{\Delta x}$$

One way of making further progress is to recall the algebraic identity (Section 4.7), or simply multiply out the right-hand side and note the cancellation:

$$b^n - a^n = (b - a)(b^{n-1} + b^{n-2}a + b^{n-3}a^2 + \cdots + a^{n-1})$$

Then apply this to the difference of the two nth powers, where $b = x + \Delta x$ and $a = x$. It works out nicely because

$$b - a = (x + \Delta x) - x = \Delta x$$

and the difference quotient is merely the second parentheses of the algebraic identity. The final step is to take the limit as Δx approaches 0. Looking at the equivalent expression,

$$\frac{\Delta y}{\Delta x} = c[(x + \Delta x)^{n-1} + (x + \Delta x)^{n-2}x + \cdots + x^{n-1}]$$

we see that each term in the brackets will approach

$$x^{n-1}$$

and that there are exactly n such terms. Therefore, the derivative of cx^n is

$$\frac{d(cx^n)}{dx} = cnx^{n-1}$$

This is a fundamental rule of the differential calculus. Notice that you "pass over" the constant c in front of the expression and differentiate x^n. When you follow the algebra, you see that as a general rule the constant c factors out of each term and enters into the derivation at each step *only* as a front constant.

An alternative derivation can be based on expanding the binomial $(x + \Delta x)^n$, canceling the x^n term, dividing by Δx, and then taking the limit. All the terms of the expansion except the first one that is left will approach zero and you will have the same result.

Example 7.6-2

If

$$y = 3x^{17}$$

then the derivative is

$$\frac{dy}{dx} = 3(17x^{16}) = 51x^{16}$$

The special case of $n = 0$ requires separate investigation. The exponent equal to zero means that the function cx^0 is a constant,

$$y(x) = c$$

Just considering what this function is, a horizontal line, and the meaning of the derivative in geometry, that is, the local slope at the point x, shows that the derivative must be zero. However, we will go through the basic four-step process. Remember $y(x) = c$ for all x. At step 2 of the four-step process, we have

$$\Delta y = y(x + \Delta x) - y(x) = c - c = 0$$

Divide by Δx to get at step 3

$$\frac{\Delta y}{\Delta x} = \frac{0}{\Delta x} = 0$$

This is always zero (for $\Delta x \neq 0$), so the limit in step 4 is also zero. Thus we have the rule that *the derivative of a constant is 0.*

Differentiation is a fundamental process that must be mastered so well that it is almost automatic. The student is apt to think that the formulas can always be looked up when needed. But consider the situation where you are teaching a child the multiplication tables. The child says, "Why must I learn them perfectly, I can always look them up or use a hand computer?" The answer is, of course, there will come a time when the child faces division and needs to know the multiplication table so well that the number 21 immediately calls up to mind 3 times 7. Indeed, when it comes to factoring polynomials and the number 12 arises, then all combinations must be already in mind,

$$12 = 12 \times 1$$

$$12 = 6 \times 2$$

$$12 = 4 \times 3$$

so that the factoring of

$$x^2 + (\text{something})x + 12$$

can be done reasonably efficiently; the algebraic sum of the two factors must be the "something."

So it is with the process of differentiation. Not only must you be able to go directly from the functions to their derivatives, but there will soon come a time when the problem is to go in the reverse direction; from the given derivative you must be able to answer rapidly and reliably the question, "Where did this derivative come from?" You must be able, almost instantly, to imagine a large number of possibilities. Therefore, you are urged to learn the process of differentiation, at each stage, so well that you can do it reliably and almost without thought.

At the moment we have two basic general rules (including the case $n = 0$, a constant):

$$\frac{dx^n}{dx} = nx^{n-1} \qquad (n \geq 0)$$

and

$$\frac{dy}{dx} = \frac{dy}{du}\frac{du}{dx}$$

You can see them working together in the simple example

$$x^{mn} = (x^n)^m$$

Differentiate both sides in their respective forms (use $u = x^n$ on the right-hand side) to get

$$mnx^{mn-1} = m(x^n)^{m-1}(nx^{n-1})$$

which after a little algebra is

$$mnx^{mn-1} = mnx^{mn-n}x^{n-1}$$

and is obviously true. You can mentally practice differentiation at odd moments using special cases of the identity

$$(x^m)^n = x^{mn}$$

EXERCISES 7.6

Differentiate on sight:

1. $1, x, x^2, x^3, x^4, x^{19}, x^1, x^n$
2. $6x^2, 4x^3, 3x^4, 2x^6, x^{12}$
3. $ax^1, cx^3, bx^n, 6, 3ax^4$

Apply the delta process to the following:

4. x^4
5. $7x^3$
6. $17x^2$
7. Show in detail that the derivatives of x^{21} and $(x^7)^3$ are the same.
8. Derive the rule for differentiating x^n using the binomial theorem.

.7 SUMS OF POWERS OF X

Now that we can differentiate individual positive powers of x we turn to the examination of sums of powers of x. As a first, simple example (note the use of prime numbers so you can follow the way things go), consider the following example.

Example 7.7-1

If

$$y(x) = 7x^3 + 5x + 11$$

We get for step 2 of the four-step rule

$$\Delta y = [7(x + \Delta x)^3 + 5(x + \Delta x) + 11] - [7x^3 + 5x + 11]$$
$$= 7[(x + \Delta x)^3 - x^3] + 5[(x + \Delta x) - x]$$

so that at step 3 (expand the first term by the binomial theorem, cancel like terms, and divide by Δx)

$$\frac{\Delta y}{\Delta x} = 7[3x^2 + 3x\Delta x + (\Delta x)^2] + 5[1]$$

Therefore, doing step 4 we get

$$\frac{dy}{dx} = 7[3x^2] + 5 = 21x^2 + 5$$

The result is the same as if we passed over the constant coefficients and differentiated each power of x term by term.

It is easy to generalize to the sum of two (or more) functions. Suppose we have two functions $u(x)$ and $v(x)$ and that

$$y(x) = au(x) + bv(x)$$

Now thinking about how the algebra in Example 7.7-1 went, we see that we can group the terms in a and b separately. Thus we are led to the corresponding expressions (at step 3)

$$\frac{\Delta y}{\Delta x} = a\frac{u(x + \Delta x) - u(x)}{\Delta x} + b\frac{v(x + \Delta x) - v(x)}{\Delta x}$$

and in the limit we will have

$$\frac{dy}{dx} = a\frac{du}{dx} + b\frac{dv}{dx} \tag{7.7-1}$$

It is evident (using mathematical induction) that the rule applies to any finite sum of terms. The rule for differentiating any finite sum of terms is: differentiate term by term and pass over (but copy down) any constants that may occur in front of the powers of x. In words, differentiation is a *linear* operation; the derivative of a sum is the sum of the derivatives. Remember that the derivative of a constant is 0.

Example 7.7-2

Given the function

$$y(x) = 5x^4 - 7x^3 + 2x - 9$$

the derivative can be written down immediately (using 7.7-1) as

$$\frac{dy}{dx} = 5(4x^3) - 7(3x^2) + 2(1) - 9(0)$$

$$= 20x^3 - 21x^2 + 2$$

Example 7.7-3

Another example of more general form (we are deliberately forcing generality on you) is the general polynomial

$$P(x) = c_n x^n + c_{n-1} x^{n-1} + \cdots + c_1 x + c_0$$

The derivative is, upon inspection,

$$\frac{dP(x)}{dx} = nc_n x^{n-1} + (n-1)c_{n-1}x^{n-2} + \cdots + c_1$$

Example 7.7-4

As an example of a composite function, consider

$$y(x) = [x^3 + x + 11]^7$$

We say to ourselves, we have some function $u(x)$ to the seventh power; therefore,

$$\frac{dy}{dx} = 7[u(x)]^6 \frac{du}{dx}$$

where $u(x) = x^3 + x + 11$, and of course, from (7.7-1),

$$\frac{du}{dx} = 3x^2 + 1$$

Therefore,

$$\frac{dy}{dx} = 7[x^3 + x + 11]^6 (3x^2 + 1)$$

Example 7.7-5

A more complex example is

$$y(x) = [5x^2 + 1]^7 + 3[x - 2]^3 - 17[x^4 - x^2 + 1]^2$$

Take it term by term. The first term is a seventh power of something. Reason as follows: the derivative is 7[something] to the sixth power times the derivative of what is inside and that is (looking at the $5x^2 + 1$) exactly $10x$. So the first term gives

$$7[5x^2 + 1]^6(10x)$$

Apply the same technique to each term, thus getting, finally,

$$\frac{dy}{dx} = 7[5x^2 + 1]^6(10x) + 3(3)[x - 2]^2(1) - 17(2)[x^4 - x^2 + 1](4x^3 - 2x)$$

$$= 70x[5x^2 + 1]^6 + 9[x - 2]^2 - 68x(2x^2 - 1)[x^4 - x^2 + 1]$$

after a little algebra.

This last example shows that after the process of formal differentiation there is often a large amount of what appears to be trivial algebra to get the final result into a decent form. There is a tendency to ignore the importance of this step, but if it is not done correctly, all the earlier work is wasted. Thus a course in the calculus consists to a great extent in learning to do algebra rapidly and reliably, a fact of life that also assures you that you will always be able to do algebra in the future.

EXERCISES 7.7

Differentiate on sight the following:

1. $x^2 + a^2$
2. $x^3 - 3x^2$
3. $x^3 + x^2 + x + 1$
4. $4x^2 - 2x + 1$
5. $x^3/3 + x^2/2 + x$
6. $ax^2 + bx + c$
7. $ax^n + bx^m$
8. $(a + 1)x^n$
9. $x^n - a^n$

Differentiate the following:

10. $3(x^2 + 1)^3 + 4(x^2 - 1)^2$
11. $(a^2 + x^2)^n + (a^2 - x^2)^m$
12. $(x + 1)^3 + 3(x + 1)^2 + 7$

7.8 PRODUCTS AND QUOTIENTS

Suppose you have a product of two functions, say

$$y(x) = (x^2 + 7)^3(x^3 - x)^2$$

or in general

$$y(x) = u(x)v(x)$$

What is the derivative? We apply the basic four-step rule: in step 1, first x gets an increment Δx; then, step 2, this induces increments in u to $u + \Delta u$, v to $v + \Delta v$, as well as in y to $y + \Delta y$; next, step 3, we take the ratio to get

$$\frac{\Delta y}{\Delta x} = \frac{[u(x) + \Delta u(x)][v(x) + \Delta v(x)] - u(x)v(x)}{\Delta x}$$

Expand the product and rearrange the terms to get

$$\frac{\Delta y}{\Delta x} = u(x)\frac{\Delta v(x)}{\Delta x} + v(x)\frac{\Delta u(x)}{\Delta x} + \frac{\Delta u(x)}{\Delta x}\Delta v(x)$$

In the last step (4), the limits of these terms, as Δx approaches 0, are

$$\frac{d(uv)}{dx} = u\frac{dv}{dx} + v\frac{du}{dx} \qquad (7.8\text{-}1)$$

In words, the derivative of a product is the first times the derivative of the second plus the second times the derivative of the first. Don't learn it in terms of u and v because in use you will be thinking about terms that may have any letters. Note the symmetry in the formula. Since $uv = vu$ (multiplication is commutative), the formula must be symmetric in u and v.

Example 7.8-1

In the example at the beginning of this section,

$$y(x) = (x^2 + 7)^3(x^3 - x)^2$$

we have

$$u(x) = (x^2 + 7)^3$$

$$v(x) = (x^3 - x)^2$$

Therefore, applying the formula for the product, we get

$$\frac{dy}{dx} = (x^2 + 7)^3[2(x^3 - x)(3x^2 - 1)] + (x^3 - x)^2[3(x^2 + 7)^2(2x)]$$

Usually, there are common powers that can be factored out; each common factor is of one lower power than in the original formula. The result is

$$\frac{dy}{dx} = (x^2 + 7)^2(x^3 - x)[2(x^2 + 7)(3x^2 - 1) + 3(x^3 - x)(2x)]$$

$$= (x^2 + 7)^2(x^3 - x)[12x^4 + 34x^2 - 14]$$

$$= 2(x^2 + 7)^2(x^3 - x)[6x^4 + 17x^2 - 7]$$

Example 7.8-2

Find the derivative of

$$y = (x^2 + a^2)^3(x^3 - x)^4$$

Apply the formula for the product to get

$$\frac{dy}{dx} = (x^2 + a^2)^3[4(x^3 - x)^3(3x^2 - 1)] + (x^3 - x)^4[3(x^2 + a^2)^2(2x)]$$

$$= (x^2 + a^2)^2(x^3 - x)^3[4(x^2 + a^2)(3x^2 - 1) + (x^3 - x)(6x)]$$

$$= (x^2 + a^2)^2(x^3 - x)^3[18x^4 + (12a^2 - 10)x^2 - 4a^2]$$

$$= 2(x^2 + a^2)^2(x^3 - x)^3[9x^4 + (6a^2 - 5)x^2 - 2a^2]$$

Naturally, products of three or more terms can occur, and we need to look at *extensions* of the formula. Consider the formula

$$y = uvw$$

We can look at this as either of the forms

$$y = (uv)w = u(vw)$$

and apply the product rule. In the first case we have

$$\frac{dy}{dx} = uv\frac{dw}{dx} + w\frac{d(uv)}{dx}$$

and to the last term we again apply the product rule

$$\frac{dy}{dx} = uv\frac{dw}{dx} + w\left[u\frac{dv}{dx} + v\frac{du}{dx}\right]$$

$$= uv\frac{dw}{dx} + uw\frac{dv}{dx} + vw\frac{du}{dx} \qquad (7.8\text{-}2)$$

Again note the necessary symmetry. Indeed, you can now easily see (and prove by induction if you wish) the formula for the product of any number of terms. The derivative of a product of n terms is the sum of n distinct terms in each of which only one factor is differentiated and all the others are copied as given. No products of derivatives occur.

For a quotient of two functions, we proceed in a similar manner as we did for the product. Suppose

$$y(x) = \frac{u(x)}{v(x)}$$

Apply the four-step rule. At step 3 you will have

$$\frac{\Delta y}{\Delta x} = \left[\frac{u(x) + \Delta u(x)}{v(x) + \Delta v(x)} - \frac{u(x)}{v(x)} \right] \frac{1}{\Delta x}$$

Get a common denominator

$$= \frac{[u(x) + \Delta u(x)]v(x) - u(x)[v(x) + \Delta v(x)]}{[v(x) + \Delta v(x)]v(x)\Delta x}$$

and then rearrange in the form

$$= \frac{v(x)\dfrac{\Delta u(x)}{\Delta x} - u(x)\dfrac{\Delta v(x)}{\Delta x}}{v(x)[v(x) + \Delta v(x)]}$$

Taking the limit, we have the formula for a quotient:

$$\frac{d(u/v)}{dx} = \frac{v(du/dx) - u(dv/dx)}{v^2} \tag{7.8-3}$$

In words, the derivative of a quotient is the denominator times the derivative of the numerator minus the numerator times the derivative of the denominator, all divided by the denominator squared.

In all these derivations there are many terms that are present before the limit is taken, and which vanish in the limit. Thus the mathematics of the limiting form is generally much simpler than is the finite form before the limit is taken. This is one reason why the algebra and other mathematics for the calculus is much simpler than is the corresponding discrete mathematics.

Example 7.8-3

Differentiate

$$y = \frac{x^2 - 1}{x^2 + 1}$$

Applying the formula for the quotient, we get

$$\frac{dy}{dx} = \frac{(x^2 + 1)2x - (x^2 - 1)2x}{(x^2 + 1)^2}$$

$$= \frac{4x}{(x^2 + 1)^2}$$

Example 7.8-4

Differentiate

$$y = \frac{ax^2 + bx + c}{x^2 + 1}$$

Using the quotient formula, we get

$$\frac{dy}{dx} = \frac{(x^2 + 1)(2ax + b) - (ax^2 + bx + c)(2x)}{(x^2 + 1)^2}$$

$$= \frac{-bx^2 + 2(a - c)x + b}{(x^2 + 1)^2}$$

EXERCISES 7.8

Differentiate both sides to check your work.

1. $(x - 1)(x + 1) = (x^2 - 1)$
2. $(x^2 - 1)(x^2 + 1) = x^4 - 1$
3. $(x^2 + 1)^2 = (x^2 + 1)(x^2 + 1)$
4. $(x^2 - 1)/(x - 1) = x + 1$
5. $(x + 1)(x + 2)(x + 3) = x^3 + 6x^2 + 11x + 6$

Differentiate the following:

6. $(x^2 + a^2)(x^2 + b^2)$
7. $(x^2 - a^2)(x^2 - b^2)(x^2 - c^2)$
8. $(x - 1)/(x + 1)$
9. $x/(x - a)$
10. $(x + a)/(x + b)$
11. $(x^3 + a^3)/(x^2 + a^2)$
12. $(a^2x^2 + b^4)/(a^2x^2 - b^4)$
13. Apply the product formula to $x^n = xxx \ldots x$
14. Apply the product rule to $x^n x^m = x^{m+n}$
15. Apply the quotient rule to $x^m/x^n = x^{m-n}$

7.9 AN ABSTRACTION OF DIFFERENTIATION

Rather than going back again and again to the basic four-step delta process, let us examine just how little we need to have so we can find the derivative of any algebraic function. What rules of differentiation suffice to generate all the rest of the rules for algebraic differentiation?

Clearly, we need both the sum formula,

$$\frac{d(u + v)}{dx} = \frac{du}{dx} + \frac{dv}{dx} \tag{7.9-1}$$

and the product formula,

$$\frac{d(uv)}{dx} = u\frac{dv}{dx} + v\frac{du}{dx} \qquad (7.9\text{-}2)$$

Both are easily extended to many terms by recognizing that the product of any two terms may be regarded as a single function. We also need the important function of a function formula:

$$\frac{dy}{dx} = \frac{dy}{du}\frac{du}{dx} \qquad (7.9\text{-}3)$$

It turns out (by trial and error) that we also need two more derivatives

$$\frac{dx}{dx} = 1 \qquad (7.9\text{-}4)$$

and

$$\frac{dc}{dx} = 0 \qquad (7.9\text{-}5)$$

both of which are easy. From these five formulas alone we will derive all the others using the methods of extension you have already learned.

First, note that applying the product formula to an expression of the form $cu(x)$ gives the rule that you pass over constants.

$$\frac{d[cu(x)]}{dx} = c\frac{du(x)}{dx} + u(x)\frac{dc}{dx} = c\frac{du(x)}{dx}$$

We next use the model of the laws of exponents (Section 3.8). We write simply

$$x^2 = xx$$

and differentiate both sides [on the right using the product formula (7.9-2) and (7.9-4)]:

$$\frac{dx^2}{dx} = x\frac{dx}{dx} + x\frac{dx}{dx}$$

$$= 2x$$

We see the basis for an induction proof (Section 2.3). Assume that the formula

$$\frac{d(x^n)}{dx} = nx^{n-1}$$

holds up to $n = m - 1$. Now take the mth case,

$$x^m = xx^{m-1}$$

and differentiate both sides:

$$\frac{dx^m}{dx} = x[(m-1)x^{m-2}] + x^{m-1}$$

$$= x^{m-1}[m - 1 + 1] = mx^{m-1}$$

The induction is complete for positive integer powers.

Next, to extend the rule to negative powers, we write

$$1 = x^n x^{-n}$$

and differentiate both sides [using the product formula (7.9-2)] to get

$$0 = x^n \frac{dx^{-n}}{dx} + x^{-n} n x^{n-1}$$

Solve for

$$\frac{dx^{-n}}{dx} = \frac{(-x^{-n} n x^{n-1})}{x^n}$$

$$= -n x^{-n-1} = \frac{-n}{x^{n+1}}$$

This is the *same formula* as for positive powers of x; you *decrease* the exponent by 1.

To attack the fractional powers, we begin by writing

$$x = x^{1/2} x^{1/2}$$

and differentiate both sides (again using the product formula). We get

$$1 = x^{1/2} \frac{dx^{1/2}}{dx} + x^{1/2} \frac{dx^{1/2}}{dx}$$

Solve for the derivative:

$$\frac{dx^{1/2}}{dx} = \frac{1}{2} x^{-1/2} = \frac{1}{2} x^{(1/2)-1}$$

Evidently, we can similarly find that for the qth root

$$\frac{dx^{1/q}}{dx} = \frac{1}{q} x^{1/q - 1}$$

We are supposing that at this point you can supply the necessary ellipsis proof. Now we have only to use the pth power of the qth root to get the derivative of $x^{p/q}$. We have, using the function of a function formula (7.9-3),

$$\frac{dx^{p/q}}{dx} = p (x^{1/q})^{p-1} \left(\frac{1}{q} x^{1/q-1} \right)$$

$$= \frac{p}{q} x^{(p/q - 1/q + 1/q - 1)}$$

$$= \frac{p}{q} x^{p/q - 1}$$

Thus we see that, using a somewhat standard pattern for the extension of ideas, we have found the derivative for all rational powers of the variable x [and by the function of function formula (7.9-3) this applies to rational powers of anything at all]. For the irrational powers such as

$$x^{\sqrt{2}}$$

we have to appeal, at this point, to the continuity of x^n as a function of n and the assumption that, if the formula holds for the everywhere dense rationals, then it also holds true for the irrationals (see Section 4.5 where this assumption 1 is explicitly stated).

We have finally to derive (again) the formula for the derivative of the quotient of two functions. We write the quotient as a product

$$y(x) = \frac{u(x)}{v(x)} = \frac{u}{v} = uv^{-1}$$

Then, applying the product formula (7.9-2),

$$dy/dx = u\left[(-1v^{-2})\frac{dv}{dx}\right] + v^{-1}\frac{du}{dx}$$

$$= -\left(\frac{u}{v^2}\right)\frac{dv}{dx} + \frac{1}{v}\frac{du}{dx}$$

We get a common denominator and have

$$\frac{dy}{dx} = \frac{[v(du/dx) - u(dv/dx)]}{v^2}$$

Thus we have *found all* the formulas we need for the moment using only five basic ones. These basic formulas used the four-step rule; the others followed logically from them.

EXERCISES 7.9

Differentiate the following:

1. $1/x$, $1/x^2$, x^{-3}, $5x^{-5}$
2. $1/(a + x)$
3. $1/(a^2 + x^2)$
4. $1/(a^2 + x^2)^2$
5. $1/(a^2 - x^2)^2$
6. $\sqrt{2x}$, $2\sqrt{x}$
7. $2x^{1/2}$, $(2x)^{1/2}$
8. $3x^{2/3}$

9. $x^{-1/2}$

10. $3x^{-1/3}$

11. $\sqrt{a + x}$

12. $(b - x)^{-1/2}$

13. $\sqrt{a^2 - x^2}$

14. $1/\sqrt{a^2 - x^2}$

15. $x/(a^2 - x^2)$

16. $\sqrt{x} + 1/\sqrt{x}$

17. $x/\sqrt{x^2 - a^2}$

18. $\sqrt{(1 + \sqrt{x})}$

19. $x^{4/3} + x^{-4/3}$

20. $(\sqrt{x} - 1)/(\sqrt{x} + 1)$

21. $[(\sqrt{x} - 1)/(\sqrt{x} + 1)]^{1/2}$

*.10 ON THE FORMAL DIFFERENTIATION OF FUNCTIONS

Just how should you approach a formula when it is to be differentiated? If it is a sum, you use the divide and conquer rule; take the terms one at a time (and try not to get lost when you finish one term but remember to go on to the next term of the sum). For example,

$$y(x) = 7x^9 + (x - 3)(x^2 + 1) + \frac{x - 7}{x^2 + 1}$$

You take one term of the sum at a time. This way the sum is broken down to a number of simpler problems. The first term goes easily, simply

$$63x^8$$

The second term is a product, the first times . . . .

$$(x - 3)\frac{d(x^2 + 1)}{dx} + (x^2 + 1)\frac{d(x - 3)}{dx}$$

Now we look at this and see (again!) the simple differentiations of sums:

$$\frac{d(x^2 + 1)}{dx} = 2x \quad \text{and} \quad \frac{d(x - 3)}{dx} = 1$$

We have to assemble the parts of the derivatives of the first two terms. Then we remember that there is a third term. This is a quotient and we start, denominator times . . . .

$$\frac{(x^2 + 1)(1) - (x - 7)(2x)}{(x^2 + 1)^2}$$

We need to assemble the results of the three terms

$$= 63x^8 + (x - 3)(2x) + (x^2 + 1) + \frac{(x^2 + 1) - 2x^2 + 14x}{(x^2 + 1)^2}$$

$$= 63x^8 + 2x^2 - 6x + x^2 + 1 + \frac{-x^2 + 14x + 1}{(x^2 + 1)^2}$$

$$= 63x^8 + 3x^2 - 6x + 1 + \frac{-x^2 + 14x + 1}{(x^2 + 1)^2}$$

Thus the process of formal differentiation is broken down to a sequence of simpler differentiations. When you begin, it is wise to write out the steps one at a time so that you do not get lost. In effect, you are trying to push the derivative sign off the right-hand side of the expression, and you finally get rid of it when you have a derivative of x with respect to x (you simply pass over constants in front of terms). You are done when there are no more derivatives to be taken. But the differentiation process is laborious, and gradually you begin to keep a mental "pushdown list" of where you are, and you start eliminating the writing down of many of the trivial steps. Of course, there is often a lot of final algebra to be done to clean up the mess that appears on the paper.

There is not a unique, single way of attacking a differentiation problem. For example, suppose you have a problem in the form

$$y = \frac{uv}{w}$$

You can look at it in many ways. Among others are

$$y = \left(\frac{u}{w}\right)v = u\left(\frac{v}{w}\right) = \frac{uv}{w} = uvw^{-1} = \frac{u}{v^{-1}w}$$

They must all give the same result. In the first case, $(u/w)v$, you have

$$\frac{dy}{dx} = \left(\frac{u}{w}\right)\frac{dv}{dx} + v\frac{d(u/w)}{dx}$$

$$= \frac{u}{w}\frac{dv}{dx} + v\left[\frac{w\dfrac{du}{dx} - u\dfrac{dw}{dx}}{w^2}\right]$$

$$= \frac{u}{w}\frac{dv}{dx} + \frac{vdu}{wdx} - \frac{uv}{w^2}\frac{dw}{dx}$$

In the second case, $u(v/w)$, you get

$$\frac{dy}{dx} = u\frac{d(v/w)}{dx} + \frac{v}{w}\frac{du}{dx}$$

$$= u\left[\frac{w\dfrac{dv}{dx} - v\dfrac{dw}{dx}}{w^2}\right] + \frac{v}{w}\frac{du}{dx}$$

$$= \frac{u}{w}\frac{dv}{dx} - \frac{uv}{w^2}\frac{dw}{dx} + \frac{v}{w}\frac{du}{dx}$$

which is the same thing with a different order of terms.

If the equations are written in the form with the original function factored out in front, then you have

$$\frac{dy}{dx} = \frac{uv}{w}\left[\frac{1}{v}\frac{dv}{dx} + \frac{1}{u}\frac{du}{dx} - \frac{1}{w}\frac{dw}{dx}\right]$$

From this you can make a good guess that the general case is

$$\frac{d\left(\dfrac{uvw}{pqr}\right)}{dx} = \frac{uvw}{pqr}\left[\frac{1}{u}\frac{du}{dx} + \frac{1}{v}\frac{dv}{dx} + \frac{1}{w}\frac{dw}{dx} - \frac{1}{p}\frac{dp}{dx} - \frac{1}{q}\frac{dq}{dx} - \frac{1}{r}\frac{dr}{dx}\right]$$

where there is a plus sign in front of the terms in the numerator and minus signs in front of the denominator terms.

Formal differentiation is a necessary process to master. We will use it constantly throughout the rest of the book. It is a matter of practice until it is almost automatic and you can concentrate on the use you are going to make of the derivative when you get it. Blind practice is not as effective as intelligent practice, especially going over a completed problem to see why and how things went, criticizing your attack, and trying the same problem another way until you understand "the way things go." You need to develop a suitable technique for this important process. From each particular case you should try to understand how other similar cases would go. You can make up as many drill problems as you think you need and check yourself by computing the derivative in two or more ways; problems that you make yourself are apt to be more effective learning experiences than those from a textbook!

EXERCISES 7.10

Differentiate the following:

1. $x^{1/3} + 2x^{2/3} + 3x^{4/3}$
2. $(ax^2 + bx + c)/(x^2 - 1)$
3. $p^m(1 - p)^{n-m}$, where the variable is p
4. $x(x - 1)(x - 2)(x - 3)$
5. $1/x(x + 1)$
6. $1/\sqrt{x}$

7. $1/x^{4/3}$

8. $-x^{-1/5}$

9. $(x - a)(x - b)/(x - c)$

10. $1/x^{-1/2}$

7.11 SUMMARY

From a formal point of view, differentiation is the basic operation of the calculus. Fundamentally, it is the four-step rule whose meaning, geometrically, is finding the slope of the local tangent line to the curve. From this rule a number of basic rules were derived, which save you from going back to all the details of the four-step rule. We found that

$$\frac{du^n}{dx} = nu^{n-1}\frac{du}{dx}$$

for all real n, not just the integers. Other rules are

$$\frac{d(au + bv)}{dx} = a\frac{du}{dx} + b\frac{dv}{dx}$$

$$\frac{d(uv)}{dx} = u\frac{dv}{dx} + v\frac{du}{dx}$$

$$\frac{d(u/v)}{dx} = \frac{v(du/dx) - u(dv/dx)}{v^2}$$

$$\frac{du}{dx} = \frac{du}{dv}\frac{dv}{dx}$$

Using these five simple rules, you can differentiate any algebraic function. This process of formal differentiation should be mastered so well that whenever you need to differentiate a function you can do so without having to think very much (and thus distract yourself from the new material at hand). We repeat this important point: voluminous practice is not as effective as a reasonable amount done thoughtfully; especially when you finish an exercise, ask yourself what other kinds could be done essentially the same way. Looking back at successes is an efficient way of mastering the tedious art of differentiation. Many short sessions are generally more effective than a few long ones.

You should not forget, while doing the formal differentiation, that the derivative is actually defined by the four-step rule, and the rules you use come from it. All the derivatives you find can be found, in principle, by the direct application of the four-step rule, but to carry out the delta process in full detail would often defeat all your energy. Thus, although at first they are troublesome to master, the simple rules for differentiation represent an enormous saving of effort.

MISCELLANEOUS EXERCISES

Find the derivatives for the following:

1. $y = (x^2 + 1)(x^2 - 1)$
2. $y = x/(x + 1)$
3. $y = x/(1 - x^2)$
4. $y = (x^4 - 1)/(x + 1)$
5. $y = [(1 - s)/(1 + s)]^{1/2}$
6. $y = x^{1/2} + a^{1/2}$
7. $y = x^2 + x(x + 1)$
8. $y = u^2 + 1, u = v^2 - 1, v = x^2 + 1$
9. $y = (x^2 + a^2)/(x^2 - a^2)$
10. $y = x + \sqrt{(x + 1)}$
11. $s = t^2/(a^2 - t^2)$
12. $y = (1 - \sqrt{x})^3$
13. $y = [a^{2/3} - x^{2/3}]^{3/2}$
14. $y = [1 + \{1 + (1 + x^2)^2\}^2]^2$
15. $y = x\sqrt{x}$
16. $y = \sqrt{\{1 + \sqrt{(1 + \sqrt{x})}\}}$
17. $y = (x + 1)/(x - 1) + (x - 1)/(x + 1)$
18. $y = 1/(x\sqrt{x})$
19. $y = x^2 + 3ax - a^2$
20. $y = \sqrt{(u + 1)}, u = \sqrt{(x + 1)}$
21. $y = \sqrt{(2at - t^2)}/t$
22. $y = (x^2 + a^2)^{3/2}$
23. $s = t^2(1 - t)^3$
24. $\theta = \frac{1}{2}\sqrt{(1 - t^2)}$
25. $x = 1/\{y^2(y + a)^2\}$

8

Geometric Applications

8.1 TANGENT AND NORMAL LINES

To a great extent, the differential calculus arose from the problem of finding the tangent line to a given curve at an arbitrary point. We have just seen how to find the slope of a curve at an arbitrary point (x, y); it is simply the derivative of the function at that point.

To find the tangent line, we could assume the general form of a straight line (Section 5.4)

$$Ax + By + C = 0 \qquad (8.1\text{-}1)$$

and impose the two conditions (using the method of undetermined coefficients) that the line pass through the point and have the given slope at that point. The resulting two equations would be linear (in the unknown coefficients) and easy to solve (there really are only two parameters in the general form of the equation). Because we now want to use the symbols x and y for points running along the tangent line, it is necessary to fix the point where the line is to be tangent; call the point on the curve (x_1, y_1). This is the point where we impose both the point and slope conditions. As an alternative to (8.1-1), we can use the more convenient point-slope form of the line, Equation (5.6-6):

$$y - y_1 = m(x - x_1) \qquad (8.1\text{-}2)$$

where, of course, m is the slope (the derivative of the curve at the point (x_1, y_1)).

Example 8.1-1

Suppose we use the simple parabola as a test case,

$$y = x^2$$

The derivative at the point (x_1, y_1) is

$$\frac{dy}{dx} = 2x_1 = m$$

Therefore, the tangent line at the point (x_1, y_1) is

$$y - y_1 = 2x_1(x - x_1)$$

In particular, at the point $(1, 1)$ on the curve, we have

$$y - 1 = 2(x - 1)$$

or $$y = 2x - 1$$

See Figure 8.1-1.

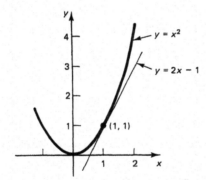

Figure 8.1-1   Tangent to a parabola

Besides asking what is the tangent line at a given point, we can ask many other questions such as: "Where does a given curve have a slope parallel to a given line?"

Example 8.1-2

Where does the parabola

$$y = x^2$$

have a tangent parallel to the line

$$y = 3x$$

(Figure 8.1-2)? The word "parallel" means "having the same slope." The slope of the given line is 3 (either from the point-slope formula or by differentiating the equation for the line). We equate this to the slope of the parabola $(2x)$

$$2x = 3$$

which tells us that, at the place where $x = \frac{3}{2}$ and (from the equation of the parabola) $y = \frac{9}{4}$, the tangent to the parabola is parallel to the given line. The actual tangent line is

$$y - \frac{9}{4} = 3\left(x - \frac{3}{2}\right)$$

or

$$y = 3x - \frac{9}{2} + \frac{9}{4} = 3x - \frac{9}{4}$$

We can check this easily. Is the slope right? Does it pass through the required point? Both the original line and the solution line have the same slope 3. Thus it remains to check that the point $(\frac{3}{2}, \frac{9}{4})$ lies on both the parabola (obviously) and on the solution line. We have, putting the coordinates of the point into the equation of the line,

$$\frac{9}{4} = 3\left(\frac{3}{2}\right) - \frac{9}{4} = \frac{9}{4}$$

as it should.

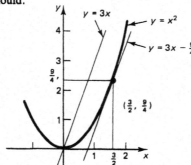

Figure 8.1-2 Parallel tangent line

Example 8.1-3

We may also ask where two curves are tangent to each other. For example, which parabola of the family (depending on the choice of a)

$$y = ax^2$$

is tangent to the line

$$y = x + 1$$

and at what point?

We require two conditions: (1) the two curves meet at a point, and (2) the two curves have the same slope at that point. In mathematical terms, we require both

(1) $$ax^2 = x + 1$$

and

(2) $$2ax = 1$$

To eliminate the a, multiply the first equation by 2 and the second equation by $-x$ and add. You get

$$0 = 2x + 2 - x$$

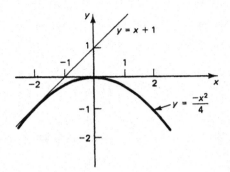

Figure 8.1-3 Parabola tangent to a line

which yields $x = -2$, $y = -1$ (from the line), and hence $a = -\frac{1}{4}$ (from the parabola). This is easily checked by sketching the two curves (Figure 8.1-3).

The *normal line* is the line perpendicular to the tangent line that passes through the point of tangency. We recall from Equation (5.5-8) that two lines are perpendicular to each other when their slopes are the negative reciprocals of each other. Thus, to find the normal line, we use the negative reciprocal of the derivative as the slope; otherwise, things are the same.

Example 8.1-4

Using the same parabola,

$$y = x^2$$

and the same point, $(1, 1)$, find the normal line. The slope of the curve at the point is $m = 2$. The negative reciprocal is $-\frac{1}{2}$. Thus the normal line is

$$y - 1 = -\tfrac{1}{2}(x - 1)$$

or

$$y = \frac{-x + 3}{2}$$

See Figure 8.1-4. Again, it is easy to check the answer.

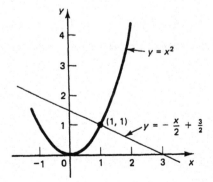

Figure 8.1-4 Normal line

EXERCISES 8.1

Find the tangent and normal lines to the curve at the indicated point:

1. $y = x^3$ at $(1, 1)$ 2. $y = x^4$ at $(1, 1)$
3. $y = 1/x$ at $(1, 1)$ 4. $y = x + 1$ at $(0, 1)$
5. $y = \sqrt{x}$ at $(1, 1)$ 6. $y = 1/\sqrt{x}$ at $(1, 1)$
7. $y = x^n$ at $(1, 1)$ 8. $y = ax^2 + bx + c$ at (x_1, y_1)
9. Show that the normal line to the semicircle $y = \sqrt{a^2 - x^2}$ passes through the center of the circle.

8.2 HIGHER DERIVATIVES—NOTATION

From a given function

$$y = y(x)$$

we derived a function, called the derivative,

$$\frac{dy}{dx}$$

This is a function of x, and therefore we can again apply the process of differentiation; we can compute

$$\frac{d}{dx}\left\{\frac{dy}{dx}\right\} = \left(\frac{d}{dx}\right)^2 y$$

Example 8.2-1

For the usual parabola

$$y = x^2$$

we have the *first* and *second* derivatives

$$\frac{dy}{dx} = 2x$$

$$\frac{d}{dx}\left\{\frac{dy}{dx}\right\} = 2$$

For the more complicated function (the upper branch of a circle)

$$y = \sqrt{1 - x^2}$$

we have

$$\frac{dy}{dx} = \frac{1}{2}(1 - x^2)^{-1/2}(-2x) = \frac{-x}{\sqrt{1 - x^2}}$$

$$\frac{d}{dx}\left\{\frac{dy}{dx}\right\} = \frac{\sqrt{1 - x^2}\,(-1) - (-x)(1/2)(1 - x^2)^{-1/2}(-2x)}{1 - x^2}$$

or $$\frac{d}{dx}\left\{\frac{dy}{dx}\right\} = \frac{-1}{(1 - x^2)^{3/2}}$$

This notation for the second derivative is clearly a nuisance. Traditionally (using the observations of Section 2.8 on the use of exponents on symbols), it follows that

$$\left(\frac{d}{dx}\right)^2 y$$

is the second derivative. The third derivative would be

$$\left(\frac{d}{dx}\right)^3 y$$

and, in general, the nth derivative would be

$$\left(\frac{d}{dx}\right)^n y$$

Remember that d/dx is a single symbol. Nevertheless, conventionally it is written as

$$\left(\frac{d}{dx}\right)^n = \frac{d^n}{(dx)^n} = \frac{d^n}{dx^n}$$

the last being clearly a convenience. Read this as a single symbol, "d to the nth over dx to the nth".

The derivative is often written in shorter forms. For example, all the following notations are in widespread use

$$\frac{dy}{dx} = Dy = D_x y = y'$$

The use of the prime notation (y prime) is especially convenient. The higher derivatives are correspondingly

$$\frac{d^2 y}{dx^2} = D^2 y = D_x^2 y = y''$$

and so on. We will often use the prime notation (y'', pronounced "y double prime," for example).

EXERCISES 8.2

Find the first and second derivatives of the following:

1. $y = \sqrt{(x^2 + 1)}$
2. $y = x^n$
3. $y = 1\sqrt{1 - x^2}$
4. $y = 1/x$

Find the third derivative of the following:

5. $y = x^3, x^n, 1/x, \sqrt{x}$

6. Show that the nth derivative of a polynomial of degree n is $a_n n!$, and that the $(n + 1)$st derivative is zero.

Find the nth derivative of the following:

7. $y = 1/x$

8. $y = \sqrt{x}$

9. $y = (a + bx)^m$

8.3 IMPLICIT DIFFERENTIATION

Example 8.3-1

The conventional form of a circle

$$x^2 + y^2 = r^2$$

requires us (at present) to solve for one *branch* or the other,

$$y = \pm\sqrt{r^2 - x^2}$$

depending on which piece of the *double-valued* function we want to use. However, we can avoid the unpleasantness of the square roots during the differentiation process by a process called *implicit differentiation*. Take the equation of the circle in the conventional form,

$$x^2 + y^2 = r^2$$

and start the four-step delta process (Section 7.5). We will get, at step 2 of the four-step process,

$$(x + \Delta x)^2 + (y + \Delta y)^2 = r^2$$

Now subtract the original equation and divide by Δx. We get

$$2x + \Delta x + 2y\frac{\Delta y}{\Delta x} + \Delta y\left(\frac{\Delta y}{\Delta x}\right) = 0$$

We then take the limit as Δx approaches 0, noting that as $\Delta x \to 0$ it also follows that $\Delta y \to 0$. We get

$$2x + 2y\frac{dy}{dx} = 0$$

This shows that we simply apply the rule for "differentiating a function of a function" to the y terms. The final result in the case of a circle is

$$y'(x) = \frac{dy}{dx} = -\frac{x}{y}$$

The value of y we use selects the branch we are on.

Example 8.3-2

Suppose we have the hyperbola

$$xy = a^2$$

Differentiate the product to get

$$x\frac{dy}{dx} + y = 0$$

or

$$y' = \frac{dy}{dx} = \frac{y}{x}$$

Example 8.3-3

Find the angle of intersection between the circle $x^2 + y^2 = 5$ and the hyperbola $xy = 2$ in the first quadrant.

    You must first find the point of intersection. To find it you (cleverly) add and subtract twice the hyperbola equation from the circle equation to get

$$x^2 + 2xy + y^2 = 5 + 4 = 9$$
$$x^2 - 2xy + y^2 = 5 - 4 = 1$$

Take the square root of both equations, from which the solutions are

$$x + y = \pm 3$$
$$x - y = \pm 1$$

Remember that the solution is to be in the first quadrant. You get, after some thinking and the use of *symmetry*,

$$x = 2, \quad y = 1 \quad \text{and} \quad x = 1, \quad y = 2$$

Check that that both points lie on both curves! From Example 8.3-1 for the circle at the point (2, 1),

$$y' = \frac{dy}{dx} = -\frac{x}{y} = -2$$

and for the hyperbola, Example 8.3-2,

$$y' = \frac{dy}{dx} = -\frac{y}{x} = -\frac{1}{2}$$

We now look back, Equation (5.5-5), for the formula for the angle between the two tangent lines. At the point of intersection we have

$$\tan \theta = \frac{-2 - (-\frac{1}{2})}{1 + (-2)(-\frac{1}{2})} = \frac{-2 + \frac{1}{2}}{1 + 1} = -\frac{3}{4}$$

from which (see Figure 8.3-1) θ is approximately $-37°$. From symmetry, the size of the angle at the second point is the same, although the sign will be different.

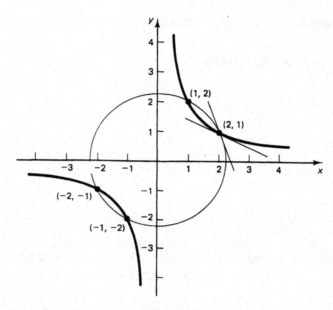

Figure 8.3-1 Angle between a circle and a hyperbola

Example 8.3-4

Suppose we have the more complex expression

$$x^2y^2 + 5y - x^4 - 5 = 0$$

To find the slope at any point, we simply differentiate according to the usual rules. The first term is a product. We have

$$x^2\left(2y\frac{dy}{dx}\right) + y^2(2x) + 5\frac{dy}{dx} - 4x^3 = 0$$

and solve for dy/dx:

$$(2x^2y + 5)\frac{dy}{dx} = 4x^3 - 2xy^2$$

or

$$y' = \frac{dy}{dx} = \frac{4x^3 - 2xy^2}{2x^2y + 5} = \frac{2x(2x^2 - y^2)}{2x^2y + 5}$$

At the particular point $(1, 1)$, which lies on the original curve, the slope is

$$y' = \frac{dy}{dx} = \frac{4 - 2}{2 + 5} = \frac{2}{7}$$

We could, of course, also compute the second derivative if we wished.

Notice how implicit differentiation lures us out of the domain where y is a single-valued function of x. Implicit differentiation is a very useful tool in many situations where we are initially given implicitly defined functions, rather than the conventional explicitly defined, single-valued function.

EXERCISES 8.3

Find dy/dx for the following:

1. $x^2/a^2 + y^2/b^2 = 1$ at the point $(a/\sqrt{2}, b/\sqrt{2})$
2. $x^3y^3 = 1$ at the point $(1, 1)$
3. $x^3 + y^3 = 2a^3$ at the point (a, a)
4. $x^2 + 3xy + y^2 = 5a^2$ at the point (a, a)
5. $\sqrt{x} + \sqrt{y} = 2$ at the point $(1, 1)$
6. $(x^2 - y^2)/(x^2 + y^2) = 1$ at the point $(1, 0)$

Find the tangent and normal lines to the following:

7. $x^3 + y^3 = 2a^3$ at (a, a)
8. $1/x + 1/y = 2$ at $(1, 1)$
9. $x^2/a^2 + y^2/b^2 = 1$ at $(a/\sqrt{2}, b/\sqrt{2})$
10. Generalize Exercise 7. Describe the result geometrically and algebraically.

8.4 CURVATURE

If we can fit a straight line (having two parameters) to a curve at a point using the two conditions of matching both point and slope at the point, then clearly we can extend the idea and fit a circle, which has three parameters, by matching at a point on the curve the three conditions of point, slope, and second derivative. By fitting a circle, we can measure the *curvature* of the function at a point. This gives us some additional information about the function.

To fit the circle, we could use the general form (6.2-3) of the circle (with D, E, and F as linear undetermined parameters); but since it is likely that we will want the geometrical properties of the circle (the center and radius), it seems better to pick the form (6.2-1):

$$(x - h)^2 + (y - k)^2 = r^2 \qquad (8.4\text{-}1)$$

In this form the unknown parameters h, k, and r occur nonlinearly, but since we believe we could solve the linear form using D, E, and F as the variables and then reduce it to the present form, *probably* we can handle this nonlinear case of undetermined coefficients (if we are careful!). After all, the solution must be the same either way we solve it.

Implicit differentiation of (8.4-1) gives

$$2(x - h) + 2(y - k)y' = 0 \tag{8.4-2}$$

Differentiating again, we get

$$2 + 2(y - k)[y''] + 2(y')^2 = 0 \tag{8.4-3}$$

The unknowns are the values of h, k, and r. The knowns are the point (x, y), slope y', and second derivative y''. Equation (8.4-3) can be solved for

$$y - k = -\frac{1 + (y')^2}{y''} \tag{8.4-4}$$

Equation (8.4-2) can be solved for

$$x - h = -(y - k)y' \tag{8.4-5}$$

These two equations give the coordinates of the center (h, k) of the circle. Put (8.4-5) in the first equation (8.4-1) to eliminate the parameter h:

$$[(y - k)y']^2 + (y - k)^2 = r^2$$

or

$$(y - k)^2[(y')^2 + 1] = r^2 \tag{8.4-6}$$

Now eliminating the $(y - k)$ from (8.4-6) and (8.4-4), we have the radius

$$r^2 = \left[\frac{1 + (y')^2}{y''}\right]^2 [1 + (y')^2]$$

$$r = \left|\frac{[1 + (y')^2]^{3/2}}{y''}\right| \tag{8.4-7}$$

The absolute value occurs because by convention the radius of a circle is a positive quantity.

This is the radius of the circle that fits *locally* the curve (matches point, slope, and second derivative). Hence r is called *the radius of curvature*. This circle is sometimes called the *osculating circle* (osculate means to kiss).

If the original curve had been a straight line, or if the particular point of the curve we were using had the second derivative equal to 0, then we could not divide by it to get the radius of curvature. We say, glibly, "The radius of curvature is infinite." Because of this possibility, we often prefer to talk about the reciprocal of the radius of curvature, and call it *the curvature*,

$$\kappa = 1/r = \left|\frac{y''}{[1 + (y')^2]^{3/2}}\right| \tag{8.4-8}$$

(κ is Greek lowercase kappa). Thus a straight line has zero curvature.

Example 8.4-1

For the circle

$$x^2 + y^2 = r^2$$

we know the center is $(0, 0)$ and the radius is r. This makes a good test of the above formulas and of their derivation. First, differentiate implicitly to get

$$2x + 2y \frac{dy}{dx} = 0$$

Divide out the factor 2, and solve for y:

$$y' = \frac{dy}{dx} = \frac{-x}{y}$$

Differentiate this again:

$$y'' = \frac{d^2y}{dx^2} = -\frac{[y - xy']}{y^2}$$

Eliminate the y' to get

$$y'' = -\frac{y + (x^2/y)}{y^2} = -\frac{x^2 + y^2}{y^3} = \frac{-r^2}{y^3}$$

We have therefore (8.4-8)

$$\kappa = \left| \frac{-r^2/y^3}{[1 + x^2/y^2]^{3/2}} \right|$$

$$= \left| \frac{-r^2/y^3}{r^3/y^3} \right| = \left| \frac{-1}{r} \right| = \frac{1}{r}$$

which is independent of x and y, as it should be for a circle. To find the center of this circle, we use formulas (8.4-4) and (8.4-5). First we find k from the equation (8.4-4):

$$y - k = -\frac{1 + (y')^2}{y^{11}}$$

$$= -\frac{1 + x^2/y^2}{-r^2/y^3}$$

$$= \left(\frac{x^2 + y^2}{y^2} \right) \left(\frac{y^3}{r^2} \right) = y$$

Therefore, $k = 0$. Similarly, we find h from (8.4-5):

$$x - h = -(y - k)y' = -y(-x/y) = x$$

And $h = 0$ as it should. The formulas pass this obvious test.

Example 8.4-2

Suppose we find the radius and center of curvature for the parabola (Figure 8.4-1)

$$y = x^2 \qquad y' = \frac{dy}{dx} = 2x \qquad y'' = \frac{d^2y}{dx^2} = 2$$

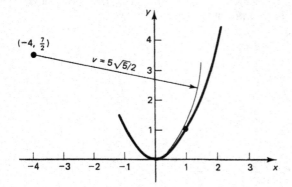

Figure 8.4-1 Osculating circles

Therefore, from (8.4-7), (8.4-5), and (8.4-4),

$$r = \frac{(1 + 4x^2)^{3/2}}{2}$$

$$h = x - \left[\frac{1 + 4x^2}{2}\right][2x] = -4x^3$$

$$k = x^2 + \frac{1 + 4x^2}{2} = \frac{1 + 6x^2}{2}$$

We can check this easily at the point $(0, 0)$. We have $r = \frac{1}{2}$, $h = 0$, and $k = \frac{1}{2}$. As another check, we can try the point $(1, 1)$. Again, see Figure 8.4-1. We get

$$r = \frac{5\sqrt{5}}{2}$$

$$h = -4$$

$$k = \frac{7}{2}$$

The distance from (h, k) to the point $(1, 1)$ must be the radius of the circle. Hence

$$(h - 1)^2 + (k - 1)^2 = 25 + \frac{25}{4} = \frac{125}{4} = r^2$$

Again, things check properly.

Generalization 8.4-3

Suppose we try to extend the above sequence of matching (1) the point, (2) the tangent line, and (3) the second derivative to find the matching circle. What curve would be sensible to fit to the function, first, second, and third derivatives? The next convenient geometric figure we know much about is the general ellipse (or hyperbola), which has six parameters. If we used this, we would have to go up to fitting the fifth derivative. We can certainly do this, because the parameters enter in linearly and the corresponding linear equations will be solvable in general (although degenerate cases of local straight lines and the like could arise). But we would not know how to interpret the result if we got it! The generalization is *probably* not worth the effort.

We see how we can fit *locally* curves from one given family of curves to another given curve at a given point; we see the essence of *local approximation* methods. "Local approximation" means "fit as many consecutive derivatives (the function is the zeroth derivative) as you can." If you try skipping some of the derivatives, you can get into trouble.

EXERCISES 8.4

1. Find the center and radius of curvature for the function $y = x^n$ at the point $(0, 0)$, and check by using $n = 2$.
2. Find the curvature of the function $y = 1/(1 + x^2)$ at $x = 0$.
3. Find the curvature of $y^2 = a^2 + x^2$ at $(0, a)$
4. Find the curvature of $x^{2/3} + y^{2/3} = a^{2/3}$ at $(a, 0)$.
5. Find the curvature of $y = a^{1/3}x^{2/3}$ at (a, a).

8.5 MAXIMA AND MINIMA

At a *local* maximum or minimum of a curve where the function has a derivative (see Figure 8.5-1), the slope must be 0. We use this observation backward when we are asked to find the local maximum or minimum of a curve (the best or worst). Simply set the derivative equal to 0 and find where this can happen (i.e., where the tangent is horizontal). It may be, as at point C in the figure, that the slope is horizontal, and that it is neither a local maximum nor a local minimum.

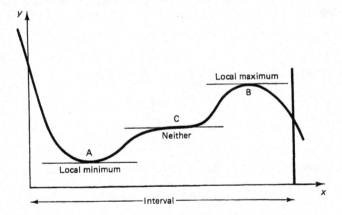

Figure 8.5-1   Local maxima and minima

Furthermore, if you are to find the maximum or minimum in an interval, say $-1 \leq x \leq 1$, it may happen that the extreme value occurs at one end of the interval, and the slope of the tangent line is not necessarily 0.

Example 8.5-1

As a simple example, consider maximizing the area of a rectangle with a given fixed perimeter. In word problems like this, you pick symbols for the quantities and write down the statements (see Figure 8.5-2). Let the rectangle have dimensions x and y. The total perimeter, P, of the rectangle is

$$P = 2x + 2y$$

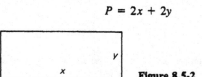

Figure 8.5-2   Rectangle

This is a condition between x and y. (Of course, we limit x and y to be nonnegative.)

The area, A, which is what we want to maximize, is

$$A = xy$$

It is easy to eliminate one of the variables using the *side condition* (the perimeter P is fixed). Thus, to eliminate y, we have from the perimeter equation

$$y = \frac{P}{2} - x$$

Put this in the area expression to get

$$A = x\left(\frac{P}{2} - x\right) = \left(\frac{P}{2}\right)x - x^2$$

We see, after a moment, that this is a parabola opening downward (consider large positive and negative x and see what happens), so there is a maximum somewhere.

The slope of the curve (= slope of the tangent line) of $A = A(x)$ is given by

$$A' = \frac{dA}{dx} = \frac{P}{2} - 2x$$

At the maximum, this derivative must be zero, so we have

$$\frac{P}{2} - 2x = 0$$

$$x = \frac{P}{4}$$

Using the perimeter equation for y, it follows that

$$y = \frac{P}{2} - \frac{P}{4} = \frac{P}{4}$$

so we have a square as expected.

It is important in learning to do mathematics to review what is done and why it was successful. The common remark that you should learn from your mistakes is misleading; there are so many ways of being wrong and so few ways of being right that it is usually better to study successes.

What did we do? First, we took the statement of the problem and converted it to symbols and then to equations. Never underestimate the power that simply naming things gives you. Next, we eliminated all but one of the variables; in this case we eliminated y. Then to find the maximum, which we had convinced ourselves was there, we differentiated and set the derivative equal to zero. We solved this equation and substituted back into the earlier relationships to get the unknowns x and y. Then we interpreted the results; $x = y$ means that we have a square. Notice that we chose a form for stating the answer that eliminates reference to both P and A; we described the nature of the solution, not the specific details.

It is worth noting that among rectangles the shape, a square, is the solution for the equivalent problem; for a fixed area, find the shape with the minimum perimeter. Why are the two problems equivalent?

Example 8.5-2

A very similar problem is that of building a rectangular-shaped fence along a river or the side of a sufficiently long barn (which saves one side of the fencing). The amount of fence is fixed, and you want the maximum amount of enclosed rectangular area. Again, we pick the variables x and y for the dimensions, y in the direction of the river or barn, and x perpendicular to y (see Figure 8.5-3). The perimeter this time is

$$P = 2x + y$$

and the area again is

$$A = xy$$

Again, eliminate the variable y; this time

$$y = P - 2x$$

and the area is

$$A = x(P - 2x) = Px - 2x^2$$

This is a parabola opening downward, so there is one maximum. To find the maximum, differentiate and equate to 0.

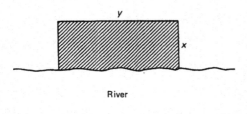

River

Figure 8.5-3   Minimum perimeter

$$A' = \frac{dA}{dx} = P - 4x = 0$$

or

$$x = \frac{P}{4}$$

Therefore,

$$y = P - 2\frac{P}{4} = \frac{P}{2}$$

The "shape" of the field is, say, the ratio of x to y:

$$\frac{x}{y} = \frac{P/4}{P/2} = \frac{P}{4}\frac{2}{P} = \frac{1}{2}$$

This expresses the answer in a form that is independent of the particular given length of fence.

Example 8.5-3

You are given a square sheet of paper or metal, and wish to cut out square corners and then fold up the sides to make a shallow tray. Find the shape of the tray with maximum volume.

Let the given square have sides of length L (if you must be specific think of $L = 17$), and let x be the side of the small corner squares you are going to cut out (see Figure 8.5-4). The volume is then (height times the square base)

$$V = x(L - 2x)^2$$
$$= L^2x - 4Lx^2 + 4x^3$$

Therefore,

$$V' = \frac{dV}{dx} = L^2 - 8Lx + 12x^2 = 0$$

gives the maximum. Factor this:

$$(L - 2x)(L - 6x) = 0$$

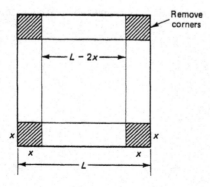

Figure 8.5-4   Shallow tray

The solution $L = 2x$ is clearly zero volume (minimum volume); it is the other solution

$$x = \frac{L}{6}$$

that you want. The bottom dimension is therefore

$$L - 2x = L - \frac{L}{3} = \frac{2L}{3}$$

So the shape, the depth to width, is

$$\frac{x}{L - 2x} = \frac{L/6}{2L/3} = \frac{1}{4}$$

Word problems are the bane of many a student's existence. One of the simplest errors is that when finally the x value of the extreme is found the student forgets that this is but a step toward the final answer, and fails to find what was asked for. There are no simple rules for solving word problems in a mechanical fashion. Indeed, in doing research it is the ability to go from vaguely formulated problems to the "proper" clear formulation that marks the great scientist from the average person. Einstein repeatedly said that he "had a nose" for smelling out the right approaches, and an examination of biographies of his life indicates the truth of this observation. The great people differ from the run-of-the-mill able people by their ability to formulate questions "properly." Thus, if you want to use the mathematics you are learning, you need to pay attention to this aspect of the course. The formulation of a real problem requires a good deal of idealization of reality; just which idealization you choose measures, in some sense, your ability to do things "right." It is the ability to select the relevant and ignore the irrelevant aspects of a problem that matters.

Example 8.5-4

What is the optimal shape of a closed cylinder? What do we mean by optimal? What can it mean? It could mean, for example, that for a given volume the surface is minimal. Or it could mean that for a fixed surface the volume is maximal. Some thought shows that in this case the two are the same question! See Figure 8.5-5.

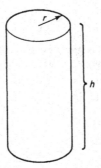

Figure 8.5-5 Optimal cylinder

Next, you need some technical knowledge, such as the area of the surface of a cylinder and its volume. The need for technical knowledge is common to all applications: *you must know the technical details of the domain of application.* The *method* of finding the optimum, however, is the same. First, introduce some notation. For the cylinder let the radius of the base be r and the height be h. Next, set up the equation for the side area of the cylinder, which is the circumference of the base times the altitude (imagine a can cut up the side and laid out flat). Therefore, the area of the side is

$$(2\pi r)h$$

To this you must add the area of the two ends:

$$2(\pi r^2)$$

Thus, the total surface area S is the sum of the two terms:

$$S = 2\pi r (h + r)$$

On the other hand, the volume V is the area of the base times the altitude,

$$V = \pi r^2 h$$

The third step is to reorganize the equations for your convenience. Suppose you decide to maximize the volume subject to a fixed area. You can eliminate h using the fixed amount of surface equation to find h. This gives

$$h = \frac{S}{2\pi r} - r$$

Therefore, the volume can be written, after a little algebra, as

$$V = \frac{rS}{2} - \pi r^3$$

This is the volume you want to maximize.

The fourth step is to differentiate with respect to the variable, which in this case is the radius r, and set the derivative equal to 0. You get

$$V' = \frac{dV}{dr} = \frac{S}{2} - 3\pi r^2 = 0$$

Therefore, solving for the radius r you have

$$r = \sqrt{\frac{S}{6\pi}}$$

Next, you need to find h (which is given by the equation used to eliminate it) in terms of r. Substitute the value of r into the equation you used to eliminate h:

$$h = \frac{S}{2\pi r} - r$$

After some algebra you will get

$$h = 2\sqrt{\frac{S}{6\pi}}$$

Is this the answer? No. Look at the question, "What is the optimal shape . . . ?" The shape must mean the ratio of the two sizes. You can take either height to diameter, D, or diameter to height. In the first case

$$\frac{h}{D} = \frac{h}{2r} = 1$$

It is clearly better to state it in terms of diameter than it is to state the result in terms of the radius. The statement height = diameter is the kind of thing you can remember and visualize; when you use the radius, you get an extra factor of 2. Often the value of a result lies in the neat statement of it, so to be an effective person you need to learn the gentle art of presenting the results in a suitable form.

As we earlier observed, in a textbook it is necessary to give very simple examples, but in practice often a single problem may require hours, days, or weeks of work. There is not time in a course to give such problems, but we can give one slightly harder problem and discuss some of the things that arise.

Case Study 8.5-5

Suppose, as in the shallow tray problem, we are again given a square, but this time are asked to form a square pyramid (including the base). We first sketch Figure 8.5-6. Before taking action, we do a little thinking. The result is that maybe Figure 8.5-7 would be a better plan. Are we sure? Well, if we had the optimal solution in the first case, we could rotate it and then have room to make the sides of the pyramid longer. Or could we rotate? Might not the corners of the base hit the sides? Well, a moment's inspection shows that the base cannot be bigger than half the original side since the flaps must at least meet at the center when folded up. Yes, we could rotate, and we need examine only the second case.

Due to the symmetry of the figure, it is probably best to pick the origin of the coordinate system at the center. (*Rule:* Try to use any symmetry you have.) We therefore

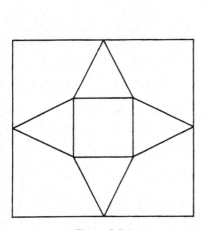

Figure 8.5-6

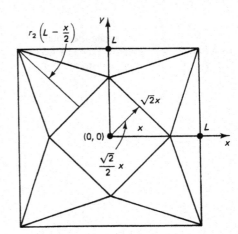

Figure 8.5-7

pick the side of the square to be $2L$. Let x be the distance out from the center to the corner of the base. The side of the base is, by the distance formula,

$$\text{side of base} = \sqrt{2}x$$

Therefore, the distance from the origin to the side is one-half of this:

$$\frac{\sqrt{2}}{2}x$$

The distance from the origin to the corner of the square is

$$\sqrt{2}L$$

so the slant height of the face to be folded up is

$$\text{slant height} = \sqrt{2}\left[L - \frac{x}{2}\right]$$

Now, looking at Figure 8.5-8, the altitude is

$$\text{altitude} = \left[\left\{\sqrt{2}L - \left(\frac{\sqrt{2}}{2}\right)x\right\}^2 - \frac{2x^2}{4}\right]^{1/2}$$

$$= \left[2L^2 - 2Lx + \frac{x^2}{2} - \frac{x^2}{2}\right]^{1/2}$$

$$= [2L(L - x)]^{1/2}$$

Finally, the volume of a pyramid is one-third of the volume of the corresponding rectangular cylinder; that is

$$V = \frac{1}{3}\left[\left(\frac{\sqrt{2}}{2}\right)x\right]^2 [2L(L - x)]^{1/2}$$

$$= \frac{\sqrt{2}}{6}x^2[L(L - x)]^{1/2} = \frac{\sqrt{2L}}{6}x^2\sqrt{L - x}$$

When we differentiate the volume with respect to the variable x and set it equal to 0, the front constant can then be divided out, so we can ignore it. The derivative of the rest, which is a product form, is

$$\frac{x^2(-1)}{2\sqrt{L - x}} + \sqrt{L - x}(2x) = 0$$

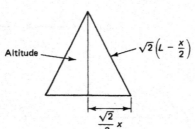

Altitude

$$\sqrt{2}\left(L - \frac{x}{2}\right)$$

$$\frac{\sqrt{2}}{2}x$$

Figure 8.5-8   Altitude of a pyramid

Multiply through by the denominator to get

$$-x^2 + 4x(L - x) = x$$

or

$$4x(L - x) = x^2$$

One solution is $x = 0$, which is of course a minimum. The other solution is

$$4L - 4x = x$$

or

$$x = \frac{4L}{5}$$

What is the altitude? We look back for the expression for it:

$$\text{altitude} = [2L(L - x)]^{1/2} = L\sqrt{\frac{2}{5}}$$

from which we can find the volume easily. The shape factor would be, probably, the ratio of the altitude to the side of the base,

$$\frac{[2/5]^{1/2}L}{[2L]} = \frac{1}{\sqrt{10}} = 0.316\ldots$$

The corresponding volume is

$$V = \frac{1}{3}[4L^2]\left[\frac{2}{5}\right]^{1/2} L = \left[\frac{4}{3}\right]\left[\frac{2}{5}\right]^{1/2} L^3$$

$$= \sqrt{\frac{32}{45}} L^3 = 0.843\ldots L^3$$

The experienced person would recognize that the answer would turn out to be linear in L for the altitude, and the volume would be proportional to the cube of L. Thus in place of the choice we made for x you could choose xL; the fraction of the edge length could be chosen as the unknown. We have ignored the question of a clever cutting up of the square and later fastening together the parts; we have used the cut and fold condition.

Example 8.5-6

A wire of length L is to be cut into two pieces. The first is formed into a square and the second into a circle. Find where to cut the wire to get the maximum enclosed area. Let

$$x = \text{length of piece for square}$$

$$L - x = \text{length for the circle}$$

The area of the square is $(x/4)^2$. The radius of the circle is $(L - x)/2\pi$. Hence the area of the circle is $\pi[(L - x)/2\pi]^2$. Therefore, the total area is

$$A = \left(\frac{x}{4}\right)^2 + \frac{(L - x)^2}{4\pi}$$

We differentiate and set equal to zero.

$$A' = \frac{x}{8} - \frac{L-x}{2\pi} = 0$$

$$x = \frac{4}{\pi}(L - x)$$

$$\left(1 + \frac{4}{\pi}\right)x = 4L$$

$$x = \frac{4\pi L}{\pi + 4}$$

But when you compute this you find that it is a minimum and that the maximum occurs for $x = 0$, at one end of the range. Once you realize this, then looking at the original function A you see that it is a parabola opening upward, and that the horizontal derivative must give the minimum, not the maximum. It is then a matter of deciding which end is the larger. The front coefficients on the two terms tells you which is best.

Aside 8.5-7

It may come as a surprise to you to realize that some reasonable sounding problems do not have, for example, a minimum. Consider Figure 8.5-9, two towns A and B, with B due east of A. You are required to depart from A in a northerly direction. Clearly, there is no shortest path that satisfies all the conditions. Only "well-formulated problems" have a unique solution (whatever well formulated may mean in general). This means that in some cases *you* have to decide if the problem posed to you makes sense or not.

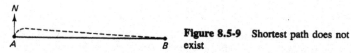

Figure 8.5-9   Shortest path does not exist

EXERCISES 8.5

See Appendix B for the common formulas from geometry.

1. Find the shape of the isoceles triangle along a river having maximum area for fixed length of the two equal legs.
2. Compare the solution of the cylindrical can problem with those in the stores. Why are there differences? What kinds of things did we neglect?
3. Find the depth of the shallow tray of maximum volume made from a piece of material with sides a and b in dimensions.
4. Optimize a tin can with no top. Compare this to the one with a top.
5. Find the maximum volume of a cone that can be put in a sphere of radius a.
6. Find the maximum volume of a right cylinder that can be put in a cone of radius R and height H.
7. A rectangle is surmounted by an equilateral triangle. Find the maximum area for a fixed exterior perimeter.

8. Find the maximum rectangle you can put in an ellipse.

9. Two spheres of radius respectively a and b have their centers c units apart ($c > a + b$). Find the point where the combined maximum surface can be seen, given that the surface area that can be seen is $A = 2\pi rh$, where h is the depth of the sector.

10. Find the largest right circular cylinder that can be put in a sphere of radius a.

11. The heat from a source varies inversely with the distance away. Given two sources of intensities a and b separated by a distance c, find the point between the sources where the intensity is least.

12. Find the largest rectangle that can be put inside the common area of two parabolas, $3y = 12 - x^2$ and $6y = x^2 - 12$.

13. A right circular cylinder is topped with a hemisphere (but no bottom). Find the shape for a fixed total surface that has maximum volume.

14. A window is surmounted by a semicircle. Find dimensions for the fixed perimeter that admits the most light.

15. A strip of metal is to be shaped into the cross section of an isoceles triangle with height y and width x. Find the shape with maximum cross-sectional area.

16. Generalize Example 8.5-6. The interesting point about this example is that the maximum occurs at the end of the interval. We therefore ask: (a) Does it depend on using a circle and a square? (b) Would any two regular figures do? (c) Would any two systems of similar convex figures do? (d) Must they be convex? (e) Could the edges of a figure shape cross itself? (f) What determines which end will have the extreme?

8.6 INFLECTION POINTS

In Section 8.4 we found the circle that best fitted a given curve at a point. As the point where we fit the local circle to a curve moves along the curve, we find that the curvature is continually changing. This osculating circle and its curvature are properties of the curve, since they do not change when the coordinate system is shifted or rotated. The curvature is an *intrinsic* property of the curve.

We now introduce a similar concept, but one that is tied to the coordinate system. This idea uses the words "concave" and "convex." Consider the simple curve

$$y = x^3$$

(see Figure 8.6-1). On the right side the curve is *concave up* (convex down), and on the left it is *concave down* (convex up). Between them there is a point where it is neither, the origin. Such a point, where the curve changes from concave up to concave down, we call an *inflection point*. Where the curve is concave up, the second derivative is positive, and where the curve is concave down, the second derivative is negative. Where it is neither concave up nor concave down, the second derivative is zero. Such a point is called an *inflection point*. That the condition $y'' = 0$ is not always an inflection point follows from an examination of the simple curve

$$y = x^4$$

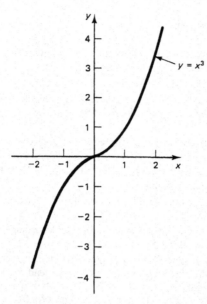

Figure 8.6-1 Inflection point

The second derivative is (see Figure 8.6-2)

$$y'' = 12x^2$$

and at the orgin this is zero, although there is no change in the convexity upward. Thus the vanishing of the second derivative is a necessary but not sufficient condition for an inflection point. At an inflection point the tangent line crosses the curve.

To get a feeling for the matter in the general case, we start with a sketch of some curve (see the top of Figure 8.6-3). Below this curve we sketch the curve of its derivative, using the simple fact that positive slopes must have derivative values that are positive and negative slopes have negative derivative values. See again Figure 8.6-3. If we now repeat the process of sketching the derived curve from a given curve,

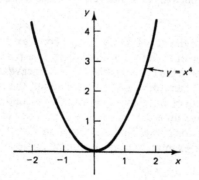

Figure 8.6-2 No inflection point

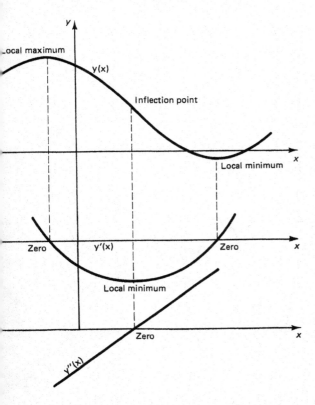

Figure 8.6-3

we are led to the second derivative curve shown at the bottom of the figure. Where the second derivative crosses the x-axis, the top curve has a change in curvature.

Example 8.6-1

Consider the simple cubic (Figure 8.6-4)

$$y = x(1 - x^2) = x - x^3$$

The first two derivatives are

$$y' = \frac{dy}{dx} = 1 - 3x^2$$

$$y'' = \frac{d^2y}{dx^2} = -6x$$

Equating the first derivative to 0 gives the positions of the local maxima and minima:

$$x = \pm \frac{1}{\sqrt{3}}$$

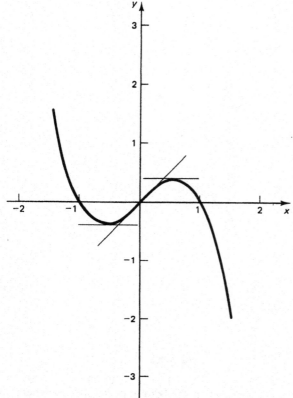

Figure 8.6-4 A cubic

Equating the second derivative to 0 gives the position of any possible inflection point:

$$x = 0$$

We need the corresponding function values at these points. For the max–min points

$$y = x(1 - x^2) = \left(\pm \frac{1}{\sqrt{3}}\right)\left(1 - \frac{1}{3}\right) = \pm \frac{2}{3\sqrt{3}} = \pm \frac{2\sqrt{3}}{9}$$

For the inflection point at $x = 0$, evidently $y = 0$. It is now easy to sketch the curve. The only additional information we might need would be the slope at the inflection point, which is exactly 1.

If we examine the idea of a local minimum closely, we see that as we go from left to right the slope passes from negative values to positive values, and this means that the second derivative must be positive. On the other hand, at a local maximum the slope passes from positive to negative, and the second derivative must be negative. Thus we have the following rule:

At a local minimum the second derivative is ≥ 0. At a local maximum the second derivative is ≤ 0.

The rule can best be remembered by Figure 8.6-5 and the soup bowls. A bowl will hold the soup, be $+$, when it is concave up and will not hold the soup, be $-$, when it is concave down.

Figure 8.6-5 Soup bowls

As we earlier saw, when the first and second derivatives are both 0 at a point, then there is no clear decision about the extreme. The curves

$$y = x^3$$

and

$$y = x^4$$

illustrate the situation clearly. In the first case we have a local horizontal slope that is neither a maximum nor a minimum, and in the second case there is indeed a local minimum; but in both cases the second derivative is 0 at the point $x = 0$. The equation

$$y = -x^4$$

also has the second derivative equal to 0 at the origin but clearly opens downward (see Figure 8.6-6).

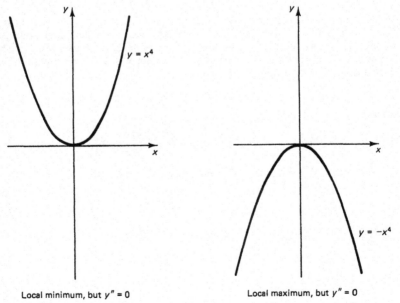

Local minimum, but $y'' = 0$ Local maximum, but $y'' = 0$

Figure 8.6-6

Another way of saying things is:

1. $y' = 0$, $y'' > 0$ implies a local minimum

2. $y' = 0$, $y'' < 0$ implies a local maximum

3. $y' = 0$, $y'' = 0$ implies no conclusion

Example 8.6-2

Find the zeros and the inflection points of

$$y = 3x^5 - 10x^3 + 7x$$

Note first that the curve has odd symmetry, $f(x, y) = f(-x, -y)$. To find the zeros, we start to factor it. There is a factor of x, and the rest is a quadratic in x^2. Therefore,

$$y = x(x^2 - 1)(3x^2 - 7)$$

The zeros are easily seen to be $x = -\sqrt{7/3}$, -1, 0, 1, and $\sqrt{7/3}$.

To find the inflection points, differentiate twice (almost directly term by term) and equate to zero:

$$y'' = 3(5)(4)x^3 - 10(3)(2)x = 0$$

Divide out the common factor 60 to leave

$$x^3 - x = x(x^2 - 1) = 0$$

whose zeros are -1, 0, and 1. The function values at these points are already known to be zero! This function has inflection points and zeros at the same three points, -1, 0, and 1 (and has a pair of other zeros). We are ready to sketch it, *except* that we have no value to give the scale in the y direction. If we put $x = 2$ in the factored form of the polynomial, we get

$$y = 2(3)(5) = 30$$

and if we use $x = \frac{1}{2}$ we get

$$y = \left(\frac{1}{2}\right)\left(-\frac{3}{4}\right)\left(-\frac{25}{4}\right) = \frac{75}{32} = 2\frac{11}{32}$$

To find the maximum and minimum, we have the equation

$$\frac{dy}{dx} = 15x^4 - 30x^2 + 7 = 0$$

(which is a quadratic in x^2), whose zeros are given by

$$x^2 = 1 \pm \sqrt{8/15}$$

$$x^2 = 1.730 \ldots, \quad 0.2697 \ldots$$

$$x = \pm 1.315 \ldots, \quad \pm 0.5193 \ldots$$

The square roots are taken to get the x values. It is less difficult to handle when you recognize that x^2 occurs frequently in the evaluation of the equation written in the form

$$y = x(3x^4 - 10x^2 + 7)$$

With a hand calculator, it is easy to place the points where the local extremes occur. The sketch is shown in Figure 8.6-7.

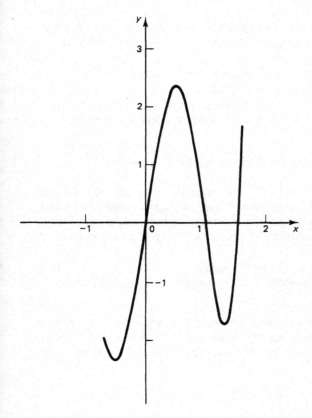

Figure 8.6-7 $y = 3x^5 - 10x^3 + 7x$

EXERCISES 8.6

1. Find the distance between the inflection points of $y = 1/(1 + x^2)$.
2. Find the slope at the inflection points of the above equation.
3. Find the zeros, maximum, minimum and inflection points of $y = x^5 - 2x^3 + x$.

8.7 CURVE TRACING

It is essential to learn to sketch the curve of a given function. It is an art to do it easily and gracefully. The most useful tool is symmetry, because each form of symmetry decreases the work by about a factor of 2. Let us review the simple rules from Section 6.11. They were based, as you remember, on invariance of the equation under substitutions of the variables.

1. Symmetry with respect to the x-axis. Typical terms that have this property are constants, y^2, y^4, . . . , $|y|$, cos y, and so on. The rule is: functions that are of even degree in y are symmetric with respect to the x-axis. Replacing y by $-y$ does not change the equation.

2. Symmetry with respect to the y-axis. Same as above except x in place of y. Replacing x by $-x$ does not change the equation.

3. Symmetry with respect to the origin. Typical terms, xy, $(xy)^2$, . . . , sin x sin y, cos x cos y, $x^2 + y^2$, . . . , and so on. Replacing both x and $-x$ and y by $-y$ does not change the equation.

Note that if you have two of the above you automatically have the third.

4. Odd symmetry. The same as above (3) ; the function on one side of the origin is the negative of the function on the other. Typical terms are odd powers of x in place of even powers, sines or tangents in place of cosines, and so on. Clearly, "symmetry" means "even symmetry" unless specifically it is said to be "odd symmetry."

5. Symmetry with respect to the 45° line. Typical terms, xy, $(xy)^2$, . . . , $x + y$, $x^2 + y^2$, . . . , and so on. In words, nothing happens when you interchange x and y in the equation.

As we said, do not try to learn all the above rules, but rather in such a situation ask, "What is behind all this complexity?" A little thought shows that it is the *invariance* of the equation under the appropriate substitutions. For example, the equation

$$x^2 + y^2 = a^2$$

is not changed by replacing x by $-x$. Hence any point (x, y) that satisfies the equation has the corresponding point $(-x, y)$. What does this mean? The same y value but opposite x values; symmetry with respect to the y-axis. Thus the mathematician learns to inspect a given equation for invariances of all kinds, and when one is found it is often easily exploitable. Once having thought your way through various invariances to their symmetries, you need only recall the method, not memorize the long list of results. Some invariances are harder to interpet because they are not symmetries. For example, the equation

$$xy + \frac{1}{xy} = 2$$

has the change of variables $x \to 1/x$ and $y \to 1/y$ that leaves the equation invariant. The meaning of the invariance is more complicated than simple symmetry. But each invariance cuts down the difficulty of plotting points by a factor of about 2.

Another tool for curve tracing is the choice of a few values of x for which it is easy to get values of y. When you are given $y = f(x)$, it is easy to compute pairs of values that satisfy the function. In cases where you cannot easily solve for either of the variables in terms of the other, you must guess at likely pairs of values.

Next, you have the maximum and minimum points. Not only do you need the x values, but also the corresponding y values (if you can get them easily). The maximum and minimum points greatly constrain the function's behavior.

Next, you have the inflection points. Again, you need the corresponding y values (if you can get them). Often the slope of the curve at the inflection points greatly aids in sketching the curve.

Finally, you have those particular values for which x or y, or both, go to infinity. When these are used, or at least some of them, you have a fairly good idea of what the curve does far away from the origin. This kind of information can help, even when you are near the origin.

The importance and usefulness of curve tracing cannot be overemphasized. On a number of occasions a simple sketch on the back of a luncheon placemat has answered important technical questions.

In curve tracing it is an art to do those things that come easily, and not to use those that are hard. All we can do is give a few examples to illustrate curve tracing, followed by some personal practice by you. Remember, the purpose of the exercises is not to get the answer but rather to learn the art.

Example 8.7-1

Sketch the curve

$$x^2 y^2 = a^4$$

Clearly, there is a scale to the figure; we really need to think of the transformation of x to ax_1 and y to ay_1. Then the a^4 would cancel out. Thus, we temporarily replace a by 1 for convenience, and in the final drawing (but not necessarily along the way) we plot in units of size a.

As for symmetry, the even powers in both variables show that we have symmetry with respect to the x-axis, the y-axis, and the origin. Thus it suffices to examine the first quadrant and then suitably duplicate it for the other quadrants. In the first quadrant we examine the curve

$$x^2 y^2 = 1$$

and, because the quantities are positive, we can take the square root of each side. We recognize the hyperbola (in the first and third quadrants)

$$xy = 1$$

There is, of course, a second equation with a minus sign. To get the scale, we note that the point $x = 1$, $y = 1$ lies on the curve. It is also easy to find the points $x = 2$, $y = \frac{1}{2}$ and $x = \frac{1}{2}$, $y = 2$ on the curve. Thus the full sketch is in Figure 8.7-1.

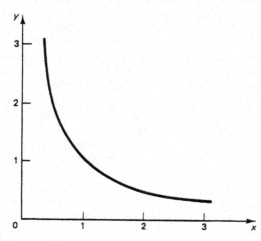

Figure 8.7-1 $x^2 y^2 = 1$

Example 8.7-2

Another example is

$$y = \frac{x^2 - 1}{x^2 + 1}$$

We see immediately that there are only even powers of x, and hence symmetry with respect to the y-axis. Next, the zeros of the function are at $x = -1$ and 1. At the point $x = 0$, the curve has the value $y = -1$.

The derivative

$$\frac{dy}{dx} = \frac{4x}{(x^2 + 1)^2}$$

gives only one value for an extreme, that is, at $x = 0$. The function value at this place is, again, -1; it looks like a local minimum. The numerator of the second derivative (which is usually all we need) is

$$(x^2 + 1)^2(4) - 4x[2(x^2 + 1)(2x)] = (x^2 + 1)[4(x^2 + 1) - 16x^2]$$

from which we deduce that at the x values $\pm 1/\sqrt{3} = \pm 0.577 \ldots$ are the inflection points. The corresponding function values are easily found to be

$$\frac{-2/3}{4/3} = -1/2$$

It remains to see what happens far from the origin. For large absolute x values, the y values approach 1 from below. There can be no large y values (apparently, but you can be fooled if you are too hasty in thinking!). A final point $x = 2$, $y = \frac{3}{5}$ helps nail down the curve.

We are ready for the sketch. At $x = 0$ we begin at $y = -1$ with $y' = 0$. It is concave up until $x = 1/\sqrt{3}$, where the y value is $-\frac{1}{2}$. Continuing, it must be concave down and pass through $x = 1$, where $y = 0$. It continues to rise slowly, passing through the point $x = 2$, $y = \frac{3}{5}$, and is limited far out by $y = 1$ (see Figure 8.7-2). Symmetry gives the curve for negative x.

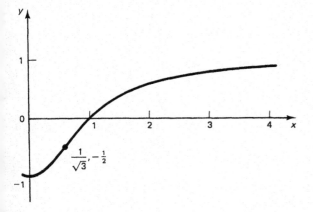

Figure 8.7-2 $y = \dfrac{x^2 - 1}{x^2 + 1}$

Example 8.7-3

Plot the family of curves $(n \geq 1)$

$$x^{2n} + y^{2n} = a^{2n}$$

Evidently, the change of variables $ax \to x$, $ay \to y$ will remove the parameter a. Hence we can act as if $a = 1$ and at the last moment put units of a on the axes.

For $n = 1$, the equation is a circle. For $n = 2$, we have

$$x^4 + y^4 = 1$$

to think about. Since numbers smaller than 1 in size when raised to a high power become smaller, it will require larger numbers than for the circle to satisfy the equation. To find out how large, we look along the $45°$ line along which $x = y$. There we find we have

$$2x^4 = 1, \qquad x = \frac{1}{\sqrt[4]{2}} = 0.84 \ldots$$

Thus we see that it looks somewhat like a circle but "bulges out" a bit.

For the next case, $n = 3$, the similar point is the solution

$$x = y = \frac{1}{\sqrt[6]{2}} = 0.9 \ldots$$

and so on. In the limit as $n \to \infty$, we have the curve approaching the square containing the original circle (see Figure 8.7-3).

Sometimes the curves you wish to understand have several parameters and cannot be plotted for any definite values without the risk of losing some information.

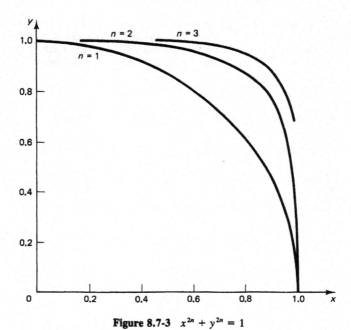

Figure 8.7-3 $x^{2n} + y^{2n} = 1$

You need to be able to plot in your imagination, and then prove, that all possible values have a given property, as the following example shows.

Abstraction 8.7-4

Can we see the previous family of curves from a unified viewpoint? If we can, then we have a great diversity mastered under a single form for reference, something you should always try to do.

The exponent $= 2$ is the familiar circle. How does the exponent 2 arise? It comes from the Pythagorean definition of the distance between two points. But suppose you defined the distance between two points as

$$d = \sqrt[n]{|x_2 - x_1|^n + |y_2 - y_1|^n} \qquad (n \geq 1)$$

You would then have "circles" about the origin $(0, 0)$ in all finite cases of the exponent.

But is that a reasonable definition of a distance? In Section 5.2 we gave the following conditions:

1. $d(p_1, p_2) \geq 0$

2. $d(p_1, p_2) = 0$ if and only if $p_1 = p_2$

3. $d(p_1, p_2) = d(p_2, p_1)$

4. $d(p_1, p_2) + d(p_2, p_3) \geq d(p_1, p_3)$

Does the generalized distance have these properties? Looking at the exponent $= 1$, called L_1 ("ell one"), we see that the distance

$$d = |x_2 - x_1| + |y_2 - y_1|$$

is simply the sum of the absolute values of the differences. It is as if you had to go along the sides and there were no diagonals when you measured distances. In this distance the circle is the square-shaped diamond shown in Figure 8.7-4.

We see from the earlier curves that as n gets larger the corresponding circles bulge out and approach the square in the limit. This limit is called L_∞ ("ell infinity," or sometimes the Chebyshev distance). The distance is given by the formula

$$d(p_1, p_2) = \text{Max}\ \{|x_1 - x_2|, |y_1 - y_2|\}$$

as you can see when you look at how the large exponents used tend to select out the largest term at the expense of all other possible terms. The corresponding root brings that term back to its original size.

It is not hard to see that by algebra alone the first three conditions for a distance are true. The fourth condition comes at present from your intuition.

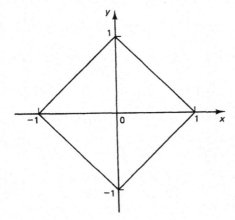

Figure 8.7-4 $|x| + |y| = 1$

Example 8.7-5

Discriminant of a cubic. One day, when I had no textbooks handy, I needed to know when a cubic

$$ax^3 + bx^2 + cx + d = 0$$

has all real roots. I recalled (and checked by carrying it out) that this could be reduced to a simpler form by the substitution

$$x = y - \frac{b}{3a}$$

which you can see eliminates the square term. The reduced cubic form (which is usually presented in the variable y) is

$$y^3 + py + q = 0$$

A typical shape of a cubic is shown in Figure 8.7-5. The wiggles are characteristic, but do not always occur. If the equation is to have three real roots, then the minimum must be a negative number, and the maximum a positive number; otherwise there will not be three crossings. These two conditions can be combined into one by calling the cubic $P(y)$ and writing

$$P(y_{min})P(y_{max}) < 0$$

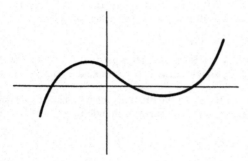

Figure 8.7-5   Cubic

What about the equality? Yes, when either of the extremes is zero, there is a double (real) root, and if both are zero, then it is clearly the triple root of the equation $y^3 = 0$. Hence we take

$$P(y_{min})P(y_{max}) \leq 0$$

as the test.

We have, therefore, to compute this quantity. The derivative of the cubic, when set equal to zero, gives the position of the extremes:

$$P'(y) = 3y^2 + p = 0$$

Hence $y_{min} = \sqrt{(-p/3)}$ and $y_{max} = -\sqrt{(-p/3)}$. We have, therefore, on setting (p must be negative for real solutions)

$$\sqrt{\frac{-p}{3}} = R$$

to compute

$$[R^3 + pR + q][-R^3 -pR + q] = q^2 - [R^3 + pR]^2$$
$$= q^2 - R^6 - 2pR^4 - p^2R^2 \leq 0$$

This becomes, on eliminating the R,

$$q^2 + \left(\frac{p}{3}\right)^3 - 2p\left(\frac{p}{3}\right)^2 + p^2\left(\frac{p}{3}\right) \leq 0$$

$$q^2 + \left(\frac{p}{3}\right)^3[1 - 6 + 9] = q^2 + 4\left(\frac{p}{3}\right)^3 \leq 0$$

This can be rewritten in the more easily remembered form

$$\left(\frac{q}{2}\right)^2 + \left(\frac{p}{3}\right)^3 \leq 0$$

which is the conventional *discriminant* of the cubic and is the condition for three real roots.

EXERCISES 8.7

1. Examine the family $x^{2/(2n+1)} + y^{2/(2n+1)} = a^{2/(2n+1)}$.

8.8 FUNCTIONS, EQUATIONS, AND CURVES

In Section 4.3 we defined a function as a relationship such that given an x (in some range) there is a unique value y corresponding to it. Generally, we deal with functions that can be expressed by equations (although not always!).

But an equation is not always a function. Thus the equation for a circle

$$x^2 + y^2 = a^2$$

is not a function, but leads to two functions

$$y = \pm\sqrt{a^2 - x^2}$$

for $|x| < a$.

When we examined the family of curves

$$|x|^n + |y|^n = a^n$$

we found various shapes depending on the exponent n. We also looked at the limit as $n \to \infty$. There we had a curve, but no equation—only the limit of an equation.

We see, therefore, that if we want to be careful we need to realize that the three concepts, function, equation, and curve, are not all exactly the same.

8.9 SUMMARY

The geometric applications of the calculus are very important. Tangent and normal lines are needed tools in many situations and lead naturally to curvature and convexity. These in turn lead to the max–min test for *local extremes in the interval*. It is not necessary to dwell on the importance of optimization in most fields since it is a natural question to ask in many situations.

The art of curve tracing is an important talent to master; in many complex situations it enables you to "see" what you are examining. Sketches are much more easily grasped by the human mind than are long tables of numbers, or even mathematical equations. It should be noted, however, that, when a parameter occurs in an equation, the equation may be more revealing than a lot of curves!

This chapter is a tool kit of useful methods for dealing with the world. Each tool should be recognized as a "chunk" of knowledge. You should review them in your mind until they are seen as units of thought, and not as individual complex patterns of details. Of course, you must also see the details when you focus on the particular tool. Mastery of a topic implies both the "chunking" and the ability, when needed, to call up the relevant details of the "chunk."

9

Nongeometric Applications

9.1 SCALING GEOMETRY

In geometric applications the units on the two axes are the same, both will be lengths, or some other common units. In such a situation a rotation of the coordinate system makes sense. Slope and angles have meaning, as has the distance between two points.

But in many situations the units on the coordinate axes are quite different. Thus you often see the number of dollars plotted against the year, distance versus time, height versus weight, and so on. There is no common unit between dollars and years; you merely pick convenient units along each axis. Furthermore, rotation, Section 6.9, which would involve adding dollars and years, makes no sense (the beginning algebra teacher said, "You can't add apples to oranges"). Nor do the words "slope," "angle," and "distance between two points" mean anything. Effectively, when you have no common unit, you are free to choose the sizes of the units as you please. *Only* those things can have any real meaning that are invariant (unchanging) under scale transformations of the form (to avoid confusion with derivatives, we are subscripting the new variables rather than using primes).

$$x = k_1 x_1$$

$$y = k_2 y_1 \tag{9.1-1}$$

These equations transform the old (x, y) coordinates to the new coordinates (x_1, y_1). The particular transformation depends on the constants k_1 and k_2 and is equivalent to stretching the two axes independently (but each uniformly). Viewed differently,

Equations (9.1-1) represent a change in the units of measurement; perhaps the change from x to x_1 is from meters to feet.

When the k_1 and k_2 are greater than 1 in size, then it is a contraction, but we will always use the word *stretching*. The k_i cannot be 0, of course, and we generally do not use negative values (which would correspond to both a stretch and a reflection).

The geometry of this situation is a special case of what is called *affine geometry*. Only things that are invariant under the above stretching transformation (9.1-1) are of interest in this scaling geometry. This is the geometry appropriate for most of the graphs you see. When $k_1 = k_2$, the transformation is merely an enlargement and is of slight interest.

The most general affine transformation allows translations and rotations of the coordinate system; thus it is of the form

$$x = a + bx_1 + cy_1$$
$$y = d + ex_1 + fy_1$$

If rotations are ruled out (and the axes are kept perpendicular), then c and e are both 0 in the above formulas for the general affine transformation. Often it is convenient to shift the origin of time to start at the first record, but it can be deceptive to subtract a fixed amount of dollars. We will restrict the class of transformations to those that only stretch the coordinate axes.

$$x = bx_1 \quad \text{and} \quad y = fy_1$$

This will be called a *scaling transformation*.

Speaking in terms of algebra, rather than geometry, *only* those expressions are acceptable that transform reasonably when the scaling transformation (9.1-1) is applied. A formula must "scale" properly to be a valid formula. This is very useful to remember and, indeed, is the basis, in a sense, of "dimensional analysis." Similar variables, such as lengths, will have the same scale factor k_i, while other variables like mass and time will have their corresponding scale factors. The translation we studied in Section 5.9 is appropriate to a geometry that uses all the same units, but it is rarely applicable to the stretch invariant situation (time is a conspicuous exception since often there is no natural origin of time).

Dimensional analysis is an example of scaling transformations. In dimensional analysis you have one or more variables (and constants) that use the same k_i when you change the units of measurement. Given any equation in the variables, it must be that each additive term scales the same (there are at least two terms since it is an equation). If we represent symbolically the length by L, use V for velocity, and T for time, then, looking at the units, we see that symbolically

$$L \neq V \neq T$$

You can have terms in these units in the form

$$\frac{L}{T} = V$$

since velocity is the distance divided by the time. In much of elementary physics the units are mass M, length L, and time T.

Dimensional analysis is a very useful tool that is widely used by experts. But it has simple applications even in geometry. If you have an area, it must depend on the square of the length, and the volume must depend on the cube of the length. For example, for the circle,

$$\text{circumference } C = 2\pi r \quad \text{and} \quad \text{area } A = \pi r^2$$

For a sphere,

$$\text{surface } S = 4\pi r^2 \quad \text{and} \quad \text{volume } V = \frac{4\pi r^3}{3}$$

It is true that in finding the volume of a box of unit height the answer will appear to depend only on the square of a length, since one length has unit size. This is one of the reasons why it is often wise to do the general case (so you can get a check on the units).

In problems that involve several variables with different units for scaling, dimensional analysis provides even more power in checking that the equations scale properly; each unit must scale properly.

Example 9.1-1

Apply a scaling transformation to dy/dx. We have

$$\frac{d(k_1 y_1)}{d(k_2 x_1)} = \frac{k_1 dy_1}{k_2 dx_1} = \left(\frac{k_1}{k_2}\right)\frac{dy_1}{dx_1}$$

where as usual we pass over any constants in the differentiation process.

EXERCISES 9.1

1. If $x = 3x_1$ and $y = 0.5y_1$, find dy_1/dx_1 and d^2y_1/dx_1^2 in terms of dy/dx and d^2y/dx^2.

2. Find the first derivative in terms of the general affine transformation.

3. Show that one affine transformation followed by another is equivalent to a single affine transformation. Show that for any affine transformation there is one that undoes it (except in degenerate cases), in short that there is generally an inverse transformation.

4. For the total surface of a cone, show that each term scales properly.

5. Show that the derivative with respect to r of the volume of a sphere is the surface, and the derivative of the area of a circle is the circumference. Apply to a cube centered at the origin. Show that the rule that the surface is the derivative of the volume is *not* generally applicable.

9.2 EQUIVALENT IDEAS

What becomes of the ideas of familiar Euclidean geometry in this scaling geometry? Since slope no longer has an absolute meaning, the interpretation of the derivative must become

$$\frac{dy}{dx} = \text{rate of change of } y \text{ with respect to } x$$

Actually, we should say "the limiting rate of change," but the shorter version is used for convenience. The interpretation of the derivative as a rate of change is an *extension* of the original interpretation as a slope.

The tangent line is still tangent to the curve after a scaling transformation and is still found the same way as before, but the angle between two lines has no fixed value since a stretch of the coordinates will alter it (unless $k_1 = k_2$).

The numerical value of the slope is no longer the same, but the sign remains (provided we use only positive stretch factors k_i). Interestingly enough, maxima and minima remain unchanged since the stretch cannot affect the slope of the horizontal line (nor a vertical line). You "see" the truth of these statements, but let us look at the mathematical details for a straight line to see how we might prove them.

Example 9.2-1

If we had the general straight line

$$Ax + By + C = 0$$

then after the transformation we would have

$$Ak_1x_1 + Bk_2y_1 + C = 0$$

and
$$\frac{dy_1}{dx_1} = \frac{-Ak_1}{Bk_2}$$

which is the slope of the original line $(-A/B)$ with the multiplier

$$\frac{k_1}{k_2}$$

and we see that the above statements are true. This is, of course, the same as we found in Example 9.1-1.

In general, the tangent circle will change its shape under a scaling transformation (circles will go into ellipses), but the sign of the second derivative will not change. Thus curvature no longer has meaning, and we must go back to the use of the second derivative and recall the soup-bowl rule: "concave up" for $y'' > 0$ and "concave down" for $y'' < 0$. The test for maximum or minimum remains the same. To see all these things intuitively, merely picture a curve on a sheet of rubber and imagine stretching it uniformly in the x direction. The general features of a curve, or curves, will remain, but the specific values of slope and curvature will alter. Remember (see Figure 9.2-1), a curve that is concave up is convex down, and conversely.

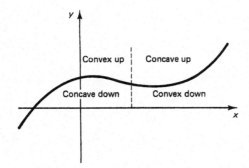

Figure 9.2-1

EXERCISES 9.2

1. Prove that a scaling transformation takes a circle into an ellipse.
2. Show that if one line has + slope and one has − slope then there is a scaling transformation that makes the lines perpendicular. But if they are both of the same sign, then there is a maximum angle that can be achieved. Find it.
3. Find the expression for the new y'' after a scaling transformation in terms of the old y''.
4.* Apply scaling transformations to the analysis of the general conic and show that you need only one circle, one equilateral hyperbola, one size of parabola, and straight lines.

9.3 VELOCITY

A very common situation in this geometry is to be given the position s (space) as a function of time t. Thus you are given the formula

$$s = f(t)$$

For example, you may be given the height in meters of a vertically thrown projectile (rock) as

$$\text{height} = y = 20 + 49t - 4.9t^2$$

where t is measured in seconds and y in meters. In this case it is convenient to use y in place of s since it is the usual vertical coordinate. At time $t = 0$, it was clearly already 20 meters high; the motion of the projectile began on top of some building. And of course the given formula is to apply only until the projectile hits the ground, $y = 0$. The curve (with time as the independent variable) is a parabola opening downward (see Figure 9.3-1).

The derivative is

$$\frac{dy}{dt} = 49 - 2(4.9)t$$

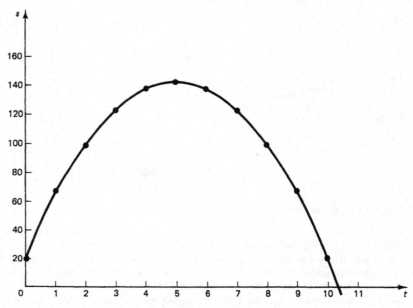

Figure 9.3-1 Projectile height versus time

and is the rate of change of position with respect to time. This is normally called *velocity*. You are going fast when your position is rapidly changing and slow when it changes slowly. Clearly, at time $t = 0$, the start, the instantaneous velocity upward was 49 meters per second and is decreasing as time goes on. Negative velocities mean that the projectile is going downward. The maximum height occurs when the velocity is 0, that is, at time (in seconds)

$$t = \frac{49}{2(4.9)} = 5$$

The position corresponding to this time of maximum height is

$$y = 20 + 49(5) - 4.9(25)$$

$$= 20 + 245 - 122.5 = 142.5 \text{ meters}$$

How shall we interpret the derivative in physical situations? It is distance per unit of time, which is velocity; of course, the derivative is the limiting value of the ratio, it is the instantaneous velocity. If we have two nearby values of position, then their difference divided by the time interval is the *average velocity* during the time interval, and in the limit we speak of the instantaneous velocity of the projectile. We do not need to get into the argument of how this would be measured in practice; it is sufficient to note that in the physical interpretation, as the interval gets shorter and shorter, we would get more and more accurate estimates of the instantaneous velocity, *except* for the simple fact that, however accurately we were measuring the position and time, we would ultimately find that we were getting such small differences as to remove all sense of accuracy. The instantaneous velocity of the mathematically defined projectile along a path approximates physical reality in some senses, but not in all. The instantaneous velocity is a mathematical concept that tends to become a physical concept in most people's minds.

There is always a gap between the mathematics and reality. Most of us believe that the world is made out of molecules, and when you try to make very, very accurate measurements, the random movement of the molecules will defeat your attempts at ultimate precision. In the modern theory of quantum mechanics, it is widely believed that you cannot, even in principle, precisely measure both the position and momentum (velocity times mass) of a particle at the same time; thus *in this interpretation* of quantum mechanics it is impossible, even theoretically, to get arbitrarily accurate measurements at the same time on certain properties of a particle. In practice, from the physical world we abstract a mathematical idealization of what is going on, and then we operate on the mathematical model. Finally, we try to interpret the mathematical results back into physical reality (see Figure 9.3-2). Surprisingly often we get

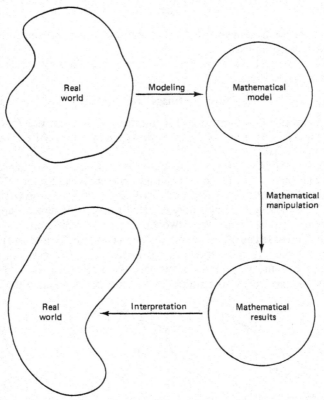

Figure 9.3-2   Mathematical modeling

useful results, but now and then we get nonsense. You need to develop your intuition about the reality of the mathematical models you see.

9.4 ACCELERATION

What about the physical interpretation of higher derivatives? The rate of change of velocity is normally called *acceleration*. In the above example we find that the second derivative

$$\frac{d^2y}{dx^2} = -9.80$$

meaning a downward acceleration of 9.80 meters per second per second. Each second the velocity changes by 9.80 meters per second. The constant in this example is merely the constant of gravity, which is conveniently taken as 980 centimeters per second per second, or in shorter form

$$-980 \text{ cm/s}^2$$

The minus sign means *downward* in the chosen coordinate system where the positive y was chosen to be upward.

Newton's second law of motion asserts that accelerations are caused by forces; indeed the law says that

$$\text{force} = (\text{mass})(\text{acceleration})$$

(when the mass is a constant). You feel the force due to the acceleration (or deceleration) whenever you speed up (or slow down) in an automobile. Thus, we need to think a bit about forces.

In the late Middle Ages, scientists (called natural philosophers in those days) found that two forces could be combined into one equivalent force. When the forces are in exactly opposite directions, it is easy to see that the *resultant* force is the difference between the two oppositely directed forces. Conversely, any force can be regarded as the difference between two forces in many different ways (just as any integer can be regarded as the difference between two positive integers in many ways).

More importantly, they found the parallelogram of forces (see Figure 9.4-1), which shows that the two forces along the two edges (represented by lines with arrowheads) are equivalent to a single force along the diagonal at the parallelogram.

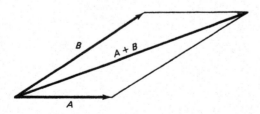

Figure 9.4-1 Parallelogram of forces

The converse, that given any force you can break it up into two forces in almost any pair of directions you choose, is a bit more startling.

Locally, there is a constant acceleration of gravity pulling things downward; this acceleration is commonly represented by the letter $g = 9.80$ m/s^2, although for convenience 980 cm/s^2 is often used in place of 9.80 m/s^2. Actually, the value differs from place to place on the surface of the earth, as well as depending on the height above the surface. This acceleration occurred in the above example of a falling body.

If we imagine firing a cannon horizontally from the top of a mountain (Figure 9.4-2), then we have the horizontal velocity v_0 in the horizontal direction. Neglecting air resistance (and the rotation and curvature of the earth), the law of inertia says that in the absence of other forces an object will continue with this velocity indefinitely.

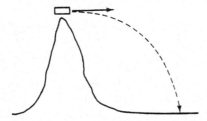

Figure 9.4-2

But the force on the mass due to gravity produces a constant acceleration in the downward direction. This gives an increasing downward velocity, which continues until the projectile hits the ground. The x position, starting at $x = 0$, is given by the formula

$$x = v_0 t$$

while the y position is (h_0 is the height of the mountain and $-g$ is the acceleration)

$$y = h_0 - \frac{1}{2}gt^2$$

If, instead of firing horizontally, we fired at some angle θ, then we could decompose the initial velocity into two components, one horizontal and the other vertical (see Figure 9.4-3). If the angle of firing is θ, then we have to add the vertical velocity component to the y direction motion. Therefore,

$$x = v_0 (\cos \theta) t$$
$$\tag{9.4-1}$$
$$y = h_0 + v_0 (\sin \theta)t - \frac{1}{2}gt^2$$

are the equations describing the motion in time. These are called *parametric equations*, where t is the *parameter*. For each t there is a pair of numbers x and y. You can plot the x and y numbers as a point in two dimensions and mark small ticks along the trajectory to indicate where the projectile is at a given time. It is clearly a parabola, as you can see if you eliminate the parameter t between the two equations.

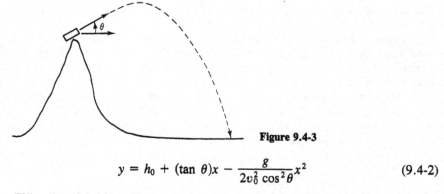

Figure 9.4-3

$$y = h_0 + (\tan \theta)x - \frac{g}{2v_0^2 \cos^2\theta}x^2 \qquad (9.4\text{-}2)$$

This general field is called *exterior ballistics* (exterior as contrasted with interior ballistics, which is what happens inside the gun barrel; the word ballistics comes from an old weapon, the ballista). Of course, in practice, you must take into consideration the effects of air resistance, which we have been neglecting as it is a messy topic. When we send vehicles into outer space, we have similar problems except that once launched and out of the earth's atmosphere there is very little air resistance; the world is almost round, so the gravitational force is directed toward the center of the earth (assuming that the earth is homogeneous, which for delicate work it is not); and there are rotational effects due to the rotation of the earth, both at takeoff and on landing. Evidently, simple mathematics will not suffice, but the general ideas will; the problems are much more complex in their details, but the mathematics to solve problems in mechanics is still the same calculus. The simulation of a space flight is a dramatic application of these same mathematical ideas.

We are forced to take simple problems as examples in a first course, but do not be deceived by them; the methods have wide applicability.

EXERCISES 9.4

1. Find the velocity and acceleration of the motion $s = t^2 - t$.
2. Using (9.4-2) with $h_0 = 0$, find the angle of maximum range.
3. Find the maximum velocity of the motion $s = 12 - t^4$, $t \geq 0$.
4. Find the maximum acceleration for the motion $s = 20t^2 - t^4$.

9.5 SIMPLE RATE PROBLEMS

Example 9.5-1

Consider the following idealized problem. The water is running at a constant rate out of a conical tank whose shape is given in the Figure 9.5-1. How fast is the surface falling when the depth is 20 units (think of meters or feet if you wish)?

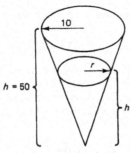

Figure 9.5-1   Conocal tank

This is a word problem; let us analyze how to approach them. First, what is the meaning of the constant rate? It means that the volume is decreasing at a constant rate; if $V = V(t)$ is the volume at any instant t, then letting the rate be R we have

$$\frac{dV}{dt} = R \quad \text{(a negative number)}$$

Next we are asked for the rate at which the surface falls. If h is the height, then we are asked to find

$$\frac{dh}{dt}$$

at a certain time (when $h = 20$).

Can we find a relationship between V and h so that after differentiating it implicitly with respect to time, t, we would have some connection with the two derivatives dV/dt and dh/dt? We need to draw a figure (Figure 9.5-2) as soon as we have any idea at all of what is going on—always draw figures when you can!

The volume of the cone of water is

$$V = \frac{1}{3}\pi r^2 h$$

where r is the radius of the surface of the liquid and h is the height. The new letter r is a nuisance to say the least. Can we get rid of it? Yes! By looking at the figure we see the similar triangles, and we write the proportion as

$$\frac{10}{50} = \frac{r}{h}$$

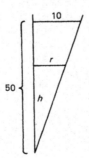

Figure 9.5-2   Schematic figure

Thus we can use $r = h/5$ and eliminate the r. We have, therefore,

$$V = \frac{\pi}{75} h^3$$

Does this seem to be right? Does the volume depend on the cube of the depth of the liquid? It certainly has the right "dimension" for a volume. If it seems right, then go ahead; and if not, then review things until you are satisfied that the equation is correct.

We are ready to differentiate with respect to time, which will generate the derivatives that were given and asked for. We have

$$\frac{dV}{dt} = \frac{\pi}{75} 3h^2 \frac{dh}{dt}$$

Solve for the requested quantity,

$$\frac{dh}{dt} = \frac{25}{\pi h^2} \frac{dV}{dt}$$

Now we put in the given quantities, $dV/dt = R$ and $h = 20$, to get

$$\frac{dh}{dt} = \frac{R}{16\pi}$$

Remember that R is a negative number if the fluid is running out (it would be positive if the tank were being filled up).

There is indeed some method in how to solve word problems! You look at what you are given and what you are asked to find. You then give them names, and try to find relationships between the quantities. If there are rates, then you take derivatives of the general equation with respect to time. *After* you take the derivatives, you assign the fixed data. Whenever possible, you draw a picture, even if it is only symbolic.

Example 9.5-2

The volume of a spherical soap bubble is increasing at a rate of 3 cm³/s. When the radius is 2 cm, how fast is the surface increasing?

We begin with the two formulas

$$V = \frac{4\pi r^3}{3} \quad \text{and} \quad S = 4\pi r^2$$

We have given $dV/dt = +3$ for all r, and we want dS/dt at $r = 2$. We could eliminate r from between the two equations, thus getting S in terms of V. There would be a lot of fractional powers to handle. An alternative is to break the problem up into two problems; first, using the volume rate, find the rate dr/dt, and then use dr/dt in the derivative of the surface equation to find dS/dt. This looks easier, so we proceed on that approach.

$$\frac{dV}{dt} = 4\pi r^2 \frac{dr}{dt} \quad \text{which implies} \quad \frac{dr}{dt} = \frac{3}{16\pi}$$

$$\frac{dS}{dt} = 8\pi r \frac{dr}{dt} = 8\pi(2)\frac{3}{16\pi} = 3$$

and we have the rate 3 cm²/s.

Example 9.5-3

Sand is poured on top of a sand pile at a rate R units per second. At one moment we observe the ratio k = height/diameter of the pile (which measures the shape and depends on the "angle of repose" of the sand). How fast is the radius changing when the height is h_1? We are deliberately being abstract to give you practice in handling general symbols. If it causes you trouble, you can assign particular numbers, work the problem, and then go over the solution and substitute the abstract values as needed.

First draw Figure 9.5-3. We have that the volume of the cone of sand is

$$V = \frac{\pi}{3} r^2 h$$

But the diameter d and the height h are connected by

$$k = \frac{h}{d} \quad \text{or} \quad h = kd = 2kr$$

so the volume is

$$V = \frac{2\pi k}{3} r^3$$

This has the right dimension, so we go ahead. We differentiate to get the needed derivatives:

$$\frac{dV}{dt} = 2\pi k r^2 \frac{dr}{dt} = R$$

Hence

$$\frac{dr}{dt} = \frac{R}{2\pi k r^2}$$

We must now get to the units that the question asked for. We have $r = h/2k$. Therefore,

$$\frac{dr}{dt} = \frac{R(2k)^2}{2\pi k h^2} = \frac{2Rk}{\pi h^2}$$

In this equation we set $h = h_1$ and we have the answer.

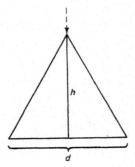

Figure 9.5-3   Sand pile

Example 9.5-4

Fermat's Principle. Fermat's principle in physics asserts that light travels from point A to point B in the minimum time (actually the extremal time). If the velocity of light in the first medium, which contains the point A, is c_1, and the velocity in the second medium, which contains the point B, is c_2, find the path from A to B. We first draw Figure 9.5-4. Let x be the point where the light ray crosses from the first to the second medium. We set up the formula for the total time:

$$T = \frac{\sqrt{a^2 + x^2}}{c_1} + \frac{\sqrt{b^2 + (d - x)^2}}{c_2}$$

To find the extreme time, we differentiate with respect to the variable x and note that the conditions at the extreme give

$$\frac{dT}{dx} = \frac{x}{c_1\sqrt{a^2 + x^2}} - \frac{d - x}{c_2\sqrt{b^2 + (d - x)^2}} = 0$$

In terms of the angles, this is

$$\frac{\sin \theta_1}{c_1} = \frac{\sin \theta_2}{c_2} \tag{9.5-1}$$

Is this a minimum? At $x = 0$ the derivative dT/dx is negative, while at $x = d$ it is positive. Yes, we have the minimum. Thus it is a simple minimum problem, but cast in the rate form of velocity.

This formula (9.5-1) is known as Snell's law and was found originally quite independently of Fermat's principle. Thus you see one principle derived from another, a common thing in a well-developed field like physics.

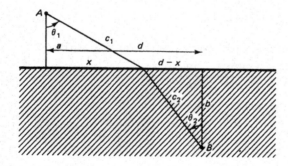

Figure 9.5-4   Fermat's principle

EXERCISES 9.5

1. Do Example 9.5-2 by the first method indicated.
2. How fast does the volume of a cylinder of height h and radius r change when the rate of change of the radius is 3? When the rate of change in the height is 3?
3. A 10-meter ladder is leaning against a wall. If the bottom is pushed in at a rate of $1/10$ m/s how fast does the top move when the distance at the bottom is 6 m from the wall?

4. How fast does the surface drop for a hemispherical tank of diameter 10 meters when the volume is decreasing at a rate of 5 m³/h? The formula for the volume of part of a sphere is $V = (\pi h^2/3)(3r - h)$.

5. Two stones are dropped in a well, one following the other. Show that the distance apart increases at a constant rate proportional to the time difference between when they were dropped.

6. Find the critical angle such that the ray of light does not emerge from the denser (slower velocity) medium.

7. If you have a mirror, then $c_1 = c_2$. Show from Fermat's principle that the angle of incidence is the angle of reflection.

.6 MORE RATE PROBLEMS

Example 9.6-1

A boat moving 20 meters per minute along a path parallel to the end of a pier is being moored as shown in Figure 9.6-1. The path of the boat is 10 meters from the end of the pier. How fast is the rope coming in when the boat is 10 meters from the front of the pier?

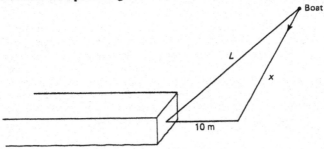

Figure 9.6-1 Mooring boat

Draw the figure and assign letters to the various quantities. The length of the rope is the hypotenuse of the triangle, and once we have that we can differentiate and get the rate. So we set out to find the length of the rope as a function of time. One side of the right triangle is 10 m; the other is the distance x from the end of the pier. We have, therefore,

$$L^2 = 100 + x^2$$

If $t = 0$ at the same time the boat is opposite the end of the pier, then the rate is to be given at $t = -1/2$. Now we have a choice. We could differentiate this equation with respect to t as it is, or we could take square roots and differentiate. To avoid the square roots as long as possible, we work with the above form. We have, on differentiating and dividing out the factor 2,

$$L\frac{dL}{dt} = x\frac{dx}{dt}$$

We have dx/dt, which is the velocity of the boat ($= -20$) and we need L *at the moment when ($t = -1/2$) we are computing the rate of change of the rope length.* It is given by

$$L^2 = 10^2 + 10^2 = 200, \qquad L = 10\sqrt{2}$$

Remember that at that moment we want the rate the value of $x = 10$ ($t = -1/2$), and dividing by the L at that moment,

$$\frac{dL}{dt} = \frac{x}{L}\frac{dx}{dt} = \frac{10}{10\sqrt{2}}(-20) = -10\sqrt{2}$$

The minus sign means that the distance is decreasing.

Example 9.6-2

For a more complex problem consider a 6-foot-tall man strolling at a rate of 3 ft/s down a path that has a light pole 20 ft high set off from the path by a distance of 5 ft. When the man is 30 ft from the position exactly opposite the pole, how fast is the tip of his shadow moving?

We need pictures to get the problem straight in our mind. A top view of the situation is given in Figure 9.6-2. It shows the path, the man, and the light pole. It is natural to take the origin of the x distance along the path as the point opposite the light pole. Remember that $dx/dt = 3$ ft/s.

This picture does not show the heights, so we need another picture, one that shows the instantaneous heights, meaning a picture in the vertical plane (which is constantly changing as the man moves along the path) through the pole and the man. We draw Figure 9.6-2, where s is the distance of the man from the pole. Notice that this figure shows that the tip of the shadow moves along a line parallel to the path of the man. Label the total length to the end of the shadow as y (had we been asked how fast the shadow is lengthening, it would be a different matter). We want to know dy/dt.

From Figure 9.6-2, we can get ds/dt as follow.

$$s^2 = 5^2 + (3t)^2$$

or, differentiating and dividing by the factor 2 again,

$$s\frac{ds}{dt} = 9t$$

$$\frac{ds}{dt} = \frac{9t}{s}$$

What is the value of t? Oh yes, t is the instant when the man is 30 feet from the pole. That must occur when $t = 10$ (we are starting time at the moment he is opposite the pole for convenience). What is s? By the right triangle, we have $s^2 = 925$. It is time to go to the other figure.

Somehow we must extract some more information connecting things together. From the similar triangles we pick, after a little thought, the ratio of similar parts as

$$\frac{14}{s} = \frac{20}{y}$$

or

$$y = \frac{20s}{14} = \frac{10s}{7}$$

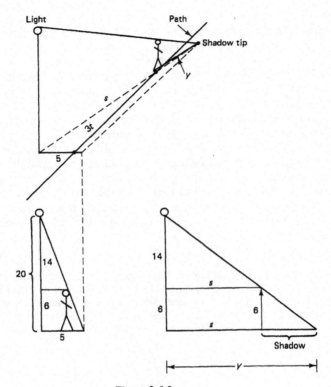

Figure 9.6-2

Now we see a possible approach to the answer. Differentiate this equation with respect to time:

$$\frac{dy}{dt} = \frac{10}{7}\frac{ds}{dt}$$

And we have only to substitute using the quantities we have found until we are done (we hope).

$$\frac{dy}{dt} = \frac{10}{7}\frac{9t}{s} = \frac{10}{7}\frac{9(10)}{\sqrt{925}} = \frac{900}{7(5\sqrt{37})}$$

$$= \frac{18}{7\sqrt{37}} = \frac{18\sqrt{78}}{259}$$

Is this the answer? What is y? What is asked for? The quantity y is the distance from the fixed light pole, and the rate of change is the velocity of the tip of the shadow. Yes, we have the answer this time.

Do not get confused with the details of the geometry and lose the simplicity of the calculus. The ability to *use* the calculus often involves the ability to draw complex situations and extract the essential features you need. This is a difficult art, especially since there is a strong tendency in elementary courses in mathematics to avoid "word problems" as being too hard. But you must master them because that is the form (or even less clearly stated) in which real problems come to you for solution.

If you have trouble with a particular word problem, then try a simpler one (have the man pass directly beneath the light in the above case); it might get you started. Once this is solved, then it is a matter of generalizing the solution to the original problem (the displacement of 5 feet from the light pole). If you cannot do the given problem, can you do one that has a part of it, and then elaborate the solution to cover the given problem? This is a very common attack. To sit and stare at a problem that you cannot understand is a waste of time. To imaginatively strip it down to a simpler problem that you think you can solve is a worthwhile use of your time and effort. It is the method that experts use when attacking a difficult new problem (Hilbert's principle).

Example 9.6-3

Two ships have paths that cross at right angles. The first ship passes the point where the paths cross at 2 P.M. and goes at a rate of 20 knots (nautical miles per hour). The second ship passes the point at 4 P.M. and goes at a rate of 15 knots. How fast are the ships separating when the time is 6 P.M.? See Figure 9.6-3 for the tracks of the ships.

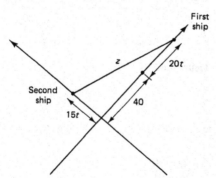

Figure 9.6-3   Two ships

We begin with the distance z between them, measuring time beginning at 4 P.M. We have

$$z^2 = (40 + 20t)^2 + (15t)^2$$

as the distance between the two ships. Differentiation with respect to time gives, at time t,

$$2z\frac{dx}{dt} = 2(40 + 20t)(20) + 2(15t)(15)$$

and at $t = 2$ (6 P.M.) we get (using the original equation to get the needed z and dividing the rate equation by 2)

$$\frac{dz}{dt} = \frac{80(20) + 30(15)}{\sqrt{80^2 + 30^2}} = \frac{205}{\sqrt{64 + 9}} = \frac{205\sqrt{73}}{73}$$

measured in knots.

Case Study 9.6-4

Van der Waals' equation. It is worth looking at a practical problem that arises in physics. The *equation of state* of a perfect gas is

$$PV = RT$$

where P is the pressure of a perfect gas, V is the volume, T is the absolute temperature, and R is the gas constant. Van der Waals (1837–1923) introduced a modified equation for a more realistic model of a gas where the gas molecules are supposed to have a finite volume. The equation is

$$\left(P + \frac{a}{V^2}\right)(V - b) = RT$$

where a and b are suitable constants whose values depend on the particular gas being studied. The term involving a comes from the slight increase in the pressure due to the fact that the finite size of the molecules causes them to go slightly less far between collisions, and the term involving b from the slight decrease in the volume available for the motion between collisions. See Figure 9.6-4 for typical curves for different values of the temperature T. Where does the ripple in the lower curves vanish?

This is a typical nongeometric problem, and therefore it is natural to consider scaling transformations of the variables. Set

$$P = c_1 p$$

$$V = c_2 v$$

$$T = c_3 t$$

Using these, you get the original van der Waals' equation in the new (lowercase) variables p, v, and t:

$$\left[c_1 p + \frac{a}{(c_2 v)^2}\right](c_2 v - b) = Rc_3 t$$

The stretch factors c_i ($i = 1, 2, 3$) can be chosen for convenience. An examination suggests first picking

$$c_2 = b$$

because this allows the b to be factored out of the left-hand side of the equation. Next pick c_1 (so that it too will factor out). This requires

$$c_1 = \frac{a}{c_2^2} = \frac{a}{b^2}$$

Finally, pick

$$c_3 = \frac{a}{bR}$$

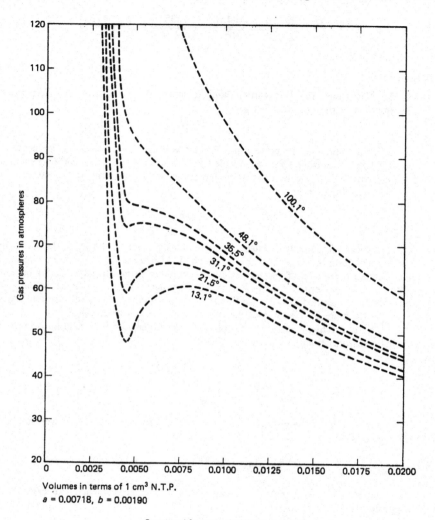

Computed from van der Waals' equation

Figure 9.6-4   Van der Waals equation of state

to remove the last constant. Thus the equation is reduced to the nice form

$$\left(p + \frac{1}{v^2}\right)(v - 1) = t$$

and you see that there is essentially only one standard van der Waals' equation to be studied, not whole families depending on the particular gases. You also see that you are interested only in $v \geq 1$. This *scaling* was early recognized and indeed was used to predict the critical temperature and pressure where the phenomenon of the vanishing of the ripple in the curve occurs. This point is called the *critical point* of the gas and plays an important role in the physics of the gas, especially for low temperatures. How can you find it on the scaled curve?

We begin by noting that for fixed t you are looking for maxima and minima, so you will need to compute the derivative of p with respect to v and set it equal to zero. First solve for p:

$$p = \frac{t}{v - 1} - \frac{1}{v^2}$$

Differentiate this and set the result equal to zero to find the equation for the maxima and minima.

$$\frac{dp}{dv} = \frac{-t}{(v - 1)^2} + \frac{2}{v^3} = 0$$

or, clearing out the fractions,

$$2(v - 1)^2 = tv^3$$

What you want to know is for which values of t this cubic in v has three real solutions and for which it has one real solution (see Figure 9.6-4). One of them is $v < 1$, and this you are not interested in. For $v \geq 1$, the left side rises from the value 0 as a quadratic power, and the right-hand side starts at t and rises like a cubic, t being positive. Since for very large values of v the cubic must be larger than the quadratic, either there are two crossings or there are none (for $v \geq 1$). The point you are looking for is the biggest t for which the equation has a solution, the maximum t. To find the maximum, differentiate the equation in t and set equal to zero. You have

$$t = \frac{2(v - 1)^2}{v^3}$$

$$\frac{dt}{dv} = \frac{-2(v^2 - 4v + 3)}{v^4} = 0$$

whose zeros are, after factoring the numerator,

$$v = 1 \quad \text{and} \quad v = 3$$

It is the $v = 3$ that you want, so you have

$$t = \frac{8}{27}$$

as the critical temperature at which the two real roots merge into one double root and then become complex. From this you can get the critical pressure,

$$p = \frac{1}{27}$$

Notice in how many ways the simple ideas of the calculus were used to get the result. This is typical of many realistic problems; they require a number of small, individually simple steps. The simple steps should be learned so well that you are later able to put them together into larger organizations (chunks) of importance.

Generalization 9.6-5

For nongeometric equations, the scaling transformation that removes as many of the arbitrary physical constants as possible leads you to the least amount of work. The transformations are the basis for any *scaling* of the problem that may be there. Sometimes physical reasons arising from the problem suggest one way of removing the constants rather than another. Not to use the scaling that is available from the scaling transformations is usually foolish, since scaling often leads you to the heart of the problem and away from the details of the particular case as it is first posed to you.

EXERCISES 9.6

1. There are two lights above a path, the first of height h_1 of intensity I_1 and the second of height h_2 and of intensity I_2. They are c units apart. If the light intensity falls off as the square of the distance, give the formula for the rate of change of the illumination if the measuring instrument moves at a rate of d units per second.

2. If a car going 60 kilometers per hour has a plane going 300 km/h fly directly over it on a bearing of 45° to the one the car is driving, how fast do they separate? (It is implied by the 45° that they are going in the same general direction.)

3. Heat radiates inversely as the square of the distance away. If two equally intense sources of heat are c units apart, how fast does the intensity change on a meter moving 5 m/min from one to the other when it is in the middle between them?

4. A light post has a light on top 6 ft high. A 4-ft child walks at a rate of 12 ft/min past it on a path 7 ft from the post and parallel to a house wall 3 ft on the other side of the child. How fast does the shadow move when the child is opposite the pole?

5. How fast does the circumference of a circle change when the area of a circle is changing at a rate R?

6. If the surface of a sphere changes at a rate of R square units per second, how fast does the volume change for a given radius?

9.7 NEWTON'S METHOD FOR FINDING ZEROS

A very common problem is to find the zeros of a function. We have carefully arranged the earlier problems so that the zeros for maximum, minimum, and inflection points are easily found. But in practice, situations often arise where this is not so. Newton devised a simple method of finding the real zeros of a function arbitrarily accurately.

The basic idea of Newton's method for finding zeros is *analytic substitution* of the tangent line for the function (see Figure 9.7-1). The argument is that the corresponding zero of the tangent line (an approximation to the curve, Section 8.4, end) would give a better local approximation to the zero than the approximation you started with. Then, using this improved approximation, we use the new tangent line and find a still better approximation. We need to translate this iterative process into a systematic method.

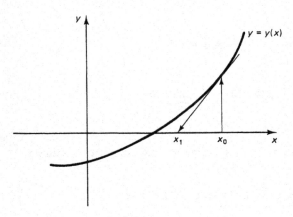

Figure 9.7-1 Newton's method

Given a function

$$y = y(x)$$

with some first approximation to the zero, call it x_0, we fit the tangent line to the curve at the point. The tangent line is, from the point-slope formula,

$$y - y_0 = \left(\frac{dy}{dx}\right)_0 (x - x_0) = y'(x_0)(x - x_0)$$

The zero of this line occurs when $y = 0$. We solve for the corresponding x value, which we label x_1.

$$x_1 = x_0 - \frac{y_0}{y_0'}$$

where, of course, the derivative is evaluated at $x = x_0$.

In view of our plan to *iterate* the formula to get a sequence of better and better estimates of the zero, it is wise to write Newton's formula in the form

$$x_{n+1} = x_n - \frac{y_n}{y_n'} \tag{9.7-1}$$

In words, the next estimate of the root is the present one minus the function value divided by the derivative at that point.

Example 9.7-1

Let us apply this method first to a case whose solution you know. We try the function

$$y = x^2 - 2$$

one of whose solutions is $\sqrt{2}$. Suppose you start with the guess $x = 1$. Then you have

$$y = -1 \quad \text{and} \quad \frac{dy}{dx} = 2x = 2$$

The next guess is (Figure 9.7-1),

$$x = 1 - \frac{(-1)}{2} = \frac{3}{2} = 1.5$$

The next guess is

$$x = \frac{3}{2} - \frac{9/4 - 2}{(3)} = \frac{3}{2} - \frac{1}{12} = \frac{17}{12}$$

$$= 1.41666\ldots$$

It is a matter of iterating the process until you get the accuracy you want. At each step you will almost double the number of digits that are correct.

TABLE OF SUCCESSIVE ESTIMATES

1.
1.5
1.416 6667
1.414 2157
1.414 2136
1.414 2136

When finding the square root of a number N, Newton's method can be arranged in a particularly neat form. The equation $y = x^2 - N$ has a zero, which is $\sqrt{N}$. Therefore, using (9.7-1),

$$x_{n+1} = x_n - \frac{y_n}{y'_n}$$

you have

$$x_{n+1} = x_n - \frac{x_n^2 - N}{2x_n}$$

$$= \frac{2x_n^2 - x_n^2 + N}{2x_n}$$

$$= \frac{x_n + N/x_n}{2}$$

In this form you see another view of why the method works; if the guess x_n is too small, then the quotient N/x_n is too large, and the average of the two numbers is a good new guess. Similarly, if the original guess is too large, then the quotient is too small, and again the average of the two numbers is a good guess.

Example 9.7-2

Suppose you want to find the solution of

$$x = \frac{1}{\sqrt{1 + x^2}}$$

Set up the function

$$y = x - \frac{1}{\sqrt{1 + x^2}}$$

whose zero is the number you want. Differentiate to get

$$\frac{dy}{dx} = 1 + \frac{x}{(1 + x^2)^{3/2}}$$

Thinking of the two graphs, $y_1 = x$ and $y_2 = 1/\sqrt{(1 + x^2)}$ (Figure 9.7-2) (remember you are looking for their intersection), you probably would guess at $x = 1$. Then

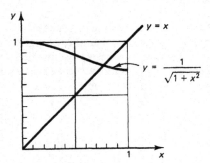

Figure 9.7-2 Finding a zero

$y = 1 - 1/\sqrt{2}$ and $dy/dx = 1 + 1/(2\sqrt{2})$. The next estimate is, therefore, 0.7836 1162. Indeed, you easily get the table.

1.0

0.7836 1162

0.7861 5122

0.7861 5138

If, instead of starting with $x = 1$, you started with the guess of $x = 0$, then the next guess is $x = 1$, and you then get the above table. Finally, if you begin with $x = \frac{1}{2}$, you get

0.5

0.6680 2248

0.7861 5644

0.7861 5138

with no further changes.

There is no unique function to use to find the zero of a given equation. In Example 9.7-2, we could have used

$$y = x\sqrt{1 + x^2} - 1$$

Again, since x is > 0, we could have squared the equation before trying to solve it; we could choose to use

$$y = x^2(x^2 + 1) - 1$$

but would have to watch for the root that is introduced by the squaring and later eliminate it.

These examples show that Newton's method usually converges rapidly. However, there are circumstances in which the method gives trouble. Figure 9.7-3 shows three such situations. The first arises when there is an inflection point between the first guess and the answer. The second shows how a local minimum can cause trouble. The third shows the slow approach when it is a multiple zero.

Let us review the method. The idea is that, given a first guess, we fit a local tangent line at that point and use the zero of the line as the next guess. Then we iterate the process, each stage getting approximately twice the number of digits correct as we have at the start of the step. Insofar as the tangent line represents the curve locally, the method is effective; but if there are serious differences between the curve and its tangent line, or if y' approaches zero, there can be trouble.

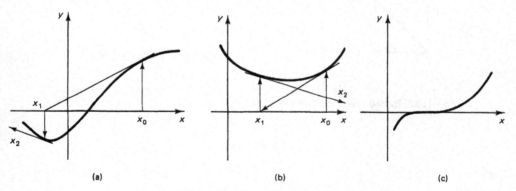

Figure 9.7-3   Failures of Newton's method

EXERCISES 9.7

1. Find the cube root of 3.
2. Find the general formula for finding the cube root of N.
3. Apply Newton's method to the suggested forms after Example 9.7-2.
4. Generalize Exercise 2 to the nth root of N.
5. Find the real root of $x^2 = 1/(1 + x)$.
6. Discuss the problem of finding all the real roots of a polynomial using Newton's method.
7.* Compare the bisection method, Exercise 3.5-9, with Newton's method. List the advantages and disadvantages of both.
8.* Extend Newton's method and fit a quadratic locally. Note there will be the question of which root of the quadratic to use.

.8 MULTIPLE ZEROS

The factor theorem that we derived in Section 2.7 shows the equivalence of a factor of a polynomial and a zero of it; if you have one, you have the other. But sometimes a polynomial may have the same factor repeated, for example,

$$P(x) = x^3 - 5x^2 + 7x - 3 = (x - 3)(x - 1)^2$$

Abusing the language, we say that the equation has a *multiple zero*, that the root $x = 1$ is a *double root*. We wish to preserve the equivalence of zeros and factors, and are therefore forced to this kind of talk. Similar remarks apply for higher multiplicities of roots.

The finding of multiple roots plays an important role in much of mathematics. We see that multiple roots can cause trouble for Newton's method for finding zeros because the derivative, which appears in the denominator will also vanish for a multiple zero. For example, in the above equation

$$\frac{dP(x)}{dx} = (x - 1)^2 + (x - 3)[2(x - 1)]$$

$$= (x - 1)[x - 1 + 2x - 6] = (x - 1)(3x - 7)$$

This shows that the first derivative also vanishes at the same point, $x = 1$. Indeed, if there is a repeated factor of order k,

$$P(x) = (x - a)^k p_{n-k}(x)$$

where $p_{n-k}(x)$ is all the rest of the factors, then the first derivative will have the form

$$\frac{dP(x)}{dx} = (x - a)^{k-1}\left[(x - a)\frac{dp_{n-k}(x)}{dx} + k p_{n-k}(x)\right]$$

and shows that the first derivative has a zero at the point $x = a$ of exactly order $k - 1$. By induction, we see that a factor of order k at $x = a$ means that the function, the first, second, . . . , $(k - 1)$th derivative will all be zero at $x = a$. It also follows by some thinking that the next derivative will not vanish at that point.

This suggests that Euclid's algorithm for common factors of a polynomial, Section 4.6, can be applied to a polynomial and its first derivative. The greatest common factor that the algorithm produces has all the repeated factors, each to one lower power than it was in the original polynomial. Note that Euclid's algorithm is a rational process (involving only the rational operations of add, subtract, multiply, and divide), so finding the polynomial containing all the repeated factors (to their appropriate degrees) of a given polynomial is a simple, although at times tedious, process.

And, of course, we can apply the same step to the greatest common factor to find the repeated factors of it. And so on. Thus, if a polynomial has a single repeated factor of highest degree, then we can find it by a rational process. If there are two different factors both of the same highest degree, we can find the corresponding quadratic equation. Thus, in a very real sense, for a polynomial of degree n, multiple roots are easier to find than many different roots of first degree.

Example 9.8-1

Find the multiple factors of

$$y = x^3 + 3x^2 - 4$$

The derivative is

$$y' = 3x^2 + 6x = 3(x^2 + 2x)$$

The coefficient 3 can be ignored, which makes the arithmetic easier to carry out. We divide the function by the derivative (using detached coefficients, and remembering to supply zero coefficients for the missing terms, just as you do when there are missing powers of 10 in a number such as four hundred and three, which is written as 403):

```
                    1   1
          ┌─────────────────
1  2  0   │ 1   3   0  -4
          │ 1   2   0
          │ ──────────
          │ 1   0  -4
          │ 1   2   0
          │ ──────────
          │    -2  -4
```

The remainder is $-2x - 4$, but can be conveniently taken as $x + 2$. We therefore divide the derivative by this

```
              1   0
        ┌──────────────
1   2   │ 1   2   0
        │ 1   2
        │ ──────
```

Hence the common factor is $x + 2$. It must occur doubly in the original function. We have only to divide this factor out of the given polynomial twice to get the remaining factor, $x - 1$. We have, therefore,

$$x^3 + 3x^2 - 4 = (x + 2)^2(x - 1)$$

as the completely factored form.

EXERCISES 9.8

1. Find the zeros of $x^4 - 5x^3 + 9x^2 - 7x + 2$.
2. Find the zeros of $x^6 - 8x^5 + 24x^4 - 34x^3 + 23x - 6$.

9.9 THE SUMMATION NOTATION

We make a digression here to develop a useful notation. We want to write the sum of a number of similar terms and do not want to write them out every time (even when using the ellipsis method of three dots). The Greek capital sigma is normally used,

$$\sum_{k=1}^{n} (x_k - x)^2$$

means

$$(x_1 - x)^2 + (x_2 - x)^2 + \cdots + (x_n - x)^2$$

This is the sum of *all* the terms of the form indicated, beginning at $k = 1$ and going through $k = n$. In general, the notation (a and b are integers and $a < b$)

$$\sum_{k=a}^{b} f(k)$$

means the sum

$$f(a) + f(a + 1) + f(a + 2) + \cdots + f(b)$$

Note that both ends of the sum are included; there are $(b - a + 1)$ terms in the sum.
  In particular,

$$\sum_{k=a}^{a} f(k) = f(a)$$

To get no terms in the sum, you write (extending the notation in a somewhat unnatural way)

$$\sum_{k=a}^{a-1} f(a) = 0$$

Another example that sometimes causes the student trouble is

$$\sum_{k=a}^{b} 1 = 1 + 1 + 1 + \cdots + 1 = b - a + 1$$

  It is easy to see that

$$\sum_{k=a}^{b} cf(k) = c \sum_{k=a}^{b} f(k)$$

since the constant c will factor out of each term in the sum. It is also easy to see that

$$\sum_{k=a}^{b} \{c_1 f(k) + c_2 g(k)\} = c_i \sum_{k=a}^{b} f(k) + c_2 \sum_{k=a}^{b} g(k)$$

and therefore Σ is a *linear operator*.
  Using this notation, the sums that occurred earlier in Section 2.2 can be written

$$S_1(n) = \sum_{k=1}^{n} k = \frac{n(n + 1)}{2}$$

$$O_1(n) = \sum_{k=0}^{n} (2k + 1) = n^2$$

$$S_2(n) = \sum_{k=1}^{n} k^2 = \frac{n(n + 1)(2n + 1)}{6}$$

$$s_n = \frac{1}{2} + \sum_{k=1}^{n} \cos kx = \frac{\sin (n + 1/2)x}{2 \sin (x/2)}$$

Similarly, the binomial coefficients can be written in the form

$$(1 + x)^n = \sum_{k=0}^{n} C(n, k)x^k$$

while the sum of a geometric progression for n terms takes the form

$$\sum_{k=0}^{n-1} ar^k = \frac{a(1 - r^n)}{1 - r}$$

It is easy to see that

$$\sum_{k=a}^{b} f(k) = \sum_{n=a}^{b} f(n)$$

and therefore the *dummy index of summation* has no real meaning beyond its immediate equation. It is often very convenient to change the index of summation in the middle of a derivation or other work, so the above fact should be clearly understood; the letter used as the index of summation is a *dummy* quantity and has only local significance.

Finally, it is often convenient to drop the first appearance of the dummy index and write

$$\sum_{n=a}^{b} f(n) = \sum_{a}^{b} f(n)$$

and we write at times, when the values of a and b are clearly evident,

$$\sum f(n)$$

9.10 GENERATING IDENTITIES

An application of differentiation that seems to have very little to do with either the slope or the rate of change is the use of differentiation to generate new identities from old identities.

Example 9.10-1

Section 2.4 developed some of the theory of the binomial coefficients. In particular, the function (we have shifted from the earlier letter x to the letter t)

$$(1 + t)^n = 1 + C(n, 1)t + C(n, 2)t^2 + \cdots + C(n, n)t^n \qquad (9.10\text{-}1)$$

was used to *generate* the binomial coefficients. Thus the function

$$(1 + t)^n$$

is called the *generating function* of the binomial coefficients. The letter t is usually used as the variable in the generating function, but when convenient other letters, such as x, are used.

From this you can get other identities by various operations, such as setting $t = 1$. The identity (9.10-1) becomes

$$\sum_{k=0}^{n} C(n, k) = 2^n$$

You can also differentiate a generating function with respect to t to get new identities. Using identity (9.10-1), you get

$$n(1 + t)^{n-1} = \sum_{k=0}^{n} kC(n, k)t^{k-1} \qquad (9.10\text{-}2)$$

When you put $t = 1$, you get

$$n2^{n-1} = \sum_{k=1}^{n} kC(n, k) \qquad (9.10\text{-}3)$$

This is a useful identity involving the binomial coefficients.

You can get further identities by further differentiation, and you can also, for example, multiply by t *before* doing the next differentiation. What is required is the imagination to see what to do to generate the identity you want.

Example 9.10-2

Find the sum

$$1 + 2x + 3x^2 + 4x^3 + \cdots, \quad \text{for } |x| < 1$$

We remember, after some thought, a similar identity for the geometric progression (assuming that $|x| < 1$):

$$\frac{1}{1 - x} = 1 + x + x^2 + x^3 + \cdots = \sum_{k=0}^{\infty} x^k$$

If we differentiate formally (without regard to whether an infinite sum can be differentiated term by term or not) with respect to x, we get

$$\frac{1}{(1 - x)^2} = 1 + 2x + 3x^2 + \cdots = \sum_{k=0}^{\infty} kx^{k-1} \qquad (9.10\text{-}4)$$

which is what is required.

If you are a bit worried about the differentiation of the infinite sum, then you can proceed as follows. Take the finite sum

$$\frac{1}{1 - x} - \frac{x^{n+1}}{1 - x} = 1 + x + \cdots + x^n = \sum_{k=0}^{n} x^k$$

and differentiate it (which is a finite sum and you know that both sides will give the same result):

$$\frac{1}{(1-x)^2} - \frac{x^{n+1}}{(1-x)^2} - \frac{(n+1)x^n}{1-x} = \sum_{k=1}^{n} kx^{k-1}$$

which becomes

$$\frac{1}{(1-x)^2} - \frac{(n+1)x^n - nx^{n+1}}{(1-x)^2} = \sum_{k=1}^{n} kx^{k-1}$$

We now see that for fixed $|x| < 1$ and increasing n, the second term on the left approaches zero; hence the series approaches the first term, which is what we had above (9.10-4). Hence this time we get the same answer in both cases.

Example 9.10-3

The purpose of this example is to show the power of the generating function approach. Consider the representation of

$$(1-x)^{1/2} = c_0 + c_1 x + c_2 x^2 + c_3 x^3 + \cdots$$

where we are supposing that x is so small that the right-hand side "converges" to a definite value for the value of x. Let us formally differentiate both sides:

$$\frac{-1}{2(1-x)^{1/2}} = c_1 + 2c_2 x + 3c_3 x^2 + \cdots$$

Now multiply both sides by $-2(1-x)$. We get

$$(1-x)^{1/2} = -2(1-x)(c_1 + 2c_2 x + 3c_3 x^2 + \cdots)$$

Since we proved that the powers of x are linearly independent (Section 4.7) for *any* finite number of terms, it is reasonable to suspect that the independence also holds for infinite sums. We are led, therefore, to equate the separate powers of x;

$$x^0: \quad c_0 = -2c_1 \qquad c_1 = -\left(\frac{1}{2}\right)c_0$$

$$x^1: \quad c_1 = -4c_2 + 2c_1 \qquad c_2 = \frac{1}{4}c_1$$

$$x^2: \quad c_2 = -6c_3 + 4c_2 \qquad c_3 = \frac{1}{2}c_2$$

$$x^3: \quad c_3 = -8c_4 + 6c_3 \qquad c_4 = \frac{5}{8}c_3$$

Now, if $x = 0$, then clearly $c_0 = \pm 1$. Taking the $+$ for the square root, we can solve for the successive coefficients of the expansion, one at a time.

$$c_1 = -\frac{1}{2}$$

$$c_2 = \frac{c_1}{4} = -\frac{1}{8}$$

$$c_3 = \frac{c_2}{2} = -\frac{1}{16}$$

If we had taken the minus sign for the square root, then each of the coefficients would change sign; hence

$$(1 - x)^{1/2} = \pm\left[1 - \frac{x}{2} - \frac{x^2}{8} - \frac{x^3}{16} - \frac{5x^4}{128} - \cdots\right]$$

The rule for computing the binomial coefficients given in Section 2.4 still applies for all fractional exponents. By Section 4.5, we assume that it will also apply for irrational numbers.

Thus the differential calculus has many uses beyond the obvious simple max–min problems and rate problems. Differentiation can be used as a formal process for obtaining new results from old, known results.

EXERCISES 9.10

1. From Example 9.10-1 multiply through by t, differentiate again, and so on, to get $\sum_{k=0}^{n} k^2 C(n, k)$.
2. Find $\sum_{0}^{\infty} k^2 x^n$. *Hint:* See Example 9.10-2 and repeat the trick of the previous exercise.
3. Find $\sum k/2^k$ and $\sum k^2/2^k$. Generalize.
4. Use the generating function $1/(1 - xt)$ to get the result (9.10-2).
5. Find the $\sum_{k=0}^{n} 2^k C(n, k)$, $n > 1$.
6. Find the $\sum_{k=1}^{n} k 2^k C(n, k)$, $n > 1$.
7. Find the $\sum_{k=1}^{n} (-1)^k k C(n, k)$, $n > 1$.

9.11 GENERATING FUNCTIONS—PLACE HOLDERS

The main purpose of this section is to explore the meaning of the dummy variable t that we used in the generating function, such as that for the binomial coefficients

$$(1 + t)^n = \sum_{k=0}^{n} C(n, k) t^k$$

We first look at a similar, familiar situation, plain old multiplication. Consider

$$
\begin{array}{r}
307 \\
742 \\
\hline
614 \\
1228 \\
2149 \\
\hline
227794 \\
\end{array}
$$

In a slightly less familiar form, we can begin the multiplication by the first digit on the left, and arrange the products in the other order

$$
\begin{array}{r}
307 \\
\underline{742} \\
2149 \\
1228 \\
\underline{614} \\
227794
\end{array}
$$

What is the role of the 0? It is a *place holder* to keep the lineup of the powers of 10, which is implicit in the whole arrangement of the multiplication; we keep the same powers of 10 in the same column.

Now suppose we consider the product

$$(1 + t)^3(1 + t)^3 = (1 + t)^6$$

We do not write the powers of t (just as we ignore the powers of 10 in arithmetic). We have

$$
\begin{array}{rrrrrrr}
1 & 3 & 3 & 1 & & & \\
1 & 3 & 3 & 1 & & & \\
\hline
1 & 3 & 3 & 1 & & & \\
& 3 & 9 & 9 & 3 & & \\
& & 3 & 9 & 9 & 3 & \\
& & & 1 & 3 & 3 & 1 \\
\hline
1 & 6 & 15 & 20 & 15 & 6 & 1
\end{array}
$$

which we immediately recognize as the binomial coefficients of order 6, as they should be. Notice how closely this resembles multiplication in arithmetic *except* that there are no carries from one column to another.

Example 9.11-1

We immediately *generalize* to the nth case:

$$(1 + t)^n(1 + t)^n = (1 + t)^{2n}$$

Let us pick out the middle column of the product, the coefficient of t^n. This is the sum of all the terms whose exponent of t in the product is n. From the top line, we get $C(n, n)C(n, 0)$; from the next line $C(n, n - 1)C(n, 1)$; from the next, $C(n, n - 2)C(n, 2)$; and so on down to the last line, $C(n, 0)C(n, n)$. The sum of all these terms is the corresponding coefficient of t^n from the expansion of order $2n$, that is, $C(2n, n)$. In the summation notation, we have

$$\sum_{k=0}^{n} C(n, n - k)C(n, k) = C(2n, n)$$

Recalling from Section 2.4 that the binomial coefficients are symmetric in the second index,

$$C(n, n - k) = C(n, k)$$

we can write the above equation as

$$\sum_{k=0}^{n} C^2(n, k) = C(2n, n)$$

In words, the sum of the squares of the binomial coefficients of order n is the single midcoefficient of the expansion of order $2n$.

In all this derivation, what is the role of the dummy letter t? It is a place holder enabling us to keep track of the various terms; it does not play the role of a number. True, the expression

$$1 + t$$

indicates that the quantity represented by t must have some properties of a number (we cannot add apples to oranges), yet it really is not a number, nor does it necessarily become one later in the above derivation. It is some kind of a generalized number having no particular size. Sometimes we do assign a numerical value to it, but often we do not.

EXERCISES 9.11

1. Find the $\Sigma (-1)^k C^2(n, k)$.
2. Find the sum of the binomial coefficients in the mth column.
3. Generalize Example 9.11-1 to the case of $(1 + t)^n(1 + t)^m$.

9.12 DIFFERENTIALS

When the student has done enough four-step derivations, the terms that are going to matter are soon recognized, and the others that will drop out are ignored. For example, suppose you have

$$y = x^4$$

At the third step of the four-step process, you get something like

$$\frac{\Delta y}{\Delta x} = \frac{x^4 + 4x^3\Delta x + \cdots - x^4}{\Delta x}$$

You simply are not interested in the terms represented by the three dots; you know that they will disappear in the limit as Δx approaches zero.

We would like to eliminate early in the derivation all the terms that will not matter. In Figure 9.12-1, you see what the derivation looks like. P is the original point (x, y), and Q is the incremented point. The point S represents the position of the point when all the terms that you want to neglect are omitted; you are on the tangent line. We cannot call this change in the y direction Δy any more because of the terms that

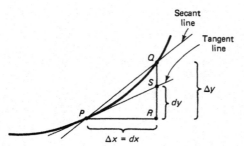

Figure 9.12-1

we dropped. A psychologically good, but logically bad, choice is conventionally made to call the quantities

$$\Delta x = dx \quad \text{and} \quad \text{the distance } RS = dy$$

Since the point S is on the tangent line, we have

$$\frac{dy}{dx} = \frac{dy}{dx}$$

where the symbol on the right is the conventional derivative. Thus on the left we think of the dx and dy as small changes running along the tangent line, while on the right the derivative is evaluated at the point on the curve. After having insisted that the symbol dy/dx is a single symbol, we now write it as a quotient of two *differentials* dy and dx.

What are these differentials dy and dx? They are symbols that are chosen to keep track of the terms that matter and to ignore those that do not; they are really place holders, much as the t was in the generating function method. They are not numbers per se (per se means by, of, or in themselves), but they have some of the attributes of numbers since we find them being added to numbers. They are not of any definite size, since they represent the terms that will still be present after the limit is taken.

The users of the calculus constantly think in terms of differentials when they set up problems from the real world. They say they think of the differentials as being "infinitely small," except that they generally do not because they also believe in the discrete nature of the physical universe to which they are applying the argument. They are trying to get rid of the terms that will not matter when they take the limit. There is a logical difficulty in the way they talk about things, but when you understand that the differentials are being used to eliminate the terms that will not matter and to concentrate on the terms that will, then what is going on becomes clear. Of course, it is implied that the person doing the derivation knows what terms will matter and those that will not. If an error is made and a needed term is omitted, then the result will, of course, be false. But experience is a very good guide, and the use of differentials saves a lot of very tedious details in the usual derivation; differentials eliminate early those terms that will vanish in the limit.

.13 DIFFERENTIALS ARE SMALL

Having insisted that differentials have no size, that they are merely place holders for the terms that will matter in taking the limit, we now turn around (again!) and let them be small and thereby use the tangent line as an approximation to the curve. There is in all this no definition of how small "small" really is. There is again an intuitive feeling that if we use (small) differentials then the terms we neglect will (near the limit) be very small; hence we will make relatively small errors when we use differentials.

Example 9.13-1

Suppose we have a rectangular shape with an area A:

$$A = xy$$

and suppose that $x = 5$ and $y = 7$. The area $A = 35$. But next we suppose that the x and y values are slightly changed (Figure 9.13-1). Let the changes be $dx = 0.01$ and $dy = 0.02$. Then the change in the area is

$$\Delta A = x\,\Delta y + y\,\Delta x + \Delta x\,\Delta y$$

Writing this in differential form, we have

$$dA = x\,dy + y\,dx + dx\,dy$$

In this equation, supposing that the differentials are small, the last term is very much smaller than the others in the equation, and we simply ignore it to get

$$dA = x\,dy + y\,dx$$
$$= 5(0.02) + 7(0.01) = 0.1 + 0.07 + 0.17$$

and the new area is

$$A = 35 + 0.17 = 35.17$$

In this approximation we see from the figure that we have taken the contributions from the two sides (rectangles) and neglected the cross-product term $dx\,dy = 0.01 \times 0.02 = 0.0002$, which in this case is a very small error indeed *when* compared to the actual area. The formula we used is simply the derivative of a product formula in the differential form.

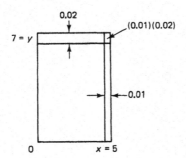

Figure 9.13-1

There is no reason for the changes to be positive; they can be either positive or negative as needed. What is required is that you realize that *the use of differentials puts you on the local tangent line and not on the curve itself.*

Example 9.13-2

Suppose you have a circle of radius r and wish to make a small change in the radius. How much does the area change? We simply write

$$A = \pi r^2$$

take differentials

$$dA = 2\pi r \, dr$$

and we have the formula telling us (approximately for small changes) the change in area for a change in the radius.

Example 9.13-3

Let us write the old derivative formulas in the new differential form. We have

$$d(u^n) = nu^{n-1} \, du$$

$$d\{au(x) + bv(x)\} = a \, du(x) + b \, dv(x)$$

$$d(uv) = u \, dv + v \, du$$

$$d\left(\frac{u}{v}\right) = \frac{v \, du - u \, dv}{v^2}$$

$$du = \left(\frac{du}{dv}\right) dv$$

These forms are only a small change from the earlier forms, where we used derivatives instead of differentials, but they are the form we will often need in the future. Hence these formulas should be learned for immediate recall at any moment.

If you want to use the idea of differentials, and not be bound to very small quantities, then a glance at Newton's method (Section 9.7) for finding zeros of a function shows that large changes can *sometimes* be accommodated by the application of an iteration process.

EXERCISES 9.13

Differentiate the following in differential form:

1. $y = 3x^4$
2. $y = \sqrt{a^2 - x^2}$
3. $y = (1 - x)/(1 + x)$
4. $y = \sqrt{x}$

5. $x^2 + y^2 = a^2$
6. $3xy^2 + 3x^2y = b^3$
7. $\sqrt{x} + \sqrt{y} = 1$
8. $(x^2 + 4)(x^2 - 6)^2 = y$
9. $y = (1 + x^2)^4$

9.14 SUMMARY

In this chapter we saw that there is an essential difference between geometric and nongeometric applications of the calculus, but that there are some equivalences between them. In particular, we saw the central role of scaling transformations in nongeometric applications.

We also saw applications to velocity and acceleration in simple and slightly complex problems, usually called *rate problems* because they involve rates (derivatives with respect to time).

Newton's method for finding the real zeros of a function is based on the idea of the *analytic substitution* of the local tangent line for the original curve, and then taking the zero of the tangent line as the new approximation to the zero of the original function. When it works, its convergence, once well started, is very rapid.

Next, we saw that the calculus is also useful in generating new identities from old ones. We can take any identity in x (or t), multiply by an arbitrary function of x, and differentiate to get a new identity. The problem is not so much that of generating identities as it is of finding how to generate the particular one you want to be. This takes a little imagination to decide where to start and how to get to where you want. This use of the calculus rests on *only* the simple fact that differentiation is a formal linear mapping of functions onto functions, and seems to have very little to do with the original concept of the limit of the ratio of $\Delta y / \Delta x$. This is a typical generalization of an idea in mathematics; something begun for one purpose is later found to have many, apparently unrelated, applications. It is an example of the universality of mathematical ideas (Section 1.3).

Finally, we looked at the delicate topic of differentials as a means of handling a sea of unnecessary details by simply eliminating them from the derivation. The concept was extended to small, local approximations along the tangent line, rather than on the curve; and insofar as the tangent line is close to the curve, their use in this manner is reasonable. The logical difficulties remain, however. The differential form is used in later parts of this book, so the rules must be learned carefully.

10

Functions of Several Variables

Up to now we have concentrated on functions of a single independent variable, and have used the notation

$$y = f(x)$$

We are often interested in phenomena that involve more than a single independent variable, and therefore we must examine functions of several variables. It is prudent to examine the case of two independent variables before examining the general case of n independent variables (Section 10.6).

The standard notation for functions of two independent variables is

$$z = f(x, y)$$

where x and y are the independent variables and z is the dependent variable. We think of the corresponding geometrical picture in three dimensions, the x and y mutually perpendicular axes horizontally, and the z variable is the third (perpendicular) direction upward. There are, unfortunately, two distinct such coordinate systems. Mathematicians generally use a left-handed one, shown in Figure 10.1-1a, and scientists and engineers often use a right-handed system, shown in Figure 10.1-1b. The names "left handed" and "right handed" come from the motion along the z-axis of a screw as you turn from the x-axis to the y-axis. It is unlikely that either group will, in the near future, abandon their system for the other, so you need to know both systems.

Plotting points in the three-dimensional (mathematical) space is straightforward; they are taken in the order (x, y, z). Thus the point P_1 with coordinates $(1, 2, 3)$ is

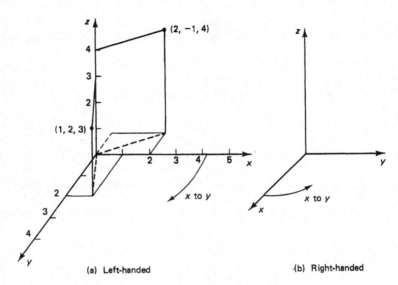

(a) Left-handed   (b) Right-handed

Figure 10.1-1

shown in Figure 10.1-1 along with the point P_2 with coordinates, $(2, -1, 4)$. There is no widely used system for numbering the eight octants (corresponding to the four quadrants), and it is customary to give the signs of the three variables x, y, and z to indicate the octant you are talking about. All three coordinates positive is the first octant.

Once you can plot points, the next thing to consider is the distance between two points. Let the three coordinates be in the same units, and let the two points be P_1, $(1, 2, 3)$, and P_2, $(2, -1, 4)$, of the previous paragraph. Think of them as the diagonal corners of a rectangular room whose sides are parallel to the coordinate axes (Figure 10.1-2). Using the Pythagorean distance, the square of the length of the diagonal on the floor is the sum of the squares of the lengths of the two sides,

$$d_1^2 = (2 - 1)^2 + (-1 - 2)^2 = 1 + 9 = 10$$

Now the vertical height of the room is measured perpendicular to the floor, and it is

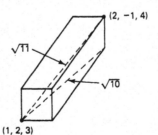

Figure 10.1-2   Distance between points

perpendicular to the diagonal line whose length we just computed. Applying the Pythagorean distance again, we find that the square of the room diagonal length is

$$d^2 = 10 + (4 - 3)^2 = 10 + 1 = 11$$

We immediately generalize; in the case of two general points, with coordinate subscripts 1 and 2, the Pythagorean distance between them is

$$d^2 = (x_2 - x_1)^2 + (y_2 - y_1)^2 + (z_2 - z_1)^2 \qquad (10.1\text{-}1)$$

Notice that (1) the distance function is symmetric in the three coordinates, (2) involves the squares of the differences of the coordinates, and (3) the points can be taken in either order (for any coordinate). In words, the square of the distance between the two points is the sum of the squares of all the differences of the corresponding coordinates of the two points.

   We next turn to linear relations among the variables. The simplest general form is

$$Ax + By + Cz + D = 0 \qquad (10.1\text{-}2)$$

Assume, for the moment, that $C \neq 0$. We can therefore write the equation as solved for z,

$$z = ax + by + c$$

where the lowercase letters are related to the uppercase letters in an obvious way: $a = -A/C$, $b = -B/C$, and $c = -D/C$. For each pair x and y, this equation gives a corresponding z value, and the triple numbers (x, y, z) represents a point. The locus of all the points that satisfy the equation is a plane. The plane

$$z = x + 2y - 3$$

is drawn in Figure 10.1-3, and all you see are the *traces* on the three coordinate axes. The plane, of course, extends indefinitely in all directions, but the figure you see is the small triangle. The figure is easiest drawn by finding the intersections of the plane

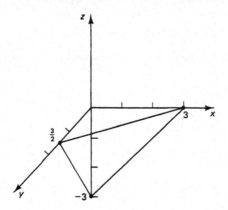

Figure 10.1-3   Plane $z = x + 2y - 3$

with the coordinate axes. To do this, you merely set each pair of variables equal to zero to find the value on the corresponding axis. Thus, in this example, to find the point on the x-axis, we set $y = z = 0$, and find that $x = 3$. Putting one letter equal to zero gives the equation of the straight line on the corresponding plane. For example, for $x = 0$ you have $z = 2y - 3$ in the y–z plane.

From your knowledge of lines in a two-dimensional plane, you see that linear equations in three variables represent planes in three dimensions without rigorously proving it. The special cases of a and/or $b = 0$ can be analyzed. For example, if both a and b are 0, then for all x and y we have $z = c$; it is a horizontal plane parallel to the x–y plane. If only one coefficient is zero, say $a = 0$, then the plane is parallel to the x-axis and cuts the y–z plane in the line (trace) defined by

$$z = by + c$$

when viewed as an equation in the y–z plane (see Figure 10.1-4).

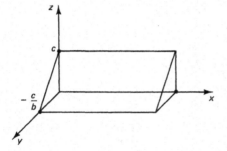

Figure 10.1-4   Plane $z = by + c$

A straight line is described by the intersection of two planes; it takes two constraints on the three variables to define a line. For example, the equations

$$z = 3 - x - y$$
$$z = 4 - 2x + 4y$$

together describe the locus of a line. The points of the line must lie on both planes at the same time.

But any point that satisfies both equations also satisfies any linear combination of the equations. For example, if you multiply the top equation by 4 and then add this to the lower equation, you eliminate the y variable. You have

$$5z = 16 - 6x$$

This is another plane through the same line defined by the two original equations. You could, instead, eliminate the x variable. Multiply the top equation by 2 and the lower by -1, and then add

$$z = 2 - 6y$$

These two derived planes equally well serve to define the line, and are somewhat preferable because you can see them more easily. The first plane is parallel to the y-axis and you see the *projection* of the line onto the z–x plane, while the second is parallel to the x-axis and you see the projection on the y–z plane (Figure 10.1-5). Thus the equations of a line are not unique; any pair of planes through the line will serve to describe it.

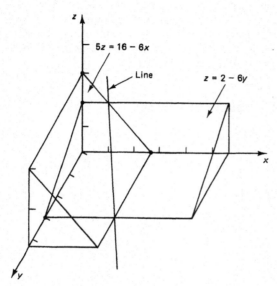

Figure 10.1-5 Two projections of a line

Generalizing the above, given the two equations that define a line, any linear combination of the equations also defines a plane through the line. This resembles the similar observation that, given two lines in two dimensions having a common point, then any linear combination of the two equations defines a line that also passes through the common point. And this probably brings to mind the warning that, just as two lines may be parallel and have no common point, so too may two planes be parallel and have no common line. Going further, you see that by picking a suitable choice of the constants you can form the linear combination that will define any line (plane) through the common point (line).

You could look at many special forms of the plane corresponding to the special forms of the line in two dimensions. The basic tool is the use of the general linear form (10.1-1) for a plane

$$Ax + By + Cz + D = 0 \qquad (10.1\text{-}3)$$

together with the method of undetermined coefficients. Note that there exists an extra parameter (just as there did in the case of a line in a plane). For example, if $D \neq 0$, then from this form by dividing by $-D$, and a little algebra, you can get the intercept form

$$\frac{x}{a} + \frac{y}{b} + \frac{z}{c} = 1 \tag{10.1-4}$$

of the plane. Inspection shows that the *two* equations

$$\frac{x - x_1}{x_2 - x_1} = \frac{y - y_1}{y_2 - y_1} = \frac{z - z_1}{z_2 - z_1} \tag{10.1-5}$$

are the equations of a line through the points P_1 and P_2, and thus represent the two-point form.

Example 10.1-1

Given three points, say $(1, 1, 1)$, $(2, -1, 3)$, and $(1, 0, 2)$, determine the corresponding plane through them. We use undetermined coefficients and substitute the values of the points into the general form for the plane (10.1-3). From the three points, we get the following three conditions on the coefficients of the plane:

$$\begin{aligned}
A + B + C + D &= 0 \\
2A - B + 3C + D &= 0 \\
A \phantom{{}- B} + 2C + D &= 0
\end{aligned} \tag{10.1-6}$$

Their solution gives the coefficients of the plane through the three points. It is easiest in this particular system to eliminate B by adding the top two equations:

$$3A + 4C + 2D = 0$$

Next, we take this with the third equation, and eliminate A by subtracting this from three times the third equation:

$$(6 - 4)C + D = 0$$

or
$$C = \frac{-D}{2}$$

From the third of the original equations (10.1-6), we have

$$A = -2C - D = D - D = 0$$

Next, from the top equation of (10.6-6), we solve for B

$$B = -A - C - D = \frac{-D}{2}$$

Therefore, the equation of the plane through the three points is

$$0x + \left(\frac{-D}{2}\right)y + \left(\frac{-D}{2}\right)z + D = 0$$

Dividing out the D (as it must) and multiplying through by 2, we get, finally,

$$y + z = 2$$

This is a plane parallel to the x-axis. It is easy to check by direct substitution that this plane contains the three points.

Generalization 10.1-2

Linear Algebra. We have just solved three equations (10.1-6) in three unknowns, A, B and C. We have progressed to the point where from a single special case we can go to the general case. In this spirit we examine the general problem of solving three linear equations in three unknowns, x, y and z (we have changed from the unknown letters A, B, and C of the plane to the more usual variables x, y, and z). Let the given equations be

$$a_1x + b_1y + c_1z = d_1$$
$$a_2x + b_2y + c_2z = d_2 \qquad (10.1\text{-}7)$$
$$a_3x + b_3y + c_3z = d_3$$

where we have adopted the usual notation of the independent variables as x, y, z, and the coefficients a_i, b_i, c_i, d_i are constants.

The *method* of solution we will use is known as *Gauss (1777–1855) elimination*. It systematically eliminates one variable at a time. We begin by supposing that at least one term in x is present, and it is no loss in generality to suppose this x is in the first equation, since we could renumber the equations if necessary. Thus we assume that $a_1 \neq 0$. Multiply the top equation by $-a_2$, and add the equation to a_1 times the second equation. We have

$$(a_1b_2 - a_2b_1)y + (a_1c_2 - a_2c_1)z = a_1d_2 - a_2d_1 \qquad (10.1\text{-}8)$$

This equation, together with the first original equation, is equivalent to the second equation, because the second equation can be reconstructed from the first equation plus the derived equation by a suitable combination.

Similarly, multiplying the top equation by $-a_3$, and the third by a_1, and then adding, you get

$$(a_1b_3 - a_3b_1)y + (a_1c_3 - a_3c_1)z = a_1d_3 - a_3d_1 \qquad (10.1\text{-}9)$$

The original system (10.1-7) is now equivalent to the first original equation plus the two derived equations, (10.1-8) and (10.1-9) (since the steps are reversible). But the two derived equations depend on only two variables y and z. Repeating the elimination process, we eliminate the y terms and get a single equation in z.

Thus the original system is now equivalent to a triangular system of equations, whose solution is called the *back solution* of the Gauss elimination process. We are *apparently* led to a unique solution by this method of systematically eliminating one variable at a time.

What can go wrong? First, at each stage the method supposes that there is at least one equation in which the next variable to be eliminated appears. It might be that this is not the case. For example,

$$x + y + 2z = 9$$
$$x + y - 3z = -6$$
$$x + y + z = 6$$

After eliminating x we will find that there is no y term in the two derived equations. The quantity $x + y$ always occurs together and it is really not two variables. Thus, we cannot hope to determine x and y separately. This situation can arise when any two (or more) variables are coupled together. This situation means that one (or more) of the variables can be picked as we please; we can have infinitely many solutions of the system *provided* the rest of the solution process does not get into a contradiction, such as $z = 6$ from one equation and $z = 7$ from another equation.

A second thing that can go wrong is that at some stage all the variables in the equation cancel out. For example,

$$x + y + z = 2$$
$$x + y + z = 3$$
$$x - y + 2z = -5$$

Evidently, when we eliminate the x at the first stage between the first two equations, we will get

$$0 = -1$$

which is a contradiction, meaning geometrically that the first two equations represent parallel, distinct planes that have no common line of intersection. If the second equation had been, for example,

$$2x + 2y + 2z = 4$$

then the elimination would have left us with

$$0 = 0$$

which means parallel coinciding planes, and hence the second equation represents no constraint on the variables other than that of the first equation. There are really only two equations defining the three variables, and one of the variables (say z if you wish) may be picked to be anything you please—leading to a *one parameter family* of solutions (a straight line).

This method of *Gauss elimination* can clearly be applied to n linear equations in n independent variables. The algebraic interpretations of what happens when you cannot go on are about the same; either a variable disappears completely from all the derived equations at that stage, or else all the variables in one equation drop out. The first shows a basic connection between the variables. The second implies a linear relationship between the involved equations; the equation that produced the row of zeros is a linear combination of the ones above. If the right-hand side is not zero, this means the equations are *inconsistent* (parallel distinct planes). If the right-hand side is equal to zero (coinciding planes), this means that at least one of the equations is *redundant*, and you have at least a one-parameter family of solutions.

EXERCISES 10.1

1. Find the plane through the three points $(1, -1, 0)$, $(0, 1, 2)$, and $(2, 1, 1)$.
2. Find the plane through the three points $(1, 1, -1)$, $(1, -1, 1)$ and $(-1, 1, 1)$.
3. Using the intercept form (10.1-4), find the plane with the intercepts 1, 2, 3.

4. Rewrite the general form of a plane (10.1-3) in a one-point form. *Ans.:* $A(x - x_1) + B(y - y_1) + C(z - z_1) = 0$.

5. What is the condition that a plane pass through the origin?

6. What is the condition that a plane be parallel to the z-axis?

7. Discuss the geometric situation where each of three planes can intersect the other two but that there is no common point of the three planes.

8. Give the condition that two planes be parallel.

9. Draw the plane $x + y + 3z = 5$.

10. Draw the plane $2x - y - z = -2$.

11. Draw the planes $x + y + z = 1$ and $x - y + z = -1$ and sketch the line.

12. Find the equations of a line through $(1, -1, 1)$ and $(1, -2, 3)$. *Hint:* Use Equation (10.1-5).

13. Solve the system of equations $x + y = 3$, $y + z = 4$, $z + x = 5$. Generalize.

10.2 QUADRATIC EQUATIONS

In Euclidean geometry (see Figure 10.2-1), the Pythagorean distance in three dimensions (all of the same kind, and not "apples plus oranges"), is given by the formula (10.1-2):

$$d^2 = (x_2 - x_1)^2 + (y_2 - y_1)^2 + (z_2 - z_1)^2$$

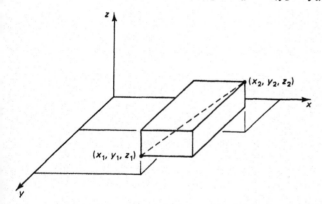

Figure 10.2-1   Distance

It is easy to recognize that if r is a fixed number then the equation

$$(x - x_1)^2 + (y - y_1)^2 + (z - z_1)^2 = r^2 \qquad (10.2\text{-}1)$$

is the equation of a *sphere* about the point (x_1, y_1, z_1) of radius r (we have set $d = r$ for convenience of calling the distance the radius of the sphere). Thus, there is a close analogy between circles in the plane and spheres in three dimensions; for example,

$$x^2 + y^2 = 1$$

is a circle of radius 1 about the origin in two dimensions (and in three dimensions this would be a cylinder of radius 1 about the z-axis). Again,

$$x^2 + y^2 + z^2 = 1$$

is a sphere of radius 1 about the origin in three dimensions. Clearly, the trivial plane $z = \frac{1}{2}$ cuts the sphere in the circle

$$x^2 + y^2 = 1 - \frac{1}{4} = \frac{3}{4} = \left(\frac{\sqrt{3}}{2}\right)^2$$

Planes clearly cut a sphere in circles, although the intersection of a tilted plane with a sphere requires some attention to prove that it is a circle.

When we go to three variables, and examine the *general equation of degree two in three variables*, we will have three-dimensional conics.

$$Ax^2 + By^2 + Cz^2 + 2Dxy + 2Exz + 2Fyz + Gx + Hy + Jz + K = 0 \quad (10.2\text{-}2)$$

By suitable rotations in each plane, the cross products xy, xz, and yz can be removed (Section 6.9*) and the general form reduced to

$$Ax^2 + By^2 + Cz^2 + Gx + Hy + Jz + K = 0 \quad (10.2\text{-}3)$$

Of course, the A, B, ... have been changed from their original values when we make the rotation, but it is convenient to use the same letters.

The complete analysis of the general case (which involves the completing of the squares of the quadratic variables to gain symmetry) will be omitted, and we will examine only a few special cases. An *ellipsoid* is of the form

$$\frac{x^2}{a^2} + \frac{y^2}{b^2} + \frac{z^2}{c^2} = 1 \quad (10.2\text{-}4)$$

Notice how this extends the idea of an ellipse, just as the distance in three dimensions extended the two-dimensional distance and the sphere extends the idea of a circle (see Figure 10.2-2). In particular, when $a = b = c$, the ellipsoid will be a sphere whose radius is the common value of the three coefficients. When only two of the coefficients are equal, it is an ellipsoid of revolution about the nonequal axis (Figure 10.2-3);

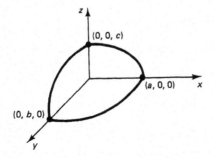

Figure 10.2-2 Ellipsoid

oblate when the equal axes are the larger, and *prolate* when they are the smaller of the axes.

For the corresponding hyperboloid, there are two cases. The *hyperboloid of one sheet* (one connected piece) occurs when there is one minus sign among the squares of the three coefficients on the left-hand side; for example,

$$\frac{x^2}{a^2} + \frac{y^2}{b^2} - \frac{z^2}{c^2} = 1 \qquad\qquad (10.2\text{-}5)$$

(see Figure 10.2-4). The *hyperboloid of two sheets* (two separate pieces) has correspondingly two minus signs,

$$\frac{x^2}{a^2} - \frac{y^2}{b^2} - \frac{z^2}{c^2} = 1 \qquad\qquad (10.2\text{-}6)$$

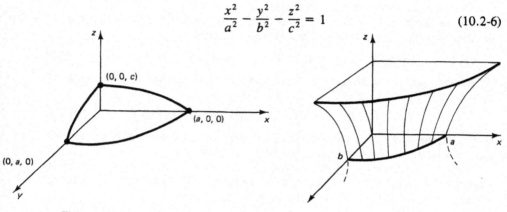

Figure 10.2-3 Prolate spheroid

Figure 10.2-4 Hyperboloid of one sheet

(see Figure 10.2-5). Three minus signs would, of course, produce an imaginary locus, since the left-hand side would then be negative (for real numbers x, y, and z, as well as real numbers a, b, and c), and the right-hand side is plus 1. Only complex (imaginary) coordinates could satisfy the equation.

In a similar way, there are paraboloids, hyperbolic paraboloids, elliptic paraboloids, and even cases when only one term is quadratic. They will occur now and then in the text as examples, but each time the meaning can be deduced as needed rather than examined now, and promptly forgotten!

Further development of the theory is now well within your abilities (chiefly completing the square on the quadratic variables and translating the coordinate system), but the number of special cases is naturally much larger (there are said to be 17 distinct cases!).

Instead, let us discuss how to draw the surfaces. Most of the time the coordinates have been chosen so that what symmetry there is occurs with respect to either the coordinate planes or coordinate axes. This is often referred to having the surface as a *central conic*. We then find the *traces* of the surface on the coordinate planes. To do this for the x–y plane, we simply set $z = 0$. The problem is thereby reduced to the

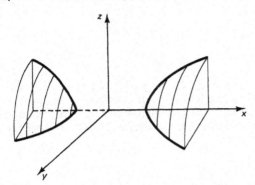

Figure 10.2-5 Hyperboloid of two sheets

two-dimensional problem of earlier chapters. Similarly, we set $y = 0$, and find the trace on the x–z plane. Finally, we set $x = 0$ and find the trace on the y–z plane. If the figure is not obvious from this, then we can try *sections* (cross sections) through the surface. For example, putting $z = 5$ gives the cross section of the surface parallel to the x–y plane at the height $z = 5$. Similar cross sections in other planes generally will fill in the surface enough for you to draw a reasonable sketch of it.

Example 10.2-1

Draw the surface

$$x^2 + y^2 - z^2 = 1$$

You immediately see that there is symmetry with respect to all three coordinate planes (replacing any variable by its negative does not change the equation; hence, if any coordinate satisfies the equation, then its negative also satisfies the equation). For $z = 0$, the trace in the x–y plane gives you a circle of radius 1. You can also see that as z gets larger the circle in the intersection plane parallel to the x–y plane grows in radius $(r^2 = 1 + z^2)$.

Next set $y = 0$. In the x–z plane you have a hyperbola with 45° asymptotes. And similarly, in the y–z plane. You are ready to draw Figure 10.2-6, which is a hyperbola of one sheet.

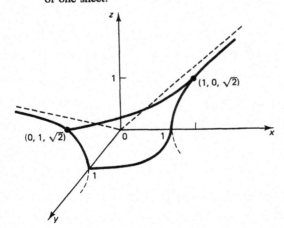

Figure 10.2-6 $x^2 + y^2 - z^2 = 1$

Example 10.2-2

Draw the surface

$$x^2 - y^2 + z = 0$$

From the squared terms you see the symmetry in the y–z and x–z planes. The trace on the x–y plane ($z = 0$) is a pair of straight lines, $y = \pm x$. The trace on the x–z plane is a parabola opening downward, while the trace on the y–z plane is a parabola opening upward. This saddle shape is hard to sketch clearly (see Figure 10.2-7). It has the general shape of a pass between two mountains.

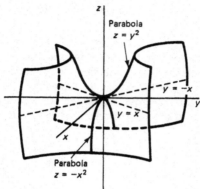

Figure 10.2-7 Hyperbolic paraboloid saddle

Example 10.2-3

Draw the surface

$$z = x^2$$

In this equation the y does not occur, so you get the same pair of (x, z) for each y. From the x^2 term you see the symmetry with respect to the y–z plane. It is a cylindrical surface along the direction of the y-axis. The trace is a parabola in the x–z plane. See Figure 10.2-8, which is a parabolic cylinder.

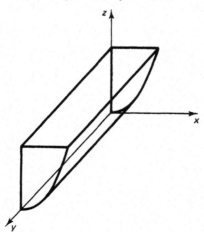

Figure 10.2-8 Cylinder $z = x^2$

EXERCISES 10.2

1. Write the equation of the sphere centered at the origin with radius 4.
2. Write the equation of the sphere with center at $(0, -3, 4)$ and radius 5.
3. Find the center and radius of $x^2 + y^2 + z^2 - 2x - 4y + 6z = 11$.
4. Find the center and radius of $x^2 + y^2 + z^2 - x + y + 3z = 0$
5. Sketch the first octant of the surface $x^2/4 + y^2 + 4z^2 = 1$.
6. Sketch $x^2 - y^2 + z^2 = 1$.
7. Sketch $x^2 + y^2 - z = 0$.
8. Find the sphere through the four points $(0, 0, 0)$, $(0, 0, 1)$, $(1, 0, 0)$, and $(0, 1, 1)$.
9. Plot $x^2/4 + y^2/9 + z^2/25 = 1$.
10. Plot $y^2 = x$.

10.3 PARTIAL DERIVATIVES

We have repeatedly used plane sections of the surface to find the nature of the surface. This amounts to assigning a fixed value for one of the variables. Suppose we fix the y variable, set $y = c$. But $y = c$ is the equation of a plane and it cuts the surface in a plane section. If we use the calculus to study the curve *in this plane*, then we will want to take derivatives. But this could be confused with the derivatives in just two variables. Thus we have to use a different notation for the letter d; we use one form of the Greek lowercase delta

$$\frac{\partial z}{\partial x}$$

when we write the symbol of the *partial derivative* of z with respect to x. Similarly, fixing the value of x, we have the partial derivative of z with respect to y.

The confusion arises in some student's minds when the choice of the fixed value, the $y = c$, is allowed to vary. Consider the function

$$z = xy^2$$

We have

$$\frac{\partial z}{\partial x} = y^2$$

$$\frac{\partial z}{\partial y} = 2xy$$

In each case you regard only the indicated variables as changing. These derivatives depend on the point where they are evaluated, the x and y coordinates, and they are the slopes in the corresponding planes parallel to the coordinate axes. Higher derivatives are similarly defined.

There are *cross* derivatives, and it is an important fact that for most functions you are apt to meet (but not for all functions) the order in which they are taken is not important.

$$\frac{\partial^2 z}{\partial x\, \partial y} = \frac{\partial^2 z}{\partial y\, \partial x} \tag{10.3-1}$$

In the above equation, $z = xy^2$, the cross partial derivative is $2y$. We will not prove this fact here, but several examples will convince you of its truth in simple cases. Pathological functions can be found for which it is not true, but for the class of functions we generally meet the cross derivatives are equal.

Example 10.3-1

For the general conic (10.2-2), there are cross-product terms in the equation; hence the cross derivatives are not identically zero. Cross derivatives can be other than 0 only when the corresponding term depends on both variables. For a term like

$$z = x^m y^n$$

we have

$$\frac{\partial^2 z}{\partial x\, \partial y} = mx^{m-1} ny^{n-1} = mnx^{m-1}y^{n-1}$$

whether we do the x or y differentiation first.

The *implicit differentiation* of Section 8.3 can now be seen in another light. Suppose you have an equation in two variables,

$$f(x, y) = 0$$

Differentiating this with respect to x, you have from basics

$$f(x + \Delta x, y + \Delta y) - f(x, y) = 0$$

But this can be written as [subtract and add the term $f(x, y + \Delta y)$]

$$\{f(x + \Delta x, y + \Delta y) - f(x, y + \Delta y)\} + \{f(x, y + \Delta y) - f(x, y)\} = 0$$

We divide this by Δx, and then take the limit as both Δx and Δy go to zero. In the first two terms, the second argument is the same, so we have a partial derivative with respect to the first variable. In the second pair, it is the first argument that is the same, so we multiply and divide by Δy and then take the limit. Thus we get

$$\frac{\partial f}{\partial x} + \frac{\partial f}{\partial y}\frac{dy}{dx} = 0$$

From this you have

$$\frac{dy}{dx} = -\frac{\partial f / \partial x}{\partial f / \partial y} \tag{10.3-2}$$

This is the same thing as you had earlier from implicit differentiation, but with a new notation.

Example 10.3-2

Using the third example in Section 8.3, we have

$$z = f(x, y) = x^2y^2 + 5y - x^4 - 5$$

Find dy/dx. The partial derivatives are

$$\frac{\partial f}{\partial x} = 2xy^2 - 4x^3$$

$$\frac{\partial f}{\partial y} = 2x^2y + 5$$

from which, using Equation (10.3-2), you have, again,

$$\frac{dy}{dx} = -\frac{2xy^2 - 4x^3}{2x^2y + 5}$$

EXERCISES 10.3

Find the first partial derivatives and the cross derivative of the following:

1. $z = x^2 + y^2$
2. $z = \sqrt{x^2 + y^2 - 1}$
3. $z = x/y$
4. $z = x^2y + y^2x$
5. $z = (x - y)/(x + y)$
6. Find the tangent plane to the sphere $x^2 + y^2 + z^2 = 1$ at the point $x = y = z = 1/\sqrt{3}$. *Hint:* Use $Ax + By + Cz + D = 0$; equate the value and the two partial derivatives of the two surfaces at the given point.

10.4 THE PRINCIPLE OF LEAST SQUARES

Up to now we have been dealing with mathematical objects having mathematical definitions; our points, lines, planes, curves, and surfaces were exact. But everyone knows that when you measure things in the real world you find that the measurements are not exact, that there is *noise* added to the *true value*, whatever "noise" and "true value" may mean! You *think* there is a true value, say x, but when you make the nth measurement, you get (ϵ is the lowercase Greek epsilon)

$$x_n = x + \epsilon_n$$

where ϵ_n is the corresponding error in the measurement. The quantity ϵ_n is called the *residual* at n. For N measurements, you have the index n running from 1 to N.

Given a number of measurements x_n, what value should you take as the *best estimate* of x? The *principle of least squares* asserts that the best estimate is the one that minimizes the sum of the squares of the differences between your estimate and the measured values. That is, you minimize the sum of all the terms

$$(x_1 - x)^2 + (x_2 - x)^2 + \cdots + (x_N - x)^2$$

as a function of x. This is exactly the same as the sum of the squares of the residuals:

$$\epsilon_1^2 + \epsilon_2^2 + \cdots + \epsilon_N^2$$

We write this more compactly as (see Section 9.9)

$$f(x) = \sum_{n=1}^{N} \epsilon_n^2 = \sum_{n=1}^{N} (x_n - x)^2$$

This compact notation carries a lot of details, and you should frequently ask yourself what the expanded version of the summation really looks like so that you will realize how much is being done with this summation notation. Being compact, it allows you to think in larger units of thought (chunks), to carry things further along, without the excessive effort of watching all the details.

This function $f(x)$ is a function of the single variable x. To find the minimum, you naturally differentiate with respect to x and set the derivative equal to zero. You get (each term in the summation differentiates the same)

$$\frac{df(x)}{dx} = -2 \sum_{n=1}^{N} (x_n - x) = 0$$

The original function was a parabola in x, so you believe that this equation gives the minimum (and you also have $d^2f/dx^2 = 2N > 0$). You have, since the sum of the terms in the parentheses of this sum can be added separately (due to the linearity of the operator Σ),

$$\sum_{n=1}^{N} x_n - \sum_{n=1}^{N} x = 0$$

The second summation is a common occurrence; the quantity x being summed over the N terms does not depend on the index n. There are N copies of x to be added, and you get for the second sum

$$Nx$$

To solve for x, divide the equation by N:

$$x = \frac{1}{N} \sum_{n=1}^{N} x_n = \frac{x_1 + x_2 + \cdots + x_N}{N}$$

which is simply the *average* of all the observations. This is also called the *mean value* of the observations. The average of all the observations is the least square estimate (according to the least squares principle).

Example 10.4-1

Given the seven observations

$$3.1, 3.3, 3.1, 3.2, 3.3, 3.1, 3.5$$

we add them to get the sum 22.6. If we now divide by the number of observations ($N = 7$), we get

$$\frac{22.6}{7} = 3.228 \ldots$$

as the least squares fit to the data.

There are a number of simple shortcuts when doing this mentally (and they are sometimes useful on a computer to save accuracy). First, note that for these data the first digit is always 3, and we can ignore it *provided* we later compensate by putting it back. Next, instead of regarding the rest of each number as a decimal, we can imagine it multiplied by 10, so we can deal with the seven integers, 1, 3, 1, 2, 3, 1, 5. The sum is 16, and the average is $\frac{16}{7} = 2.28 \ldots$. This is to be divided by 10 to compensate for making the numbers into integers, and then the initial digit 3 that we set aside is to be added. We get the same result as before:

$$3 + \left(\frac{1}{10}\right)\left(\frac{16}{7}\right) = \frac{21 + 1.6}{7} = \frac{22.6}{7}$$

Abstraction 10.4-2

The above simplification of the arithmetic is a general principle. Given N observations x_n, they can each be written as

$$x_n = a + (x_n - a) = a + \frac{1}{b}[b(x_n - a]$$

where a and b are convenient numbers; typically, b is a power of 10 and a is near the mean of the observations.

When we take the average (mean) of the N observations x_n, we get

$$\frac{1}{N}\sum_{n=1}^{N} x_n = \frac{1}{N}\sum_{n=1}^{N} a + \frac{1}{Nb}\sum_{n=1}^{N} b(x_n - a)$$

$$= a + \frac{1}{b}[\text{average of } b(x_n - a)]$$

This is an extremely useful result not only for mental arithmetic, such as computing grade averages in your head, but also in large computing machines to keep accuracy.

EXERCISES 10.4

Find the least squares estimate of the following data:

1. 5.1, 5.3, 5.2, 5.0, 4.9
2. $-3.7, -3.9, -3.5, -3.6, -3.6$
3. $-5.1, -4.7, -5.2, -4.8, -5.3, -5.3, -4.9$

10.5 LEAST SQUARES STRAIGHT LINES

Often you have N observations y_n, each observation depending on a corresponding x_n. In mathematical notation you are given the N data points

$$(x_n, y_n) \qquad (n = 1, \ldots, N)$$

The x_n values are *assumed* to be correct, and the errors are in the y_n values. From a look at a plot of some data (Figure 10.5-1), the data seem to lie on a straight line. Suppose you try to fit a straight line to the data as best you can. Using the principle of least squares, you pick that line for which the sum of the squares of the differences (residuals) between the data at x_n and the value on the line at x_n is minima. Thus the sum of the squares of the residuals, the squares of the differences between the y values on the line, and the given values at x_n is to be made minimal.

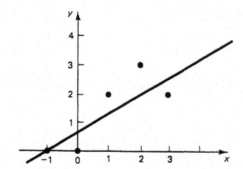

Figure 10.5-1 Least squares line

To get things in more mathematical terms, you want to fit the straight line

$$y = mx + b$$

to the N data points (x_n, y_n). The value on the line at x_n is the corresponding $y(x_n)$ value from the line; that is, $y(x_n) = mx_n + b$. Thus the difference between the observed data y_n and the value from the line $y(x_n) = mx_n + b$ is

$$y_n - y(x_n) = y_n - (mx_n + b) = \epsilon_n$$

where the error ϵ_n is the *residual* at x_n. The error is considered to be *noise* in the original measurements y_n. According to the principle of least squares, the sum of the squares of the residuals is to be minimized; that is,

$$\sum_{n=1}^{N} \epsilon_n^2 = \sum_{n=1}^{N} (y_n - mx_n - b)^2$$

This is a function of two variables m and b. This often comes as a surprise to the student: the constants of the line have become the variables! But consider the problem, "Which line is best?" The particular line you choose depends on the choice of the parameters m and b. They are the variables, and the x and y have been summed over all their values. It is necessary to belabor this point; the variables of the problem are the coefficients of the line. You are searching for the best line.

As a function of m and b, you can write it as

$$f(m, b) = \sum_{n=1}^{N} (y_n - mx_n - b)^2 \tag{10.5-1}$$

and you see that $f(m, b)$ is a paraboloid in the independent variables m and b. If you set the partial derivatives of $f(m, b)$ with respect to each of these variables (m and b) equal to 0, you will be led to the minimum. You have to differentiate each term in the sum, one by one. You get

$$\frac{\partial f}{\partial b} = 2 \sum_{n=1}^{N} (y_n - mx_n - b)(-1) = 0$$
$$\frac{\partial f}{\partial m} = 2 \sum_{n=1}^{N} (y_n - mx_n - b)(-x_n) = 0 \tag{10.5-2}$$

Since the right-hand sides are all 0, you can ignore the -2 factors that always arise. Next break up the sum of the terms of (10.5-2) into sums over each term:

$$\sum_{n=1}^{N} y_n = m \sum_{n=1}^{N} x_n + b \sum_{n=1}^{N} 1$$
$$\sum_{n=1}^{N} x_n y_n = m \sum_{n=1}^{N} x_n^2 + b \sum_{n=1}^{N} x_n \tag{10.5-3}$$

These equations are called the *normal equations* because they normally occur, and with familiarity you learn to write them down directly, if you do the least square fitting of polynomials often enough! If you don't, then you remember the derivation of the normal equations: set up the sum of the squares of the residuals; then differentiate with respect to each of the parameters of the curve you are fitting and set equal to zero, and finally rearrange the equations.

When you compute the sums, which refer to the measured data, you will have two linear equations in the unknowns m and b to be solved. When solved, you will know the values of m and b for the least squares straight line that fits the data.

Example 10.5-1

Suppose you have the data given in the x and y columns of Table 10.5.1. The choice of very simple integers allows you to follow easily what is going on. In practice, of course, the numbers can be very awkward and require at least a hand calculator to carry out the arithmetic easily. To get the sums, it is best to make a table like Table 10.5-1.

TABLE 10.5-1

n	x_n	y_n	x_n^2	$x_n y_n$	
1	-1	0	1	0	
2	0	0	0	0	
3	1	2	1	2	
4	2	3	4	6	
5	3	2	9	6	
	5	7	15	14	Sums of columns

The normal equations (10.5-3) become, when these numbers are substituted for the sums,

$$7 = 5m + 5b$$

$$14 = 15m + 5b$$

(the 5 in the $5m$ term in the top line arises because the sum of the x_n is 5, and the 5 in the $5b$ because there are exactly 5 data points; while the 5 in the bottom equation is again the sum of the x_n). To eliminate b, subtract the top equation from the bottom:

$$7 = 10m$$

or

$$m = \frac{7}{10} = 0.7$$

Next, you get from the top equation after substitution for m and some algebra,

$$b = \frac{7}{10} = 0.7$$

The equation of the least squares line fitting the given points is

$$y = \frac{7}{10}(x + 1)$$

In Figure 10.5-1 you can inspect the results where both the data and the line appear. It is only by chance that the line goes through a given point. It does look like the proper choice since any changing of position vertically, or the slope, or any combination of the two, does not appear to reduce the sum of the squares of the errors (the vertical distance between the point and the line, *not* the perpendicular distance as you might suppose). Remember what the residuals are and what you minimized!

Generalization 10.5-2

Let us generalize the method. You are given N data points (x_n, y_n) and you have a *model*, in this case the straight line ($y = mx + b$). You want the least squares fit of the model to the data. You form the sum of the squares of the residuals and set out to minimize this as a function of the parameters of the model (the coefficients of the straight line). When the coefficients appear in the model in a linear fashion, as the m and b do for the straight line, then following through the partial differentiation to find the minimum, you see that the parameters will *always* occur linearly in the normal equations. You solve these linear equations, and if you care about the result, you plot the original data and the computed model (the straight line). Finally, you *think* carefully whether or not you would be willing to act on the results.

Example 10.5-3

We could go on fitting specific data by the model of a straight line to give you more exercise; instead, we will only partially abstract the problem in another direction and suppose (a very common situation) that the data at a set of N *equally spaced values* of x_n are simply $y_1, y_2, \ldots, y_N$. The table for the sums is:

n	x_n	y_n	x_n^2	$x_n y_n$
1	1	y_1	1	y_1
2	2	y_2	4	$2y_2$
3	3	y_3	9	$3y_3$
		$\cdots$		
N	N	y_N	N^2	Ny_N

where we have assumed that the x values were at the integers. From Table 2.5-1 we know the sum of the first N integers and the sum of the squares. They are

$$\sum n = \frac{N(N+1)}{2}, \qquad \sum n^2 = \frac{N(N+1)(2N+1)}{6}$$

We have, putting the summations involving y on the right-hand side,

$$m \sum n + b \sum 1 = \sum y_n$$

$$m \sum n^2 + b \sum n = \sum n y_n$$

where we have abbreviated the notation but what is meant is clear. The sums on the left-hand side are given immediately above, so we have only to solve for m and b to have the least squares straight line fitting the given data. In any specific case, we generally take the trouble to plot the points and the computed straight line to see if we have made any arithmetic mistakes along the way. You should also see if you like the result of assuming the criterion of least squares; when the result obtained is not reasonable and pleasing, then maybe it was the criterion of least squares that was inappropriate! It is generally a good idea to plot the residuals to see if there is any strange pattern in them.

Fitting a least squares line may be regarded as trying to remove the "noise of the measurement" from the data, "smoothing the data" if you prefer those words. It rests directly on the *assumption* that the best fit is that which minimizes the sum of the squares of the errors in the measurements y_n (and which also assumes that the x measurements are exactly right). Since least squares is so widely used, we had better examine this assumption in more detail. There is a saying that scientists and engineers believe that it is a mathematical principle and that mathematicians believe that it is a physical principle. It is neither; it is just an assumption. And although it can be derived from other assumptions (as we will in Section 15.9), in the final analysis it is an assumption, and not a fact of the universe. What is a fact is that when the parameters of the model occur in a linear fashion then the resulting equations to be solved are linear in the unknown coefficients, and hence are easily solved. This is what accounts for much of the popularity of the method.

What is wrong with the least squares is easily stated: it puts too much weight on large errors. A single bad data point can greatly distort the solution. Thus, given 100 observations, a single errror of size 10 (whose square is 100) has greater effect than 99 errors each of size 1. As a result of this, it is customary to inspect the data for *outliers* before applying least squares methods (and to inspect again, via the residuals, after the fitting).

EXERCISES 10.5

1. Examine the theory of fitting a straight line when there are an odd number $2N = 1$ of points symmetrically spaced about the origin.

Find the least squares line through the following data:

2. $(0, 2)$, $(1, 1)$, $(2, 2)$, $(3, 0)$, $(4, 1)$.
3. $(-2, 2)$, $(-1, 1)$, $(0, 0)$, $(1, -1)$, $(2, -2)$
4. $(-2, y_{-2})$, $(-1, y_{-1})$, $(0, y_0)$, $(1, y_1)$, $(2, y_2)$
5. In Example 10.5-3, suppose the data were at $x = 0, 1, \ldots, N - 1$. What changes would be necessary?
6. Carry out Example 10.5-3.

10.6 n-DIMENSIONAL SPACE

In Section 10.5, the fitting of a straight line to some data takes place in two-dimensional space, the space of the coefficients of the straight line. Suppose we had tried to fit a parabola,

$$y(x) = ax^2 + bx + c$$

which has three coefficients, to some data; we would have to fit the data in the space of the three coefficients. The data (x_n, y_n) are in a two-dimensional space, but the fitting of the coefficients a, b, and c takes place in a three-dimensional space.

Now, suppose we tried to fit a polynomial having four coefficients. I say we would have to do the fitting in a four-dimensional space of the coefficients (parameters). And you see where we are headed.

As a specific example, suppose that the design of something, say an airplane engine, depended on 17 different parameters: the diameter of the rotor, the length of the blades, the number of rings of blades, and so on. The design would take place in 17-dimensional space. Of course, the engine is built in 3-dimensional space, but the design takes place in the space of the parameters, one dimension for each parameter. Do we mean a "real physical space"? No, what we mean is that you must give 17 different numbers to describe a single design. Each design is an ordered array of the 17 specific numbers. It is natural to regard this array of 17 numbers,

$$(x_1, x_2, x_3, \ldots, x_{17})$$

as a point in space. (We are now using x_m as the name of the mth variable.) We can imagine holding all but one of the variables fixed and varying that one, let it be the ninth. With the other 16 held fixed, we have the concept of the partial derivative and can imagine searching for the optimum value of x_9.

In going from three- to four-dimensional space, the student often struggles with the concept, trying to imagine, perhaps, if time is the fourth dimension. This is needless worrying and is the reason we have jumped so rapidly to 17-dimensional space. With 17 dimensions there is simply no question of asking if the space is "real" or not. Careful thinking will show you that we are being formal when we say that two variables led to two-dimensional space, that three variables is equivalent to three dimensions. The modeling of the mathematical formulation to common experience is so close that the student feels comfortable with the two- and three-dimensional mathematical models, but they are only models after all! Similarly, given 17 parameters (coefficients), we think of them as being modeled by a 17-dimensional space; each particular set of the 17 parameters is a point in the space, just as before each pair (or triple) of numbers was a point in two- or three-dimensional space. You have plotted three-dimensional figures on a two-dimensional plane, and not let that bother you very much.

Let us, for the moment, return to the example of N measurements of the same thing. We could adopt an alternative view and think of these N numbers from the measurements as a single point in N-dimensional space if we wished. This is quite a different space than the space of the *coefficients* of the line we fitted to the set of pairs of observations (x_n, y_n). The idea of an n-dimensional space is very fundamental to many mathematical ways of looking at things. It is a very useful concept, which requires gradual mastery since it can often provide a simple, familiar way of looking at a new situation that involves N numbers or M parameters. Both views lead to the concept of an n-dimensional space; one is that it is an N-dimensional space of the data, and the other that it is an M-dimensional space of the coefficients.

In the method of undetermined coefficients, we set up the criterion we want to meet, at present the least squares criterion, and find the coefficients by some means—in least squares by solving the normal equations. It is important to note that in least squares it is the linearity of the coefficients, not the terms they multiply, which assures you that the resulting normal equations will be linear in the unknowns.

The N measurements of the experiment often have very different kinds of dimensions (feet, weight, etc.), and we therefore are in a nongeometric space where rotations are meaningless. But it is convenient to keep the language of space when talking about the design. There is little meaning for slant distances between points in such a space, but there is meaning to distances between values of the same parameter. We can measure along the coordinate axes, but not diagonals (if we wish to keep much sense in what we are doing).

On the other hand, often the values in the space have a common unit of measurement, such as N measurements of the same thing. In such a space the Pythagorean distance between points can be very useful. We can say how far one set of N measurements that you may have made are from another set of N measurements of the same thing I may have made; after all, we do not expect them to coincide, and the distance between them gives some indication of the consistency between our measurements. It is dangerous, although unfortunately all too common, to use the Pythagorean distance when the units are not comparable. Least squares is related to the Pythagorean distance, but the distance function implies homogeneity of the units that are being added together to get the total distance. Sometimes a suitable scaling can make them the same. For example, in the special relativity theory the three distance units are made comparable to the time unit by using ict in place of t (c being the velocity of light and $i = \sqrt{-1}$) because the dimension of ct is (distance per unit time) $\times$ (time) = distance, and is therefore comparable with the distance units of the other three variables. The $i = \sqrt{-1}$ is not so easily explained!

The idea of n-dimensional space is simply a large generalization of our ideas of one-, two-, and three-dimensional spaces. When you think carefully about it, then the lower-dimensional spaces are seen to be no more "real" than the higher-dimensional spaces. It is just that you feel that the physical experiences from which you abstracted your ideas of the spaces are close by; they are easy for you to visualize. But the mathematical spaces of one, two, and three dimensions exist only in the mind, and as such are no more, nor less, "real" than the higher-dimensional mathematical spaces. There is an inevitability of higher-dimensional space, and you might as well get used to the idea.

Example 10.6-1

Suppose we tried to fit the special cubic (one parameter, $M = 1$)

$$y = ax^3$$

to the following five data sets, $(-2, -3)$, $(-1, 0)$, $(0, 0)$, $(1, 2)$, $(2, 4)$, using the principle of least squares. We first write the residuals

$$\epsilon_n = y_n - ax_n^3$$

and sum their squares over all the data. We have

$$\sum_{n=-2}^{2} (y_n - ax_n^3)^2 = f(a)$$

Differentiate with respect to the variable a and set the derivatives equal to zero to find the minimum:

$$\frac{df(a)}{da} = -2 \sum_{n=-2}^{2} (y_n - ax_n^3)x_n^3 = 0$$

From this we get

$$\sum y_n x_n^3 = a \sum x_n^6$$

or

$$a = \frac{\sum yx^3}{\sum x_n^6} = \frac{52}{130} = \frac{26}{65}$$

We now plot the result, $y = 26x^3/65$, in Figure 10.6-1. It looks reasonable.

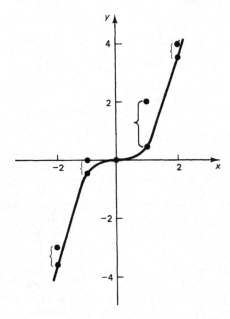

Figure 10.6-1 Least squares cubic

Example 10.6-2

To illustrate the use of a higher-dimensional space, we will fit a cubic (having four parameters, coefficients) to some symbolic data (x_n, y_n) for $n = 1, \dots, N$, using the criterion of least squares to choose the particular cubic from the family of all possible cubics. In mathematical symbols, we want to fit

$$y = c_0 + c_1 x + c_2 x^2 + c_3 x^3$$

to the data. "Least squares" means that we set up the sum of the squares of the residuals as a function of the parameters c_0, c_1, c_2, and c_3.

$$\sum \epsilon_n^2 = F(c_0, c_1, c_2, c_3)$$

But ϵ_n is the difference between the observed data y_n and the corresponding value $y(x_n)$ of the cubic (the model). Hence

$$F(c_0, c_1, c_2, c_3) = \sum [y_n - (c_0 + c_1 x_n + c_2 x_n^2 + c_3 x_n^3)]^2$$

where we have clearly shown that the variables are the coefficients of the equation. To find the minimum, we differentiate $F(c_0, c_1, c_2, c_3)$ with respect to each parameter in turn, equate each to zero, divide out the -2, and finally transpose things to get the normal equations (c_0, then c_1, etc.). To avoid a lot of writing, we use the notation

$$S_k = \sum x_n^k = \sum_{n=1}^{N} x_n^k$$

The normal equations are, in this notation,

$$S_0 c_0 + S_1 c_1 + S_2 c_2 + S_3 c_3 = \sum y$$

$$S_1 c_0 + S_2 c_1 + S_3 c_2 + S_4 c_3 = \sum xy$$

$$S_2 c_0 + S_3 c_1 + S_4 c_2 + S_5 c_3 = \sum x^2 y$$

$$S_3 c_0 + S_4 c_1 + S_5 c_2 + S_6 c_3 = \sum x^3 y$$

We need to make a table (corresponding to Table 10.5-1) of values, but with many more columns. We need the sums of the powers of the observed x_n up through the sixth power, and the sums of the data multiplied by the powers of x up through the third power. With even a small programmable hand calculator (let alone a large computer), this is not a great effort. Once they are substituted into the equations, we have four equations in four variables to solve. Again, from generalization 10.1-2, it is tedious, but not difficult, to do.

It is important that you understand what is going on; then when you face a practical problem you will know what to do even though it may take time. Normally, it takes many hours to gather some data, at times even years, and the effort to fit a least squares *model* to the data is trivial in comparison. Data gathering is expensive and data processing is relatively cheap. Indeed, routines for curve fitting of the above type are in standard statistical packages available on most computers. You know what the routine should be doing. If you take the trouble to plot the results (and the residuals) and examine it with a jaundiced eye, then you can catch most errors in either the way you used the statistical package or else in *your* application of the criterion of least squares. The final criterion is simply, "Are the results reasonable?" If not, why not? Is it the data, an error in arithmetic or algebra, or is it the peculiarity of the least squares method that puts great emphasis on large errors?

Example 10.6-3

To make sure we understand the general case, we will fit a cubic ($M = 4$) to nice integer data ($N = 7$). The equation is

$$y = c_0 + c_1 x + c_2 x^2 + c_3 x^3$$

The data are in the first two columns of Table 10.6-1.

TABLE 10-6.1

x	y	x^2	x^3	x^4	x^5	x^6	xy	$x^2 y$	$x^3 y$
-3	1	9	-27	81	-243	729	-3	9	-27
-2	1	4	-8	16	-32	64	-2	4	-8
-1	2	1	-1	1	-1	1	-2	2	-2
0	1	0	0	0	0	0	0	0	0
1	0	1	1	1	1	1	0	0	0
2	1	4	8	16	32	64	2	4	8
3	0	9	27	81	243	729	0	0	0
0	6	28	0	196	0	1588	-5	19	-29

The normal equations are, therefore,

$$7C_0 + 0C_1 + 28C_2 + 0C_3 = 6$$

$$0C_0 + 28C_1 + 0C_2 + 196C_3 = -5$$

$$28C_0 + 0C_1 + 196C_2 + 0C_3 = 19$$

$$0C_0 + 196C_1 + 0C_2 + 1588C_3 = -29$$

Due to the symmetry of the data points x_i, these equations fall into two systems of two equations each. The first and third equations are one pair:

$$7c_0 + 28c_2 = 6$$

$$28c_0 + 196c_2 = 19$$

To solve them, multiply the top equation by -4 and add to the bottom equation to get

$$(196 - 112)c_2 = 19 - 24$$

$$84c_2 = -5 \quad \text{or} \quad c_2 = \frac{-5}{84}$$

From the top of the pair, we have

$$7c_0 = 6 - 28\left(\frac{-5}{84}\right) = 6 + \frac{5}{3} = \frac{23}{3}$$

$$c_0 = \frac{23}{21}$$

Similarly, we solve the second pair of equations, the second and fourth equations.

$$28c_1 + 196c_3 = -5$$

$$196c_1 + 1558c_3 = -29$$

Multiply the top of these two equations by -7 and then add to the lower equation to get

$$c_3 = \frac{1}{36} \quad \text{and} \quad c_1 = \frac{-47}{126}$$

The solution is, therefore,

$$y = \frac{23}{21} - \frac{47}{126}x - \frac{5}{84}x^2 + \frac{1}{36}x^3$$

$$= \frac{1}{252}[276 - 94x - 15x^2 + 7x^3]$$

and is shown in Figure 10.6-2.

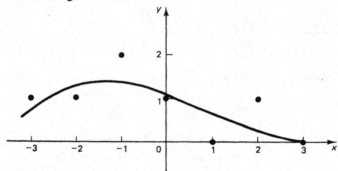

Figure 10.6-2   Cubic fit

Abstraction 10.6-4

Linear Algebra. The purpose of this abstraction is to show, using some conventional mathematical ways of thinking, that the normal equations that arise in polynomial fitting can always be solved in principle because they are linearly independent.

The method for proving that the normal equations have a solution is to show that the corresponding *homogeneous* set of equations (i.e., with the right-hand sides set equal to zero), have no solution other than all 0's. We study, therefore, the homogeneous equations in $M + 1$ variables c_i, with $S_m = \Sigma x_k^m$

$$S_0c_0 + S_1c_1 + \cdots + S_Mc_M = 0$$

$$S_1c_0 + S_2c_1 + \cdots + S_{M+1}c_M = 0$$

$$\cdots$$

$$S_Mc_0 + S_{M+1}c_1 + \cdots + S_{2M}c_M = 0$$

as the set of equations to be solved. Multiply the top equation by c_0, the second by $c_1, \ldots$, the last by c_M, and add. You will get the squares of each c_i plus twice the cross products $c_i c_j$. When you follow out the details and write the S_i as sums, you will find that you have

$$\sum_{n=1}^{N} \{c_0 + x_n c_1 + \cdots + x_n^M c_M\}^2 = 0$$

The quantity in the braces is a polynomial of degree M, say $P_M(x)$. This polynomial of degree M is evaluated at $N > M + 1$ points (more points than parameters or else it is not least squares). If the sum of the squares is to be zero, then each term of the sum must be zero since each term is nonnegative. Therefore, the expression in the braces must be zero for all N values of n. But a polynomial of degree M can only vanish at M distinct points. Said in other words, the powers of x are linearly independent; hence no linear relationship can exist among the terms other than with all zero coefficients. Therefore, the polynomial is identically zero; all its coefficients are zero. Thus the system of homogeneous equations has only the all zero solution.

We next apply the following general theorem, which we will prove shortly (note the change in notation):

Theorem.   If n homogeneous equations

$$\sum_{j=1}^{n} a_{i,j} x_j = 0 \qquad (i = 1, \ldots, n)$$

have only the solution of all zeros, then the nonhomogeneous equations

$$\sum_{j=1}^{n} a_{i,j} x_j = b_i \quad (i = 1, \ldots, n)$$

have a single solution.

From this theorem it follows that the *complete system of normal equations* as originally given has a unique solution, and the system of equations can be solved easily (in principle). There can be no trouble because the process cannot lead to multiple or inconsistent solutions. (Note that there can be roundoff trouble, however.)

It remains to prove this important theorem. When we passed from the n original nonhomogeneous equations to the n homogeneous equations, we moved each plane (corresponding to its equation) parallel to itself so that it finally passed through the origin. (The partial derivatives on each plane remained the same.) The theorem assumes that these homogeneous equations have only the single solution, which is the origin. We have now to get back to the solution of the original nonhomogeneous equations.

We first observe that the original n equations could have no solution, one solution, or infinitely many solutions. Suppose for the moment that the original nonhomogeneous equations had two *different* solutions; call them Y and Z. Each capital letter is the whole set of n numbers that constitutes the solution. Then the difference

$$U = Y - Z$$

(this is a symbolic equation indicating differences of solution values taken term by term, $u_i = y_i - z_i$, $i = 1, \ldots, n$) will satisfy the homogeneous equations. To see this, we merely take any equation of the original set, say the first one,

$$a_{1,1} x_1 + a_{1,2} x_2 + \cdots + a_{1,M} x_M = b_1$$

put the Y values into this equation, then put the Z values into this same equation, and finally subtract. You see the U values on the left side and the cancellation of the b_1 on the right, leaving a homogeneous equation with the same coefficients. Since the homogeneous equations were assumed to have only the all zero solution, it follows that the nonhomogeneous equations cannot have two distinct solutions.

It remains to show that the nonhomogeneous equations have exactly one solution when the homogeneous equations have a single solution (which must be all zeros). Each equation defines a plane (sometimes called *hyperplane*). Consider one of these n equations, say the last one. The other $n - 1$ planes intersect in a line. This line *pierces* (intersects) the last plane in one point by the assumption that there is a single unique solution. Thus the $n - 1$ equations cannot have more than a line in common or else the intersection with the last plane would be more than a point, or else no points at all (in case the line is parallel to the chosen plane). Now we move the last plane parallel to itself back to its original position. The unique intersection will persist, although the values of the coordinates may change.

In a similar fashion, we move each plane back to its original position. In each case the movement may change the coordinates of the intersection, but cannot destroy the fact of a unique intersection.

For those who prefer algebraic to geometric arguments, we consider the same question in terms of the Gauss elimination (Generalization 10.1-2) applied to these equations. We will come down to a triangular system if all goes right, and if anything goes wrong and we do not get the triangular system, then the equations are either inconsistent (no solution) or else redundant (many solutions). But the Gauss reduction method operates *only* on the coefficients on the left-hand side of the equations and carries the corresponding right-hand sides along. The assumption that the homogeneous equations have a unique solution indicates that all the stages of the elimination went along properly and nothing went wrong. Supplying the appropriate right-hand sides in the triangular system, we get a unique back solution to the original system of equations. Thus we see that again the algebra and geometry agree, that each illuminates the other.

EXERCISES 10.6

1. Fit a quadratic $y = ax^2$ to the data of Example 10.6-1.
2. Write out the normal equations for fitting a fifth-degree equation as in Example 10.6-2.
3. Outline the proof that you can always solve the normal equations in the case of polynomial fitting.

10.7 TEST FOR MINIMA

We have been assuming that a minimum was found by equating the $M + 1$ partial derivatives to zero (corresponding to the $M + 1$ parameters $c_0, c_1, \ldots, c_M$). Since we were forming a quadratic expression in the parameters (which is the surface in the $M + 1$ dimensions we are exploring for a minimum), we see that there is a unique minimum, and that at this minimum point the partial derivatives must be all zero (or else we could move in some direction to decrease the value of the sum of the squares of the residuals). We are effectively asserting that for this surface the minimum occurs where for each plane section *with all other variables held fixed* there is a minimum in this plane. Thus to find the minimum we compute the partial derivative in the plane and equate the partial derivatives to zero to define the unique minimum. This is true in this special case, but you should be wary of this approach in general, as the following classic example shows.

Example 10.7-1

Consider the simple z surface with two independent variables x and y:

$$z = (y - x^2)(y - 2x^2) = y^2 - 3x^2y + 2x^4$$

Let us look at a plane section made by a plane through the z-axis. Such planes have the form

$$y = cx$$

On this plane we have

$$z = c^2x^2 - 3cx^3 + 2x^4$$

and we have

$$\frac{dz}{dx} = 2c^2x - 9cx^2 + 8x^3 = 0$$

at the origin. The second derivative at $x = 0$ is

$$\frac{d^2z}{dx^2} = 2c^2 \geq 0$$

Thus, except when $c = 0$, this is a guaranteed minimum on this plane. For $c = 0$, we have the cross section in the plane

$$z = 2x^4$$

and on this plane we know that the origin is a local minimum,. Finally, we take the plane that goes along the y-axis ($x = 0$). We have

$$z = y^2$$

and again it is a local minimum. To summarize at this point, *every* plane section containing the z-axis gives a local minimum. Yet when we look at the picture from above (Figure 10.7-1), we see the x–y plane and the two curves along which $z = 0$; that is,

$$y = x^2 \quad \text{and} \quad y = 2x^2$$

We also see that between the two parabolas the function must take on negative values! Note that the factor $y - x^2$ is positive and $y - 2x^2$ is negative in the curved region between the two parabolas. The origin is not a local minimum; there are negative values of the function in the immediate neighborhood, no matter how small, of the origin.

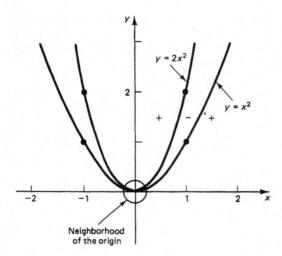

Figure 10.7-1

Thus we see that plausible arguments are not always sufficient. It seems to be so reasonable that: "If for every possible plane section through the z-axis we have a local minimum, then it ought to be a minimum for the function." Yet it is not. We are not in a position to give a rigorous mathematical proof that the minima we find by least squares are the true minima, but the nature of the surface, quadratic in each parameter, makes one feel relatively safe, even in the face of the above example.

10.8 GENERAL CASE OF LEAST-SQUARES FITTING

An examination of what we have done, using least squares fitting of polynomials, shows that only at few places in the derivation did we depend on the fact that the functions we were using to approximate the data were polynomials. The method depended rather on the fact that the unknown parameters occurred linearly in the formula.

Example 10.8-1

Suppose we want to fit some data by a linear combination of the functions $(M = 2)$ x and $1/x$. The original data $(N = 8)$ and the working computations are shown in Table 10.8-1. The numbers are simple, the curve is crude, and normally given noisy data you have at least ten times the number of data points as you have coefficients to be fitted. All of this is neglected to produce an example where you can compute the numbers easily.

We begin with the formula we want:

$$y = Ax + \frac{B}{x}$$

Suppose we have the curve; then the errors (difference between the given y_n and the value from the fitted line $Ax_n + B/x_n$) are the residuals,

TABLE 10.8-1

n	x	y	xy	y/x	x^2	$1/x^2$	$\dfrac{Ax + B}{x}$	ϵ
1	0.5	3	1.5	6.000	0.25	4.000	2.80	0.20
2	1.0	2	2.0	2.000	1.00	1.000	1.80	0.20
3	1.5	1	1.5	0.667	2.25	0.444	1.64	−0.64
4	2.0	1	2.0	0.500	4.00	0.250	1.69	−0.69
5	2.5	2	5.0	0.800	6.25	0.160	1.82	0.18
6	3.0	2	6.0	0.667	9.00	0.111	2.00	0.00
7	3.5	2	7.0	0.571	12.25	0.082	2.20	−0.20
8	4.0	3	12.0	0.750	16.00	0.063	2.42	0.58
	18.0	16	37.0	11.955	51.00	6.110	← Column totals	

$$\epsilon_n = y_n - \left(Ax_n + \frac{B}{x_n}\right)$$

The principle of least squares says that we should minimize the sum of the squares of these residuals (summed over all the given data); that is, we minimize

$$f(A, B) = \sum_{n=1}^{8} \left\{y_n - \left(Ax_n + \frac{B}{x_n}\right)\right\}^2$$

To do this, we differentiate with respect to each of the variables (A and B) and set the corresponding derivatives equal to zero.

$$2 \sum \left\{y_n - \left(Ax_n + \frac{B}{x_n}\right)\right\}\{-x_n\} = 0$$

$$2 \sum \left\{y_n - \left(Ax_n + \frac{B}{x_n}\right)\right\}\left\{\frac{-1}{x_n}\right\} = 0$$

After a slight rearrangement, we get

$$A \sum x_n^2 + B \sum 1 = \sum y_n x_n$$

$$A \sum 1 + B \sum 1/x_2^n = \sum y_n/x_n$$

The two coefficients $\sum 1$ arise because $x(1/x) = 1$. From the Table 10.8-1, we have all the coefficients, when we remember that $\sum 1 = 8$ in this case because there are eight data points.

These are the normal equations for this problem, and after substituting from the totals of the appropriate columns we have

$$51A + 8B = 37$$

$$8A + 6.11B = 11.955$$

To eliminate A, multiply the top equation by -8 and the lower by 51 and then add the two equations. As a result we have

$$B = 1.267 \quad \text{and} \quad A = 0.527$$

We use these to compute the values on the curve as well as the residuals ϵ_n in the last column. Figure 10.8-1 shows both the original data and the computed least squares curve. The fit is, of course, crude as you would expect from the crudeness of the data and having only two coefficients to use in fitting the data.

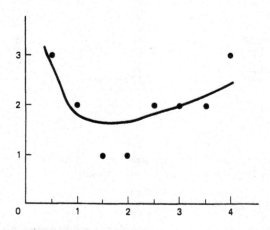

Figure 10.8-1 $\quad y = 0.527x + \dfrac{1.267}{x}$

EXERCISES 10.8

1. Write the normal equations for fitting N data points (x_n, y_n) by three functions $u_1(x)$, $u_2(x)$, and $u_3(x)$.
2. Generalize to fitting the data by M functions $u_m(x)$, $m = 1, 2, \ldots, M$.

10.9 SUMMARY

The first idea we introduced is that of a function of several variables, and along with this comes the idea of an n-dimensional space. This is the space in which the set of n values are imagined as a point, and the corresponding equations are surfaces (hypersurfaces if you prefer that word). We looked briefly at both linear and quadratic surfaces in three dimensions as natural extensions of what we earlier did in two dimensions. The surfaces studied are useful later.

Then we introduced the central idea of *partial derivatives*, which are simply derivatives of one variable with respect to another, when all the other variables are held constant. We found that implicit differentiation had a new look in this notation. The general theory of partial differentiation (say in thermodynamics) is very complex, and we have looked only at the simplest parts.

The principle of least squares was introduced, and its virtues (easy to solve such problems) as well as its main fault, too much emphasis on an outlier (error?), were discussed. We fitted a point to a set of data (measured on the same thing) and found that the least squares fit was simply the mean of the values. We then approached fitting straight lines, and later polynomials, to given data, and then finally general functions to sets of data. As long as the coefficients to be determined enter linearly into the formula, and the functions you are combining are linearly independent, then in principle the least squares solution can be found. The emphasis on "in principle" is because when there are many functions the solution may not be well determined, in the sense that small changes in the input numbers will produce large changes in the output (coefficients).

But do not confuse the quality of the least squares fit, which is measured by the sum of the squares of the residuals, with the uncertainty in the coefficients you are determining. It can often happen that the quality of the fit of the curve to the data is good, but that the particular coefficients are ill determined by the data. You merely have a poorly designed experiment *if* you wanted to determine the coefficients accurately.

The principle of least squares is widely used in many fields to fit data. It is often regarded as "smoothing the data" in some sense, eliminating the "noise of the measurements." The ease of solution (we are not speaking of the arithmetic labor) makes it very popular. Furthermore, as you will see in later chapters, there is a good deal of theory connected with least squares, and the background of theory gives a better basis for understanding the results than methods that lack the corresponding rich background.

11

Integration

11.1 *HISTORY*

The integral calculus was actually developed before the differential calculus. It arose from the problem of finding areas and volumes defined by mathematical expressions. In the early stages each particular area or volume problem was solved by a suitable trick. The great progress occurred when systematic methods were introduced and the *fundamental theorem of the calculus* was discovered. This theorem says that differentiation and integration are inverse processes one to the other (much as multiplication and division are inverse processes).

Experience seems to show that when presenting the calculus for the first time it is preferable to put the differential calculus before the integral calculus (which is what we have done). Nevertheless, mathematically it is the other way around; the rigorous approach is easiest from the integration side. We will base our approach to the integral calculus on the idea of area, and then extend, *generalize* if you prefer, the idea to a broader context. It is necessary, therefore, to first review your beliefs about area.

11.2 *AREA*

You emerge from elementary Euclidean geometry with the belief that the area of a square is proportional to the square of the side. For example, you can see directly that a square 10 by 10 is composed of 100 congruent unit squares (Figure 11.2-1). It is conventional to take the constant of proportionality as being 1. Hence, the measure of the area of a square is the square of the side.

Next, you believe that the area of a finite sum of nonoverlapping areas is the sum of the individual areas, in short that area is *additive* for finite sums of areas. Along the way you also agree that area is to be measured by a positive number, with 0 being the area of "nothing."

You also believe that congruent figures must have the same area; you would be logically embarrassed if you did not, since congruent figures are interchangeable as far as area is concerned.

Finally, and this is an assumption, too, you probably believe that every figure has a unique area. This means, for example, that you are assured, when you compute it, that the area of a triangle is independent of which side you pick for the base.

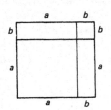

Figure 11.2-1
A 10 by 10 square

Figure 11.2-2
Square $(a + b)^2$

Looking at Figure 11.2-2, you can see that the algebraic identity

$$A = (a + b)^2 = a^2 + 2ab + b^2$$

corresponds to a geometric identity. This identity says that the whole square is the sum of the two squares (of area a^2 and b^2), plus the two, congruent rectangles. The two congruent rectangles have necessarily the same area. From this you deduce that the area of a single rectangle of sides a and b must be

$$A = ab$$

The diagonal of a rectangle (Figure 11.2-3) divides it into two congruent triangles, and therefore each of these right triangles must have the area

$$A = \frac{ab}{2}$$

Now any triangle, right or otherwise, can have a perpendicular dropped from a vertex to the base, which cuts the given triangle into two right triangles (Figure 11.2-4). You can either make sure that the vertex chosen is the largest angle, or else

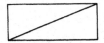
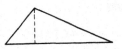

Figure 11.2-3 Rectangle as two triangles

Figure 11.2-4 Triangle

discuss the fact that, if the perpendicular (the altitude) falls outside the triangle then you are talking about the difference between the areas of the two triangles. Either way, from the assumptions you come fairly directly to the formula for the area of a general triangle:

$$A = \frac{(\text{base})(\text{altitude})}{2}$$

Next, any plane closed figure that does not cross itself and is bounded by a finite number of straight lines may be decomposed into triangles (Figure 11.2-5). At any convex angle (less than 180° as seen from the inside of the region, and there is always at least one such angle), you can, as in Figure 11.2-6, reduce the number of sides by 1 when you draw the third side of the triangle since what is left to triangulate has one less side. The triangulation is not unique (but you can see that it can always be done). Since you assumed that the area is unique, it does not matter just how the triangle decomposition is done; the resulting area will be the same for a given figure (again, having a finite number of sides).

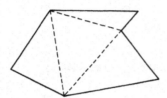   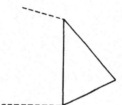

Figure 11.2-5   **Figure 11.2-6**

It is when you face the area of a circle, for example, that you need to think. It soon becomes evident that some new definition, an *extension* of the old definition, of area *must be made* for regions bounded by other than straight lines. No matter how you choose to extend the definition, *it should be consistent with the earlier definition*. In short, it is necessary to make an extension of area to new shapes, a standard situation in creating new mathematics.

EXERCISE 11.2

1. Draw a right triangle and drop a perpendicular from the right angle to the hypotenuse. This divides the original triangle T_1 into two triangles T_2 and T_3. The three triangles are similar. Derive Pythagoras' theorem.

11.3 THE AREA OF A CIRCLE

The simplest figure that has curved sides is the circle. While Euclid's *Elements* proves that the areas of circles are to each other as the squares of their respective sides, nowhere does he discuss the number π; nowhere does he get the measure of the area.

The historical approach to finding the area of a circle is to inscribe (inscribe means draw inside) a regular polygon in the circle and then compute the area of the polygon. This area is taken to be a lower bound on the area of the circle. Archimedes by a clever device doubled the number of sides of the inscribed polygon (see Figure 11.3-1). Then he again doubled the number of sides, and so on. He then took the *limit* of the area of the sequence of inscribed polygons to be the area of the circle.

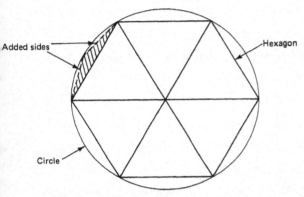

Figure 11.3-1

There were endless arguments as to whether the polygon became the circle in the limit or not. There was confusion between (1) the area of the inscribed polygons approaching the area of the circle, and (2) various properties of the perimeter (straight line segments) approaching a constantly curving circle. We have learned to avoid such questions and confine ourselves to the question, "Does the *area* of the inscribed polygons get arbitrarily close to the area of the circle?"

But we are not as sure as we might be when we merely use this approach of inscribing regular polygons; instead, we propose to *also* circumscribe (circum-, around) a regular polygon around the outside of the circle (Figure 11.3-2) and again let the number of sides be doubled indefinitely. If the two areas, that of the inscribed

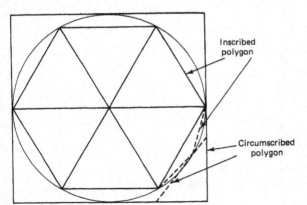

Figure 11.3-2

and that of the circumscribed polygons, approach *the same number*, then it seems reasonable to take this common number as the area of the circle. Note that we are actually *defining* what we mean by the area of a figure with curved sides.

This approach is a tedious piece of special algebra and trigonometry to carry out for a circle. What is important is both the plan for finding a common limit and the realization that a new definition is required to deal with areas bounded by curved lines. Instead of finding the area of a circle, we will start finding the areas under the parabolas (a generalization of $y = x^2$),

$$y = cx^k \qquad (k > 0)$$

and later (Examples 11.9-4 and 16.7-1) find the area of a circle in a simple fashion.

11.4 AREAS OF PARABOLAS

Example 11.4-1

Given the family of parabolas

$$y = cx^k \qquad\qquad (11.4\text{-}1)$$

the simplest case, $k = 1$, is a straight line. This forms a triangle whose area we already know (we can use this case as a check on what we are doing). We have

$$y = cx$$

To estimate the area under the curve, above the x-axis, and to the left of the vertical line $x = a$, we first inscribe (see Figure 11.4-1) a sequence of narrow rectangles, and then examine the limit as the number of rectangles approaches infinity (gets arbitrarily large). Take the width of each rectangle to be

$$\Delta x = \frac{a}{N}$$

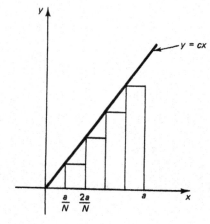

Figure 11.4-1

where N = the number of rectangles. From the figure we have the inscribed area $A_I(N)$ [$A_I(N)$ depends on N, of course].

$$A_I(N) = \Delta x \left[c\frac{0}{N} + c\frac{a}{N} + c\frac{2a}{N} + \cdot\cdot\cdot + c\frac{(N-1)a}{N} \right]$$

But this can be simplified by factoring ac/N from each term:

$$A_I(N) = \Delta x \left[\frac{ac}{N}(0 + 1 + 2 + 3 + \cdot\cdot\cdot(N-1)) \right] \qquad (11.4\text{-}2)$$

We know the sum of the consecutive integers from Section 2.3 ($N - 1 = n$ in the formula), and we also can eliminate the $\Delta x = a/N$. We get the area of the inscribed polygons:

$$A_I(N) = \frac{a}{N}\left(\frac{ac}{N}\right)\frac{(N-1)N}{2} \qquad (11.4\text{-}3)$$

Rearranging this, we have

$$A_I(N) = \frac{a^2 c}{2}\frac{N-1}{N} = \frac{a^2 c}{2}\left[1 - \frac{1}{N}\right]$$

and this approaches the limit A_I:

$$A_I = \frac{a^2 c}{2} = \frac{a(ac)}{2} \qquad (11.4\text{-}4)$$

as N approaches infinity. We recognize that this is the correct answer for a triangle. But let us persevere in our approach and compute the area of the circumscribed set of rectangles.

For the circumscribed set of rectangles, we get exactly the same thing (see Figure 11.4-2), *except* that the sequence of rectangles begins with the height proportional to 1 and goes on to height proportional to N, instead of beginning with the 0 and going to $N - 1$. We have, therefore [compare with Equation (11.4-2)], the circumscribed area

$$A_C(N) = \Delta x \left[\frac{ac}{N}(1 + 2 + 3 + \cdot\cdot\cdot + N) \right] \qquad (11.4\text{-}5)$$

and the corresponding sum is [compare with Equation (11.4-3)]

$$A_C(N) = \frac{a}{N}\left(\frac{ac}{N}\right)\frac{N(N+1)}{2}$$

Rearranging this, we have

$$A_C(N) = \frac{a^2 c}{2}\left[\frac{N+1}{N}\right] = \frac{a^2 c}{2}\left[1 + \frac{1}{N}\right] \qquad (11.4\text{-}6)$$

which again approaches (11.4-4):

$$A_C = \frac{a^2 c}{2}$$

Thus the limit of both the inscribed and the circumscribed areas is the same. We have tested the approach on a known result and found that it works correctly.

When you consider the difference between the inscribed and circumscribed sums, (11.4-5) and (11.4-6), you see that it is

$$A_C(N) - A_I(N) = \frac{a^2c}{2}\left[\frac{2}{N}\right] = \frac{a^2c}{N}$$

and that as $N \to \infty$ the difference must approach zero.

Example 11.4-2

The next case is $k = 2$,

$$y = x^2$$

We have chosen to ignore the front constant c (since from Example 11.4-1 you see that it will simply factor out of everything, and remain in front of the final answer). From Figure 11.4-3 you have the inscribed area under the curve, above the x-axis, and to the left of the line $x = a$. The area is

$$A_I(N) = \Delta x\left[\left(\frac{0a}{N}\right)^2 + \left(\frac{1a}{N}\right)^2 + \cdots + \left(\frac{(N-1)a}{N}\right)^2\right] \qquad (11.4\text{-}7)$$

After factoring out the common factors and replacing Δx by its proper value (a/N) in terms of N, you have [compare with (11.4-2)]

$$A_I(N) = \frac{a}{N}\left(\frac{a^2}{N^2}\right)[0^2 + 1^2 + 2^2 + 3^2 + \cdots + (N-1)^2]$$

The circumscribed area is correspondingly [compare with (11.4-5)]

$$A_C(N) = \frac{a}{N}\left(\frac{a^2}{N^2}\right)[1^2 + 2^2 + 3^2 + \cdots + N^2] \qquad (11.4\text{-}8)$$

(the difference being again the end terms). For the inscribed area, from Example 2.3-2, we have, on substituting for the sum of the squares of the consecutive integers $(n = N - 1)$,

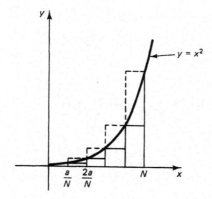

Figure 11.4-2

$$A_I(N) = \frac{a^3}{N^3}\left[\frac{(N-1)N(2N-1)}{6}\right] = \frac{a^3}{6}\left[\left(1-\frac{1}{N}\right)\left(2-\frac{1}{N}\right)\right]$$

and for the circumscribed area ($n = N$),

$$A_C(N) = \frac{a^3}{N^3}\left[\frac{N(N+1)(2N+1)}{6}\right] = \frac{a^3}{6}\left[\left(1+\frac{1}{N}\right)\left(2+\frac{1}{N}\right)\right]$$

It is easy to see that as $N \to \infty$ both expressions approach

$$\frac{a^3}{3}$$

Thus we take $a^3/3$ as the area under the second-degree parabola. From these two examples (plus just plain thinking), we are inclined to accept the definition of the area as being this common limit of the areas of the inscribed and circumscribed polygons.

Generalization 11.4-3

We naturally turn next to the general case of

$$y = x^k \qquad (k = \text{positive integer})$$

(see Figure 11.4-4). Again, we will find the area under the curve, above the x-axis, and to the left of the vertical line $x = a$. We proceed slowly. For the inscribed area, we will get

$$A_I(N) = \Delta x\left[\left(\frac{0a}{N}\right)^k + \left(\frac{1a}{n}\right)^k + \cdots + \left(\frac{(N-1)a}{N}\right)^k\right]$$

and, factoring out the common quantities while also replacing Δx by a/N, we get for the inscribed area [compare with (11.4-2) and (11.4-7)]

$$A_I(N) = \frac{a^{k+1}}{N^{k+1}}[0^k + 1^k + 2^k + \cdots + (N-1)^k]$$

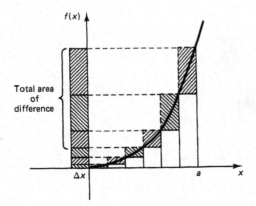

Figure 11.4-3

Using the sum of the kth powers of the consecutive integers from Generalization 2.5-2, we have

$$A_I(N) = \frac{a^{k+1}}{N^{k+1}}\left[\frac{(N-1)^{k+1}}{k+1} + \frac{(N-1)^k}{2} + \cdots\right]$$

which is a polynomial of degree $k + 1$ in $N - 1$ ($N =$ the number of rectangles). We rearrange this:

$$A_I(N) = a^{k+1}\left[\frac{(1 - 1/N)^{k+1}}{k+1} + \frac{(1 - 1/N)^k}{2N} + \cdots\right]$$

As N goes to infinity, all the terms in the bracket except the first approach zero, and we get the limit,

$$A_I = \frac{a^{k+1}}{k+1} \tag{11.4-9}$$

For the circumscribed sum, we will have one more term in the sum of the kth powers. This approximation produces an N in place of $N - 1$ in the sum:

$$A_C(N) = \frac{a^{k+1}}{N^{k+1}}\left[\frac{N^{k+1}}{k+1} + \frac{N^k}{2} + \cdots\right]$$

but otherwise it is the same. In the limit the result is the same. Thus the circumscribed polygons and inscribed polygons both lead to the same limit (of the two approximate areas):

$$A_C = \frac{a^{k+1}}{k+1}$$

Looked at in another way (Figure 11.4-4), the *difference* between the areas of the sums for the circumscribed rectangles and the inscribed rectangles is the single term

$$A_C(N) - A_I(N) = \frac{a^{k+1}N^k}{N^{k+1}} = \frac{a^{k+1}}{N}$$

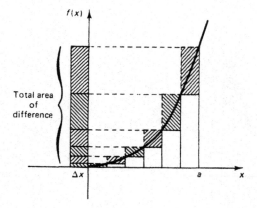

Figure 11.4-4

Therefore, as N approaches infinity, the difference between the circumscribed and inscribed areas must approach 0; both the upper and the lower sums *must* approach the same limit, and thus they serve to define the area under the curve $y = x^k$ out to $x = a$. These are often called the *Riemann* (1826–1866) *upper and lower sums*.

EXERCISES 11.4

1. Carry out the details for the function $y = x^3$.
2. Carry out the details for the function $y = x^4$.

1.5 AREAS IN GENERAL

We now turn to the case of a general function and the area under it (assuming that the function is above the x-axis). Thus we are given a function

$$y = f(x)$$

We assume that this is composed of a finite number of pieces each of which is monotone increasing or monotone decreasing (not strictly monotone, but only monotone). See Figure 11.5-1. These are the famous Dirichlet (1805–1859) conditions. Since they permit discontinuities and other reasonable behavior, we are allowing a broad enough class of functions to meet most elementary needs when modeling the real world.

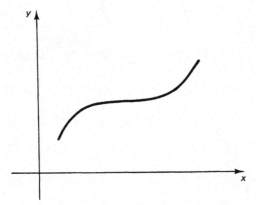

Figure 11.5-1 Monotone increasing function

We consider the area under one piece of this function, and argue that if we can find this area then, because the area is additive for any finite number of pieces, we can find the area under the whole function. Suppose we begin at $x = a$ and go to $x = b$, where $a < b$. Figure 11.5-2 shows one piece of the function. For convenience, we

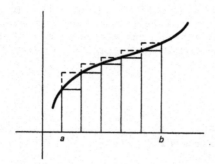

Figure 11.5-2

assume that in this interval $f(x)$ is monotone increasing. The changes for a monotone decreasing function are trivial.

The difference between the upper sum $A_U(N)$ and the lower sum $A_L(N)$ for equally spaced rectangles will be (Figure 11.5-3 shows these differences projected into a single column on the left)

$$A_U(N) - A_L(N) = [f(b) - f(a)]\Delta x$$

But

$$\Delta x = \frac{b - a}{N}$$

Therefore, the difference between the upper and lower sums is

$$A_U(N) - A_L(N) = [f(b) - f(a)]\frac{b - a}{N}$$

which must approach zero in the limit as $N \to \infty$. Thus the method of upper and lower sums defines a common limit to associate with the concept of the area under the continuous monotone curve $y = f(x)$ between the two limits a and b.

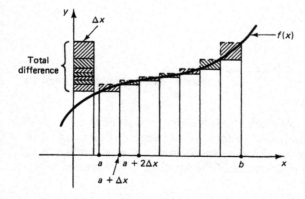

Figure 11.5-3

To *generalize* the idea of the approximating upper and lower sums of a monotone continuous function, we see first that we need not require that all the rectangles have the same width; they can be of any convenient widths, *provided* the widest one approaches 0. The projection of all the differences multiplied by their widths onto one column will give a *bound*:

$$A_U(N) - A_L(N) \leq |f(b) - f(a)| \text{ (widest interval)}$$

(see Figure 11.5-4). Thus we have some flexibility in picking the interval widths. Any set of rectangles will do, *provided* the widest interval approaches 0 in the limit. The difference between the upper and lower sums will be bounded by

$$A_U(N) - A_L(N) \leq |f(b) - f(a)| \text{ (widest interval)}$$

The upper and lower sums will therefore approach a common limit, which we call the area under the curve. Further thought shows that the upper and lower sums need not use the same intervals.

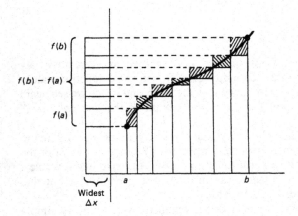

Figure 11.5-4

As a further generalization, we see that we can estimate the sum of the rectangles for a monotone continuous piece of the function by choosing the height of the *i*th rectangle corresponding to *any* value θ_i (Greek lowercase theta) of the function in the *i*th interval:

$$x_{i-1} \leq \theta_i \leq x_i$$

The estimate for the sum (which in the limit is to approach the area) is now

$$\sum_{i=1}^{N} f(\theta_i)(x_i - x_{i-1})$$

where the maximum difference $x_i - x_{i-1}$ approaches zero in the limit. Hence (Figure 11.5-5), we have for a monotone increasing continuous function

$$A_L(N) = \text{lower sum} \leq \text{this sum} \leq \text{upper sum} = A_U(N)$$

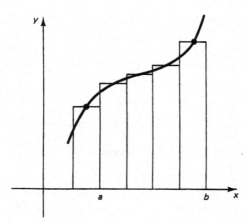

Figure 11.5-5

In mathematical symbols (using the notation of Section 9.9),

$$\sum_{i=1}^{N} f(x_{i-1}) \, \Delta x_i \le \sum_{i=1}^{N} f(\theta_i) \, \Delta x_i \le \sum_{i=1}^{N} f(x_i) \, \Delta x_i$$

For a monotone decreasing function, the inequalities of this last equation are obviously in the opposite direction. Due to the continuity of $f(x)$, the three function values $f(x_{i-1}), f(\theta_i)$, and $f(x_i)$ all approach the same value in the limit, and hence we see again that the middle sum must also approach the common sum of the two ends as N approaches infinity.

The upper and lower sums are called Riemann sums because he was the first to make general the idea of an area mathematically rigorous. The common limit of the upper and lower sums is called the *integral* (integrate means to make into a whole). Since his time the concept of an integral has been further generalized, but that development lies beyond the needs of this course.

From the areas of the monotone pieces of the function, we obtain, by simple addition, the area under the whole curve (consisting of a finite number of the pieces of monotone continuous functions).

Thus we see that the limit of the sum can be defined using any arbitrary intervals (as long as the largest approaches 0 as N approaches infinity), and we can pick any typical value in the ith interval as the height of the corresponding rectangle. We are not limited to equal spacing and picking particular values in each interval, nor must the intervals chosen for the upper and lower sums be exactly the same.

Further thinking about what we have done shows that actually we are computing the *limit of a sum;* the word "area" was only a colorful way of referring to the problem. Integration is actually the computation of the limit of a sum that is chosen in a rather flexible manner; it is not the rigid thing we began with. This is very typical in mathematics; an idea that arises in a specific context is gradually generalized until the original idea is merely a very special case.

We assumed that the two limits between which we were computing the area were different. It is a natural extension of the idea of area (and the generalizations we have made from it) to say that if

$$b = a \qquad \text{then the integral (area)} = 0$$

A further extension is that if $b < a$ then the integral (the limit of the sum) is to be negative, since the Δx_i will be negative. Similarly, if the function is below the x-axis and the Δx_i is positive, we would naturally call the integral negative. Finally, it follows that if $b < a$ and $f(x) < 0$ then the integral would again be positive, since it would be the limit of a sum of positive terms.

Notice that we began by claiming that area was measured by a positive number or 0, and now we are admitting negative areas. The contradiction is only apparent; we still feel that areas are nonnegative but that at times it is convenient to call an area negative. We made the positive area convention when we first examined areas. Now, due to the generalizations we have made, from our original idea of area to the idea of a limit of a sum, it is convenient to allow negative areas; otherwise, we would be forced to limit ourselves to functions that were nonnegative and sums where $b \geq a$. We could easily fix up the contradiction if we wished, but that would later force us into a lot of circumlocutions when we wanted to discuss problems where the areas (sums) cancel. We have only to keep in mind the apparent contradiction and watch for any foolishness that may emerge when we are careless. In a sense, we have extended the idea of area, and this generalization needs to be watched to see all the consequences.

We now introduce the usual notation. The limit of the sum is suggested by the elongated S (Sum) that begins the symbol. The letters a and b at the bottom and top of this elongated S are the limits of the range we are using. The next piece is the name of the function $f(x)$. Finally, we have the name of the variable with respect to which we are summing, the differential dx. We write it in the differential form as

$$\int_a^b f(x)\, dx$$

This is called the *integral from a to b of the function $f(x)$ with respect to the variable x*. In words, "the integral from a to b of $f(x)\, dx$."

The integral can be thought of as an operator, like d/dx, but in the form

$$\int_a^b \ldots dx$$

where the . . . is the place for the name of the function to be *integrated* with respect to the variable x.

If b lies between a and c, then it is clear that (see Figure 11.5-6)

$$\int_a^b \ldots dx + \int_b^c \ldots dx = \int_a^c \ldots dx$$

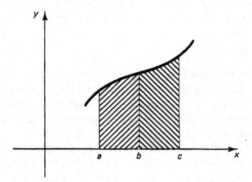

Figure 11.5-6

This is merely the statement that adjacent nonoverlapping intervals add properly for sums as well as for areas. Due to our use of the algebraic sign of areas, it is still true when b is not between a and c (provided the integrals exist). With all this formal apparatus, we can now find sums and areas by a uniform method, rather than a collection of artificial special techniques.

We remind you again, we generalized our primitive ideas about areas until we found that a suitable limit of a sum is the integral of the function. Areas are only a special application of the idea of an integral.

11.6 THE FUNDAMENTAL THEOREM OF THE CALCULUS

The great discovery that made the calculus a "calculus" (a routine process) is that it is sensible to ask, "What is the derivative of the integral?" We now think of the area under a function $y = y(x)$ (or any of the extensions of this idea) as depending on the upper limit of the range of the integration (set $b = x$):

$$\int_a^x f(x)\, dx$$

Before going on, it is convenient to remove a possible source of confusion. The x in the above integrand is a dummy variable in the sense that if we used any other letter for the variable x, say the letter t in

$$f(x)\, dx$$

then in its place we would have

$$f(t)\, dt$$

and we would have the integral

$$\int_a^x f(t)\, dt \qquad\qquad (11.6\text{-}1)$$

The area (or, more generally, the limit of the sum) would be the same. The reason is simply that the expression says that the variable of summation goes from the lower limit of integration a to the upper limit x. This change of the dummy variable of integration is exactly the same as we have for summation:

$$\sum_{n=1}^{N} n^k = \sum_{m=1}^{N} m^k$$

Whether we use n or m makes no difference in the answer. If we do not make this change in notation for the integral (11.6-1), we will become confused with the different meanings of x. In the old form it was both the variable of summation and the value we used as the upper limit in the final sum. In the new form we are summing and then taking the limit of a set of function values times the corresponding interval widths, depending on the dummy variable t, and using x as the upper value.

We now ask, "What is the derivative of this integral?" What is

$$\frac{d}{dx} \int_a^x f(t)\, dt \qquad (11.6-2)$$

Intuitively, what are we asking? For a continuous curve $f(x)$, when $f(x)$ is a large number, the area is increasing rapidly, and where $f(x)$ is small, the area is increasing slowly. The rate of growth of the area under a continuous curve is exactly the height of the curve at that point.

Due to the importance of this result, we need to back up our intuition with some rigor; in the past we have occasionally found that our intuition led us astray.

The question of what is the derivative with respect to the upper limit of the integral is a basic question, and we therefore go back to the basic definition of a derivative, the four-step process of Section 7.5. At step 4 we are to take the limit of the difference quotient:

$$\frac{\int_a^{x+\Delta x} f(t)\, dt - \int_a^x f(t)\, dt}{\Delta x}$$

But since this is the difference between two partially overlapping ranges of integration (Figure 11.6-1), the result must be merely over the range that is not common to both, that is, x to $x + \Delta x$. (Remember that Δx can be either positive or negative.) Therefore, we have

$$\frac{\int_x^{x+\Delta x} f(t)\, dt}{\Delta x}$$

If we think of $f(t)$ as being a positive, continuous, monotone increasing function in this interval (decreasing makes trivial changes), then (going back to the sum definition of the integral) we have the bounds on the function in the interval Δx. If Δx is positive, then

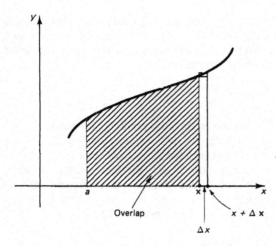

Figure 11.6-1

lower bound $= f(x)$ and   upper bound $= f(x + \Delta x)$

where Δx is the length of the integration interval. Because $f(x)$ is continuous, as Δx approaches 0 we have the same upper and lower bounds. And similarly for Δx negative. Thus we conclude that the derivative of the integral with respect to the upper limit of integration x is simply the function being integrated, the *integrand*. Be sure you see the inevitability of this result. *Think* about what is happening as the derivative of the integral is taken, how the rate of growth of the area under a continuous curve must be exactly the height of the curve at that point. The mathematics this time matches our intuition.

This is the fundamental theorem of the calculus,

$$\frac{d}{dx} \int_a^x f(t) \, dt = f(x)$$

(where a is some fixed value of x); *the derivative with respect to the upper limit of the integral of a function is the function itself*. Thus differentiation undoes integration. We see the truth of this in the integration of x^k, Generalization 11.4-3. The integral with the upper limit x [use x in place of a in the formula (11.4-4)] is

$$\int_0^x t^k \, dt = \frac{x^{k+1}}{k + 1}$$

Differentiating this with respect to x, we get the original integrand,

$$\frac{d}{dx} \frac{x^{k+1}}{k + 1} = \frac{(k + 1)x^k}{k + 1} = x^k$$

as the theorem requires. Thus integration is the *inverse* operation to differentiation, just as division is the inverse to multiplication. In both cases (division and integration), it comes down to a guess process. The question in integration is, "What function is this the derivative of?"

While the derivative of the integral is the original function, it is not true in general that the integral of the derivative of a function is the original function (the two operations do not "commute"). The reason for this is simple. Consider two functions that differ by a constant. Since the derivative of a constant is zero, their derivatives are the same. Therefore, given only the derivative, we cannot know which one was the original function. We know the integral only to within an additive constant; there is an arbitrary additive constant to be tacked onto the integration process when we think of integration as the *antiderivative*.

Example 11.6-1

Given that the derivative of a polynomial is

$$\frac{dy}{dx} = 3x^5 - 4x^3 + 2x + 7$$

we find that the antiderivative is (where C is some constant)

$$y = \frac{3x^6}{6} - \frac{4x^4}{4} + \frac{2x^2}{2} + 7x + C$$

After some algebra, this is simply

$$y = \frac{x^6}{2} - x^4 + x^2 + 7x + C$$

You can always check an integration by differentiating the answer to recover the original expression.

Thinking about the matter (upper and lower sums, the limit, and the continuity in each interval), we see that the sum of two functions added together is the same as the sum of the two individual sums added together; integration is a *linear operator*, just as differentiation was.

$$\int_a^x \{Af(t) + Bg(t)\} \, dt = A \int_a^x f(t) \, dt + B \int_a^x g(t) \, dt$$

We also see from the fundamental theorem of the calculus that since we can differentiate any powers, not only integral powers, we can also integrate them. It is convenient to write the formula for integration in terms of the antiderivative (any function whose derivative is the given function) of a single power of x as

$$\int x^n \, dx = \frac{x^{n+1}}{n + 1} + C$$

for *any* n other than $n = -1$. We have deliberately omitted the limits of integration for the antiderivative, and have supplied the missing constant C that is arbitrary when we handle the antiderivative.

We clearly cannot do the case $n = -1$ using this formula, both because it requires a division by zero and because no differentiation of a power of x leads to the exponent -1. We will take up the problem of its integration in Chapter 14.

We need some notation. It is very convenient to write an antiderivative of $f(x)$ as $F(x)$. We have, therefore,

$$\int f(x)\,dx = F(x) + C$$

To get from the antiderivative to the integral between the limits of an interval, we merely note that the integral over an interval of no length

$$\int_a^a f(x)\,dx = 0$$

so that we must have for the special value $x = a$

$$\int_a^a f(x)\,dx = 0 = F(a) + C$$

From this we see that

$$C = -F(a)$$

or in general

$$\int_a^x f(t)\,dt = F(x) - F(a) \tag{11.6-3}$$

where $F(x)$ is any antiderivative of $f(x)$. Notice how the additive constant C in the antiderivative $F(x)$ disappears in the final result.

It is customary to call the antiderivative the *indefinite integral,* as contrasted with the integral with limits, which is called the *definite integral.*

Besides polynomials and sums of arbitrary powers of x (excluding $n = -1$), we can integrate many different expressions when we recall (Section 7.6) the function of a function formula for differentiation:

$$\frac{dy}{dx} = \frac{dy}{du}\frac{du}{dx} \tag{11.6-4}$$

Example 11.6-2

What is the antiderivative of

$$\frac{dy}{dx} = 2x(1 + x^2)$$

We identify the u of formula (11.6-4) with

$$u = 1 + x^2$$

Then the $du/dx = 2x$, and we have the form

$$\frac{dy}{dx} = u\frac{du}{dx}$$

which integrates into

$$y = \frac{u^2}{2} + C = \frac{1}{2}(1 + x^2)^2 + C$$

It is *always* easy to check the antiderivative by the simple process of direct differentiation of the answer, and in this case we see that we have the correct antiderivative.

The constant of integration for the indefinite integral is not unique as the following example shows: one C is the other $C + \frac{1}{2}$.

Example 11.6-3

Given the function to integrate

$$\frac{dy}{dx} = x + 1$$

we can think of it as the sum of two terms and get the answer

$$y = \frac{x^2}{2} + x + C$$

or we can think of it in the form

$$u = x + 1$$

and get

$$y = \frac{(x + 1)^2}{2} + C$$

We see that the letter C is not the same in the two cases.

Let us solve this same problem in the differential notation. We have

$$dy = (x + 1)\, dx$$

Upon integrating both sides, we get

$$y = \int (x + 1)\, dx$$
$$= \frac{x^2}{2} + x + C$$

The result is, of course, the same.

The lack of uniqueness of the constant of integration can cause confusion for the beginner when the results obtained are not those in the book.

Example 11.6-4

What is the antiderivative of

$$\frac{dy}{dx} = \frac{x}{\sqrt{1 - x^2}}$$

We write this in the differential form

$$y = \int \frac{x}{\sqrt{1 - x^2}} \, dx$$

Again, how shall we pick the u? The radical is awkward, so we pick

$$u^2 = 1 - x^2$$

because this form will get rid of the radical. We get

$$2u \, du = -2x \, dx \quad \text{or} \quad dx = -\frac{u}{x} du$$

We now have

$$y = \int \frac{x}{u} \left(\frac{-u}{x} \, du \right)$$
$$= -u + C = -\sqrt{1 - x^2} + C$$

Differentiation confirms this result.

Example 11.6-5

Consider

$$\frac{dy}{dx} = x^3 (1 + x^4)^5$$

We see that if we take the $u = 1 + x^4$ then

$$du = 4x^3 \, dx$$

If only we had an extra 4 to make the front term $4x^3$, we would have the proper du. Evidently, we merely multiply and divide by the 4 to get this. Thus we have, finally,

$$y = \frac{1}{4} \int u^5 \, du$$

and this upon integration becomes

$$y = \frac{1}{4} \left(\frac{u^6}{6} \right) + C = \frac{(1 + x^4)^6}{24} + C$$

Differentiation verifies that we are right and indicates how the integration process reverses the differentiation steps.

There are no infallible rules for picking the substitution of a variable u for the original variable x. The fact is that the ability to chose a good substitution depends on your familiarity with the differentiation process; you must "see" how the verification by differentiation would go, how the terms would combine, cancel, and so on. If you are having trouble at this point, perhaps the best thing to do is practice some comparable differentiations until you have a feeling for how they go and can see how to make them go backward, beginning with the function, computing the derivative, and then imagine starting there and trying to reconstruct the original function. The differentiation process will show you, when you inspect it, exactly how to go backward as the integration process requires. It is probably much better to try a few problems and to carefully analyze what happens in each case, and why it happens, than it is to merely do a lot of problems with no careful examination of how the process goes. Familiarity is necessary so that you will learn to reverse the process of differentiation. Differentiation must be so automatic that you can see almost immediately how various choices will work out. In this, it is like factoring a quadratic such as

$$x^2 - 4x - 12$$

You must see all the various factorings of 12 and select the one for which the difference of the factors is 4.

Furthermore, there is no unique way of doing an integration; there are often various substitutions all of which will work (but some may be easier than others). Finding the antiderivative is a guess process whose truth is easily verified by a final differentiation *if* there is any doubt in the correctness of the result. There is no excuse for accepting a result that does not differentiate back to the given derivative.

Example 11.6-6

Find the antiderivative of

$$\frac{dy}{dx} = x^3\sqrt{1 - x^4}$$

This is the same as

$$y = \int x^3\sqrt{1 - x^4}\, dx$$

We see the x^4 under the radical and out in front the x^3, which is what we need for the derivative of x^4. Therefore, we pick the u as

$$u = 1 - x^4$$

which leads to

$$du = -4x^3\, dx$$

We are lacking a -4, so we supply it, and compensate in the denominator to get

$$y = \frac{-1}{4}\int u^{1/2}\, du$$

which integrates into

$$y = \frac{-1}{4}\left(\frac{u^{3/2}}{3/2}\right) + C$$

or

$$y = \frac{-1}{6}(1 - x^4)^{3/2} + C$$

Rather than doing a long list of exercises, you are well advised, after doing a few, to make up your own exercises. At first, as we earlier suggested, you pick the function, differentiate it, and then imagine starting with the derivative to integrate. If you get stuck, then the differentiation process will show you exactly how to do the integration. Later, you should start with the function to integrate. This will require you to pick your problems with some care so that they can be integrated. In doing this, you will find more clearly exactly what is involved in the integrals you can do. What evidence there is on the matter seems to show that your active involvement in the process teaches you more in less time than does the passive doing of problems that have been selected for you. As we have repeatedly said, we can only coach you; you must do the learning, and active participation in the learning process is generally much more effective than is passive following of what you are told to do. The best mathematicians generally find their own problems, work them out, and then go over them again and again, examining alternative strategies, until they have a sure mastery of the situation. There is an emotional involvement when you do your own problem that is usually missing when you merely do problems that others suggest.

EXERCISES 11.6

Find the antiderivative of the following:

1. $y' = x^3$
2. $y' = x + x^2 - x^3$
3. $y' = a + bx + cx^2$
4. $y' = x^{1/2}$
5. $y' = 3 + x^{-1/2}$
6. $y' = c - \sqrt{x}$
7. $y' = x^{3/2} - x^{1/2} + x^{-1/2}$
8. $y' = x(a^2 + x^2)^4$
9. $y' = x\sqrt{x^2 - a^2}$
10. $y' = (x^3 - x)^3(3x - 1)$
11. $y' = \sqrt{x - 1}$
12. $y' = 1/\sqrt{x + a}$
13. $y' = \sqrt{x} - 1/\sqrt{x}$
14. $y' = \sqrt{x}(1 + x^{3/2})^3$
15. $y' = (a^2 + x^2)^3(3x)$

11.7 THE MEAN VALUE THEOREM

Before we go too far, it is necessary to examine the above argument closely. We glibly said that since the derivative of a constant was 0 it follows that all we have to do to get the general antiderivative is add an arbitrary constant C to any particular antiderivative we find. But are *you certain* that no function other than a constant can go into 0 when you differentiate?

The point is not trivial. The role of the additive constant in integration is central. You are claiming that when you differentiate then all functions differing only by a constant must go into the same function, and *only* those. You are claiming that there is no function other than a constant whose derivative is identically zero.

You need some kind of mathematical proof of this important observation. Given a function that is continuous in a closed interval $a \le x \le b$, denoted by $[a, b]$, and that has a continuous derivative in the open interval $a < x < b$, denoted by (a, b), which is identically zero (in principle you cannot compute the derivative at the ends of the interval), then the function must be a constant. (A more complicated proof would not require the continuity of the derivative.) How to prove this kind of a statement is your problem.

Naturally, you begin by thinking about what you mean by a function being a constant. You mean that at any two points a and b (let them be the ends of the interval) you have

$$f(b) = f(a)$$

But you have learned from observation that it pays to try to prove not that two things are equal, but rather that their difference is zero. A trivial change in the problem, but one whose importance professional mathematicians well recognize. Therefore, the statement you want to prove is

$$f(b) - f(a) = 0$$

The statement you are given is

$$\frac{df(x)}{dx} = 0$$

at all points in the interior of the interval. Can you equate these two expressions? No, the dimensions are wrong, the resulting equation would not be invariant under a scaling transformation; hence it would be a meaningless expression. The derivative is the quotient of the scaling factors for $f(x)$ and for x, so you try

$$\frac{f(b) - f(a)}{b - a} = \frac{df(x)}{dx}$$

What value x? Certainly not all values! Is there at least one such x? If there is, then from knowing that the derivative is always 0 you would have the proof that the function is a constant, that $f(b) = f(a)$.

Some pictures of the situation are needed (see Figure 11.7-1). Yes, it does look as if it is a true statement and avoids the examples where there is a corner and hence no derivative at some point. The point where the equality holds is conventionally labeled by the Greek letter θ (lowercase theta).

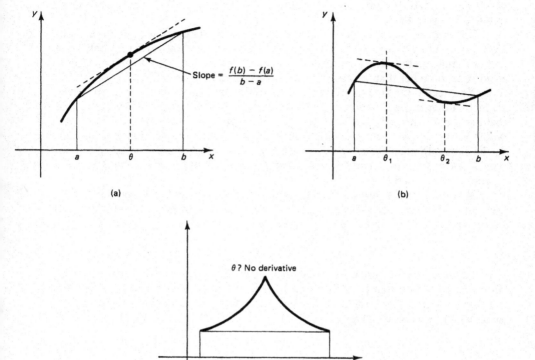

Figure 11.7-1

Mean Value Theorem.   If a function $f(x)$ is continuous in a closed interval $[a, b]$ and has a continuous derivative in the open interval (a, b), then there is at least one value of x, labeled θ, such that

$$\frac{f(b) - f(a)}{b - a} = \frac{df(\theta)}{dx} \qquad (a \le \theta \le b)$$

This theorem states geometrically that the slope of the secant line through the end points of the interval is equal to the slope of the curve for at least one interior point θ of the curve.

Can you simplify this theorem before you try to prove it? It seems that if you subtracted out the secant line through the end points you would have an easier theorem to prove. The secant line through the two end points is (it is a linear equation so you need only check this for $x = a$ and $x = b$)

$$f(x) - f(a) = \frac{f(b) - f(a)}{b - a}(x - a)$$

using the point-slope form (or two-point form if you prefer) of the line. Remove this line by subtracting it from the function. Now the theorem you want to prove is:

Rolle's Theorem.   If a function is continuous in the closed interval $[a, b]$ and vanishes at both ends, and has a continuous derivative in the open interval (a, b), then there is at least one value of x, $x = \theta$, such that

$$\frac{df(\theta)}{dx} = 0$$

where $a < \theta < b$.

Now look at the corresponding Figure 11.7-2. If you suppose that the function is positive in the interval, then when it first rises the derivative (slope or rate of change) is positive, and when the function returns to zero, as it must before or at the other end of the interval, the slope must be negative. Thus the derivative must change sign somewhere. But the derivative is assumed to be continuous, and recalling our assumptions about the existence of numbers, you recall (Section 4.4, Assumption 2) that there will exist a number where the function takes on the value 0. If there are several zeros of $f(x)$ in the interval, then it suffices to use any one to show the existence of a place in the interval where the derivative is zero. If the function is negative, the argument is easily modified to fit this situation.

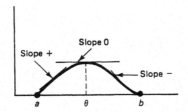

Figure 11.7-2

In the trivial case for $f(x) = 0$ for all values, it is easily seen that the derivative does vanish for at least one point in the interval.

From Rolle's theorem you have the proof of the mean value theorem. And from that you get, via the assumption that the derivative was identically zero, that

$$f(b) - f(a) = 0$$

This says that any continuous function in a closed interval $[a, b]$ whose derivative is continuous in the open interval (a, b) and is always zero must be a constant. There is no other admissible function, no matter how peculiar it may be, that meets the conditions. Therefore, it is sufficient to add the *additive constant* C to the antiderivative to get the most general function whose derivative is the given function.

You need, now, to examine the way you combined the individual pieces (each of which is a piecewise monotone continuous function) into the complete function over the whole range. First, there are only a finite number of such pieces. Second, there is a bit of trouble at the ends. At a discontinuity (see Figure 11.7-3) of the function you have the question of the value of the function. For example, suppose between

$$-1 \leq x < 0, \qquad f(x) = x - 1$$
$$0 < x \leq 1, \qquad f(x) = x^2$$

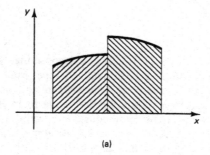

(a)

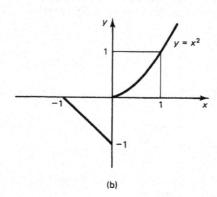

(b)   **Figure 11.7-3**

When you are working in the first interval and substitute in the upper limit of the range, $x = 0$, you want to use the value appropriate to the interval, that is, $f(0) = -1$, while for the lower limit of the second integral you naturally want to use $f(0) = 0$. Thus you want to have two distinct values of the function at the same place! This contradicts the claim of handling only single-valued functions. The trouble is clear; you should take the limits appropriate for each interval, but this requires a lot of words throughout the rest of the text. We leave it to the student to supply the limit arguments for handling this essentially trivial point.

We remind you again, the constant C need not be the same in various methods of integration of the same function, as shown in Example 11.6-3.

We proved the mean value theorem for functions with a continuous derivative in the open interval and remarked that a proof could be given with merely the assumption of a derivative in the open interval. The difference is so small that you do not often meet a function that has only the weaker condition and not the stronger, although such functions can be "made up." We are generally concerned to handle the needs that arise in practice, and do not attempt to prove the weakest theorems we could.

EXERCISES 11.7

1. Prove Rolle's theorem using only the existence of the derivative. *Hint:* Pick a point where the function takes on its extreme value in the closed interval. Argue that the difference quotient on one side must be ≥ 0 and on the other must be ≤ 0; hence, since the limit exists, the limit must be 0.
2. Using the function $y = x(1 - x)^2$, show that in the interval $[0, a]$ the corresponding value $\theta(a)$ is not a continuous function of a, for a as large as 3.

11.8 THE CAUCHY MEAN VALUE THEOREM

We will later need a generalization of the mean value theorem. It is probably best to deal with it now, because it uses the same kind of techniques we just used.

Consider two functions $f(x)$ and $g(x)$. We wish to prove that there exists a θ such that

$$\frac{f'(\theta)}{g'(\theta)} = \frac{f(b) - f(a)}{g(b) - g(a)}$$

supposing that both functions satisfy the same kinds of conditions that were needed in the mean value theorem, plus the condition that $g(b) \neq g(a)$. We cannot apply the mean value theorem to the numerator and denominator separately because then, in general, the two theta values would not be the same, as is asserted in the theorem we are to prove.

To apply Rolle's theorem (which is the main tool), we construct the function

$$\Phi(x) = f(x) - f(a) - \frac{f(b) - f(a)}{g(b) - g(a)}[g(x) - g(a)]$$

so that $\Phi(a) = \Phi(b) = 0$. We conclude that there is at least one point θ in the interval at which the derivative $\Phi'(x)$ vanishes. At this point we have

$$f'(\theta) = \frac{f(b) - f(a)}{g(b) - g(a)}[g'(\theta)]$$

Now if $g'(x)$ does not vanish in the interval, then we have

$$\frac{f'(\theta)}{g'(\theta)} = \frac{f(b) - f(a)}{g(b) - g(a)}$$

This is the extension we need. If $g(x) = x$, then this extension reduces to the original mean value theorem.

Cauchy Mean Value Theorem. If $f(x)$ and $g(x)$ are continuous in the closed interval $[a, b]$ and have continuous derivatives in the open interval (a, b), and if $g'(x) \neq 0$ [hence $g(b) \neq g(a)$] in the interval, then

$$\frac{f(b) - f(a)}{g(b) - g(a)} = \frac{f'(\theta)}{g'(\theta)}$$

for some θ in the interval $a < \theta < b$.

11.9 SOME APPLICATIONS OF THE INTEGRAL

At this point you can integrate comparatively few functions, so we must pick the problems carefully. Later, in Part III, you will greatly increase the class of functions you can integrate. We can, however, at this point give some flavor of the richness and variety of the applications of integration.

Example 11.9-1

Find the area under the curve

$$y = x - x^3$$

between $x = 0$ and $x = 1$ (see Figure 11.9-1). You want the integral

$$\int_0^1 (x - x^3)\, dx = \left(\frac{x^2}{2} - \frac{x^4}{4}\right)\Bigg|_0^1$$

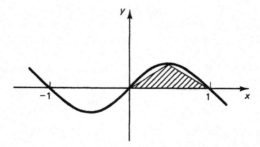

Figure 11.9-1

On the right-hand side, the bar at the end of the line corresponds to the earlier limits on the integral sign. From (11.6-3) we substitute the upper limit into the integrated function and subtract the value of it at the lower limit to get

$$\int_0^1 (x - x^3) \, dx = \frac{1}{2} - \frac{1}{4} - [0 - 0] = \frac{1}{4}$$

Does the answer seem reasonable? The area of a triangle inside is $\frac{3}{16}$ which is less than the computed area by about the right amount.

Example 11.9-2

Find the area under the curve

$$y = x^{1/2}$$

as a function of the upper limit (Figure 11.9-2). Remember to use a dummy variable of integration to avoid confusion with the upper limit x:

$$\int_0^x t^{1/2} \, dt = \frac{t^{3/2}}{3/2} \Big|_0^x$$

$$= \frac{2}{3} x^{3/2}$$

Is this reasonable? We can test it for $x = 1$. A bounding rectangle gives an area of 1, while an enclosed triangle gives an area of $\frac{1}{2}$, and the triangle should be closer to the area, which is $\frac{2}{3}$; and it is!

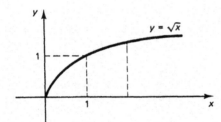

Figure 11.9-2

Example 11.9-3

Find the area under the curve

$$y = \frac{x}{(a^2 + x^2)^{1/2}}$$

(see Figure 11.9-3) from the origin out to an arbitrary position x. You want the integral

$$\int_0^x \frac{t}{(a^2 + t^2)^{1/2}} \, dt$$

This is not a simple power of t, so you need to think of what function of t you can call $u(t)$. Some thought suggests trying $u = a^2 + t^2$ because you have

$$du = 2t \, dt$$

and there is already a t in the numerator. For the moment you lack the factor 2, so you put the 2 in the numerator and compensate by a factor $\frac{1}{2}$ out in front (for the same reason we could pass over constants when differentiating). You have, therefore,

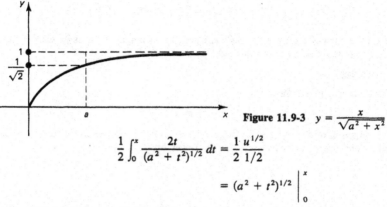

Figure 11.9-3 $y = \dfrac{x}{\sqrt{a^2 + x^2}}$

$$\frac{1}{2} \int_0^x \frac{2t}{(a^2 + t^2)^{1/2}}\, dt = \frac{1}{2} \frac{u^{1/2}}{1/2}$$

$$= (a^2 + t^2)^{1/2} \Big|_0^x$$

to be evaluated between the limits 0 and x. You get, finally, the area (upper limit $t = x$ minus the lower limit $t = 0$):

$$A = (a^2 + x^2)^{1/2} - a$$

A more suggestive way of writing this is

$$A = [\{1 + (x/a)^2\}^{1/2} - 1]a$$

since this shows the answer in a form where x is related to the parameter a. In particular, at $x = a$ you have the area

$$A = (\sqrt{2} - 1)a = (0.414 \ldots)a$$

Does it seem to be a reasonable area? Look at the figure and estimate the area to check the answer. The enclosed triangle has an area of $a/2\sqrt{2} = (0.35 \ldots a)$, while the bounding rectangle has area of $a/\sqrt{2} = (0.707 \ldots a)$. All this suggests that in the original problem we should have set $x = at$ and worked with the corresponding integral $y = t/\sqrt{1 + t^2}$.

Example 11.9-4

The number π. The transcendental number π arises in two ways: the circumference of a circle, $C = 2\pi r$, and the area of a circle, $A = \pi r^2$. The first arises from the theorem that linear measurements of similar figures (circles in this case) are to each other as any other corresponding linear measurement (say radii). Thus

$$\frac{C_1}{C_2} = \frac{r_1}{r_2}$$

from which

$$C = 2\pi r$$

for some constant 2π.

For areas the theorem states that areas of similar figures are to each other as the squares of their corresponding linear measurements. Hence

$$A = \pi r^2$$

Are these two occurrences of the symbol π the same number? To prove that they are, we divide the area of the circle into concentric rings of thickness Δr_i (see Figure 11.9-4), choose any convenient radius r_i in the ring of width Δr_i, form the sum

$$\sum_{i=1}^{n} 2\pi r_i \, \Delta r_i$$

and finally take the proper limit to find

$$\int_0^r 2\pi r \, dr = \pi r^2$$

Thus the two occurrences of π are both the same number.

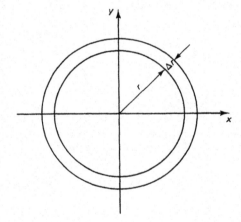

Figure 11.9-4   Area of a circle

Example 11.9-5

Find the formula for the volume of a cone. We think of the cone as upright with the vertex at the origin. The side that lies in the x–y plane would have the equation

$$y = \frac{h}{r}x$$

where h is the height and r is the radius of the cone. We imagine the cone cut into horizontal slices of thickness Δy and radius x. The volume is, therefore,

$$V = \pi \int_0^h x^2 \, dy$$

$$= \pi \int_0^h \left(\frac{ry}{h}\right)^2 dy$$

$$= \pi \frac{r^2}{h^2}\left(\frac{y^3}{3}\right)\bigg|_0^h = \frac{\pi r^2 h}{3}$$

which is the known answer.

Example 11.9-6

Suppose we rotate the parabola

$$y = x^2$$

about the x-axis (see Figure 11.9-5). If we think of the fundamental process of integration, that is, the limit of a sum of terms, we see that the volume out to $x = 1$ can be approximated by a series of flat discs of radius y and thickness Δx. The element of volume is, therefore,

$$\pi y^2 \, \Delta x$$

We need no longer fill in all the details of upper and lower sums since clearly their difference will be bounded by the flat disc of maximum radius multiplied by the maximum thickness Δx, and in the limit the corresponding upper and lower sums will pass over to

$$\int_0^1 \pi y^2 \, dx$$

But $y = x^2$, so we have

$$\int_0^1 \pi x^4 \, dx = \frac{\pi}{5}$$

as the volume. Does this answer seem reasonable? The volume of an enclosing cone would be $\pi/3$ (from Exercise 11.9-5). Yes, it seems reasonable.

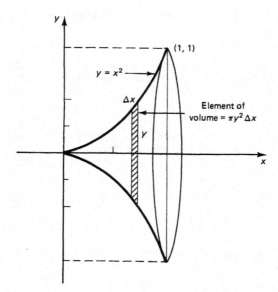

Figure 11.9-5   Parabola rotated about x-axis

Example 11.9-7

Suppose we use the same parabola as in the previous problem except we rotate it about the y-axis and ask for the corresponding volume. We note (Figure 11.9-6) how the shaded area is being rotated. Here the element of volume is the anchor ring whose inner radius is x and whose outer radius is 1. The element of volume is, therefore,

$$\pi(1 - x^2)\,\Delta y$$

and the integral (remember we are integrating in the y direction and $x^2 = y$)

$$\int_0^1 \pi(1 - y)\,dy = \pi\left(y - \frac{y^2}{2}\right)$$

$$= \pi\left(1 - \frac{1}{2}\right) = \frac{\pi}{2}$$

As a check, we could compute the volume of the inside and compare the sum of the two with the volume of the cylinder (see the exercises).

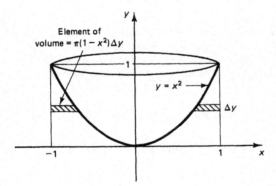

Figure 11.9-6 Parabola rotated about y-axis

Example 11.9-8

Find the volume of a sphere of radius a. Draw a picture of a cross section (Figure 11.9-7). The volume of a slice is

$$\pi y^2\,\Delta x$$

By symmetry, we have that the total volume is

$$2\int_0^a \pi y^2\,dx = 2\pi\int_0^a (a^2 - x^2)\,dx$$

which easily integrates into

$$2\pi\left(a^2 x - \frac{x^3}{3}\right) = 2\pi[a^3]\left(1 - \frac{1}{3}\right) = \frac{4}{3}\pi a^3$$

as it should.

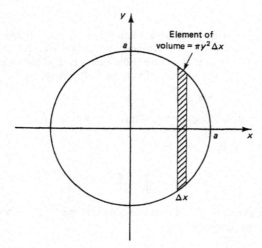

Figure 11.9-7 Volume of a sphere by discs

Example 11.9-9

We have used discs to estimate the volume of a solid of revolution. Sometimes it is easier to use cylinders as the elements of volume (Figure 11.9-8). For the sphere in Example 11.9-8, we now have that the volume of the elementary cylinder is the surface, $2\pi rh$, times the thickness Δx. Using only the upper half and doubling to get the total volume, we have

$$2\int_0^a 2\pi yx \, dx = 4\pi \int_0^a x\sqrt{a^2 - x^2} \, dx$$

$$= \frac{4\pi}{3}\{-1(a^2 - x^2)^{3/2}\} \Big|_0^a = \frac{4}{3}\pi a^3$$

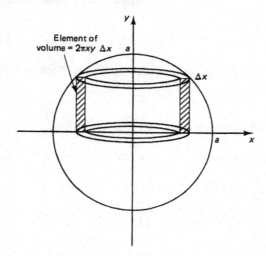

Figure 11.9-8 Volume of a sphere by cylinders

The use of a disc or a cylinder depends on the integration to be done; sometimes one is much easier to use than the other.

Example 11.9-10

Just as we could differentiate identities to get new identities (Section 9.9), so too we can integrate them provided we watch the evaluation of the additive constant. For example, suppose you wanted the sum of

$$S = \frac{C(n, 0)}{1} + \frac{C(n, 1)}{2} + \frac{C(n, 2)}{3} + \cdots + \frac{C(n, n)}{n + 1}$$

You could start with the basic generating function

$$(1 + t)^n = C(n, 0) + C(n, 1)t + C(n, 2)t^2 + \cdots + C(n, n)t^n$$

and integrate it. Naturally, you integrate between $t = 0$ and $t = 1$. On the right you get the desired sum expression, and on the left you get, after the integration,

$$S = \frac{(1 + t)^{n+1}}{n + 1} \Big|_0^1$$

Putting in the limits, you get

$$S = \frac{2^{n+1} - 1}{n + 1}$$

as the desired sum.

EXERCISES 11.9

Given the following function and limits compute the corresponding area:

1. $y = x - x^3$, $(0, 1)$
2. $y = x^2 - x^4$, $(0, 1)$
3. $y = 3x^5 - 10x^3 + 7x$, $(-1, 1)$
4. $y = 1 - \sqrt{x}$, $(0, 1)$
5. $y = 1/\sqrt{x}$, $(1, 4)$
6. $y = x^{1/2} - x^{3/2} + x^{5/2}$, $(0, 1)$
7. $y = x/\sqrt{a^2 - x^2}$, $(0, a)$
8. $y = x\sqrt{a^2 - x^2}$, $(0, a)$
9. $y = 1/\sqrt{1 + x} + \sqrt{1 + x}$, $(0, 1)$

Rotate the following curves about the x-axis and find the corresponding volume.

10. Exercise 1
11. Exercise 2
12. Exercise 4

13. Exercise 6
14. Exercise 8

Rotate the following about the y-axis and find the corresponding volume.

15. Exercise 1
16. Exercise 2
17. Exercise 5
18. Exercise 4
19. Given the curve $y = x^n$, for $n > 0$, and the interval $(0, 1)$, (a) find the area under the curve and above the x-axis, (b) rotate about the x-axis, and (c) rotate about the y-axis.
20. Find the value of $\Sigma_{k=0}^n (-1)^k C(n, k)/(k + 1)$.
21. Evaluate $\Sigma_{k=0}^n C(n, k)/(k + 2)(k + 1)$.

11.10 INTEGRATION BY SUBSTITUTION

One of the two powerful methods of finding integrals of functions is substituting one variable for another, also called *change of variables*. We have used the method a number of times. It is based on the fundamental relation

$$dy = \frac{dy}{du} du$$

which is the basis for much of differentiation; hence it is reasonable to believe that it could play a similar part in integration. For the purposes of integration, we prefer to write this in the differential form

$$dx = \frac{dx}{dt} dt$$

when we change from the variable x to the variable t. Hence if $x = g(t)$, then

$$\int f(x)\ dx = \int f(g(t))g'(t)\ dt$$

We often use this in the reverse direction; we recognize that we have some function of a variable multiplied by the derivative of that variable.

Example 11.10-1

Suppose you have the indefinite integral

$$\int x\sqrt{1 + x^2}\ dx$$

If you set

$$1 + x^2 = t$$

then using the differential notation you have

$$2x \, dx = dt$$

and in the t variable you have

$$\int \frac{1}{2} t^{1/2} \, dt = \left(\frac{1}{2}\right) \frac{t^{3/2}}{3/2} + C = \frac{(1 + x^2)^{3/2}}{3} + C$$

Differentiation will convince you that this is the right answer.

The careful student may worry about the substitution

$$1 + x^2 = t$$

being multiple valued. You have only to think of the substitution as a notational change and in each appearance of t write $1 + x^2$ to see that the result is safe at this step. You do need to think about the point when you substitute in the two limits.

Case Study 11.10-2

Find the indefinite integral

$$J = \int \sqrt{1 + \sqrt{x}} \, dx$$

Even the experienced person finds it difficult to imagine what function would differentiate into this integrand, $\sqrt{1 + \sqrt{x}}$.

Square roots are generally awkward to handle, so the most obvious substitution is

$$x = u^2$$

$$dx = 2u \, du$$

Therefore, the integral is

$$J = \int \sqrt{1 + u} \, (2u) \, du = 2 \int u \sqrt{1 + u} \, du$$

which shows some progress in making the integrand more tractable. To get rid of the remaining square root, we next try

$$1 + u = v^2$$

$$du = 2v \, dv$$

and we have

$$J = 2 \int (v^2 - 1) v \, 2v \, dv = 4 \int (v^4 - v^2) \, dv$$

This integral is easy to do. We get

$$J = 4 \left[\frac{v^5}{5} - \frac{v^3}{3} \right] + C$$

It remains to substitute back twice, once to get from v to u, and then to go from u to get to the original x variable. We have first

$$J = 4v^3\left[\frac{v^2}{5} - \frac{1}{3}\right] + C = 4(1 + u)^{3/2}\left(\frac{1 + u}{5} - \frac{1}{3}\right) + C$$

$$= 4(1 + u)^{3/2}\left(\frac{u}{5} - \frac{2}{15}\right) + C$$

Finally, going back to the x variable we get

$$J = 4(1 + \sqrt{x})^{3/2}\left(\frac{\sqrt{x}}{5} - \frac{2}{15}\right) + C$$

as the indefinite integral. This can also be written in the form

$$J = \frac{4}{15}\sqrt{1 + \sqrt{x}}[3x + \sqrt{x} - 2] + C$$

Direct differentiation will show that this is the correct answer. Of course, now that we have the result we see that the two substitutions could be combined into one change of variable

$$x = [v^2 - 1]^2$$

This is the typical polishing up mathematicians do when presenting the final result; and if you can see far enough ahead, you can likewise do the integral with one substitution.

Example 11.10-3

In the previous case study, suppose we had been asked for the definite integral

$$\int_0^1 \sqrt{1 + \sqrt{x}}\, dx$$

Using the answer, we merely put in the limits to get

$$4\left\{\left[\left(1 + 1\right)^{3/2}\left(\frac{1}{5} - \frac{2}{15}\right)\right] - \left[1^{3/2}\left(\frac{-2}{15}\right)\right]\right\}$$

$$= 4\left\{2\sqrt{2}\left(\frac{1}{15}\right) + \frac{2}{15}\right\} = \frac{8(\sqrt{2} + 1)}{15} = 1.287\ldots$$

A sketch of the integrand shows that it lies between 1 and $\sqrt{2}$ so that the answer must lie there too.

However, had we started with the definite integral problem there is another approach. Once we made each substitution, we could have also brought the limits of the integral to the new variable, and thus saved all the back substitution to get back to the variable x. The start would have been

$$\int_0^1 \sqrt{1 + \sqrt{x}}\, dx$$

The substitution

$$x = u^2$$

moves the limits in the variable x from 0 to 1 to the same range 0 to 1 for the variable u, so we have

$$2 \int_0^1 u\sqrt{1 + u}\ du$$

The substitution

$$1 + u = v^2$$

moves the limits 0 to 1 in u to the range 1 to $\sqrt{2}$ in the variable v, so we have

$$4 \int_1^{\sqrt{2}} (v^4 - v^2)\ dv$$

whose integration gives, as before,

$$4v^3\left(\frac{v^2}{5} - \frac{1}{3}\right)\Bigg|_1^{\sqrt{2}}$$

and substituting in the limits we get

$$\left[4\left(2\sqrt{2}\right)\left(\frac{2}{5} - \frac{1}{3}\right)\right] - 4\left[\frac{1}{5} - \frac{1}{3}\right] = 4\left\{\left(2\sqrt{2}\right)\left(\frac{1}{15}\right) + \frac{2}{15}\right\}$$

$$= \frac{8(\sqrt{2} + 1)}{15}$$

as before. A comparison shows that probably in this case the moving of the limits to the new variables is preferable to transforming the variables back to the old limits. Sometimes it is easier to do the back substitution of the variables; there is no fixed rule in this matter.

EXERCISES 11.10

Integrate the following:

1. $\int (1 + x^2)^3 x\ dx$
2. $\int [(x^{2/3} + 1)^2 x^{-1/3}\ dx]$
3. $\int_0^1 (t^{3/2} - 1)^3 \sqrt{t}\ dt$
4. $\int_0^1 \theta\sqrt{1 - \theta^2}\ d\theta$
5. $\int_0^1 \theta(1 - \theta^2)^k\ d\theta,\ k \neq -1$
6. $\int_0^1 (1 + \sqrt{x})/(\sqrt{x})\ dx$
7. $\int (1 + \sqrt{t})\sqrt{t}\ dt$
8. Show that for *any* integrable function $f(x)$

$$\int_0^a f(x)/\{f(x) + f(a - x)\}\ dx = a/2$$

Hint: Substitute $a - x = t$ and combine with original integral.

9. Show that $\int_0^1 f(x^2)x \, dx = (\frac{1}{2}) \int_0^1 f(u) \, du$
10. Show that $\int_0^1 f(\sqrt{x}) \, dx = 2 \int_0^1 f(u)u \, du$
11. Show that $\int_0^a f(\sqrt{a^2 - x^2})x \, dx = \int_0^a f(u)u \, du$
12. Show that $\int f(\sqrt{g(x)})g'(x) \, dx = 2 \int uf(u) \, du + C$
13. Show that $\int_{1/a}^a f(x + 1/x)(1 - 1/x^2) \, dx = 0$
14. Show that for $n > 0 \int_0^1 x^n \, dx + \int_0^1 x^{1/n} \, dx = 1$

11.11 NUMERICAL INTEGRATION

Many integrals that naturally arise cannot be integrated in any finite form in terms of commonly used functions (cannot in the same sense that $\sqrt{2} \neq p/q$, and not in the sense of ignorance). Still, you believe that the corresponding problem, *when* presented as an area, has some definite value. Thus, when all other integration methods fail, *numerical integration* is used to estimate the corresponding area. If the integral involves one or more parameters, you *try* to eliminate them by suitable scale transformations, and if this does not work, then you face the numerical evaluation for each parameter value of interest.

For example, given the integral

$$\int_0^1 \sqrt{a^2 - x^2} \, dx$$

the substitution $x = at$ gives

$$\int_0^1 a\sqrt{1 - t^2}\, a \, dt = a^2 \int_0^1 \sqrt{1 - t^2} \, dt$$

and we have an integral free of the parameter a.

The central idea behind numerical integration is *analytic substitution*: for the function you cannot handle, you approximate it locally by one you can handle. This is the same idea that is behind Newton's method (Section 9.7, where you replaced, locally, the curve by the tangent line, and then found the zero of the tangent line as a new approximation to the zero of the function.

Given

$$\int_0^1 f(x) \, dx$$

suppose we approximate $f(x)$ by a straight line through the two end points, that we approximate the area by a trapezoid. The approximation to the function is the line

$$f(x) = f(0) + [f(1) - f(0)]x$$

That this is the desired line can be seen from the fact that it passes through the two points $(0, f(0))$ and $(1, f(1))$. We therefore integrate this approximation as if it were the function

$$J = \int_0^1 \left\{ f(0) + [f(1) - f(0)]x \right\} dx$$

$$= f(0)x + \left. \frac{\{f(1) - f(0)\}x^2}{2} \right|_0^1$$

$$= f(0) + \frac{1}{2}\{f(1) - f(0)\}$$

$$= \frac{1}{2}\{f(0) + f(1)\}$$

which is the known area of a trapezoid (see Figure 11.11-1).

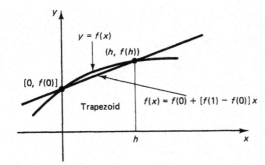

Figure 11.11-1   Trapezoid area

This is a crude approximation to the function, and you promptly think of breaking up the interval $[0, 1]$ into, say, N subintervals, and then using this same trapezoid approximation in each interval (see Figure 11.11-2). The fact that the integral was defined to be the limit of a sum suggests inverting this definition and estimating the integral as a sum of small elements. If we write $h = 1/N$, then we have for an estimate of the area J the *composite trapezoid rule*

$$J = \frac{h}{2}\{f(0) + f(h)\} + \frac{h}{2}\{f(h) + f(2h)\} + \cdots + \frac{h}{2}\{f((N-1)h) + f(Nh)\}$$

$$= h\left\{ \frac{f(0)}{2} + f(h) + f(2h) + \cdots + f((N-1)h) + \frac{f(Nh)}{2} \right\}$$

It is easy to generalize to an arbitrary interval (a, b). Again using N subintervals, we have

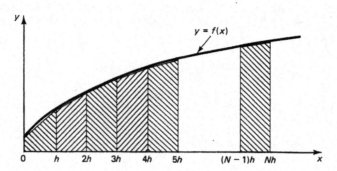

Figure 11.11-2   Trapezoid rule

$$J = h\left\{\frac{f(a)}{2} + f(a + h) + f(a + 2h) + \cdots + \frac{f(b)}{2}\right\} \tag{11.11-1}$$

where $h = (b - a)/N$. In words, the area estimate for the trapezoid rule is the sum of all the equally spaced $N + 1$ integrand values except that the first and last have coefficients $\frac{1}{2}$, and the whole sum is multiplied by the length of a subinterval $(b - a)/N$.

It can be shown that the error in the interval can be expressed by

$$\frac{- h^3 f''(\theta)}{12}$$

(for some θ in the interval), and hence for the whole interval

$$E = \frac{-(b - a)h^2 f''(\theta)}{12} \tag{11.11-2}$$

(although the complete proof is not easy). Thus we have the following rule: halving the interval reduces the trapezoid rule error by a factor of 4.

An alternative approximation, still using straight lines, is the midpoint formula. We again divide the interval $[a, b]$ into N subintervals, but this time we use the tangent line at the midpoint of the interval. From the Figure 11.11-3 we see that by tilting the tangent line until it is horizontal the area under it in the subinterval will not change. The *composite midpoint formula* is, therefore,

$$h\left\{f\left(a + \frac{h}{2}\right) + f\left(a + \frac{3h}{2}\right) + f\left(b + \frac{5h}{2}\right) + \cdots + f\left(a - \frac{h}{2}\right)\right\}$$

$$\tag{11.11-3}$$

It is simply the sum of the samples of the function taken at the midpoints of each interval multiplied by the length of a subinterval.

It can be shown (and you can see that it is approximately true from Figure 11.11-4) that the error per interval is half that of the trapezoid rule and with the opposite sign:

$$\frac{h^3 f''(\theta)}{24}$$

Therefore, the total error is approximately

$$E = \frac{(b - a)h^2 f''(\theta)}{24} \qquad (11.11\text{-}4)$$

We see that the midpoint formula has approximately half the error as the trapezoid rule, has simpler coefficients, and requires one less evaluation of the function. Furthermore, the error is of the opposite sign.

Midpoint error

Trapezoid error

Midpoint

Midpoint

Figure 11.11-3 Midpoint integration

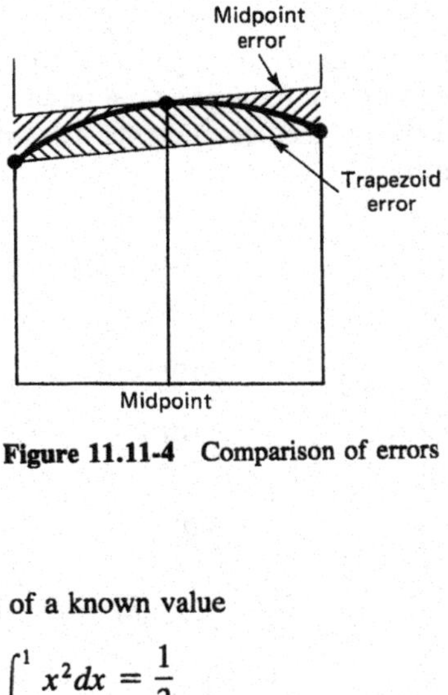

Figure 11.11-4 Comparison of errors

Example 11.11-1

As a check case we find the area of a known value

$$\int_0^1 x^2 dx = \frac{1}{3}$$

Let us take $N = 4$ so that we can follow the arithmetic easily. The trapezoid rule gives

$$J = \frac{1}{4}\left[\frac{1}{2}\, 0^2 + \left(\frac{1}{4}\right)^2 + \left(\frac{2}{4}\right)^2 + \left(\frac{3}{4}\right)^2 + \frac{1}{2}\,(1)^2\right]$$

$$= \frac{1}{4}\left[\frac{1}{16}\{1 + 4 + 9 + 8\}\right] = \frac{22}{64} = \frac{11}{32} = 0.344 \ldots$$

The midpoint formula gives

$$J = \frac{1}{4}\left[\left(\frac{1}{8}\right)^2 + \left(\frac{3}{8}\right)^2 + \left(\frac{5}{8}\right)^2 + \left(\frac{7}{8}\right)^2\right]$$

$$= \frac{1}{4}\frac{1}{64}[1 + 9 + 25 + 49] = \frac{84}{256} = \frac{21}{64} = 0.327\ldots$$

The true answer is, of course, $\frac{1}{3}$, and the errors are

$$\text{trapezoid error} = 0.011$$

$$\text{midpoint error} = -0.006$$

and, as predicted, the midpoint has half the error and the error is of the opposite sign. The error formulas could be used to estimate the error since the second derivative is exactly 2 for all θ.

To get more accuracy, you could take many more intervals (N much larger), but that involves more arithmetic, and leads in extreme cases to increased roundoff error in the arithmetic done. Instead, there is an alternative approach to get more accuracy. We could use a better local approximation to the function than straight lines. Thus we next look at local parabolas through three equally spaced points.

To find the formula for a parabola through three equally spaced points, we use the method of undetermined coefficients. Let the given x values be $-h$, 0, and h. For the function we have

$$y = A + Bx + Cx^2$$

and we have the three equations (see Figure 11.11-5)

$$f(-h) = A - Bh + Ch^2$$

$$f(0) = A$$

$$f(h) = A + Bh + Ch^2$$

We get as solutions

$$A = f(0)$$

$$B = \frac{1}{2h}\{f(h) - f(-h)\}$$

$$C = \frac{1}{2h^2}\{f(-h) - 2f(0) + f(h)\}$$

The integral from $-h$ to h of this parabola is

$$2Ah + \frac{2}{3}Ch^3$$

and substituting in the values of the coefficients we get, after some algebra,

$$\frac{h}{3}\{f(-h) + 4f(0) + f(h)\}$$

To convert this into a composite formula, we note that translating a parabola

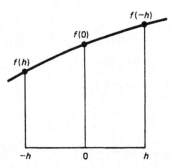

Figure 11.11-5 Derivation of Simpson's formula

does not change the quadratic nature of the curve; hence the same formula applies after translation. The composite formula, when you note that the ends of each interval (except the very first and last) occur twice, is

$$\frac{h}{3}\{f(a) + 4f(a + h) + 2f(a + 2h) + 4f(a + 3h) + \cdots + f(b)\}$$

For the inner points, the odd numbered get weight 4 and the even ones get weight 2, while the end ones get weight 1. The pattern of weights is

$$\frac{h}{3}\{1, 4, 2, 4, 2, 4, \ldots, 4, 2, 4, 1\}$$

The error term of the composite formula has the form

$$\frac{-(b - a)h^3 f^{(4)}(\theta)}{90}$$

Notice that this formula is exact for cubics, although we only started with quadratics in our approximation. The reason is easy to see from Figure 11.11-6, where we see that the cubic has a canceling error in the double interval. This formula is called *Simpson's formula,* is exact for cubics, and requires an even number of subintervals.

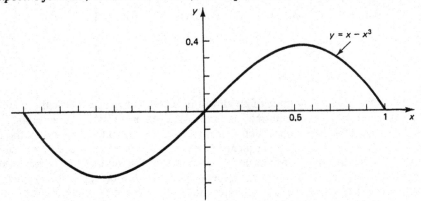

Figure 11.11-6 Typical cubic

Example 11.11-2

As a check on Simpson's formula, suppose we apply it to a cubic

$$y = x^3$$

from 0 to 1. We pick $N = 2$, meaning four subintervals, and $h = \frac{1}{4}$.

$$\frac{1}{12}\left\{ 1(0)^3 + 4\left(\frac{1}{4}\right)^3 + 2\left(\frac{2}{4}\right)^3 + 4\left(\frac{3}{4}\right)^3 + 1(1)^3 \right\}$$

$$= \frac{1}{12}\frac{1}{64}\{4(1) + 2(8) + 4(27) + 64\} = \frac{192}{(12)(64)}$$

$$= \frac{1}{4}$$

which is the correct answer.

Example 11.11-3

Suppose we use the known integral (which at present you cannot integrate)

$$\int_0^1 \frac{1}{1 + x^2}\,dx = \frac{\pi}{4}$$

We will again use Simpson's formula with a crude spacing, $N = 2$ (meaning four intervals, $h = \frac{1}{4}$). We have the table

x	$1/(1 + x^2)$	Weight	Product
0	1.00000	1	1.0000
$\frac{1}{4}$	0.94118	4	3.7647
$\frac{2}{4}$	0.80000	2	1.6000
$\frac{3}{4}$	0.64000	4	2.5600
$\frac{4}{4}$	0.50000	1	0.5000

The sum is 9.4247, and this is to be multiplied by $h/3 = \frac{1}{12}$. The result is

$$\text{Simpson} = 0.78539$$

$$\text{true} = 0.78540$$

which is remarkably close.

We have given only three of an almost infinite number of numerical integration formulas. Each may be programmed as a process, even on a small programmable hand calculator, with the particular function left as a subroutine to be supplied by the user in the specific case. The calling sequence gives the range numbers a and b, the number of intervals N, and the name (or location) of the description of the function to be integrated. Thus, these formulas are general methods of wide applicability *once* you have written the basic program. Unquestionably, Simpson's method is the most widely used in practice. The other two are inferior for most functions (but there are functions where the midpoint formula gives a better answer). You can see how you could derive many other integration formulas if you wished; simply find another approximation method and integrate it. Given noisy data, for example, you might choose to fit by least squares.

EXERCISES 11.11

Compute the following areas using the indicated methods:

1. $(1 + x^3)^{1/2}$, $(0, 1)$, trapezoid and midpoint, $N = 4$.
2. $(1/x)$, $(1, 2)$, trapezoid and midpoint, $N = 6$.
3. $1/x$, $(1, 2)$ Simpson, $N = 3$
4. $1/(1 + x^2)$, $(0, 1)$ trapezoid and midpoint, $N = 4$
5. $1/(a^2 + x^2)$, $(0, a)$, reduce to computable form
6. $1/x^2$, $(1, \infty)$, midpoint, any N you want

11.12 SUMMARY

We introduced the idea of an integral as a limiting method for finding areas of regions with curved edges. For continuous functions, the upper and lower sums were found to converge to the same limit as the largest interval size approached zero. The idea of area was then generalized to be the limit of any sum of elements, including discs, rings, and cylinders, or any other regular shapes.

We proved the fundamental theorem of the calculus, that the derivative of the integral is the integrand. This is the tool we use to find areas; we find an antiderivative (the indefinite integral) of the integrand and then substitute in the limits of integration.

Not only did we find various limits of sums; we also used the operation of integration to find new identities from old ones.

Finally, we looked briefly at the problem of what you can do when you cannot integrate the given function in terms of known functions. We gave three methods of integration, although there are many others. Day in and day out it is Simpson's formula that is used more than all others put together.

12

Discrete Probability

12.1 INTRODUCTION

The role of probability in science and engineering is rapidly increasing, not only in areas of applications but also in the very foundations, such as in quantum mechanics. It is essential that you master this way of thinking. To do so, you need to first learn the language of probability.

Probability has a long and confused history. People generally have an intuitive idea of what probability is, but in practice they often hold contradictory ideas at the same time; and certainly they are often confused about the logical relationship between the probability of an event occurring on a single trial and its frequency of occurrence in a sequence of many trials. To correct some of these logical confusions, it is necessary to start from the simplest elements of the subject.

In this chapter we will examine simple, discrete situations involving probability, and in the next chapter we will examine continuous situations, which are generally the more useful ones in science and engineering. The theory of probability arose mainly from gambling situations, which are typically discrete events such as the toss of a coin, the roll of a die (singular of dice), the draw of a card, or the turn of a roulette wheel, and they still provide the best intuitive approach to the subject. But, as noted, there are often confusions in the mind of the beginner, so please be patient with the following rather pedantic approach to the subject.

2.2 TRIALS

We begin with the concept of the toss of a "well-balanced coin," the roll of a "perfect die," or the draw of a card from a "well-shuffled deck" (each is clearly an idealization of a common situation in the real world). Consider, for example, the toss of a coin. Each toss is called a *trial*, and the trials are supposed to occur under identical conditions. We do not mean exactly identical, but rather under conditions that merely seem to us to be identical so far as the coin toss is concerned. Absolute identity in space and time is not possible. The identical trials produce the outcomes that we regard as *random*. By "random" we mean, at the moment, merely "unpredictable." We cannot say which of the two outcomes, heads or tails, will occur on any particular trial (toss). We are further assuming that there is no wear and tear on the coin during the repeated tosses, and that each trial is equivalent to any other trial, although the outcomes may be different.

The first thing we do is make a list of the possible elementary outcomes of a trial. Each possible outcome we regard as a point in a *sample space* (a common mathematical approach to a set of items). Thus we have a space of outcomes (*events*) corresponding to a trial. In simple problems, like those above, there are a finite number of possible outcomes in the sample space of outcomes. For a coin, the outcomes are H and T; for a die, the outcomes (the top face) are 1, 2, 3, 4, 5, 6; for the draw of a card from a deck (using the standard deck of spades, hearts, diamonds, and clubs), 1S, 2S, . . . , 13C. An event may involve a number of the elementary outcomes. For example, consider the event that the outcome of the roll of a die is an odd number on the upper face of the die, a member of the set $\{1, 3, 5\}$.

Sample spaces of events are often presented in terms of set theory and Venn diagrams (Figure 12.2-1). Set theory has been taught until the typical student is weary of it, so we will assume that it is familiar. Venn diagrams are apt to be misleading to the unwary. You would probably lay out the sample space of a random draw of a card from a deck in four rows of 13 items each:

$$1S, \ 2S, \ 3S, \ . \ . \ . \ , \ 13S$$

$$1H, \ 2H, \ 3H, \ . \ . \ . \ , \ 13H$$

$$1D, \ 2D, \ 3D, \ . \ . \ . \ , \ 13D$$

$$1C, \ 2C, \ 3C, \ . \ . \ . \ , \ 13C$$

The set of odd spades $\{1S, 3S, 5S, 7S, 9S, 11S, 13S\}$ is scattered and not in one nice connected piece.

We would like to define an *elementary event* exactly, but recall that the Greeks thought that the world might be made of indecomposable units called *atoms*, as did science for a long time. Later it was thought that the atoms were composed only of electrons and protons. Still later, further divisions of the elementary particles were assumed. Regularly, what the best minds of one generation considered as elementary

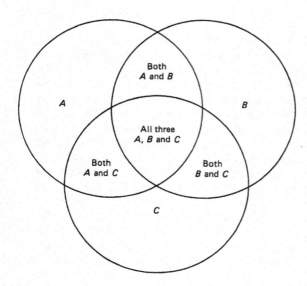

Figure 12.2-1 A Venn diagram

later was found to be composed of smaller elements. Yet we need the concept of an elementary *event*. We "feel" that the individual cards drawn from a deck are elementary units that cannot be further decomposed. However, we must leave the definition of an elementary event as an intuitive concept.

The above examples were sample spaces consisting of a finite number of elementary events. But if the trial is a sequence of tosses of a coin *until* a head appears, then we have the sample space of possible outcomes (events)

H

TH

TTH

TTTH

. . .

This sample space of outcomes (*n* tails followed by a head) is infinite, $(T)^n H$, for all nonnegative integers n.

Students are sometimes bothered by questions like, "What happens if the coin lands on its edge?" or "What happens if the trial is not completed because the coin is lost?" The answers are simple; we do not regard them as possible outcomes (events) of the trials. It is customary to consider only heads (H) and tails (T) as the two possible outcomes of a toss of a coin. Similarly, the roll of a die has six possible outcomes, and the draw of a card has 52 possible outcomes. We simply do not, generally, consider the outcome that a person dropped dead before the trial was completed. We only consider a selected list of outcomes from the class of all theoretically conceivable outcomes of a trial.

Next, we face the problem of assigning a *probability* to each outcome (event). The founders of probability theory spoke of "equally likely events" and said that the probability of a compound event, such as the face of the die being either 1 or 2, was the ratio of the total number of favorable elementary outcomes to the total of all possible elementary outcomes. For example, the probability of drawing a spade from a deck of cards is the ratio of

$$\frac{\text{number of spades}}{\text{total number of cards}} = \frac{13}{52} = \frac{1}{4}$$

The probability of a 1 or a 2 on the die is the ratio

$$\frac{2}{6} = \frac{1}{3}$$

The founders of probability theory regarded each possible outcome (event) of the trial of drawing a card from a well-shuffled deck as being equally likely. Similarly, they regarded each outcome of a roll of a perfect die as equally likely. There was confusion when they began to think about compound events like the sequence of tosses until an H appears. Each event in the sample space is not equally likely; only H and T are assumed to be equally likely on a single toss. This kind of distinction is not easy for the beginner to grasp. In this approach to probability it is necessary to decide which are the equally likely events and which are not equally likely events.

The definition of the probability (for the outcomes of the equally likely events) is the ratio of the successes to the total number of possible outcomes. If the given set of outcomes (compound events or simple events) exhausts (but does not duplicate) all possible events of the list of possible outcomes, then the total probability of all these outcomes must be exactly equal to 1. Thus the probability of any particular event is a real number between 0 and 1.

But what is the meaning of "equally likely" if not the assertion of a probability? There is a circularity in the very definition. Probability is an intuitive idea and cannot be based on other than a somewhat circular definition. Therefore, we will examine a bit more carefully why they chose equally likely events as the basic element of the theory.

The toss of a well-balanced coin seems to have equally likely outcomes (what else does "well balanced" mean?). The basic rule of avoiding contradictions in mathematics (and in science generally) requires that we assign the same weight (probability) to things (events) that we find indistinguishable (except for the *label* H or T, which we *assume* has no effect on the outcome of the trial). If we did not assign the same probability to each, then we would face the situation that we could interchange the indistinguishable outcomes and thus find different probabilities for the same outcomes of the same trial.

Thus the fundamental tool in assigning probabilities is the basic symmetry of situations, physical, geometric, logical, or algebraic symmetry. Just as congruent figures have the same area, so too in probability theory symmetric situations must have the same probability. Otherwise, we face a logical contradiction.

Unfortunately, experience in physics has shown that sometimes our intuition can be wrong, and that nonintuitive equally likely events are at the basis of the world (as we now understand it).

Applying this symmetry rule to a well-balanced coin, the two possible outcomes are equivalent; hence each outcome must have the probability of $\frac{1}{2}$ (since the total probability is 1). For a perfect die, each outcome must have the probability of $\frac{1}{6}$, and for the well-shuffled deck, each card has the probability of being drawn of $\frac{1}{52}$. The words "well balanced," "perfect," and "well shuffled" mean that the possible outcomes are all equivalent, and hence have the same probability. Thus in a symmetric situation the probability of each simple outcome is 1/(total number of possible simple outcomes).

Probabilities are sometimes stated in the form of *odds*. For equally likely events the odds of "three to two" implies that the ratio of successes to the total number is $\frac{3}{5}$, and the unfavorable ratio is $\frac{2}{5}$. The odds of 1 to the $\sqrt{2}$ is less easy to explain!

There is often a basic confusion in the student's mind. We are saying that the mathematically conceived events are equally likely; we are not asserting that the actual physical coins have, or have not, this property. We are modeling reality with the mathematics, but we are not handling reality itself. Common observation shows that a typical coin has something like a 50–50 chance (probability of $\frac{1}{2}$) of being either an H or a T. But a carefully performed experiment using a real coin would probably show some bias, a slight lack of exactly $\frac{1}{2}$ for the ratio of the outcomes. We are clearly applying, in this experiment, the concept that probability somehow determines the ratio of the outcomes of many trials. This is *not* part of the basic definition of probability. One of the main things we must at least partially justify is this connection, which is in your mind. With respect to the conventional die, the very fact that one face has 6 hollowed out spots and the opposite side has only 1 suggests that the center of gravity is not exactly in the middle of the die, and that we will, from this lack of symmetry, find a lack of symmetry in the corresponding outcome frequencies from many trials. But the bias is so slight in practice that the mathematically perfect die does closely approximate the behavior of a real die.

To repeat the main point, we are modelling reality, we are assigning mathematical probabilities to the outcomes in the list of possible outcomes of a single trial. The relationship between the mathematical model and the actual situation is subject to investigation, but lies outside the theory. In modern physical theory, such as quantum mechanics, there is an increasing acceptance that we can never know, *even in principle*, exactly what something is, that we must settle for probable values. (*Note:* It is fashionable in some circles to assert that if in principle we cannot know something then it does not exist!)

12.3 INDEPENDENT AND COMPOUND EVENTS

The concept of *independent events* is also an intuitive idea. We feel that, in the combined trial of a toss of a coin and the roll of a die, the outcome of the roll of the die is *independent* of the outcome of the toss of the coin. Actually, we are simply asserting that in the mathematical model they are independent, that the outcome of the coin toss does not influence the outcome of the roll of the die. Therefore, the possible outcomes (events) can be listed as

$$\text{H1} \quad \text{H2} \quad \text{H3} \quad \text{H4} \quad \text{H5} \quad \text{H6}$$

$$\text{T1} \quad \text{T2} \quad \text{T3} \quad \text{T4} \quad \text{T5} \quad \text{T6}$$

There are 12 possible events, and due to the apparent symmetry each should have the same probability of $\frac{1}{12}$. If the two events, the outcomes of the roll of the die and the toss of the coin, are taken as being independent, then we see that the probability of any combined outcome is the product of the probabilities of its individual outcomes. This is a general rule: *for independent events, the probability of the compound event is the product of the probabilities of the simple events.* In the above case, we have

$$\frac{1}{2}\frac{1}{6} = \frac{1}{12}$$

This follows from the feeling that the list of the 12 events of the trial are all equally likely since the outcomes of the individual parts are equally likely. We do not prove that the two events are independent; rather we assume that they are because we can see no immediate interconnection between them.

This sample space of events is often called the *product space.* The space of possible events consists of all possible products, one factor from each of the basic sample spaces H, T and 1, 2, . . . , 6. Note that the corresponding probabilities assigned to the events are the corresponding products of the probabilities of the individual independent events.

But some events are not independent. If we draw a card at random from a deck and return it, shuffle, and then draw again, we feel that the two drawings are independent events. For example, drawing a 4H and a 7S (H = hearts, S = spades) on the two independent draws would be realized by either ("&" means followed by)

$$\text{4H \& 7S} \quad \text{or} \quad \text{7S \& 4H}$$

since the order of drawing them is supposed not to matter. In this case we have, for each order of the two drawn cards,

$$\left(\frac{1}{52}\right)^2$$

so the total probability of this event is

$$P = \frac{2}{(52)^2}$$

On the other hand, if we do not replace the first card before drawing the second card from the deck, then the second drawing is from a deck of 51 cards. The probability of 4H & 7S is

$$\frac{1}{52} \frac{1}{51}$$

and is the same as for 7S & 4H. The total probability of the *compound* event of drawing two cards without replacement is therefore

$$P = \frac{2}{(52)(51)}$$

which is not the same as when the card is replaced. Why the difference? We easily see that one of the possible outcomes of the replacement trial is that the same card is drawn both times, while in the nonreplacement case this is impossible ($p = 0$). Thus the ratio of the successes to the total number of equally likely outcomes is different in the two cases; in the second case there are fewer possible equally likely outcomes than in the first case.

The probabilities of compound events are based on the probabilities of the simple (elementary) events. The probability of HH in the toss of a coin (assuming independent tosses) is the product of the probabilities of the individual events. The sample space of events is

$$HH \quad \frac{1}{4}$$

$$HT \quad \frac{1}{4}$$

$$TH \quad \frac{1}{4}$$

$$TT \quad \frac{1}{4}$$

This is merely the product space (for independent events) H = $\frac{1}{2}$, T = $\frac{1}{2}$, taken two times. Note that the compound event of an H and T *in either order* is $\frac{1}{2}$.

It is the general rule (in the mathematical model) that probabilities of independent events are to be multiplied to get the probability of both events happening. This is what we mean by a product space of independent events. If the events in the original spaces are all equally likely, then the events in the product space of independent events are also equally likely.

Example 12.3-1

What is the probability of exactly three heads in four independent tosses of a coin? The sample space consists of any sequence of H and T, and each sequence of the four has the same probability (because we are assuming the independence of the tosses). The probabilities of these 2^4 simple events are each

$$\left(\frac{1}{2}\right)^4$$

This is the probability of each event in the basic sample space of events where each event is the result of four tosses (or four coins tossed at the same time, since both lead to the same product space). How many ways can three heads occur in the four positions, the other position being filled wtih a T? We first list all 16 equally likely possible outcomes of four tosses:

HHHH	HHHT	HHTT	HTTT	TTTT
	HHTH	HTHT	THTT	
	HTHH	HTTH	TTHT	
	THHH	THHT	TTTH	
		THTH		
		TTHH		

We have put in the first column all heads, in the second column those events with 1 T, in the third those with 2 T's, in the fourth those with 3 T's, and in the fifth column those with 4 T's. We see (from the complete listing) that there are exactly 4 successes out of the total 16 possible outcomes. The ratio gives the probability

$$\frac{4}{16} = \frac{1}{4}$$

There is an alternative approach to the problem. We note that the probability of getting one of the two faces of the coin is exactly $\frac{1}{2}$. The four tosses are supposed to be independent, so the probability of any given sequence is the product

$$\frac{1}{2}\frac{1}{2}\frac{1}{2}\frac{1}{2} = \frac{1}{16}$$

Furthermore, there are only four ways of getting one T (3 H's):

$$\text{THHH} \qquad \text{HTHH} \qquad \text{HHTH} \qquad \text{HHHT}$$

And each being of probability $\frac{1}{16}$, we get

$$\frac{1}{16} + \frac{1}{16} + \frac{1}{16} + \frac{1}{16} = \frac{1}{4}$$

as the probability of three H's.

Example 12.3-2

We now consider the trial of a coin toss until a head appears. We regard each single toss as independent of any other preceding or succeeding toss. Each toss has probability of $\frac{1}{2}$ to be either an H or a T. The compound events, which make up the events of our sample space of runs of tosses until a head, have the corresponding probabilities

$$\text{H} \qquad \frac{1}{2}$$

$$\text{TH} \qquad \frac{1}{4}$$

$$\text{TTH} \qquad \frac{1}{8}$$

$$\text{TTTH} \qquad \frac{1}{16}$$

$$\ldots \qquad \ldots$$

In general, we have

$$(\text{T})^n\text{H}, \qquad \left(\frac{1}{2}\right)^{n+1}$$

for all $n = 0, 1, 2, \ldots$.

Do all these probabilities total up to exactly 1 as they should? We have an infinite geometric progression with

$$a = \frac{1}{2} \quad \text{and } r = \frac{1}{2} < 1$$

so the sum is [Equation (7.2-3)] the total probability

$$S = \frac{a}{1 - r} = \frac{1/2}{1 - (1/2)} = 1$$

The *expectation* of a variable is defined to be the average over the sample space of the variable you are examining. Thus the expected length of run is the weighted average of the run lengths; each run length is weighted by the corresponding probability of the occurrence of that run length. This is not, as you may think, the average over repeated trials. You think that the two are the same (one average is over the probability space and one is from the average of a long sequence of trials), but we have not proved in what sense these can be the same. You associate probability with long-term ratios of successes to total trials, but that is not how we started to define probability. We got the basic probabilities from symmetry, not from ratios of repeated trials.

The concept of *expectation* as an average over the sample space is indeed what you "expect" from repeated trials, but you need to be careful. There are traps awaiting the careless person. For example, if you call a head 1 and a tail 0, then the expected value of a single trial is

$$\left(\frac{1}{2}\right)1 + \left(\frac{1}{2}\right)0 = \frac{1}{2}$$

which is *not* any single outcome. You may "expect" to get $\frac{1}{2}$ from the toss of a coin, but in fact you must get either a 0 or a 1. Similarly, for the toss of n coins (or n tosses of a single coin) you expect $n/2$ heads, but if n is odd, this is impossible; and in any case you may, due to chance fluctuations, get values far from $n/2$.

Example 12.3-3

(Previous example continued) What is the expected number of tosses of a coin before an H appears? If we say that the event H has length 1, the event TH has length 2, and so on then to get the expected length of run until the first H turns up we have to sum the lengths multiplied by their corresponding probabilities; that is

$$\text{expectation} = 1\left(\frac{1}{2}\right) + 2\left(\frac{1}{2}\right)^2 + 3\left(\frac{1}{2}\right)^3 + \cdots$$

How can we sum this? Equation (9.10-2) was something like it. Let us look at it. Example 9.10-2 showed that

$$\frac{1}{(1-x)^2} = 1 + 2x + 3x^2 + 4x^3 + \cdots$$

If we multiply this identity through by x,

$$\frac{x}{(1-x)^2} = x + 2x^2 + 3x^3 + 4x^4 + \cdots$$

and then set $x = \frac{1}{2}$, we would have our result. Therefore, the

$$\text{expectation} = \frac{1/2}{[1-(1/2)]^2} = \frac{1/2}{1/4} = 2$$

This seems like a reasonable result. (You can always approximately test it by tossing a coin and finding that the average number of tosses needed is approximately two.) If you are dubious about handling infinite series in this formal way, then you can take a finite segment of the geometric series, compute the answer, and then let the number of terms in the series approach infinity.

From these simple examples, we see the need for more elaborate mathematics to count the possible outcomes to be listed in the event space (suppose you asked for 3 H's out of 10 tosses; then you would have a space of 1024 outcomes, which would be unpleasant to list in all detail). We also see the necessity for taking a closer look at the use of the calculus in generating identities from known identities. We now turn to the first of these, after looking at one final example.

Example 12.3-4

There is a famous puzzle based on two coins in two drawers of a box. The problem is that there is a box with two drawers. In one drawer there are two gold coins and in the other drawer there is a gold and a silver coin. Given that you opened at random one drawer, picked at random one coin in that drawer, and you found that it was gold, what is the probability that the other is silver? The false reasoning is that the other coin is either a gold or a silver coin and therefore each has probability $\frac{1}{2}$. The correct reasoning goes as follows. The words "at random" mean that both the drawer and the coin have equal probabilities (remember the words "at random")

$$G_1G_2 = \frac{1}{4}, \qquad G_2G_1 = \frac{1}{4}, \qquad GS = \frac{1}{4}, \qquad SG = \frac{1}{4}$$

The actual trial that found the gold coin eliminated the fourth possibility; there are only three possible outcomes left and each has the same probability, now $\frac{1}{3}$ (since the sum must be 1). Of these three equally likely outcomes, only one is successful (given that you have already seen the gold coin), and the probability is therefore $\frac{1}{3}$. This shows the necessity of watching what are the actual equally likely cases, and the advantage of imagining an experiment that might be done to test out the result. This is an example of *conditional probability* (see Section 12.9); the probability is *conditioned* by having found a gold coin in the first drawer.

12.4 PERMUTATIONS

There is a simple rule for combining independent choices. If the first selection can be made in n_1 ways, the second independent choice in n_2 ways, and the third independent choice in n_3 ways, then the number of ways you can choose a set consisting of one from each is

$$N = n_1 n_2 n_3$$

This rule (generalized in an obvious way, which we need not state) is very useful in solving problems involving the selection of items. It is simply the product space of the individual independent sample spaces (choices).

Example 12.4-1

Suppose we have four items, and label them A, B, C, and D. In how many ways can we arrange them? We make a systematic list.

ABCD	BACD	CABD	DABC
ABDC	BADC	CADB	DACB
ACBD	BCAD	CBAD	DBAC
ACDB	BCDA	CBDA	DBCA
ADBC	BDAC	CDAB	DCAB
ADCB	BDCA	CDBA	DCBA

There are 24 cases. We may analyze them in the following way. The first letter could be picked in any of four ways (the four columns). The second letter could be picked in any of three ways, and finally the third letter could be picked in any of two ways; the last letter is forced to be the one left over. Thus we have the product

$$4 \times 3 \times 2 \times 1 = 24$$

as the number of ways of selecting the four items.

Generalization 12.4-2

Suppose we have a collection of n distinct items. In how many ways can we arrange them when we regard different orderings as being different? Evidently, we can choose the first item in any of n ways, the second item as one of the $n - 1$ remaining items, the next item from one of the $n - 2$ items, and so on, down to the last item, which is the one that is left. As we think about it, we see that there will be a total of

$$n(n - 1)(n - 2) \ldots (2)1 = n!$$

distinct ways of selecting the n items. These are called *permutations,* because if we permute (interchange) items we have distinct selections. In mathematical notation we have

$$P(n, n) = n!$$

Rule: The number of permutations of n things taken n at a time is $n!$.

Example 12.4-3

Suppose from the same four items A, B, C, and D we select only two items (independently). We can get

$$AB, \ BA, \ AC, \ CA, \ AD, \ DA, \ BC, \ CB, \ BD, \ DB, \ CD, \ DC$$

which are 12 arrangements (permutations). We could reason as follows; the first letter could be picked in any of four ways, and the second letter in any of three ways, which makes a total of

$$4 \times 3 = 12$$

ways.

Generalization 12.4-4

If, from a collection of n distinct things, we select only r of them, then we have

$$P(n, r) = n(n - 1)(n - 2) \ldots (n - r + 1)$$

different permutations since we can select the first in n ways, the second in $n - 1$ ways, the third in $n - 2$ ways, on down to the rth, which is selected from the $n - r + 1$ things that are left at that time. If we multiply both numerator and denominator by $(n - r)!$, we get

$$P(n, r) = \frac{n!}{(n - r)!} \tag{12.4-1}$$

which is the number of permutations of n things taken r at a time. *Rule: The number of permutations of n things taken r at a time is n factorial divided by $n - r$ factorial.*

Example 12.4-5

The birthday problem. We wish to know the probability that one or more persons in a room will have the same birthday. It is natural to assume that the individuals have birthdays that are independent (that both of a known pair of twins, for example, are not there at the same time), that birthdays are uniformly probable over all the days (which is slightly wrong), and that there are exactly 365 days in a year (which is definitely wrong by a very small amount). The errors in the assumptions are all fairly small.

We begin by inverting the problem (a very common thing in probability) and compute the probability that no two persons have the same birthdate. We can do this in one of two ways. If we ask, "How many ways can we select the k people, one at a time, from the 365 days with no duplicates?" the answer is that this is the number of permutations of n things taken k at a time:

$$P(365, k)$$

This is the number of successes. Next, how many sets of k (with duplicates) can you select? This is

$$365^k$$

and is the total number of equally likely elements in the sample space. The ratio of successes to the total possible is the desired probability

$$Q(k) = \frac{P(365, k)}{365^k}$$

The second approach is to deal in the probabilities from the start. We can select the first person with probability (of a duplicate birthday) $365/365 = 1$ of success, the next with probability $364/365$, the next with probability $363/365, \ldots$, and the kth with probability $(365 - k + 1)/365$. Reasoning that these are independent choices, then the probability that no two persons have the same birthdate is the product

$$Q(k) = \frac{[365(364)(363) \ldots (365 - k + 1)]}{365^k}$$

What was asked for was the probability $P(k) = 1 - Q(k)$ of a duplicate, which is

$$P(k) = 1 - Q(k) = 1 - \frac{365!}{(365 - k)! \, (365)^k} \qquad (12.4\text{-}2)$$

It surprises many people that when $k = 23$ the probability $P(k)$ is greater than $1/2$. We give a short table to show how $P(k)$ depends on k.

k	$Q(k)$	$P(k) = 1 - Q(k)$
5	0.9729	0.0271
10	0.8831	0.1169
15	0.7471	0.2529
20	0.5886	0.4114
25	0.4313	0.5687
30	0.2937	0.7063
35	0.1856	0.8144
40	0.1088	0.8912
45	0.0590	0.9401
50	0.0296	0.9704

A pair of interesting numbers are $P(22) = 0.4757$ and $P(23) = 0.5073$. With 23 people in a room (and the assumed independence), there is better than a 50% chance of a duplicate birthday.

Why is this true? You tend to forget that with k people there are $k(k - 1)/2$ possible pairs to have duplicate birthdays; you tend to think of yourself having a birthday in common with someone else, forgetting that this also applies to everyone else in the room.

Example 12.4-6

Biased coin. How can you get an unbiased choice from a coin that is biased? Simple! If you toss the coin twice, you have the sample space with the probabilities (where p = probability of an H and $q = 1 - p$)

Outcome	Probability
HH	p^2
HT	pq
TH	qp
TT	q^2

We notice that HT and TH have the same probabilities. Thus we call one outcome (say HT) the event H and the other outcome (say TH) the event T, and if either of the other two cases occurs, we simply try again. We can get an unbiased decision from a biased source.

It is interesting to compute the expected number of double tosses to get a decision. You get a decision with probability $P = 2pq$, and you fail to get a decision with probability $Q = 1 - P$. Thus you have the equation (one trial = a double toss)

$$\text{expected no. of trials} = (1)P + (2)QP + (3)Q^2P + \cdots$$

$$= \frac{P}{(1 - Q)^2} = \frac{1}{P} = \frac{1}{2pq}$$

If the coin is unbiased, $p = \frac{1}{2}$, you can expect two (double) tosses, but the more biased the coin is, the longer you will have to toss until you get a decision.

EXERCISES 12.4

1. In how many ways can you select seven items from a list of ten items?
2. In how many ways can you go to work if you have a choice of two cars, four routes, and seven parking spaces?
3. In how many ways can you select (independently) a shirt, jacket and shoes if you have nine shirts, three jackets, and four pairs of shoes?
4. What is the probability that two faces are the same when you roll two dice?

5. What is the probability of a total of 7 on the faces of two dice thrown at random?

6. What is the probability of each sum on the faces of two dice?

7. Assuming that each month has 30 days, what is the probability of two people having the same birthdate (in the month) when there are k people in the room? Where is the 50% chance point?

8. If I put k marbles in ten boxes, what is the probability of no box having more than one marble?

9. Write out all the permutations of two items selected from a set of five items.

10. If the amount of change in a pocket is assumed to be uniformly distributed from 0 to 99 cents, how many people must be in a room until it is at least a 50% chance that two will have the same amount of change?

11. How many three-letter arrangements can you make of the form consonant, vowel (a, e, i, o, u), consonant?

12. Criticize the proposal that 50 people use the last three digits of their home phone number as a private key for security purposes with the requirement that no two have the same key.

12.5 COMBINATIONS

Frequently, the order in which the numbers from the selected set occur is of no importance. Thus, when dealing cards in a card game, the order in which you receive the cards is often of no importance. When order is neglected, the sets are called *combinations*.

If we start with the number of permutations $P(n, r)$, how many of them are the same when we ignore the order? There are exactly $r!$ orderings of r things, so $r!$ different permutations all go into the same combination. Thus the number of combinations of n things taken r at a time is

$$C(n, r) = \frac{P(n, r)}{r!} = \frac{n!}{(n - r)!\, r!}$$

These are exactly the binomial coefficients of Section 2.4. (We saw them in the table of heads and tails in Example 12.3-1; the number of items in the columns were 1, 4, 6, 4, 1.) Therefore, we are justified in using the same notation for the quite different appearing things; they are the same numbers. You see the central role that the binomial coefficients play in the theory of combinations. This is an illustration of the fact that mathematical functions devised for one purpose often have applications to very different situations, the universality of mathematics (Section 1.3).

Example 12.5-1

This observation supplies an alternative proof of the binomial theorem (Section 2.4). We consider

$$(a + b)^n = (a + b)(a + b)(a + b) \ldots (a + b)$$

Let us multiply out the first stage (keeping the order of the factors)

$$(a + b)a + (a + b)b = aa + ba + ab + bb$$

The next stage gives

$$aaa + baa + aba + bba + aab + bab + abb + bbb$$

At the nth stage, the number of times we will have

$$a^k b^{n-k}$$

is the number of ways a can occur in k places of the n possible places. This is $C(n, k)$ times. Thus it will give rise, when these terms are all combined, to the term

$$C(n, k)a^k b^{n-k}$$

The whole expansion will be

$$(a + b)^n = \sum_{k=0}^{n} C(n, k)a^k b^{n-k}$$

which is the binomial theorem. Thus we have given a "counting" proof of the binomial theorem in place of the earlier inductive proof.

This also means that what we learn in one of the situations can be applied to the other. For example, we showed that

$$C(n, r) = C(n, n - r)$$

In the new situation, this says that the number of combinations of n things taken r at a time is the same as the number of combinations of n things taken $n - r$ at a time. The truth of this is seen when you count the combinations of those that are not taken; they match, one by one, those that are taken. What you take is uniquely defined by what you leave.

Another example of this transfer of knowledge is the binomial identity

$$C(n, 0) + C(n, 1) + C(n, 2) + \cdots + C(n, n) = 2^n$$

which is the same statement we proved in Example 2.3-4 of counting the number of subsets (proper and improper) of a set of n items.

Example 12.5-2

How many different hands of 13 cards are possible? There are conventionally 52 cards in a deck, and in a hand of cards the order of the cards is of no importance; so we have

$$C(52, 13) = \frac{52!}{13! \, 39!}$$

which is a very large number. We will leave to later how to estimate such numbers. We note, however, that it can be written as

$$C(52, 13) = \frac{(52)(51) \cdots (40)}{(13)(12) \cdots (1)}$$

$$= \binom{52}{13}\binom{51}{12} \cdots \binom{40}{1}$$

$$= 6.350 \ldots \times 10^{11}$$

which is fairly easy to compute on a hand calculator.

EXERCISES 12.5

1. In two independent draws of a card from a deck, what is the probability of getting a duplicate?
2. How many different bridge deals can be made, regarding the various positions as independent?
3. What is the ratio of the number of combinations to the number of permutations of n things taken k at a time?
4. How many sets of k items can be selected from n items?
5. How many hands of seven cards are possible?
6. How many teams of 5 players can be selected from a class of 20 students?
7. What is the probability of seeing 7 heads in the toss of 12 coins?
8. Given n items, show that the number of combinations involving an odd number of selected items equals the number of combinations in which an even number are selected. *Hint:* $(1 - 1)^n = 0$.
9. What is the probability $P(k)$ of a run of length k for rolling a die until a duplicate number appears?

12.6 DISTRIBUTIONS

Example 12.6-1

Now that we have the mathematical tools and notation, we are ready to explore questions such as the number of possible outcomes, say, of ten tosses of a (mathematical) coin. There are

$$2^{10} = 1024$$

possible outcomes. How many have 0, 1, 2, . . . , 10 heads? These are simply the numbers

$$C(10, k) \qquad (k = 0, 1, \ldots, 10)$$

These eleven numbers give the *distribution of the number of heads* that can occur in ten tosses of a coin (see Figure 12.6-1). They can be computed sequentially using the rule of Section 2.4. They are

$$1, 10, 45, 120, 210, 252, 210, 120, 45, 10, 1$$

which add up to the required 1024 cases in total. Each of the 1024 events is equally

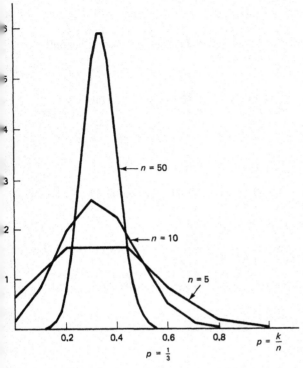

Figure 12.6-1

likely, *provided* both the original individual H and T were equally likely and the tosses were independent. Therefore, to get the probabilities, we merely divide the above numbers by the total number of equally likely possible outcomes. Thus the probability of k heads is

$$\text{Prob\{of exactly } k \text{ heads in ten trials}\} = \frac{C(10, k)}{2^{10}}$$

This is the *distribution* of the probability of k heads in ten trials and the total probability is 1. The sum of all the probabilities is 1 since the sum of the binomial coefficients of order n is 2^n.

Suppose we have a *binary event* (a trial with only two possible outcomes) and one of the outcomes has probability p. The other outcome naturally has probability $1 - p$. Conventionally, we write

$$q = 1 - p \qquad (12.6\text{-}1)$$

as the *complementary* probability to p. Let a "success" have probability p; then the "failure" has probability q. For example, items coming down a production line may be "good" or "bad." A medical bandage may be sterile or not. A single trial may succeed or fail. When interviewing people, a single trial (asking a person a question) may yield a response or no response. In an experiment, a mouse may or may not catch a disease. These yes–no situations in which the trials are independent are very common in life, and are called *Bernoulli trials*, since Jacques (James) Bernoulli (1654–1705) first investigated the theory extensively.

Example 12.6-2

For n Bernoulli trials, what is the distribution of the possible outcomes? There are $n + 1$ possible outcomes *if* we ignore the order in which the events occur. In how many ways can exactly k successes occur? There are $C(n, k)$ ways of this happening. Next, what is the probability of each of these individual events? The probability of exactly k successes followed by $n - k$ failures must be

$$p^k(1 - p)^{n-k}$$

This is the probability for any *preassigned* sequence of k successes and $n - k$ failures. Thus the *distribution* of the probability of k successes is given by the function

$$C(n, k)p^k(1 - p)^{n-k} \qquad (12.6\text{-}2)$$

which is the number of different ways it can occur times the probability of the single event. In Figure 12.6-1, there is a plot of this for $p = \frac{1}{3}$ and for both $n = 5$ and $n = 10$. In both cases the total probability is, of course, exactly 1. This can be verified mathematically, because they are the individual terms obtained by expanding the binomial

$$(p + q)^n = 1$$

remembering that $p + q = 1$. *Note:* This result (12.6-2) agrees with the result of Example 12.6-1 when $p = \frac{1}{2}$.

Example 12.6-3

Frequency of occurrence in Bernoulli trials. We can now make a connection between the probability p of a success in a single trial and the probability $P(k)$ of k successes in n independent trials. By (12.6-2), the probability distribution function for k successes is

$$P(k) = C(n, k)p^k q^{n-k}$$

For what value(s) of k is this a maximum? This is a discrete maximization problem. We write

$$P(k + 1) = \frac{n!}{(k + 1)!(n - k - 1)!}p^{k+1}q^{n-k-1}$$

$$= \frac{n!}{k!(n - k)!}\left(\frac{p}{q}\right)\left[\frac{n - k}{k + 1}\right]p^k q^{n-k}$$

$$= P(k)\left[\frac{n - k}{k + 1}\right]\left(\frac{p}{q}\right)$$

The function $P(k)$ is increasing when

$$\frac{(n - k)p}{(k + 1)q} > 1$$

$$np - kp > (k + 1)(1 - p) = k + 1 - kp - p$$

$$np - q > k$$

If the equality holds, then there are two terms of the maximum size.

Thus we see that the distribution function $P(k)$ for k successes in n trials (derived from the product space of single trials) rises for values $k < np - q$ and falls for values $k > np - q$. Figure 12.6-1 and a little familarity with the binomial coefficients show the increasing peakedness of the distribution function $P(k)$ as n gets large. The peak is near $p = k/n$. The number of successes divided by the total number of trials is near p.

Distributions of many other kinds of events can be computed. The above distributions are simply a grouping of the original terms that had the same probabilities into a single term. This term is exactly k successes out of the n trials; that is why the number of times something can occur is so often needed.

Given a distribution, it is common to ask, "What is the *expected value* of this distribution?" (Section 12.3). This is the same thing as the *average* over the sample space. The average of what? Of the number of successes, presumably. Thus we are asked to compute the expression (the expected value)

$$E\{K\} = \sum_{k=0}^{n} kC(n, k)p^k(1 - p)^{n-k} \tag{12.6-3}$$

where k is the number of successes. The notation requires some explanation. The letter k is the dummy index of summation and could be changed to any other letter we pleased. But we have the letter K on the left-hand side. What is the meaning of K? It gives a clue, via the variable of summation, to what is being computed. It is k times the probability of k successes that goes into the sum. If we had written

$$E\{K^2\}$$

this would mean

$$E\{K^2\} = \sum_{k=0}^{n} k^2C(n, k)p^k(1 - p)^{n-k} \tag{12.6-4}$$

the sum of the square of k multiplied by the corresponding probability from the distribution we are averaging over.

Example 12.6-4

To find a more compact form for representing the expected value (12.6-3),

$$E\{K\} = \sum_{k=0}^{n} kC(n, k)p^k q^{n-k}$$

we have to think how we can create the appropriate identity. The $C(n, k)$ come from the generating function (2.4-1):

$$(1 + t)^n$$

Looking at our expression (12.6-3), we finally think that starting with

$$(q + pt)^n \tag{12.6-5}$$

would be a good idea since it looks something like what we want. We have

$$(q + pt)^n = \sum_{k=0}^{n} C(n, k) p^k q^{n-k} t^k$$

If we differentiate this identity with respect to t, we will have the correct coefficient in front of each binomial coefficient. We now set $t = 1$:

$$n(q + p)^{n-1} p = \sum_{k=1}^{n} k C(n, k) p^k q^{n-k}$$

The difference in the ranges of summation between this and the one we started with (which began with 0) has no effect since the omitted term is multiplied by 0. *Now*, using the fact that $p + q = 1$, we have, finally,

$$E\{K\} = np = \sum_{k=1}^{n} k C(n, k) p^k q^{n-k} \tag{12.6-6}$$

In words, the expected value of the binomial distribution is the number of trials times the probability of a success.

We regard $E\{\ \}$ as *the expectation operator* (just as we regarded $d\{\ \}/dx$ and $\int \ldots dx$ as operators), operating on a function to be supplied in the braces. The expectation is a linear operator. The function operated on is to be multiplied by the corresponding probability of occurring and then summed over all possible cases that can occur. The denominator of the averaging is the sum of the probabilities, and, since this is always 1, it can be ignored.

We have the simple rules (A and B are constants)

$$E\{Af(K) + Bg(K)\} = AE\{f(K)\} + BE\{g(K)\}$$

$$E\{A\} = A$$

When computing an average (expected value), it is important to ask, "Whose average?" The following example will illustrate this point.

Example 12.6-5

Suppose there is a chemistry department that has

No. of Classes	Size of Class	Total Students
10 (freshman)	30	300
10 (sophomore)	20	200
10 (jr.–sr.)	10	100

What is the average class size? If you ask the faculty, they will say that the average class size is 20, since

$$\frac{10(30) + 10(20) + 10(10)}{10 + 10 + 10} = \frac{600}{30} = 20$$

But note that half the students are in a class of size 30. If we ask the average size as seen by the students, we get

$$\frac{30(300) + 20(200) + 10(100)}{300 + 200 + 100} = \frac{14,000}{600}$$

$$= 23\frac{1}{3}$$

Now suppose that the chemistry department decides to combine the 10 freshman classes into one large lecture, and then use the 9 free lecturers to add 9 jr.-sr. courses of size 10 (thus increasing the number of advanced students by 90). You now have:

No. of Classes	Size of Class	Total Students
1 (freshman)	300	300
10 (sophomore)	20	200
19 (jr.–sr.)	10	190

The faculty would now say that the average class size is

$$\frac{1(300) + 10(20) + 19(10)}{1 + 10 + 19} = \frac{690}{30} = 23$$

but the students would see the average class size as

$$\frac{300(300) + 200(20) + 190(10)}{300 + 200 + 190} = \frac{90,000 + 4000 + 1900}{690}$$

$$= \frac{95,900}{690} = 139^{-}$$

Thus you see how different the service offered looks from the point of view of the people offering the service (23) and those getting it (139). This explains the well-known phenomenon that the airlines claim that many flights are fairly empty, but the customers see mainly the crowded flights (after all, no passenger has ever flown on a completely empty flight!). This same phenomenon applies to crowded nursing homes, jails, transportation systems, and so on. Notice how small fluctuations from average as they grow in size are rapidly transformed to very great differences in the quality of service as seen by the customer.

Generalization 12.6-6

Find the general formula for computing the phenomenon in Example 12.6-5. We begin with the distribution n_i units of size s_i. The server computes

$$\frac{\Sigma\, n_i s_i}{\Sigma\, n_i}$$

the consumer computes

$$\frac{\sum n_i s_i^2}{\sum n_i s_i}$$

These are very different things! They are the same only if all the sizes s_i are the same. Once there is a reasonable variability in the sizes, then rapidly the customer tends to be in the big groups and sees a degradation in the quality of service. Thus there is a simple rule: to improve the quality of the service, you should work hard on the worst cases and not spend a lot of time on the slightly bad cases.

EXERCISES 12.6

1. Find the maximum in Example 12.6-3 by differencing the successive terms $P(k + 1) - P(k)$.
2. Apply the theory of Generalization 12.6-6 to the class sizes you currently have.
3. Compare the case of three classes, each of size s, with three with sizes $s - k$, s, and $s + k$.
4. What is the expected value of the distribution of the number of rolls of a die until a 6 appears?
5. What is the distribution of the number of double tosses of a biased coin until you get a matching pair? Find the expected value.

12.7 MAXIMUM LIKELIHOOD

In the above probability problems we assumed that we had the initial probabilities. In statistics we invert things, and we start with the results of the observations. Assume that we have made n Bernoulli trials and have seen exactly k successes. What is a reasonable value to assign to p? Of course, any value of p might have given the result of k successes, but some are very unlikely to have done so.

In this situation we need a rule to decide which value of p we should pick from all the possible values of p available. Bernoulli suggested that we take the p that is "most likely." True, the outcome of the n trials is a random event, and p could be number $0 \leq p \leq 1$, but what is a *reasonable* choice for the unknown value of p? Bernoulli argued that *in the absence of other information, you should choose the value of p that is "most probable," most likely to give rise to what you have observed, the one with the maximum likelihood. How could you seriously propose to choose a very unlikely value of p?* It is, of course, implied that the distribution of outcomes has a single, sharp peak, and not two or more approximately equal sized peaks. In this latter case, the concept of picking the most likely probability has much less sense. Maximum likelihood applies mainly to single, sharp peak situations.

Example 12.7-1

Find the value of p for the maximum of the function $P(k)$ for k successes in n Bernoulli trials,

$$P(k) = C(n, k)p^k(1 - p)^{n-k}$$

To find the value of p, you differentiate and set the derivative equal to zero. In this maximizing process, the leading constant coefficient $C(n, k)$ has no effect, so you can ignore it and maximize, as a function of p, the expression

$$f(p) = p^k(1 - p)^{n-k}$$

The derivative is

$$f'(p) = p^k(n - k)(1 - p)^{n-k-1}(-1) + (1 - p)^{n-k}kp^{k-1} = 0$$

Divide out the common factor $p^{k-1}(1 - p)^{n-k-1}$ to get

$$-(n - k)p + k(1 - p) = 0$$

Solve for p:

$$p = \frac{k}{n}$$

This is exactly what you would expect! If all you know is that there were k successes out of n Bernoulli trials, you would guess that the probability is the ratio of the successes to the total number of trials, k/n.

Beware! This is not *the* probability; it is your "best guess" at it, it is the "most likely value." Due to the method used, the result must be a rational number, but we saw in Section 3.6 that rational numbers have decimal representations that are periodic, so it is very "unlikely" that the "true" probability is a rational number. It is your estimate that is a rational number. Probabilists deal with *given* probabilities; statisticians *infer* likely values for probabilities and do not deal with the certainties of the probabilists. Furthermore, you might have seen a very unlikely set of trial outcomes and be very far from an accurate estimate of p.

But we have done more. We have made another connection between the idea of probability (based on symmetries in the product space of repeated independent trials) and the idea of the frequency of occurrence (based on repeated trials). It is the principle of maximum likelihood which says that what we observe should be the most likely event in the product space of events. It is true that this is only an assumption, but in the one use (so far) of this principle the result agrees with common sense (often meaning uncommon sense!).

One verification does not establish a principle as being infallible. Indeed, no finite number of verifications can do it. Maximum likelihood must remain a very plausible principle, and it is not a mathematically derived result. We will, in Section 13.6, derive a similar result that connects the probability with the frequency of occurrence.

The Bernoulli assumption of maximum likelihood is that you should take the value of the unknown parameter (in this case the value of p) that is "most likely" to give rise to the observations you have seen. It seems hard to argue that you should arbitrarily pick any other value of p, that you should pick an unlikely value—one less likely to give rise to what you have seen—but this is not a proof that the maximum likelihood principle always gives the "best" value for an unknown parameter. It remains a very plausible rule, a very useful rule, but arbitrary in the final analysis.

Example 12.7-2

Proofreading. It is a common experience to have two different people read a table of numbers to find the errors that may be in the table. Suppose the first reader finds n_1 errors (wrong digits), the second reader finds n_2 errors, and that when the two lists of errors are examined there are n_3 errors in common. Can we make any reasonable guess at the remaining errors in the table? Answer: not if we do not make some assumptions about the process of finding the errors. The assumptions we make must be very simple since we lack much data on the subject. A realistic model would require extensive measurements of both the kinds of errors and the habits of the readers. We therefore make a rather idealistic model.

We will assume that there is an unknown number N of errors in the table and that each is as easily found as any other. Second, we will assume that the two readers find errors independently of each other. These are ideal assumptions to be sure, but they are plausible, and they enable us to make a guess at the number N.

From Example 12.7-1, we see that a reasonable assumption for the probability of the first reader finding an error is

$$p_1 = \frac{n_1}{N}$$

and for the second reader we have correspondingly

$$p_2 = \frac{n_2}{N}$$

Now what is the probability of both readers finding the same error? Since we assumed independence in the reader's abilities to find errors, it is

$$p_1 p_2 = \frac{n_3}{N}$$

Eliminate p_1 and p_2. You get

$$\frac{n_1 n_2}{N^2} = \frac{n_3}{N}$$

and, supposing that n_3 is not zero, you have an estimate of the total number N of errors in the table:

$$N = \frac{n_1 n_2}{n_3}$$

Is this reasonable? We try some extreme cases. If both readers found exactly the same errors, then $n_1 = n_2 = n_3 = N$. This means that all the errors were found and that both readers were perfect, $p_i = 1$. Yes, this is reasonable. At the other extreme, suppose there were no common errors. Then N is infinite! This means that the readers were very bad at finding errors since they found none in common. The result is, in a sense, reasonable, although it is unrealistic since the table must have been finite to begin with (although we made no claim as to the size).

There were many assumptions in getting this formula. First, you are guessing at the values of the p_i. Second, the expected number of errors may be a fraction, but the number observed must be an integer. Finally, you are dealing with probabilities, and you may have seen an unlikely example of the readers' abilities. The above result is thus more understandable.

Another exception to this reasoning occurs if we assume that either reader found no errors. Say $n_2 = 0$. This means, of course, that $n_3 = 0$. If n_1 is not zero, then we have removed only those errors the first reader found, and we have no real estimate of the original number of errors in the table. If neither reader found any errors, we are inclined to assume that the table is perfect, but it could be that both readers were no good; we simply cannot choose between these two cases without outside information such as the known performance of the readers in other situations.

We have not yet found the answer to the original question, the number of errors left in the table. It is easy to see that

$$N - n_1 - n_2 + n_3$$

is a reasonable answer, since the first two subtractions each remove the common errors, and we should add the common errors back to get a reasonable result. Again, we test the formula by common sense. When you go through the exercise, you find that the formula is at least plausible, although certainly not perfect.

This is a good example of the use of probability in common life. The assumptions that must be made to get anywhere (without a large number of measurements) are often somewhat unrealistic, yet the result is a "better than nothing" guide. If we find that the estimated number of errors is larger than we care to tolerate, then we will have to get further readers to search and make further measurements on the common errors between each of the readers. On the other hand, if the estimated number of errors left is quite low, we are inclined to let the table go and concentrate on other things. The case of three readers is much more complex as there will be four unknowns (p_1, p_2, p_3, N) and seven conditions, which suggests a least square approach.

12.8 THE INCLUSION–EXCLUSION PRINCIPLE

If there had been three readers in the above example of proofreading, then we would have had to subtract out the errors found by all three individually, add back those that were found by each pair, and then subtract out again those that were common to all three. This is a special case of the general *inclusion–exclusion principle*, where we alternately add and subtract the larger and larger commonality. Consider a particular item that is common to m sets. If we add all the times it occurs in a set A_i, we will have $m = C(m, 1)$ of them and this is too many. Next we subtract all the times it occurs in two sets $A_i A_j$, $-C(m, 2)$, and then add all the times it occurs in three sets $A_i A_j A_k$, $C(m, 3)$, and so on. As a result we have counted the item

$$C(m, 1) - C(m, 2) + C(m, 3) - \cdots + (-1)^{m-1} C(m, m)$$

But the sum of *all* the binomial coefficients of order m [we neglected $C(m, 0)$] with alternating sign is zero since

$$(1 - 1)^m = 0$$

Hence the above sum is exactly 1, and we have finally entered the point a net of 1 time. Thus the above formula gives an alternative way of counting a set.

Example 12.8-1

There is a famous problem in probability that can be phrased in many ways. One way is to imagine n letters and n envelopes addressed separately, and then the letters are put in the envelopes at random. What is the probability that at least one letter will be in the right envelope? Again, if you number a set of items 1 to n, mix them up, and count down from the top, what is the probability that the rth item will occur at the rth count?

To apply the above formula, we first note that the probability that a letter is in its proper envelope is $1/n$. We designate the event that the ith letter is in the ith envelope by A_i. There are exactly n such trials (one for each letter i); hence we have for the first term

$$\sum[\Pr\{A_i\}] = n\left(\frac{1}{n}\right) = 1$$

For a pair of letters, the probability that the first is in its envelope and that the second is in its is $1/n(n - 1)$, and there are $C(n, 2)$ such pairs. This gives the second term

$$\sum[\Pr\{A_i A_j\}] = \frac{C(n, 2)}{n(n - 1)} = \frac{1}{2!}$$

correspondingly, we have for triples,

$$\sum[\Pr\{A_i A_j A_k\}] = \frac{C(n, 3)}{n(n - 1)(n - 2)} = \frac{1}{3!}$$

This continues through the n possible terms; and when they are added with alternating sign, we get for the probability that at least one letter is in its correct envelope

$$p_n = 1 - \frac{1}{2!} + \frac{1}{3!} - \frac{1}{4!} + \cdot \cdot \cdot + \frac{(-1)^{n-1}}{n!} \qquad (12.9\text{-}1)$$

In view of the rapid rate at which the individual terms approach 0, it is evident that for even reasonable sized n the value of p_n is close to its limiting value ($n = \infty$) of $0.63212 \ldots$. By $n = 7$, the correct value and the limiting value agree to four decimal places. Thus whether there are 10 letters or 10,000, the probability of at least one letter being in its correct envelope is about the same. Correspondingly, the complementary probability that no envelope has its proper letter is $0.36788 \ldots$. We will see later (Section 20.8) that the number called $e = 2.71828 \ldots$ is involved in the limit of P_n as $n \to \infty$; that is, $P_n \to 1 - 1/e$.

This is an example of the famous *Problème de recontres*, or matching problem, whose solution was given by Montmort (1678–1719) in 1708.

EXERCISES 12.8

1. *Work out the details for a theory for three proofreaders, given the number of errors of each, each set of errors common to two, and the number common to all three.

2. Show that the number of integers $\leq N$ that are not divisible by the primes 2, 3, 5, and 7 is

$$N - \Sigma\left[\frac{N}{i}\right] + \Sigma\left[\frac{N}{ij}\right] - \Sigma\left[\frac{N}{ijk}\right] + \Sigma\left[\frac{N}{ijkl}\right]$$

where $[N/i]$ means (here) greatest integer not greater than N/i, and the sum is over all given primes, all unlike pairs of primes, and so on. If $N = 210n$, show that this number is $48n$.

3. Suppose n balls are placed at random in m boxes. Let A_i be the event that box i is empty, for $i = 1, 2, \ldots, m$. Show that (a) $P(A_i) = (m - 1)^n/m^n$, $P(A_iA_j) = (m - 2)^n/m^n$ for $i \neq j$, and so on. Hence we have $S_k = C(m, k)(m - k)^n/mn$, where S_k is the sum of all the terms of single, double, and so on, occupancy. (b) Show that the probability of no empty boxes is $\Sigma(-1)^kC(m, k)(m - k)^n/m^n$.

2.9 CONDITIONAL PROBABILITY

We have seen a couple of examples of situations where the probability was *conditional* on what happened. There are the original probabilities, often called *a priori* (prior) probabilities, and then there are the probabilities *a posteriori* (posterior means after), after some event has happened. For the two drawers and the gold and silver coins (Example 12.3-4) the original distribution of four possible outcomes was equally likely (a priori), but, having seen a gold coin, the distribution was now conditional on that, and one outcome was excluded. Similarly, when we draw a card from a deck, and then draw another card without replacing the first, the probabilities are conditional on what was first drawn. If the first card was a ten of diamonds, then the probability of another ten of diamonds is 0, and the probability of any other named card is 1/51 (not 1/52).

Example 12.9-1

Information. Suppose you draw one card from a deck, do not look at it, and then draw a second card. What is the probability that the second card is 10D? If the first card drawn had been the 10D, then the probability would be zero; but if it had not been the 10D, then the probability would be 1/51. We need to combine these two probabilities with their probability of occurrence, that is, 1/52, and 51/52. Thus using these conditional probabilities we get

$$\frac{1}{51}0 + \frac{51}{52}\left(\frac{1}{51}\right) = \frac{1}{52}$$

This is the same as if the first card had not been drawn. We learned nothing (got no information) from the first draw, so it is not surprising that the probability was not affected. This is a general principle, if nothing is learned, you can ignore the event; but it is sometimes a subtle question whether you have, or have not, learned anything.

Often, in a sequence of trials, what happens at one stage can affect the later events. This is the concept of *conditional probability*. We use the notation

$$p(x \mid y)$$

to mean "the probability of x given that y has occurred." The vertical bar is the condition sign, and what follows is assumed to have already happened (a typical inversion of notation!).

Example 12.9-2

Bayes' Relationship. There is the original sample space of all possible pairs of events $p(x, y)$. We see that conditional probability is effectively defined by the equation

$$p(x, y) = p(y)p(x|y)$$

In words, the probability of the pair x and y occurring is equal to the probability of y occurring followed by the conditional probability of x given that y has occurred. But supposing we look first at the event x; then this can also be written as

$$p(x, y) = p(x)p(y|x)$$

Since these two probabilities are the same, we have

$$p(x, y) = p(y)p(x|y) = p(x)p(y|x)$$

or, if $p(x) \neq 0$,

$$p(y|x) = \frac{p(y)p(x|y)}{p(x)}$$

This is a special case of the famous *Bayes' relation*. It relates the two conditional probabilities; it is the device that enables you to reverse the arguments in the conditional probabilities. More generally, if y is composed of independent sets y_i, then since

$$p(x) = \sum [p(y_i)p(x|y_i)]$$

we get, similarly,

$$p(y_j|x) = \frac{p(y_j)p(x|y_j)}{p(x)}$$

The equation is above reproach, but there is great controversy over its use.

The failure to recognize such conditional effects can be serious in science, as the following case study shows.

Case Study 12.9-3

ESP. There is a widespread belief that some people have extrasensory perception (ESP), meaning that, for example, they can foretell the random toss of a coin or in some way influence the outcome of the toss of the coin. Among the believers in ESP, there is also the belief that this ability comes and goes, that sometimes you have it and sometimes you do not.

To test the presence of ESP, the following experiment is proposed. We will make ten trials in all. If the first five show that ESP is present, we will go on; but if it is not present, then there is no use wasting time on this test. How shall we decide if it is present or not? If the first five trials show three or more successes, then we will assume that ESP is present and go on; otherwise, we will abort the test.

Let us analyze the experiment from the point of view that there is no such effect as ESP, that calling the correct head or tail is solely a matter of luck.

What is the sample space of the first five trials? There are six possibilities, 0, 1, 2, 3, 4, or 5 successes. We have the following table:

Successes	Probability
0	$\frac{1}{32}$
1	$\frac{5}{32}$
2	$\frac{10}{32}$
3	$\frac{10}{32}$
4	$\frac{5}{32}$
5	$\frac{1}{32}$

The expected number of successes is $\frac{5}{2}$. But *conditional* on achieving at least three correct guesses, the expected value is

$$\frac{3(10/32) + 4(5/32) + 5(1/32)}{(10/32) + (5/32) + (1/32)} = \frac{55}{16}$$

which is about $\frac{7}{2}$ rather than $\frac{5}{2}$. Thus the conditioning has influenced the expected number of right guesses (as you would expect).

For the second five trials, we expect

$$\frac{5}{2} = \frac{40}{16}$$

successes. The sum of the two expected values is

$$\frac{55}{16} + \frac{40}{16} = \frac{95}{16}$$

which is very close to 6 ($16 \times 6 = 96$).

Without this *optional stopping* after the first five trials, the expected number of successes in the ten trials is five. By optional stopping and saying that ESP is not present in the current run of trials (and thus not counting the trials that contained a lot of failures), we have raised the expected number of successes from five to almost six.

If I am allowed to proceed this way, then I can make so many runs that the experiment will appear to show significantly better results than chance, that there is such a thing as ESP. These arguments in no way prove that there is no such thing as ESP, but are designed to alert you to any experiments where there is optional stopping;

even in a physics lab sometimes a run is aborted because . . . . Again, this is not an argument that every run that is started must be completed no matter how obvious it is that an error has occurred; it is a warning to be very careful when you engage in optional stopping. You may find what you set out to find, and not what the experiment was supposed to reveal about the world.

12.10 THE VARIANCE

Given a distribution in the random variable k, we have introduced the *expectation* of k, $E\{K\}$, as one measure of what statisticians call "typical values." Another typical value is the *median*, which is the value (or average of two adjacent values) for which there are as many items above as there are below. A third typical value is the *mode*, which is the most fashionable, the most frequently occurring value (if there is one). All these are summarized in Figure 12.10-1.

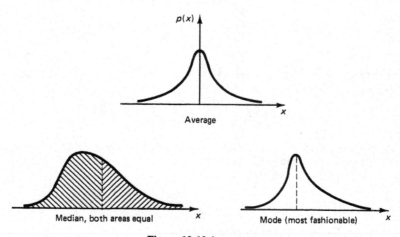

Figure 12.10-1

We all know that there is usually a spread about the average (see Figure 12.10-2), that some distributions are centered close to their average value and some are spread out. To measure the spread, we introduce the *variance* about the mean as the *average of the squares of the distance from the mean*. In mathematical symbols, using μ (Greek lowercase letter mu) as the expected value of the distribution,

$$\mu = \text{expected value of the distribution} = E\{K\}$$

$$= \sum_{k=0}^{n} k\, p(k) \tag{12.10-1}$$

then we have for the variance (σ^2) (Greek lowercase sigma)

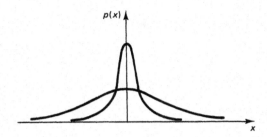

Figure 12.10-2

$$V\{K\} = E\{(K - \mu)^2\} = \sum_{k=0}^{n} (k - \mu)^2 p(k) = \sigma^2 \qquad (12.10\text{-}2)$$

where $p(k)$ is the probability of the k occurring.

Example 12.10-1

What is the variance of a Bernoulli distribution? The variance $V\{K\}$ is defined as the second moment about the mean, and the mean of n Bernoulli trials is $\mu = np$. Thus

$$V\{K\} = \sum_{k=0}^{n} (k - np)^2 C(n, k) p^k q^{n-k}$$

The term $(k - np)^2$ can be expanded into three terms (we temporarily drop the range of summation for convenience):

$$V\{K\} = \sum k^2 C(n, k) p^k q^{n-k} - 2np \sum k C(n, k) p^k q^{n-k}$$

$$+ (np)^2 \sum C(n, k) p^k q^{n-k} \qquad (12.10\text{-}3)$$

The middle sum we have just done in (12.6-6), and it is simply np. The sum in the last term is obviously equal to 1 (it is the sum of all the probabilities). Thus it combines with the middle term to give, finally,

$$V\{K\} = \sum k^2 C(n, k) p^k q^{n-k} - (np)^2 \qquad (12.10\text{-}4)$$

To attack this summation term, we proceed along the lines we did to get the average value. We start with the generating function (12.6-5):

$$(q + pt)^n = \sum_{k=0}^{n} C(n, k) p^k q^{n-k} t^k$$

If we (1) differentiate this with respect to t, we will get a coefficient k; then if we (2) multiply through by t, we will be back with the same power of t ready to differentiate again; so we (3) differentiate again with respect to t to get a second factor of k; and then (4) set $t = 1$ to get rid of the generating variable t; and, finally, (5) recall that $p + q = 1$. We have therefore to compute

$$\frac{d}{dt}\left\{t\frac{d}{dt}(q + tp)^n\right\} = \frac{d}{dt}\left\{t\frac{d}{dt}\sum C(n, k)p^k q^{n-k} t^k\right\}$$

Doing steps (1) and (2), we get

$$\frac{d}{dt}\{tpn(q + tp)^{n-1}\} = \frac{d}{dt}\left\{\sum kC(n, k)p^k q^{n-k} t^k\right\}$$

We now differentiate again to get for step (3)

$$\{tp^2 n(n - 1)(q + tp)^{n-2} + pn(q + tp)^{n-1}\} = \sum k^2 C(n, k)p^k q^{n-k} t^{k-1}$$

We now carry out step (4), which requires us to set $t = 1$. Finally, we do step (5), put $p + q = 1$, to get the sum we want:

$$(np)^2 - np^2 + pn = (np)^2 + np(1 - p)$$

$$= (np)^2 + npq = \sum k^2 C(n, k)p^k q^{n-k}$$

Now going back to the formula for the variance of the Bernoulli distribution (12.10-4), we have

$$V\{K\} = [(np)^2 + npq] - (np)^2 = npq = \sigma^2 \qquad (12.10\text{-}5)$$

This is a very nice, compact formula for the variance, and goes along with the expected value $E\{K\} = np$.

Example 12.10-2

For the binomial distribution, which probability p gives the maximum variance? The variance of the binomial distributon is

$$V\{K\} = npq = np(1 - p) = n(p - p^2)$$

To find the maximum, you differentiate this, set the derivative equal to zero, and solve the resulting equation. The details are

$$\frac{dV\{K\}}{dp} = n(1 - 2p) = 0$$

Since $n \neq 0$, we have

$$p = \frac{1}{2}$$

Thus the maximum variance occurs when $p = 1/2$. A plot of the function

$$f(p) = p(1 - p)$$

is given in Figure 12.10-3 and confirms the result. You could use the test

$$V'' = -2n$$

to prove that this is a maximum. Thinking about symmetry and the meaning of the symbols shows why this must be true.

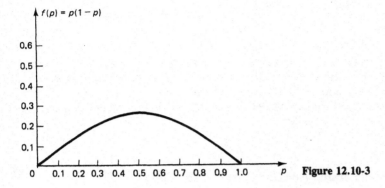

Figure 12.10-3

Generalization 12.10-3

A general property of the variance. The mean of a distribution of $p(k)$ is defined by

$$\mu = E\{K\} = \sum kp(k) \tag{12.10-6}$$

and the variance is defined by

$$\operatorname{Var}\{K\} = V\{K\} = E\{(K - \mu)^2\} = \sum (k - \mu)^2 p(k)$$

We are, of course, supposing that the sums exist. Expand the square term and spread out the sum into three sums (as we did in Example 12.10-1):

$$V\{K\} = \sum k^2 p(k) - 2\mu \sum kp(k) + \mu^2 \sum p(k)$$

The middle sum is by definition μ, so the term is $-2\mu^2$. The last sum by definition is 1, and when multiplied by μ^2 it combines with the middle term to give

$$V\{K\} = \sum k^2 p(k) - \mu^2 = \sum k^2 p(k) - \left[\sum kp(k)\right]^2 \tag{12.10-7}$$

When you compare the simplicity of this derivation in the general case with the top of Example 12.10-1 for the specific Bernoulli distribution, you see the advantage of using the general case of a distribution $p(k)$, rather than the details of the particular case at hand. Abstraction often makes things easier and more compact and in the long run perhaps much clearer, because the details are removed and you are looking at the essentials of the situation.

Equation (12.10-7) is of great use in the theory, although you should be wary of using it in numerical work with real data; the roundoff may cause you trouble. The formula says, "The sum of the squares minus the square of the sum of the first powers, each weighted by the probability distribution, gives the variance."

The *m*th *moment* of a distribution is defined by

$$E\{K^m\} = \sum k^m p(k) \tag{12.10-8}$$

Thus the expected value is the first moment, and the variance is the second moment (about the origin) minus the square of the first moment. The third moment about the mean is called *skewness*, and the fourth is called *kurtosis* (see Figure 12.10-4). Higher moments than this are not usually named. Generally, the higher moments are taken about the mean (as is the variance)

$$E\{(K - \mu)^m\} = \sum (k - \mu)^m p(k) \tag{12.10-9}$$

and are called the *m*th *central moments* of the distribution $p(k)$.

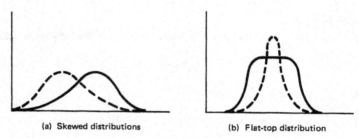

(a) Skewed distributions   (b) Flat-top distribution

Figure 12.10-4   Examples of distributions

EXERCISES 12.10

1. Compute the third moment of the Bernoulli distribution.
2. Given the geometric distribution $q, pq, p^2q, p^3q, \ldots$, find the first three moments.
3. In Exercise 2, find the mean and variance.

12.11 RANDOM VARIABLES

We begin with a sample space of events (outcomes of a trial). We can do no numerical computation with the names of the outcomes of the trials; we need numerical values to assign to the events. It is natural to assign the numerical value of the face of the die to the corresponding outcome. In principle, we could assign any numerical values to any of the faces, but then there might be difficulties in later interpretations.

For a coin, you are apt to assign 0 to one face and 1 to the other face. The expected value would then be

$$0\left(\frac{1}{2}\right) + 1\left(\frac{1}{2}\right) = \frac{1}{2}$$

But you could assign -1 to one face and 1 to the other. The expected value would then be

$$-1\left(\frac{1}{2}\right) + 1\left(\frac{1}{2}\right) = 0$$

Thus the assignment of numerical values to the outcomes of a trial is not unique. The values you assign should be made with an eye toward the later interpretations you are going to make, but for the beginner this is hardly a useful remark!

Again, it is natural to assign the number k to the length of a run of k tosses of a coin until a success. Generally, it soon becomes evident how to assign numbers to the events; indeed, it is so natural that often the student is unaware that this is being done. But it is a logical step that must be taken with an eye toward the final results. With only names for the outcomes, we can do little further mathematics; with numbers properly assigned, we can find many interesting results.

We have been using the notation

$$E\{K\}$$

which looks as if it were a function of K, but on the right-hand side of the equation the dummy variable k of summation occurs. Therefore, the result does not depend on k or K. We need to introduce the corresponding idea behind the notation.

The outcome of a trial cannot be known definitely. It can be any of the possible outcomes; but we need a notation to refer to the trial. For this purpose, we introduce the concept of a *random variable* (it is *not* a variable in the usual sense). We use capital letters for random variables (we used K before) to correspond to the lowercase index of the outcomes k. Thus we wrote

$$E\{K\} = \sum_{k=1}^{n} kp(k)$$

as the expected value of the distribution $p(k)$, and, in general,

$$E\{f(K)\} = \sum_{k=1}^{n} f(k)p(k)$$

as the expected value of the combination $f(K)$ of the random variable K. For example,

$$E\{K^2\} = \sum_{k=0}^{n} k^2 p(k)$$

For the general values of the random variable x_i (not necessarily integers), we write (supposing, as usual, that the sums exist and are absolutely convergent, Section 20.5)

$$E\{X^m\} = \sum x_i^m p(x_i)$$

As before, we have the moments

$$E\{1\} = \sum p(x_i) = 1$$

$$E\{X\} = \sum x_i p(x_i) = \mu$$

$$E\{(X - \mu)^2\} = \sum (x_i - \mu)^2 p(x_i) = \sigma^2$$

We have introduced the notation of X as the *random variable* whose possible values are the x_i. We cannot say which outcome will occur, and we need a notation to talk about the trial. Thus X is the name of the trial, and the various x_i's are the possible values of the outcome. The $p(x_i)$ are the probabilities of the outcomes. Each time we form a sum, we weight the numerical value assigned to the outcome by its probability of occurrence. It is a matter of rereading much of this chapter to put the results in terms of random variables. Thus you can write, if you wish,

$$p_k = p(k) = p\{X = k\}$$

The idea is simple, once understood, and we have delayed its introduction until it would be comparatively easy to grasp. Sometimes you see the word "stochastic" where we have used "random," but as far as the author can determine there is no difference among the two schools of notation in what they do with random (stochastic) variables. The values of outcomes you are interested in are multiplied by their probabilities and are then summed over all those of interest.

12.12 SUMMARY

The purpose of this chapter is to introduce the concepts of probability so that they can be used in later sections of the book. They also illustrate the use of the calculus, as well as some of the methods of mathematics. It is not expected that on this first encounter with probability you have mastered all the finer parts of it. You will get constant reviews of the material in the rest of the book. The next chapter will review the ideas in terms of continuous probability.

The main ideas are a trial, especially a Bernoulli trial where the trials are the same and independent of each other; the concept of the sample space of possible outcomes; the assignment of the probability to the events in the sample space; the selection of some subpart of the space; and the computation of various averages over the sample space. Sometimes the probabilities are conditioned on earlier events, and the notion of conditional probability is necessary.

The idea of a random variable was introduced so that we can speak of a trial X whose outcomes are the x_i, but which x_i occurs is unknown. Typically, we have dealt with averages of a random variable over all the possible outcomes, but sums over subsets can be used when appropriate.

13

Continuous Probability

13.1 PROBABILITY DENSITY

Given a continuous probability distribution in the interval $0 \leq x \leq 1$, what is the probability of picking a given point at random with equal likelihood for all points? It is zero! If it were any finite number, then each of the infinite number of points in the interval would have the same probability, and the total probability would be infinite, not 1. This is the *same* situation as we had when, in Chapter 4, we thought of the line segment as being composed of points that have no length.

To get around this logical difficulty, we adopt the standard strategy. We first consider the *cumulative distribution*

$$\text{Prob}\{X \leq x\} = P(x)$$

which is the probability that an observed value of the random variable X is less than some given x. If $P(x)$ has a derivative, then let this be denoted by

$$\frac{dP(x)}{dx} = p(x)$$

where $p(x)$ is called the *probability density*. Evidently, we have

$$P(x) = \int_{-\infty}^{x} p(t) \, dt$$

with $P(\infty) = 1$. We generally use the entire range $-\infty$ to ∞ when we are dealing with the theory. We have earlier assumed that the range of integration was finite. We will take up the use of an infinite range in Section 14.9; for the present we will use only probability distributions that are zero outside a finite range. When the range is finite, we simply define the probability $p(x) = 0$ for values outside the range.

To examine the meaning of the probability density function $p(x)$ a bit closer, we go back to the basic definition of a derivative (Section 7.5) and set up the third step of the four-step process:

$$\frac{\Delta P(x)}{\Delta x} = \frac{\text{Prob } \{X \leq x + \Delta x\} - \text{Prob } \{X \leq x\}}{\Delta x}$$

Since this is the difference of two partially overlapping integrals, we have

$$\frac{\Delta P(x)}{\Delta x} = \frac{\text{Prob } \{x < X \leq x + \Delta x\}}{\Delta x}$$

When we take the limit, we get the probability density. By this technical device we get around the fact that the continuous line is composed of points that have no length and that the probability of a random variable having any particular value (of getting any point) is 0, while the probability that X falls in an interval is in general not 0. It is a mathematical device that is widely used. The cumulative probability distribution is very useful for mathematically deriving things, but you tend to think in terms of the probability density. In a sense we have repeated the derivation of the fundamental theorem of the calculus.

"Random" means not predictable, but in continuous variable probability (as in discrete probability) "random" without modification means "uniform probability" unless otherwise stated.

Example 13.1-1

Consider the situation when we are given a number such as $\pi = 3.1415926535 \ldots$. In practice, we are forced to round it off to k decimal digits. For the ensemble of all numbers between 1 and 10, the roundoff error in the kth decimal digit is (very closely) uniformly distributed from $-\frac{1}{2}$ to $\frac{1}{2}$ of the last place kept. The probability of an error of size x is $p(x) = 1$ for $(-\frac{1}{2} \leq x < \frac{1}{2})$ and 0 elsewhere. Hence (Figure 13.1-1) the

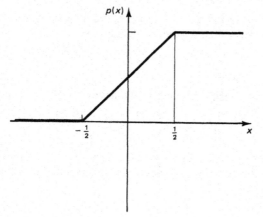

Figure 13.1-1 Probability of roundoff

cumulative probability distribution $P(x)$ for roundoff is

$$P(x) = x + \frac{1}{2}, \qquad \text{for } -\frac{1}{2} \leq x < \frac{1}{2}$$

where we have adopted the usual (computing machine) rule of rounding up for the half-values. Note that $P(x) = 0$ for $x < -\frac{1}{2}$, and $P(x) = 1$ for $x \geq \frac{1}{2}$.

We can compute both the mean and the variance of this continuous distribution by using the calculus in place of the finite summations of the previous chapter. You simply replace the sums by integrals. We have, for the expected value of the random variable X,

$$E\{X\} = \mu = \int_{-\infty}^{\infty} x \, p(x) \, dx$$

$$= \int_{-1/2}^{1/2} x \, dx$$

$$= \frac{x^2}{2} \Big|_{-1/2}^{1/2} = 0$$

which is obvious since the integrand is odd over a symmetric interval about 0. The variance is, therefore,

$$V\{X\} = \sigma^2 = \int_{-1/2}^{1/2} (x - 0)^2 \, dx$$

$$= \frac{x^3}{3} \Big|_{-1/2}^{1/2} = \frac{1}{3}\left[\left(\frac{1}{8}\right) - \left(\frac{-1}{8}\right)\right]$$

$$= \frac{1}{12}$$

In words, the variance of a random variable that is uniformly distributed from $-\frac{1}{2}$ to $\frac{1}{2}$ is $\frac{1}{12}$.

13.2 A MONTE CARLO ESTIMATE OF PI

In the abstract, the concept of a *random number* is the same as that of a random variable. Hence a random number x_i between 0 and 1 is a particular realization of the random variable X betwen 0 and 1 (the sample space). This means that each number has the same probability of being selected as any other number in the interval. As a result, we assign the probability density function $p(x)$ as uniform in the interval. Since the total probability must be 1, it follows that $p(x) = 1$ in the interval and 0 outside. For continuous distributions, the uniform distribution plays the same role as does symmetry for finite, discrete distributions.

Example 13.2-1

Now consider two independent random numbers X and Y, both in the interval $0 \leq x < 1$, and consider plotting their realizations (x_i, y_i) in the unit square (see Figure 13.2-1).

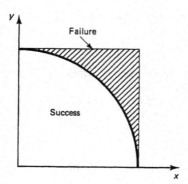

Figure 13.2-1   Monte Carlo for $\pi/4$

What is the probability that the random point will fall inside the arc of the unit circle? Because of the assumed uniform distribution in the two variables, the probability is the ratio of the two areas, the ratio of the area of the quadrant of the circle to the area of the unit square. This ratio is exactly $\pi/4$.

Experiment 13.2-2

Monte Carlo estimate of π. Using k digit numbers (which for reasonably large k introduces a fine granularity), suppose we select a random point in the unit square and count it a success when it falls within the circle and a failure when it falls outside. (The probability that the random variable falls exactly on the perimeter is zero, so we can treat the circumference as either a success or a failure.)

If we make many trials, then our intuition suggests that the ratio of the successes of the Bernoulli trials to the total number of trials should be approximately $\pi/4$. Most computers, even the programmable hand-held computer used below, have a random-number generator. Typically, the generator has the ability to produce a sequence of *pseudorandom numbers* n_i based on the formula

$$n_i = rn_{i-1}$$

where only the last k digits of the $2k$ digit product are kept. There are two choices to be made in setting up such a generator of pseudorandom numbers, the *multiplier r* and the *seed* n_0. For a binary computer, when these two numbers are properly chosen, the generator will produce $\frac{1}{4}$ of all the 2^k possible k digit binary numbers in some order, without replacement, and then repeat the cycle. For a k decimal digit random-number generator, there are, of course, 10^k numbers possible, and there are other minor differences in the details of the random-number generator.

The sequence is not random in the true sense since it is predictable; but when viewed as a stream of numbers, it has many of the properties of a sequence of random numbers, because the numbers pass most reasonable tests for randomness.

A simple program to compute a Monte Carlo estimate of π based on Example 13.2-1 produced the following results. (It is, of course, a test of the random-number generator and not a validation of the number pi.) The test is an interesting example of the use of a random process to estimate a completely determinate number. Such methods are called *Monte Carlo methods* after the famous gambling place.

Trials	Experimental number of: Successes	Pi (est.)	Theoretical number of successes	$\sqrt{npq}$
100	82	3.28	78.5	4.11
200	160	3.20	157.1	5.81
300	245	3.2667	235.6	7.11
400	323	3.23	314.2	8.21
500	407	3.256	392.7	9.18
600	482	3.2133	471.2	10.1
700	557	3.1829	549.8	10.9
800	636	3.18	628.3	11.6
900	707	3.1422	706.9	12.3
1000	788	3.152	785.4	13.0

Are these results reasonable? Do they seem to you (intuitively) to differ too much or too little from the expected value $n\pi/4$ (column 4)? The variance is the average of the square of the difference from the expected value, so the square root of the variance measures the deviation you might expect from the mean. It seems reasonable to suppose that in practice we can measure this deviation by the square root of the variance, which is called the *standard deviation*. We know from Section 12.8 that the square root of the variance of n trials of a Bernoulli process is

$$\sigma = \sqrt{npq}$$

We have $p = \pi/4$, so we have

$$\text{variance} = n\left(\frac{\pi}{4}\right)\left(1 - \frac{\pi}{4}\right) = (0.168548 \ldots)n$$

In the above table the last column gives the standard deviations.

The variance is the "expected value" of the square of the deviations from the mean; hence for repeated trials we "expect" to have a mean square deviation from the expected value of the variance. The mean square accentuates large values more than small ones, so the values of the number of successes should fall, on the average, within the range

$$np \pm \sqrt{npq}$$

and this is what we see. Note that for the average this is divided by n:

$$p \pm \frac{\sqrt{pq}}{\sqrt{n}}$$

From this approximation we see that to get one more decimal digit of accuracy requires 100 times as many trials. Of course, in this case of estimating π by the above method we do not really believe that enough trials will get arbitrary accuracy; rather we believe that pushed far enough the pseudorandom number generator will produce systematic errors, and these will prevent us from getting an arbitrarily accurate estimate.

EXERCISES 13.2

For the following probability density distributions in the indicated intervals, show that it is a probability distribution and compute the mean and variance.

1. $p(x) = 6x(1 - x)$, $0 \le x \le 1$
2. $p(x) = 1 + x$ for $-1 \le x \le 0$ and $1 - x$ for $0 \le x \le 1$
3. $p(x) = 1/4$, $-2 \le x \le 2$
4. $p(x) = (3/4)(1 - x^2)$, $-1 \le x \le 1$
5. Test your random-number generator for 1000 numbers by computing the 0th, 1st, 2nd, and 3rd moments. Why do you expect larger errors on the higher moments?
6. Divide the range of numbers into tenths, and count the number that fall in each interval for 1000 random numbers. Is the result reasonable?
7. In Example 13.2-1, suppose the radius of the circle is 8/10; what is the probability of a random point falling inside the circle?

13.3 THE MEAN VALUE THEOREM FOR INTEGRALS

Discrete distributions often do not have a value corresponding to the computed expected value. For the toss of a coin, the expected value is $\frac{1}{2}$ and for the roll of a die it is $\frac{7}{2}$, neither of which can occur on any single trial. The "expected value" cannot occur for some discrete distributions. But for a continuous distribution it is reasonable to expect that there is some value where the average (the expected value) occurs.

We will prove a more general case than this; we will take a *weighted average* of the values of the continuous function $f(x)$, using nonnegative weights $g(x) \ge 0$, and show that the average value will occur somewhere. Thus, in place of a probability density function $p(x)$ we are using any nonnegative integrable function $g(x)$ as weights. We wish to prove

$$\int_a^b f(x)g(x)dx = f(\theta) \int_a^b g(x)dx$$

for some θ in the interval $a < \theta < b$.

To prove this theorem, first consider the values of $f(x)$ in the closed interval $a \le x \le b$. From Section 4.5, assumption 3, we see that there is a number where the function takes on its largest value; call this largest function value F. Similarly, there is a place where it takes on its smallest value (algebraically); call this function value f. We have

$$f \le f(x) \le F \quad (a \le x \le b)$$

Multiply these inequalities by the nonnegative weight $g(x)$ to get

$$fg(x) \le f(x)g(x) \le Fg(x)$$

and then integrate (sum) to get

$$f \int_a^b g(x) \, dx \le \int_a^b f(x) g(x) \, dx \le F \int_a^b g(x) \, dx$$

Now consider the constructed function of t:

$$\Phi(t) = f(t) \int_a^b g(x) \, dx - \int_a^b f(x) g(x) \, dx$$

(look at where the t appears). At the value of t where

$$f(t) = f, \quad \text{then } \Phi(t) \le 0$$

$$f(t) = F, \quad \text{then } \Phi(t) \ge 0$$

Hence, since $f(t)$ is continuous, then $\Phi(t)$ is continuous (see Section 4.5, assumption 2), and there is a value θ such that $\Phi(\theta) = 0$, which means that

$$\int_a^b f(x) g(x) \, dx = f(\theta) \int_a^b g(x) \, dx$$

Mean Value Theorem for Integrals.   If $f(x)$ is continuous in the closed interval $[a, b]$ and $g(x)$ is of constant sign, then there exists a number θ such that the above equation is satisfied (we, of course, assume that the integrals exist).

Note that if $g(x) > 0$ for at least some values then $\int g(x) \, dx > 0$. In this case you can divide both sides by the integral on the right. You then see that this is simply stating that there is a weighted average value of $f(x)$, labeled $f(\theta)$, for any set of nonnegative weights $g(x)$.

The mean theorem for integrals is related to the earlier mean value theorems of Section 11.7. If we write

$$\int f(x) \, dx = F(x)$$

and set $g(x) = 1$, then the above theorem is simply

$$F(b) - F(a) = F'(\theta)(b - a)$$

To get the Cauchy form of the mean value theorem (Section 11.8), we write, instead,

$$\int f(x) g(x) \, dx = F(x)$$

$$\int g(x) \, dx = G(x)$$

and we have

$$F(b) - F(a) = [G(b) - G(a)] \frac{F'(\theta)}{g(\theta)}$$

But

$$g(\theta) = G'(\theta)$$

and we have the Cauchy form of the mean value theorem (in a slightly different notation):

$$\frac{F(b) - F(a)}{G(b) - G(a)} = \frac{F'(\theta)}{G'(\theta)}$$

This illustrates again the fact that mathematical results may be obtained in many different ways. The alternative approaches illustrate different mathematical methods, but note the close similarity of the underlying method of constructing a function in each proof.

EXERCISES 13.3

1. Find the θ for $f(x) = x^k$ and $g(x) = 1$, $0 \le x \le 1$.

2. Write an essay comparing the integral approach of this section to the earlier results.

13.4 THE CHEBYSHEV INEQUALITY

There is a famous inequality named after the Russian mathematician Chebyshev (1821–1894) that is easy to derive and is of great theoretical usefulness. We begin by considering a probability density distribution $p(x)$, and the second moment about the origin

$$\sigma^2 = \int_{-\infty}^{\infty} x^2 p(x) \, dx$$

Since $p(x)$ is nonnegative, if we remove a small region about the origin, then we have not increased the integral. Therefore, we have (Figure 13.4-1)

$$\sigma^2 \ge \int_{|x| \ge \epsilon} x^2 p(x) \, dx$$

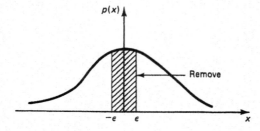

Figure 13.4-1   Chebyshev inequality

If next, in place of x^2, we take its smallest value, ϵ^2, then the integral can only further decrease:

$$\sigma^2 \geq \epsilon^2 \int_{|x| \geq \epsilon} p(x) \, dx$$

$$\geq \epsilon^2 \text{ Prob } \{|X| \geq \epsilon\}$$

We rearrange this to get

$$\text{Prob } \{|X| \geq \epsilon\} \leq \frac{\sigma^2}{\epsilon^2}$$

This is the desired inequality. A review of the derivation shows that for a typical distribution we threw away a lot at each step, although there are distributions that are close to the equality. Such a distribution has most of its probability just outside the values of $\pm\epsilon$.

EXERCISE 13.4

1. Derive the Chebyshev inequality for a discrete distribution.

13.5 SUMS OF INDEPENDENT RANDOM VARIABLES

Suppose we have two independent (see Section 12.3 for the meaning of independent) random variables X_1 and X_2 from the corresponding probability density distributions $p_1(x)$ and $p_2(x)$, with the corresponding mean values μ_1 and μ_2. Thus we think of each pair (x_1, x_2) as a point in the plane (the product space). The expected value for the product of the deviations of each from its mean is

$$E\{(X_1 - \mu_1)(X_2 - \mu_2)\}$$

Since they are independent random variables, we think of the product space of pairs of points (x_j, x_k), each point being in the product space of the two distributions. Each point has the probability of the product $p_1(x_j)p_2(x_k)$ of the corresponding probabilities. When we compute the expected value, we sum over the sample space of these products of probabilities. But the sum can be found by summing the rows and then summing the column of row sums, or the other way around, summing the columns and then summing the sums in the row. Either way, we have for the expected value of the product

$$E\{(X_1 - \mu_1)(X_2 - \mu_2)\} = E\{X_1 - \mu_1\}E\{X_2 - \mu_2\}$$

But

$$E\{X_i - \mu_i\} = \mu_i - \mu_i = 0$$

so the expected value of the product is $0 \times 0 = 0$.

By mathematical induction, what applies to a pair of independent random variables also applies to any finite product of independent random variables. The proof will have the same form, only it will be a bit more elaborate.

We now turn to the important topic of sums of independent random variables. Suppose for simplicity we compute the average (mean) of n independent random variables x_i from the same distribution $p(x)$ (called IID for independent identically distributed variables):

$$S_n = \frac{X_1 + X_2 + \cdots + X_n}{n}$$

To get the expected value $E\{S_n\}$, we simply multiply the values of each random variable X_i by the probability density function $p(x)$ and integrate over all the probability range. This will be, since integration is a linear operator (the integral of a sum is the sum of the integrals), the sum of the expected values, that is, the sum of μ taken n times divided by n, which is simply μ, the common expected value. From this it follows immediately that

$$E\{S_n\} = \mu$$

In words this says, the expected value of the average of n random variables from the same distribution is the expected value of any one random variable.

Next we look at the variance of the average S_n. We have

$$V\{S_n\} = E\{(S_n - \mu)^2\}$$

$$= \frac{1}{n^2} E\left\{\sum_{k=1}^{n}(X_k - \mu) \sum_{j=1}^{n}(X_j - \mu)\right\}$$

where we have brought μ into each term (written $\mu = n\mu/n$), and used different indices of summation so that when we multiply them together we can keep track of the various terms. We also brought out from the expectation operator the denominator n^2. Upon multiplication of the two sums, we will get terms of the form

$$E\{X_k - \mu)(X_j - \mu)\}$$

which we just saw above were zero if $j \neq k$, since X_k and X_j are assumed to be independent random variables. But when we have $j = k$, we have

$$E\{(X_k - \mu)^2\}$$

which is the variance of each term. There are a total of exactly n such terms when $i = j$. Hence when we remember the $1/n^2$ in front, we have in words,

The variance of the average of n independent random variables selected from the same distribution $p(x)$ is the variance of any single variable divided by n.

In mathematical symbols we have

$$V\{S_n\} = \frac{\sigma^2}{n}$$

Averaging reduces the variance *provided* the variance is finite.

EXERCISES 13.5

1. Repeat the derivation using Equation (12.10-7).
2. Carry out the derivation for random variables from a discrete distribution.

3.6 THE WEAK LAW OF LARGE NUMBERS

We are now ready to examine the weak law of large numbers, which puts into precise mathematical form your general experience that the average of repeated trials is close to the expected value of a single trial, that 1000 tosses of a coin will generally give around 500 H and 500 T. This law gives more details than the general discussion in Example 12.6-3, where we discussed only Bernoulli trials; here we have any continuous distribution (provided the variance of the distribution is finite). This "law" connects these two concepts of expected value: the expected value averaged over the sample space, and the expected value averaged over a long run of trials. This form is called "weak" because there is a stronger form called the strong law of large numbers.

You may wonder how there can be laws about random events, since on the face of it *random* is the opposite of *predictable* (law). We will be predicting the average, the long-run behavior. As you will see, we will have allowed for the obvious fact that it is possible, although not probable in any reasonable sense, that out of 1000 tosses of a well-balanced coin you will get 1000 H. This last probability is easily calculated to be

$$\left(\frac{1}{2}\right)^{1000}$$

which is around 10^{-300}. There are less than $\pi \times 10^9$ seconds in your probable lifetime, so you can expect to spend roughly 3×10^{290} lifetimes of trials at a rate of one trial (1000 tosses) per second to get such an outcome.

To get the desired result, we have only to use the Chebyshev inequality (Section 13.4)

$$\text{Prob}\,\{|X| \ge \epsilon\} \le \frac{\sigma^2}{\epsilon^2}$$

with the result of Section 13.5.

$$V\{S_n\} = \frac{\sigma^2}{n}$$

We identify the random variable X in the Chebyshev inequality with random variable $S_n - \mu$. The variance (which is independent of the mean) is now the variance of any single variable divided by n. We have, therefore,

$$\text{Prob } \{|S_n - \mu| \geq \epsilon\} \leq \frac{\sigma^2}{n\epsilon^2}$$

Theorem.   The probability that the mean of n observations of the same random variable (having a finite variance) differs from the mean of the random variable by more than ϵ can be made arbitrarily small by making n sufficiently large.

Since this law is so widely misunderstood, it is necessary to make a number of remarks (always assuming that the distribution has a finite variance). First, it is the probability that is small; the law does not say that the difference must be small. The statement allows the unusual event of 1000 heads in 1000 tosses to occur; it keeps it under the statement that the probability can be made sufficiently small for large enough n (the number of tosses in this case).

Second, the law does not say that after "a run of bad luck there must be a run of good luck." What it says is that any run of events, no matter how peculiar, will *probably* be "swamped" by later events, not compensated. The model of the random *independent* tosses says that the coin has no memory to remember past behavior so that it could not possibly compensate. There is nothing in the law that says that a peculiar run is later compensated. In other words, because the law asserts that it is the probability that is small, it does not prevent unusual events from occurring; due to the independence of the trials a run of good luck can be as likely followed by a run of good, indifferent, or bad luck as any other run of events. The model assumes that the trials are independent of each other, that there is no memory.

Third, the law does not say that the number of heads, for example, must come close to the expected number; on the contrary, the number will (probably) slowly, like the $\sqrt{n}$, deviate from the expected number. It is the average that will come close to the expected value (probably). The difference in these two statements is that in the second there is an n in the denominator.

Abstractly, how is it that we have finally connected the probability of a single trial with the outcomes of runs of trials? It is simply that from the original space of the probability distribution $p(x)$ we moved to the n-dimensional product space of n trials and in this product space examined the distribution of probabilities. When you consider that high powers of probabilities less than 1 become small numbers, but that high powers of small probabilities become very small, then you see that the relatively large probabilities are the dominant numbers in the product space; hence the distribution of probabilities is bunched up, and this accounts for the law of large numbers. Almost all the probability in the product space is "near" the expected value, "near" as measured by the standard deviation. Most of the probability values very far from the expected value are very small, even when added together. These are the intuitive reasons why we find that almost all the probability is near the expected value of the

n trials, and that the sum of all the other outcomes is arbitrarily small, *provided* we raise the dimension n of the product space sufficiently high. We identified the outcome of a run of n trials with a point in the corresponding n-dimensional product space of single trials, and it is hard to believe that this is not a reasonable thing to do.

We illustrated the law in Example 12.6-2, where we picked the probability $p = \frac{1}{3}$ and computed the set of numbers

$$P(k) = C(n, k)p^k(1 - p)^{n-k}$$

which gives the probability distribution of exactly k successes in n trials. In Figure 12.6-2 you can see the deviation of the average from the expected value $\frac{1}{3}$. You see how the distribution gets narrower and narrower.

3.7 EXPERIMENTAL EVIDENCE FOR THE MODEL

How well is this probability model realized in practice? Numerous experiments have been performed and recorded. If you try to make such a test, you need to be careful to see that you have a device that has constant probabilities for each outcome and that the trials are indeed independent. Gambling places (when honest) try to maintain these two properties and thus can be used as sources of data. During the First and Second World Wars, apparently there was idle time for some soldiers because records of dice tosses were kept and found to follow the computed values from the ideal model of unbiased, independent trials. Similarly, roulette wheels are regularly tested by the gambling houses for wear and tear that might destroy the equally probable properties of the wheel. The very design, as well as the way they are operated, makes the trials quite independent; there seems to be little chance for "influencing" the outcomes.

Numerous trials of coins have been made, and most of them show that the model is realized rather closely, although as you would suspect any particular coin is not perfect. The difficulty is that when a larger than expected deviation occurs you are left in the position of wondering if it is due to the coin or to one of the unlikely fluctuations that the law allows to occur. After all, in the model unlikely events should occur with their predicted probabilities, and if the model represents reality very closely, then unlikely events should also occur in real life. In delicate tests, the wear of the coin and the tossing mechanism must be allowed for.

In this day and age of computers with random-number generators, many people try computing simulations of random processes, and as indicated in Experiment 13.2-2 with respect to the computation of π, the simulation is a test of the random-number generator rather than a test of the theory. Nevertheless, simulations on computers are often useful for showing how things go. You can see the law of large numbers in action when you compute the probabilities of various numbers of H's and T's for n tosses and then plot the results. The plots showing the law of large numbers were not a Monte Carlo simulation but rather a display of the theoretical probabilities. If you want to do a simulation, you have only to get a computer, even a small programmable hand computer, and write the appropriate program. A few thousand trials will be useful, but don't make the mistake of running the machine all night (the power consumption is slight after all and the wear and tear on the transistors is negligible), thinking that you are doing much other than testing the random-number generator. We will use Monte Carlo tests many times to check our derivations to see if we have missed an algebraic sign or a factor of 2. Such tests cannot prove that a formula is right, but it can strengthen your belief that you have not made a serious mistake.

EXERCISES 13.7

1. Test a local random-number generator to see if the mean and variance are reasonable.
2. Compute the average of 12 numbers from the flat distribution ($0 \leq x \leq 1$). Show that the variance is 1 while the mean is 6.

13.8 EXAMPLES OF CONTINUOUS PROBABILITY DISTRIBUTIONS

We now take up a few examples of the use of continuous probability distributions. We are greatly limited in the problems we can do since we are limited at present by the class of functions we can integrate. Many more examples will be given later when other functions can be integrated.

Example 13.8-1

Given a unit square, what is the probability that a random point is within $\frac{1}{4}$ of a unit from an edge? The sample space is the unit square. The word "random" implies a uniform distribution of probability; hence we see that $p(x) = p(y) = 1$. We draw Figure 13.8-1. The area of the center square (failures) is $(\frac{1}{2})(\frac{1}{2}) = \frac{1}{4}$. Hence the required probability is

$$P = 1 - \frac{1}{4} = \frac{3}{4}$$

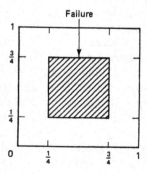

Figure 13.8-1   Probability of being within one-quarter of an edge

Example 13.8-2

Roots of a quadratic. What is the probability that the quadratic equation

$$x^2 + bx + c = 0$$

has real roots? To make sense of the question, we need to define the ranges and probability distributions permitted for the coefficients. Suppose we restrict the coefficient probability distributions to be uniform and $|b| \leq 1$, $|c| \leq 1$. The sample space is, therefore, the 2×2 square centered about the origin (Figure 13.8-2). It is reasonable to assume that the coefficients are selected independently and at random in their intervals;

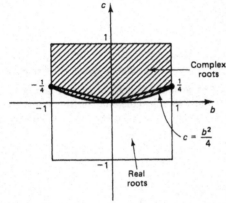

Figure 13.8-2   Roots of a quadratic

for each coefficient the probability density $p(b) = p(c) = \frac{1}{2}$. What is the dividing line between the region of real and the region of complex coefficients? The condition

$$b^2 - 4c = 0$$

that the discriminant be zero, divides the two regions. The shaded region is the region of complex roots. The probability density integrated over the whole region must be 1. Since $p(b, c) = p(b)p(c) = \frac{1}{4}$, the successes have the probability

$$R = 2\left(\frac{1}{4}\right) + 2 \int_0^1 \frac{1}{4}\left(\frac{b^2}{4}\right) db$$

$$= \frac{1}{2} + 2\left(\frac{1}{16}\right)\left(\frac{1}{3}\right) = \frac{1}{2} + \frac{1}{24} = \frac{13}{24}$$

From the figure you see that this is a reasonable answer. If you cut out the shaded figure, you find the upper bound on the area of $\frac{9}{16}$.

Example 13.8-3

A stick is broken at random in two places. What is the probability that the three pieces will form a triangle?

     Let the stick have unit length since the units will not matter in the final answer. Next, let the first break be at the point x and the second at the point y. The sample space is the unit square (Figure 13.8-3). The words "at random" mean that the probability is uniform over the sample space. From the total area, we conclude that $p(x) = p(y) = 1$.

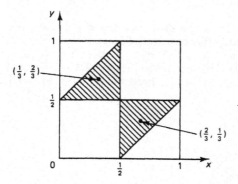

Figure 13.8-3   Broken stick problem

We need the dividing lines between the regions of success and failure. Clearly, when $x = \frac{1}{2}$, we have a triangle with no area, and therefore the line $x = \frac{1}{2}$ is a boundary. Similarly, for $y = \frac{1}{2}$. The other degenerate triangle occurs when the difference between y and x is $\frac{1}{2}$:

$$y - x = \frac{1}{2}$$

or

$$y = x + \frac{1}{2}$$

and this is shown in the figure. Similarly, for $x - y = \frac{1}{2}$ we have a boundary. Now, to determine which regions are successes, we take the equilateral triangle formed when $x = \frac{1}{3}$, $y = \frac{2}{3}$ and plot the corresponding point in the sample space. And similarly for $x = \frac{2}{3}$, $y = \frac{1}{3}$. These points are also shown and define the shaded regions. We have only to convince ourselves that all other regions do not give a triangle. When both x and y are small, or both large, it is impossible for a triangle to exist, and that accounts for the two squares. The two remaining triangles can be also eliminated easily. Inspection shows that the total shaded area is $\frac{1}{4}$; hence the probability that the three pieces form a triangle is $\frac{1}{4}$.

EXERCISES 13.8

1. What is the probability of a random point in a unit circle being within one-quarter of the edge?

2. What is the probability of a point in an equilateral triangle with sides = 2 being within one-third of an edge?

3. Repeat Example 13.8-2 for the equation $x^2 + 2bx + c = 0$ with $|b| \leq |c| \leq 1$.

4. Repeat Example 13.8-2 with $0 \leq b, c \leq 1$.

5. A stick is broken at random into two pieces. What is the probability that the longer piece is at least twice as long as the shorter piece?

6.* Find the expected value of the kth moment of the distribution between two random points selected in the unit interval.

3.9 BERTRAND'S PARADOX

For a finite, discrete sample space of events (outcomes), the word "random" means (unless otherwise stated) that the probabilities of the various outcomes are all equal. When the discrete space is infinite, as in the number of tosses until a head, then we cannot assign to each outcome the same probability, and we have to think a bit more. For continuous distributions, the meaning of "random" and "uniformly random" can be much more subtle. This is perhaps best illustrated by the famous Bertrand's paradox.

Example 13.9-1.

Bertrand's paradox asks the question, "What is the probability that a random chord of a circle has a length greater than the side of an inscribed equilateral triangle?" We will give three different answers to this question, depending on how the "random" is chosen.

The first approach is based on the idea that a "random chord" should be taken to mean that the distance from the center of the circle to the chord is uniformly distributed. To make the assessment of this probability easy, we introduce as a "scaling factor" an equilateral triangle with one side parallel to the chord in question (Figure 13.9-1). There is then a uniform distribution along the perpendicular bisector of the chord, and it is easy to see from the figure that the required probability is $\frac{1}{2}$.

The second approach is to take "random chord" to mean that the end points of the chord are uniformly distributed along the circumference. Equivalently, one end point is held fixed and the other uniformly selected along the circumference. In this case we introduce as the "scaling factor" an equilateral triangle with one vertex at the fixed point (Figure 13.9-2). From the figure the probability is clearly $\frac{1}{3}$.

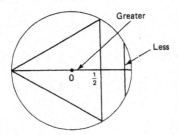

Figure 13.9-1 Chord perpendicular to diameter

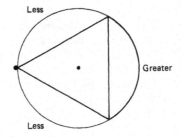

Figure 13.9-2 Uniform along circumference

In the third approach, "random chord" is taken to mean that the midpoint of the random chord should be uniformly distributed over the circular disc. In this case only points inside a circle of half the original radius lead to chords that are longer (Figure 13.9-3), and the probability is clearly $\frac{1}{4}$.

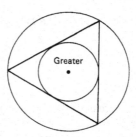

Figure 13.9-3   Center of chord inside circle

Which is the correct answer? In a purely mathematical sense, each is correct; they are merely different interpretations of the meaning assigned to the words "random chord." But to the average student this is unsatisfactory, because they can imagine situations where they are going to act on some computed probability; to be told that there is no answer, or that which is true depends on how you happen to interpret "random chord," leaves a lot to be desired.

One approach to resolving this dilemma is to look more closely at how *you* would do an equivalent experiment. There is a tendency to think of the circle as fixed and the chord as the random element, but the equivalent situation (apparently!) of the line being fixed and the circle chosen at random has the great advantage that the selection of the random circle involves only the choice of the center. We therefore think of a plane surface, say the floor of a room, with a line segment drawn across the middle of the floor, with the ends not too close to the walls. We next pick a circle of radius r much less than the size of the room (and the distance from the line segment to the walls). Now imagine the circle "tossed at random." We assume that the center lies uniformly in this area and *only count* those cases where there is an intersection of the circle with the line. We are using conditional probability, conditioned on there being an intersection

We get two regions (see Figure 13.9-4), the shaded one where the center of the circle is close enough so that the chord is longer than the side of the equilateral triangle we are using for the scale of measurement, and the nonshaded one where there is still an intersection. It is evident immediately that the ratio of successes to total is the ratio of the corresponding areas, and this ratio is near $\frac{1}{2}$. The longer the line segment is with respect to the diameter of the circle we toss, the closer we are to $\frac{1}{2}$.

This gives us a bit more confidence in the $\frac{1}{2}$ value as we have invoked a reasonable invariance; a random line with respect to a fixed circle should agree with

Line
segment

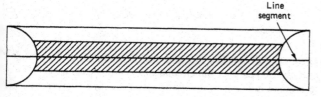

Figure 13.9-4   Region of chords

a random circle with respect to a fixed line. "The experiment and result seem plausible," is all that can be said. Basically, it is a matter of how you decide the uniform distribution is to be selected, and Bertrand's paradox is a clear reminder that this is not as easy as you may at first think it is. Experiments to get the two other solutions to Bertrand's paradox seem to be much more artificial. In any particular case, it is probably best to talk with the consumer of the result before deciding on this random element. This is vital in many simulations that are done on computers; various tacit assumptions on the part of the programmer will give, at times, quite different results. If you are to act on the results, you had best be very careful to check on this point; everyone may think that they understand what is meant by "at random," but a close examination may reveal quite different concepts when you are dealing with continuous distributions. The most dangerous part is that you act without ever thinking about the question!

3.10 SUMMARY

As you have seen, the topics of continuous probability and area are very closely related to each other. If the continuous probability is not uniform, then the equivalent problem is "weighted area" where a different amount of weight is assigned to various points in the region. The probability density function $p(x)$ measures this variability in weighting the elements of the sample space.

We estimated π (which is a fixed constant) by a random process; this is called the Monte Carlo method, and consists of finding a problem involving random elements whose answer is the desired result. When the methods of mathematics fail us, we often turn to a Monte Carlo version of the problem to get a feel for what is happening, and sometimes even to get design parameters. We will use Monte Carlo estimates on occasions to confirm our analytical calculations, and in particular to check that we have the right area or volume and have not made an error of a factor of 2. The Monte Carlo method is very useful for getting approximate guesses as to what is going on in some situation, but sometimes it is used to get moderately accurate answers. You saw that to get slightly increased accuracy greatly increases the amount of labor that must be done; roughly, another decimal place means 100 times as many trials.

We also looked at the Mean Value Theorem for Integrals, and saw that it included the earlier mean value theorems as special cases. Hence it is a fundamental relationship. We put it in the form of an average value which is the way it is often used.

We next turned to the averages of random variables, looking toward the weak law of large numbers, which is a central concept. The law is widely misunderstood; hence a good deal of effort was spent in trying to correct the usual conceptual errors. In particular, the point that the coin, die, or roulette wheel has no memory seems to finally bring a realization that the theorem does not imply compensation for runs of good or bad luck. Along the way we needed the important Chebyshev inequality.

We applied the ideas of probability to a few situations and gave a clear warning of the difficulty of interpreting the concept of "at random." We will later give other examples of this difficulty. Unfortunately, in practice this awkward problem of the interpretation of what is to be taken as "uniform" will not go away, and you need to think carefully about what it means to you, how in the future cases you may face *you* will decide how to choose "at random." You need to develop your intuition before the day you must decide for yourself without the help of some book or other outside source of answers.

PART III

THE TRANSCENDENTAL FUNCTIONS AND APPLICATIONS

The Logarithm Function

We have developed the calculus in terms of the simple algebraic functions, and this has limited the class of problems you can do. Part III of the text is devoted to extending the range of functions you can handle, and consequently the range of problems you can do. The main ideas and applications will be repeated in terms of the new functions. Thus there is a great deal of repetition of ideas in this part, the spiral reinforcement of what you have just learned. It is often said that you learn each mathematics course when you use it regularly in the next course. We deliberately and regularly use old ideas, at first reminding you where they came from but gradually assuming that you know them.

We also introduce some new mathematical ideas. The first one is how to study a situation for which you have no already prepared solution. For example, in Part II you had the indefinite integral

$$y = \int x^n \, dx = \frac{x^{n+1}}{n+1} + C$$

for all $n \neq -1$, but when $n = -1$ this formula cannot apply (note the indicated division by 0). This integral

$$y = \int \frac{dx}{x}$$

is a new function.

The integrand does not exist for $x = 0$, so it is convenient to start the definite integral at $x = 1$. We will call this the ln function (pronounced "the ell en function," which will soon turn out to be a logarithm function). It is defined by

$$y = \ln x = \int_1^x \frac{dt}{t}$$

It is often convenient to write this as

$$y = \ln x = \int \frac{dx}{x}$$

when no confusion can arise (it amounts to dropping the constant of integration C).

We know from the existence of the integral (Section 11.5) that since $1/x$ is continuous for $x \neq 0$, the ln x function exists. Are you sure you cannot find some function whose derivative is $1/x$? It is evident, after some trial differentiations, that fractional exponents always remain fractional exponents, so you are confined to integer exponents. After some trials, and finding that you cannot find such a function, you are naturally driven to the idea that there is no rational function whose derivative is $1/x$, that the ln x is not a rational function in x. In mathematical symbols, this is saying

$$\ln x = \int \frac{dx}{x} \neq \frac{N(x)}{D(x)}$$

where $N(x)$ and $D(x)$ are polynomials of some degrees and the fraction is in lowest terms (meaning no common factors in both the numerator and denominator). You wish to show the nonequality is true, that there can be no two polynomials $N(x)$ and $D(x)$ with the required property. As for most impossibility theorems, the proof is by contradiction.

You may recall from Section 4.6 that polynomials are very much like the integers, and this reminds you that you have seen a similar sort of impossibility problem for the $\sqrt{2}$ in the example of Section 3.4. There you had

$$\sqrt{2} \neq \frac{p}{q}$$

as the impossibility; here you have

$$\ln x \neq \frac{N(x)}{D(x)}$$

What did you do there? You first assumed that there was such an equality, and then squared both sides to get rid of the operation of square root; this suggests that you assume the equality and then *differentiate* both sides to get rid of the integration. The result is

$$\frac{1}{x} = \frac{D(x)N'(x) - N(x)D'(x)}{D^2(x)}$$

Continuing to model this proof on the earlier one, you remove the fractions

$$D^2(x) = x[D(x)N'(x) - N(x)D'(x)]$$

Next was the divisibility argument. Clearly, $D(x)$ is divisible by x; indeed it may be divisible by x several times. Let k be the highest power of x that divides $D(x)$; that is, write

$$D(x) = x^k D_0(x), \qquad \text{where } D_0(0) \neq 0, \qquad k > 0$$

$D_0(0) \neq 0$ says that there is no factor x in $D_0(x)$. The result is

$$x^{2k}D_0^2(x) = x^{k+1}D_0(x)N'(x) - xN(x)[x^k D_0'(x) + kx^{k-1}D_0(x)]$$

Now x^k can be divided out to get

$$x^k D_0^2(x) = xD_0(x)N'(x) - xN(x)D_0'(x) - kN(x)D_0(x)$$

Here each term *except* the last one is divisible by x, so the last one must be, too. But $D_0(x)$ is not divisible by x by its definition; therefore, $N(x)$ must be divisible by x, the same sort of contradiction as before. You assumed that the original expression is in lowest form, and then you found that there is a common factor! Hence the ln x is not a rational function. The proof is a good example of the extension of a method of proof to a similar situation, once you have seen some of the similarities between rational numbers and rational functions.

The function ln x will turn out to be the logarithm of x to the base $e = 2.7182818284 \ldots$, which is a transcendental number. Hence *now* is the time for a brief review of the topic of logarithms.

If y is the logarithm of x to the base b, we write this as

$$y = \log_b x$$

By definition (recall that a logarithm is the power to which the base must be raised to give the original number), it follows that

$$x = b^y$$

Let us take the log to the base a of this equation. We get

$$\log_a x = y \log_a b$$

and, since by definition $y = \log_b x$, we have

$$y = \log_b x = \frac{\log_a x}{\log_a b}$$

This says, *any system of logs is proportional to any other system.* If you have a table of logs to any base (the common logs use base 10), then dividing these entries in the table by the value (from the same table) of the log of the new base, you get the corresponding table in the new base.

If you set $x = a$ in the above equation, you get the useful relationship

$$\log_b a = \frac{1}{\log_a b}$$

A pair of numbers that are needed in converting from base 10 to base e and back are

$$\log_{10} e = 0.4342944819 \ldots \quad \text{and} \quad \log_e 10 = 2.302585092 \ldots$$

The three relationships you need involving logs are (for any base)

$$\log(xy) = \log x + \log y$$

$$\log \frac{x}{y} = \log x - \log y$$

$$\log x^y = y \log x$$

In the last equation, y is any real number. In general, when taking logs, you must start with positive numbers, since the logs of negative numbers are not defined in terms of real numbers. The base is conventionally a positive number greater than 1.

Remember that these are properties of logs, and that we have not yet shown that the integral

$$\int \frac{dx}{x}$$

is a logarithm, let alone the actual base involved.

14.2* ln x IS NOT AN ALGEBRAIC FUNCTION

It is slightly less easy to prove that $\ln x$ is not an algebraic function. In mathematical terms, this statement means that there is no polynomial in x and y of the form

$$P(x, y) = 0$$

such that

$$y = \ln x = \int \frac{dx}{x}$$

is a solution. The conics of Chapter 6 were algebraic functions where the polynomial was of degree 2. The above assertion is that there can be no polynomial of any finite degree for which $y = \ln x$ is a solution. In more detailed mathematical notation this statement is as follows: There can be no polynomial in y with polynomial coefficients in x for which

$$P(x, y) = P_n(x)(\ln x)^n + P_{n-1}(x)(\ln x)^{n-1} + \cdots + P_0(x) = 0$$

for all x in some interval [where the $P_k(x)$ are polynomials in x of arbitrary degrees]. The proof is, as usual, by contradiction. We assume that there is such an equation, and that the degree n is the lowest that is possible in $y = \ln x$. There cannot be more than one polynomial of lowest degree that has the $\ln x$ as a solution, since if there were two then we could eliminate the highest power of $\ln x$ (by the multiplication of each equation by the coefficient of the highest power of the other followed by a subtraction of the two equations). As a result we would have a still lower degree polynomial (degree in $\ln x$ of course), which is not possible since we assumed that we had the lowest.

Using the calculus, we can differentiate this equation. Remember that by the fundamental theorem of the calculus, Section 11.6,

$$\frac{dy}{dx} = \frac{d\{\int dx/x\}}{dx} = \frac{1}{x}$$

We obtain another equation

$$P_n'(x) \ln^n x + P_n(x) n (\ln x)^{n-1} \frac{1}{x} + \cdot \ \cdot \ \cdot$$

of the same degree in $\ln x$. Next, by using suitable multipliers and a subtraction, we could eliminate the highest power of $\ln x$ to get a lower-degree polynomial and hence a contradiction, *provided* the resulting equation is not identically zero! We must take care of this possibility. This suggests an alternative approach to the theorem; we should divide through by $P_n(x)$ *before* the differentiation. You then get, after differentiation,

$$(\ln x)^{n-1} \frac{1}{x} + \left[\frac{P_{n-1}(x)}{P_n(x)} \right]' (\ln x)^{n-1} + \cdot \ \cdot \ \cdot = 0$$

The coefficient of the highest power of $\ln x$ is

$$\frac{1}{x} + \left[\frac{P_{n-1}(x)}{P_n(x)} \right]'$$

and is exactly what we just showed (Section 14.1) could not be identically zero. Since the derived polynomial for which $\ln x$ is a solution is of lower degree than the one you assumed was of lowest degree, you have a contradiction.

Thus the function $\ln x$ is definitely something new, and it is not an algebraic expression in some disguise. We have shown that not only is $\ln x$ linearly independent of the powers of x, but that any finite number of integer powers of $\ln x$ are linearly independent, using polynomials in x as coefficients of the powers of $\ln x$. This will be useful later in the book.

EXERCISES 14.2

1. Carry out the first indicated proof in this section.
2. Prove that $\ln x \neq N(x)/D(x)$ using an argument based on the degrees of $N(x)$ and $D(x)$.

14.3 PROPERTIES OF THE FUNCTION ln x

We will investigate this new function using the techniques that have been developed in a field called *functional analysis*. First, you got rid of the arbitrary constant of the indefinite integral by taking the definite integral form with a fixed lower limit. You did not pick 0 as the lower limit, since at $x = 0$ the integrand $1/x$ does not exist. The next natural choice is at $x = 1$. You have, therefore, the definition (again)

$$\ln x = \int_1^x \frac{dt}{t}$$

where it is convenient to use the dummy variable t to avoid confusion with the upper limit of x. Suppose you replace t by a new variable

$$t = y^{1/n} \qquad dt = \left(\frac{1}{n}\right) y^{1/n-1} \, dt$$

for any real $n \neq 0, 1$. The integral becomes

$$\ln x = \int_1^{x^n} \frac{1}{n} \frac{y^{1/n-1} \, dy}{y^{1/n}} = \frac{1}{n} \int_1^{x^n} \frac{dy}{y}$$

Hence we have the *functional equation*

$$\ln x = \left(\frac{1}{n}\right) \ln (x^n)$$

(Normal equations define sets of numbers; functional equations define whole functions.)

Inspection of the functional equation shows that it has the solution

$$\ln x = k \log_{10} x$$

for any $k \neq 0$.

Are there any other solutions to this functional equation? Suppose that $g(x)$ is any solution; hence

$$g(x) = \left(\frac{1}{n}\right) g(x^n)$$

and consider the difference

$$g(x) - k \log_{10} x = \left(\frac{1}{n}\right) [g(x^n) - k \log_{10}(x^n)]$$

For any $x_0 \neq 1$, set

$$k = \frac{g(x_0)}{\log_{10}(x_0)}$$

Then at $x = x_0$ the left-hand side is zero; hence

$$g(x_0^n) - k \log_{10}(x_0^n) = 0$$

For any real $z > 0$, there is an n such that

$$x_0^n = z$$

namely,

$$n = \frac{\log_{10}(z)}{\log_{10}(x_0)}$$

Thus for all real $z > 0$ we have shown that

$$g(z) = k \log_{10}(z)$$

and the solution is essentially unique (within a constant k). The constant k determines the base of the system of logarithms.

The next step is, therefore, to determine the base. What is it that you know? The main property of logs that you know which involves the base b is

$$\log_b b = 1$$

If you integrate numerically, using Simpson's formula (Section 11.11) from 1 to 3 by double steps of $\frac{1}{10}$, you can get an estimate of where the area under the curve $1/x$ from 1 to x has the value 1. At this point, x will equal the logarithm base. Table 14.3-1 (see also Figure 14.3-1) shows that the base, commonly called e, is approximately

$$e = 2.718$$

TABLE 14.3-1 $\int_1^x \frac{dt}{t}$

x	Simpson's approximation	
2.5	0.91629	
2.6	0.95551	Linear interpolation
2.7	0.99325	gives 2.71855 . . .
2.8	1.02962	
2.9	1.06471	
3.0	1.09861	

A much more accurate value is

$$e = 2.71828\ 18284\ 59045 \ldots$$

We will later investigate this number e more closely. It is conventional in mathematics to write logs to the base e as $\ln x$. Logs to the base e are called *natural logarithms*. They are also called *Naperian logs* after John Napier (1550–1617), who

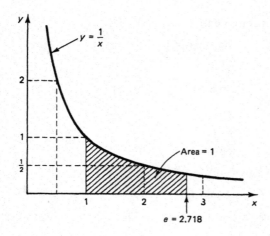

$e = 2.718$

Figure 14.3-1 Estimate of e

discovered logarithms, although his were *not* to the base e in spite of what you may read in some books. We have already used the notation $\log_b x$ for other bases, with $\log x$ meaning base 10.

EXERCISES 14.3

Write the value of the following in terms of ln 2 and ln 3:

1. ln 9
2. ln 12
3. ln (3^3)
4. ln $\sqrt{2}$
5. $(\ln \sqrt{2})^2$
6. ln 24
7. 3 ln $\sqrt{3}$
8. ln $(3/2)^3$
9. 4 ln $1/4$
10. $\sqrt{\ln e}$

What are the following?

11. $\ln_2 4$
12. $\ln_4 8$
13. $(\ln \sqrt{2})^2$
14. $\ln_{10} 100$
15. $\ln_a \sqrt{a}$

4.4 AN ALTERNATIVE DERIVATION—COMPOUND INTEREST

Since we have regularly preached going back to basics, how would the direct derivation of the derivative of ln x go? We apply the four-step rule (Section 7.5) to the $\log_a x$ function. At the third step we have

$$\frac{\Delta y}{\Delta x} = \frac{\log_a (x + \Delta x) - \log_a x}{\Delta x}$$

$$= \frac{\log_a [(x + \Delta x)/x)]}{\Delta x}$$

$$= \frac{\log_a [1 + (\Delta x/x)]}{\Delta x}$$

$$= \frac{1}{x} \log_a \left(1 + \frac{\Delta x}{x}\right)^{x/\Delta x}$$

We need to examine the limit of

$$\left(1 + \frac{\Delta x}{x}\right)^{x/\Delta x} \tag{14.4-1}$$

as $\Delta x \to 0$. Set $x/\Delta x = t$. The limit is now the same as

$$\lim [1 + t]^{1/t} \tag{14.4-2}$$

as $t \to \infty$. We could instead set $t = 1/u$ and have the limit

$$\lim \left[1 + \frac{1}{u}\right]^{u} \tag{14.4-3}$$

as $u \to 0$. The three limits are clearly equivalent. From the following table of numbers, it *seems* as if the limit is the same number $e = 2.71828 \ldots$ as before.

n	$\left(1 + \frac{1}{n}\right)^n$
1	2
2	2.25
4	2.44141
8	2.56578
16	2.63793
32	2.67699
64	2.69734
128	2.70774
1,024	2.71696
8,192	2.71812
65,536	2.71826
524,288	2.71828
4,194,304	2.71828

Thus we again have

$$\frac{d(\log_a x)}{dx} = \frac{1}{x} \log_a e \tag{14.4-4}$$

For $a = e$, this becomes

$$\frac{d(\ln y)}{dx} = \frac{1}{x} \tag{14.4-5}$$

The three limits occur frequently.

Example 14.4-1

Compound interest. This limit e occurs in compound interest problems. Suppose you have an interest rate r of 6%, 12%, or 18% as given in the headings of the columns of Table 14.4-2. The left-hand column has the number of compounding periods. The first line gives the growth per year on 1 unit of money (or other quantity such as biological size). The entry in the body of the table is simply the total amount, interest plus original amount for one period. The second line is for compounding each quarter, the rate per quarter is one-quarter as much, but you get the interest four times *plus* the interest on the interest. The formula is, for rate r,

$$\left[1 + \frac{r}{4}\right]^4 = 1 + 4\left(\frac{r}{4}\right) + 6\left(\frac{r}{4}\right)^2 + \cdots \tag{14.4-6}$$

The first two terms account for the principal and the original simple interest, and the later terms are the interest on the interest due to the compounding.

Next, suppose it is compounded monthly. The effective rate is $r/12$ per month, and it is compounded 12 times. Finally, consider compounding daily. The daily rate is $r/365$, and it is done 365 times.

TABLE 14.4-2

n	6%	12%	18%
1	1.06	1.12	1.18
4	1.06136	1.12551	1.19252
12	1.06168	1.12683	1.19562
365	1.06183	1.12747	1.19716
Estimate	1.06184	1.12750	1.19722

We see that the numbers approach limits, and we see that well before we are at 365 we are close to the position where we could use the number e. We have, corresponding to (14.4-2), the limit

$$\left[1 + \frac{r}{n}\right]^n = \left\{\left[1 + \frac{r}{n}\right]^{n/r}\right\}^r \to e^r \tag{14.4-7}$$

Using this, estimates are written at the bottom of the columns of the table. These values correspond to *continuous compounding*.

This is another example of the fact that, although the phenomenon you are studying may be discrete (a finite number of fixed sized events), nevertheless the continuous model of the calculus provides a very convenient tool for getting good approximations.

EXERCISES 14.4

1. Show that $\int_{1/10}^{10} 1/x \, dx = 2 \ln 10$.
2. Show that, for any function $f(x)$ and $a > 1$, $\int_{1/a}^{a} f(x)/\{f(x) + f(1/x)\} \, dx/x = \ln a$. *Hint:* Use the reciprocal substitution.
3. Compute the estimate in Table 14.4-2 for 8% interest.
4. Find a formula for the doubling period at any given rate of compounding.
5. Make a table of the doubling periods and compare with the following classic rule: Divide 72 by the rate (expressed as a percent) to get the number of years to double.

14.5 FORMAL DIFFERENTIATION AND INTEGRATION INVOLVING ln x

We need to look at the differentiation and integration of this new function. As the first few examples show, there are often ways of simplifying the expression before differentiation.

Example 14.5-1

Compute

$$\frac{d(\ln 17x)}{dx} = \frac{d(\ln 17 + \ln x)}{dx} = \frac{1}{x}$$

$$\frac{d(\ln x^n)}{dx} = \frac{d(n \ln x)}{dx} = \frac{n}{x}$$

Find dy/dx where

$$y(x) = \frac{A(x)B(x)}{C(x)D(x)}$$

Take the ln of both sides before differentiating (and use the function of a function idea):

$$\frac{d(\ln y)}{dx} = \frac{d\{\ln A(x) + \ln B(x) - \ln C(x) - \ln D(x)\}}{dx}$$

We see that each term is to be differentiated separately, and this makes things much easier.

$$\frac{y'}{y} = \frac{A'}{A} + \frac{B'}{B} - \frac{C'}{C} - \frac{D'}{D}$$

It is important to note that if x is replaced by $-x$ in the integral, then the dx is replaced by $-dx$, and the quantity

$$\frac{dx}{x}$$

does not change! Thus in truth

$$\int \frac{dx}{x} = \ln|x| + C$$

We simply cannot tell which sign to use when we are given the derivative. Usually, the problem indicates which sign to use, but as protection we must remember that all we get is the absolute size, not the algebraic size.

Example 14.5-2

Differentiate

$$y = \ln(1 + x^2)$$

We get immediately ($U = 1 + x^2$; note that we have shifted from the lower case u to the upper case U)

$$\frac{dy}{dx} = \left(\frac{dy}{dU}\right)\left(\frac{dU}{dx}\right)$$

$$= \left(\frac{1}{1 + x^2}\right)\left(\frac{d(1 + x^2)}{dx}\right)$$

$$= \frac{1}{1 + x^2}(2x) = \frac{2x}{1 + x^2}$$

Example 14.5-3

Differentiate

$$y = \ln(1 + \sqrt{x})$$

This gives

$$\frac{dy}{dx} = \frac{1}{1 + \sqrt{x}} \frac{d(1 + \sqrt{x})}{dx} = \left(\frac{1}{2\sqrt{x}}\right)\frac{1}{1 + \sqrt{x}} = \frac{1}{2(\sqrt{x} + x)}$$

Example 14.5-4

Differentiate

$$y = x^2 \ln x - \frac{x^2}{2}$$

This gives

$$\frac{dy}{dx} = x^2\left(\frac{1}{x}\right) + 2x \ln x - x = 2x \ln x$$

Example 14.5-5

Integrate

$$\int \frac{dx}{1 + x}$$

If we think of the denominator as $U = 1 + x$, then the $dU = dx$, and we have $(1/U)\,dU$ to integrate. The result is

$$\int \frac{dx}{1 + x} = \ln|1 + x| + C$$

Example 14.5-6

Integrate

$$\int \frac{x\,dx}{1 + x^2}$$

This time we pick $U = 1 + x^2$ and need $dU = 2x\,dx$. We lack a factor 2, so we supply it inside the integral and compensate by a factor of $\frac{1}{2}$ in front. Thus

$$\int \frac{x\,dx}{1 + x^2} = \frac{1}{2}\ln(1 + x^2) + C$$

Differentiation will verify it easily.

Example 14.5-7

Integrate

$$\int \frac{dx}{a^2 - x^2}$$

Here we appear to be in trouble since there is no x in the numerator for the derivative of the denominator. After some thought, we notice that $a^2 - x^2 = (a - x)(a + x)$, and that maybe the term came from the sum of two terms of the form

$$\frac{1}{a + x} \quad \text{and} \quad \frac{1}{a - x}$$

We do not know the actual numerators, so we apply the method of undetermined coefficients and write

$$\frac{1}{a^2 - x^2} = \frac{A}{a + x} + \frac{B}{a - x}$$

with A and B the undetermined coefficients. Cleared of fractions,

$$1 = A(a - x) + B(a + x)$$
$$= (-A + B)x + (B + A)a$$

From the linear independence of the powers of x, we may equate like powers of x. We get from the x terms

$$A = B$$

Then from the constant terms we get

$$2aA = 1$$

Hence

$$\frac{1}{a^2 - x^2} = \frac{1}{2a}\left[\frac{1}{a + x} + \frac{1}{a - x}\right]$$

It is easy to check this result by adding the two terms together. The integral is, therefore,

$$\int \frac{dx}{a^2 - x^2} = \frac{1}{2a}\int \frac{dx}{a + x} + \frac{1}{2a}\int \frac{dx}{a - x}$$

$$= \frac{1}{2a}[\ln|a + x| - \ln|a - x|] + C$$

$$= \frac{1}{2a}\ln\left|\frac{a + x}{a - x}\right| + C$$

This can be verified by differentiating.

Example 14.5-8

Integrate

$$\int \frac{\ln x}{x}\,dx$$

Set $U = \ln x$, and we have the form $U\,dU$, which leads to $U^2/2$:

$$\int \frac{\ln x}{x}\,dx = \frac{1}{2}(\ln x)^2 + C$$

Example 14.5-9

Integrate

$$\int \frac{dx}{x \ln x}$$

The $\ln x$ in the denominator looks difficult to handle. How could it have gotten there from a differentiation? Well, a log does that, so we finally think of

$$\ln|\ln x| + C$$

as the answer. Differentiation checks our guess.

Example 14.5-10.

Integrate

$$\int \frac{x^2\,dx}{1 + x^3}$$

This lacks a 3 in the numerator of being just right, $d\dot{U}/U$. Thus the integral must be

$$\frac{1}{3}\ln|1 + x^3| + C$$

which is easily verified by differentiation.

EXERCISES 14.5

Differentiate the following on sight:

1. $y = \sqrt{\ln x}$
2. $y = x^2 \ln x$
3. $y = \ln(x^3 - 3x)$
4. $y = x^n \ln x - x^n/n$

Integrate the following:

5. $\int 1/(1 + x)\, dx$
6. $\int 1/(a + bx)\, dx$
7. $\int 1/(a + bx)^2\, dx$
8. $\int x/(1 + x^2)\, dx$
9. $\int x/(a^2 + x^2)\, dx$
10. $\int x/(a^2 + x^2)^2\, dx$
11. $\int (\ln x)^2/x\, dx = |\ln x|^3/3 + C$
12. $\int (\ln_3 x)/x\, dx$
13. $\int (\ln_a x)/x\, dx$
14. $\int_1^e \dfrac{(1 - x)^2}{x}\, dx$
15. $\int (1/x)(1 + \ln x)\, dx$
16. $\int 1/\{x(1 + \ln x)\}\, dx$

14.6 APPLICATIONS

The curve $y = \ln x$ rises smoothly but increasingly slowly as you go from $x = 1$ ($y = 0$) to the right (Figure 14.6-1). Each time you go a distance $e = 2.71828\ldots$ times as far out, you rise by another unit; the curve becomes very flat, but continues to rise indefinitely (since $y' = 1/x > 0$ for $x > 0$). There is no upper bound of the $\ln x$ function. The second derivative is $-1/x^2$ and is always negative, so the curve is always concave downward. Since $\ln 1/x = -\ln x$, the values going to the left of $x = 1$ head toward minus infinity as $x \to 0$.

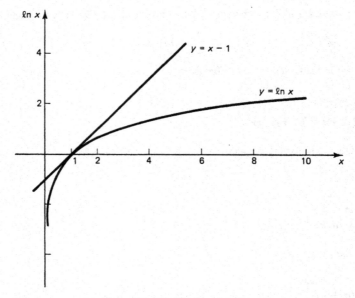

Figure 14.6-1   The ln x function

Example 14.6-1

Graph the function

$$y = x \ln x \qquad\qquad (14.6\text{-}1)$$

(which occurs frequently). To study it near the origin where one factor is 0 and the other is arbitrarily large, we compute Table 14.6-1.

TABLE 14.6-1

x	$x \ln x$
0.1	-0.23026
0.01	-0.04605
0.001	-0.00691
0.0001	-0.00092

We see why we think that the limit is zero. We will later prove in a more mathematical setting (Section 15.5) that the limit is zero. At the origin, the zero from the term x dominates the infinity from the term ln x. Thus the function

$$y = x \ln x$$

has a pair of zeros (Figure 14.6-2), one at $x = 0$ and the other at $x = 1$, and between them the function is negative. Where does the minimum occur? Differentiate, set equal to zero, and solve.

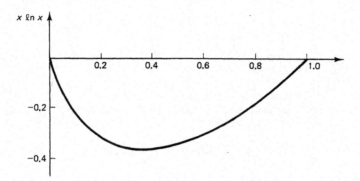

Figure 14.6-2 $y = x \ln x$

$$y' = \frac{dy}{dx} = x\left(\frac{1}{x}\right) + (\ln x)(1) = 1 + \ln x = 0$$

or $\ln x = -1$

Hence the minimum occurs at $x = 1/e$. At this point

$$y = \left(\frac{1}{e}\right) \ln \frac{1}{e} = \frac{-1}{e}$$

Since $y'' = 1/x > 0$, this indicates that the curve is always (for $x > 0$) convex upward.

Example 14.6-2

The *entropy function*, which is defined for any discrete probability distribution p_i, is given by the expression

$$H_2(P) = H(p_1, p_2, \ldots, p_n) = \sum_{i=1}^{n} p_i \log_2 \frac{1}{p_i} \qquad (14.6\text{-}2)$$

where $\Sigma p_i = 1$ and P represents the probability distribution, or if you wish the random variable. This function occurs in many places, particularly in *information and coding theory;* and as is customary there, we have taken 2 as the base of the logarithm system (also to give you practice in other bases than e and 10). In words, the entropy of a distribution is the probability-weighted average of the negative logarithms of the probabilities of the distribution.

It is natural to ask for the set of probabilities p_i for which the entropy function $H_n(P)$ is maximum. By a study of the way each individual term of the function behaves (see Figure 14.6-2), and recognizing that since $\Sigma p_i = 1$, then any increase in one term must decrease some other term, you soon come to the conclusion that the symmetric case must give the maximum, all the

$$p_i = \frac{1}{n}$$

Alternatively, you can eliminate one of the probabilities, say p_n, by using the relation

$$p_1 + p_2 + \cdots + p_n = 1 \tag{14.6-3}$$

You get for $H_2(P)$

$$-\sum_{i=1}^{n-1} p_i \log_2 p_i - (1 - p_1 - p_2 - \cdots - p_{n-1})\log_2 (1 - p_1 - \cdots - p_{n-1})$$

and when you differentiate with respect to p_i (for each i $= 1, \ldots, n - 1$), you get for each derivative equated to zero [recall (14.4-4) and write $1 - p_1 - p_2 - \cdots - p_{n-1} = p_n$ after the differentiation]

$$-p_i \frac{1}{p_i} \log_2 e - \ln_2 p_i + p_n \frac{1}{p_n} \log_2 e + \ln_2 p_n = 0$$

Thus

$$\log_2 e + \ln_2 p_i = \log_2 e + \ln_2 p_n$$

and each $p_i = p_n$ (because the log function is monotone and therefore there is only one solution for a given value of the function). Hence, since the sum of all the p_i is 1, it follows that each $p_i = 1/n$. What is the maximum value? Put $p_i = 1/n$ in each term and you get n terms, each of which is

$$\frac{1}{n} \log_2 n$$

so the sum of all n terms gives the maximum

$$\text{Max}\{H_2(P)\} = \log_2 n \tag{14.6-4}$$

Example 14.6-3

The Gibbs inequality (used extensively in information theory and probability). Given two discrete probability distributions

$$\sum p_i = 1 \quad \text{and} \quad \sum q_i = 1$$

show that

$$\sum p_i \ln\frac{q_i}{p_i} \leq 0 \tag{14.6-5}$$

To prove this we need a simple lemma.

Lemma. For all $x > 0$

$$\ln x \leq x - 1 \tag{14.6-6}$$

and the equality holds only at $x = 1$.

The proof is easy; we simply fit the tangent line to the curve $y = \ln x$ at the point $x = 1$ (Figure 14.6-1) and note that the curve is below the tangent line. The derivative of $\ln x$ is $1/x$, and at $x = 1$ this is 1. The tangent line, therefore, is

$$y - 0 = 1(x - 1)$$

The second derivative of $\ln x$ is $-1/x^2$, which is always negative, and this leads to the statement of the lemma (14.6-6).

To prove the main result, we simply apply this lemma to each term of the sum (14.6-5):

$$\sum p_i \ln \frac{q_i}{p_i} \le \sum p_i \left\{ \frac{q_i}{p_i} - 1 \right\} \le \sum q_i - \sum p_i = 1 - 1 = 0$$

For the equality condition to hold, it must hold for each term of the sum; that is, each $q_i = p_i$ for all i. This is known as the Gibbs (1839–1903) inequality. The inequality applies for any base $b > 1$.

EXERCISES 14.6

1. Sketch $H_2(P) = p \log_2 (1/p) + (1 - p) \log_2 1/(1 - p)$ when the distribution has only two probabilities, p and $1 - p$. Note carefully the slope at $p = 0$ and $p = 1$.
2. From the Gibbs inequality, show directly that a maximum of the entropy function is $\ln n$. *Hint:* Set $q_i = 1/n$ for all i.
3. Prove from Exercise 2 that the maximum entropy occurs when all the p_i are equal.

4.7 INTEGRATION BY PARTS

One of the most useful tools for integration is the method of *integration by parts*. It is easily derived from the formula for the product of two functions (in differential form)

$$d(uv) = u \, dv + v \, du$$

We merely take the indefinite integral of both sides (and neglect the constant C of integration). We get

$$uv = \int u \, dv + \int v \, du$$

or in the standard form (where we shift to the conventional uppercase letters)

$$\int U \, dV = UV - \int V \, dU \qquad (14.7\text{-}1)$$

The lost constant of integration can be associated with the second integral if you are worried about it.

The reason that the method is so important is that when you see an awkward function under the integral sign then calling it U means that you will later see it only as the derivative (which is often simpler than the function!). We will illustrate this observation by a series of examples.

Example 14.7-1.

Consider the integral

$$\int \ln x \, dx$$

At first you are probably stunned by the appearance of the ln x. You probably do not know an integral of it. But the above remark suggests that you can get rid of the ln x if you pick $U = \ln x$ and the rest as the dV. You have

$$U = \ln x, \qquad dV = dx$$

$$dU = \frac{1}{x} dx, \qquad V = x$$

(It is usually convenient to pick the constant of integrating C when integration dV as 0, as we have done, but occasionally other choices are useful.) Using the formula for integration by parts (14.7-1), you have for the given integral

$$\int \ln x \, dx = x \ln x - \int x \frac{1}{x} \, dx$$

$$= x \ln x - x + C$$

You can verify this easily by differentiating the answer. You get

$$\frac{d(x \ln x - x + C)}{dx} = x \frac{1}{x} + \ln x - 1 = \ln x$$

which is indeed the integrand.

Example 14.7-2

Next consider the integral

$$\int x \ln x \, dx$$

How would you proceed? Pick U as $x \ln x$? That would still leave a ln x after the differentiation to get the dU. This could be done by another application of integration by parts, but perhaps

$$U = \ln x, \qquad dV = x \, dx$$

$$dU = \frac{1}{x} dx, \qquad V = \frac{x^2}{2}$$

would be better. So you try it. You get

$$\int x \ln x \, dx = (\ln x)\frac{x^2}{2} - \int \frac{x^2}{2} \frac{1}{x} \, dx$$

and the second integral can now be done. You get, finally,

$$\int x \ln x \, dx = \frac{x^2}{2} \ln x - \frac{x^2}{4} + C$$

which you then verify by differentiating back to check that the integrand comes out.

Generalization 14.7-3

Generalize the previous two examples. *One* possible generalization is

$$\int x^n \ln x \, dx$$

To get rid of the unwanted ln x function, you pick it as

$$U = \ln x$$

That leaves the rest as the dV:

$$dV = x^n dx, \qquad V = \frac{x^{n+1}}{n+1}$$

You have

$$dU = \frac{1}{x} \, dx$$

and therefore

$$\int x^n \ln x \, dx = \frac{x^{n+1}}{n+1} \ln x - \int \frac{x^n}{n+1} \, dx$$

$$= \frac{x^{n+1}}{n+1} \ln x - \frac{x^{n+1}}{(n+1)^2} + C$$

Again you verify by direct differentiation. With this simple check there is no excuse for giving a wrong answer; the only possible excuse is that you could not get started on the problem correctly! Note that this generalization provides a check on the two previous examples, $n = 0$ and $n = 1$. The formula is true for all real $n \neq -1$.

Example 14.7-4

Consider

$$\int (\ln x)^2 \, dx$$

Obviously, you pick $U = (\ln x)^2 = \ln^2 x$, and $dV = dx$. You get

$$\int \ln^2 x \, dx = x \ln^2 x - \int x \left\{ 2 \ln x \frac{1}{x} \right\} dx$$

The x terms cancel in the second integrand, and you have

$$\int \ln^2 x \, dx = x \ln^2 x - 2 \int \ln x \, dx$$

You are now reduced to an earlier case, Example 14.7-1, which will yield to another integration by parts. Thus you have, finally,

$$\int \ln^2 x \, dx = x \ln^2 x - 2x \ln x + 2x + C$$

To check the answer, you differentiate the result. You get from differentiating the right-hand side

$$\ln^2 x + x(2 \ln x)\frac{1}{x} - 2 \ln x - 2x\frac{1}{x} + 2$$

and all the terms cancel, as they should, except the first one. This kind of cancellation is very typical of results obtained by repeated integration by parts.

Example 14.7-5

An example that usually bothers the student the first time it occurs is

$$\int \frac{\ln x}{x} \, dx$$

This is the missing case $n = -1$ in Generalization 14.7-3. There are many ways of doing it. There is the direct way used in Section 14.5 (not integration by parts)

$$U = \ln x, \qquad dU = \frac{1}{x} \, dx$$

which gives, as before,

$$\int \frac{\ln x}{x} \, dx = \frac{\ln^2 x}{2} + C$$

If you try integration by parts, you get

$$U = \ln x, \qquad dV = \frac{dx}{x}$$

$$dU = \frac{1}{x} \, dx, \qquad V = \ln x$$

Notice the advantage of the fixed pattern for writing the terms you are going to use. The use of the same technique every time frees the mind for more important things. When you do the integration by parts, you get

$$\int \frac{\ln x}{x} \, dx = (\ln x)(\ln x) - \int \frac{\ln x}{x} \, dx$$

The integral on the right end is the same as the original integral! A moment's thought shows that you can transpose it to the left-hand side, divide through by 2 (and remember to attach the constant C) to get

$$\int \frac{\ln x}{x} \, dx = \frac{(\ln x)^2}{2} + C$$

as you should.

There are often several ways of doing the same integral. We have apparently neglected to handle the absolute values in doing the integration; this leaves you a problem to discuss for your own peace of mind. The results must, of course, be the same within an additive constant C. Note that $\ln 6x + C$ and $\ln x + C$ are the same (although the two values of the constant C differ by the number $\ln 6$).

EXERCISES 14.7

Integrate the following:

1. $\int x^3 \ln x \, dx$
2. $\int P(x) \ln x \, dx$, where $P(x)$ is an arbitrary polynomial. *Hint:* write $P(x) = \sum_0^N c_k x^k$.
3. $\int \ln^n x \, dx$
4. Carry out one stage of the general case $J(m, n) = \int x^m \ln^n x \, dx$ to get a *reduction formula* lowering the degree of n to $n-1$. $J(m, n) = x^{m+1}(\ln^n x)/(m+1) - n/(m+1)J(m, n-1)$.
5. Discuss why any polynomial in x and $\ln x$ can be integrated in principle.
6. In the above exercises, when can the implied condition of integer exponents be relaxed?
7. Why do we not use the derivative of a quotient formula as the basis for two corresponding integration formulas?

14.8* THE DISTRIBUTION OF NUMBERS

The widely used *scientific notation* for representing numbers places the decimal point *after* the first nonzero digit and follows the number by an appropriate power of 10; for example

$$\pi = 3.14159 \ldots \times 10^0$$

The velocity of light is approximately 299,792.5 kilometers per second, and is written as

$$c = 2.997\,925 \times 10^5$$

Likewise $1/1000$ is written as

$$1 \times 10^{-3}$$

The number part is often carelessly called the *mantissa* in analogy with the mantissa of logarithms.

You might think that these mantissas are uniformly distributed between 1 and 10, but simple experiments do not bear this out. Rather, for most (but not all) sources of numbers the probability density is close to

$$p(x) = \frac{1}{x \ln b} \tag{14.8-1}$$

where b is the number base, in the usual case 10, but in computers it may be 2, 8, or 16. This distribution is often called the *reciprocal distribution* because $p(x)$ is proportional to the reciprocal of x. The cumulative probability distribution for the random variable X is $(1 \le x < b)$, $P(X \le x) = P(x)$

$$P(x) = \int_1^x \frac{dt}{t \ln b} = \frac{\ln x}{\ln b} \tag{14.8-2}$$

In computers it is usual to place the decimal point *before* the first digit, thus making the exponent 1 greater. The difference is a trivial nuisance, but it is probably too late to get the computer people to reform. In any case, the density distribution is essentially the same, $(\ln bx)/\ln b$. We will use the conventional scientific notation and the interval (sample space) $1 \le x < 10$.

It is easy to verify that the total probability, using base 10, is

$$\int p(x) \, dx = \int_1^{10} \frac{dx}{x \ln 10}$$

$$= \frac{\ln x}{\ln 10} \Big|_1^{10} = 1$$

The cumulative distribution for the random variable X is $P(x) = \ln x/\ln 10 = \log_{10} x$. What is the probability of a random number beginning with a given decimal digit? For the digit N, it is the difference

$$P(N + 1) - P(N) = \frac{\ln(N + 1) - \ln N}{\ln 10} = \log_{10}\left(1 + \frac{1}{N}\right)$$

As an experimental test of the distribution, we took the 50 physical constants listed in a standard table and counted the number of entries beginning with each decimal digit. The result is given in Table 14.8-1,

TABLE 14.8-1

Digit	Observed	Theoretical	Difference
1	16	15	1
2	11	9	2
3	2	6	-4
4	5	5	0
5	6	4	2
6	4	3	1
7	2	3	-1
8	1	3	-2
9	3	2	1
	50	50	

which seems to verify the model for physical constants. The distribution also applies to such widely diverse sources of numbers as the binomial coefficients, the factorials, the powers of any number whose $\log_{10}$ is not a rational number, and so on.

The nonuniform distribution of the numbers has a number of consequences. For example, the effect of roundoff propagating through a sequence of multiplications depends, on the average, on the distribution of the leading digits, since if we have two numbers with corresponding small errors

$$x_1 + \epsilon_1 \quad \text{and} \quad x_2 + \epsilon_2$$

and form their product, we get

$$x_1 x_2 + x_1 \epsilon_2 + x_2 \epsilon_1 + \epsilon_1 \epsilon_2$$

Neglecting the last term as being too small to matter, we see that the expected value of the x_i gives the effective multiplier of each propagated error for randomly chosen numbers.

Example 14.8-1

Find the expected value of a number from the reciprocal distribution. We have

$$E\{X\} = \int_1^{10} x \left\{ \frac{1}{x \ln 10} \right\} dx$$

$$= \frac{1}{\ln 10} \int_1^{10} dx = \frac{9}{\ln 10} = 3.90865 \ldots$$

which is a good deal less than the 5.5 that is given by the model of a uniform distribution.

EXERCISES 14.8

1. Make a table of the distribution of the first digit of the function 2^n for $n = 1, \ldots, 25$.
2. Same for $y = 3^n$.
3. Same for $y = 5^n$.
4. Compute the distribution of leading digits for the base 8.
5. Compute the expected propagation coefficient for base b and examine the results for the bases 2, 4, 8, 10, and 16.
6. Find the variance of the reciprocal distribution.

14.9 IMPROPER INTEGRALS

Up to this point we have insisted that the integral have two properties. First, that the range of integration, typically from a to b, be finite. Second, that in the range of integration the integrand be finite. Without these two conditions the derivations we made in connection with the development of the concept of the integral would not be valid.

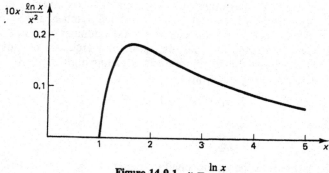

Figure 14.9-1 $y = \dfrac{\ln x}{x^2}$

However, frequently problems arise where one or the other (or both) of these conditions is violated. As an example of an infinite range, consider the area under the curve

$$y = \frac{\ln x}{x^2}$$

from $x = 1$ to $x = \infty$ (see Figure 14.9-1). An example of an infinite integrand is the area below the curve

$$y = -\ln x$$

above the x-axis and between $x = 0$ and $x = 1$ (see Figure 14.9-2). We treat both cases by the following very reasonable device. We ask, "What is the integral when the range is chosen to be slightly less?" and then we examine the answer as the range

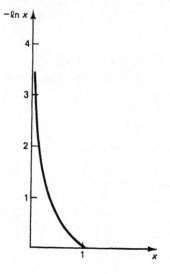

Figure 14.9-2 $y = -\ln x$

is increased to the limit, either (1) for infinite range or (2) to where the integrand becomes infinite.

This process of handling the infinite by dropping back to the finite, working out the result, and then taking the limit is a basic approach, which we have already used many times. When in doubt, it is a sound rule: "Do the finite case, and then examine how it behaves as you approach infinity." If the result appears strange, then you are in a position to see whether it is your intuition or an error that bothers you.

Example 14.9-1

We first consider the case of an infinite range. In this example we begin by considering the integral over the finite range

$$J(N) = \int_1^N \frac{\ln x}{x^2}\, dx$$

(Figure 14.9-1). When we get the answer in terms of N, we will *then* let N approach infinity. To do the integration, we use integration by parts with

$$U = \ln x, \qquad dV = \frac{dx}{x^2}$$

$$dU = \frac{1}{x}\, dx, \qquad V = \frac{-1}{x}$$

The indefinite integral becomes

$$\int = (\ln x)\left(\frac{-1}{x}\right) - \int \left(-\frac{1}{x}\right)\left(\frac{1}{x}\right)\, dx$$

$$= \frac{-\ln x}{x} - \frac{1}{x} + C$$

We have now to put in the limits 1 and N. We get the definite integral

$$J(N) = -\frac{\ln N}{N} + 0 + \left(\frac{-1}{N}\right) + 1$$

We are now ready to let N approach infinity. The first term, $-(\ln N)/N$, requires attention. As before, the denominator N dominates the slower growing numerator $\ln N$, and the quotient goes to 0. The final answer is, therefore,

$$J(\infty) = 1$$

The area under the curve $y = (\ln x)/x^2$ from $x = 1$ to infinity is exactly 1.

Example 14.9-2

An example of the infinite integrand is the area between

$$y = -\ln x$$

the x-axis, and in the interval $x = 0$ to $x = 1$ (see Figure 14.9-2). We have to creep down to the origin, so we start with the integral (with $a > 0$):

$$J(a) = \int_a^1 (-\ln x)\, dx = -(x \ln x - x)\big|_a^1$$

$$= -[0 - 1] + [a \ln a - a]$$

We now let a approach 0. Again the a dominates the $\ln a$ as a approaches 0. The final result is

$$J(0) = 1$$

There is one unit of area below the $-\ln x$ curve and above the x-axis while lying between $x = 0$ and $x = 1$.

Example 14.9-3

Suppose the curve $y = -\ln x$ for $(0 < x \le 1)$ is rotated about the y-axis to form a volume of rotation (see Figure 14.9-3). What is the volume? Again, there is an improper integral to evaluate. From the figure, the elements (of volume) we are summing are cylinders, each of volume

$$2\pi xy\, dx$$

Thus we have (and we eliminate, *except* in our minds, all the details of letting the lower value approach 0)

$$\int_0^1 \{-2\pi x (\ln x)\}\, dx = -2\pi \left\{ \frac{x^2}{2} \ln x - \frac{x^2}{4} \right\}\Bigg|_0^1$$

We get

$$\int = \frac{\pi}{2}$$

for the volume of rotation. The direct substitution of the value 0 masks the careful process of examining the value obtained. This can at times lead to troubles, for example, when terms from two different places both cancel before putting in the limit. You should be prepared to follow the longer path of carefully filling in all the details in any doubtful situation.

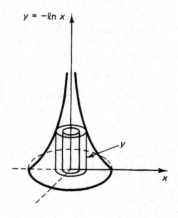

Figure 14.9-3 Method of cylinders

Example 14.9-4

The same problem can be done using discs instead of cylinders (Figure 14.9-4). We have the element of volume

$$\pi x^2 \, dy$$

so we have the integral

$$\int_0^\infty \pi x^2 \, dy$$

We will change the variable of integration from dy to dx using the equation

$$y = -\ln x, \qquad dy = -\frac{1}{x} \, dx$$

The limits, of course, also change. The infinity for y becomes $x = 0$, and $y = 0$ becomes $x = 1$. We have, therefore (remember the minus sign),

$$\int_0^1 \pi x^2 \left(\frac{1}{x}\right) dx = \pi \int_0^1 x \, dx = \frac{\pi x^2}{2} \bigg|_0^1$$

$$= \pi/2$$

as before.

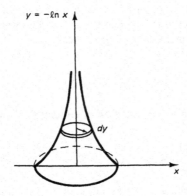

$y = -\ell n\ x$

dy

Figure 14.9-4   Method of discs

Example 14.9-5

Sometimes (as in Example 14.9-4) a change of variable will eliminate an improper integral. Another example is, given the integral

$$J = \int_1^\infty \frac{1}{x^2} \, dx$$

the change of variable $x = 1/y$, $dx = -1/y^2 \, dy$ will give

$$J = \int_1^0 y^2 \left(\frac{-1}{y^2}\right) dy = \int_0^1 dy = 1$$

EXERCISES 14.9

Evaluate the following:

1. $\int_1^\infty \frac{1}{x^2} \, dx$ and $\int_0^1 \frac{1}{\sqrt{x}} \, dx$

2. $\int_1^0 \frac{1}{x^{3/2}} \, dx$ and $\int_0^1 \frac{1}{x^{2/3}} \, dx$

3. $\int_1^\infty \frac{1}{x^{1+a}} \, dx$ and $\int_0^1 \frac{1}{x^{1-a}} \, dx \ (1 > a > 0)$

4. Find the area under $H(P) = p \ln 1/p + (1 - p) \ln 1/(1 - p)$ between 0 and 1.

5. Integrate $\int_0^\infty \frac{\ln (x + 1)}{(x + 1)^2} \, dx$

6. Eliminate the improper integral $J = \int_1^\infty \frac{1}{x^{5/2}} \, dx$ by the substitution $x = 1/y$.

14.10 SYSTEMATIC INTEGRATION

As you approach the problem of integration, it usually appears as a vast morass of special cases, some of which can be done by one trick, some by another, and some which cannot be done within the framework of a given set of functions. For example, the integral

$$\int \frac{1}{\ln x} \, dx$$

cannot be done in closed form (a finite combination of elementary functions). On the other hand, the integral

$$\int \frac{dx}{x \ln x} = \ln |\ln x| + C$$

But we have preached the necessity of the larger view, the uniformization of special tricks, the need for abstraction to master the sea of details, and the corresponding hopelessness of going down the path of special tricks for the endless number of special cases. It is necessary, therefore, to look a bit more abstractly at the problem of integration of functions involving the function $\ln x$. In particular, what broad classes of integrals can be integrated in closed form? There will still be special isolated cases that require tricks to integrate, but at least we will know some of the general cases we can hope to handle.

From the Generalization 14.7-3 that we have already done

$$\int x^n \ln x \, dx$$

we now see that the integral

$$\int P(x) \ln x \, dx$$

where $P(x)$ is any polynomial of degree n in x, can be integrated. We see that the *form* of the result is

$$\int P(x) \ln x \, dx = Q(x) \ln x + R(x) + C$$

If we apply the fundamental theorem of the calculus to this, we have (differentiate)

$$P(x) \ln x = Q'(x) \ln x + Q(x)\frac{1}{x} + R'(x)$$

But we proved (Section 14.1) the important result that the $\ln x$ cannot be represented by a rational function (indeed we proved in Section 14.2* that the powers of $\ln x$ are linearly independent when the coefficients are polynomials). Therefore, we may equate all the terms in $\ln x$ to zero and the terms free of $\ln x$ to zero. We get

$$P(x) = Q'(x) \quad \text{and} \quad 0 = \left(\frac{1}{x}\right)Q(x) + R'(x)$$

from which we get

$$Q(x) = \int_0^x P(s) \, ds$$

$$R(x) = -\int \frac{Q(x)}{x} \, dx + C$$

We picked the lower limit in the first integral to be zero so that in the next integrand $Q(x)/x$ is a polynomial.

Example 14.10-1

Integrate

$$\int (3x^2 + 2) \ln x \, dx$$

We try the form

$$(q_1 x + q_2 x^2 + q_3 x^3) \ln x + r_1 x + r_2 x^2 + r_3 x^3$$

Upon differentiating both expressions, we get

$$(3x^2 + 2) \ln x = (q_1 + 2q_2 x + 3q_3 x^2) \ln x$$
$$+ (q_1 + q_2 x + q_3 x^2) + r_1 + 2r_2 x + 3r_3 x^2$$

Equating the coefficients, one by one, of the two ln terms, we get

$$q_1 = 2, \qquad q_2 = 0, \qquad q_3 = 1$$

Now equating the terms without the ln function, we get, in sequence,

$$r_1 = -q_1 = -2, \qquad r_2 = \frac{-q^2}{2} = 0, \qquad r_3 = \frac{-q_3}{3}$$

The integral is, therefore,

$$\int (3x^2 + 2) \ln x \, dx = (x^3 + 2x) \ln x + \left(\frac{-x^3}{3} - 2x \right) + C$$

Upon differentiation, this easily checks.

This method of integration is a general one, but it is usually neglected in the standard textbooks. It is a powerful method worth remembering. First, you guess at the form (as you often do in the method of undetermined coefficients). Second, you apply the fundamental theorem of the calculus (differentiate both sides). Third, you equate the coefficients of the linearly independent functions to zero, thus getting as many conditions as there are unknowns in the form you assumed. As in the method of undetermined coefficients, if you guess the form wrong, you will find out that you are wrong and often get a clue as to where you were wrong. You can then either include more terms as suggested by the failure, or else set about proving that it cannot be done in the general form you are trying (see Sections 14.1 and 14.2*).

The approach depends on your understanding of what terms can possibly lead to the integrand you are given and what terms simply cannot be involved. Thus it should be clear that radicals when differentiated lead to radicals and, therefore, cannot be involved in a problem where the integrand is rational. You can fill in the details of the argument to convince yourself that if you try such a term it will end up with a zero coefficient (when you equate such linearly independent terms to zero).

Evidently, the method that assumes the form is often easier to carry out than the equivalent method of direct integration (using repeated integration by parts in this case).

The method is easily extended to

$$\int P(x) \ln^2 x \, dx$$

and then you see that the entire class

$$\int P(x, \ln x) \, dx$$

can be done, where now $P(x, \ln x)$ is the polynomial in two variables x and $\ln x$.

Why can these integrals be done? One partial explanation is that the ln x function arises from the integration of a rational function $1/x$, and therefore when we either integrate by parts or differentiate in the above method, the ln function effectively disappears.

EXERCISES 14.10

1. Write out the form you would assume for the integration of $P(x, \ln x)$.
2. Carry out the details for the integrand $P(x) \ln^2 x$.
3.* Following the pattern of Section 14.2*, show that $\int 1/\ln x \, dx \neq N(x, \ln x)/D(x, \ln x)$, where $N(x, \ln x)$ and $D(x, \ln x)$ are polynomials.

14.11 SUMMARY

We introduced and studied a new function $\ln x$. It was defined by an integral, which we proved could not be done in a closed rational (or algebraic) form. We further showed that the powers of $\ln x$ were linearly independent, and this is useful in the later stages of the systematic integration of integrals with powers of $\ln x$.

We also examined the use of the new function in a number of old and new applications, including compound interest, the distribution of numbers, and the entropy function.

The powerful method of integration by parts was introduced; it is simply the inverse of the derivative of a product. The method enables you to replace a function by its derivative in the new integral, which is often, when combined with the integral of the other part of the integrand, such that you can carry out an integration. Integration by parts and the method of changing variables are two of the most powerful methods of integration available.

We also *extended* the idea of an integral to improper integrals by the usual method of examining the finite case and letting it approach infinity. In general, it is not the details but the methods used to develop new applications that are important.

Finally, we introduced the obvious "guess at the solution" method as another illustration of the usefulness of the method of undetermined coefficients. If you guess the correct form, then from the fundamental theorem of calculus you can equate the derivative of your guess to the original integrand. If the result is an identity, for a suitable choice of the undetermined parameters you introduced into the guessed form, then you have the integral. If it is not an identity, then you can guess again, guided by the failure to fit, or else abandon that approach.

15

The Exponential Function

15.1 *THE INVERSE FUNCTION*

Given a point in the Euclidean plane, say (3, 5), then the *inverse point* has the coordinates interchanged (inverted), that is, the point (5, 3). Similarly, the inverse of $(-2, 4)$ is $(4, -2)$, and in general the point (x, y) has it inverse as (y, x). From Figure 15.1-1, you see that the point and its inverse are symmetrically located with respect to the 45° line; the 45° line bisects perpendicularly the line segment joining the two

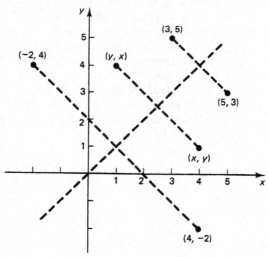

Figure 15.1-1 Inverse points

points. Hence the 45° line is a line of symmetry. Note that the inverse of the inverse of a point is the original point.

When every point of one curve (function) is the inverse point of another curve (function), then the first curve (function) is said to be the inverse curve (function) of the second. For example, given

$$y = x^2, \qquad \text{for } x \geq 0$$

we interchange x and y to get the inverse curve (function)

$$x = y^2$$

Solving this for y, we have the inverse curve (function)

$$y = \sqrt{x}$$

Figure 15.1-2 shows the two curves. You see the reason for the condition of non-negative x in the first equation; it is to avoid the double values that would arise if we had allowed the full parabola.

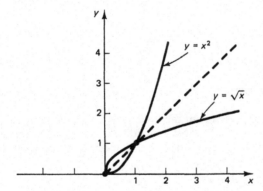

Figure 15.1-2   Inverse curves

This brings up a difficult point when finding the inverse function: in general, how can you limit the range to avoid multiple values in the inverse function? For strictly monotone functions there is no problem, as the function must be single valued and hence the inverse function is also single valued, but in general it is a problem that must be faced for each inverse function.

The $\ln x$ function is strictly monotone, $y' \neq 0$, and hence it has an inverse function. To find the inverse of the function

$$y = \ln x$$

we interchange the letters

$$x = \ln y = \log_e y$$

To solve this for y, we need to remember a property of the ln function; x is the power to which the base e is raised to get the number y. We get

$$y = e^x$$

This is the *exponential function*. It is shown in Figure 15.1-3. Note that e^x is always > 0 for all finite x. As $x \to \infty$, so does $e^x \to \infty$; and as $x \to -\infty$, $e^x \to 0$. Also, $e^0 = 1$.

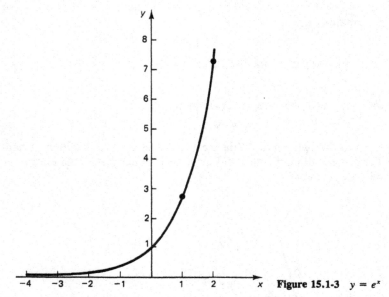

Figure 15.1-3 $y = e^x$

It is worth noting that, since the inverse of the inverse of a point is the original point, so too the inverse function of the inverse function is the original function (provided you watch the limitations on the range that is necessary to make the functions single valued). Thus we have both

$$y = e^{\ln y}$$

and

$$y = \ln e^y$$

15.2 THE EXPONENTIAL FUNCTION

The exponential function plays a central role in the calculus, probability, and statistics, as well as in mathematics generally. We therefore need to investigate both its derivative and its integral. You know that

$$y = e^x \tag{15.2-1}$$

is equivalent to

$$x = \ln y$$

You know how to differentiate this last equation

$$\frac{dx}{dy} = \frac{1}{y}$$

What relation is this to dy/dx? Go back to fundamentals and recall that the derivative is the limit of a difference quotient. It is clear that (barring a division by zero)

$$\frac{\Delta y}{\Delta x} = \frac{1}{\Delta x/\Delta y}$$

and that when you take the limit you will get

$$\frac{dy}{dx} = \frac{1}{dx/dy} \tag{15.2-2}$$

This equation requires that neither of the derivatives be 0. If they are not zero, then sufficiently close to the limit neither of the difference quotients will be zero. Therefore, you have for the inverse of the ln x function

$$\frac{dy}{dx} = \frac{1}{d(\ln y)/dy} = \frac{1}{1/y} = y = e^x$$

The derivative of the function is the function itself! This accounts for much of the importance of the exponential function $y = e^x$.

More generally, you have from the chain rule for differentiation,

$$\frac{d(e^u)}{dx} = e^u \frac{du}{dx} \tag{15.2-3}$$

From this it follows immediately that the corresponding integral is

$$\int e^u \, du = e^u + C \tag{15.2-4}$$

As a result, you have a new pair of formulas to learn.

The particular case

$$\int e^{ax} \, dx = \frac{1}{a} e^{ax} + C \tag{15.2-5}$$

is very useful and should also be memorized.

Because there are frequently a lot of symbols in the exponent, an alternative notation is often used. We write

$$y = e^u = \exp u \tag{15.2-6}$$

The function $y = \exp x$ "grows exponentially" as you go to the right; and it "decays exponentially" as you go to the left. The graph is shown in Figure 15.1-3. Since $y' = \exp x > 0$, the curve always rises as you go to the right. Furthermore, since $y'' = \exp x > 0$, the curve is always convex upward. If you shift the coordinate system left or right, you merely multiply the exponential function by a constant, as is evident from

$$e^x = e^h e^{x-h}$$

Thus the exponential function has a "constant shape" everywhere; it is the "size" that changes, not the "shape."

Example 15.2-1

Find the derivative of

$$y = e^{-3x}$$

You have directly from the formula (15.2-3)

$$y' = e^{-3x}(-3) = -3e^{-3x}$$

Example 15.2-2

Find the derivative of

$$y = \exp\left(\frac{-x^2}{2}\right)$$

You get immediately from (15.2-3)

$$y' = \exp\left(\frac{-x^2}{2}\right)\frac{d(-x^2/2)}{dx} = \exp\left(\frac{-x^2}{2}\right)[-x] = -x\exp\left(\frac{-x^2}{2}\right)$$

Example 15.2-3

Find

$$\int e^{-3x}\,dx$$

Using $-3x$ as your U in (15.2-4), you see that you lack a -3 in the integrand for the corresponding dU. When you supply this in both numerator and denominator, you get for the integral

$$\int e^{-3x}\,dx = \frac{-e^{-3x}}{3} + C$$

You can use (15.2-5) directly if you wish. This can be checked by differentiation.

Example 15.2-4

Integrate

$$\int x\exp\left(\frac{-x^2}{2}\right)\,dx = -\exp\left(\frac{-x^2}{2}\right) + C$$

where $U = -x^2/2$ and $dU = -x\,dx$. Check by differentiating the result.

Example 15.2-5

Integrate

$$\int xe^{-x}\, dx$$

The form suggests integration by parts. One choice, which gets rid of the x term (you cannot hope to get rid of an exponential by either integration or differentiation) in the resulting integral, is

$$U = x, \qquad dV = e^{-x}\, dx$$
$$dU = dx, \qquad V = -e^{-x}$$

From this you get

$$\int xe^{-x}\, dx = -xe^{-x} + \int e^{-x}\, dx$$
$$= -xe^{-x} - e^{-x} + C$$

This is easily checked by direct differentiation.

Example 15.2-6

Evaluate the improper integral

$$\int_0^\infty \frac{1}{1 + e^x}\, dx$$

At first glance it is a bit baffling to start doing anything. The e^x in the denominator requires a corresponding e^x in the numerator for the derivative, but there isn't one there! Suppose you multiplied both numerator and denominator by e^{-x}. You would have

$$\int_0^\infty \frac{e^{-x}}{e^{-x} + 1}\, dx = -\ln(e^{-x} + 1)\Big|_0^\infty$$

When you put in the upper limit, you get the $\ln 1 = 0$; and for the lower limit you get $\ln(1 + 1) = \ln 2$. Thus you have the answer

$$\int_0^\infty \frac{1}{1 + e^x}\, dx = \ln 2$$

EXERCISES 15.2

Differentiate the following:

1. $\exp(x - 1/x)$
2. $\exp(x^n)$
3. $x^2 \exp(-x^2)$
4. $x^n \exp(-x)$

5. $1/(1 + e^{-x})$

6. $\exp(ax)/(1 - \exp(bx))$

7. $x^n \exp(-x^2/2)$

Find the maximum of the following:

8. $x^n e^{-x}$

9. $e^x/(1 + e^x)$

10. $x \exp(-x^2/2)$

Find the inflection points of the following:

11. $y = \exp(-x^2/2)$

12. $x^n \exp(-x)$

Integrate the following:

13. $\displaystyle\int x^2 \exp(x^3)\, dx$

14. $\displaystyle\int x \exp(x)\, dx$

15. $\displaystyle\int x^2 \exp(x)\, dx$

16. $\displaystyle\int x^3 \exp(-x^2/2)\, dx$

17. Show that writing $\int dx = \int e^x/e^x\, dx$ gives the right answer. *Hint*: $\ln e^x = x$.

18. Show that writing $e^x = (e^{x/n})^{(n-1)} e^{x/n}$ gives the correct answer for the $\int e^x\, dx$.

15.3 SOME APPLICATIONS OF THE EXPONENTIAL FUNCTION

The exponential function $y = \exp x$ is of very common occurrence, and perhaps the negative exponential function

$$y = e^{-x}$$

is of even more common occurrence.

Example 15.3-1

The area under the negative exponential leads to an improper integral (Figure 15.3-1). We have

$$\int_0^\infty e^{-x}\, dx = -e^{-x}\Big|_0^\infty = 0 - (-1) = 1$$

Thus the area is finite.

Example 15.3-2

Find the volume inside the surface of revolution obtained by rotating

$$y = e^{-x} \qquad (x > 0)$$

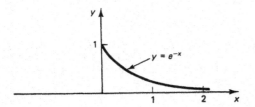

Figure 15.3-1   Area under e^{-x}

about the x-axis (Figure 15.3-2). We have for the element of volume (a disc)

$$\pi y^2 \, \Delta x$$

so the total volume is the improper integral

$$V = \int_0^\infty \pi e^{-2x} \, dx = -\left(\frac{\pi}{2}\right) e^{-2x} \Big|_0^\infty = \frac{\pi}{2}$$

Example 15.3-3

If we rotate the area under $y = e^{-x}$ between 0 and 1 about the y-axis, what is the volume? See Figure 15.3-3. The element of volume we have chosen is the cylinder

$$2\pi xy \, dx$$

(we have used dx in place of Δx since it is widely used this way; no real confusion should arise by doing so). The desired volume is, therefore (using the elements of volume in the form of cylinders),

$$V = 2\pi \int_0^1 xy \, dx$$

This is the same as [using the notation of (15.2-6)]

$$V = 2\pi \int_0^1 x \exp(-x) \, dx = 2\pi[-x \exp(-x) - \exp(-x)] \Big|_0^1$$

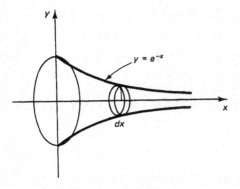

Figure 15.3-2   Volume about the x-axis

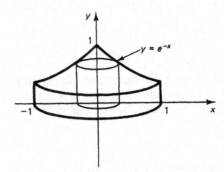

Figure 15.3-3   Volume about the y-axis

$$= 2\pi(-e^{-1} - e^{-1} + 1) = 2\pi\left(1 - \frac{2}{e}\right)$$

Is this a reasonable answer? The enclosing cylinder has volume π and the volume of the pedestal is π/e. The numbers π, $2\pi(1 - 2/e)$, and π/e seem to be about the right relative sizes.

EXERCISES 15.3

1. Find the volume generated by rotating $y = xe^{-x}$ about the x-axis (for $x \geq 0$).
2. Given the probability distribution $p(x) = (x^n/n!)\, e^{-x}$ ($0 \leq x \leq \infty$) find the mean and variance.
3. Find the volume generated by rotating the area under $y = xe^{-x}$ between 0 and 1 about the y-axis.
4. Write a short essay on how you would handle $\int a^x\, dx$.

15.4 STIRLING'S APPROXIMATION TO n!

The expression $n!$ occurs frequently in mathematics, especially in probability and statistics. It is often awkward to use in mathematical expressions, and the Stirling (1692–1770) approximation

$$n! = n^n e^{-n} \sqrt{2\pi n} \qquad (15.4\text{-}1)$$

is very useful. Actually, DeMoivre (1667–1754) published it first in 1730. We will derive it in two stages. In this section, using a powerful method for getting bounds on finite sums, we will get the useful bounds

$$n^n e^{-n} e^{7/8} \sqrt{n} < n! < n^n e^{-n} e \sqrt{n} \qquad (15.4\text{-}2)$$

Then in Section 16.6 we will obtain the estimate of the coefficient as $\sqrt{2\pi}$. Note that

$$e^{7/8} = 2.39888 \ldots < \sqrt{2\pi} = 2.50663 \ldots < e = 2.71828 \ldots$$

We will use numerical integration to get the bounds. The original definition of an integral was as the limit of a sum; in a typical mathematical style, we will invert this idea and use the integral to get an approximation to a sum.

The expression $n!$ is a product of n factors, and products are generally hard to handle. This suggests starting with

$$\ln n! = \sum_{k=1}^{n} \ln k$$

Next, the sum suggests an integral, and we are led to consider the trapezoid rule (11.11-1) applied to

$$\int_1^n \ln x \, dx = n \ln n - n + 1$$

We have done this integral before. Since the ln x curve is concave downward, the trapezoid rule will give a lower bound on the integral (Figure 15.4-1). The trapezoid rule gives, using unit sized intervals,

$$\frac{1}{2} \ln 1 + \ln 2 + \ln 3 + \cdots + \ln(n - 1) + \frac{1}{2} \ln n < n \ln n - n + 1$$

Since ln $1 = 0$ we ignore the first term. To fix up the last term on the left, we add $(1/2) \ln n$ to both sides. We have

$$\ln n! = \sum_{k=1}^n \ln k < n \ln n - n + 1 + \frac{1}{2} \ln n$$

Take the exponential function of both sides (which is the antilog) and remember that ln x is monotone increasing, so the inequality is preserved. We get

$$n! < n^n e^{-n} e \sqrt{n}$$

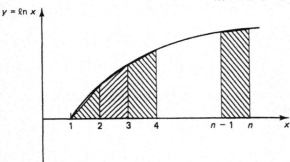

Figure 15.4-1   Trapezoid rule

The second stage is to get an upper bound on the integral (a lower bound on the sum). Figure 15.4-2 shows the midpoint formula (11.11-3) applied at the integer points to intervals of width 1. There is a first triangle to be taken care of. The slope of ln x at $x = 1$ is 1, so the tangent line at $x = 1$ provides an upper bounding triangle. The area of the triangle is clearly $\frac{1}{8}$. There is also a last half-interval to be accounted for. We simply use the maximum height, ln n, times the width $1/2$ to get a bound $\frac{1}{2} \ln n$. Thus the bounding area is (remember ln $1 = 0$)

$$\frac{1}{8} + \sum_{k=1}^{n-1} \ln k + \frac{1}{2} \ln n > n \ln n - n + 1$$

which becomes (transpose the $1/8$ and add $\frac{1}{2} \ln n$ to both sides)

$$\sum_{k=1}^n \ln k > n \ln n - n + \frac{7}{8} + \frac{1}{2} \ln n$$

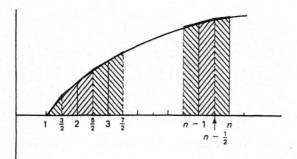

$$1 \quad \tfrac{3}{2} \quad 2 \quad \tfrac{5}{2} \quad 3 \quad \tfrac{7}{2} \qquad\qquad n-1 \uparrow n$$
$$n - \tfrac{1}{2}$$

Figure 15.4-2   Midpoint rule

Taking antilogs, we get

$$n! > n^n e^{-n} e^{7/8} \sqrt{n}$$

As an aside, we note that the midpoint rule has typically half the error of the trapezoid rule, and the errors are of opposite sign. Hence a good estimate of the value would be the weighted average

$$\frac{2e^{7/8} + e}{3} = 2.50534 \ldots$$

which is close to the true value 2.50663 . . . .

How good is Stirling's approximation to $n!$ (with the standard $\sqrt{2\pi}$ coefficient)? Table 15.4-1 shows the first ten values.

TABLE 15.4-1

n	Stirling	True	Stirling/true
1	0.92214	1	0.92214
2	1.91900	2	0.95950
3	5.83621	6	0.97270
4	23.50618	24	0.97942
5	118.01916	120	0.98349
6	710.07818	720	0.98622
7	4,980.3958	5,040	0.98817
8	39,902.395	40,320	0.98964
9	359,536.87	362,880	0.99079
10	3,598,695.6	3,628,800	0.99170

A study of the table shows that, although the ratio of the two values approaches 1 as $n \to \infty$, still the difference gets continually larger. In this respect it resembles the law of large numbers (Section 13.6), where the average approached the expected value, but the deviations from the expected value tended to get larger and larger with increasing number of trials.

Example 15.4-1

Estimate the middle (maximum) binomial coefficient of even order. Its importance derives partly from the identity (Example 9.11-1)

$$C(2n, n) = \sum_{k=0}^{n} C^2(n, k)$$

The maximum size of a binomial coefficient is of interest in its own right. We have

$$C(2n, n) = \frac{(2n)!}{n! \, n!}$$

Using Stirling's approximation for each factorial, we get

$$C(2n, n) = \frac{(2n)^{2n} e^{-2n} \sqrt{2\pi} \sqrt{2n}}{n^n e^{-n} \sqrt{2\pi} \sqrt{n} \; n^n e^{-n} \sqrt{2\pi} \sqrt{n}}$$

$$= \frac{2^{2n} \sqrt{2}}{\sqrt{2\pi} \sqrt{n}} = \frac{2^{2n}}{\sqrt{\pi n}}$$

Note that $2^{2n} = \sum_{k=0}^{2n} C(2n, k)$, and hence the maximum coefficient is $1/\sqrt{\pi n}$ times the sum of all the coefficients of order $2n$.

Generalization 15.4-2

We see from the preceding that an estimate of the sum of a finite series

$$\sum_{k=1}^{n} \ln k$$

can be found by the adroit use of the trapezoid and midpoint rules for numerical integration. The estimate of the sum is by the corresponding integral

$$\int_1^n \ln x \, dx$$

both from below and above. Then by rearranging things we get the corresponding bounds on the sum. The method does not depend on the particular sum to be approximated.

The errors in the approximation are the differences between the approximate curves and the continuous integrand. If the curve is uniformly convex (or concave) and is fairly smooth, then usually the errors are small even for the unit interval you are forced to use. Evidently, it is a general method and puts few requirements on the function (which was $\ln k$).

Example 15.4-3

The *harmonic series*

$$H_n = \sum_{k=1}^{n} \frac{1}{k}$$

occurs frequently in mathematics. The sum suggests the corresponding integral

$$\int_1^n \frac{dx}{x} = \ln n$$

Since this goes to infinity as $n \to \infty$, it is better to study the difference

$$h_n = H_n - \ln n$$

From Figure 15.4-3, we see that the trapezoid rule will give an upper bound (the curve $y = 1/x$ is concave up)

$$\frac{1}{2}1 + \frac{1}{2} + \frac{1}{3} + \frac{1}{4} + \cdots + \frac{1}{2}\frac{1}{n} \geq \ln n$$

or

$$h_n = \sum_{k=1}^{n} \frac{1}{k} - \ln n > \frac{1}{2} + \frac{1}{2n}$$

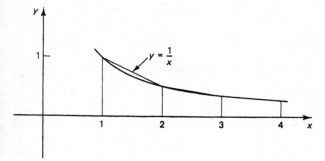

Figure 15.4-3   Trapezoid bound

For the lower bound, we use the midpoint rule (see Figure 15.4-4):

$$\frac{1}{2}\left(\frac{1}{3/2}\right) + \frac{1}{2} + \frac{1}{3} + \cdots + \frac{1}{n-1} + \frac{1}{2}\frac{1}{n} \leq \ln n$$

Hence

$$h_n = \sum_{k=1}^{n} \frac{1}{k} - \ln n < \frac{2}{3} + \frac{1}{2n}$$

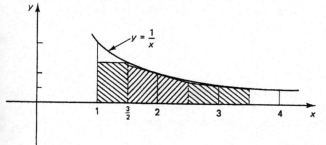

Figure 15.4-4   Midpoint bound

and we have the bounds

$$\frac{1}{2} + \frac{1}{2n} < h_n < \frac{2}{3} + \frac{1}{2n}$$

EXERCISES 15.4

1. Using the interval $\frac{3}{2} \leq x < n - \frac{1}{2}$, get bounds on h_n.
2. Find $h_n - h_{n-1}$, and using the mean value theorem, show that it is negative; hence h_n is monotone decreasing. Therefore, $h_n \to \gamma = 0.577 \dots$, which is Euler's constant.
3. Show that $h_{2n} - h_n = \sum\limits_{k=n+1}^{2n} \frac{1}{k} - \ln 2 \to 0$; hence $\sum\limits_{n+1}^{2n} \frac{1}{k} \to \ln 2$.
4. Find bounds on $\sum \frac{1}{k^2}$.
5. Find bounds on $m_n = \sum \sqrt{k} - (\frac{2}{3})n^{3/2} \to C\sqrt{n}$.
6. Study $G_n = \sum 1/(n \ln n)$.

15.5 INDETERMINATE FORMS

We have met a few cases where we faced the product of two terms, one growing toward infinity and the other toward zero, and needed to know which factor of the product dominated the other. For example, we looked at

$$x \ln x \qquad \text{as } x \to 0$$

and saw from the Table 14.6-1 that the x factor approaching zero dominates the $\ln x$ factor going to minus infinity.

Many *indeterminate forms*, as they are called, can be written

$$\frac{0}{0}$$

which is the first form we attack. We are extending the *missing value* method of Section 7.4.

Given two functions $f(x)$ and $g(x)$, both approaching 0 as x approaches a, what value shall we assign to

$$\lim_{x \to a} \frac{f(x)}{g(x)}$$

Apply the Cauchy mean value theorem (Sections 11.8 and 13.3) to the quotient $f(x)/g(x)$ separately. The theorem states that under suitable conditions there exists a value θ such that

$$\frac{f(b) - f(a)}{g(b) - g(a)} = \frac{f'(\theta)}{g'(\theta)}$$

But we are assuming that $f(a) = g(a) = 0$, so, on writing b as the variable x, we have

$$\frac{f(x)}{g(x)} = \frac{f'(\theta)}{g'(\theta)}$$

As x approaches the limit a, the variable θ is caught between x and a and must (due to continuity) also approach a. The two limits will therefore be the same. This result is known as *L'Hopital's rule* (1661–1704). The name is pronounced "low-pea-tahl." The method applies to indeterminate forms of the form $0/0$. If the resulting *ratio* of derivatives is still indeterminate, then we can apply the theorem again—and again if necessary!

Example 15.5-1

Suppose you want to know the value to assign to the limit as x approaches 0 of

$$\frac{e^x - 1}{x}$$

It is of the form $0/0$, so we apply L'Hopital's rule, which calls for the *separate* differentiation of the numerator and the denominator. We get

$$\frac{e^x}{1}$$

which clearly approaches 1 in the limit.

Example 15.5-2

As a more complex case, consider

$$\frac{e^x - 1 - x}{x^2}$$

as x approaches 0. The first application of L'Hopital's rule gives

$$\frac{e^x - 1}{2x}$$

and the next application of the rule gives

$$\frac{e^x}{2}$$

so the limit as $x \to 0$ is clearly $\frac{1}{2}$.

Example 15.5-3

Consider

$$\frac{\sqrt{1 + x^2} - 1}{x^2}$$

as $x \to 0$. It is of the form $0/0$, so you apply L'Hopital's rule (differentiate both numerator and denominator separately). You get

$$\frac{x/\sqrt{1 + x^2}}{2x}$$

At this point you can either persist with applying the rule or else note that the common factor x will divide out, leaving the equivalent form

$$\frac{1}{2\sqrt{1 + x^2}}$$

and in the limit you will get

$$\frac{1}{2}$$

The rule also applies to the indeterminate form

$$\frac{\infty}{\infty}$$

To see this, we first reduce the situation to the earlier case of $0/0$. If $f(x)$ and $g(x)$ both $\rightarrow \infty$ as $x \rightarrow a$, we write

$$\frac{f(x)}{g(x)} = \frac{1/g(x)}{1/f(x)}$$

and have $0/0$ form. We apply L'Hopital's rule to get

$$\frac{f(x)}{g(x)} = \frac{-g'(\theta)/g^2(\theta)}{-f'(\theta)/f^2(\theta)}$$

Multiply through by $g(\theta)/f(\theta)$ to get

$$\frac{f(x)}{g(x)} \frac{g(\theta)}{f(\theta)} = \frac{f(\theta)}{g(\theta)} \frac{g'(\theta)}{f'(\theta)}$$

As $x \rightarrow a$, the left side approaches 1, since whatever $f(x)/g(x)$ approaches so too must $f(\theta)/g(\theta)$. Hence by rearranging we get

$$\frac{f'(a)}{g'(a)} = \frac{f(a)}{g(a)}$$

All of this requires that the things we multiplied and divided by were not zero.

Example 15.5-4

Consider

$$\lim_{x \to 0} \{x \ln x\}$$

Write it in the form

$$\frac{\ln x}{1/x}$$

It is now in the form ∞/∞. Separate differentiation of numerator and denominator gives

$$\frac{1/x}{-1/x^2} = -x$$

which approaches 0, as was strongly suggested by Table 14.6-1.

Example 15.5-5

Consider $x^n e^{-x}$ as x approaches infinity. This is an indeterminate form if you write it as

$$\frac{x^n}{e^x}$$

By L'Hopital's rule, differentiating numerator and denominator each separately, you get

$$\frac{nx^{n-1}}{e^x}$$

which is still of the same form of ∞/∞. But you see some progress because the degree of the numerator is less. You apply the rule again and get

$$\frac{n(n-1)x^{n-2}}{e^x}$$

This indicates (and you could prove it by induction if you wished) that after n stages you will have

$$\frac{n!}{e^x}$$

which will clearly approach 0 as x approaches infinity. Thus, in words, as x approaches infinity the exponential dominates any finite power of x.

EXERCISES 15.5

1. For $a > 0$, examine the limit $x^a \ln x$ as x approaches 0.
2. Find the limit $x^n \exp(-x^2/2)$ as x approaches infinity.
3. Find the limit of $(e^x - e^{-x})/x$ as $x \to 0$.
4. Find the limit of $(e^x - 1 - x - x^2/2 - x^3/3!)/x^4$ as $x \to 0$.
5. Find the limit of $(e^{-x} - 1 + x - x^2/2 + x^3/3!)/x^4$ as $x \to 0$.

15.6 THE EXPONENTIAL DISTRIBUTION

The probability density

$$p(x) = ae^{-ax}$$

for $0 \le x < \infty$ and 0 elsewhere, is very common. This is called the *exponential distribution*. We need to look at many of its properties.

Example 15.6-1

Find the expected value and variance of the exponential distribution. First, check that the total probability is 1. You have

$$\int_0^\infty ae^{-ax}\,dx = \frac{ae^{-ax}}{-a}\bigg|_0^\infty = 0 - (-1) = 1$$

as it should. Next, using integration by parts, the mean is

$$\mu = E\{X\} = \int_0^\infty x\,ae^{-ax}\,dx = x\{-e^{-ax}\}\bigg|_0^\infty - \int_0^\infty -e^{-ax}\,dx$$

$$= \int_0^\infty e^{-ax}\,dx = \frac{1}{a}$$

To get the variance, you first need the second moment:

$$E\{X^2\} = \int_0^\infty x^2\,ae^{-ax}\,dx = x^2\left(\frac{-1}{a}\right)ae^{-ax}\bigg|_0^\infty - \int_0^\infty 2x(-e^{-ax})\,dx$$

The integrated term is zero at both limits and the integral is the same one just done for the mean (except for the factor $2/a$); hence

$$E\{X^2\} = \frac{2}{a^2}$$

Thus, from the basic formula for the variance (12.10-7), you have

$$V\{X\} = E\{X^2\} - E^2\{X\} = \frac{2}{a^2} - \left(\frac{1}{a}\right)^2 = \frac{1}{a^2}$$

EXERCISES 15.6

1. Find the third moment about the mean of the exponential distribution.
2. Find the fourth moment about the mean of the exponential distribution.

15.7 RANDOM EVENTS IN TIME

Many phenomena occur at random instants in time. For example, the decay of a radioactive atom, the time the next telephone call arrives at the central office, when the next "job" is submitted to the central computing facility, the time of the next death in a hospital, and the time of the next failure of a machine. We need to develop the mathematical tools for thinking about such situations.

We will use a standard approach: examine the finite case first, and then let the number of intervals approach infinity.

Consider a fixed interval of time t. Let the interval be divided into n equal-sized subintervals. If in each subinterval the probability of an event occurring is p, then the probability of exactly k events in the n intervals is

$$P(k) = C(n, k)p^k(1 - p)^{n-k}$$

Since we are using time in this problem, we will change the notation from $P(k)$ to

$$P_k(t) = C(n, k)p^k(1 - p)^{n-k}$$

Now fix $n = n_0$ and $p = p_0$. If we further subdivide each of the n_0 intervals into m equal parts, then we have

$$p = \frac{p_0}{m} \quad \text{and} \quad n = n_0 m$$

(as in the case of compound interest, Section 14.4). We have for the finer subdivision

$$P_k(t) = C(mn_0, k)\left(\frac{p_0}{m}\right)^k\left(1 - \frac{p_0}{m}\right)^{mn_0-k}$$

It is now a matter of reorganizing this expression before we take the limit as $m \to \infty$. We have

$$P_k(t) = \frac{(mn_0)(mn_0 - 1) \ldots (mn_0 - k + 1)}{k!} \frac{p_0^k}{m^k}\left(1 - \frac{p_0}{m}\right)^{mn_0}\left(1 - \frac{p_0}{m}\right)^{-k}$$

$$= n_0\left(n_0 - \frac{1}{m}\right)\left(n_0 - \frac{2}{m}\right) \ldots$$

$$\left(n_0 - \frac{k - 1}{m}\right)\left(\frac{p_0^k}{k!}\right)\left[\left(1 - \frac{p_0}{m}\right)^{-m/p_0}\right]^{-n_0 p_0}\left[1 - \frac{p_0}{m}\right]^{-k}$$

Let $m \to \infty$ (remember that k is fixed). We get

$$P_k(t) \to n_0^k\left[\frac{p_0^k}{k!}\right]\exp(-n_0 p_0)[1]$$

$$= \frac{(n_0 p_0)^k}{k!}\exp(-n_0 p_0)$$

But $n_0 p_0$ is the expected number of successes in the original interval t. Hence we now write

$$n_0 p_0 = at$$

where a is the average rate of events. We then have

$$P_k(t) = \frac{(at)^k}{k!}\exp(-at)$$

In particular, the case of no successes is

$$P_0(t) = e^{-at}$$

For a shift in time, $t = t_0 + t'$, we have

$$P_0(t) = P_0(t_0)P_0(t')$$

The probability of no successes in time t_0 multiplied by the probability of no success in time t' is the probability of no successes in the sum of the two times.

We see that we have a constant rate of *failure*. The probability of a failure at any instant is *independent* of any past failures. This, you recognize, is a common situation; you are dealing with a source of independent events and the occurrence of one event does not influence the occurrence of another.

In practice, failures of a computer, or a component of a computer, for example, has a "bathtub" rate of failure (Figure 15.7-1). At first there is a "shake-down" stage of failures, then there is a long period of random, independent failures, and finally at the end of the life of the computer there is a rise in the rate due to the aging of the whole system. But the middle, flat part is very long in comparison with the two ends. Thus the above model provides a good guide to machine maintenance. However, in the author's experience the probability of a failure in a computer is usually much greater shortly after preventative maintenance than it was before! Thus, if you have a situation of a constant failure rate, it is not a bad rule to "leave well enough alone."

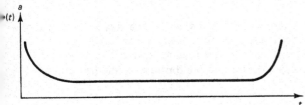

Figure 15.7-1   Failure rate

People have difficulties when thinking about such constant-rate models. They expect, usually, too even a rate and are surprised by the bunching of failures. The probability of k failures in time t is

$$P_k(t) = \frac{(at)^k}{k!} e^{-at}$$

In particular, the probability of one failure is at times the probability of no failures. The ratio of exactly two failures in time t divided by the probability of a single failure in the same interval is

$$\frac{P_2}{P_1} = \frac{at}{2}$$

If t is picked to be the time you expect a single failure ($at = 1$), then the probability of two failures in this interval is only half that of one failure, while zero and one failure are equally probable!

The probability distribution of the frequencies

$$P_0(t), \ P_1(t), \ P_2(t), \ \ldots$$

gives a convenient test for the homogeneous mixing of two materials. Suppose raisins were mixed in some dough and you wished to know if they were uniformly spread around (or bacteria in a solution). Simply take equal-sized samples and count the number of samples with 0, 1, 2, . . . raisins. The counts should obey the above distribution when you have used at as the probable number per sample.

One of the events, 0, 1, 2, . . . must occur. Hence we have

$$\sum_{k=0}^{\infty} P_k(t) = 1$$

We will later prove this relationship mathematically.

EXERCISE 15.7

1. Compute the ratio for $t = 1/a$ of observing k failures to $k - 1$ failures.

15.8 POISSON DISTRIBUTIONS

If we shift our attention and ask how the probability of k events changes in time, we see that we are faced with a probability density problem. But we need to be careful! To get the probability density in standard notation, we need to replace k by n and use the density $\Delta(at) = a\Delta t$. This produces an extra a in the formula for the probability density:

$$p_n(t) = \frac{a^{n+1}t^n e^{-at}}{n!}$$

for the range $0 \le x \le \infty$ and $a > 0$. The case $n = 0$ is the exponential distribution of Section 15.6; hence this distribution is a generalization of it.

Example 15.8-1

Check that for every n the total probability of $p_n(x)$ is 1. You need to evaluate

$$\int_0^{\infty} p_n(t) \, dt = 1$$

It is convenient to change variables (to get rid of the letter a) by setting

$$at = x \quad \text{then} \quad a \, dt = dx$$

The integral becomes the *Poisson* (1781–1840) *distribution*

$$\frac{1}{n!} \int_0^{\infty} x^n e^{-x} \, dx$$

Integration by parts is the obvious thing to do.

$$U = x^n, \qquad dV = e^{-x} \, dx$$
$$dU = nx^{n-1} \, dx, \qquad V = -e^{-x}$$

Dropping, for the moment only, the $n!$ term,

$$\int_0^\infty x^n e^{-x}\, dx = x^n(-e^{-x}) \Big|_0^\infty - \int_0^\infty nx^{n-1}(-e^{-x})\, dx$$

The integrated term drops out at both ends (when $n > 1$), and you have the *reduction formula*

$$\int_0^\infty x^n e^{-x}\, dx = n \int_0^\infty x^{n-1} e^{-x}\, dx$$

$$= n!$$

When you restore the $n!$ term you dropped, you get the result that the total probability is 1, as it should be.

Example 15.8-2

Sketch the curve of the Poisson distribution. To find the maximum, you differentiate and set the derivative equal to zero. As usual, you can ignore the constants $n!$ and the power of a. The result is

$$nt^{n-1}e^{-at} - at^n e^{-at} = 0$$

from which ($e^{-at} \neq 0$)

$$t = \frac{n}{a}$$

is the t coordinate where the maximum occurs. The solutions $t = 0$ (for $n > 1$) give the minimum. The $t = n/a$ value must be put into the function to get the maximum value. This gives

$$p(n/a) = \frac{a^{n+1}[n/a]^n e^{-n}}{n!} = \frac{an^n e^{-n}}{n!}$$

To form some idea of this number we use the Stirling approximation (15.4-1) for the factorial, and get

$$p\left(\frac{n}{a}\right) = \frac{a}{\sqrt{2\pi n}}$$

The inflection points require setting the second derivative equal to zero. The result is

$$n(n-1)t^{n-2}e^{-at} - 2nt^{n-1}ae^{-at} + t^n a^2 e^{-at} = 0$$

from which it follows that

$$(at)^2 - 2n(at) + n(n-1) = 0$$

gives the positions of the inflection points. The solutions of this quadratic are

$$at = \{n \pm \sqrt{n^2 - n(n-1)}\} = n \pm \sqrt{n}$$

These two t values are symmetric about the location of the maximum. The evaluation of the function at the inflection points is tricky, and for any particular case you can evaluate it numerically.

EXERCISE 15.8

1. Find the mean and variance of the Poisson distribution of order n.

15.9 THE NORMAL DISTRIBUTION

Another distribution that frequently occurs is the *normal distribution*,

$$p(x) = \frac{1}{\sqrt{2\pi}} \exp\left(\frac{-x^2}{2}\right) \tag{15.9-1}$$

where we have again written "exp(. . .)" for the equivalent $e^{(\cdots)}$ to avoid piling exponents on top of exponents. The factor $\frac{1}{2}$ in the exponent is common usage. We will later prove (Example 19.4-4) that the total probability is exactly 1; that is,

$$\int_{-\infty}^{\infty} \frac{1}{\sqrt{2\pi}} \exp\left(\frac{-x^2}{2}\right) dx = 1 \tag{15.9-2}$$

For the moment we will use this whenever we need it. It is not possible to find a nice, simple, closed form for the indefinite integral of the integrand

$$\exp\left(\frac{-x^2}{2}\right)$$

although, as we just said, the definite integral over the whole x-axis will be found later.

The general case of the normal distribution with mean μ and variance σ^2 is given by the formula

$$N(\mu, \sigma) = \frac{1}{\sigma\sqrt{2\pi}} \exp\left\{\frac{-(x - \mu)^2}{2\sigma^2}\right\} \tag{15.9-3}$$

To verify this statement, we need to compute the central moments $k = 0, 1, 2$:

$$\mu_k = \frac{1}{\sigma\sqrt{2\pi}} \int_{-\infty}^{\infty} (x - \mu)^k \exp\left\{\frac{-(x - \mu)^2}{2\sigma^2}\right\} dx$$

It is usually best to introduce immediately the *normalized variable*

$$(x - \mu)/\sigma = t$$

The central moments become

$$\mu_k = \frac{1}{\sigma\sqrt{2\pi}} \int_{-\infty}^{\infty} \sigma^k t^k \exp\left\{\frac{-t^2}{2}\right\} \sigma \, dt$$

The σ's cancel in the exponent, and we have for the central moments of the distribution

$$\mu_k = \frac{\sigma^k}{\sqrt{2\pi}} \int_{-\infty}^{\infty} t^k \exp\left\{\frac{-t^2}{2}\right\} dt$$

For $k = 0$, this integral is 1 [from (15.9-2)]; for $k = 1$, the integrand being odd and the interval being symmetric about the origin gives 0; and for $k = 2$, we need to apply integration by parts, picking

$$U = t, \qquad dV = t \exp\left\{\frac{-t^2}{2}\right\} dt$$

The rest follows automatically:

$$\mu_2 = \frac{\sigma^2}{\sqrt{2\pi}} \left[-t \exp\left\{\frac{-t^2}{2}\right\} \Big|_{-\infty}^{\infty} + \int_{-\infty}^{\infty} \exp\left\{\frac{-t^2}{2}\right\} dt \right]$$

The integrand term vanishes at both ends, and the integral is exactly $\sqrt{2\pi}$ (15.9-2), so you have

$$\mu_2 = \sigma^2$$

As expected, we find that the variance is σ^2. Thus the general form (15.9-3) exhibits the mean and variance as stated.

The cumulative distribution of the normal distribution is given by

$$\Phi(x) = \frac{1}{\sqrt{2\pi}} \int_{-\infty}^{x} \exp\left(\frac{-x^2}{2}\right) dx$$

and is very useful in practice. It is easily seen from symmetry that

$$\frac{1}{\sqrt{2\pi}} \int_{-\infty}^{0} = \frac{1}{2}$$

Hence we have

$$\Phi(x) = \frac{1}{2} + \frac{1}{\sqrt{2\pi}} \int_{0}^{x} \cdots$$

This cannot be integrated in closed form, but can be integrated by numerical methods (Section 11.11). We used Simpson's formula with spacing 0.025 on a programmable hand calculator to get Table 15.9-1.

Example 15.9-1

What percent of the normal distribution is within $k\sigma$ of the mean? Table 15.9-1 gives the distribution in units of σ about the mean of 0. Thus 1σ reaches from -1 to $+1$, and for $k\sigma$ we have from $-k\sigma$ to $+k\sigma$. We want to know the total area within the interval $t\sigma$. To get the amount above the mean, we subtract $\frac{1}{2}$ from the table value, and then we double that to get the total amount to account for both sides. Thus the formula is

$$2\Phi(k\sigma) - 1$$

TABLE 15.9-1   ERROR FUNCTION

x	Function value
0.0	0.50000
0.1	0.53983
0.2	0.57926
0.3	0.61791
0.4	0.65542
0.5	0.69146
0.6	0.72575
0.7	0.75804
0.8	0.78814
0.9	0.81594
1.0	0.84134
1.1	0.86433
1.2	0.88493
1.3	0.90320
1.4	0.91924
1.5	0.93319
1.6	0.94520
1.7	0.95543
1.8	0.96407
1.9	0.97128
2.0	0.97725
2.5	0.99379
3.0	0.99865
3.5	0.99977
4.0	0.99997

We get the following short table of values:

$\pm\sigma$	Amount in the interval
1	0.68268
2	0.95450
3	0.99730
4	0.99994

EXERCISES 15.9

1. Show that the integral of $\exp(-x^2/2)$ cannot be a polynomial times the exponential. *Hint:* Use the highest power of x.
2.* As in Exercise 1, but use a rational function in place of the polynomial.

15.10 NORMAL DISTRIBUTION, MAXIMUM LIKELIHOOD, AND LEAST SQUARES

Suppose we have a set of n independent observations x_i from a normal distribution

$$p(x) = \frac{1}{\sigma\sqrt{2\pi}} \exp\left\{\frac{-(x-\mu)^2}{2\sigma^2}\right\}$$

where σ is supposed known and μ is the unknown parameter of the distribution. This says that the probability density of seeing x_i is this function $p(x_i)$. The probability density of seeing all the independent observations is, therefore, the product

$$p(x_1)p(x_2) \ldots p(x_n)$$

This is

$$\left(\frac{1}{\sigma\sqrt{2\pi}}\right)^n \exp\left\{-\sum_{i=1}^{n} \frac{(x_i-\mu)^2}{2\sigma^2}\right\}$$

This is the *likelihood* (probability density) of seeing the set of independent observations x_i, assuming that the n independent observations come from the same normal distribution.

If we believe in the principle of maximum likelihood (Section 12.7), then we ask what value of μ is "best" (assuming that σ is known)? The maximum occurs, obviously, when the exponent is a minimum, that is, when

$$\sum_{n=1}^{n} (x_i - \mu)^2$$

is least. But this is exactly the principle of least squares (Section 10.4)! We have derived the principle of least squares from the assumption of independent observations from a normal distribution with a given σ but unknown mean μ, together with the principle of maximum likelihood.

Of course, the principle of least squares may be applicable, or not, when the assumptions of the derivation do not apply. At least the principle of least squares is consistent with (1) independent observations from a normal distribution of known variance, plus (2) the maximum likelihood principle.

15.11 THE GAMMA FUNCTION

Euler (1707–1783) defined the gamma function as

$$\Gamma(n) = \int_0^\infty e^{-x}x^{n-1}\,dx \qquad (15.11\text{-}1)$$

(Γ is the uppercase Greek gamma). This integral defines the function for all $n > 0$. It is easy to get a reduction formula for $\Gamma(n)$ by using integration by parts:

$$U = x^{n-1}, \qquad\qquad dV = e^{-x}\, dx$$

$$du = (n - 1)x^{n-2}\, dx, \qquad V = -e^{-x}$$

Therefore,

$$\Gamma(n) = -x^{n-1}e^{-x}\Big|_0^\infty + (n - 1)\int_0^\infty e^{-x}x^{n-2}\, dx$$

The integrated term drops out at both limits for $n > 1$, and we have the recurrence relation

$$\Gamma(n) = (n - 1)\Gamma(n - 1) \tag{15.11-2}$$

with

$$\Gamma(1) = \int_0^\infty e^{-x}\, dx = 1$$

It follows that, for the positive integers,

$$\Gamma(n) = (n - 1)! \tag{15.11-3}$$

For other values of $n > 0$, the integral exists and serves to *extend* the definition of the factorial function. In particular,

$$\Gamma\left(\frac{1}{2}\right) = 2\int_0^\infty e^{-x}x^{-1/2}\, dx$$

To get rid of the radical, set $x = t^2$; then $dx = 2t\, dt$, and

$$\Gamma\left(\frac{1}{2}\right) = 2\int_0^\infty \exp\{-t^2\}t\left(\frac{1}{t}\right)\, dt$$

$$= \int_{-\infty}^\infty \exp\{-t^2\}\, dt$$

To compare this with the normal distribution (15.9-1), we need to set $t = u/\sqrt{2}$. This gives

$$\Gamma\left(\frac{1}{2}\right) = \frac{1}{\sqrt{2}}\int_{-\infty}^\infty \exp\left\{\frac{-u^2}{2}\right\}\, du = \sqrt{\pi} \tag{15.11-4}$$

This is the main noninteger value of the gamma function that occurs in practice. Table 15.11-1 gives the gamma function for other values should the need arise.

The table was computed by a programmable hand calculator using the following method.

$$\int_0^\infty e^{-x}x^{n-1}\, dx = \int_0^1 \cdots + \int_1^\infty \cdots$$

In the second integral, we use the reciprocal substitution $x \to 1/x$.

TABLE 15.11-1

$\Gamma(x)$	$(1 \leq x \leq 2)$
x	$\Gamma(x)$
1.0	1.00000
1.1	0.95135
1.2	0.91817
1.3	0.89747
1.4	0.88726
1.5	0.88622
1.6	0.89352
1.7	0.90864
1.8	0.93138
1.9	0.96177
2.0	1.00000

$$\int_0^1 e^{-x} x^{n-1}\, dx + \int_0^1 \frac{e^{-1/x}}{x^{n+1}}\, dx$$

The integrand is not very much like a polynomial since the derivative is infinite at the origin. Instead, the integral was computed for values between 2 and 3 and then reduced by the recurrence formula (15.11-2) backward. We used, also because of the nonpolynomial behavior, the midpoint formula with 100 points.

A plot of the gamma function is shown in Figure 15.11-1.

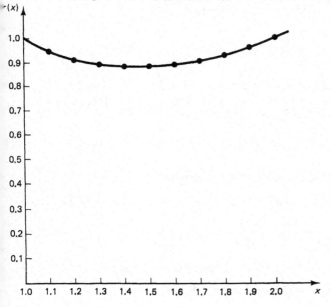

Figure 15.11-1 $\Gamma(x)$ $(1 \leq x \leq z)$

EXERCISES 15.11

1. Express the central moments of the normal distribution in terms of the gamma function.
2. Estimate the minimum of the gamma function in the interval $1 \leq x < 2$ by fitting a parabola to three adjacent points and finding the minimum of the parabola.

15.12 SYSTEMATIC INTEGRATION

In Section 14.9 we examined some of the kinds of integrands we could expect to do in closed form, which involved the $\ln x$ function. The method was based on the linear independence of the powers of $\ln x$. We proved (Section 14.2*) that the polynomial

$$P(x, \ln x) \neq 0$$

cannot be identically zero in any interval in x. Thus the powers of the $\ln x$ are linearly independent over any interval. Set $x = e^t$; then $\ln x = t$, and we have

$$P(e^t, t) \neq 0$$

(for all t in an interval) for any polynomial in the two variables. Thus the power of e^t

$$1, e^t, e^{2t}, \cdots, e^{nt}$$

are linearly independent, using coefficients that are polynomials, and we have the basis for a similar investigation of what can be integrated easily.

Consider the integral

$$\int e^{ax} P(x) \, dx \qquad (15.12\text{-}1)$$

where, as usual, $P(x)$ is a polynomial in x. Since under differentiation an exponential does not disappear, it likewise cannot appear when integrating; it must be there all the time! Thus the result of the integration must be of the form

$$\int e^{ax} P(x) \, dx = Q(x)e^{ax} + C \qquad (15.12\text{-}2)$$

Differentiate both sides of this and divide out the exponential:

$$P(x) = aQ(x) + Q'(x) \qquad (15.12\text{-}3)$$

This gives the conditions on $Q(x)$. In detail, let

$$P(x) = p_0 + p_1 x + p_2 x^2 + \cdots + p_n x^n$$

A little thought suggests that $Q(x)$ is of the same degree, so we write

$$Q(x) = q_0 + q_1 x + q_2 x^2 + \cdots + q_n x^n$$

therefore, equating like powers of x (we are here using the linear independence of the powers of x), we have from (15.12-3)

$$p_n = aq_n$$

$$p_{n-1} = aq_{n-1} + nq_n$$

$$\cdots$$

$$p_0 = aq_0 + q_1$$

Solving this triangular set of equations for the q_i, we have

$$q_n = \frac{1}{a}p_n$$

$$q_{n-1} = \frac{1}{a}(p_{n-1} - nq_n)$$

$$q_{n-2} = \frac{1}{a}[p_{n-2} - (n-1)q_{n-1}]$$

$$\cdots$$

$$q_0 = \frac{1}{a}(p_0 - q_1)$$

It is evident that if we have the sum of a number of terms with different exponentials then each must be treated separately, since the powers of exponentials are linearly independent. The result is, of course, the same as you would get by repeated integration by parts.

A similar examination of integrals of the form

$$\int \exp\left(\frac{-x^2}{2}\right) P(x)\, dx$$

will show you that when $P(x)$ has *only* odd powers you can do it in closed form; but when there are even powers, you will always come down to the integral

$$\int \exp\left(\frac{-x^2}{2}\right) dx$$

which cannot be done in closed form, and effectively defines a new function (see Table 15.9-1). You can see this intuitively when you think of doing the integral by parts; one x is needed for the derivative of the exponent in the dV, and what is left is reduced one further power by the differention in the dU. Thus one step of integration by parts reduces the power of x by 2. You can do the first power, but you cannot do the zeroth power of x in closed form. But note that all the cases of powers of x reduce to a single table (15.9-1) of function values that you will need.

Yes, integration is not easy, but the method of undetermined coefficients we have used is a powerful guide; it depends on your understanding of the process of differentiation so that you can guess intelligently what form to assume. It is only in this "backward" way that much sense can be made of what you can and cannot integrate in closed form and what the answer will be when you can. It is not an exact method, but it does reduce the confusion of the direct approach a little.

EXERCISES 15.12

1. Integrate $\int x^3 e^{-x} \, dx$.
2. Integrate $\int x^4 e^{-x} \, dx$.
3. Integrate $\int x^2 e^{-ax} \, dx$.
4. Develop the theory for integrating $\int P(x^2) x \exp(-x^2/2) \, dx$.

15.13 SUMMARY

This chapter introduced the concept of the inverse function. The inverse function of the ln x function is the exponential function

$$y = e^x$$

We repeated the pattern of Chapter 14 and developed the differentiation, integration, and applications of this new function to several new situations.

The applications we made of the exponential function were specialized, but it is hoped that they were more interesting than the usual routine drill problems. Again, it is the method of approach more than the results that is important. The examples showed the range of applications you can make once a reasonable class of functions is mastered. The next chapter will complete the study of the elementary functions and free us from needing to carefully select the problems we study.

16

The Trigonometric Functions

6.1 REVIEW OF THE TRIGONOMETRIC FUNCTIONS

We first review the trigonometric functions carefully because we are going to use them extensively in the calculus. It is necessary for you to overlearn the many relationships between these functions so that in the middle of doing a calculus problem you are not distracted by the inability to recall easily the needed information. You are advised to learn by rote the many things that are reviewed here (remember the usefulness of flash cards). There is no royal road to mathematics, and the longer you delay in mastering the details the longer you will be held up in learning to use the calculus and probability.

The first novelty is that in the calculus we *always* use the radian measure of angles; it is the natural measure to use in the calculus. Radian measure is usually defined as the ratio of the length of the arc of the circle, centered at the vertex of the angle, to the length of the radius (see Figure 16.1-1). We have, therefore,

$$\text{Measure of angle} = \frac{\text{arc length}}{\text{radius}} \tag{16.1-1}$$

Thus the radian measure of an angle is a pure, dimensionless number (a dimensionless number does *not* depend on the units used). Since the circumference of a circle is $2\pi r$, we have that 360° corresponds to 2π radians. One radian is clearly $180°/\pi = 57.295779 \ldots °$, and one degree is $\pi/180 = 0.017543 \ldots$ radians.

Trigonometry rests on the theorem in geometry that similar triangles have the same ratios of corresponding sides (Figure 16.1-2). Consequently, the ratio of any two

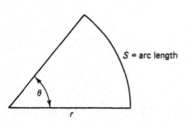

Figure 16.1-1 Angle in radian measure

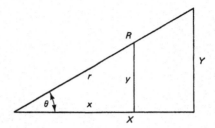

Figure 16.1-2 Similar triangles

sides of a triangle does not depend on the *size* of the triangle but only on the *shape*. Trigonometry is further simplified by tabulating only the ratios for right triangles. The six possible ratios (dimensionless numbers) for right triangles are defined by Figure 16.1-3.

$$\sin \theta = \frac{y}{r}, \qquad \cos \theta = \frac{x}{r}, \qquad \tan \theta = \frac{y}{x}$$

$$\csc \theta = \frac{r}{y}, \qquad \sec \theta = \frac{r}{x}, \qquad \text{ctn } \theta = \frac{x}{y}$$

$$(16.1\text{-}2)$$

The second line is simply the reciprocal of the first line. The "co" in the *cofunction* means "any function of $\pi/2 - \theta$ (the *complementary angle) is the cofunction." Thus sine and cosine are each the complementary function of the other. The sine, cosecant, tangent and cotangent are all odd functions, meaning that $f(-\theta) = -f(\theta)$. The other two, cosine and secant, are even functions, meaning $f(-\theta) = f(\theta)$.

Next, the idea of an angle is extended to any size. Figure 16.1-4 shows the angle in the second quadrant. Any real number, positive, negative, or zero, no matter how big or small, can be an angle. The corresponding graphs of the trigonometric functions are periodic with period 2π (or shorter)(see Figure 16.1-5). You have only to imagine the angle increasing at a constant rate, and imagine the corresponding ratio as time goes on, to see why the graphs appear as they do.

Certain angles occur repeatedly, and corresponding values of the trigonometric functions should be learned. At 0 and $\pi/2$ the values are obvious from their graphs, and we have written "∞" when we have a division by 0. The other angles arise from special triangles. For $\pi/4$ you have the 45° triangle (Figure 16.1-6) with both sides

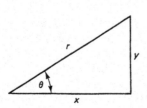

Figure 16.1-3 Basic triangle

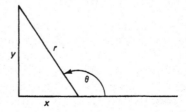

Figure 16.1-4 Extension of angle

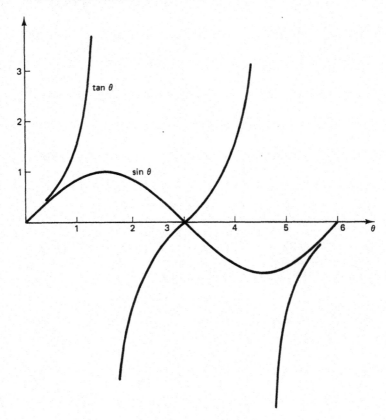

Figure 16.1-5   Sin θ and tan θ

equal to 1 and the hypotenuse equal to $\sqrt{2}$. The other special triangle is the 30°–60° triangle with one side equal to 2; it arises from bisecting one angle of an equilateral triangle of side equal to 2 (Figure 16.1-7). The angles are, in radian measure, $\pi/6$ and $\pi/3$. We thus have the Table 16.1-1

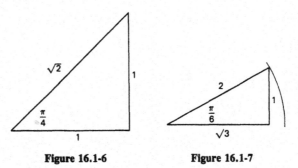

Figure 16.1-6   **Figure 16.1-7**

TABLE 16.1-1

Angle	sin	cos	tan
0	0	1	0
$\dfrac{\pi}{6}$	$\dfrac{1}{2}$	$\dfrac{\sqrt{3}}{2} = 0.8660$	$\dfrac{1}{\sqrt{3}} = 0.5771$
$\dfrac{\pi}{4}$	$\dfrac{\sqrt{2}}{2} = 0.7071$	$\dfrac{\sqrt{2}}{2} = 0.7071$	1
$\dfrac{\pi}{3}$	$\dfrac{\sqrt{3}}{2} = 0.8660$	$\dfrac{1}{2}$	$\sqrt{3} = 1.7321$
$\dfrac{\pi}{2}$	1	0	∞

One easy way to remember these numbers is to note that the entries in the sine column are $\sqrt{0}/2$, $\sqrt{1}/2$, $\sqrt{2}/2$, $\sqrt{3}/2$, and $\sqrt{4}/2$.

Next come the identities among the different functions. The circle, which we imagine the point (x, y) running around, has the equation

$$x^2 + y^2 = r^2$$

Divide this equation in turn by r^2, x^2, and y^2 to get, using (16.1-2), the three *basic identities:*

$$\cos^2 \theta + \sin^2 \theta = 1$$

$$1 + \tan^2 \theta = \sec^2 \theta \qquad (16.1\text{-}3)$$

$$\text{ctn}^2 \theta + 1 = \csc^2 \theta$$

To handle the solution of triangles, we draw a typical (obtuse) triangle in Cartesian coordinates (see Figure 16.1-8). If we apply Pythagoras's theorem to the corresponding right triangle, we get (in the figure x is negative)

$$(a - x)^2 + y^2 = c^2$$

Expand the binomial and recall that

$$x = b \cos C \quad \text{and} \quad y = b \sin C$$

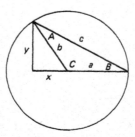

Figure 16.1-8

We have

$$a^2 - 2ab \cos C + (b \cos C)^2 + (b \sin C)^2 = c^2$$

Using $\sin^2 C + \cos^2 C = 1$ (16.1-3), we get *the law of cosines:*

$$a^2 + b^2 - 2ab \cos C = c^2 \tag{16.1-4}$$

Note that if $C = \pi/2$ then $\cos C = 0$, and you have the Pythagorean theorem as a special case of the law of cosines.

The *law of sines* is derived by dropping a suitable perpendicular in the general triangle and then expressing this distance as the sine of the opposite angles (see Figure 16.1-9). Eliminating h, we have

$$\frac{\sin A}{a} = \frac{\sin C}{c} \tag{16.1-5}$$

from which, by symmetry, it follows that this is also equal to $(\sin B)/b$.

To get the *addition formulas* of trigonometry, we consider Figure 16.1-10. For convenience, we pick a unit circle. We first compute the length of the side $PQ = z$. We have

$$\begin{aligned}
z^2 &= (x_Q - x_P)^2 + (y_Q - y_P)^2 \\
&= x_Q^2 + y_Q^2 + x_P^2 + x_Q^2 - 2(x_Q x_P + y_Q y_P) \\
&= 1 + 1 - 2(\cos A \cos B + \sin A \sin B)
\end{aligned}$$

If we apply the law of cosines to this same triangle with side z, we have a second expression for the same thing:

$$z^2 = 1 + 1 - 2 \cos (B - A)$$

Equating these two expressions, we find, after some algebra, that

$$\cos (B - A) = \cos A \cos B + \sin A \sin B = \cos (A - B) \tag{16.1-6}$$

This is one of the four addition formulas for the sum and difference of two angles.

To get the next addition formula, merely change (since the identity is true for

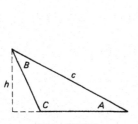

Figure 16.1-9

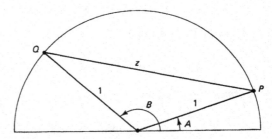

Figure 16.1-10

all angles) B to $-B$, and remember that $\cos(-B) = \cos B$, while $\sin(-B) = -\sin B$. We get

$$\cos(A + B) = \cos A \cos B - \sin A \sin B \qquad (16.1\text{-}7)$$

Next, in (16.1-6), replace B by $\pi/2 - B$. We get (on interchanging the order of the terms and using the complementary angle)

$$\cos(\pi/2 - \{B + A\}) = \sin(A + B) = \sin A \cos B + \cos A \sin B \qquad (16.1\text{-}8)$$

In this formula we replace B by $-B$ to get the last of the four addition formulas:

$$\sin(A - B) = \sin A \cos B - \cos A \sin B \qquad (16.1\text{-}9)$$

The *double-angle formulas* follow easily by putting $A = B$ in (16.1-8) and (16.1-7):

$$\sin 2A = 2 \sin A \cos A \qquad (16.1\text{-}10)$$

$$\cos 2A = \cos^2 A - \sin^2 A = 2 \cos^2 A - 1 = 1 - 2 \sin^2 A \qquad (16.1\text{-}11)$$

In (16.1-11), write $A/2$ in place of A, and solve for the sine and cosine to get the *half-angle formulas:*

$$\sin \frac{A}{2} = \sqrt{\frac{1 - \cos A}{2}}$$
$$\cos \frac{A}{2} = \sqrt{\frac{1 + \cos A}{2}} \qquad (16.1\text{-}12)$$

where, of course, you need to be careful as to which quadrant you are in since we have neglected to attach the $\pm$ sign to the radicals.

To get the *sum formulas,* we first add (16.1-8) and (16.1-9):

$$\sin(A + B) + \sin(A - B) = 2 \sin A \cos B \qquad (16.1\text{-}13)$$

Now change variables by setting

$$A + B = x$$

$$A - B = y$$

Solve these equations for A and B:

$$A = \frac{x + y}{2}, \qquad B = \frac{x - y}{2}$$

And substitute into equation (16.1-13):

$$\sin x + \sin y = 2 \sin \frac{x + y}{2} \cos \frac{x - y}{2} \qquad (16.1\text{-}14)$$

Similarly, we can get

$$\sin x - \sin y = 2 \cos \frac{x + y}{2} \sin \frac{x - y}{2}$$

$$\cos x + \cos y = 2 \cos \frac{x + y}{2} \cos \frac{x - y}{2} \qquad (16.1\text{-}15)$$

$$\cos x - \cos y = -2 \sin \frac{x + y}{2} \sin \frac{x - y}{2}$$

Finally, the *product formulas* are found from (16.1-6), (16.1-7), (16.1-8), and (16.1-9) by suitable additions and subtractions:

$$\cos x \cos y = \frac{1}{2}[\cos (x + y) + \cos (x - y)]$$

$$\sin x \cos y = \frac{1}{2}[\sin (x + y) + \sin (x - y)] \qquad (16.1\text{-}16)$$

$$\sin x \sin y = \frac{1}{2}[-\cos (x + y) + \cos (x - y)]$$

A useful formula for tan $(A \pm B)$ can be found as follows:

$$\tan (A \pm B) = \frac{\sin (A \pm B)}{\cos (A \pm B)}$$

$$= \frac{\sin A \cos B \pm \cos A \sin B}{\cos A \cos B \mp \sin A \sin B}$$

Now divide both numerator and denominator by cos A cos B to convert terms to tangents.

$$\tan (A \pm B) = \frac{\tan A \pm \tan B}{1 \mp \tan A \tan B} \qquad (16.1\text{-}17)$$

The inverse trigonometric functions are widely used in the calculus and require examination. From Section 15.1, we see the necessity for defining the inverse function so that only one value arises, called the *principal value*. An examination of the sine curve shows that it goes from -1 to $+1$ as the angle goes from $-\pi/2$ to $\pi/2$. The inverse function of the sine is sometimes written as

$$\sin^{-1} x$$

but this notation can be confused with $(\sin x)^{-1} = 1/\sin x$, and therefore we will always use the safer notation

$$\text{Arcsin } x$$

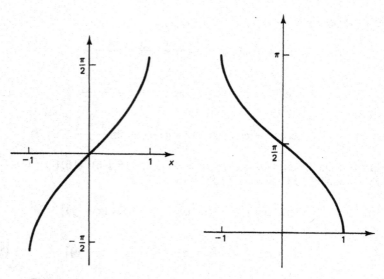

Figure 16.1-11   Arcsin x **Figure 16.1-12**   Arccos x

to mean the inverse function (Figure 16.1-11), and use

$$\text{Arcsin } x$$

to mean the *principal value*.

Correspondingly, we need an Arccos x function. The interval in which the cosine covers the full range only once is conveniently chosen as the interval from 0 to π. Finally, for the Arctan x we have the interval $-\pi/2$ to $\pi/2$. The sketches are shown in Figures 16.1-12 for the Arccos x and 16.1-13 for the Arctan x.

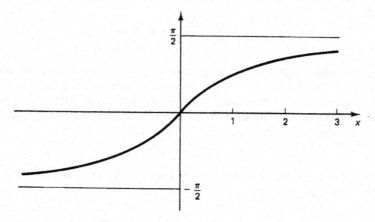

Figure 16.1-13   Arctan x

6.2 A PARTICULAR LIMIT

Before we take up the differentiation of the trigonometric functions, we develop a special limit.

$$\lim_{\theta \to 0} \frac{\sin \theta}{\theta} = 1 \qquad (16.2\text{-}1)$$

To prove this formula, we can use the area rather than the circumference (since from Example 11.9-4 they are somewhat equivalent). We look at Figure 16.2-1 and compare the three areas, the area of the inner triangle, the area of the sector of the circle, and the area of the outer triangle. These areas are

$$\frac{1}{2} \sin \theta \cos \theta < \frac{\theta}{2} < \frac{1}{2} \tan \theta \qquad (16.2\text{-}2)$$

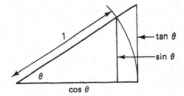

Figure 16.2-1   A particular limit

The area $\theta/2$ of the sector follows from the symmetry of the circle plus the fact that the area of the circle is π while the central angle is 2π radians. For the moment, consider only $\theta > 0$ to get

$$\cos \theta < \frac{\theta}{\sin \theta} < \frac{1}{\cos \theta}$$

Now take the reciprocals of each term (thus reversing the inequalities):

$$\frac{1}{\cos \theta} > \frac{\sin \theta}{\theta} > \cos \theta$$

Finally, as $\theta \to 0^{+}$, both end terms approach 1; hence, by continuity, the quantity between them must also approach 1. The proof is for positive θ; but if θ is replaced by $-\theta$ then $(\sin -\theta)/(-\theta) = (\sin \theta)/\theta$, and the proof goes much the same.

This limit depends crucially on the use of radian measure for the angle and is the reason we use radian measure throughout higher mathematics.

16.3 THE DERIVATIVE OF SIN x

To find the derivative of $y = \sin x$, we go back to fundamentals, to the four-step delta process (Section 7.5). At the third step, we have

$$\frac{\Delta y}{\Delta x} = \frac{\sin (x + \Delta x) - \sin x}{\Delta x}$$

We apply the top trigonometric identity of (16.1-15) for the difference of the two sines to get (note the 2 in the denominator of the denominator)

$$\frac{\Delta y}{\Delta x} = \cos\left(x + \frac{\Delta x}{2}\right)\left[\frac{\sin\left(\dfrac{\Delta x}{2}\right)}{\left(\dfrac{\Delta x}{2}\right)}\right]$$

As $\Delta x \to 0$, so also does $\Delta x/2 \to 0$, and the limit of the term on the right is 1 [from 16.2-1)]. At the same time, $x + \Delta x/2 \to x$. In the limit we have

$$\frac{dy}{dx} = \frac{d(\sin x)}{dx} = \cos x$$

One crude way to see the truth of this is to inspect the slope of the sine curve at each x value; the slope of the sine curve is the value of the cosine curve at the same x value for all x!

More generally, we have

$$\frac{d(\sin u)}{dx} = \cos u \frac{du}{dx} \tag{16.3-1}$$

Examples 16.3-1

We have (sight differentiation)

$$\frac{d(\sin 2x)}{dx} = (\cos 2x)(2) = 2\cos 2x$$

$$\frac{d(\sin x^2)}{dx} = (\cos x^2)(2x) = 2x\cos x^2$$

$$\frac{d(\sin^2 x)}{dx} = 2(\sin x)(\cos x) = \sin 2x$$

16.4* AN ALTERNATIVE DERIVATION

We found in Section 14.3 the derivative of the $\ln x$ function by considering it as an integral, and later in Section 14.4 used the direct delta process. It is natural to ask if we need to apply the direct delta process [which used the special limit of $(\sin \theta)/\theta$ of Section 16.2] in order to find the derivative of $\sin x$. Could we not find it, as we did for the $\ln x$ function, by some method, perhaps directly from the integral? The answer is yes, as the following shows.

Consider finding the area in Figure 16.4-1. It is

$$J = \int_0^x y\, dt = (\text{area of triangle}) + (\text{area of sector})$$

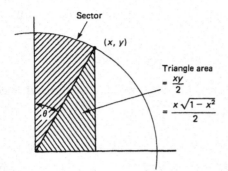

Sector

(x, y)

Triangle area
$= \dfrac{xy}{2}$

$= \dfrac{x\sqrt{1 - x^2}}{2}$

θ

Figure 16.4-1   Alternative derivation

Using for convenience the unit circle

$$x^2 + y^2 = 1$$

the integral becomes (note the dummy variable t)

$$J = \int_0^x \sqrt{1 - t^2}\, dt$$

Integrate by parts, choosing

$$U = \sqrt{1 - t^2} \qquad\qquad dV = dt$$

$$dU = -t\sqrt{1 - t^2}\, dt, \qquad V = t$$

Hence

$$J = t\sqrt{1 - t^2}\,\Big|_0^x + \int_0^x \frac{t^2}{\sqrt{1 - t^2}}\, dt$$

The integrand of the second integral can be written as

$$\frac{1 - (1 - t^2)}{\sqrt{1 - t^2}} = \frac{1}{\sqrt{1 - t^2}} - \sqrt{1 - t^2}$$

The original integral can now be written

$$J = x\sqrt{1 - x^2} + \int_0^x \frac{1}{\sqrt{1 - t^2}}\, dt - J$$

Solve for J:

$$J = \frac{x}{2}\sqrt{1 - x^2} + \frac{1}{2}\int_0^x \frac{dt}{\sqrt{1 - t^2}}$$

The integrated term is exactly the area of the triangle, and hence the integral on the right is the area $\theta/2$ of the sector. We have (upon doubling each side of the equation)

$$\theta = \int_0^x \frac{dt}{\sqrt{1 - t^2}}$$

From the fundamental theorem of the calculus, the derivative of the integral is the integrand:

$$\frac{d\theta}{dx} = \frac{1}{\sqrt{1 - x^2}}$$

This in turn means that, since from the figure $x = \sin\theta$, the radical is $\cos\theta$, and we have

$$\frac{dx}{d\theta} = \sqrt{1 - x^2} = \cos\theta$$

(in the first quadrant). We have the derivative of $\sin\theta$ is $\cos\theta$ (without having used the limit of Section 16.2). The extension to the other quadrants follows easily from the addition formulas.

16.5 DERIVATIVES OF THE OTHER TRIGONOMETRIC FUNCTIONS

The derivatives of the other trigonometric functions follow immediately. For the function $\cos x$, once you remember that

$$\text{fcn}\left(\frac{\pi}{2} - \theta\right) = \text{cofcn}\,\theta \qquad (16.5\text{-}1)$$

you have

$$\sin\left(\frac{\pi}{2} - \theta\right) = \cos\theta \quad \text{and} \quad \cos\left(\frac{\pi}{2} - \theta\right) = \sin\theta$$

Using these,

$$\frac{d(\cos x)}{dx} = \frac{d\left\{\sin\left(\frac{\pi}{2} - \theta\right)\right\}}{dx} = \cos\left(\frac{\pi}{2} - x\right)(-1) = -\sin x$$

where the cofunction rule is applied again in the last step.

The derivative of the tangent is (using the quotient form)

$$\frac{d(\tan x)}{dx} = \frac{d\left\{\dfrac{\sin x}{\cos x}\right\}}{dx} = \frac{\cos x \cos x - \sin x(-\sin x)}{\cos^2 x}$$

$$= \frac{\cos^2 x + \sin^2 x}{\cos^2 x}$$

$$= \sec^2 x$$

The derivative of the secant follows from

$$\frac{d(\sec x)}{dx} = \frac{d\left\{\dfrac{1}{\cos x}\right\}}{dx} = -\cos^{-2} x(-\sin x) = \sec x \tan x$$

The other two derivatives can be found directly or from the cofunction relationship. In the later case you replace each function by its cofunction and add a minus sign (coming from the differentiation of the $\pi/2 - \theta$) to the result. You get the formulas

$$\frac{d(\text{ctn } x)}{dx} = -\csc^2 x$$

$$\frac{d(\csc x)}{dx} = -\csc x \text{ ctn } x$$

Thus we have the six basic formulas

$$\frac{d(\sin u)}{dx} = \cos u \frac{du}{dx}$$

$$\frac{d(\cos u)}{dx} = -\sin u \frac{du}{dx}$$

$$\frac{d(\tan u)}{dx} = \sec^2 u \frac{du}{dx} \qquad (16.5\text{-}2)$$

$$\frac{d(\text{ctn } u)}{dx} = -\csc^2 u \frac{du}{dx}$$

$$\frac{d(\sec u)}{dx} = \sec u \tan u \frac{du}{dx}$$

$$\frac{d(\csc u)}{dx} = -\csc u \text{ ctn } u \frac{du}{dx}$$

These formulas should be memorized carefully so they are available for immediate recall when we get to the topic of integration.

Example 16.5-1

Sight differentiate

$$\frac{d(\cos 3x)}{dx} = -\sin 3x(3) = -3 \sin 3x$$

$$\frac{d(\tan^2 x)}{dx} = 2 \tan x(\sec^2 x) = \frac{2 \sin x}{\cos^3 x}$$

Differentiate

$$\frac{d\left\{\dfrac{\sin^2 x}{1 + \cos x}\right\}}{dx} = \frac{(1 + \cos x)2 \sin x \cos x - \sin^2 x(-\sin x)}{(1 + \cos x)^2}$$

$$= \frac{\sin x[2 \cos x + 2 \cos^2 x + \sin^2 x]}{(1 + \cos x)^2}$$

$$= \frac{\sin x[2 \cos x + \cos^2 x + 1]}{(1 + \cos x)^2}$$

$$= \frac{\sin x[1 + \cos x]^2}{(1 + \cos x)^2} = \sin x$$

This surprising result comes from not first simplifying the expression before differentiation. If you simplify first, then instead of the above mess of algebra and trigonometry you have

$$\frac{d\left\{\dfrac{1 - \cos^2 x}{1 + \cos x}\right\}}{dx} = \frac{d\{1 - \cos x\}}{dx} = \sin x$$

EXERCISES 16.5

Differentiate the following:

1. $\cos 2x$, $\cos x^2$, $\cos^2 x$
2. $\tan^2 x$
3. $\sec^2 x$

Check the following identities by differentiating both sides and comparing the results:

4. $\sin^2 x + \cos^2 x = 1$
5. $1 + \tan^2 x = \sec^2 x$
6. $\text{ctn}^2 x + 1 = \csc^2 x$
7. $\sin (a + x) = \sin a \cos x + \cos a \sin x$
8. $\cos (a + x) = \cos a \cos x - \sin a \sin x$
9. $\sin x = \tan x(\cos x)$
10. $\sin 2x = 2 \sin x \cos x$
11. $\cos 2x = \cos^2 x - \sin^2 x$
12. $\tan 2x = (2 \tan x)/(1 - \tan^2 x)$

Find the derivatives of the following:

13. $\sin x/(1 + \cos x)$
14. $(1 + \cos x)/\sin x$
15. $(1 + \sin x)$
16. $\tan x \sec x$
17. $\operatorname{ctn} x \csc x$
18. $(1 - \sin x)/(1 + \sin x)$
19. $(1 - \cos x)/(1 + \cos x)$
20. $\tan^2 x - \sec^2 x$

16.6 INTEGRATION FORMULAS

The six differentiation formulas lead directly to the following integration formulas:

$$\int \cos u \, du = \sin u + C$$

$$\int \sin u \, du = -\cos u + C$$

$$\int \sec^2 u \, du = \tan u + C$$

$$\int \csc^2 u \, du = -\operatorname{ctn} u + C$$

$$\int \sec u \tan u \, du = \sec u + C$$

$$\int \csc u \operatorname{ctn} u \, du = -\csc u + C$$

It is natural to wonder about the missing formulas, such as

$$\int \tan u \, du = \int \frac{\sin u}{\cos u} \, du = -\ln |\cos u| + C$$

$$= \ln |\sec u| + C$$

$$\int \operatorname{ctn} u \, du = \int \frac{\cos u}{\sin u} \, du = \ln |\sin u| + C$$

$$\int \sec u \, du = \, ?$$

For the integral of sec x, we need to know a trick (which was not easy to find the first time). Multiply and divide by the same quantity sec u + tan u:

$$\int \sec u \left\{ \frac{\sec u + \tan u}{\sec u + \tan u} \right\} du = \int \frac{\sec^2 u + \sec u \tan u}{\sec u + \tan u} du$$

$$= \ln |\sec u + \tan u| + C$$

Correspondingly,

$$\int \csc u \, du = -\ln |\csc u + \operatorname{ctn} u| + C$$

From these basic formulas for integration, we can find numerous other integrals. The main tools are as follows:

1. Rearrangement of the integrand (in these cases this means extensive use of the trigonometric identities of Section 16.1).
2. Recognition of a basic formula in some disguise. For example,

$$\int \frac{\cos x}{\sqrt{\sin x}} dx$$

Set $U = \sin x$, then $dU = \cos x \, dx$, and the integral has the form

$$\int U^{-1/2} \, dU = 2U^{1/2} + C = 2\sqrt{\sin x} + C$$

When used systematically, this becomes change of variables of Section 11.9.
3. Integration by parts, Section 14.7, which is very powerful, especially in getting rid of undesired terms.
4. Change of variables is mainly a matter of trial and error. It is often the only practical approach to difficult integrals.

As the following examples show, these tools can be combined in many ways to accomplish an integration. The examples also show that integration is a complex, difficult art that requires your best efforts. Each example of integration you do should be carefully studied so you gradually learn the art of integration; examine your successes as well as your failures! Ask yourself what other integrals could you do using the same method (or a similar method)?

Example 16.6-1

$\int \cos 3x \, dx$. You see immediately that $3x$ is the U, so that dU is $3 \, dx$ and you need a factor 3. Thus you get

$$\int \cos 3x \, dx = \frac{1}{3} \int \cos 3x \, 3 \, dx = \frac{\sin 3x}{3} + C$$

Similar integrals are $\cos ax$, $\cos (ax + b)$, and $\sin (ax + b)$. Do these in your head.

Example 16.6-2

$\int \sin^2 x \, dx$. Here you use the trigonometric identity for the half-angles (16.1-12) applied to the angle x:

$$\sin^2 x = \frac{1 - \cos 2x}{2}$$

Hence

$$\int \sin^2 x \, dx = \frac{x}{2} - \frac{1}{4} \sin 2x + C$$

$$= \frac{x}{2} - \frac{1}{2} \sin x \cos x + C$$

Differentiation will show how the terms combine to reverse the steps you used. Evidently, $\cos^2 x$ could be done in a similar way.

Example 16.6-3

$\int \cos^3 x \, dx$. Here you need to see that $\cos^2 x = 1 - \sin^2 x$. Hence

$$\int \cos^3 x \, dx = \int (1 - \sin^2 x) \cos x \, dx$$

$$= \sin x - \frac{1}{3} \sin^3 x + C$$

The integral of $\sin^3 x$ would be similar. Integrals with odd powers of either sin or cos could be done similarly.

Example 16.6-4

$\int \tan^2 x \, dx$. Here you need to notice that $\tan^2 x = \sec^2 x - 1$ and that $\sec^2 x$ is the derivative of $\tan x$. Hence you have

$$\int \tan^2 x \, dx = \int (\sec^2 x - 1) \, dx = \tan x - x + C$$

How about $\text{ctn}^2 x$?

Example 16.6-5

$\int x \sin x \, dx$. Clearly, the trouble is in the factor x, and to get rid of it calls for integration by parts.

$$U = x, \qquad dV = \sin x \, dx$$
$$dU = dx, \qquad V = -\cos x$$

Hence

$$\int x \sin x \, dx = -x \cos x + \int \cos x \, dx$$

$$= -x \cos x + \sin x + C$$

Direct differentiation will serve to check this result. Similar integrals are $x \cos x$ and $x^2 \sin x$.

Example 16.6-6

$\int \sin x/(1 + \cos x)\, dx$. You need to notice that (except for the sign) the numerator is the derivative of the denominator. Hence

$$\int \frac{\sin x}{1 + \cos x}\, dx = -\ln (1 + \cos x) + C$$

Absolute values are not needed in this example.

Example 16.6-7

$J = \int e^{-ax} \sin bx\, dx$. This integral occurs frequently and requires integration by parts twice, which will make the same integrand appear again. It does not matter which way we pick the U and dV. We pick

$$U = \sin bx, \qquad dV = e^{-ax}\, dx$$

$$dU = b \cos bx\, dx, \qquad V = -\left(\frac{1}{a}\right)e^{-ax}$$

Hence

$$J = -\left(\frac{1}{a}\right)e^{-ax} \sin bx + \frac{b}{a} \int e^{-ax} \cos bx\, dx$$

We apply integration by parts again, keeping the same order (lest we merely undo the step we just did):

$$U = \cos bx, \qquad dV = e^{-ax}\, dx$$

$$dU = -b \sin bx\, dx, \qquad V = -\left(\frac{1}{a}\right)e^{-ax}$$

$$J = -\left(\frac{1}{a}\right)e^{-ax} \sin bx + \frac{b}{a}\left\{-\left(\frac{1}{a}\right)e^{-ax} \cos bx - \left(\frac{b}{a}\right)J\right\}$$

Notice that the original integral J has reappeared. Multiply by a^2, and clean up the expression a bit:

$$a^2 J = -ae^{-ax} \sin bx - be^{-ax} \cos bx - b^2 J$$

Transpose the $b^2 J$ term and solve for J:

$$J = e^{-ax}\left\{\frac{-a \sin bx - b \cos bx}{a^2 + b^2}\right\} + C$$

Differentiation will verify this integral. The integral of $e^{-ax} \cos bx$ would go much the same.

$$\int e^{-ax} \cos bx\, dx = e^{-ax}\left\{\frac{b \sin bx - a \cos bx}{a^2 + b^2}\right\} + C$$

Example 16.6-8

$\int \sec^3 x \, dx$. This is even more complex than the last one, but again it is one of common occurrence. After a number of fruitless trigonometric substitutions, you will hit upon trying $\sec^2 x = 1 + \tan^2 x$.

$$J = \int \sec x\{1 + \tan^2 x\} \, dx$$

$$= \int \sec x \, dx + \int \sec x \tan^2 x \, dx$$

$$= \ln |\sec x + \tan x| + \int \tan x(\sec x \tan x) \, dx$$

Apply integration by parts to this integral:

$$U = \tan x, \qquad dV = \sec x \tan x \, dx$$
$$dU = \sec^2 x \, dx, \qquad V = \sec x$$

$$J = \ln |\sec x + \tan x| + \sec x \tan x - \int \sec^3 x \, dx$$

and the integral is J again. Solve for J and get

$$J = \frac{1}{2} \ln |\sec x + \tan x| + \frac{1}{2} \sec x \tan x + C$$

Again, differentiation will check this integration. Note that the device of getting the original integral back again and solving for it has occurred three times. This indicates that it is more than an isolated trick.

Example 16.6-9

Buffon's needle. Buffon (1707–1788), a French naturalist, suggested the following Monte Carlo determination of π. Imagine a flat plane ruled with parallel lines having unit spacing. Drop at random a needle of length $L < 1$ on the plane. What is the probability of the needle crossing a line?

"At random" will be taken to mean that the center of the needle is uniformly spaced from the nearest line. It is no loss in generality to assume that it lies on the positive side, $0 \le x \le 1/2$. We also assume that the angle the needle makes with the line, θ, is uniform, $0 \le \theta \le \pi$ (Figure 16.6-1). A crossing occurs (see Figure 16.6-2) when the coordinates fall in the shaded region of the sample space. The bounding line between the two regions, success and failure, is

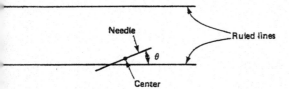

Needle

Ruled lines

θ

Center

Figure 16.6-1 Buffon's needle

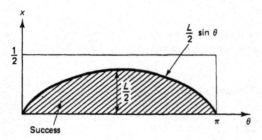

Figure 16.6-2   Buffon's needle

$$x = \frac{L}{2} \sin \theta$$

The probability assignment is uniform in the sample space. Thus the probability of a crossing is the ratio of the shaded region to the whole sample space. The area of the shaded space is

$$\int_0^\pi \frac{L}{2} \sin \theta \, d\theta = \frac{L}{2}(-\cos \theta) \Big|_0^\pi = L$$

The area of the whole space is $(1/2)\pi = \pi/2$. Hence the probability of a crossing is

$$\frac{2L}{\pi}$$

Again, we have a Monte Carlo method of estimating the known number π. As a test, we chose $L = 0.8$ and ran the cases as shown in Table 16.6-1. Thus the estimate of π is given by

$$\frac{1.6 \text{ (total trials)}}{\text{successes}}$$

We notice that the first 100 trials were not close to the correct answer, but the following trials are gradually "swamping" the first unlikely run.

TABLE 16.6-1

Successes	Total	Estimate of π
43	100	3.7209
87	200	3.6782
142	300	3.3803
192	400	3.3333
246	500	3.2520

EXERCISES 16.6

Integrate the following:

1. $\int \sin (ax + b) \, dx$
2. $\int \cos^2 x \, dx$

   3. $\int \sin^3 x \, dx$
   4. $\int \text{ctn}^2 x \, dx$
   5. $\int x^2 \sin x \, dx$
   6. $\int \sqrt{\tan x}/\cos^2 x \, dx$
   7. $\int \sec^2 x/(1 + \tan x) \, dx$
   8. $\int e^{ax} \cos bx \, dx$
   9. $\int \csc^3 x \, dx$
   10. Show that $\int_0^\infty 1/(1 + e^x) \, dx = \ln 2$.
   11. Integrate $\int \sec x \tan x/(1 + \sec x) \, dx$.
   12. Derive the integral for csc x.
   13. Generalize Exercise 5 and get a reduction formula for it.
   14. Do the Buffon needle problem in case $L = 1$.
   15. Analyze the Buffon needle problem in case $L > 1$.
   16. You are on a river bank. You walk in a random direction away from the river for one mile. At this point, you choose a new random direction and start walking another mile. What is the probability you will reach the river?

16.7 SOME DEFINITE INTEGRALS OF IMPORTANCE

Definite integrals are usually found from the corresponding indefinite integrals. In the process of substituting in the limits, often the integrated term will vanish at either or both ends, thus making the final formula much simpler.

Example 16.7-1

In Section 11.3, we said that we would later find the area of a circle. We found the area in Example 11.9-4, but we will do it again by a different method. Let the circle be defined by

$$x^2 + y^2 = a^2$$

The area is four times the area of one quadrant

$$A = 4 \int_0^a \sqrt{a^2 - x^2} \, dx$$

Set $x = a \sin \theta$; then $dx = a \cos \theta \, d\theta$, and we have

$$A = 4 \int_0^{\pi/2} a \cos \theta \, a \cos \theta \, d\theta$$

$$= 4a^2 \int_0^{\pi/2} \cos^2 \theta \, d\theta$$

$$= 4a^2 \int_0^{\pi/2} \frac{1 + \cos 2\theta}{2} \, d\theta$$

$$= \frac{4a^2}{2}\left\{\theta + \frac{\sin 2\theta}{2}\right\}\Big|_0^{\pi/2}$$

$$= 2a^2\left(\frac{\pi}{2}\right) = \pi a^2$$

which is the correct answer.

Example 16.7-2

The Wallis (1616–1703) integrals occur so often in practice that they are worth special attention. They are

$$W_n = \int_0^{\pi/2} \cos^n x \, dx = \int_0^{\pi/2} \sin^n x \, dx$$

where n is a nonnegative integer. The equality of the two integrals is immediately obvious from their graphs (or else you can set $x = \pi/2 - x'$ to get the cofunction, the interchange of the limits, and the minus sign from the differentiation). We plan to first get a *reduction formula* for W_n. We write, for $n \geq 2$,

$$W_n = \int_0^{\pi/2} \cos^{n-2} x (1 - \sin^2 x) \, dx$$

$$= W_{n-2} - \int_0^{\pi/2} \sin x (\cos^{n-2} x \sin x) \, dx$$

Apply integration by parts:

$$U = \sin x, \qquad dV = \cos^{n-2} x (-\sin x) \, dx$$

$$du = \cos x \, dx, \qquad V = \frac{\cos^{n-1} x}{n - 1}$$

Hence

$$W_n = W_{n-2} + \frac{\sin x \cos^{n-1} x}{n - 1}\Big|_0^{\pi/2} - \frac{1}{n - 1}\int_0^{\pi/2} \cos^n x \, dx$$

The integrated piece vanishes at both ends since $n \geq 2$. Multiply through by $n - 1$.

$$(n - 1)W_n = (n - 1)W_{n-2} - W_n$$

and solve for W_n.

$$W_n = \frac{n - 1}{n} W_{n-2} \tag{16.7-1}$$

which is the required reduction formula. Each application of the formula reduces the index n by 2. To complete the cases, we need to compute the two bottom ones, $n = 0$ and $n = 1$. We have

$$W_0 = \int_0^{\pi/2} dx = \frac{\pi}{2}$$

$$\tag{16.7-2}$$

$$W_1 = \int_0^{\pi/2} \cos x \, dx = 1$$

For example, repeated use of (16.7-1) ending with one application of (16.7-2) gives

$$W_6 = \left(\frac{5}{6}\right)\left(\frac{3}{4}\right)\left(\frac{1}{2}\right)\left(\frac{\pi}{2}\right) = \frac{5\pi}{32}$$

$$W_7 = \left(\frac{6}{7}\right)\left(\frac{4}{5}\right)\left(\frac{2}{3}\right)(1) = \frac{16}{35}$$

Rule. To write out the answer, you begin with n in the denominator, $n - 1$ in the numerator, $n - 2$ in the denominator, $n - 3$ in the numerator, . . . , until you are down to 2. If the 2 ends in the denominator, you include a factor $\pi/2$, and if the 2 ends in the numerator you are done.

It is slightly more complex to handle the generalized Wallis integrals of the form

$$W_{m,n} = \int_0^{\pi/2} \cos^m x \sin^n x \, dx$$

but corresponding reduction formulas for either exponent can be found by the same method. As you can see, there are no new ideas involved, only a great deal more tedious algebra.

Reduction formulas are similar to mathematical induction downward; therefore, it is not surprising that reduction formulas play a prominent role in dealing with classes of formulas.

Example 16.7-3

Stirling's formula again. In Section 15.4 we found bounds on $n!$:

$$n^n e^{-n} \sqrt{n} e^{7/8} \leq n! \leq n^n e^{-n} \sqrt{n} e$$

We asserted that for large n the proper coefficient of the approximation is

$$n! = n^n e^{-n} \sqrt{2\pi n}$$

To find this coefficient, we use the Wallis integral. First notice that $|\cos x| \leq 1$. Therefore, for the range $0 \leq x \leq \pi/2$,

$$\cos^{2n-1} x \geq \cos^{2n} x \geq \cos^{2n+1} x$$

Hence the Wallis integrals have the same relationship:

$$W_{2n-1} \geq W_{2n} \geq W_{2n+1}$$

Divide by W_{2n-1}. We get

$$1 \geq \frac{W_{2n}}{W_{2n-1}} \geq \frac{W_{2n+1}}{W_{2n-1}}$$

We now use the rule for the Wallis integrals. This gives

$$1 \geq \left[\frac{(2n-1)(2n-3)\ldots(1)}{(2n)(2n-2)\ldots(2)}\frac{\pi}{2}\right]\left[\frac{(2n-1)(2n-3)\ldots(1)}{(2n-2)(2n-4)\ldots(2)}\right] \geq \frac{2n}{2n+1}$$

Rearranging this we can write it in the form (where we supply a needed factor $2nn$ in the numerator and denominator)

$$1 \geq \left[\frac{(2n-1)(2n-3) \ldots (1)}{(2n)(2n-2) \ldots (2)} \right]^2 \frac{\pi}{2}(2n) \geq \frac{2n}{2n+1}$$

If we take the positive square root of each term, we preserve the inequalities. We next insert in the numerator and denominator of the middle term a copy of the denominator. This gives

$$1 \geq \frac{(2n)!}{[2^n n!]^2} \sqrt{\pi n} \geq \left\{ \frac{2n}{2n+1} \right\}^{1/2} \tag{16.7-3}$$

Write Stirling's approximation in the form (with the coefficient C undetermined)

$$n! = n^n e^{-n} \sqrt{n}\, C$$

We substitute this into each appearance of a factorial in (16.7-3). This gives

$$1 \geq \frac{(2n)^{2n} e^{-2n} \sqrt{2n}\, C}{2^{2n} n^{2n} e^{-2n} n C^2} \sqrt{\pi n} \geq \left\{ \frac{2n}{2n+1} \right\}^{1/2}$$

Almost everything cancels out to leave

$$1 \geq \frac{\sqrt{2\pi}}{C} \geq \left\{ 1 - \frac{1}{2n+1} \right\}^{1/2}$$

Now as $n \to \infty$ the quantity in the middle is squeezed toward 1; hence

$$C = \sqrt{(2\pi)}$$

as we asserted.

Numerous other definite integrals occur in practice and are worth looking at.

Example 16.7-4

Probably the most common definite integrals are

$$\int_0^{2\pi} \cos mx \cos nx \, dx = \begin{cases} 0, & \text{for } m \neq n \\ \pi, & \text{for } m = n \neq 0 \\ 2\pi, & \text{for } m = n = 0 \end{cases}$$

$$\int_0^{2\pi} \cos mx \sin nx \, dx = 0$$

$$\int_0^{2\pi} \sin mx \sin nx \, dx = \begin{cases} 0, & \text{for } m \neq n \\ \pi, & \text{for } m = n \neq 0 \\ 0, & \text{for } m = n = 0 \end{cases}$$

The proofs are all the same in the sense that they are based on the same group of trigonometric identities (16.1-16). For the top integral, we use

$$\cos mx \cos nx = \frac{1}{2}\{\cos (m+n)x + \cos (m-n)x\}$$

The integration of this gives, if $m \neq n$,

$$\frac{1}{2}\left\{ \frac{\sin (m + n)x}{m + n} + \frac{\sin (m - n)x}{m - n} \right\}$$

and at the two ends both terms vanish, being sines of multiples of 2π. For the case $m = n \neq 0$, the second term will be $\cos 0x = 1$, which integrates into an x. The limits and the $1/2$ in front combine to give just π. In the final case, when both $m = n = 0$, the original integrand is simply 1 and the integral is, of course, 2π. The other two integrals may be found similarly, something that the reader should do.

Example 16.7-5

The integral for $a > 0$

$$J = \int_0^\infty e^{-ax} \sin bx \, dx$$

Its indefinite integral is given in Example 16.6-7. Hence

$$J = e^{-ax} \left[\frac{-a \sin bx - b \cos bx}{a^2 + b^2} \right] \Bigg|_0^\infty$$

and it is a matter of substituting in the limits. At the upper limit, ∞, the exponential dominates everything and gives 0. At the lower limit the sine vanishes and the cosine is 1; so we have

$$J = \int_0^\infty = \frac{b}{a^2 + b^2}$$

Similarly (for $a > 0$),

$$\int_0^\infty e^{-ax} \cos bx \, dx = e^{-ax} \left[\frac{b \sin bx - a \cos bx}{a^2 + b^2} \right] \Bigg|_0^\infty$$

$$= \frac{a}{a^2 + b^2}$$

Example 16.7-6

Another definite integral of common occurrence is

$$I_n = \int_{-\infty}^\infty x^n \exp \left(\frac{-x^2}{2} \right) dx]$$

for n a positive integer. Integration by parts (when $n \geq 2$), using one x for the integration of the exponential and the resulting x^{n-1} for the U, gives

$$I_n = -x^{n-1} \exp \left(\frac{-x^2}{2} \right) \Bigg|_{-\infty}^\infty + (n - 1) \int_{-\infty}^\infty x^{n-2} \exp \left(\frac{-x^2}{2} \right) dx$$

The integrated part vanishes at both limits since the exponential dominates any power of x. Thus you have the reduction formula

$$I_n = (n - 1)I_{n-2}$$

The index drops by two each step, and as a result you need the two bottom cases, $n = 0$ and $n = 1$. For $n = 1$,

$$I_1 = \int_{-\infty}^{\infty} x \exp \left(\frac{-x^2}{2}\right) dx = -\exp \left(\frac{-x^2}{2}\right) \Big|_{-\infty}^{\infty} = 0$$

since the exponential approaches zero more rapidly than any power of x. We could also note that the integrand is an odd function over a symmetric interval about the origin; hence the integral must be zero. The other case, $n = 0$, we will later show (Example 19.4-4) is

$$I_0 = \int_{-\infty}^{\infty} \exp \left(\frac{-x^2}{2}\right) dx = \sqrt{2\pi}$$

EXERCISES 16.7

1. $\int_0^{\infty} x^n e^{-x}\, dx$, $n =$ a positive integer
2. $\int_0^{\pi} \cos mx \cos nx\, dx$
3. $\int_0^{\pi} \cos mx \sin nx\, dx$
4. Equations in Example 16.7-4 for the range $-\pi \leq x \leq \pi$
5. $\int_{-a}^{a} (a^2 - x^2)^{n/2}\, dx$
6. $\int_{-a}^{a} x^n / \sqrt{a^2 - x^2}\, dx$
7. $\int_{-a}^{a} x^n \sqrt{a^2 - x^2}\, dx$
8.*Find the formulas for the generalized Wallis integrals $W_{m,n}$.

16.8 THE INVERSE TRIGONOMETRIC FUNCTIONS

The inverse trigonometric functions occupy a surprisingly large role in the calculus, and we therefore need to examine them carefully. The mathematical tool is the same in each case.

Consider the direct function

$$y = \tan x$$

$$\frac{dy}{dx} = \sec^2 x = 1 + \tan^2 x = 1 + y^2$$

The *inverse* function has the x and y interchanged:

$$x = \tan y$$

or, since we need to stick with single-valued functions,

$$y = \text{Arctan } x$$

$$\frac{dy}{dx} = \frac{1}{\dfrac{dx}{dy}} = \frac{1}{1 + x^2}$$

Hence we have the general formula

$$\frac{d(\text{Arctan } u)}{dx} = \frac{1}{1 + u^2}\frac{du}{dx} \tag{16.8-1}$$

Note that (like the logarithm) the derivative of the Arctangent is an algebraic function.
Next consider the direct function

$$y = \sin x$$

$$\frac{dy}{dx} = \cos x = \pm \sqrt{1 - y^2}$$

The inverse function is

$$x = \sin y \quad \text{or} \quad y = \text{Arcsin } x$$

$$\frac{dy}{dx} = \frac{1}{\dfrac{dx}{dy}} = \frac{\pm 1}{\sqrt{1 - x^2}}$$

and for the principal value of the Arcsin we need the plus sign. Hence we have the general formula

$$\frac{d(\text{Arcsin } u)}{dx} = \frac{1}{\sqrt{1 - u^2}}\frac{du}{dx} \tag{16.8-2}$$

Again the derivative is an algebraic function.
The Arccos x goes similarly, except that we must pick the minus sign for the slope of the principal value; hence

$$\frac{d(\text{Arccos } x)}{dx} = \frac{-1}{\sqrt{1 - u^2}}\frac{du}{dx} \tag{16.8-3}$$

The other three inverse functions are of less importance but are given for the sake of completeness.

$$\frac{d(\text{Arcctn } u)}{dx} = \frac{-1}{1 + x^2}\frac{du}{dx}$$

$$\frac{d(\text{Arcsec } u)}{dx} = \frac{1}{|u|\sqrt{u^2 - 1}}\frac{du}{dx} \tag{16.8-4}$$

$$\frac{d(\text{Arccsc } u)}{dx} = \frac{-1}{|u|\sqrt{u^2 - 1}}\frac{du}{dx}$$

The corresponding integrals are

$$\int \frac{1}{1+u^2} \, du = \text{Arctan } u + C = -\text{Arcctn } u + C$$

$$\int \frac{1}{\sqrt{1-u^2}} \, du = \text{Arcsin } u + C = -\text{Arccos } u + C \qquad (16.8\text{-}5)$$

$$\int \frac{1}{|u| \sqrt{1-u^2}} \, du = \text{Arcsec } u + C = -\text{Arccsc } u + C$$

It is important to notice that the inverse trigonometric functions arise from the integration of algebraic functions, as does the logarithm (which is the inverse of the exponential function). As the calculus of the algebraic functions was gradually developed, these inverse trigonometric functions were bound to arise.

Example 16.8-1

Show that

$$\int_{-\infty}^{\infty} \frac{1}{1+x^2} \, dx = \pi$$

Integration leads immediately to

$$\text{Arctan } x \, \Big|_{-\infty}^{\infty} = \frac{\pi}{2} - \left(\frac{-\pi}{2}\right) = \pi$$

Note that the integral of a simple algebraic function leads to the transcendental number π, which arose originally in connection with the circumference (or area) of a circle.

Example 16.8-2

Integrate

$$\int \frac{dx}{a^2+x^2}$$

If we set $x = at$, we get $dx = a \, dt$, and the integral becomes

$$\frac{1}{a^2} \int \frac{1}{1+t^2} a \, dt = \frac{1}{a} \text{Arctan } t + C$$

$$= \frac{1}{a} \text{Arctan } \frac{x}{a} + C$$

Example 16.8-3

Show that for $0 < x < 1$

$$\text{Arcsin } x = \text{Arccos } \{\sqrt{1-x^2}\}$$

We can prove this in many ways. One useful approach is to observe that (from the mean value theorem) *if* (1) *two differentiable functions have the same derivative, and* (2) *have one point in common, then they are the same function.* Hence we differentiate both sides:

$$\frac{1}{\sqrt{1-x^2}} = \left(\frac{-1}{\sqrt{1-(1-x^2)}}\right)\frac{1}{2}\frac{1}{\sqrt{1-x^2}}(-2x)$$

$$= \frac{1}{\sqrt{x^2}}\frac{x}{\sqrt{1-x^2}} = \frac{1}{\sqrt{1-x^2}}$$

For $x = 0$, we have

$$\text{Arcsin } 0 = 0 = \text{Arccos } 1$$

and we have completed the proof.

EXERCISES 16.8

Find the following:

1. $\int_0^1 \text{Arctan } x \, dx$
2. $\int \text{Arcsin } x \, dx$
3. $\int x^n \text{Arctan } x \, dx$
4. Show that $\int_0^\infty [(\text{Arctan } x)/1 + x^2) \, dx] = \pi^2/8$
5. Show by differentiation that $\text{Arctan } x = \text{Arcctn } 1/x$.

16.9 PROBABILITY PROBLEMS

We are now in a position to examine a number of interesting points that arise in doing probability problems.

Example 16.9-1

A particle is moving in a sinusoidal motion. What is the probability density distribution for finding the particle in the range -1 to 1? It is no restriction if we assume that the motion is given by

$$y = \sin t$$

We see immediately that a random sample from the range $-\pi/2 \le t \le \pi/2$ will be a typical range in which to sample and will give the correct answer. We are selecting a random time (uniformly) in the interval, so we have the probability of falling in the interval dt of

$$\frac{dt}{\pi}$$

But

$$t = \text{Arcsin } y$$

and therefore

$$\frac{dt}{dy} = \frac{1}{\sqrt{1 - y^2}}$$

From this we get

$$\frac{dt}{\pi} = \frac{1}{\pi} \frac{dy}{\sqrt{1 - y^2}}$$

The probability density distribution in y is given by

$$p(y) = \frac{1}{\pi\sqrt{1 - y^2}}$$

If this seems strange then you have only to look at Figure 16.1-12 to see that most of the time the particle is near the ends of the range and that it rushes through the middle of the interval. If you are shooting at this target, then aiming at the middle (mean) position is poor policy.

Case Study 16.9-2

A paradox. Lewis Carroll (1832–1898), author of *Alice in Wonderland* and other children's stories, was an English mathematician whose real name was Charles Dodgson. He proposed the following problem. Pick three points at random in the plane and form a triangle. What is the probability that the triangle will be obtuse?

His solution was as follows. We pick the coordinate system with the x-axis along the longest side of the triangle with the origin at the end having the shortest side and the length of the unit chosen as the longest side. Figure 16.9-1 shows the sample space and the region of obtuse triangles. It is a matter of computing the two areas and then taking their ratio. For the whole region we have

$$2 \int_{1/2}^{1} y \, dx = 2 \int_{1/2}^{1} \sqrt{1 - x^2} \, dx$$

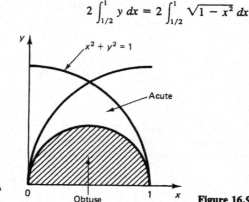

Figure 16.9-1 Base as longest side

Set $x = \sin \theta$; then $dx = \cos \theta \, d\theta$, and the area is

$$2 \int_{\pi/6}^{\pi/2} \cos^2 \theta \, d\theta = \frac{\pi}{3} - \frac{\sqrt{3}}{4}$$

The region of obtuse triangles is a semicircle of radius $1/2$, so the region of successes has an area of $\pi/8$. The ratio is

$$\frac{\pi}{8} \Big/ \left(\frac{\pi}{3} - \frac{\sqrt{3}}{4}\right) = 0.639 \ldots$$

Suppose we doubt this solution, and chose the coordinate system along the second longest side, with the origin at the vertex with the shortest side. Figure 16.9-2 shows the situation. The other vertex must lie inside the unit circle about the origin and outside the unit circle drawn about the point $x = 1$. Only to the left of the vertical line through the origin are the triangles obtuse. With slight trouble, we get $\pi/4$ for the area of the quadrant. For the area of the small sector, we have

$$\int_0^{1/2} \{\sqrt{1 - x^2} - \sqrt{1 - (1 - x)^2}\}\, dx$$

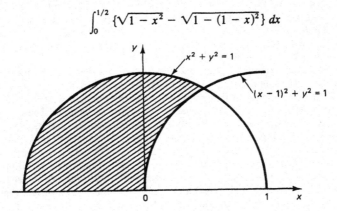

Figure 16.9-2   Second longest side

In the first integral we use $x = \sin\theta$, and in the second we use $1 - x = \sin\theta$. We get

$$\int_0^{\pi/6} \cos^2\theta\, d\theta - \int_{\pi/6}^{\pi/2} \cos^2\theta\, d\theta$$

In each we use the half-angle substitution to get

$$\frac{\theta}{2} + \frac{\sin 2\theta}{4}\Bigg|_0^{\pi/6} - \frac{\theta}{2} - \frac{\sin 2\theta}{4}\Bigg|_{\pi/6}^{\pi/2} = \frac{\sqrt{3}}{4} - \frac{\pi}{12}$$

which gives, finally,

$$\frac{\text{area success}}{\text{total admissible area}} = \frac{\pi/4}{\pi/6 + \sqrt{3}/4} = 0.821 \ldots$$

What is wrong? The first objection that might be raised is the choice of the different coordinate systems. But when you think about it, we assumed that area was independent of coordinate translation and rotation, and you believe that you could find the areas using any coordinate system, although it might be a lot more trouble to do it in some peculiar coordinate systems than in others. We believe that areas are independent of the translation and rotation of the coordinate system, and the scaling does not matter since we are taking ratios of the areas. No, the error is not in the coordinate system rotation and translation.

Maybe the calculus is wrong. But an inspection of the figures shows that the ratios of the computed areas are about correct, and the difference is too large to be explained that way.

We are reduced to thinking hard. As usual, let us go back to a simple problem with the infinite in it. Suppose I ask you to pick a random positive integer. What is its expected value? It is infinite! I mean by those words that you can name any fixed number n you want, no matter how large, and agree that the words "almost all" will be any fraction less than 1 that you care to name; then almost all positive integers are larger than the number n you named. The same sort of argument will apply to choosing any point along a line.

Next we ask, What is the distance between any two random points on an infinite line? The answer is "infinite," if we similarly understand the word to mean larger than any preassigned number. So when we scaled the units by the length of the longest side, or the second longest side, what were we doing?

The assignment of a uniform probability distribution to the infinite plane is dubious to say the least. Suppose we try to approach the infinite by the usual finite case first. We take a square and imagine that we have solved the problem with a uniform probability assignment and have found the probability of an obtuse triangle. Then we let the side of the square go to infinity. The probability will not change. Next we take an inscribed circle (which cuts off the corners of the square), again with a uniform probability assignment. The probability of an obtuse triangle in the circle will not be the same! We will get a different value for the infinite plane when we let the radius go to infinity. As a more extreme case, consider taking a long, low rectangle, which favors obtuse triangles. Then let the rectangle go to infinity while keeping its shape. Still a different answer.

We begin to realize that no matter how seductive the words may sound, or how many famous people may have fallen into the trap before us, the fact is that we often do not know when a sentence is meaningful and when it is not. The random triangle in the plane seemed reasonable when we started, and apparently seemed reasonable to generations of mathematicians, but now we do not believe that the words say anything definite. Too much more must be said before we think that the problem is well defined. We have a rising standard of rigor, and we probably have not reached the ultimate of rigor; so we need always to keep in mind that maybe we are talking nonsense no matter how fancy the mathematics we use may be.

Example 16.9-3

The Cauchy distribution, or improper integrals revisited. The Cauchy probability distribution density is given by

$$p(x) = \frac{1}{\pi(1 + x^2)} \qquad (-\infty, \infty)$$

The total probability is

$$\int_{-\infty}^{\infty} p(x)\, dx = \frac{1}{\pi} \int_{-\infty}^{\infty} \frac{dx}{1 + x^2}$$

$$= \frac{1}{\pi} \operatorname{Arctan} x \Big|_{-\infty}^{\infty} = 1$$

Nothing was said about how we went from the finite limit to the infinite limit; the upper and lower limits were done separately (in our heads).

For probability distributions, the convergence of the absolute value of the integral of the random variable is required. For the first moment, this means

$$\int_{-\infty}^{\infty} |x|\, p(x)\, dx$$

must be finite. For the Cauchy distribution, this gives

$$\frac{1}{\pi} \int_{-\infty}^{\infty} \frac{|x|}{1 + x^2}\, dx = \frac{2}{\pi} \int_0^{\infty} \frac{x}{1 + x^2}\, dx$$

$$= \frac{2}{\pi} \ln (1 + x^2) \Big|_0^{\infty}$$

and the first moment does not exist no matter whether we let the two limits go to infinity together or separately.

If we ignore the absolute value in the first moment (the expected value) of the Cauchy distribution, then we have

$$E\{X\} = \frac{1}{\pi} \int_{-\infty}^{\infty} \frac{x}{1 + x^2}\, dx$$

$$= \frac{1}{2\pi} \ln (1 + x^2) \Big|_{-\infty}^{\infty}$$

We need to pause and consider matters because the direct substitution would give an infinity at each limit.

If we were to use the limits $-a$ and a and then let $a \to \infty$, we would have

$$\lim \left\{ \frac{1}{2\pi} \ln \frac{1 + a^2}{1 + a^2} \right\} = \frac{1}{2\pi} \lim \{\ln 1\} = 0$$

This is a reasonable answer since the integrand is odd and the interval of integration is even. However, there is a bit of a worry nevertheless, and we look further.

If we let the upper and lower limits of integration approach infinity *separately*, we would have at the finite stage

$$\frac{1}{2\pi} \ln \left\{ \frac{1 + b^2}{1 + a^2} \right\}$$

where b is the finite upper limit and $-a$ is the finite lower limit. Now if we suppose that b is proportional to a, say $b = ka$, then we have, on dividing numerator and denominator by a^2,

$$\frac{1}{2\pi} \ln \left\{ \frac{1/a^2 + k^2}{1/a^2 + 1} \right\}$$

and this approaches, as $a \to \infty$,

$$\frac{1}{2\pi} \ln k^2 = \frac{\ln k}{\pi}$$

We can get any limit we please for the value of the expected value of X.

We face a decision; shall we demand using independent limits for the general improper integral when both ends require attention, or shall we permit the simultaneous, equal approach to the limit? The conventional decision is to demand independent limits, but in some fields of mathematics the simultaneous approach is used and is called the *Cauchy principal value*. It is foolish to make up the definitions and rules of mathematics without an understanding of the consequences that will follow in your field of interest.

In this case it is known that the Cauchy distribution has peculiar properties. For example, if we wish to estimate the "center" of the distribution

$$p(x) = \frac{1}{\pi[1 + (a - x)^2]}$$

(estimate a), then the average of n observations is no better than one observation! This result is certainly nonintuitive.

We do not want to commit ourselves universally on what to do in all cases when you have both limits of an integral approaching infinity, one plus and one minus infinity; nor do we want to say that always the absolute value of the random variable must converge, since it is conceivable that in some circumstance you might have an ordering of the random variable, but it seems highly unlikely! You have been warned again of the dangers of not thinking carefully about meanings behind the rules and formulas of mathematics. New applications often require new interpretations of old things. To be ready to create the new things, it is necessary that you understand not only the formal results of past mathematics, but also what assumptions were made, why they were made, and how the assumptions were combined; otherwise, you will be trapped in past thinking. Neither the applications nor formal mathematics is the final judge of definitions and assumptions you make; the decision to make is a blend of the two, involving both what you need and what you can do.

16.10 SUMMARY OF THE INTEGRATION FORMULAS

It is worth assembling the useful integration formulas in one place for easy reference. We make small changes in the notation for convenience.

1. $\int (au + bv - cw)\, dx = a \int u\, dx + b \int v\, dx - c \int w\, dx$
2. $\int U\, dV = UV - \int V\, dU$
3. $\int f(x)\, dx = \int f(g(t))\, g'(t)\, dt$

4. $\int u^n \, du = \dfrac{u^{n+1}}{n+1} + C$

5. $\int \dfrac{1}{u} \, du = \ln |u| + C$

6. $\int e^u \, du = e^u + C$

7. $\int \sin u \, du = -\cos u + C$

8. $\int \cos u \, du = \sin u + C$

9. $\int \tan u \, du = \ln |\sec u| + C$

10. $\int \operatorname{ctn} u \, du = \ln |\sin u| + C$

11. $\int \sec u \, du = \ln |\sec u + \tan u| + C$

12. $\int \csc u \, du = -\ln |\csc u + \operatorname{ctn} u| + C = \ln \left| \tan \dfrac{u}{2} \right| + C$

13. $\int \sec^2 u \, du = \tan u + C$

14. $\int \csc^2 u \, du = -\operatorname{ctn} u + C$

15. $\int \sec u \tan u \, du = \sec u + C$

16. $\int \csc u \operatorname{ctn} u \, du = -\csc u + C$

17. $\int \dfrac{1}{a^2 + u^2} \, du = \dfrac{1}{a} \arctan \dfrac{u}{a} + C$

18. $\int \dfrac{1}{(a^2 - u^2)^{1/2}} \, du = \arcsin \dfrac{u}{a} + C$

19. $\int \dfrac{1}{u^2 - a^2} \, du = \dfrac{1}{2a} \ln \left| \dfrac{u-a}{u+a} \right| + C$

20. $\int \dfrac{1}{a^2 - u^2} \, du = \dfrac{1}{2a} \ln \left| \dfrac{u+a}{u-a} \right| + C$

21. $\int \dfrac{1}{(u^2 + a^2)^{1/2}} \, du = \ln |u + \sqrt{u^2 + a^2}| + C$

22. $\int (a^2 - u^2)^{1/2} \, du = \dfrac{u}{2} \sqrt{a^2 - u^2} + \dfrac{a^2}{2} \arcsin \dfrac{u}{a} + C$

23. $\int (u^2 \pm a^2)^{1/2} \, du = \dfrac{u}{2} \sqrt{u^2 \pm a^2} + \dfrac{a^2}{2} \ln (u + \sqrt{u^2 \pm a^2}) + C$

The last five integrals have been added to round out the general forms and are easily derived.

EXERCISES 16.10

Derive the following formulas from Section 16.10.

1. Formula 19
2. Formula 20

3. Formula 21
4. Formula 22
5. Formula 23

Integrate the following:

6. $\int 5x/\sqrt{5 - 2x^2}\, dx$
7. $\int 5x/(5 - 2x^2)\, dx$
8. $\int x \cos 3x\, dx$
9. $\int (x + 3)/(x^2 + 4x + 3)\, dx$
10. $\int 1/(x^2 - 4x + 4)\, dx$
11. $\int (e^{2x} - 2 + e^{-2x})\, dx$
12. $\int 1/(e^x + e^{-x})\, dx$
13. $\int \ln (1 + \sqrt{x})\, dx$
14. $\int x^3/(x^2 + 1)\, dx$
15. $\int x^3/\sqrt{(x^2 + 1)}\, dx$
16. $\int e^{2x} \sin 3x\, dx$
17. $\int \sin^4 (y/2)\, dy$
18. $\int \{(\arcsin x)/(1 - x^2)\}^{1/2}\, dx$
19. $\int (1 + \tan t)^3\, dt$
20 $\int (\sin \theta)/(1 - \cos \theta)^3\, d\theta$
21. $\int e^{-t} \cos at\, dt$
22. $\int \sin 2\theta \sin 3\theta\, d\theta$
23. $\int 1/\sqrt{a^2 - x^2}\, dx$
24. $\int x/\sqrt{a^2 - x^2}\, dx$
25. $\int x^2/\sqrt{a^2 - x^2}\, dx$
26. $\int x \ln (1 - x)\, dx$
27. $\int \arcsin \sqrt{x}\, dx$
28. $\int x^2 e^{-x}\, dx$
29. $\int \cos t \cos 2t\, dt$
30. $\int x^2/\sqrt{x^2 + a^2}\, dx$
31. $\int (\sqrt{x} - 1/\sqrt{x})^2\, dx$
32. $\int \sin^3 x\, dx$
33. $\int_0^{\pi/2} \cos^4 x\, dx$
34. $\int_0^\infty \dfrac{dx}{a^2 + x^2}$
35. $\int_0^\infty xe^{-x}\, dx$
36. $\int_{-\infty}^\infty x^3 \exp (-x^2/2)\, dx$
37. $\int_0^\infty \dfrac{x}{(a^2 + x^2)^2}\, dx$

38. $\int_0^1 \left(\sqrt{x} + \frac{1}{\sqrt{x}} \right) dx$

39. $\int_0^1 x \ln \left(\frac{1}{x} \right) dx$

6.11 SUMMARY

We first reviewed, in a highly condensed form, the elements of trigonometry. We then examined the problem of finding the derivative of $y = \sin x$. This was first based on an important limit $(\sin \theta)/\theta$ as $\theta \to 0$, but later we found the derivative directly from the fundamental theorem of the calculus. In both cases it was necessary to use radian measure to get a formula free of extraneous constants.

We then examined the derivatives and integrals of the various trigonometric functions, as well as their inverses. The inverses all came from the integration of albebraic expressions and show the inevitability of the inverse trigonometric functions arising in many problems.

The next chapter will complete, for a beginning course, the class of functions whose derivatives and integrals you need to know. As a result, you will be ready to tackle a large class of applications of mathematics. We will, at last, have disposed of the more formal details of much of advanced mathematics. It is unfortunate, in a way, that such routine material must be learned before you can get to the interesting, useful applications of mathematics, but there seems to be no way around mastering a minimum of technique if you are to use mathematics, whether to create more mathematics or apply it to other fields.

17

Formal Integration

17.1 PURPOSE OF THIS CHAPTER

The purpose of this chapter is to investigate the systematic integration of functions. You have seen some samples of the variety of functions you can integrate, and perhaps suspect that there are many you cannot integrate. You are right in this assumption; in a sense it is unlikely that a truly random integrand can be integrated in any finite form. However, a surprising number of the integrals that arise in practice can be done in closed form.

If you write any finite expression (that has a derivative) and differentiate it, then you will have the corresponding integration problem where the answer can be written in closed form. Thus there are potentially an infinite number of different integrals that you can do, and it is necessary to surmount the sea of special cases to obtain a larger look at the topic of integration. Unfortunately, since integration is an inverse process, there is no simple answer to the problem of integration; but some general observations can be made.

You may, of course, think that an extensive table of integrals will suffice to meet your needs, but the next two sections should convince you that there are integrands you could handle, but that are not likely to be in any integral table, no matter how extensive. Furthermore, although you may not like it, there are occasional errors in the integral tables. The obvious check that differentiation gives the original integrand usually works, but there are often subtle things to watch for. For example, the obvious

$$\int_{-3}^{3} \left(\frac{1}{x}\right) dx = \ln|x| \Big|_{-3}^{3} = 0$$

simply ignores the fact that the ln function does not give the correct answer; you have either to stay with the plus or the minus branch and not jump over the infinite value of the integrand at $x = 0$. In an important problem you should go through all the details to convince yourself (and others) of the validity of the result before you risk your reputation on it. Finally, it has often been observed that those who cannot integrate without a table cannot integrate with a table.

A table of integrals is very convenient to have handy, and a small table has been included in Appendix A. Larger tables such as [G–R] are available. The table can often suggest things to do even when your particular integral is not there.

The tools for integration are the ones listed in Section 16.6: (1) rearrangement of the integrand, (2) recognition of the form of the integrand using a simple change of variables, (3) integration by parts, and (4) significant changes of variable. We will examine these in turn.

We are concerned in this chapter with the systematic integration of classes of integrands. You already know that you can integrate polynomials and sums of powers of the variable, integral or not. We next turn to the class of rational functions in x. The method is based on the rearrangement of the integrand using the method of *partial fractions* as the main tool.

17.2 PARTIAL FRACTIONS—LINEAR FACTORS

We have on several occasions, for example in Section 14.5, broken up an expression into a sum of terms. Now we examine the method when applied to rational functions.

Example 17.2-1

Consider the following expression:

$$\frac{x + 2}{x^2 - 5x + 6}$$

The denominator can be factored into two factors, $x - 3$ and $x - 2$. This suggests that the expression might be written in the form, with unknown coefficients A and B,

$$\frac{x + 2}{x^2 - 5x + 6} = \frac{A}{x - 2} + \frac{B}{x - 3}$$

since when the two terms on the right are added they would lead to the denominator on the left.

The next step is to clear of fractions by multiplying both sides by the original denominator. We get

$$x + 2 = A(x - 3) + B(x - 2)$$

This is an identity in x; and since the powers of x are *linearly independent* (Section 4.7), we may equate the like powers on both sides (x terms and constant terms).

$$1 = A + B$$

$$2 = -3A - 2B$$

To solve these equations, we multiply the top equation by 3 and add both equations:

$$5 = B$$

To eliminate the B, we multiply the top equation by 2, and add.

$$4 = -A$$

Hence we have

$$\frac{x + 2}{x^2 - 5x + 6} = \frac{-4}{x - 2} + \frac{5}{x - 3}$$

This is easily checked by adding the two terms on the right. If we are presented with the original expression as an integrand, then we can easily integrate the equivalent form on the right.

$$\int \frac{x + 2}{x^2 - 5 + 6}\, dx = -4 \ln|x - 2| + 5 \ln|x - 3| + C$$

Example 17.2-2

Consider the following expression:

$$\frac{x^2 - 4x + 4}{x(x - 1)(x - 2)}$$

An examination of the denominator suggests that the expression can be written in the form

$$\frac{A}{x} + \frac{B}{x - 1} + \frac{C}{x - 2}$$

because the fractions, when they are added, have the common denominator of the original expression. The numerators have the *undetermined coefficients* A, B, and C. We simply equate the two expressions and then multiply through by the common denominator $x(x - 1)(x - 2)$ to get

$$x^2 - 4x + 4 = A(x - 1)(x - 2) + Bx(x - 2) + Cx(x - 1)$$

$$= (A + B + C)x^2 + (-3A - 2B - C)x + 2A$$

This is an identity in the variable x, and hence the coefficients of each power can be equated. Thus we have the three equations

$$1 = A + B + C$$

$$4 = 3A + 2B + C$$

$$4 = 2A$$

These equations determine the undetermined coefficients of the *partial fraction* expansion of the original rational function. From the third equation, we get immediately $A = 2$. The other two equations are now

$$B + C = -1$$

$$2B + C = -2$$

Subtract the top equation from the bottom equation to get $B = -1$, and hence $C = 0$. Therefore,

$$\frac{x^2 - 4x + 4}{x(x - 1)(x - 2)} = \frac{2}{x} - \frac{1}{x - 1}$$

We notice there is no term with the denominator $(x - 2)$. A careful look shows the original numerator was $(x - 2)^2$, and hence the factor $x - 2$ can be canceled. We actually had the original expression

$$\frac{x - 2}{x(x - 1)}$$

Thus if we are careless we still get the right answer, but thereby have to do more work than necessary.

The method clearly requires the degree of the numerator to be less than the degree of the denominator. If this is not true, then you divide the numerator by the denominator until the remainder is of lower degree than the degree of the denominator (the same as you do with integers).

Example 17.2-3

Given

$$J = \int \frac{x^3 - 4x + 5}{x^2 - x - 2}\, dx$$

we divide out the integral part of the quotient as follows:

$$
\begin{array}{r}
x + 1 \\
x^2 - x - 2 \overline{)x^3 + 0x^2 - 4x + 5} \\
\underline{x^3 - x^2 - 2x} \\
x^2 - 2x + 5 \\
\underline{x^2 - x - 2} \\
- x + 7
\end{array}
$$

(This closely resembles the way you divide two decimal numbers. We have kept the powers of x in the same column, just as you keep the powers of 10 in a column when dividing numbers. The process may require supplying missing zero coefficients.) From this division we can write

$$\frac{x^3 - 4x + 5}{x^2 - x - 2} = x + 1 + \frac{-x + 7}{x^2 - x - 2}$$

and the integral part, $x + 1$, is easily integrated to $x^2/2 + x$, leaving the fraction whose numerator is of degree at least one less than the degree of the denominator. The fractional part is

$$\frac{7-x}{(x-2)(x+1)} = \frac{A}{x+1} + \frac{B}{x-2}$$

Repeating the earlier partial fraction process, we get, in turn,

$$7 - x = A(x-2) + B(x+1)$$
$$= (A+B)x + (-2A+B)$$

Hence, equating coefficients of like powers of x, we get

$$7 = -2A + B$$
$$-1 = A + B$$

from which $A = -8/3$, and $B = 5/3$. The partial fraction decomposition is

$$\frac{1}{3}\left[\frac{5}{x-2} - \frac{8}{x+1}\right]$$

The complete integral is, therefore,

$$J = \frac{x^2}{2} + x + \frac{5}{3}\ln|x-2| - \frac{8}{3}\ln|x+1| + C$$

Do not confuse the constant of integration C with an undetermined parameter C. This result is easily checked by differentiation and combining terms.

Generalization 17.2-4

We immediately *generalize* to the case $N(x)/D(x)$, where the degree of $N(x)$ is less than the degree of $D(x)$. *We assume that the denominator can be factored* into real linear and real quadratic factors. The fundamental theorem of algebra states that we can do this although in practice we may not know how to do it, especially when there are letters (parameters) in the denominator. Therefore, this is an assumption that the method always makes, but fortunately in practice this factorization can usually be done. The idea that a computer could be used to find the factors sufficiently accurately (the zeros of the denominator are equivalent to the factors) does not apply when there are parameters in $D(x)$.

The method of partial fractions is based on the linear independence of the powers of x and leads to a set of corresponding linear equations in the undetermined coefficients. This method (as in the previous examples) always works, whether or not there are repeated (multiple) factors. There will always be just the right number of equations to determine the coefficients.

We first look at the details of the case of all linear factors.

Example 17.2-5

Consider the fraction with a repeated linear factor

$$\frac{x^2 + x + 1}{x(x-2)^2}$$

We try the form

$$\frac{A}{x} + \frac{B}{x - 2} + \frac{C}{(x - 2)^2}$$

It is necessary to allow for *every* possible factor; thus repeated factors require a term of each possible degree. When we multiply through by the common denominator, we get

$$x^2 + x + 1 = A(x - 2)^2 + Bx(x - 2) + Cx$$

$$= (A + B)x^2 + (-4A - 2B + C)x + 4A$$

from which we get the three equations (the coefficients of x^2 and x and the constant terms)

$$1 = A + B$$

$$1 = -4A - 2B + C$$

$$1 = 4A$$

whose solution is $A = 1/4,$ $B = 3/4$, and $C = 14/4$. The integral of the original integrand is therefore

$$\frac{1}{4}\left[\ln|x| + 3\ln|x - 2| - \frac{14}{x - 2}\right] + C$$

Note that the simple factors lead to logarithms and that the multiple linear factors lead to algebraic terms.

The method we used requires the solution of a set of simultaneous linear equations, which, for denominators of high degree, can be tedious. There is a simpler, *alternative approach* for solving them by use of special values of x. This method gets most, or all, of the coefficients directly.

Example 17.2-6

Given the expression

$$\frac{N(x)}{D(x)} = \frac{x^2 + x + 3}{x(x - 1)(x - 2)}$$

We try the partial fraction form

$$\frac{x^2 + x + 3}{x(x - 1)(x - 2)} = \frac{A}{x} + \frac{B}{x - 1} + \frac{C}{x - 2}$$

Multiply through by the denominator $D(x)$ to get the indentity

$$N(x) = (x^2 + x + 3) = A(x - 1)(x - 2) + Bx(x - 2) + Cx(x - 1)$$

If we now set $x = 0$ (the zero from the first fraction), we get

$$N(0) = 3 = A(-1)(-2) = 2A \to A = \frac{3}{2}$$

If we set $x = 1$ (the zero from the second fraction), we get

$$N(1) = 5 = B(-1) \to B = -5$$

Finally, we set $x = 2$ (from the third fraction) to get

$$N(2) = 9 = C(2)(1) = 2C \rightarrow C = \frac{9}{2}$$

Thus, without solving a set of simultaneous equations, we have the integral

$$\int \frac{x^2 + x + 1}{x(x-1)(x-2)} dx = \frac{3}{2} \int \frac{dx}{x} - 5 \int \frac{dx}{x-1} + \frac{9}{2} \int \frac{dx}{x-2}$$

$$= \frac{3}{2} \ln |x| - 5 \ln |x-1| + \frac{9}{2} \ln |x-2| + C$$

For a simple first-order linear factor, this method of evaluating the coefficients by using the zeros of the denominator works and produces the coefficients of the corresponding terms one by one. The rule is (once the degree of the numerator has been reduced to less than the degree of the denominator): set up the *partial fraction* decomposition of the given rational fraction, multiply through by the denominator, and then evaluate both sides at each of the zeros of the denominator. Each result is the value of the corresponding coefficient.

What about the multiple factors such as in Example 17.2-5 with the denominator $x(x-2)^2$? We will take a third power so that the general case is obvious from the example.

Example 17.2-7

Integrate

$$\int \frac{x+1}{x(x-2)^3} dx$$

The partial fraction decomposition to try is

$$\frac{x+1}{x(x-2)^2} = \frac{A}{x} + \frac{B}{x-2} + \frac{C}{(x-2)^2} + \frac{D}{(x-2)^3}$$

Multiply through by the denominator:

$$x + 1 = A(x-2)^3 + Bx(x-2)^2 + Cx(x-2) + Dx \qquad (17.2\text{-}1)$$

To get A, set $x = 0$ (the corresponding zero):

$$1 = A(-2)^3 = -8A \rightarrow A = \frac{-1}{8}$$

We can get the coefficient D by setting $x = 2$.

$$N(2) = 3 = 2D \rightarrow D = \frac{3}{2}$$

There are many ways of getting B and C. One of the simplest is to return to the method of equating like terms on both sides.

$$x + 1 = A(x^3 - 6x^2 + 12x - 8) + B(x^3 - 4x^2 + 4x) + C(x^2 - 2x) + Dx$$

We will need only two equations since only two undetermined parameters remain. Using the third-degree terms will involve only the unknown B:

$$0 = \frac{-1}{8} + B \rightarrow B = \frac{1}{8}$$

Then, using the second-degree terms, we have

$$0 = \frac{6}{8} - \frac{4}{8} + C \rightarrow C = \frac{-1}{4}$$

We see that in this case there is a way of picking the powers so that each equation has one new unknown (a triangular system of equations), and the system that we have to solve is easy. For several different multiple zeros, the problem is a bit more complex but is greatly reduced by an adroit combination of the use of zeros and the equating of like powers.

The integral is, therefore,

$$\int \frac{x+1}{x(x-2)^2}\, dx = \frac{1}{8} \int \left[\frac{-1}{x} + \frac{1}{x-2} - \frac{2}{(x-2)^2} + \frac{12}{(x-2)^3} \right] dx$$

$$= \frac{1}{8} \left[-\ln|x| + \ln|x-2| + \frac{2}{x-2} - \frac{6}{(x-2)^2} \right] + C$$

A third, *alternative* method for handling multiple zeros is simply to differentiate, one or more times, the identity (17.2-1) that you used in equating coefficients. Then, when you put the value of the multiple root into each successive derivative, you will find that you get one more coefficient determined. Thus you can get the coefficients one at a time by this method. The first differentiation gives

$$1 = 3A(x-2)^2 + 2Bx(x-2) + B(x-2)^2 + Cx + C(x-2) + D$$

At $x = 2$, this gives

$$1 = 2C + D = 2C + \frac{3}{2} \rightarrow C = \frac{-1}{4}$$

A second differentiation gives

$$0 = 6A(x-2) + 2Bx + 2B(x-2) + 2B(x-2) + C + C$$

and for $x = 2$ we get

$$0 = 4B + 2C \rightarrow B = \frac{-C}{2} = \frac{1}{8}$$

as before.

The basic method of partial fractions for linear (first degree) factors of the denominator is clear; you use the zeros of the denominator to get the coefficients of the highest power of the corresponding factor. Using the method of equating like powers of the variable on both sides gives equations that determine the missing coefficients that arise from multiple factors. As an alternative, you differentiate the identity and then substitute in the value of the mulitple root.

EXERCISES 17.2

Find the partial fraction decomposition of the following:

1. $(x - 1)/(x^2 + 5x + 6)$
2. $x/(x^2 + 3x + 2)^2$
3. $x^2/(x + 1)(x + 2)(x + 3)$
4. $x/(x + a)^2$
5. $1/x(x + a)^3$
6. $1/x^2(x + b)^2$

17.3 QUADRATIC FACTORS

The theory of quadratic factors is slightly more complicated, but the same main ideas apply. From Section 4.8, which discusses the complex roots of a quadratic equation, you know that either there are real roots, and hence linear factors, or else the roots are complex conjugates of each other and we have a corresponding real quadratic factor. Note that in the method of partial fractions, after multiplying through by the denominator, we have to evaluate the numerator $N(x)$ at the zeros of the corresponding factor we are using. This suggests that we digress and look at the general problem of evaluating a polynomial of degree n,

$$P_n(x)$$

at the zeros of a quadratic (or other degree equation), which we take in the form

$$x^2 - px - q = 0$$

to resemble the linear factor $x - a$. Let a be a root of this quadratic equation. We want to find $P_n(a)$. All we know about a is contained in the equation

$$a^2 = pa + q$$

If we multiply this equation by a^{n-2}, we have an expression of a^n in terms of lower powers of a. In this way each higher power of a can be reduced to lower powers until we have only a first power and a constant. The easiest method of doing this reduction is to divide the polynomial by the quadratic factor to get a quotient, labeled $P_{n-2}(x)$, and a remainder $r_1x + r_0$. Thus we have

$$P_n(x) = P_{n-2}(x)(x^2 - px - q) + r_1x + r_0$$

At a root of the quadratic, we have clearly

$$P_n(a) = r_1a + r_0$$

which closely resembles the remainder theorem of Section 4.7.

Example 17.3-1

If

$$P_4(x) = x^4 + 2x^3 + 3x^2 + 4x + 5$$

$$Q(x) = x^2 - x - 2$$

then, using detached coefficients (just as when writing numbers, which are actually sums of powers of 10), we merely keep track of the positions and arrange each column to have the same power of x. We get

$$
\begin{array}{r}
 1 \quad\;\; 3 \quad\;\; 8 \\
1\;\; -1\;\; -2\,)\;\; 1 \quad 2 \quad 3 \quad 4 \quad 5 \\
\underline{1 \quad -1 \quad -2 } \\
3 \quad\;\; 5 \quad\;\; 4 \\
\underline{3 \quad -3 \quad -6 } \\
8 \quad 10 \quad\;\; 5 \\
\underline{8 \quad -8 \quad -16} \\
18 \quad 21
\end{array}
$$

In the mathematician's standard form, this is

$$P_4(x) = (x^2 + 3x + 8)(x^2 - x - 2) + 18x + 21$$

If we write either root of $Q(x) = 0$ as a, then

$$P_4(a) = 18a + 21$$

we now solve the quadratic $Q(x) = 0$ for the two values of a and have the answer.

In general, we have the result (a remainder theorem, Section 4.7)

$$P_n(a) = r_1 a + r_0$$

This method of evaluating a polynomial at a root of another polynomial avoids all the awkward square roots and complex numbers until the very last step, and hence is very useful in practice. This *method* of reducing high powers of a root of a quadratic (or other polynomial) to lower powers is used in such diverse fields as electrical network theory and coding theory (to design error-correcting codes suitable for signaling through noisy channels).

Example 17.3-2

Integrate

$$\int \frac{x^3 + 1}{(x^2 + 1)(x^2 + x + 1)}\, dx$$

The degree of the numerator is less than the degree of the denominator, so we write (note that the degree of the undetermined numerators of the quadratic factors is 1, and there are two coefficients in each)

$$\frac{x^3 + 1}{(x^2 + 1)(x^2 + x + 1)} = \frac{Ax + B}{x^2 + 1} + \frac{Cx + D}{x^2 + x + 1}$$

In the usual method, we multiply through by the denominator:

$$x^3 + 1 = (x^2 + x + 1)(Ax + B) + (x^2 + 1)(Cx + D)$$

In this simple problem, we can equate coefficients of like powers and solve the resulting four equations easily. We get at the first stage

$$\begin{aligned}
x^3 + 1 = [Ax^3 &+ Ax^2 + Ax \\
&+ Bx^2 + Bx + B \\
Cx^3 \quad &\quad\quad + Cx \\
&+ Dx^2 \quad\quad + D]
\end{aligned}$$

Next, equating the coefficients of like terms, we get the four equations

(1) $1 = A \quad\quad + C$

(2) $0 = A + B \quad\quad + D$

(3) $0 = A + B + C$

(4) $1 = \quad\quad B \quad + D$

Subtract equation (4) from (2): → $A = -1$. From (1): → $C = 2$. From (3): → $B = -1$. Finally, from (4): → $D = 2$. Hence

$$\int = \int \left[-\frac{x + 1}{x^2 + 1} + \frac{2x + 2}{x^2 + x + 1} \right] dx$$

We rewrite the first term in the integrand in the form

$$\int \left[\frac{-x}{x^2 + 1} - \frac{1}{x^2 + 1} \right] dx = -\frac{1}{2} \ln (x^2 + 1) - \text{Arctan } x$$

The second term in the integral can then be written in the form

$$\frac{2x + 2}{x^2 + x + 1} = \frac{2x + 1}{x^2 + x + 1} + \frac{1}{x^2 + x + 1}$$

The first of these integrates into

$$\ln (x^2 + x + 1)$$

The second requires completing the square (Section 6.3).

$$x^2 + x + 1 = \left(x + \frac{1}{2} \right)^2 + \frac{3}{4}$$

whose integral is

$$\frac{2}{\sqrt{3}} \text{Arctan} \left[\frac{x + 1/2}{\sqrt{3}/2} \right] = \frac{2}{\sqrt{3}} \text{Arctan} \left[\frac{2x + 1}{\sqrt{3}} \right]$$

Assembling all the terms, we have

$$\int = -\frac{1}{2} \ln (x^2 + 1) - \text{Arctan } x + \ln (x^2 + x + 1) + \frac{2}{\sqrt{3}} \text{ Arctan } \frac{2x + 1}{\sqrt{3}} + C$$

To illustrate the shorter, alternative method of getting the coefficients of a partial fraction decomposition (using the same integrand), we begin with $a^2 + a + 1 = 0$ (a is a zero of this factor). This is the same as

$$a^2 = -a - 1$$

We reduce $a^3 + 1$ (without the formal dividing out, which is mainly useful in larger problems).

$$a^3 + 1 = a(a^2) + 1 = a(-a - 1) + 1$$
$$= -a^2 - a + 1 = (a + 1) - a + 1 = 2$$

Next we reduce

$$a^2 + 1 = (-a - 1) + 1 = -a$$

Hence, we have for the partial fraction expansion, after multiplying by the denominator,

$$2 = (Ca + D)(-a) = -Ca^2 - aD = C(a + 1) - aD$$
$$= (C - D) a + C$$

from which, since 1 and a are linearly independent (a is a complex number), $C = 2$ and $D = C = 2$. The process for the other factor begins with the choice $a^2 + 1 = 0$ or

$$a^2 = -1$$

Then for this root a,

$$a^3 + 1 = a(-1) + 1 = -a + 1$$

and

$$a^2 + a + 1 = (-1) + a + 1 = a$$

Hence

$$-a + 1 = (Aa + B)a = Aa^2 - Ba = -A + Ba$$

from which we get

$$A = -1 \quad \text{and} \quad B = -1$$

The rest is the same. It is only in complicated problems that the power-reducing method is more efficient; but the concept is of significant value and widely useful.

The case of multiple quadratic factors is merely messy and *involves no new ideas* (we presented things carefully so this would be true). We will merely outline the steps. But first we summarize the method of partial fractions.

Summary: The method of partial fractions requires first dividing the numerator by the denominator until the degree of the numerator is less than the degree of the denominator. We then *assume* that the denominator can be factored into real linear and real quadratic factors. Next, write out a form with undetermined coefficients, one term for each possible degree that can arise from multiple factors, and with a linear form in the numerators of the quadratic terms. Then multiply through by the denominator. We then determine the coefficients either (1) by the slow but sure method of equating like coefficients, (2) by the proper choice of various values, typically zeros of the factors, or (3) by differentiation of the polynomial identity and then evaluation of the equation at the proper value. In every case we are led to terms of the forms

$$\frac{1}{(x - a)^n} \quad \text{and} \quad \frac{Ax + B}{(x^2 - px - q)^n}$$

where the discriminant of the quadratic is negative ($p^2 + 4q < 0$). For the first term, if $n \neq 1$, you get, upon integrating,

$$\frac{-1}{(n - 1)(x - a)^{n-1}}$$

and for $n = 1$ you get

$$\ln |x - a|$$

The quadratic terms require some manipulation. The numerator can be written in the form

$$Ax + B = \frac{A}{2}(2x - p) + \frac{pA}{2} + B$$

so that the first numerator term is a constant times the derivative of the denominator, and the integration of this term can be done; the case $n = 1$ leads to $\ln (x^2 - px - q)$ and otherwise to a suitable power, as in the case of the linear terms. It is the constant numerator term that requires attention. You need to complete the square (Section 6.3) and get the denominator in a form equivalent to

$$\frac{1}{(x^2 + a^2)^n}$$

The case $n = 1$ leads to an Arctan form $(1/a) \text{Arctan} (x/a)$. The remaining cases require a reduction formula that can be found as follows:

$$\frac{1}{(x^2 + a^2)^n} = \frac{1}{a^2}\left[\frac{a^2 + x^2 - x^2}{(x^2 + a^2)^n}\right]$$

$$= \frac{1}{a^2}\left[\frac{1}{(x^2 + a^2)^{n-1}} - \frac{x^2}{(x^2 + a^2)^n}\right]$$

We write

$$I_n = \int \frac{dx}{(x^2 + a^2)^n} = \frac{1}{a^2} I_{n-1} - \frac{1}{a^2} \int \frac{x^2}{(x^2 + a^2)^n} \, dx$$

In the integral we pick $U = x$ and $dV = x/(x^2 + a^2)^n \, dx$, thus reducing the integral to one lower index. The result is

$$I_n = \frac{1}{2(n-1)a^2} \left[\frac{x}{(x^2 + a^2)^{n-1}} + (2n - 3) I_{n-1} \right] \qquad (17.3\text{-}1)$$

Because rational functions occur frequently, it is important to understand that essentially they can be integrated. The "essentially" depends on your ability to factor the denominator. The presence of multiple zeros in the denominator can be detected and isolated by applying the Euclidean algorithm (Section 4.6) to the function and its derivative, since all multiple factors must appear in the greatest common factor of the function and its derivative (Section 9.8). Usually, problems that make use of partial fractions are comparatively simple, but now and then a real nasty one comes along and requires careful attention.

EXERCISES 17.3

Find the partial fraction decomposition of the following:

1. $x^3/(x^2 + 1)^2$
2. $x^3/(x^2 + 1)$
3. $x^2/(x^2 + a^2)$
4. $x^2/(a^2 + x^2)^2$
5. $x/(x^2 + a^2)^2$
6. $x/(x^2 + 1)(x^2 + 2)(x^2 + 3)$
7. $1/x(x^2 + a^2)^3$
8. $x^3/x^2(x + 1)^2$
9. Fill in the details of the last derivation in this section.
10.* Ostrogradski–Hermite method. The algebraic part of the integration of a rational function can be found by algebraic methods, leaving the simple real linear and quadratic factors to be found for the transcendental part. Write

$$\int \frac{N(x)}{D(x)} \, dx = \frac{P(x)}{Q(x)} + \int \frac{N_1(x)}{D_1(x)} \, dx$$

where $Q(x)$ is the GCD$\{D(x), D'(x)\}$(Section 4.6). Write $D(x) = Q(x)D_1(x)$ and $D'(x) = Q(x)T(x)$. Differentiate the equation and multiply through by $D(x)$ to get $N = D_1P' - PQ'D_1/Q + N_1Q$. Differentiate $D = QD_1$ to remove the indicated division by Q. In the resulting identity, $N = D_1P' - P(T - D_1') + N_1Q$, there are enough coefficients to determine both $P(x)$ and $N_1(x)$. Fill in the details.

17.4 RATIONAL FUNCTIONS IN SINE AND COSINE

The next large class of integrals that can be done is that of rational functions in sines and cosines. Note that a rational function in the trigonometric functions is a rational function in sine and cosine. For example,

$$\frac{\sin x}{\tan x + \cos x} = \frac{\sin x \cos x}{\sin x + \cos^2 x}$$

The following method shows how this type of integral is handled, but this does not mean that in any particular case the method is best; it is merely a guarantee that it can be done if you follow the method.

The method depends on a change of variable (which is not obvious until you see why it works). Given the function that is rational in $\sin x$ and $\cos x$, we make the tangent half-angle substitution

$$\tan \frac{x}{2} = z \qquad (17.4\text{-}1)$$

We must write $\sin x$, $\cos x$, and dx in terms of z. To do this, we must express each of these in terms of $\tan (x/2)$, using suitable trigonometric identities. We start with

$$\sin x = 2 \sin \frac{x}{2} \cos \frac{x}{2} = 2 \frac{\tan (x/2)}{\sec^2 (x/2)}$$

$$= 2 \frac{\tan (x/2)}{1 + \tan^2 (x/2)}$$

$$= \frac{2z}{1 + z^2} \qquad (17.4\text{-}2)$$

Next

$$\cos x = 2 \cos^2 \frac{x}{2} - 1 = \frac{2}{\sec^2 (x/2)} - 1$$

$$= \frac{2}{1 + z^2} - 1$$

$$= \frac{1 - z^2}{1 + z^2} \qquad (17.4\text{-}3)$$

Finally, to get the dx expression, we solve the transformation (17.4-1) for x:

$$x = 2 \text{ Arctan } z$$

from which it follows, by differentiation, that

$$dx = \frac{2 \, dz}{1 + z^2} \qquad (17.4\text{-}4)$$

Thus any integrand that is rational in sines and cosines will become a rational function in z when the substitutions (17.4-2, 17.4-3, and 17.4-4) are used.

When you finally have the answer, it is often necessary to convert back from the z variable to the original x variable. The tan $(x/2)$ is seldom a convenient form for the answer, so the following alternative substitutions are useful:

$$z = \tan \frac{x}{2} = \frac{\sin x}{1 + \cos x} = \frac{1 - \cos x}{\sin x} \qquad (17.4\text{-}5)$$

Again, this method assures us that we can do it, but often there are far simpler ways of integrating a particular case.

Example 17.4-1

Integrate

$$\int \cos x \, dx$$

which we know is $\sin x + C$. We now compare this with the proposed substitution method.

$$\int \cos x \, dx = \int \frac{1 - z^2}{1 + z^2} \frac{2 \, dz}{1 + z^2}$$

We need not apply the method of partial fractions directly to this integrand, since we can write

$$\frac{2(1 - z^2)}{(1 + z^2)^2} = \frac{2[2 - (1 + z^2)]}{(1 + z^2)^2}$$

$$= \frac{4}{(1 + z^2)^2} - \frac{2}{1 + z^2}$$

The first term, using the reduction formula (17.3-1), integrates into

$$4\left(\frac{1}{2}\right) \left\{ \frac{z}{1 + z^2} + \int \frac{dz}{1 + z^2} \right\}$$

and the integral exactly cancels the integral from the second term. Hence we have

$$\int = \frac{2z}{1 + z^2} + C = \sin x + C$$

as we must since the answer is unique. We see, however, that at times the tangent half-angle method is more laborious than necessary.

Example 17.4-2

Integrate

$$\int \frac{dx}{a + \sin x} \qquad (a > 1)$$

Using the substitutions (17.4-3) and (17.4-4), this becomes

$$\int \left\{ \frac{2/(1 + z^2)}{a + [2z/(1 + z^2)]} \right\} dz = 2 \int \frac{1}{az^2 + a + 2z} \, dz$$

$$= 2 \int \frac{1}{a[z^2 + (2/a)z + (1/a^2)] + [a - (1/a)]} \, dz$$

$$= \frac{2}{a} \int \frac{1}{[z + (1/a)]^2 + [1 - (1/a^2)]} \, dz$$

$$= \frac{2}{a} \left\{ \frac{1}{\sqrt{1 - (1/a^2)}} \text{ Arctan} \left[\frac{z + (1/a)}{\sqrt{1 - (1/a^2)}} \right] \right\} + C$$

$$= \frac{2}{\sqrt{a^2 - 1}} \text{ Arctan} \left(\frac{az + 1}{\sqrt{a^2 - 1}} \right) + C$$

$$= \frac{2}{\sqrt{a^2 - 1}} \text{ Arctan} \left[\frac{a \tan (x/2) + 1}{\sqrt{a^2 - 1}} \right] + C$$

The nature of the result (since the answer is unique within a constant C) shows that there cannot be some simple method of getting it; at times the tangent half-angle method is necessary.

Example 17.4-3

Using the tangent half-angle method, we again examine the integral

$$\int \sec x \, dx = \int \frac{1}{\cos x} \, dx = \int \frac{2 \, dz}{1 - z^2}$$

$$= \int \left\{ \frac{1}{1 - z} + \frac{1}{1 + z} \right\} dz$$

$$= \ln \left| \frac{1 + z}{1 - z} \right| + C$$

But using the fact that $\tan \pi/4 = 1$, we write this as

$$\int \sec x \, dx = \ln \left| \frac{\tan (\pi/4) + \tan (x/2)}{1 - \tan (\pi/4) \tan (x/2)} \right| + C$$

$$= \ln \tan \left| \frac{x}{2} + \frac{\pi}{4} \right| + C$$

This is an alternative form to the one we found in Section 16.6:

$$\int \sec x \, dx = \ln |\sec x + \tan x| + C$$

Case Study 17.4-4

The integral

$$I = \frac{1}{\pi} \int_0^\pi \frac{e^2 \sin^2 x}{(1 + e \cos x)^2} \, dx$$

occurs in quantum mechanics (among other places). Even before doing all the details of the method given in this section, we see that the partial fraction step may be messy. Is there anything else we could do? The sin x in the numerator goes with the cos x in the denominator, so we pick integration by parts:

$$U = e \sin x, \qquad dV = \frac{e \sin x}{(1 + e \cos x)^2} \, dx$$

$$dU = e \cos x \, dx, \qquad V = \frac{1}{1 + e \cos x}$$

Hence

$$I = \frac{1}{\pi} \frac{e \sin x}{1 + e \cos x} \Big|_0^\pi - \frac{1}{\pi} \int_0^\pi \frac{e \cos x}{1 + e \cos x} \, dx$$

The integrated piece vanishes at both ends. The new integral seems to be simpler than the one we started with, so we continue. Writing the numerator in the form

$$-e \cos x = 1 - (1 + e \cos x)$$

will further simplify the integral.

$$I = \frac{1}{\pi} \int_0^\pi \frac{dx}{1 + e \cos x} - \frac{1}{\pi} \int_0^\pi dx$$

and apparently we have made further progress. We now apply the method of this section to the first integral and integrate the second directly. The new limits of the first integral go to the range $(0, \infty)$ since the transformation is $z = \tan (x/2)$. We clear the denominator of fractions and get

$$I = \frac{1}{\pi} \int_0^\infty \frac{2}{1 + z^2 + e(1 - z^2)} \, dz - 1$$

$$= \frac{2}{\pi} \int_0^\infty \frac{1}{(1 + e) + z^2(1 - e)} \, dz - 1$$

$$= \frac{2}{\pi(1 - e)} \int_0^\infty \frac{1}{[(1 + e)/(1 - e)] + z^2} \, dz - 1$$

$$= \frac{2}{\pi(1 - e)} \frac{1}{\sqrt{(1 + e)/(1 - e)}} \, \text{Arctan} \, \sqrt{\frac{1 - e}{1 + e}} \, z \, \Big|_0^\infty - 1$$

$$= \frac{2}{\pi\sqrt{1 - e^2}} \left\{ \frac{\pi}{2} - 0 \right\} - 1$$

$$= \frac{1}{\sqrt{1 - e^2}} - 1$$

Does this answer seem reasonable? (1) It is positive and the original integrand was positive. (2) It approaches 0 as e approaches 0, as it should. (3) It is an even function of e as was the original integrand (since replacing e by $-e$ and changing the independent variable x to $\pi - x'$ leaves the integral the same). (4) As e approaches 1, the original integrand becomes strongly infinite as does the answer. Thus the answer seems reasonable.

This case study shows that often a number of different devices can be combined to get the answer. Reality is seldom as nice as are the typical exercises in a textbook.

EXERCISES 17.4

Integrate the following:

1. $\int 1/(a \sin x)\, dx$
2. $\int \cos x/(a + \cos x)\, dx, |a| < 1$
3. $\int \cos x/(a + \cos x)\, dx, |a| > 1$
4. $\int \sin x/(a + b \cos x)\, dx$

17.5 POWERS OF SINES AND COSINES

Powers of sines and cosines occur frequently. We saw the powers in the definite integrals called Wallis formulas (Section 16.7). For indefinite integrals, we need to examine the problem again. We consider the integrals

$$\int \sin^m x \cos^n x\, dx$$

If either (or both) m or n is an odd integer, regardless of the other; the use of the identity

$$\sin^2 \theta + \cos^2 \theta = 1$$

makes the problem immediately integrable.

Example 17.5-1

Integrate

$$\int \sin^5 x \cos^{3/2} x\, dx$$

We write

$$\sin^5 x = (1 - \cos^2 x)^2 \sin x$$
$$= (1 - 2 \cos^2 x + \cos^4 x) \sin x$$

and the integral is

$$\int (\cos^{3/2} x - 2 \cos^{7/2} x + \cos^{11/2} x) \sin x\, dx$$

$$= -\frac{2}{5} \cos^{5/2} x + \frac{4}{9} \cos^{9/2} x - \frac{2}{13} \cos^{13/2} x + C$$

For integer exponents, the main cases we need to examine further are those where both exponents are even integers. We have the following three identities from (16.1-11) and (16.1-10), which are useful to reduce the effective degree of the integrand:

$$\sin^2 \theta = \frac{1 - \cos 2\theta}{2}$$

$$\cos^2 \theta = \frac{1 + \cos 2\theta}{2} \qquad (17.5\text{-}1)$$

$$\sin \theta \cos \theta = \frac{1}{2} \sin 2\theta$$

Example 17.5-2

Consider

$$J = \int \sin^4 x \cos^2 x \, dx$$

Write the integrand as

$$(\sin x \cos x)^2 \sin^2 x = \left(\frac{1}{2} \sin 2x\right)^2 \frac{1 - \cos 2x}{2}$$

$$= \frac{1}{8} \{\sin^2 2x - \sin^2 2x \cos 2x\}$$

$$= \frac{1}{8} \left\{\frac{1 - \cos 4x}{2} - \sin^2 2x \cos 2x\right\}$$

This can now be integrated into

$$J = \frac{1}{8} \left\{\frac{x}{2} - \frac{\sin 4x}{8} - \frac{\sin^3 2x}{6}\right\} + C$$

$$= \frac{x}{16} - \frac{\sin 4x}{64} - \frac{\sin^3 2x}{48} + C$$

The form of this answer was quite satisfactory in the days when multiplication was awkward and looking up values in a trigonometric table was easier, but the reverse is true now with computers and hand calculators readily available. Instead, we need to convert the multiple angles into powers of the fundamental angle x (Section 21.5) or addition formulas in Section 16.1. As an alternative, we will give another approach based on *undetermined coefficients*. It is necessary to guess the proper form of the answer. We do the same problem again.

Example 17.5-3

Integrate

$$\int \sin^4 x \cos^2 x \, dx$$

The integrand is even and of total degree 6. We therefore expect the integral to be odd and the highest degree term to be of the same degree. Since integration of sines and cosines either does not change the total degree or else drops it by two, we guess the form

$$\int \sin^4 x \cos^2 x \, dx = \sin x \{A \cos^5 x + B \cos^3 x + C \cos x\}$$

$$+ \text{ a constant term} \qquad (17.5\text{-}2)$$

We now differentiate both sides:

$$\sin^4 x \cos^2 x = \cos x \{A \cos^5 x + B \cos^3 x + C \cos x\}$$

$$+ \sin x \{5A \cos^4 x + 3B \cos^2 x + C\}(-\sin x)$$

Arrange both sides in powers of $\cos x$ (use $\sin^2 x + \cos^2 x = 1$):

$$\cos^6 x - 2 \cos^4 x + \cos^2 x = A \cos^6 x + B \cos^4 x + C \cos^2 x$$

$$+ 5A \cos^6 x + 3B \cdot \cos^4 x + C \cos^2 x - 5A \cos^4 x - 3B \cos^2 x - C$$

Since the $\cos x$ (like the $\sin x$) is a transcendental function, the powers of $\cos x$ are linearly independent (Section 4.7); we must have the like powers on each side the same. Thus we can equate coefficients of like powers of $\cos x$.

$$1 = 6A$$
$$-2 = -5A + 4B$$
$$1 = \qquad - 3B + 2C$$
$$0 = \qquad\qquad\qquad C$$

and we see that we have four equations in three unknowns. Thinking a bit about the matter, we see that we overlooked the necessity of including in our guessed solution a term in x, say Dx. If we put this into the form (17.5-2), we then get in place of the bottom equation

$$0 = -C + D$$

We solve these equations one after another:

$$A = \frac{1}{6}$$

$$B = \frac{1}{4} - 2 + 5A = \frac{1}{4}\left[-2 + \frac{5}{6}\right] = \frac{-7}{24}$$

$$C = \frac{1}{2}(1 + 3B) = \frac{1}{2}\left(1 - \frac{21}{24}\right) = \frac{3}{48}$$

$$D = C = \frac{3}{48}$$

The solution is, therefore

$$\frac{\sin x}{48}[8 \cos^5 x - 14 \cos^3 x + 3 \cos x] + \frac{3x}{48} + C$$

This formula is much more convenient for many purposes than is the first solution. The method of undetermined coefficients is a powerful method for finding integrals when you have some idea of the form of the answer; even small errors in your guess will be caught and are often easily corrected (as shown above). It is easy (though tedious) to show that the two answers are the same.

EXERCISES 17.5

1. Show that the two answers are the same in Example 17.5-3.

Integrate the folowing:

2. $\int \sin^2 x \, dx$
3. $\int \sin^a x \cos^3 x \, dx$
4. $\int \cos^4 x \, dx$
5. $\int \sin^2 x \cos^4 x \, dx$ both ways
6. $\int \cos^6 x \, dx$
7. $\int \sin^2 x \cos^2 x \, dx$ as many different ways as you can
8. $\int \sin^4 y \cos^4 y \, dy$
9. $\int \sin x / \cos^2 x \, dx$

17.6 INTEGRATION BY PARTS— REDUCTION FORMULAS

We earlier saw (Section 14.7) that integration by parts is very useful. It is expecially useful for finding reduction formulas, which are one of the bases of systematic integration.

Example 17.6-1

Integrate

$$I_n = \int \tan^n x \, dx \qquad (n \geq 2)$$

We have immediately

$$I_n = \int \tan^{n-2} x (1 + \sec^2 x) \, dx$$

$$= I_{n-2} + \int \tan^{n-2} x \sec^2 x \, dx$$

$$= \frac{\tan^{n-1} x}{n - 1} + I_{n-2}$$

Since this formula drops the index by two each time, we need the two bottom cases:

$$I_1 = \ln |\sec x| + C$$

$$I_0 = x + C$$

Example 17.6-2

Consider

$$J_n = \int x^n e^{-ax} \, dx$$

$$= \frac{x^n e^{-ax}}{-a} + \frac{n}{a} \int x^{n-1} e^{-ax} \, dx$$

$$= \frac{-x^n e^{-ax}}{a} + \frac{n}{a} J_{n-1}$$

We need the single special case $n = 0$,

$$J_0 = \frac{-e^{-ax}}{a} + C$$

If we had the limits $(0, \infty)$ for the improper integral, then these two formulas would become

$$J_n = \frac{n}{a} J_{n-1}, \qquad J_0 = \frac{1}{a}$$

and we can apply them directly to get (for the improper integral)

$$J_n = \frac{n!}{a^{n+1}}$$

This is the familiar $\Gamma(n + 1)$ function (Section 15.11).

Example 17.6-3

Evaluate the $\beta(m, n)$ integral (β is Greek lowercase beta):

$$\beta(m, n) = \int_0^1 x^m (1 - x)^n \, dx = \beta(n, m)$$

as can be seen by replacing x by $1 - x$ in the integral. Suppose that the integers m and n satisfy $m \geq n > 0$. Integrate by parts using $U = (1 - x)^n$ and $dV = x^m \, dx$.

$$\beta(m, n) = \frac{x^{m+1}(1 - x)^n}{m + 1} \bigg|_0^1 + \frac{n}{m + 1} \int_0^1 x^{m+1}(1 - x)^{n-1} \, dx$$

The integrated part vanishes at both ends of the range of integration, and we have

$$\beta(m, n) = \frac{n}{m + 1} \beta(m + 1, n - 1)$$

with

$$\beta(m, 0) = \int_0^1 x^m \, dx = \frac{1}{m + 1}$$

Therefore,

$$\beta(m, n) = \frac{n}{m + 1}\beta(m + 1, n - 1) = \left(\frac{n}{m + 1}\right)\left(\frac{n - 1}{m + 2}\right)\beta(m + 2, n - 2)$$

$$= \frac{n(n - 1)(n - 2) \cdots (1)}{(m + 1)(m + 2)(m + 3) \cdots (m + n)}\beta(m + n, 0)$$

$$= \frac{n!\, m!}{(m + n + 1)!}$$

Example 17.6-4

Evaluate the integral $\int (\ln x)^n \, dx$. Using integration by parts with the $dV = dx$, we have the reduction formula

$$L(n) = \int (\ln x)^n \, dx = x\,(\ln x)^n - n \int (\ln x)^{n-1} \, dx$$

$$= x\,(\ln x)^n - nL(n - 1)$$

$$L(0) = x + C$$

EXERCISES 17.6

Find the reduction formulas for the following:

1. $\int \text{ctn}^n \theta \, d\theta$
2. $\int \sec^n t \, dt$
3. $\int 1/x^n(1 - x) \, dx$
4. $\int_0^1 x^n/(1 - x) \, dx$
5. $\int_0^1 x\,(\ln x)^n \, dx$
6. $\int f'(x) \ln x \, dx$
7. If $f(x) = f(1 - x)$, show that $I = \int_0^1 f(x) \cos^2 \pi x/2 \, dx = \frac{1}{2} \int_0^1 f(x) \, dx$
8. $\int e^{-ax}P(x) \, dx = e^{-ax}[P(x)/a + P'(x)/a^2 + \cdots]$

17.7 CHANGE OF VARIABLE

We have used a change of variable in a number of circumstances (Section 11.10). It is a very powerful tool for integrating functions, but unfortunately its use depends on insight into what will and what will not work. The variety of substitutions is so great that we can only suggest a few by means of examples. You simply try various ones and note why they did you little good, and from this sometimes a much better substitution arises.

Example 17.1-1

Integrate

$$J = \int_0^a \frac{dx}{\sqrt{x(x + a)}} \qquad (a > 0)$$

Set $x = a \tan^2 t$ (note how it removes the radical). Then $dx = 2a \tan t \sec^2 t\, dt$, and the integral becomes

$$J = \int_0^{\pi/4} \frac{2a \tan t \sec^2 t\, dt}{a \tan t \sec t}$$

$$= 2 \int_0^{\pi/4} \sec t\, dt = 2 \ln \left| \sec t + \tan t \right| \Big|_0^{\pi/4}$$

$$= 2 \ln (\sqrt{2} + 1)$$

Note that the integral does not depend on the parameter a, although it appears to in the initial formulation.

Example 17.7-2

Integrate

$$J = \int_{-1}^1 \sqrt{\frac{1 + x}{1 - x}}\, dx$$

The substitution $x = \cos t$ followed by multiplying both numerator and denominator by the numerator will remove the radical and produce

$$J = \int_\pi^0 \left(\frac{1 + \cos t}{\sin t} \right)(-\sin t)\, dt$$

$$= \int_0^\pi (1 + \cos t)\, dt$$

$$= (t + \sin t) \Big|_0^\pi = \pi$$

An alternative substitution, which also removes the radical, is

$$\frac{1 + x}{1 - x} = z^2$$

Solve for x:

$$1 + x = (1 - x)z^2$$

or

$$x = \frac{z^2 - 1}{z^2 + 1}$$

$$dx = \frac{(z^2 + 1)2z - (z^2 - 1)2z}{(z^2 + 1)^2}\, dz = \frac{4z\, dz}{(z^2 + 1)^2}$$

$$J = 4 \int_0^\infty \frac{z^2}{(z^2 + 1)^2}\, dz$$

This suggests the next substitution, $z = \tan t$:

$$J = 4 \int_0^{\pi/2} \frac{\tan^2 t \sec^2 t}{\sec^4 t}\, dt$$

$$= 4 \int_0^{\pi/2} \sin^2 t\, dt$$

$$= 4 \int_0^{\pi/2} \frac{1 - \cos 2t}{2}\, dt$$

$$= 2\left(\frac{\pi}{2}\right) = \pi$$

as before. There are often a number of different transformations that will work.

Example 17.7-3

Integrate

$$I_n = \int_0^a \frac{x^n}{\sqrt{a - x}}\, dx \qquad (a > 0)$$

Set $x = a \sin^2 t$; then $dx = 2a \sin t \cos t\, dt$, and the integral becomes

$$I_n = \int_0^{\pi/2} \frac{a^n \sin^{2n} t (2a \sin t \cos t)}{\sqrt{a} \cos t}\, dt$$

$$= 2a^{n+1/2} \int_0^{\pi/2} \sin^{2n+1} t\, dt = 2a^{(2n+1)/2} W_{2n+1}$$

where the W_{2n+1} is the Wallis number from Section 16.7.

Example 17.7-4

Integrate

$$\int \frac{dx}{x^{1/2} + x^{2/3}}$$

The radicals are both multiples of the sixth root, so we try

$$x = z^6, \qquad dx = 6z^5\, dz$$

$$\int = \int \frac{6z^5}{z^3 + z^4}\, dz = 6 \int \frac{z^2}{1 + z}\, dz$$

$$= 6 \int \left(z - 1 + \frac{1}{z + 1}\right) dz$$

$$= 6\left\{\frac{z^2}{2} - z + \ln |z + 1|\right\} + C$$

$$= 6\left\{\frac{x^{1/3}}{2} - x^{1/6} + \ln |x^{1/6} + 1|\right\} + C$$

Example 17.7-5

Integrate

$$\int \sin \sqrt{x}\, dx$$

The radical immediately suggests $x = t^2$, and we have

$$\int \sin t\, 2t\, dt$$

Apply integration by parts with the $U = t$. We have

$$\int = 2t(-\cos t) + 2 \int \cos t\, dt$$

$$= -2\sqrt{x} \cos\sqrt{x} + 2 \sin\sqrt{x} + C$$

You can check this easily by differentiation.

Example 17.7-6

Integrate (an important integral in probability theory)

$$J = \int_0^P \frac{dt}{\sqrt{t(1-t)}}$$

Set $t = \sin^2 x$ to eliminate the radical. From this we get, neglecting limits for the moment,

$$J = \int \frac{2 \sin x \cos x}{\sin x \cos x}\, dx = 2 \int dx$$

$$= 2x + C = 2 \operatorname{Arcsin} \sqrt{t} + C$$

We now put the limits into the indefinite integral to get

$$J = 2 \operatorname{Arcsin} \sqrt{p}$$

EXERCISES 17.7

Integrate the following using change of variable methods.

1. $\displaystyle\int_0^a \frac{1}{\sqrt{x(a-x)}}\, dx$

2. $\displaystyle\int_1^1 \sqrt{\frac{1-x}{1+x}}\, dx$

3. $\int_0^a x^n/(a+x)\, dx$

4. $\int 1/x^{1/2} + x^{1/4})\, dx$

5. $\int \dfrac{\tan \sqrt{x}}{\sqrt{x}}\, dx$

6. $\int f(\ln x)\, dx$, where $f(x)$ is an arbitrary function of x

7. $\int x^2 \arcsin x\, dx$

8. $\int \arctan x\, dx$

9. $\int x \ln x\, dx$

10. $\int \sqrt{x(a - x)}\, dx$

11. $\int 1/(x^{1/3} + x^{3/4})\, dx$

7.8 QUADRATIC IRRATIONALITIES

There are three trigonometric substitutions that are useful in removing quadratic irrationalities of the general form

$$\sqrt{ax^2 + bx + c}$$

when the integrand is of the form of a rational function in x and the square root

$$R(x, \sqrt{ax^2 + bx + c})$$

The method is first to complete the square, and then to remove from under the radical sign the coefficient of x by a suitable shift in the coordinates. You reduce the square root to one of the following forms:

Form	Use	Radical becomes	dx becomes
$\sqrt{a^2 - x^2}$	$x = a \sin \theta$	$a \cos \theta$	$a \cos \theta\, d\theta$
$\sqrt{a^2 + x^2}$	$x = a \tan \theta$	$a \sec \theta$	$a \sec^2 \theta\, d\theta$
$\sqrt{x^2 - a^2}$	$x = a \sec \theta$	$a \tan \theta$	$a \sec \theta \tan \theta\, d\theta$

The purpose of this table is not to have you memorize it; it is to let you check yourself on the mental process of reconstructing the transformations. If you see an $a^2 - x^2$, you ask, "Some constant minus something squared? How about $a^2 - a^2 \sin^2 \theta$? Yes, that gives an $a^2 \cos^2 \theta$ and removes the radical. I will need to use $a \sin \theta$. The form $a^2 + x^2$ suggests 1 plus the $\tan^2 \theta$, and $x^2 - a^2$ suggests $\sec^2 \theta - 1 = \tan^2 \theta$ and each eliminates the radical." You need to learn the *pattern* of thought that square roots of quadratics can be removed via completing the square, followed by a suitable trigonometric substitution. The method can be remembered, the details should not, and the substitutions should therefore be reconstructed as you need them.

In each case there are alternative substitutions that will remove the radical just as effectively, but they reverse the direction of the integration and hence are a bit messier. They are, in order, $a \cos \theta$, $a \operatorname{ctn} \theta$, and $a \csc \theta$. It is probably better to generally use the same substitution than it is to use alternative ones at random.

Using these substitutions, you are reduced to a rational function in sin and cos and can apply the method of Section 17.4, although it may not be necessary if you see how to do the integral by some other convenient method.

Example 17.8-1

Integrate

$$\int x^2 \sqrt{a^2 - x^2} \, dx = \int a^2 \sin^2 \theta \, a \cos \theta \, a \cos \theta \, d\theta$$

$$= a^4 \int (\sin \theta \cos \theta)^2 \, d\theta$$

$$= \frac{a^4}{4} \int \sin^2 2\theta \, d\theta$$

$$= \frac{a^4}{4} \int \frac{1 - \cos 4\theta}{2} \, d\theta$$

$$= \frac{a^4}{4} \left[\frac{\theta}{2} - \frac{\sin 4\theta}{8} \right] + C$$

It is a matter of details to get this back to the original variable x in case that is necessary. See Figure 17.8-1 for the substitutions from $\sin \theta$ and $\cos \theta$ to the variable x.

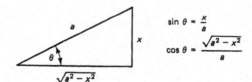

$$\sin \theta = \frac{x}{a}$$

$$\cos \theta = \frac{\sqrt{a^2 - x^2}}{a}$$

Figure 17.8-1

Example 17.8-2

Integrate

$$\int \frac{1}{\sqrt{a^2 + x^2}} \, dx = \int \frac{1}{a \sec \theta} \, a \sec^2 \theta \, d\theta$$

$$= \int \sec \theta \, d\theta$$

$$= \ln |\sec \theta + \tan \theta| + C$$

$$= \ln \left| \frac{\sqrt{a^2 + x^2}}{a} + \frac{x}{a} \right| + C$$

$$= \ln |\sqrt{a^2 + x^2} + x| + C$$

See Figure 17.8-2 for the substitutions to get to x.

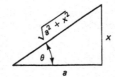

Figure 17.8-2

Example 17.8-3

Integrate

$$\int \frac{x^2}{\sqrt{x^2 - a^2}}\, dx = \int \frac{a^2 \sec^2 \theta}{a \tan \theta}\, a \tan \theta \sec \theta\, d\theta$$

$$= a^2 \int \sec^3 \theta\, d\theta$$

which we have done before (Example 16.5-8) and we need not repeat it here. The back substitution is fairly easy.

Example 17.8-4

Integrate

$$J = \int \frac{dx}{x\sqrt{x^2 + x + 1}} = \int \frac{dx}{x\sqrt{(x + \frac{1}{2})^2 + \frac{3}{4}}}$$

Use the change of variable

$$x + \frac{1}{2} = \frac{\sqrt{3}}{2} \tan \theta$$

to get, after some cancellation,

$$J = 2 \int \frac{\sec \theta}{\sqrt{3} \tan \theta - 1}\, d\theta$$

$$= 2 \int \frac{1}{\sqrt{3} \sin \theta - \cos \theta}\, d\theta$$

$$= 4 \int \frac{1}{2\sqrt{3}\, z - 1 + z^2}\, dz$$

where we have made the half-angle substitution and cleaned up the result a bit. Again we complete the square in the denominator

$$J = 4 \int \frac{1}{(z + \sqrt{3})^2 - 4}\, dz$$

and apply partial fractions to get

$$J = \int \left[\frac{1}{z + \sqrt{3} - 2} - \frac{1}{z + \sqrt{3} + 2} \right] dz$$

$$= \ln \left| \frac{z + \sqrt{3} - 2}{z + \sqrt{3} + 2} \right| + C$$

The path back to the original variables is messy to say the least. The purpose of this example is to show that persistence will get you through many times; and when you finally get the result in a nice form

$$J = \ln \left| \frac{(2 + x - 2\sqrt{x^2 + x + 1}}{x} \right| + C$$

you can always differentiate it both (1) to check the result and (2) to find a suggestion for a more direct method of integration.

Note that in each case to get back to the original variable you needed to draw a triangle and that the missing side of the triangle is exactly the radical you faced (possibly each side of the triangle is multiplied by a constant). This is a convenient check on your work.

EXERCISES 17.8

1. Write out your rules for integrating functions with a quadratic irrationality.
2. $\int_{-1}^{1} (1 - x^2)^{n/2} \, dx$
3. $\int_{-1}^{1} \frac{x^n}{\sqrt{1 - x^2}} \, dx$
4. $\int x^n/(x^2 + a^2) \, dx$
5. $\int x/(x^2 + a^2)^2 \, dx$
6. Rotate $x^{2/3} + y^{2/3} = a^{2/3}$ about the x-axis and find the volume inside.
7. Using the methods of this section, derive the integration formulas for arcsin x and arctan x.

17.9 SUMMARY

Formal integration is an important process. Unfortunately being, in practice, the inverse of differentiation, it is essentially a guess process. This chapter has shown that some classes of integrands that you can recognize are solvable in closed form. Let us list the forms you can easily recognize.

1. Polynomials $P(x)$
2. Sums of powers of x
3. $P(x) (\ln x)^m$
4. $\exp (ax)P(x)$
5. Rational functions $N(x)/D(x)$ [if you can factor $D(x)$]
6. Rational functions in sin x and cos x.
7. Quadratic irrationalities $R(x, \sqrt{a^2 + bx + c})$

These are the easily recognized types of integrands you can handle. It is not hard to make up other classes, but generally they are too specialized to be worth examining *unless* they tend to arise in the field you are currently pursuing. Again, it is worth mentioning that the methods given above will get the results, but sometimes involve a great deal of labor. Often you can recognize shortcuts that are worth the effort.

In Section 16.6 (and mentioned in Section 17.1), we gave some rules for an integrating strategy. They are as follows:

1. Rearrange the integrand, if necessary, to get the basic formulas you recognize. This involves some imaginative algebra or trigonometry at times and relies heavily on your almost instantaneous recognition of the basic formulas.

2. Use simple substitutions so you have a basic formula. For example,

$$J = \int \frac{x}{1 + x^2} \, dx$$

suggests the substitution $1 + x^2 = U$. Hence $dU = 2x \, dx$, and you lack a factor 2; so you write

$$J = \frac{1}{2} \int \frac{dU}{U} = \frac{1}{2} \ln |U| + C$$

$$= \frac{1}{2} \ln |1 + x^2| + C$$

Success here depends on both the rapid recognition of the various forms for integration and also the automatic recognition of derivatives; $a \tan x$ should suggest immediately $a \sec^2 x$. When this is combined with the rearrangements mentioned in rule 1, you see that it is necessary to have the formulas of differential and integral calculus memorized thoroughly, as well as to be skilled in algebra and trigonometric rearrangements.

3. Integration by parts is a very powerful tool. It can eliminate terms in $\ln x$, $\arctan x$, $\arcsin x$, and powers of x when the parts are properly chosen. When the integrand has an exponential term, integration by parts can sometimes be used twice to get the original integrand back (after some manipulation), and hence the integral on the right-hand side can be combined with the original integral to get the answer. This works best on e^{ax}, $\sin bx$, $\cos bx$, and the like terms, but has wider applications especially in advanced mathematics. Integration by parts is especially useful in getting rid of the inverse trigonometric functions.

4. Change of variables. A few systematic rules for change of variables have been given in this chapter, but it is an art that is developed by practice and by studying why some transformations work in some situations and others do not. One obvious way of using the change of variables is to get rid of radicals, since only a few basic integration formulas have radicals. Given a radical, you set the

quantity inside equal to a suitable power of the new variable, thus getting rid of the radical. Sometimes using the inverse function will get rid of an awkward term. If it is a definite integral, sometimes the limits will give you clues as to the change of variable to use. Unfortunately, the method often requires a sequence of transformations and the beginner soon gets discouraged. You need to have a sense when one integrand is (probably) easier to do than another. Know the basic forms well, and learn how the various types of transformations get rid of the corresponding awkward terms. The tangent half-angle substitution is a good example of a widely used set of transformations. So is the set of trigonometric substitutions of Section 17.8. But when these fail, you are forced to create your own transformation to solve the particular problem. It is an art that you learn from experience plus carefully and thoughtfully examining both the successes and failures you meet. Good luck!

5. Integral tables. We have appended a short table of commonly occurring integrals. Much larger tables are readily available, such as [G–R]. These are often useful. Browsing through a table may suggest things to try even if you do not find your particular integral in the table. One of the main difficulties with using a table is that you need to recognize your particular case among the whole set of general forms given in the table, and this is often not easy. Thus you should think of one or more general forms for which your integral is a particular case *before* you start looking in the table.

18

Applications Using One
Independent Variable

Now that we have a reasonable ability to integrate various functions, we can return to the applications of the ideas of the calculus. This time we do not need to pick the applications carefully, as we did earlier, so that the resulting integrals can be done by very limited tools. It is worth repeating that the calculus was developed, to a great extent, as a method for solving problems that arose in mechanics and related fields; but we will look only briefly at these applications.

We will first review a few of the kinds of applications that we have done, using our greater range of functions, and then turn to new applications. Thus we are partially reinforcing the earlier learning.

Calculus, probability, and statistics may each be studied for their own interest. For example, we looked at how little we had to initially derive (by the Δ process) in order to find all the derivatives we needed. Similarly, we looked slightly at the bases of both probability and statistics. On the other hand, these three fields may be looked at in terms of the problems they allow us to solve, the applications we can make. Both views are legitimate ones to take. They tend in practice to supplement each other.

18.2 WORD PROBLEMS

When we apply mathematics we form a model of the situation (see Figure 9.3-2). Mathematics cannot be applied directly to the real world; it can only be applied to the model we form of reality. In practice, the model we form rests on intuitive bases, partly arising from experience and partly from fairly direct understanding of the world and the model. The purpose of this chapter is to increase your experience in using the mathematics that you already have learned (and to reinforce this learning).

The application of mathematics implies *word problems*. These problems are often difficult for the student to handle. Real-life problems are usually much more difficult than are textbook problems. The first, and usually the most difficult, step in real-life problems is to recognize that there is a problem to solve. The next most difficult step is to formulate the problem properly. When these two steps have been taken, you are at the stage of doing word problems.

The usual advice to the student is, "Understand the problem first before you try to solve it." Unfortunately, these words are of little practical value since the main difficulty is that rereading the problem often produces no further understanding. There is no magic solution to the dilemma, but there are some helpful things you can do.

Read the problem and find out what is being asked. Give it a name, typically x; but if you see that you will need a coordinate system, maybe you should save x for the corresponding coordinate. It is surprising how powerful is the single step of naming things; it gives you a feeling of mastery over the situation, it is the first step in any science. Then reread the problem assigning names to the various items, a, y, c, and so on. Along with the naming process, you make a picture of the situation. Together, the two processes of naming and drawing a picture often produce an increased understanding of the problem, at least enough to allow you to go forward. It is quite common that you have to abandon one or more of the figures and redraw them as your understanding of the problem progresses. Sometimes you will also have to relabel things a bit. With luck, after several iterations of this process, you will have a fairly clear idea of the problem.

However, some problems do not permit drawing useful pictures, as the following classic example shows. In this problem, you merely rewrite the original statements of the problem in equivalent, more useful, forms, a characteristic feature of mathematics.

Example 18.2-1

If a hen and a half lays an egg and a half in a day and a half, how long does it take for 6 hens to lay 8 eggs?

The problem is designed to be confusing! Therefore, we read it carefully, translating the words to more common usage as we go. "If a hen and a half lays an egg and a half . . ." means the same as "If one hen lays one egg . . . ." Now, "If one hen lays one egg in a day and a half . . ." means "One hen lays 2/3 of an egg per day . . . ." Hence, 6 hens will lay $6(2/3) = 4$ eggs in one day. From this it follows that to lay 8 eggs will take $8/4 = 2$ days. The answer is, therefore, two days.

Once the textbook problem is understood and properly formulated, it is a matter of applying the techniques you have just learned, or possibly slightly extending them, to get the answer. Of course, the solution may also depend on things that you learned long ago in this course or earlier courses. But usually in textbooks the mathematical tools required are related to the material just covered.

In real life this sorting out of the problems according to the tools needed for their solution does not occur. Even when you have a clear formulation of a problem you may still not know what mathematical tools will be needed to solve it. Giving you carefully selected word problems is about as close as an elementary textbook can come to real-world situations.

The calculus not only opens the door on the solution of word problems, but you have also arrived at the stage where applications to new kinds of questions can be made easily. Hence many of the new applications in this chapter should be viewed as merely larger-than-usual word problems.

18.3 REVIEW OF APPLICATIONS

The applications of derivatives were to (1) curves, especially curve tracing, (2) maxima and minima problems, (3) rate problems, (4) generating functions, (5) least squares, (6) maximum likelihood, and finally (7) indeterminate forms. The applications of integration included the computation of (1) areas, (2) volumes of rotation, and (3) probability. We do a few examples for purposes of review.

Example 18.3-1

Given the curve

$$y = \cos x \qquad (18.3\text{-}1)$$

is there a normal line that passes through the origin? See Figure 18.3-1. We know immediately that we need the derivative

$$y' = -\sin x$$

Figure 18.3-1   Normal line

At the point (x_1, y_1) the normal line has the negative reciprocal slope of the tangent line; hence the equation of the normal line is

$$y - \cos x_1 = \frac{1}{\sin x_1}(x - x_1)$$

If this line is to pass through the origin, then the point $(0, 0)$ must lie on the line

$$-\cos x_1 = \frac{-x_1}{\sin x_1} \qquad (18.3\text{-}2)$$

or $\qquad\qquad\qquad\qquad \sin 2x_1 = 2x_1 \qquad\qquad\qquad\qquad$ (18.3-3)

The only solution of this equation is $x_1 = 0$. This line, $x = 0$, is a vertical line through the origin.

Example 18.3-2

Sketch the *normal distribution*

$$y = \frac{1}{\sqrt{2\pi}} \exp\left(\frac{-x^2}{2}\right)$$

The curve will later be shown (Example 19.4-4) to have an integral equal to 1; thus $y(x)$ can represent a probability distribution. First, since x occurs only in even powers, there is symmetry about the y-axis. Next, we compute $y'(x)$ and $y''(x)$. We get

$$y'(x) = -x \exp\frac{(-x^2/2)}{\sqrt{2\pi}}$$

$$y''(x) = \frac{(x^2 - 1) \exp(-x^2/2)}{\sqrt{2\pi}}$$

We see that $y(0) = 1/\sqrt{2\pi}$, $y'(0) = 0$, and $y''(0) = -1/\sqrt{2\pi}$. The inflection points occur where $y'' = 0$ and these are $x = \pm 1$. The value of the function $y(1) = 1/\sqrt{2\pi e}$. The slope $y'(\pm 1) = \mp 1/\sqrt{2\pi e}$ at the inflection points is also useful for plotting the curve. With this tied down, and the knowledge that as $x \to \infty$ the curve rapidly approaches 0, we can draw Figure 18.3-2.

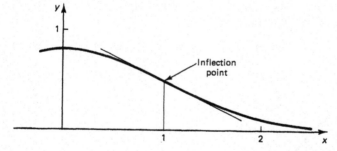

Figure 18.3-2 Error curve

The applications of integrals is much more varied, and we give only a few examples.

Example 18.3-3

Find the maximum area of the ellipse

$$\frac{x^2}{a^2} + \frac{y^2}{b^2} = 1$$

when $a + b = 1$. We must first find the area of the ellipse. We have (from symmetry)

$$A = 4\int_0^a y\, dx = 4b\int_0^a \sqrt{1 - \frac{x^2}{a^2}}\, dx$$

In this we set $x = a \sin \theta$. We get

$$A = 4b \int_0^{\pi/2} \cos \theta \, a \cos \theta \, d\theta = 4ab \int_0^{\pi/2} \cos^2 \theta \, d\theta$$

$$= 2ab \int_0^{\pi/2} (1 + \cos 2\theta) \, d\theta = \pi ab$$

We now know that the area of the ellipse is πab; hence we have the simple problem of finding the maximum of a product when the sum $a + b = 1$. We have the area

$$A = \pi a(1 - a) = \pi(a - a^2)$$

$$A' = \pi(1 - 2a) = 0$$

from which it follows that $a = \frac{1}{2} = b$, and we have a circle whose area is $A = \pi/4$.

Example 18.3-4

Find the volume of one arch of $y = \sin x$ between 0 and π when the curve is rotated about the x-axis. The volume is given by (see Figure 18.3-3), using discs as shown,

$$V = \int_0^\pi \pi y^2 \, dx$$

$$= 2\pi \int_0^{\pi/2} \sin^2 x \, dx$$

$$= \pi \int_0^{\pi/2} (1 - \cos 2x) \, dx$$

$$= \frac{\pi^2}{2}$$

which is less than the enclosing cylinder, π^2, by a reasonable amount. A lower bound of an enclosed double cone is easily found to be $\pi^2/3$.

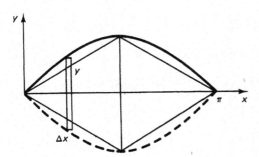

Figure 18.3-3

Example 18.3-5

Find the volume generated by the area under $y = \sin x$, from 0 to π, and above the x-axis, when rotated about the y-axis (see Figure 18.3-4). The volume is, using the element of volume as a cylinder (as shown),

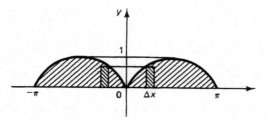

Figure 18.3-4

$$V = \int_0^{\pi} 2\pi x y \, dx$$

$$= 2\pi \int_0^{\pi} x \sin x \, dx$$

We naturally apply integration by parts to remove the x factor.

$$V = 2\pi \left[x(-\cos x) \Big|_0^{\pi} + \int_0^{\pi} \cos x \, dx \right]$$

$$= 2\pi [\pi] = 2\pi^2$$

The volume of the enclosing cylinder is π^3.

Example 18.3-6

What is the variance of the distribution

$$p(x) = \frac{1}{2} e^{-|x|}$$

First, we need to check that it is a distribution. We have, due to the symmetry and the obvious convergence of the integral,

$$\int_{-\infty}^{\infty} p(x) \, dx = 2 \int_0^{\infty} p(x) \, dx$$

$$= \int_0^{\infty} e^{-x} \, dx = -e^{-x} \Big|_0^{\infty} = 1$$

Again, by symmetry, the first moment is zero; hence the variance is

$$\text{Var} \{X\} = 2 \int_0^{\infty} (x - 0)^2 \left(\frac{e^{-x}}{2} \right) dx = \Gamma(3) = 2!$$

EXERCISES 18.3

1. Find the tangent line to $y = x^n$ at $x = 1$.
2. Find the tangent line to $x^2 + y^2 - 4x + 3 = 0$ through the origin.
3. Find the equations giving the positions of the max–min and inflection points for the curve $y = \exp(-ax) \cos bx$.

4. Find the formula giving the shape of the maximum volume for a cylindrical block fitting inside one arch of $y = \sin x$ rotated about the x-axis.

5. Given the curve $(0, \infty)$ $y = \exp(-ax) \sin bx$ $(a > 0)$, find the area under the square of the curve.

6. Rotate the curve $y = x^n \exp(-x)/n!$ about the x-axis. Find the volume from 0 to infinity. (*Hint*: Remember the gamma function.) Use Stirling's formula to approximate the answer for large n.

7. Sketch $y = \exp(-x^4/4)$.

8. Given $y = \ln x$ $(0 < x \leq 1)$ rotated about the y-axis. Find the rate at which the surface falls when $x = 1/2$ and the rate of decrease in the volume is 4 units per second.

9. Approximate $y = \sin x$ $(0, \pi/2)$ by a straight line in a least squares sense.

10. Show that $\int_{-\infty}^{\infty} 1/(1 + x^4)\, dx] = \sqrt{2\pi}/2$.

8.4 ARC LENGTH

We now turn to new applications that use the same ideas as we developed earlier. These applications are slightly more difficult to do than the above word problems. However, you need practice in applying what you have learned to situations that have not been previously explained in detail to you.

A topic that will arise, sooner or later, is that of the length of a curve. Just as we extended the idea of area from triangles and rectangles to curved regions, so too we now *extend* our ideas of the length of a straight line segment to curved lines. The basic definition of length is that of a line segment. The extension to a circle in geometry comes from taking the limit of secant lines through points on the circle as the largest interval approaches zero length. This is analogous to what was done for areas. Intuitively, the length of the circumference of a circle comes from the feeling of rolling the circle along a plane surface and measuring how far the rim goes while completing a full revolution of the circle. For curves other than convex ones, we cannot do this. We can, of course, imagine a string that is glued to the edge and later unwound. This is very much like using short secant lines. As with areas, we propose to use a general approach, and not one tied to a particular curve.

Consider a curve (Figure 18.4-1) with the short secant lines drawn in. We see many small triangles each with the property that the horizontal distance is Δx_i, the vertical distance is Δy_i and the hypotenuse is Δs_i. This is the elementary triangle. We have, dropping the subscripts,

$$(\Delta s)^2 = (\Delta x)^2 + (\Delta y)^2 \tag{18.4-1}$$

If we divide this by $(\Delta x)^2$ and take the limit, we get the relationship between the derivatives:

$$\left(\frac{ds}{dx}\right)^2 = 1 + \left(\frac{dy}{dx}\right)^2 \tag{18.4-2}$$

assuming that they exist. We could, instead, have divided by the $(\Delta y)^2$ term to get

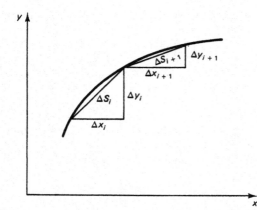

Figure 18.4-1 Arc length

$$\left(\frac{ds}{dy}\right)^2 = \left(\frac{dx}{dy}\right)^2 + 1 \tag{18.4-3}$$

We could also have divided by $(\Delta t)^2$ to get

$$\left(\frac{ds}{dt}\right)^2 = \left(\frac{dx}{dt}\right)^2 + \left(\frac{dy}{dt}\right)^2 \tag{18.4-4}$$

These three formulas, when integrated, enable us to find the length of a curve. The first formula (18.4-2) becomes

$$s = \int_{x_0}^{x_i} \sqrt{1 + \left(\frac{dy}{dx}\right)^2} \, dx \tag{18.4-5}$$

for the arc length along the curve $y = y(x)$ from x_0 to x_1.

It is always wise to check any extension by trying it out on known cases. Thus we try the formula on the straight line and the circle before using it more generally.

Example 18.4-1

Find the length of the line $y = mx$ from the origin to $x = 2$. We have $y' = m$, and formula (18.4-5) becomes

$$s = \int_0^2 \sqrt{1 + m^2} \, dx = \sqrt{1 + m^2} \int_0^2 dx$$
$$= 2\sqrt{1 + m^2}$$

The base of the triangle is 2, the height is $2m$, and we have, therefore, the hypotenuse we just computed.

Example 18.4-2

Find the circumference of a circle of radius a. We represent the circle in the form

$$x^2 + y^2 = a^2$$

We use the form for the arc length (18.4-5) and compute (using symmetry) four times the length in the first quadrant:

$$s = 4 \int_0^a \sqrt{1 + (y')^2} \, dx$$

We must find y'. By implicit differentiation of the equation for the circle, we get

$$y' = \frac{-x}{y}$$

and the formula for the arc length becomes (using the equation of the circle)

$$s = 4 \int_0^a \sqrt{1 + x^2/y^2} \, dx = 4 \int_0^a \frac{a}{y} \, dx$$

$$= 4a \int_0^a \frac{1}{\sqrt{a^2 - x^2}} \, dx = 4a \, \text{Arcsin} \, \frac{x}{a} \bigg|_0^a$$

$$= 4a\left(\frac{\pi}{2}\right) = 2\pi a$$

which is what it should be.

We conclude that the definition we made for the length of a curve is reasonable. It assumes that at every point the derivative of the curve exists. At corners, we simply break up the curve into pieces each of which has a derivative in the interval (although the ends do not have a two-sided derivative). This is simply our usual (in this course) piecewise decomposition of curves into a finite number of pieces. We are now free to apply it to other functions, but we should still keep an eye on the results to be sure that they are reasonable. Beware of positive and negative lengths that cancel!

Example 18.4-3

Consider the function $y = \ln x$. Suppose we want to measure its arc length, from $x = 1$. We have

$$y' = \frac{1}{x}$$

and therefore we have

$$s = \int_1^{x_1} \sqrt{1 + \frac{1}{x^2}} \, dx$$

This integral is one of those we can hope to do, so we persist with

$$s = \int_1^{x_1} \frac{\sqrt{x^2 + 1}}{x} \, dx$$

Make the usual transformation

$$x = \tan \theta$$

to get rid of the radical (we could also use $1 + x^2 = u^2$ instead if we wished). We have (dropping the limits for the moment)

$$s = \int \frac{\sec \theta}{\tan \theta} \sec^2 \theta \, d\theta$$

$$= \int \sec \theta \left(\frac{1 + \tan^2 \theta}{\tan \theta} \right) d\theta$$

$$= \int \frac{\sec \theta}{\tan \theta} \, d\theta + \int \sec \theta \tan \theta \, d\theta$$

$$= \int \csc \theta \, d\theta + \sec \theta$$

$$= \sec \theta - \ln \{\csc \theta + \text{ctn } \theta\}$$

$$= \sqrt{1 + x^2} - \ln \left\{ \frac{\sqrt{1 + x^2}}{x} + \frac{1}{x} \right\}$$

between the limits 1 and x_1. The result of some algebra is, on dropping the subscript on the upper limit,

$$s = \sqrt{1 + x^2} - \sqrt{2} - \ln \left\{ \frac{\sqrt{1 + x^2} + 1}{x} \right\} + \ln \{\sqrt{2} + 1\}$$

$$= \sqrt{1 + x^2} - \sqrt{2} + \ln \left\{ \frac{(\sqrt{2} + 1)x}{\sqrt{1 + x^2} + 1} \right\}$$

At $x = 1$, we of course get 0. For large x we find that the increase in arc length is mainly from the first term and grows like x; and, since the ln curve flattens out as x gets large, this is what we expect.

From this simple example we see the nature of the problem: the formula for arc length often leads to integrands that are hard, or impossible, to integrate in closed form. It is rare that arc length problems are nicely integrated. Thus for arc length problems we are often forced to numerical integration (Section 11.11). There is a collection of nice problems that have been found over the years and occupy the exercises in the standard textbooks. Few are of real importance beyond their use as exercises.

Example 18.4-4

Find the length of the four-cusp hypocycloid

$$x^{2/3} + y^{2/3} = a^{2/3}$$

We need $y' = -(y/x)^{1/3}$ for formula (18.4-5):

$$s = 4 \int_0^a \sqrt{1 + (y')^2} \, dx = 4 \int_0^a \sqrt{1 + (y/x)^{2/3}} \, dx$$

$$= 4 \int_0^a \left(\frac{a}{x}\right)^{1/3} dx = (4a^{1/3})\frac{3}{2}x^{2/3} \Big|_0^a$$

$$= 6a$$

Does this seem reasonable? If we look at the circle of radius a and compare it to the four-cusp hypocloid (Figure 18.4-2), we see that the arc lengths should be much the same. The circle has $2\pi a$ length, which is close to $6a$. Yes, the answer seems to be reasonable.

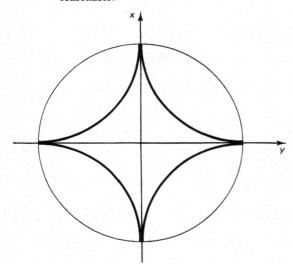

Figure 18.4-2   Four-cusp hypocloid

Example 18.4-5

Lest you think that the length of a curve is a very simple idea, we give the following example of a sequence of curves each having a length that is the same fixed number; hence the sequence has this same limit. But the sequence aproaches a curve that has a different length!

Consider the line segment from 0 to 1. It has length 1. Erect an equilateral triangle on it (see Figure 18.4-3) whose two sides have a combined length of 2. Now "fold" the peak down to the line, making the path the edges of two equilateral triangles whose combined length is still 2. Now we fold down the two peaks to get the next curve of the sequence of curves. And so on. Each time the length of the broken path is 2, but the sequence of curves approaches the line segment. The limiting broken line has length 2, but it approaches the line of length 1.

You see what is wrong. Among other things, the limiting curve no longer has a derivative in the whole interval. In some sense the broken line is not approximating the straight line, although it is getting closer and closer, as measured by the maximum distance away, to the straight line. When we picked secant lines, then in the limit the secant lines approached the *direction* as well as the position of the curve. This is the

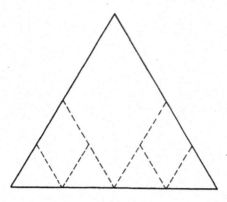

Figure 18.4-3 Arc length

essential difference between what we have been doing to approximate the arc length and this example.

EXERCISES 18.4

1. In Example 18.4-3 make the $1 + x^2 + u^2$ substitution.
2. Find the arc length of $y^2 = 4x^3$ from $x = 0$ to $x = x_1$.
3. Find the arc length of $x^2 = 9y^3$.
4. Find the length of the curve $y = \ln \cos x$ from $x = 0$.

18.5 CURVATURE AGAIN

We studied curvature in Section 8.4 by fitting a local circle to the curve and using the reciprocal of the radius of the circle as a measure of the curvature at that point. The formula for the curvature kappa was

$$\kappa = \frac{|y''|}{[1 + (y')^2]^{3/2}}$$

Now that we have the concept of arc length, an alternative definition of curvature suggests itself; how fast does the angle of the tangent line change as a function of the arc length? This definition appears to be independent of the coordinate system and hence to measure something intrinsic to the curve itself. (When we studied the ellipse in Section 6.5, we had two different measures of the deviation of an ellipse from a circle, the eccentricity e and the ratio r of the two major axes of the ellipse.)

From Figure 18.5-1 we have

$$y' = \tan \theta$$

or

$$\theta = \arctan y'$$

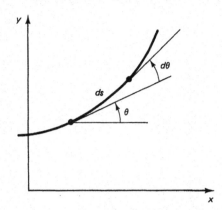

Figure 18.5-1 Curvature

The rate of change of the angle θ with respect to the arc length s is simply

$$\frac{d\theta}{ds} = \frac{1}{1 + (y')^2} \frac{dy'}{ds}$$

$$= \frac{1}{1 + (y')^2} \left(\frac{dy'}{dx}\right)\left(\frac{dx}{ds}\right)$$

Since $dy'/dx = y''$ and

$$\frac{ds}{dx} = \sqrt{1 + (y')^2}$$

we have

$$\frac{d\theta}{ds} = \frac{y''}{[1 + (y')^2]^{3/2}}$$

which is the *same* formula (almost) as before. This time the two definitions lead to essentially the same formula for computing the curvature and are therefore equivalent.

18.6 SURFACES OF ROTATION

It is natural to pass from arc length to the surface S of a volume generated by rotating a curve around an axis. If we look at a picture of what an element of arc length does as it is rotated about a line, we see that it is closely approximated by the surface of a local tangent cone. In Figure 18.6-1, where the rotation is about the y-axis, we see that the surface is reasonably approximated by

$$\Delta S = 2\pi x \Delta s$$

where S is the surface generated and s is the arc length.

It is reasonable to test this idea on a cylinder (the result is obvious) and on a sphere.

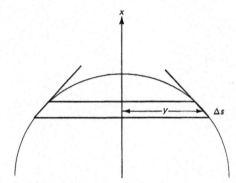

Figure 18.6-1   Surface area

Example 18.6-1

Find the surface of a sphere of radius a. We rotate the circle

$$x^2 + y^2 + a^2$$

about the x-axis (see Figure 18.6-1). By implicit differentiation we have $y' = -x/y$. The formula for the surface, therefore,

$$S = 2\pi \int y \, ds = 2\pi \int_{-a}^{a} y\sqrt{1 + (y')^2} \, dx$$

$$= 2\pi \int_{-a}^{a} y\sqrt{1 + (x/y)^2} \, dx$$

$$= 2\pi \int_{-a}^{a} y\frac{\sqrt{y^2 + x^2}}{y} \, dx$$

$$= 2\pi \int_{-a}^{a} a \, dx = 2\pi a(2a) = 4\pi a^2$$

This checks our known result and gives us confidence in the method for finding the area of surfaces of revolution. A few other check cases are left as exercises.

Example 18.6-2

Suppose the curve $y = \ln x$ is rotated about the y-axis. Find the surface generated by the segment between $x = 1$ and $x = x_1$ (see Figure 18.6-2). The formula for the surface is

$$S = 2\pi \int_{1}^{x_1} x \, ds = 2\pi \int_{1}^{x_1} x\sqrt{1 + \frac{1}{x^2}} \, dx$$

$$= 2\pi \int_{1}^{x_1} \sqrt{1 + x^2} \, dx$$

We naturally set $x = \tan \theta$; hence $dx = \sec^2 \theta \, d\theta$, and

$$S = 2\pi \int \sec^3 \theta \, d\theta$$

which is an integral we have done (Example 16.5-8). Thus we have

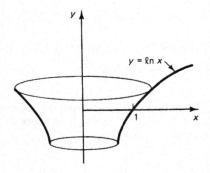

Figure 18.6-2

$$S = \pi[\ln|\sec\theta + \tan\theta| + \sec\theta\tan\theta]$$

to convert back to the variable x. We get

$$S = \pi[\ln|\sqrt{1 + x^2} + x| + x\sqrt{1 + x^2}]\Big|_{1}^{x1}$$

$$= \pi[\ln(x_1 + \sqrt{1 + x_1^2})] + x_1\sqrt{1 + x_1^2} - \ln(1 + \sqrt{2}) - \sqrt{2}]$$

Example 18.6-3

Find the surface of the volume formed by rotating the curve $y = \sin x$ about the x-axis from $x = 0$ to $x = \pi$ (Figure 18.6-3). We have immediately

$$S = 2\pi\int y\, ds = 2\pi\int_0^\pi \sin x(1 + \cos^2 x)^{1/2}\, dx$$

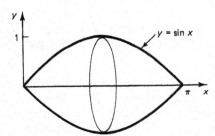

Figure 18.6-3

Set $\cos x = t$; then $dt = -\sin x\, dx$ and we have, on changing the limits,

$$S = 2\pi\int_{-1}^{1}\sqrt{1 + t^2}\, dt = 4\pi\int_0^1\sqrt{1 + t^2}\, dt$$

Now set $t = \tan z$; then $dt = \sec^2 z\, dz$, and the integral becomes

$$S = 4\pi\int_0^{\pi/4}\sec^3 z\, dz$$

We know (from the integral table if no place else) that this is

$$S = 2\pi\{\ln|\sec z + \tan z| + \sec z \tan z\}\Big|_0^{\pi/4}$$

$$= 2\pi\{\ln(\sqrt{2} + 1) + \sqrt{2}\} = 2\pi(2.295\ldots) = 4\pi(1.147\ldots)$$

which is slightly more than the surface of an enclosed sphere, 4π.

Example 18.6-4

Find the volume generated by rotating the area above $y = -\exp(-x)$ and below the positive x-axis about the y-axis (see Figure 18.6-4). We choose the cylinder method as shown (and use $-y$ to get a positive volume):

$$V = 2\pi \int_0^\infty x(-y)\, dx$$

$$= 2\pi \int_0^\infty xe^{-x}\, dx$$

$$= 2\pi[-xe^{-x} - e^{-x}]\Big|_0^\infty$$

$$= 2\pi$$

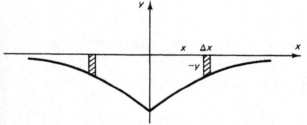

Figure 18.6-4

Note that the volume is finite, but that the surface, being at least as large as the whole plane, must be infinite! As they say, "You can fill the container full of paint, but you cannot paint the whole surface!"

This last remark shows some of the dangers of the immediate identification of the results of the mathematical model with the real world. The difficulty is that in mathematics the infinite surface generated by the rotating curve does not make a volume; it merely bounds a volume at best. You tend to think of the coat of paint as having some thickness (volume) and are led astray by the confusion between the mathematical and real models.

This is an interesting example, so we ask, "Can we make up another example of this type, this time having a container with finite maximum diameter, but having (necessarily when restricted to our class of functions) an infinite depth?"

Example 18.6-5

We begin with the reasonable choice of a curve to rotate about the y-axis as $(0 < x \leq 1)$

$$y = 1 - \frac{1}{x^a}$$

where we must require $a > 0$ to have the shape shown in Figure 18.6-5. The volume is given by

$$V = \pi \int_{-\infty}^{0} x^2 \, dy$$

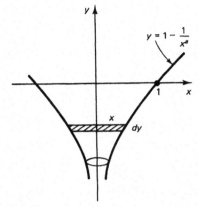

$$y = 1 - \frac{1}{x^a}$$

Figure 18.6-5

We compute dy, put that in the integral, and change the limits to the variable x:

$$V = \pi \int_0^1 x^2 \, ax^{-a-1} \, dx$$

$$= \pi a \int_0^1 x^{1-a} \, dx$$

$$= \pi a \frac{x^{2-a}}{2-a} \Big|_0^1 = \frac{\pi a}{2-a}$$

provided we make $2 - a > 0$ so that the improper integral will have the lower limit vanish. Thus the volume is finite when

$$0 < a < 2$$

and we cannot have the equality at either end, as you can show by considering those cases.

Next, we compute the surface area. We get

$$S = 2\pi \int x \, ds = 2\pi \int_0^1 x\sqrt{1 + a^2 x^{-2(a+1)}} \, dx$$

$$= 2\pi \int_0^1 \sqrt{x^2 + a^2 x^{-2a}} \, dx$$

This integral is not easy to do. But all we need are bounds, so we think as follows: in the range ($0 \le x \le 1$) the factor x^2 is not more than 1 while the other factor is large, especially near $x = 0$. Hence we drop the x^2 term and write

$$\sqrt{x^2 + a^2 x^{-2a}} > ax^{-a}$$

and use this in the integrand for a lower bound on the integral. We get

$$S > \int_0^1 ax^{-a}\, dx = \frac{2\pi a}{1 - a}$$

provided $1 - a > 0$. For $a < 1$, it is easy to get a bound on the other side (since $x^2 < 1$ and $a^2 < 1$):

$$x^2 + a^2 x^{-2a} < 2x^{-2a}$$

which will change the answer by a multiplicative factor of $a^{-1}\sqrt{2}$. Thus the surface area is infinite for $1 < a < 2$, while for this range the volume is finite. Both the surface and the volume are finite for $0 < a < 1$.

EXERCISES 18.6

1. Check the formula for the surface of a cone.
2. Check the formula for the surface of a cylinder.
3. Find the surface generated by rotating $y = \exp(-x)$ $(0 \le x \le 1)$ about the x-axis.
4. Find the total surface of the four-cusp hypocycloid $x^{2/3} + y^{2/3} = a^{2/3}$ when rotated about the x-axis.

18.7 EXTENSIONS

When finding areas and volumes, we have taken rectangular strips, discs, and cylinders as our elements to sum. In the limit the sum passes over into the integral. Similarly, we took short line segments and strips of cones to find lengths of lines and surfaces of volumes of rotation.

But the integral is a limit of a sum, as was proved in great generality; although we began with areas, the elements in the sum were not tied to any specific quantities. This suggests that we can use any shaped sections we want in finding a volume (and other quantities). They can be similar shapes or not, as the following examples show.

Example 18.7-1

Two circular cylinders of the same diameter have their axes intersecting at right angles; find the common volume. The main problem is to visualize and draw the figure. Looking from the top of the intersecting cylinders, we get Figure 18.7-1, where, after some thought, we see the two lines of intersection meeting at right angles.

How to cut the common volume up into computable parts is the problem. After some thought we see that squares parallel to the top will lead us to elements of volume for which we can write a formula. Looking at the end view down one cylinder, we see that a side of a square is $2x$; hence the element of volume is

$$(2x)^2\, dy$$

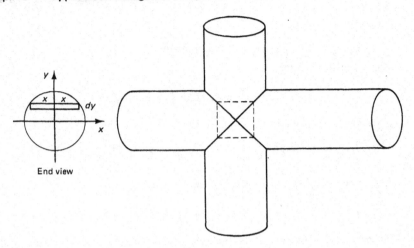

Figure 18.7-1

where the variables are related by the equation

$$x^2 + y^2 = a^2$$

The volume we want is, therefore,

$$V = 2 \int_0^a 4x^2 \, dy = 8 \int_0^a (a^2 - y^2) \, dy = 8 \left[a^3 - \frac{a^3}{3} \right] = \frac{16a^3}{3}$$

Does the answer seem reasonable? The enclosing cube has volume $8a^3$, which makes the answer seem reasonable.

Example 18.7-2

A volume has a base that is circular,

$$x^2 + y^2 = a^2$$

and the top is a line above the y-axis at a height h. The cross sections perpendicular to the y-axis are isosceles triangles with their bases on the circle. Find the volume (Figure 18.7-2).

The elements of volume are triangles with heights h, bases $2x$, and thickness dy,

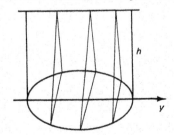

Figure 18.7-2

so we have

$$dV = hx \, dy$$

to sum. This leads to the integral

$$V = 2 \int_0^a hx \, dy = 2h \int_0^a \sqrt{a^2 - y^2} \, dy$$

Put $y = a \sin \theta$ to get rid of the radical.

$$V = 2h \int_0^{\pi/2} a \cos \theta \, a \cos \theta \, d\theta = 2a^2 h \int_0^{\pi/2} \cos^2 \theta \, d\theta$$

and we recognize a Wallis integral. Hence the answer is

$$V = 2a^2 h \left(\frac{\pi}{4}\right) = \frac{\pi a^2 h}{2}$$

Thus the volume is one-half of the enclosing cylinder. A little thought about a cylinder, and the parts cut off, shows that this is indeed the answer.

Example 18.7-3

Find the volume of the ellipsoid

$$\frac{x^2}{a^2} + \frac{y^2}{b^2} + \frac{z^2}{c^2} = 1$$

We cut the ellipsoid by planes parallel to the x–y axes. The section is

$$\frac{x^2}{a^2} + \frac{y^2}{b^2} = 1 - \frac{z^2}{c^2}$$

where z is a constant for the moment. We put this in the standard form for an ellipse:

$$\frac{x^2}{a^2[1 - (z^2/c^2)]} + \frac{y^2}{b^2[1 - (z^2/c^2)]} = 1$$

The area of the ellipse is (see Example 18.3-3) π times the product of the two semiaxes; hence the element of volume is

$$dV = \pi ab \left(1 - \frac{z^2}{c^2}\right) dz$$

and the volume is given by

$$V = 2\pi ab \int_0^c \left(1 - \frac{z^2}{c^2}\right) dz$$

$$= 2\pi ab \left(c - \frac{c}{3}\right) = \frac{4}{3} \pi abc$$

When $a = b = c$, we get the volume of a sphere; and when we think about it, the formula is just what it must be.

EXERCISES 18.7

1. The cylinder $x^2 + y^2 = a^2$ above the x–y plane is cut off by the plane $z = x + a$. Find the volume. Why is the answer obvious?

2. A volume has a base that is circular, $x^2 + y^2 = a^2$, and cross sections are right triangles of sides h with the hypotenuse in the circle. Find the volume.

3. Do Example 18.7-2 when the base is an ellipse.

4. A volume has a base that is circular of radius a. Cross sections in a vertical plane perpendicular to the x-axis are parabolas centered above the x-axis with the top along a line c units above the x-axis. Find the volume.

5. A solid has square cross sections perpendicular to the x-axis with the bases on an ellipse. Find the volume.

18.8 DERIVATIVE OF AN INTEGRAL

It frequently happens that the integrand of an integral depends on a parameter c. In mathematical notation, we have

$$F(c) = \int_a^b f(x, c)\, dx \qquad (18.8\text{-}1)$$

An example was the $\Gamma(n)$ function (Section 15.11), where, since n can be any real, positive number, we write c in place of n:

$$\Gamma(c) = \int_0^\infty x^{c-1} e^{-x}\, dx \qquad (18.8\text{-}2)$$

where now c is a real number greater than 0. This gives a reasonable extension of the notion of the factorial to all positive real numbers.

It is natural to ask about the derivative of $F(c)$ with respect to c in (18.8-1). To answer the question, we go back to fundamentals and write out the details of the four-step process. At the third step, we have

$$\frac{F(c + \Delta c) - F(c)}{\Delta c} = \frac{\int_a^b f(x, c + \Delta c)\, dx - \int_a^b f(x, c)\, dx}{\Delta c}$$

$$= \int_a^b \frac{f(x, c + \Delta c) - f(x, c)}{\Delta c}\, dx$$

By the mean value theorem (Section 11.7) applied to the function $f(c)$, the integrand has the value

$$\frac{\partial f(x, \theta)}{\partial c} \Delta c$$

for some θ in the interval from c to $c + \Delta c$, provided there is a continuous finite derivative of $f(x, c)$ with respect to c in the whole interval $[a, b]$. The Δc cancels out of numerator and denominator. Now when we let $\Delta c \to 0$, the left side approaches the partial derivative, and the integrand approaches its partial derivative. Hence we have the formula

$$\frac{\partial F(c)}{\partial c} = \int_a^b \frac{\partial f(x, c)}{\partial c} \, dx \qquad (18.8\text{-}3)$$

provided $\partial f/\partial c$ is finite and continuous in the finite interval $[a, b]$. Why all these conditions? Some come from the hypotheses of the mean value theorem, and some come from the fact that we want a bound on the final integrand that is uniform in the interval lest at some place the values become unbounded. Finally, unless the interval is finite, we may be facing a situation where we have an error that is zero times an infinity, and the indeterminate value may not be zero. In loose words, you can differentiate under an integral sign with respect to a parameter provided the derivative is continuous in the interval of integration. In practice, you had best avoid improper integrals in both the original integral and in the derivative integral, or else be prepared to examine things much more closely than we have done.

Example 18.8-1

Consider the known integral

$$\int \frac{dx}{x^2 + c} = \frac{1}{\sqrt{c}} \arctan \frac{x}{\sqrt{c}} + C$$

If we differentiate the left side with respect to c, we get

$$-\int \frac{dx}{(x^2 + c)^2}$$

We can, of course, differentiate again and again, obtaining the integrals

$$\int \frac{dx}{(x^2 + c)^{n+1}} = \frac{(-1)^n}{n!} \frac{\partial^n \{1/\sqrt{c} \arctan x/\sqrt{c}\}}{\partial c^n}$$

The simple formula (18.8-3) allows us to generate new integrals from old ones (which contain a parameter) by the *direct* process of differentiation with respect to the parameter. Indeed, there is an art of changing variables to introduce the proper parameter into the integral so that the differentiation will lead you to the integral you want to solve. It is a powerful method for generating a table of integrals.

Example 18.8-2

Consider the integral

$$\int_0^\infty e^{-ax} \, dx = \frac{1}{a}$$

Differentiating once with respect to the parameter a and multiplying through on both sides by (-1), we get

$$\int_0^\infty xe^{-ax}\,dx = \frac{1}{a^2}$$

Differentiating again we get (again multiply by -1)

$$\int_0^\infty x^2 e^{-ax}\,dx = \frac{2!}{a^3}$$

and in general we get from the $(n-1)$th derivative

$$\int_0^\infty x^{n-1} e^{-ax}\,dx = \frac{(n-1)!}{a^n}$$

which is the $\Gamma(n)$ function of Section 15.11 and (18.8-2). What was obtained there by integration by parts is here found by simple differentiation. We ignored the warning about improper integrals, because in this case it is easy to start with a finite range, get the answer, and then let the range go to infinity.

Example 18.8-3

Consider the indefinite integral

$$\int \sin ax\,dx = \frac{-1}{a}\cos ax + C$$

Differentiate with respect to a:

$$\int x\cos ax\,dx = \frac{1}{a^2}\cos ax + \frac{1}{a}x\sin ax + C$$

where we have added the constant of integration, which vanished when we differentiated, but is needed for indefinite integrals. Again, what you might have obtained by integration by parts is found by differentiation.

Since the method is clearly so powerful, we should look a bit closer and answer the question of partial differentiation when the parameter can also occur in the limits.

Example 18.8-4*

Suppose we have the integral

$$F(c) = \int_{a(c)}^{b(c)} f(x,\,c)\,dx$$

and want to differentiate with respect to c. One suspects how the formula will go, but to be safe we will go through the details of the four-step process. It is only algebra with nothing really new in the whole process, just the use of old ideas in a new situation.

$$\frac{F(c+\Delta c) - F(c)}{\Delta c} = \frac{\int_{a(c+\Delta c)}^{b(c+\Delta c)} f(x,\,c+\Delta c)\,dx - \int_{a(c)}^{b(c)} f(x,\,c)\,dx}{\Delta c}$$

But this can be written as the difference of two integrals over the common range plus the pieces over the extra ranges:

$$\frac{F(c + \Delta c) - F(c)}{\Delta c} = \int_{a(c)}^{b(c)} \left\{ \frac{f(x, c + \Delta c) - f(x, c)}{\Delta c} \right\} dx$$

$$+ \int_{b(c)}^{b(c + \Delta c)} \frac{f(x, c + \Delta c)}{\Delta c} dx$$

$$- \int_{a(c)}^{a(c + \Delta c)} \frac{f(x, c + \Delta c)}{\Delta c} dx$$

In the first integral we apply the mean value theorem (Section 11.7), and in the second two we apply the mean value theorem for integrals [Section 13.3: the average over an interval is some value of the integral in the interval with the $g(x) = 1$]:

$$\frac{F(c + \Delta c) - F(c)}{\Delta c} = \int_{a(c)}^{b(c)} \frac{\partial f(x, c + \theta)}{\partial c} dx$$

$$+ f(\theta_1, c + \Delta c) \left\{ \frac{b(c + \Delta c) - b(c)}{\Delta c} \right\}$$

$$- f(\theta_2, c + \Delta c) \left\{ \frac{a(c + \Delta c) - a(c)}{\Delta c} \right\}$$

with θ_1 in $[b(c), b(c + \Delta c)]$ and θ_2 in $[a(c), a(c + \Delta c)]$. We now take the limit as $\Delta c \to 0$.

$$\frac{\partial F(c)}{\partial a} = \int_{a(c)}^{b(c)} \frac{\partial f(x, c)}{\partial c} dx + f(b, c) \frac{\partial b(c)}{\partial c} - f(a, c) \frac{\partial a(c)}{\partial c} \qquad (18.8\text{-}4)$$

This is more or less what you would expect from the function of a function rule for differentiation, plus the rule for differentiation of an integral with respect to its limits. If we write the indefinite integral of $f(x, c)$ as $F(x, c)$, then the integral would be

$$F(b(c), c) - F(a(c), c)$$

and you see the derivative emerging. It is futile to try to remember the formula in detail; it is easy to remember if you ask, "What does it say?" It says that you differentiate with respect to the parameter wherever it occurs; in the integrand you copy the result with the partial derivative (as before), but for the parameter in the limits, remembering the fundamental theorem of the calculus, you get the integrand at the limits *times the derivatives of the limits*; the upper limit of course is positive and the lower is negative. Thus it includes the fundamental theorem of the calculus that the derivative of an integral with respect to the upper limit is the integrand.

EXERCISES 18.8

1. Given the integral $1/a(\sqrt{2\pi}) \int_0^\infty \exp(-x^2/a^2) \, dx = 1$, differentiate with respect to a once to get a new integral.
2. Use the formula for $\int_0^1 dx/\sqrt{a^2 - x^2}$ to get new integrals.

3. The integral leading to the arctangent seems difficult to differentiate repeatedly. Note that, using complex numbers, $1/(c^2 + x^2)$ can be factored into complex linear factors and you are led formally to the ln of complex numbers, whose formal derivatives are easily found.
4. Discuss introducing a parameter by a change of variables and then using it to get new integrals.

8.9* MECHANICS

We earlier noted that the calculus was developed in terms of mechanics. It is mainly the words that are changed between mechanics and statistics, as we will now show.

We have already discussed velocity (Section 9.3) and acceleration (Section 9.4). The next application is to the center of gravity. At first we imagine a solid of uniform density (which for convenience we can take as 1) and ask where all the mass could be put so that it is equivalent to being at one single point. This is the *same question* as the mean value of a distribution. We can take the origin of the coordinate system at any place we please and measure the distance to the typical unit of mass from that point. This must be divided by the total mass (in probability theory the total probability was 1, so it was ignored). The formula for the x coordinate of the center of gravity is

$$\bar{x} = \frac{\iiint xm \, dV}{\iiint m \, dV}$$

We can think of $dV = dx \, dy \, dz$, and m as the element of mass. We will then be summing an element to get a row, the rows to get a plane, and the planes to get the volume. This will be studied in the next chapter; meanwhile we will only use elements of volume in the cases where we can get a suitably shaped element of volume directly. There are similar formulas for the other coordinates of the center of mass. This is also called the *center of gravity* if you prefer. Notice that the mass or density will drop out, and the result is independent of units used except for the distance unit. We can therefore drop the quantity m from the formula for the center of gravity if we wish, *provided* it is homogeneous material.

If the density is not uniform, then we merely have a weighted average in both the numerator and denominator to compute.

Example 18.9-1

Find the center of mass of a uniform hemisphere of radius a. We first pick the coordinate system at the center of the sphere. We have the equation

$$z = \sqrt{a^2 - x^2 - y^2}$$

and use the upper half-plane as the hemisphere. The center in x and y lie on the axis of symmetry, and it is only the z coordinate that is needed. The formula, using discs as in Figure 18.9-1, is

$$\bar{z} = \frac{\int_0^a \pi x^2 z \, dz}{2\pi a^3/3}$$

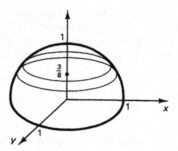

Figure 18.9-1 Hemisphere

where we have used the fact that we know the volume of the hemisphere. When we recognize that we are using the circular symmetry found by rotating the circle

$$x^2 + z^2 = a^2$$

about the z-axis, we can use the fact that

$$x^2 = a^2 - z^2$$

The integral becomes

$$\bar{z} = \int_0^a (a^2 - z^2)z \, dz \left(\frac{3}{2a^3}\right)$$

$$= \left(\frac{3}{2a^3}\right)\left(\frac{a^2 z^2}{2} - \frac{z^4}{4}\right)\Bigg|_0^a$$

$$= \left(\frac{3}{2a^3}\right)a^4\left(\frac{1}{2} - \frac{1}{4}\right) = \frac{3a}{8}$$

This seems like a reasonable result; the average of the mass is slightly below the midpoint on the z-axis.

Another mechanical concept is the *moment of inertia* about a line through the center of mass. This is exactly the same as the variance (about the mean) in statistics, again remembering to divide by the total mass. In place of the first moment used for the center of gravity, we use the second moment about the axis of rotation.

Example 18.9-2

Moment of inertia of a sphere. The moment of inertia of a sphere rotating about, say, the z-axis can be found by cutting the sphere up into cylinders (Figure 18.9-2) and noting that all points on the cylinder have the same distance from the axis of rotation. Thus the ratio of the second moment divided by the volume (mass) gives the moment of inertia about the z-axis:

$$I_z = \frac{\int_0^a x^2 (2z)m(2\pi x) \, dx}{\int_0^a (2z)m(2\pi x) \, dx}$$

$$= \frac{4\pi \int_0^a x^3 \sqrt{a^2 - x^2} \, dx}{4\pi \int_0^a x \sqrt{a^2 - x^2} \, dx}$$

Set $x = a \sin \theta$, $dx = a \cos \theta \, d\theta$, to get

$$I_z = \frac{\int_0^{\pi/2} a^5 \sin^3 \theta \cos^2 \theta \, d\theta}{\int_0^{\pi/2} a^3 \sin \theta \cos^2 \theta \, d\theta}$$

$$= a^2 \frac{W_3 - W_5}{W_1 - W_3} = \frac{2a^2}{5}$$

Another concept that is often used in mechanics is the center of pressure, say on a dam. We picture the cross section of the water against the dam as a parabola (Figure 18.9-3). The pressure of the water is proportional to the mass in a column above the element shown. Thus the pressure is the first moment about the top of the water. Nothing else is new.

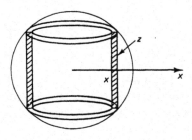

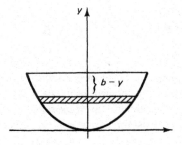

Figure 18.9-2 Moment of inertia of a sphere

Figure 18.9-3 Pressure on a dam

Example 18.9-3

Using the parabolic shaped dam

$$y = ax^2$$

we have the total pressure as

$$p = \int_0^b k(b - y) \, 2x \, dy$$

where k is some suitable constant (see Figure 18.9-3). We can either substitute for the x values the corresponding values in y or else change from y to x as the independent variable. We do the latter.

$$p = 2k \int_0^{\sqrt{b/a}} (bx - ax^3) \, 2ax \, dx$$

$$= 4ak \int_0^{\sqrt{b/a}} (bx^2 - ax^4) \, dx$$

$$= 4ak \left[\frac{bx^3}{3} - \frac{ax^5}{5} \right] \Bigg|_0^{\sqrt{b/a}}$$

$$= 4ak \left[\frac{b}{3} \left(\frac{b}{a} \right)^{3/2} - \frac{a}{5} \left(\frac{b}{a} \right)^{5/2} \right]$$

$$= 4ak\left(\frac{b^{5/2}}{a^{3/2}}\right)\left[\frac{1}{3} - \frac{1}{5}\right]$$

$$= \frac{8k}{15}\frac{b^{5/2}}{a^{1/2}}$$

You see that, if you understand the physical concepts, then it is a matter of using the same sort of calculus as you have been doing to get the answers. Of course, the details may be messier, but the calculus ideas are the same. If you do not understand the field of application, then it all becomes hazy to say the least! To use mathematics successfully, you need to know what you are talking about, both the mathematics and the field of application.

EXERCISES 18.9

1. Find the center of gravity of a cone.
2. Find the center of gravity of half an ellipsoid of revolution.
3. Find the moment of inertia of a cylinder rotated about the axis of symmetry.
4. Find the moment of inertia of a cone rotated about its axis of symmetry.
5. Find the force on a semicircular dam filled to a height h.

18.10* FORCE AND WORK

Force is another topic that arises in mechanics. Forces play a central role in mechanics. Any force can be broken up into a pair of forces in different directions, such that the original force is the diagonal of the parallelogram formed by the two derived forces (Figure 18.10-1). See Section 9.4.

Newton's law

$$f = ma$$

connects mass with weight. In a gravitational field, such as the earth, the weight of an object is the mass times the gravitational force, usually labeled g.

Force applied through a distance is called *work*, the force times the distance being the measure of the amount of work. If the force is constant and the motion is in a straight line, then the computation is easy. But if (1) the mass varies with time,

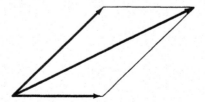

Figure 18.10-1 Parallelogram of forces

as in a rocket using up its fuel, (2) the force varies with position, or (3) the motion is along a curved line and the motion is not necessarily in line with the force, then the calculus is necessary. Notice that work can be positive or negative, meaning you put in work or get it out of the system studied. It is also true that the amount of mass moved, as well as the distance it is moved per unit time, may vary as the task proceeds.

Example 18.10-1

The force exerted by a spring, when displaced from its equilibrium position (of no force), is proportional to its displacement (extension) x (Figure 18.10-2). This is known as Hooke's (1635–1703) law. Let the constant of proportionality be k; that is,

$$f = kx$$

If the end is moved from $x = 0$ to $x = a$, how much work is done?

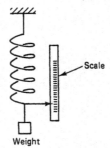

Weight   **Figure 18.10-2**   Spring scale

We have to compute

$$W = \int_0^a f(x)\,dx$$

$$= \int_0^a kx\,dx = \frac{ka^2}{2}$$

Example 18.10-2

How much work is required to lift the water (one cubic foot of water weighs 62.5 pounds) in a cone shaped as in Figure 18.10-3 to a level 2 feet above the rim? We take an element of volume $\pi x^2\,dy$, and since the weight is 62.5 lb/ft^3, we merely multiply by 62.5, and integrate it when multiplied by the distance the element of volume is moved; that is, $2 + (10 - y)$. We have

$$W = 62.5 \int_0^{10} (12 - y)\pi x^2\,dy$$

We need to get x in terms of y, and from the figure $x = 3y/10$. Hence we have

$$W = (62.5)\pi\left(\frac{3}{10}\right)^2 \int_0^{10} (12y^2 - y^3)\,dy$$

$$= (62.5)\frac{9\pi}{100}\left(4y^3 - \frac{y^4}{4}\right)\bigg|_0^{10}$$

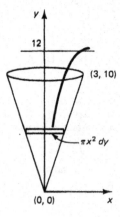

Figure 18.10-3 Work

$$= (62.5)\frac{9\pi}{100}(4000 - 2500)$$

$$= (62.5)(9\pi)(40 - 25) = (62.5)135\pi$$

where we have kept the weight separate so you can follow it easily.

Example 18.10-3

Show that the work done in emptying a volume is the same as moving the center of mass through the distance. We simply write out the two formulas, noting how the density term arises, and see the equivalence. Thus there is little new in the matter of emptying volumes of liquid.

EXERCISES 18.10

1. Find the work done in emptying a hemispherical bowl when the water is lifted to a height h.

2. How much work is done when a hemispherical bowl is drained through a hole in the bottom?

3. How much work is done in bailing out a 1-foot cube of water when the water is lifted just over the upper edge of the cube?

18.11* INVERSE SQUARE LAW OF FORCE

Many phenomena, for example the force of gravity, the force between electrostatic charges, and light intensity, have an inverse square law

$$f = \frac{k}{d^2}$$

where d is the distance involved. For example, the force of gravity is given by

$$f = \frac{Gm_1 m_2}{d_2}$$

where G is a universal constant (depending on the units used to measure things), m_1 and m_2 are the two attracting masses, and d is the distance between them.

Example 18.11-1

Suppose you have wire of uniform density in the shape of a semicircle of radius a. What is the force on a unit of mass located as shown in Figure 18.11-1?

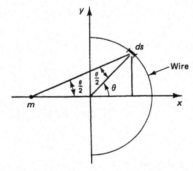

Figure 18.11-1   Semicircular wire

Each element ds along the circumference has mass proportional to $a\, ds$. The component of force along the axis is multiplied by $\cos \theta/2$ (since the angle is half the central angle). The distance comes from the law of cosines:

$$d^2 = a^2 + a^2 + 2aa \cos \theta = 2a^2(1 + \cos \theta)$$

The element of force due to the element of mass along the circumference has some constant of proportionality, say k, and is given by

$$df = \frac{ka \cos (\theta/2)}{2a^2(1 + \cos \theta)} d\theta$$

The total force is, therefore,

$$f = \int_{-\pi/2}^{\pi/2} \frac{ka \cos (\theta/2)}{2a^2(1 + \cos \theta)} d\theta$$

We use the half-angle formula on the denominator and symmetry to get

$$f = \frac{k}{a} \int_0^{\pi/2} \sec \frac{\theta}{2} \frac{d\theta}{2}$$

$$= \frac{k}{a} \ln\left\{ \sec \frac{\theta}{2} + \tan \frac{\theta}{2} \right\} \Big|_0^{\pi/2}$$

$$= \frac{k}{a} \ln\{\sqrt{2} + 1\} = \frac{k \ln(\sqrt{2} + 1)}{a}$$

Example 18.11-2

What is the attraction of a rod of length $2a$ on a point opposite its center and at a distance h? We draw Figure 18.11-2. Remember that the force is a vector, and by symmetry we see that only the component in the direction perpendicular to the rod will matter. The square of the distance to an element of mass of the rod is

$$x^2 + h^2$$

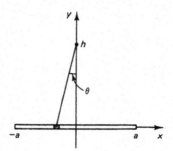

Figure 18.11-2   Attraction of rod

We must take the cosine component of the force:

$$\frac{h}{\sqrt{x^2 + h^2}} \frac{1}{x^2 + h^2} = \frac{h}{(x^2 + h^2)^{3/2}}$$

and sum this up for each element dx. We have (k is the density of the rod)

$$f = k \int_{-a}^{a} \frac{h}{(x^2 + h^2)^{3/2}} \, dx$$

Set $x = h \tan \theta$; then $dx = h \sec^2 \theta \, d\theta$, and

$$f = 2k \int_{0}^{\arctan a/h} \frac{h}{h^3 \sec^3 \theta} h \sec^2 \theta \, d\theta$$

$$= \frac{2k}{h} \int_{0}^{\arctan a/h} \cos \theta \, d\theta$$

$$= \frac{2k}{h} \sin\{\arctan a/h\}$$

From Figure 18.11-2, we see that the answer is

$$f = \frac{2k}{h} \frac{a}{\sqrt{a^2 + h^2}} = \frac{2ka}{h\sqrt{a^2 + h^2}}$$

which seems to be reasonable.

EXERCISES 18.11

1. Find the force exerted by a wire ring on an element of mass located on the axis of the circle at a distance b away.
2. Using the result of Exercise 1, find the force of a disc on an element of mass on its axis.
3. Find the increase in velocity of a point mass in falling through a gravitational field from b to a. Let b approach infinity. This is also the "escape velocity" and can be evaluated from the gravitational force at the surface of a planet and the radius of the surface. For the earth, it is about 6.9 miles per second (neglecting all such details as air drag, rotational effects, etc.).

8.12 SUMMARY

In this chapter, now that we have a larger class of functions we can handle, we first reviewed a few of the standard applications of the calculus. Then we examined a number of new, but at this point not essentially difficult, applications. They depended mainly on a knowledge of the field of application. When you know your field well, then it is not as difficult as it may at first seem to make new applicaitons of the calculus to your field. It is necessary to practice converting word problems to mathematical formulas and the reverse, from the mathematics to get back to the words. By practicing these two steps, you prepare yourself for when the time comes for you to contribute to the advancement of knowledge.

Most of the material should be learned not in terms of the results, but in terms of the approach. Thus the elementary triangle, from which we derived the arc length and later the formulas for surface of revolution, is the basic concept and can hardly be called sophisticated or difficult. It is interesting that the two apparently different notions of curvature should turn out to be the same thing (almost).

Section 18.7, devoted to extensions, is based on the generalization we derived of the fundamental theorem that connected the summation of elements of any general form with the corresponding integral. It is another example of learning what the general formula says and then being willing to apply the generality, rather than being limited to the specific cases you began with. Thus you see, again, the value of generalization and the power that comes when you have investigated a result to understand essentially what it does and does not depend on. The author is convinced that (1) only by making these generalizations your own can you hope to make significant contributions to your chosen field, and (2) that it is often only this step that significantly advances a field. The world is rapidly changing, and the applications of the calculus, probability, and statistics will also probably be greatly extended and changed within your lifetime; so if you are to play a part in this process, it is wise to master not only the details, but the underlying assumptions that produced the current version of mathematics.

The differentiation of an integral is a natural question and generalizes the fundamental theorem that the derivative of the integral with respect to the upper limit gives the integrand at that point. We investigated the problem in some detail to familiarize you with the processes of abstraction, generalization, and extension. These are the heart of mathematics; these are what the book is really about.

The specific applications to mechanics, force, work, and inverse square law are interesting applications of mathematics and have shaped mathematics extensively; yet they are only the applications, not the essence of mathematics. We avoided the applications to probability and statistics in this chapter, but similar examples could be found easily.

19

Applications Using Several
Independent Variables

19.1 FUNDAMENTAL INTEGRAL

The original approach to the integral was based on (1) dividing the interval into a large number of parts, (2) constructing an estimate of the area for each interval, (3) adding all the elements of area together, and then (4) taking the limit as the width of the largest interval approached zero. This simple concept was later generalized to cover the process of summation generally. Thus the idea of integration is independent of the idea of area. The integral is the limit of a suitable sum, regardless of the meaning of the elements entering into the sum.

In the approach to the problem of several independent variables, we begin with the case of two independent variables. If we can do this case, it will probably be easy to extend to three dimensions, to four, and so on, until we understand the case of n independent variables. We have to consider not only independent variables in geometry, but also independent random variables as we did when we examined the sum of n independent identically distributed (IID) random variables in the law of large numbers.

We first consider the problem of finding the volume under a continuous surface

$$z = f(x, y)$$

above some region in the x–y plane. This is analogous to the area under a curve and above an interval of a line. For example, we might ask for the volume under the surface

$$z = xy \tag{19.1-1}$$

where x and y are confined to some region R in both variables (above the unit square $0 \leq x, y \leq 1$ is one such example).

We divide up the region R in the x–y plane (Figure 19.1-1) into small rectangles $\Delta x_i \Delta y_j$ and, using the height $z = f(x, y)$ at some point (x_i, y_j) in the rectangle $\Delta x_i \Delta y_j$ as an estimate of the average height in the rectangle, we sum all these elements of volume

$$f(x_i, y_j)\Delta x_i \Delta y_j = f(x_i, y_j)\Delta A_{i,j} \tag{19.1-2}$$

to get an estimate of the volume we want. We then take the limit of the sum, as we let the size of the rectangles decrease such that the largest dimension of any rectangle approaches zero. This is the natural extension of the idea of an integral to two independent variables. We write this as

$$\int_R f(x, y) \, dA \tag{19.1-3}$$

where R is the region in the x–y plane, and dA is used to remind us of the element of area in the x–y plane.

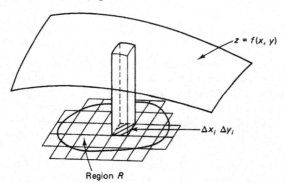

Region R **Figure 19.1-1**

It is important to see that if we sum all the rectangular elements of area $\Delta A_{i,j} = \Delta x_i \Delta y_j$ in the y direction, and then sum these sums, we will get the same result as if we first summed in the x direction, and then summed these sums. Each element of area will be included once and only once. Thus the area dA in Figure 19.1-1 can be written in either of the two forms:

(1)
$$\int_R f(x, y) \, dA = \int_a^b \int_{c(x)}^{d(x)} f(x, y) \, dy \, dx$$

where $c(x)$ and $d(x)$ describe the limits in the y direction of the inner summation for that x, and a and b are the limits in the x direction (see Figure 19.1-2); or the alternative of summing first in the x direction (Figure 19.1-3), which gives

(2)
$$\int_R f(x, y) \, dA = \int_c^d \int_{a(y)}^{b(y)} f(x, y) \, dx \, dy$$

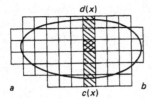

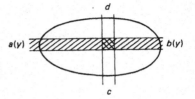

Figure 19.1-2   **Figure 19.1-3**

where this time it is the $a(y)$ and $b(y)$ that are the limits of the summation in the x direction (for that particular y), and c and d are the limits in the y direction. Thus the *double integral* of a continuous function is equivalent to the *iterated integral* in either order. The student is left the exercise of constructing upper and lower sums and showing that for continuous functions they approach each other in the limit, and thus squeeze this estimate to the same limit. You must also have the upper sum contain the region R, while the lower sum must lie entirely within the region R. As you let the maximum size of an element of area approach zero, the differences of the contributions along the edge will also approach 0. The region R is thus supposed to have a finite length to its boundary. Improper integrals can, of course, arise and require investigation, but we will not do so here.

Example 19.1-1

Given Equation (19.1-1) of the surface $z = xy$, and given the unit square as the region R above which you want the volume, you have the iterated integral

$$J = \int_0^1 \int_0^1 xy \, dA = \int_0^1 \int_0^1 xy \, dy \, dx$$

To evaluate this integral, we start by evaluating the inner integral (remember that y is the variable and therefore x is a constant in *this* inner integration). The inner integration followed by the outer integration gives

$$J = \int_0^1 \frac{xy^2}{2} \Big|_0^1 dx = \int_0^1 \frac{x}{2} dx = \frac{x^2}{4} \Big|_0^1 = \frac{1}{4}$$

Example 19.1-2

The previous example (19.1-1) of an integral over a square did not involve variable limits in the inner integral. Let us find the volume of a sphere of radius a by this new method (as a check on the method). Using symmetry, we will find the area in one octant only and then multiply by the factor 8 to get the total volume. Let the surface of the sphere in the first octant be defined by

$$z = \sqrt{a^2 - x^2 - y^2}$$

What is the region in the x–y plane under the surface? It is the region in the first quadrant of the x–y plane $(z = 0)$ inside the circle

$$x^2 + y^2 = a^2$$

See Figure 19.1-4. If we decide to use the inner integral with respect to dy, then we have

$$V = 8 \text{ volume of octant} = 8 \int_0^a \int_0^{\sqrt{a^2-x^2}} z \, dy \, dx$$

$$= 8 \int_0^a \int_0^{\sqrt{a^2-x^2}} \sqrt{a^2 - x^2 - y^2} \, dy \, dx$$

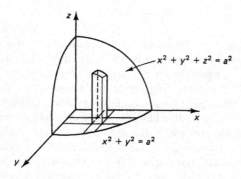

Figure 19.1-4   Volume of a sphere

Because the integration in the inner integral is with respect to y, the variable x is viewed as a constant. Think of $a^2 - x^2 = b^2$ as a constant. The inner integral is, therefore,

$$\int_0^b \sqrt{b^2 - y^2} \, dy$$

which suggests the substitution

$$y = b \sin \theta$$

The inner integral becomes

$$\int_0^b \sqrt{b^2 - y^2} \, dy = \int_0^{\pi/2} b \cos \theta \, b \cos \theta \, d\theta$$

$$= b^2 \int_0^{\pi/2} \frac{1 + \cos 2\theta}{2} \, d\theta = \frac{b^2 \pi}{4}$$

Now putting this in the outer integral, and remembering that $b^2 = a^2 - x^2$, we have (integrating and substituting in the limits in one step)

$$V = 8 \left(\frac{\pi}{4} \right) \int_0^a (a^2 - x^2) \, dx$$

$$= 2\pi \left(a^3 - \frac{a^3}{3} \right) = \frac{4}{3} \pi a^3$$

as the volume of the sphere. Thus we see that the method, in this check case, gives the correct answer.

Example 19.1-3

If the coefficients a, b, and c are selected independently and at random from the interval $[0, 1]$ what is the probability that the roots of the quadratic

$$ax^2 + 2bx + c = 0$$

are real?

First, the sample space of three random variables is a three-dimensional cube of side 1. Thus the probability density $p(a, b, c) = 1$. The condition for real roots is, of course, the discriminant (for the above form of the quadratic)

$$b^2 - ac \geq 0$$

To describe the surface that divides the region of real roots from the region of complex roots, we need to pick two of the coefficients as the independent variables and the third as the dependent variable. Since the variables a, b, and c can take on the value 0, we want to avoid any divisions. Thus we think of the independent variables as being a and c and the dependent variable as b. The whole sample space is the unit cube. The admissible region is above the surface

$$b^2 = ac$$

which divides the whole sample space into two regions.

What does this surface look like? After some thinking we note that $a = b = c$ is always a solution. The points along the line from the origin to the point $(1, 1, 1)$ all lie on the surface. Next we ask ourselves what the surface looks like for a fixed constant b. This is a plane parallel to the a–c plane. The curve in that plane (of height b) is an equilateral hyperbola through the point $a = b$ and $c = b$. For a or c equal to 1, the other value is b^2. We can draw these cross sections easily as shown in Figure 19.1-5. It is not easy to draw the isometric figure in this case, but a sketch is given in Figure 19.1-5.

The point with values $b = 1$, $a = c = 0$ is a "success." Hence the probability we want is (again) the volume of the region above the surface and inside the unit cube (which

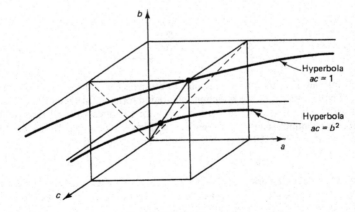

Figure 19.1-5

has unit volume), with $p(a, b, c) = 1$. This total probability is the integral

$$P = \int_0^1 \int_0^1 (1 - \sqrt{ac})\, da\, dc$$

$$= 1 - \int_0^1 \int_0^1 \sqrt{ac}\, da\, dc$$

But a and c are now independent dummy variables, so this is the same as the product of two integrals, each of exactly the same form. Hence we have

$$P = 1 - \left\{ \int_0^1 \sqrt{x}\, dx \right\}^2 = 1 - \left(\frac{2}{3} \right)^2 = 1 - \frac{4}{9} = \frac{5}{9}$$

Does this seem reasonable? We inspect the region in the figure and try to decide if it is believable. A bounding figure [the cube minus a tetrahedron with vertex $(1, 1, 1)$] would have volume $5/6$, and the answer seems reasonable.

We can make a further check, and do a Monte Carlo (Section 13.2) on it by (1) taking the coefficients a, b, and c as uniform random numbers each in the interval $[0, 1]$, (2) forming, say, 100 such quadratics, and then (3) computing the sign of the discriminant. Table 19.1-1 gives the results (from a programmable hand calculator) and convinces us that the computation agrees reasonably with experience. The theoretical value $= 0.5555 \ldots$. Notice that all the values tend to be on one side; there is not the alternate crossing of the true value that one might expect. An explanation of why a Monte Carlo tends to be most of the time on one side rests on a result (beyond this text) known as the Arcsine law, which indicates that, for example, in tossing coins one opponent or the other will be ahead, probably, most of the time.

TABLE 19.1-1

Successes	Total trials	Ratio
57	100	0.57
123	200	0.615
183	300	0.61
300	500	0.6
576	1000	0.576

Instead of using random samples to estimate the volume, we could use a regular grid of points, based on the midpoint integration formula (in each of the three variables, Section 11.11). The fault with this is that a lot of points will fall right on the surface (unlike the random sample case), and we will have trouble classifying them fairly. The result of 1000 ($10 \times 10 \times 10$) trials using the regular grid was

561 successes → 0.561 estimate of volume

Thus both numerical methods of checking the reasonableness of the result have their faults, but both tend to confirm the computation of the probability of a random polynomial of the given form having real roots.

Example 19.1-4

Suppose we were to take the quadratic in the previous problem in the form (drop the factor 2 in the middle coefficient)

$$ax^2 + bx + c = 0$$

What would be the answer? The surface that divides the two regions this time is

$$b^2 - 4ac = 0$$

This cuts the top surface $b = 1$ in the curve

$$ac = \frac{1}{4}$$

Therefore, the region for integration is no longer the whole unit square but only over the area bounded by $ac = 1/4$. The integral (see Figure 19.1-6) is therefore

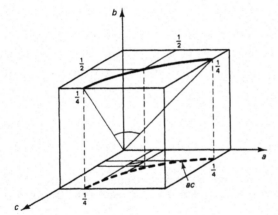

Figure 19.1-6 $b^2 = 4ac$

$$P = \int_0^{1/4}\!\!\int_0^1 \{1 - 2\sqrt{ac}\} \, da \, dc + \int_{1/4}^1\!\!\int_0^{1/4c} \{1 - 2\sqrt{ac}\} \, da \, dc$$

The two intervals $[0, 1/4]$ and $[1/4, 1]$ have different upper limits for their inner integrals.

We first do the two inner integrals

$$\int_0^1 \{1 - 2\sqrt{ac}\} \, da = 1 - 2\sqrt{c}\left(\frac{2}{3}\right) = 1 - \frac{4\sqrt{c}}{3}$$

$$\int_0^{1/4c} \{1 - 2\sqrt{ac}\} \, da = \frac{1}{4c} - 2\sqrt{c}\left(\frac{2}{3}\right)\left(\frac{1}{4c}\right)^{3/2}$$

$$= \frac{1}{4c} - \frac{1}{6c} = \frac{1}{12c}$$

These are then put into their outer integrals, and the results added to get

$$P = \int_0^{1/4} \left\{ 1 - \frac{4\sqrt{c}}{3} \right\} dc + \int_{1/4}^1 \frac{1}{12c} dc$$

$$= \frac{1}{4} - \left(\frac{8}{9}\right)\left(\frac{1}{8}\right) + \frac{1}{12} \int_{1/4}^1 \frac{dc}{c}$$

$$= \frac{1}{4} - \frac{1}{9} + \frac{1}{12} \ln c \Big|_{1/4}^1$$

$$= \frac{5}{36} + \frac{1}{12}(2 \ln 2) = \frac{5 + 6 \ln 2}{36} = 0.2544 \ldots$$

This is quite different from the first answer and shows how the answer to the question of the probability of a real root can depend on the details of the model assumed. The result of a corresponding Monte Carlo experiment is as follows:

Successes	Total trials	Ratio
28	100	0.28
60	200	0.3
85	300	0.28333 ...
113	400	0.2825
139	500	0.278

True answer = 0.2544 . . . .

Since multiplying the original quadratic equation through by a constant N is equivalent to increasing the range of the coefficients a, b, and c to $[0, N]$, the probabilities in Examples 19.1-3 and 19.1-4 will remain the same for any range $0 < x < N$.

EXERCISES 19.1

1. From Example 8.7-5, the discriminant of the reduced cubic $y^3 + py + q = 0$ is $(q/2)^2 + (p/3)^3 \le 0$. If p and q are chosen uniformly at random in the interval $-1 \le p$, $q \le 1$, what is the probability of the equation having three real roots?
2. Find the volume inside the paraboloid $z = 1 - x^2 - y^2$ and above the xy-plane.
3. Find the volume in the first octant below $z = x + y$ and inside the cylinder $y = 1 - x^2$.
4. The base of a cylinder lies between $y = x$ and $y = x^2$ ($0 \le x \le 1$), and the top is cut off by $z = x^2y^2$. Find, in two different ways, the volume in the first octant.

9.2 FINDING VOLUMES

From the previous few examples we see that the central difficulty in setting up such problems is that of visualizing the three-dimensional surfaces involved. This is not a trivial problem; even experts have their troubles in doing this. Therefore, let us examine the resources we have available.

1. There is symmetry of various kinds to be exploited as best you can (Section 6.11).
2. We can draw the intersections of the surface on the three coordinate planes (if there are any intersections). These are called the *traces* of the surface. They often give very valuable indications as to the general shape of the surface.
3. We can set the value of any one variable equal to a constant, and then draw the plane figure in the corresponding plane parallel to the corresponding coordinate plane. These are called *cross sections*. There are thus three possible ways of cross-sectioning the surface to get an idea of what it looks like.
4. There is the tedious method of plotting points that lie on the surface, triples of values that satisfy the equation for the surface.

But the truth is, in many cases there is no easy way of visualizing surfaces, and often the isometric sketch is very difficult to draw even when you have all the above information available. Go in the corner of a room, use the two walls and the floor as the coordinate planes in the first octant (or other octant if necessary), and simply go to work to combine the various pieces of information you have obtained. This is often the only way you can convince yourself that you indeed have the proper picture of the surface. If you do not get the surface correct, then the integrals you set up will not be correct, and it is therefore very unlikely that the answer you finally get after all the work of integration will be correct. Do not try to hurry through this vital step of getting straight in your mind the picture of the surfaces involved.

We will at the beginning draw the figures of the surfaces involved rather carefully, but will deliberately gradually draw more and more the hasty sketches you are likely to generate. It is necessary that you learn the art of getting the figure straight before you go ahead with the problem.

Example 19.2-1

Interchange of order of integration. Evaluate

$$J = \int_0^1 \int_x^1 \exp\left(y^2\right) \, dy \, dx$$

Using Figure 19.2-1, we interchange the order of integration to get

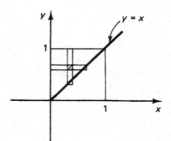

Figure 19.2-1

$$J = \int_0^1 \int_0^y \exp(y^2) \, dx \, dy$$

$$= \int_0^1 y \exp(y^2) \, dy = \frac{1}{2} \exp(y^2) \Big|_0^1$$

$$= \frac{e-1}{2} = 0.85914 \ldots$$

Example 19.2-2

Evaluate

$$\int_0^a \int_0^{\sqrt{a^2-x^2}} (x+y) \, dy \, dx = \int_0^a \left\{ x\sqrt{a^2-x^2} + \frac{1}{2}(a^2-x^2) \right\} dx$$

$$= -\frac{1}{3}(a^2-x^2)^{3/2} + \frac{1}{2}\left(a^2 x - \frac{x^3}{3} \right) \Big|_0^a$$

$$= \frac{a^3}{3} + \frac{a^3}{2} - \frac{a^3}{6} = \frac{2a^3}{3}$$

Example 19.2-3

Interchange of order of integration. Evaluate the integral, where $p(x)$ is a probability density,

$$\int_0^1 \int_0^x p(y) \, dy \, dx$$

Using Figure 19.2-2, we interchange the order of integration to get

$$\int_0^1 p(y) \int_y^1 dx \, dy = \int_0^1 p(y)(1-y) \, dy$$

$$= E\{1\} - E\{Y\}$$

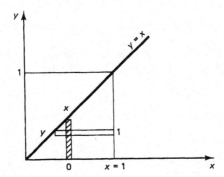

Figure 19.2-2   Interchange of limits

Example 19.2-4

Two points are chosen uniformly and at random on a line of unit length. What is the expected distance between them?

Let the two points be at x and y. The sample space is therefore the unit square with a uniform probability assignment $p(x) = p(y) = 1$. The diagonal (in Figure 19.2-3) is $x = y$ and has 0 distance between the two points. The points a distance y away from this line lie on a line parallel to this diagonal. The length of such a line is

$$\sqrt{2}(1 - y)$$

and the width of the element of area is

$$dw = dy/\sqrt{2}$$

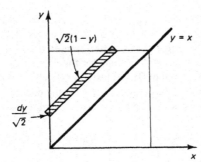

Figure 19.2-3   Two random points

Hence, using symmetry, we have the total area as a check:

$$2 \int_0^1 (1 - y) \, dy = 2\left(y - \frac{y^2}{2}\right)\Bigg|_0^1 = 2\left(1 - \frac{1}{2}\right) = 1$$

To get the expected value of the random variable D, which is the distance between the points, we compute

$$E\{D\} = 2 \int_0^1 y(1 - y) \, dy = 2\left(\frac{y^2}{2} - \frac{y^3}{3}\right)\Bigg|_0^1 = \frac{1}{3}$$

Does this seem like a reasonable answer? If one point fell at the end, then the average distance to the other point would be $\frac{1}{2}$; while if it fell in the middle, then the average would be $\frac{1}{4}$. Yes, the answer seems reasonable.

EXERCISES 19.2

1. Find the volume beneath $z = 1 - (x^2 + y^2)$ and above the x–y plane.
2. Find the volume between $z = x^2 + y^2$ and $z = 2(x^2 + y^2)$ and above $x^2 + y^2 = a^2$.
3. Find volume between $z = x^2 + y^2$ and $z = 2 - (x^2 + y^2)$.
4. Find volume between $x^2 - y^2 - z^2 = 1$ and $x = 2$.
5. Generalize Example 19.2-3.
6. Find the volume in the first octant bounded by $az = xy$, $x^2 = 4ay$, $x - y = a$.
7. Given a region in the x–y plane bounded by $y = x, y = 0, x = 1$, write the integral below the surface $z = f(x, y)$ in two ways. Discuss the interchange of the order of integration in this case.
8. Solve Example 19.2-4 using $\int_0^1 \int_0^1 |y - x| \, dy \, dx$.
9. Generalize Example 19.2-3 to n iterated integrals.

19.3 POLAR COORDINATES

Polar coordinates are often used instead of rectangular coordinates. In polar coordinates in the plane you are given how far and in what direction to go. Thus the pair of numbers r and θ, the distance and the angle (in radians), names the point (r, θ). See Figure 19.3-1. In polar coordinates the constant values for the coordinates are straight lines through the origin ($\theta = c$) and concentric circles ($r = c$) about the origin.

The advantages of polar coordinates are connected with the ease of representing both lines through the origin and concentric circles. The disadvantages are several as you will see.

From Figure 19.3-2 it is easy to see that the rectangular coordinates (x, y) and

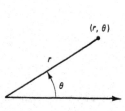

Figure 19.3-1 Polar coordinates

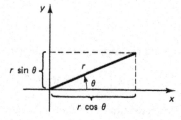

Figure 19.3-2

the polar coordinates (r, θ) are related by the equations

$$x = r \cos \theta$$

$$y = r \sin \theta \qquad\qquad (19.3\text{-}1)$$

and

$$r^2 = x^2 + y^2$$

$$\theta = \arctan \frac{y}{x} \qquad\qquad (19.3\text{-}2)$$

This second form is the conventional way of representing the arctangent, but in practice it is usually better to think of the arctangent as a function of two variables, arctan (x, y), so the value will fall in the proper quadrant. In the conventional way, you customarily assume that the angle is in the first or fourth quadrants. [Consider the point in rectangular coordinates $(-1, -1)$, convert it to the polar coordinates $r = \sqrt{2}$, $\theta = \pi/4$, the wrong point! When you convert back to rectangular coordinates, you have the rectangular point $(1, 1)$]. If you take both values, x and y, into consideration, you can place the point in its proper quadrant.

Consider the circle $r = a$ and the circle $r = -a$ (Figure 19.3-3). The two circles appear to be the same; all points have the distance a from the origin, but algebraically they have no common solution! There is an ambivalence about whether or not you may let r take on negative values. Some people believe that you should restrict $r \geq 0$, and some do not. There is no easy solution to this dilemma. If you do not so restrict r, then each point will have the representations (r, θ) and $(-r, \theta + \pi)$, which is awkward in practice.

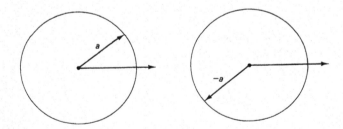

Figure 19.3-3

The restriction to nonnegative r does not completely solve the problem of multiple representations, however. Consider the spiral

$$r = \theta$$

(Figure 19.3-4). Next consider the spiral

$$r = \theta + 2\pi$$

Again, the two spirals give the same geometric picture, but have no algebraic solutions in common. Again, we have lost the fundamental property of the Cartesian coordinate

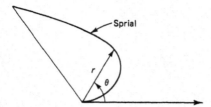

Figure 19.3-4 $r = \theta$ and $r = \theta + 2\pi$

system, that the algebra and geometry coincide exactly. To combat this difficulty, some people believe in restricting θ to a total angle of 2π, but others do not. It would appear that in practice what you should do depends on the problem at hand and not on some abstractly laid down rule. You should *listen* to each problem and then decide what is proper in that case.

How do straight lines and other curves appear in polar coordinates? You have only to take an equation in rectangular coordinates, and substitute (19.3-1)

$$x = r \cos \theta \quad \text{and} \quad y = r \sin \theta$$

to convert to polar coordinates.

Example 19.3-1

The general straight line has the equation

$$Ax + By + C = 0$$

This becomes

$$Ar \cos \theta + Br \sin \theta + C = 0$$

or

$$r = \frac{-C}{A \cos \theta + B \sin \theta} = \frac{1}{a \cos \theta + b \sin \theta} \tag{19.3-3}$$

where a and b are defined by $a = -A/C$ and $b = -B/C$ (supposing that $C \neq 0$). If $C = 0$, then the line passes through the origin and

$$A \cos \theta + B \sin \theta = 0$$

$$\tan \theta = -\frac{A}{B}$$

$$\theta = -\text{Arctan}\frac{A}{B}$$

Example 19.3-2

Consider the circle (Figure 19.3-5)

$$(x - a)^2 + y^2 = a^2$$

In polar coordinates, this leads to

$$r^2 \cos^2 \theta - 2ar \cos \theta + a^2 + r^2 \sin^2 \theta = a^2$$

which simplifies into

$$r^2 - 2ar \cos \theta = 0 \tag{19.3-4}$$

and leads to the point $r = 0$, plus the circle in polar coordinates

$$r = 2a \cos \theta \tag{19.3-5}$$

Notice that in going from 0 to 2π the circle is drawn twice (if *you* believe in negative r).

Similarly, we have that

$$r = 2a \sin \theta \tag{19.3-6}$$

is a circle.

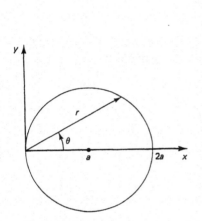

Figure 19.3.5 $r = 2a \cos \theta$

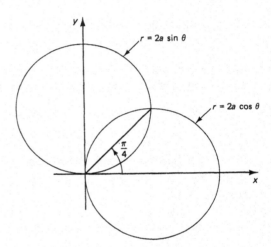

Figure 19.3-6 Intersection of two circles

Example 19.3-3

Now consider the intersections of these two circles, (19.3-5) and (19.3-6). See Figure 19.3-6. We equate the two expressions, thus eliminating r. The result is, after some algebra,

$$\sin \theta = \cos \theta$$

which has solutions $\pi/4$ and $5\pi/4$. But we see that for the second solution r is negative, and perhaps we want to exclude this double counting of the same point. However, no matter how you look at it, the origin is a common point of the two curves in Figure 19.3-6, but is not an algebraic solution! The trouble is that the origin has the coordinates $(0, \theta)$ for all θ! The origin has an infinity of different names. Thus, even when both r and θ are suitably restricted, there is trouble with the multiple representation of points in polar coordinates. This forces a constant attention to this question when converting from algebra to geometry and back; you must stop and do some checking *in every problem* to see that the multiple representations in polar coordinates have not led you astray. Still, the assets of polar coordinates are enough to make them well worth the trouble in many problems, especially those with some circular symmetry. Notice that it was the dividing out of the r term in Example 19.3-2 in going from (19.3-4) to (19.3-5) (which lost the $r = 0$ solution) that is the villain; if you do not divide out the r, then the equations of the circles are

$$r^2 = 2ar \cos \theta \quad \text{and} \quad r^2 = 2ar \sin \theta$$

and you will find the common point $r = 0$.

The conversion from rectangular to polar coordinates has been covered. We need to consider the reverse conversion, from polar to rectangular coordinates. The substitutions are [Equation (19.3-2)]

$$r = \sqrt{x^2 + y^2}$$

provided you believe in only positive r values, and

$$\theta = \arctan (x, y)$$

where, again, we think of the quadrant that the point lies in and do not exclude the second and third quadrants, as is often done when the definition

$$\theta = \text{Arctan} \frac{y}{x}$$

is used.

The following examples show some of the typical curves that occur in polar coordinates.

Example 19.3-4

Plot $r = a \sin 2\theta$. We have that at $\theta = 0, r = 0$. As $\theta \rightarrow \pi/4$, $\sin 2\theta$ approaches 1. In the next eighth of a circle it goes back to 0, then to -1, then to 0, and so on. The result is Figure 19.3-7.

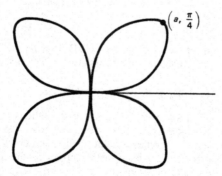

$$\left(a, \frac{\pi}{4} \right)$$

Figure 19.3-7 $r = a \sin 2\theta$

Example 19.3-5

Plot $r^2 = a^2 \cos 2\theta$. This is much as in the previous exercise except that for negative values of $\cos 2\theta$ the r becomes imaginary. We have Figure 19.3-8.

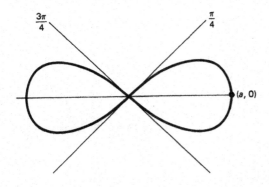

$\frac{3\pi}{4}$

$\frac{\pi}{4}$

$(a, 0)$

Figure 19.3-8 $r^2 = a^2 \cos 2\theta$

EXERCISES 19.3

Plot the following curves:

1. $r = a \sin 3\theta$
2. $r = a(1 - \cos \theta)$
3. $r = a \cos 4\theta$
4. $r = b - a \cos \theta$, for $b \geq a > 0$, and for $0 < b < a$
5. $r^2 = a^2 \sin 2\theta$
6. $r = a \sec^2 \theta/2$

19.4 THE CALCULUS IN POLAR COORDINATES

We now look at the application of the ideas of the calculus to functions given in polar coordinates. The first topic is naturally the tangent to a curve. It is immediately evident that

$$\frac{dr}{d\theta}$$

is the rate of change of r with respect to θ. The local extremes, maxima and minima, of the distance from the origin to the curve occur when this derivative is 0 (or occur at the ends of the interval in θ). If tangent (or normal) lines are desired, then it is a matter of determining the unknown parameters in the equation of a line, (19.3-3).

Arc length is another topic to examine in polar coordinates. Figure 19.4-1 shows the *elementary triangle* in polar coordinates. We have

$$ds^2 = (dr)^2 + (r \, d\theta)^2 \qquad\qquad (19.4\text{-}1)$$

We need to check this formula for reasonableness.

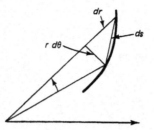

Figure 19.4-1 Elementary triangle

Example 19.4-1

Find the circumference of a circle. If we take the circle in the form

$$r = a \tag{19.4-2}$$

then $dr/d\theta = 0$, and we have

$$ds = r\, d\theta = a\, d\theta$$

to be integrated around the whole central angle of 2π. This clearly gives the right answer. If instead of (19.4-2) we assume the circle in the form (Figure 19.4-2)

$$r = 2a \cos \theta \tag{19.4-3}$$

we have (19.4-1)

$$\left(\frac{ds}{d\theta}\right)^2 = \left(\frac{dr}{d\theta}\right)^2 + r^2$$

$$= (2a)^2\{\sin^2 \theta + \cos^2 \theta\} = (2a)^2$$

Hence

$$s = \int_{-\pi/2}^{\pi/2} 2a\, d\theta = 2\pi a$$

as it should.

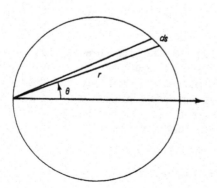

Figure 19.4-2 Arc length of a circle

Area is the next topic to examine. We divide up a region into elementary pieces of size dr by $r\,d\theta$ (see Figure 19.4-3). Thus the element of area is

$$r\,dr\,d\theta \qquad\qquad (19.4\text{-}4)$$

which we use in the integrals to get the area.

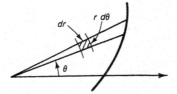

Figure 19.4-3   Element of area

Example 19.4-2

We naturally check this rule by computing the area of a standard region whose area we know. Consider the circle (Figure 19.4-2).

$$r = 2a \cos \theta$$

The area is using (19.4-4)

$$
\begin{aligned}
A &= \int_{-\pi/2}^{\pi/2}\int_0^{2a\cos\theta} r\,dr\,d\theta \\
&= \int_{-\pi/2}^{\pi/2} \frac{(2a \cos \theta)^2}{2}\,d\theta \\
&= \int_{-\pi/2}^{\pi/2} a^2(1 + \cos 2\theta)\,d\theta = \pi a^2
\end{aligned}
$$

This is the correct answer, and we now have more confidence in the formula for the area in polar coordinates.

Example 19.4-3

Find the area common to the two circles

$$r = 2a \cos \theta \quad \text{and} \quad r = 2a \sin \theta$$

Looking at Figure 19.4-4, we see that the common area is symmetrical about the line $\theta = \pi/4$. Hence the area is twice that up to this line of symmetry:

$$
\begin{aligned}
A &= 2\int_0^{\pi/4}\int_0^{2a\sin\theta} r\,dr\,d\theta = \int_0^{\pi/4} (2a \sin \theta)^2\,d\theta \\
&= 4a^2 \int_0^{\pi/4} \frac{1 - \cos 2\theta}{2}\,d\theta = 2a^2\left\{\theta - \frac{\sin 2\theta}{2}\right\}\Big|_0^{\pi/4} \\
&= (2a^2)\left(\frac{\pi}{4} - \frac{1}{2}\right) = \frac{a^2(\pi - 2)}{2}
\end{aligned}
$$

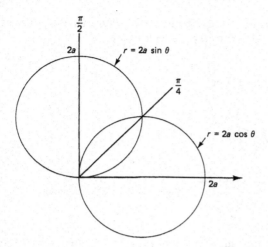

Figure 19.4-4   Common area

Example 19.4-4

There is one particular integral of great interest that can be solved in polar coordinates by a trick,

$$J = \int_{-\infty}^{\infty} \exp\left(\frac{-x^2}{2}\right) dx$$

We cannot do this by means of the indefinite integral because there is no closed form for such an integral. The trick is to first compute the square of the integral and then convert the product to a double integral:

$$J^2 = \int_{-\infty}^{\infty} \exp\left(\frac{-x^2}{2}\right) dx \int_{-\infty}^{\infty} \exp\left(\frac{-y^2}{2}\right) dy$$

$$= \int_{-\infty}^{\infty} \int_{-\infty}^{\infty} \exp\left\{\frac{-(x^2 + y^2)}{2}\right\} dy\, dx$$

Because of the suggestive $x^2 + y^2$, we naturally convert to polar coordinates:

$$J^2 = \int_{0}^{2\pi} \int_{0}^{\infty} \exp\left\{\frac{-r^2}{2}\right\} r\, dr\, d\theta$$

$$= \int_{0}^{2\pi} -\exp\left\{\frac{-r^2}{2}\right\}\bigg|_{0}^{\infty} d\theta$$

$$= \int_{0}^{2\pi} 1\, d\theta = 2\pi$$

Hence

$$J = \sqrt{2\pi}$$

as we long ago promised we would prove (Sections 15.9 and 16.7). Thus, this definite integral can be evaluated in closed form, even though the indefinite integral cannot be represented in closed form.

At this point the student should be alert enough to be a bit bothered by the use we made of the rule for converting from rectangular to polar coordinates for the above *improper* integral. If so, then it is necessary (as always for improper integrals) to go back to the finite case, evaluate it, and then let the range approach infinity.

Example 19.4-5

We begin with the definition

$$J(N) = \int_{-N}^{N} \exp\left(\frac{-x^2}{2}\right) dx$$

and form $J^2(N)$ using the variables x and y as before. Next, looking at Figure 19.4-5, we see that we can get the bounds (the integration in θ can be done in each case and gives 2π):

$$2\pi \int_0^N \exp\left(\frac{-r^2}{2}\right) r\, dr < J^2(N) < 2\pi \int_0^{\sqrt{2}N} \exp\left(\frac{-r^2}{2}\right) r\, dr$$

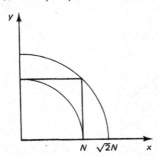

Figure 19.4-5

Doing the r variable integrations, we get

$$2\pi\left\{1 - \exp\left(\frac{-N^2}{2}\right)\right\} < J^2(N) < 2\pi\{1 - \exp(-N^2)\}$$

As $N \to \infty$, the two ends of this running inequality squeeze the middle between them to the common value 2π, as we got earlier. Thus the careless handling of the improper integral gave the right answer this time, and we got further exercise in how to deal with improper integrals: go back to the finite situation, and then at the last stage take the limit.

EXERCISES 19.4

1. Find the circumference and area of the circle $r = 2a \sin\theta$.
2. Find the area of one loop of $r = a \sin 2\theta$.
3. Find the area of one loop of $r = a \cos n\theta$.

19.5* THE DISTRIBUTION OF PRODUCTS OF RANDOM NUMBERS

In Section 14.8* we studied the distribution of physical constants and looked at a rough calculation of the distribution of the product of two numbers taken from a flat distribution. If we want to study this question further, we must find a formula for the product of any two numbers x and y taken from arbitrary probability distributions $f(x)$ and $g(y)$.

Example 19.5-1

Product of two independent random numbers. We begin, as usual, with a picture of the sample space. For the external representation of *floating* point numbers from a computing machine, the range is from $1/10$ to 1, but in customary scientific notation the range is 1 to 10 (which we shall use). Figure 19.5-1 shows the range in the two variables. While humans use the base 10 most of the time, machines often use the bases 2, 8, and 16. It is convenient to make the derivation independent of the base by writing numbers in base b. (Besides, we want to force you to think and work abstractly.) As usual, if the generality of the base bothers you, think of it as some specific number, say 10. Then when you understand the specific case, repeat the argument with the general base b.

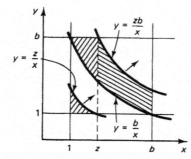

Figure 19.5-1 Cumulative probability distribution for the product $z = xy$

In scientific notation, when a product such as $3 \times 4 = 12$ arises, it requires a shift of the decimal place and writing 12 as 1.2×10^1. We are concerned with the leading digit, which is the single nonzero digit of the number in front of the decimal point when the number is in the scientific notation. Thus some products are shifted after the multiplication and some are not. Where does this shift occur? In the Figure 19.5-1 it occurs above the curve

$$xy = b \quad \text{or} \quad y = \frac{b}{x}$$

Thus all the numbers above the curve are effectively brought below the curve by the shift.

It is generally easier to work with the cumulative distribution than with the probability density, so we set up the formula for the condition that the product be less than z, where

$$z = xy$$

For fixed z the boundaries in the two regions are

$$z = xy \quad \text{and} \quad z = \frac{xy}{b}$$

There are three regions to be integrated over. We take the probability density as $f(x)g(y)$, since the two factors are assumed to be chosen independently from their corresponding probability density distributions, and we integrate this product over the three regions

$$\Pr\{Z < z\} = \int_1^z \int_1^{z/x} f(x)g(y)\, dy\, dx + \int_1^z \int_{b/x}^b f(x)g(y)\, dy\, dx + \int_z^b \int_{b/x}^{bz/x} f(x)g(y)\, dy\, dx$$

Let $G(y)$ be the cumulative distribution of $g(y)$; that is,

$$G(y) = \int_1^y g(t)\, dt$$

where of course $G(1) = 0$ and $G(b) = 1$. Then

$$\Pr\{Z < z\} = \int_1^z f(x)\left\{ G\!\left(\frac{z}{x}\right) - G(1) + G(b) - G\!\left(\frac{b}{x}\right) \right\} dx$$

$$+ \int_z^b f(x)\left\{ G\!\left(\frac{bz}{x}\right) - G\!\left(\frac{b}{x}\right) \right\} dx$$

$$= \int_1^z f(x)\left\{ G\!\left(\frac{z}{x}\right) + 1 - G\!\left(\frac{b}{x}\right) \right\} dx$$

$$+ \int_z^b f(x)\left\{ G\!\left(\frac{bz}{x}\right) - G\!\left(\frac{b}{x}\right) \right\} dx$$

Let us differentiate this with respect to z, recalling all the rules for differentiating an integral with a parameter from Equation (18.7-3). We get

$$h(z) = f(z)\left\{ G(1) + 1 - G\!\left(\frac{b}{z}\right) \right\} - f(z)\left\{ G(b) - G\!\left(\frac{b}{z}\right) \right\}$$

$$+ \int_1^z f(x)\left\{ g\!\left(\frac{z}{x}\right)\!\left(\frac{1}{x}\right) \right\} dx$$

$$+ \int_z^b f(x)\left\{ g\!\left(\frac{bz}{x}\right)\!\left(\frac{b}{x}\right) \right\} dx$$

$$= \int_1^z f(x)\left\{ g\!\left(\frac{z}{x}\right)\!\left(\frac{1}{x}\right) \right\} dx$$

$$+ \int_z^b f(x)\left\{ g\!\left(\frac{bz}{x}\right)\!\left(\frac{b}{x}\right) \right\} dx \qquad (19.5\text{-}1)$$

This is the formula for the distribution of the product of two factors each from its own distribution.

An examination of the formula (19.5-1) leaves us a bit uneasy since we know that $xy = yx$, so there should be some symmetry between $f(x)$ and $g(y)$. To check that the formula is correct, we try to transform it into the equivalent form with $f(x)$ and $g(y)$ interchanged. In the first integral, we replace x by z/x to get

$$\int_1^z f(x)g\left(\frac{z}{x}\right)\left(\frac{1}{x}\right)\,dx = \int_z^1 f\left(\frac{z}{x}\right)g(x)\left(\frac{x}{z}\right)\left(\frac{-z}{x^2}\right)\,dx$$

$$= \int_1^z f\left(\frac{z}{x}\right)g(x)\left(\frac{1}{x}\right)\,dx$$

as we should. Similarly, in the second integral we use the replacement of x with bz/x to get

$$\int_z^b f(x)g\left(\frac{bz}{x}\right)\left(\frac{b}{x}\right)\,dx = \int_b^z f\left(\frac{bz}{x}\right)g(x)\left(\frac{bx}{bz}\right)\left(\frac{-bz}{x^2}\right)\,dx$$

$$= \int_z^b g(x)f\left(\frac{bz}{x}\right)\left(\frac{b}{x}\right)\,dx$$

Thus, although the formula does not look symmetric, we find that it is indeed symmetric in the two variables.

Example 19.5-2

The persistence of the reciprocal distribution. It is worth looking at the question of what happens when one of the distributions is the reciprocal distribution. Suppose

$$g(y) = \frac{1}{y \ln b}$$

Then we have, using (19.5-1),

$$h(z) = \int_1^z f(x)\left(\frac{x}{z \ln b}\right)\frac{1}{x}\,dx + \int_z^b f(x)\left(\frac{x}{bz \ln b}\right)\frac{b}{x}\,dx$$

$$= \frac{1}{z \ln b}\left\{ \int_1^z f(x)\,dx + \int_z^b f(x)\,dx \right\}$$

$$= \frac{1}{z \ln b}$$

since the sum of the two integrals must be 1. In words, if one factor of the product is from the reciprocal distribution, then the product has the reciprocal distribution. We see the stability of the reciprocal distribution; once it arises it tends to persist indefinitely through multiplications (and divisions as the exercises will show).

Example 19.5-3

The probability of shifting. As we noted earlier (Example 19.5-1), when two numbers x and y in the standard scientific notation are multiplied, then the decimal (binary) point of the product $z = xy$ may not be in the right place. The process of shifting a product in a computer takes time, and the question arises whether the scientific or usual computer notation for floating-point numbers will save machine time on the average. Figure 19.5-2 shows the region for no shifting (in the scientific notation). The probability of selecting a given pair of numbers from the reciprocal distribution for the independent choices of the two factors is

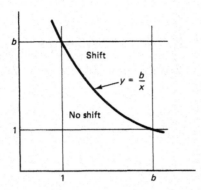

Figure 19.5-2

$$p(x)p(y) = \frac{1}{x \ln b}\frac{1}{y \ln b}$$

Hence the probability of the product lying in the no-shift region is

$$
\begin{aligned}
P &= \int_1^b \int_1^{b/x} \frac{1}{x \ln b}\frac{1}{y \ln b}\, dy\, dx\\[2mm]
&= \frac{1}{(\ln b)^2}\int_1^b \frac{1}{x}\ln y\,\Big|_1^{b/x}\, dx\\[2mm]
&= \frac{1}{(\ln b)^2}\int_1^b \frac{1}{x}\{\ln b - \ln x\}\, dx\\[2mm]
&= \frac{1}{(\ln b)^2}\left[\ln b \ln x - \frac{(\ln x)^2}{2}\right]\Big|_1^b\\[2mm]
&= \frac{1}{(\ln b)^2}\left[(\ln b)^2 - \frac{(\ln b)^2}{2}\right] = \frac{1}{2}
\end{aligned}
$$

Thus, it is a matter of indifference between the two notations *provided* you assume that the numbers come from the reciprocal distribution. For other assumed distributions, you will not generally get $p = 1/2$, and one or the other design of the arithmetic unit would be preferable.

EXERCISES 19.5

1. Discuss the probability of shifting for a product if you assume a uniform probability distribution rather than the reciprocal distribution. Give the details for the number bases 2, 4, 8, 10, and 16.
2. Find the distribution of the quotient of two numbers corresponding to Example 19.5-1 for the interval $1/b \le x < 1$.
3. Show the persistence of the reciprocal distribution for quotients as in Example 19.5-2.
4. If a number is from a reciprocal distribution, is its reciprocal from a reciprocal distribution?

19.6 THE JACOBIAN

We have evaded a common, although difficult topic, the effects of a change of
variables in an integral. For one independent variable, we used the idea of a function
of a function and merely supplied the derivative of the old variable with respect to the
new variable. Thus, given

$$I = \int_a^b f(x)\, dx$$

and the change of variable

$$x = g(t)$$

where $g(t)$ is some function of t, we write

$$I = \int_c^d f\big(g(t)\big) \frac{dg(t)}{dt}\, dt$$

where c is the value

$$a = g(c)$$

and correspondingly d is defined by

$$b = g(d)$$

To be even more concrete (a highly recommended habit), consider the specific
case of the integral

$$\int_0^1 x\, dx$$

and set $x = e^{-t}$. You get

$$\int_\infty^0 e^{-t}(-e^{-t})\, dt = \int_0^\infty e^{-2t}\, dt = \frac{1}{2}$$

Let us analyze this transformation a bit more carefully. The region near $x = 1$
is going into the region near $t = 0$. But the region near $x = 0$ is being stretched out
to infinity! When computing the area under the curve, it is necessary to compensate
for this enormous stretching out by multiplying the integrand (the curve under which
we are finding the area) by a suitable factor, thus making it enormously smaller, by
multiplying by $\exp(-t)$. Thus the derivative of the transformation locally compensates for the local stretching out (or compression) of the coordinate system. Be sure
you see intuitively why the derivative exactly compensates for the stretching.

For example, in the trivial case of a linear change of scale

$$x = kt$$

the interval $[0, 1]$ becomes $[0, 1/k]$. If we suppose k is less than 1, then this is a
stretching of the interval, and we get for the Jacobian

$$dx = k\, dt$$

The integrand is multiplied by the small number k to compensate for the stretching.

We now turn to two dimensions. In going from rectangular to polar coordinates, for example, you tend to think of the picture drawn earlier to get the element of area in polar coordinates, but a closer look at the general transformation (the use of the capital letters does *not* mean random variables here)

$$x = X(u, v), \qquad y = Y(u, v)$$

requires compensation *when* the figure is drawn in the Cartesian (rectangular) coordinates u and v of the u–v plane. In polar coordinates we are doing the integration in the Cartesian coordinates (r, θ), as shown in Figure 19.6-1. Thus we need some compensation. A look at how the area is distorted shows that the distortion is independent of θ, but increases as r becomes larger. Thus, the compensation that we used, multiply by r, seems reasonable. Hence the rule $r\, dr\, d\theta$ is what you use.

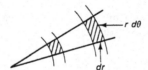

Figure 19.6-1

The general case is too complex to present a complete proof, and we will merely give the formula for the weighting. It is the *Jacobian* (named after C. G. J. Jacobi, 1804–1851)

$$\begin{vmatrix} X_u & X_v \\ Y_u & Y_v \end{vmatrix}$$

(X and Y are not random variables.) This determinant has the value $X_u Y_v - X_v Y_u$.

Example 19.6-1

For polar coordinates (where u is now r and v is now θ), the formulas give

$$x = X(r, \theta) = r \cos \theta$$
$$y = Y(r, \theta) = r \sin \theta$$

The Jacobian determinant becomes

$$J = \begin{vmatrix} \cos \theta & -r \sin \theta \\ \sin \theta & r \cos \theta \end{vmatrix} = r$$

which is exactly r as it should be, since $r(\cos^2 \theta + \sin^2 \theta) = r$.

Example 19.6-2

As a check on the formula for the Jacobian, suppose we reduce it to a single variable. The determinant is the single derivative we have always been using. Thus it is a reasonable formula, especially as it seems to have the formal symmetry that enables us to extend it to more variables.

EXERCISES 19.6

1. Show that the Jacobian gives the right result when we make the transformation $x = hu$, $y = kv$.
2. Show that the Jacobian gives the proper answer for a general linear transformation.
3. Find the Jacobian for the transformation $x = u^2 + v^2$, $y = 2uv$.
4. Show that the Jacobian gives the right result if $x = f(u)$, and $y = g(v)$ is the transformation.

19.7 THREE INDEPENDENT VARIABLES

The case of three independent variables is very common. Indeed, we can easily convert some of our earlier examples of finding the volume under a surface to three-dimensional problems by simply breaking up the volume into elementary cubes, and then summing them in one direction, summing these sums in the second direction, and finally summing these in the third direction. Thus the integral we would be led to when we took the limit as the largest dimension of any cube approached zero,

$$\int_R dV = \iiint dz\, dy\, dx$$

would be the same as the indicated iterated integral.

If we do the integration in the z variable first, we will find that we have what we wrote before, the difference between the values at the upper surface and the lower surface: the height of the column. There are $3! = 6$ different orders of selecting the variables of integration. Often one order is much easier than the others, and hence you need to think *before* you select the order.

Of course, we could integrate some function, say a three-dimensional probability distribution, or a density of material, or anything else that depended on the position, over the volume. We would have

$$\int_R f(x, y, z)\, dV = \iiint f(x, y, z)\, dz\, dy\, dx$$

Example 19.7-1

Find the volume in the first octant under the plane

$$\frac{x}{a} + \frac{y}{b} + \frac{z}{c} = 1$$

with $a, b, c > 0$.

First draw Figure 19.7-1. We set up the integral from the inside out, the range of z, then the range of y, then the range of x. We get

$$V = \int_0^a \int_0^{b[1-(x/a)]} \int_0^{ca[1-x/a-y/b]} dz\, dy\, dx$$

$$= c \int_0^a \int_0^{b[1-(x/a)]} \left(1 - \frac{x}{a} - \frac{y}{b}\right) dy \, dx$$

$$= cb \int_0^a \left\{\left(1 - \frac{x}{a}\right) - \frac{x}{a}\left(1 - \frac{x}{a}\right) - \frac{1}{2}\left(1 - \frac{x}{a}\right)^2\right\} dx$$

$$= cb \int_0^a \left(1 - \frac{x}{a}\right)\left\{1 - \frac{x}{a} - \frac{1}{2}\left(1 - \frac{x}{a}\right)\right\} dx$$

$$= \frac{bc}{2} \int_0^a \left(1 - \frac{x}{a}\right)^2 dx = \frac{bc}{2}\left(\frac{-a}{3}\right)\left(1 - \frac{x}{a}\right)^3\bigg|_0^a$$

$$= \frac{abc}{6}$$

Notice that the last integrand is the area of a slice. The answer is obviously correct, since it is a pyramid with a triangular base and has the right dimensions.

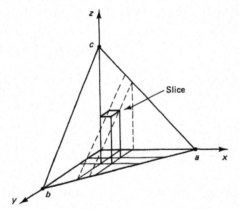

Figure 19.7-1

As an alternative approach, we get rid of the parameters at the start by setting

$$x = au, \qquad y = bv, \qquad z = cw$$

We get immediately (check using the corresponding three-dimensional Jacobian, Section 19.6)

$$V = abc \int_0^1 \int_0^{1-u} \int_0^{1-u-v} dw \, dv \, du$$

which is much easier to carry out. Indeed, we have that the integral is now a constant independent of the three parameters; we have found a more fundamental result.

Example 19.7-2

Find the volume inside

$$x^{2/3} + y^{2/3} + z^{2/3} = a^{2/3}$$

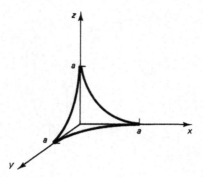

Figure 19.7-2

We draw Figure 19.7-2 as best we can and note its general shape and symmetry. We will do the first octant and then multiply by 8 to get the total volume. Again, beginning with the inner integral, we construct the formula

$$V = 8 \int_0^a \int_0^{(a^{2/3}-y^{2/3})^{3/2}} \int_0^{(a^{2/3}-x^{2/3}-y^{2/3})^{3/2}} dz\, dy\, dx$$

We do the inner integration with respect to z immediately, and then attack that integral setting $a^{2/3} - x^{2/3} = b^{2/3}$ to save messy notation and fix our attention on the variables. We have

$$\int_0^b (b^{2/3} - y^{2/3})^{3/2}\, dy$$

Set $y = b \sin^3 \theta$. You get

$$\int_0^{\pi/2} b(1 - \sin^2 \theta)^{3/2} 3b \sin^2 \theta \cos \theta\, d\theta = 3b^2 \int_0^{\pi/2} \cos^4 \theta \sin^2 \theta\, d\theta$$

$$= 3b^2\{W_4 - W_6\}$$

where the W_n are the Wallis integrals. The answer for the middle integral is, therefore,

$$3b^2 \left\{ \left(\frac{3}{4}\right)\left(\frac{1}{2}\right)\frac{\pi}{2} \right\}\left\{ 1 - \frac{5}{6} \right\} = \frac{3\pi}{32}(a^{2/3} - x^{2/3})^3$$

This goes into the outer integral, multiplied by 8:

$$\frac{3\pi}{4} \int_0^a (a^{2/3} - x^{2/3})^3\, dx$$

We could do this integral another way, but let us use the same method. Set $x = a \sin^3 \theta$ to handle this step.

$$\frac{3\pi}{4} \int_0^{\pi/2} a^2 \cos^6 \theta\, 3a \sin^2 \theta \cos \theta\, d\theta = \frac{3\pi}{4} a^3 (W_7 - W_9) = \frac{12\pi}{315} a^3$$

EXERCISES 19.7

1. Find the volume of a sphere by the method of this section.
2. Find the volume inside $x^{1/3} + y^{1/3} + z^{1/3} = a^{1/3}$.

9.8 OTHER COORDINATE SYSTEMS

In three dimensions there are other useful coordinate systems beyond the rectangular one. For example, we could use polar coordinates in place of the x and y variables and keep the z variable. Thus the transformation would be

$$x = r \cos \theta$$
$$y = r \sin \theta \qquad (19.8\text{-}1)$$
$$z = z$$

(Figure 19.8-1). We can either directly realize that the Jacobian of the transformation will be r, or else we can compute the corresponding determinant

$$\begin{vmatrix} \cos \theta & -r \cos \theta & 0 \\ \sin \theta & r \cos \theta & 0 \\ 0 & 0 & 1 \end{vmatrix}$$

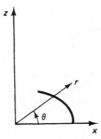

Figure 19.8-1 Cylindrical coordinates

Since the bottom row of the determinant has a single 1, it reduces immediately to the corresponding two-dimensional case. This coordinate system is called *cylindrical coordinates*, since the shapes that are easily handled are cylinders. The element of volume is

$$r \, dr \, d\theta \, dz$$

The full equivalent of polar coordinates in three dimensions is called *spherical coordinates*. In the notation of Figure 19.8-2, we have (ρ = lowercase Greek rho)

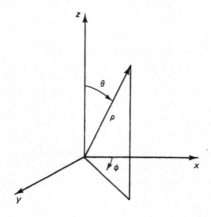

Figure 19.8-2

$$x = \rho \cos \phi \sin \theta$$
$$y = \rho \sin \phi \sin \theta \qquad\qquad (19.8\text{-}2)$$
$$z = \rho \cos \theta$$

We see immediately that $x^2 + y^2 + z^2 = \rho^2$. It is customary in three dimensions to use ρ rather than r (as in polar coordinates) for the distance from the origin. The element of volume is (from the Jacobian, Section 19.5)

$$\rho^2 \sin \theta \, d\rho \, d\theta \, d\phi$$

Example 19.8-1

Volume of a sphere. Find the volume of a sphere using spherical coordinates.

We first decide on the ranges. The angle θ goes from 0 to π (from pole to pole), ϕ from 0 to 2π (around the equator), and ρ from 0 to a (the radius of the sphere). The element of volume is

$$\rho^2 \sin \theta \, d\rho \, d\theta \, d\phi$$

The ρ variable integrates to the $a^3/3$. The θ variable integrates the sine into a minus cosine, and the two limits produce a factor of 2. The ϕ integration gives 2π. As a result, we have

$$(2\pi)(2)\frac{a^3}{3} = \frac{4\pi}{3}a^3$$

as we should. The simplicity of spherical coordinates, when you have spherically symmetric problems, should be obvious.

19.9 n-DIMENSIONAL SPACE

Suppose we have a design problem that involves 17 parameters, say a space vehicle, a nursing home, or a governmental structure of organization. Each particular design is represented by the particular choices made for each parameter. Suppose we agree on the order in which we will take the parameters. Then $(\dot{x}_1, x_2, \ldots, x_{17})$ describes a particular design. The very form suggests regarding this array of numbers as a point in 17-dimensional space, just as we regarded (x, y, z) as a point in three-dimensional space.

The reason for going to 17-dimensional space immediately is to get over the step of going from three to four dimensions. The student often gets confused about the reality of four-dimensional space and about the possibility of time being the fourth dimension and, if the student is a bit knowledgeable about special relativity theory, whether it is imaginary time that is meant. Suppose we have 17 parameter variables; we mean to think of them as an array; as a point in a corresponding 17-dimensional space.

The student is apt to worry about the "reality" of the 17-dimensional space. If you are so bothered, then ask yourself about the reality of one, two, and three dimensions. You have strong intuitive perceptions about them, but actually you have been handling not the physical space but the corresponding mathematical arrays of one, two, or three numbers. You have been dealing with the mathematical space, not the physical space of your intuitions. Similarly, we are dealing with the mathematical space. In the final analysis, in spite of all the color words like distance, sphere, and the like, we are dealing with ordered arrays of numbers (or symbols).

For example, since you believe in the Pythagorean distance in one, two, and three dimensions, where distance between points is measured by the sum of the squares of the differences of the coordinates, you are apt to accept that the distance between the point X with coordinates $(x_1, x_2, \ldots, x_n)$ and the point Y with coordinates $(y_1, y_2, \ldots, y_n)$ should be defined by the formula

$$d^2 = (x_1 - y_1)^2 + (x_2 - y_2)^2 + \cdots + (x_n - y_n)^2$$

(This may remind you of least squares, Sections 10.4, 10.6, and 15.10.) The formula certainly agrees with that used for the first three dimensions and has, when you think through the derivation in three dimensions, a plausibility about it.

If you fix the distance as a constant r, then all the points that lie on a *sphere* of radius r about the origin are described by the formula

$$x_1^2 + x_2^2 + \cdots + x_n^2 = r^2$$

Example 19.9-1

Suppose we ask for the volume of an n-dimensional sphere. To take advantage of the symmetry of the problem, we break up the volume of the sphere into shells of thickness Δr and then sum the shells, take the limit, and end up with an integral.

A little thinking about cubes and spheres in n dimensions will convince you that the volume of a sphere of radius r has the form

$$V_n(r) = C_n r^n$$

where C_n is some constant depending on n. Our problem is to find C_n.

Recall the trick, in an example of Section 19.4, that was used to find the integral of the exponential function. We obtained the result equivalent to (note that there the exponent was $-x^2/2$ and here it is $-x^2$, so a transformation is needed to convert that result to this notation)

$$I = \sqrt{\pi} = \int_{-\infty}^{\infty} \exp{(-x^2)}\, dx$$

If we now take the product of n of these integrals, we will get

$$I^n = \pi^{n/2} = \int_R \exp{\{-(x_1^2 + x_2^2 + \cdots + x_n^2)\}}\, dV$$

Naturally, we think of the appropriate coordinate system, a series of shells. For a sphere the surface of a shell, after a little thought, must be the rate of change of the volume as a function of the radius. In mathematical notation, this is

$$\frac{dV_n(r)}{dr} = nC_n r^{n-1} = S$$

Thus we have

$$I^n = \pi^{n/2} = \int_0^{\infty} (nC_n r^{n-1}) \exp{(-r^2)}\, dr$$

Last time we coped with this form of integration by the substitution $r^2 = t$. This gets us to the gamma function $\Gamma(x)$ (Section 15.11). We have, in more detail,

$$\pi^{n/2} = nC_n \int_0^{\infty} t^{(n-1)/2} \exp{(-t)} \frac{1}{2\sqrt{t}}\, dt$$

$$= \frac{n}{2} C_n \int_0^{\infty} \exp{(-t)} t^{(n/2)-1}\, dt$$

This integral is $\Gamma(n/2)$. Thus we have, solving for the coefficient C_n,

$$C_n = \frac{\pi^{n/2}}{(n/2)\Gamma(n/2)} = \frac{\pi^{n/2}}{\Gamma[(n/2)+1]}$$

Remembering what the gamma function is for various values, we get the Table 19.9-1 for C_n, from which we see that the volume of a unit sphere, or equivalently the coefficient C_n of r^n, comes to a maximum at $n = 5$ and falls off rather rapidly toward 0 as n approaches infinity.

Another property of high-dimensional spheres is that "almost all the volume is on the surface." We write the formula for the volume of a shell of thickness ϵ divided by the total volume to get the relative amount in the shell.

$$\frac{\text{shell}}{\text{volume}} = \frac{C_n r^n - C_n (r - \epsilon)^n}{C_n r^n}$$

$$= 1 - \left(\frac{1 - \epsilon}{r}\right)^n$$

and as $n \to \infty$ this approaches 1; hence almost all the volume is on the surface, *provided* you raise the dimension of the space suitably high.

TABLE 19.9-1

n	C_n
1	$2 = 2.00000\ldots$
2	$\pi = 3.14159\ldots$
3	$\dfrac{4\pi}{3} = 4.18879\ldots$
4	$\dfrac{\pi^2}{2} = 4.93480\ldots$
5	$\dfrac{8\pi^2}{15} = 5.26379\ldots$
6	$\dfrac{\pi^3}{6} = 5.16771\ldots$
7	$\dfrac{16\pi^3}{105} = 4.72477\ldots$
8	$\dfrac{\pi^4}{24} = 4.05871\ldots$
$\ldots$	$\ldots$
$2k$	$\dfrac{\pi^k}{k!} = \longrightarrow 0$

Example 19.9-2

Find the volume under the surface $\sum_1^n x_i/a_i = 1$ and in the first 2^nth. This is a generalization of the three-dimensional tetrahedron (Example 19.7-1). Rather than try to set up the general integral, we attack it recursively (the idea of mathematical induction if you prefer). We see that if we make the substitution $x_i = a_i t_i$ then the constants will factor out (or apply the Jacobian), and we will have the simpler problem

$$\sum_{i=1}^n t_i = 1$$

We think that if we knew the answer for $n - 1$ dimensions then integrating this, suitably scaled, would give us the answer for the nth case.

We conjecture that the answer in $n - 1$ dimensions is $1/(n - 1)!$. Hence the answer for dimensional space is, since the volume is proportional to the nth power of the side,

$$\int_0^1 \frac{(1 - x)^{n-1}}{(n - 1)!}\, dx = -\frac{1}{(n - 1)!}\left.\frac{(1 - x)^n}{n}\right|_0^1 = \frac{1}{n!}$$

and we have done the induction step. For $n = 1$ dimension, we have, of course, the value 1 to back up our conjecture. We can check it against the next two cases to verify the result.

This recursive method of using the answer for the next lower case and integrating up to the nth case, as well as verifying it for some starting case, has the same pattern of logic as has mathematical induction and is very useful in such problems. Unfortunately, it is hard to describe most of the problems that arise of this form, so we will (as we have) use what appear to be "made up" problems to familiarize you with the method of finding integrals in n dimensions. Usually, there is this kind of regularity in the problem, or else you could not require the answer in an arbitrary dimensional space.

Such problems cannot be as simple as those we began with. As noted before, you should by this time have developed your "intellectual muscles" from earlier work and be prepared to take on significantly harder problems that take a lot more effort and time. Any reasonable course must not only develop new material, it must deliberately raise the level of difficulty to continue developing your abilities.

EXERCISES 19.9

1. Given the line from the origin to the point $(1, 1, \ldots, 1)$ in n dimensions, show that the cosine of the angle between it and any of the coordinate axes is $\cos \theta = 1/\sqrt{n}$. Hence show that for sufficiently high dimension the diagonal line is almost perpendicular to all the coordinate axes.

2. Choose two random points with coordinates $(\pm 1, \pm 1, \ldots, \pm 1)$. Using the law of cosines, show that the cosine of the angle between them is $(1/n) \sum_1^n (\pm 1)$. Using the law of large numbers, show that any two of the 2^n such lines chosen at random are almost certainly almost perpendicular, that in n-dimensional space there are almost 2^n almost perpendicular lines!

3. Find the volume inside $\sum_{i=1}^n x_i^{2/3} = a^{2/3}$.

4. Find the volume inside $\sum_{i=1}^n x^{1/p} = a^{1/p}$.

5. Generalize Exercises 3 and 4.

19.10 PARAMETRIC EQUATIONS

In general, we have been using a function of a single variable in the form

$$y = f(x)$$

However, if we write

$$x = a \cos t$$

$$y = a \sin t$$

we have a pair of equations in terms of the *parameter* t. For each value of the parameter t we get a pair of values (x, y), which corresponds to a point on the locus defined by the equations. In this particular example, as t goes from 0 to 2π, the point goes around the circle of radius a centered about the origin. (See Figure 19.10-1). Sometimes the parameter can be eliminated from the defining equations. In the above case we simply square the two equations and add, thus eliminating the parameter t. There is a loss of information, however. In the original parametric form, assuming that t represents time, we have not only the trajectory (orbit) of the quantity defined by

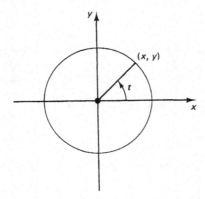

Figure 19.10-1

(x, y), we also have the time when it was at various points along the path. For example, the parametric equations

$$x = a \cos t^2$$

$$y = a \sin t^2$$

also describe the same circle, but for the corresponding time t the points occur at different places. Thus, there is not a unique parameterization of a curve *unless* you require that the value of the parameter coincide point by point in each of the two trajectories.

Many other curves can be written in parametric form. For example, the parabola

$$y = x^2$$

can be written in the form

$$x = t^n$$

$$y = t^{2n}$$

for any n not 0. If n is positive, then if we start at $t = 0$ the parabola begins at the origin and goes forward to the right. If n is an even number, then the part of the parabola in the third quadrant is lost. If n is negative, then the parabola begins, for $t = 0$, out at infinity and comes in toward the origin as t increases.

19.11 *THE CYCLOID—MOVING COORDINATE SYSTEMS*

The motion of a point on the rim of a wheel, as it rolls along a smooth surface, describes a curve known as the *cycloid*. The cycloid plays an important part in many situations, and is customarily written in parametric form.

Example 19.11-1

We could find the equations for the cycloid by a straight approach to the problem; we prefer to use a little less direct, but easier, approach. We think of the motion of the point on the rim as a combination of a rotation of the wheel followed by a uniform translation of the center of the wheel. We propose to *construct* the equations from simple known parts.

The polar coordinates of the stationary wheel must be

$$x = a \cos t$$

$$y = a \sin t$$

We would like to have the bottom of the wheel on the x-axis. Thus we raise y by an amount a and write the equations as

$$x = a \cos t, \qquad y = a + a \sin t$$

But this starts with the point halfway up to the highest point, and it would be nice to start with the point on the axis. We need a shift in the t variable by a quadrant; that is, we write the equations as

$$x = a \sin t, \qquad y = a - a \cos t$$

At $t = 0$, we now have $x = 0$ and $y = 0$. Now looking at the rotation we see that it is counterclockwise, and it would be nice to have the wheel rotate to the right (meaning the rotation is clockwise). This requires replacing t by $-t$ to give, finally, the equations we want for the stationary wheel:

$$x = -a \sin t, \qquad y = a - a \cos t$$

Now that we have the wheel rotating to our satisfaction, we need to make the translation to the *moving coordinate system*. This requires y to stay the same, but x to have the proper amount added. What is the proper amount? The wheel moves in one rotation exactly $2\pi a$ to the right, so we add the proper amount to the x coordinate, at, and we have

$$x = a(t - \sin t), \qquad y = a(1 - \cos t)$$

These are the parametric equations of the cycloid (Figure 19.11-1), where this time the

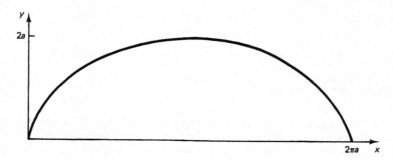

Figure 19.11-1   Cycloid

parameter has physical meaning, that is, the time when the chosen point on the rim is in a given position.

Example 19.11-2.

Find the slope at any point on the cycloid. To do this, we remember that

$$\frac{dy}{dx} = \frac{\dfrac{dy}{dt}}{\dfrac{dx}{dt}}$$

Hence

$$\frac{dy}{dx} = \frac{\sin t}{1 - \cos t}$$

At $t = 0$, this is an indeterminate form, and we need L'Hopital's rule (Section 15.5) to evaluate the limit at $t \to 0$. Thus, we differentiate both numerator and denominator separately to get

$$\frac{\cos t}{\sin t} = \operatorname{ctn} t$$

which gives the value infinity at $t = 0$. At time $t = \pi$, meaning half a rotation, we have

$$\frac{dy}{dx} = \frac{0}{1 + 1} = 0$$

The derivative is perhaps better written in the form

$$\frac{dy}{dx} = \frac{2 \sin\left(\dfrac{t}{2}\right)\cos\left(\dfrac{t}{2}\right)}{2 \sin^2\left(\dfrac{t}{2}\right)} = \operatorname{ctn} \frac{t}{2}$$

since the periodicity of the derivative is more clearly seen in this form. Furthermore, the slope is more easily evaluated.

EXERCISES 19.11

1. Find the area under one arch of the cycloid.
2. Find the volume inside the surface formed by rotating the cycloid about the x-axis.
3. Find the volume generated by rotating the cycloid about the y-axis.
4. Find the equation of motion of a point on the rim of a wheel with radius b and rotation rate R_1 when the center of this wheel is on the rim of a second wheel of radius $a > b$ and rotating with a rate R_2. This is called a *hypercycloid*.
5. The same as Exercise 4 except that $b > a$. This is called a *hypocycloid*.
6. Find the conditions on a, b, R_1, and R_2, for which retrograde motion (as seen from the origin of the coordinate system) can occur.

19.12 ARC LENGTH

The basic differential triangle of Section 18.4 leads to
$$(ds)^2 = (dx)^2 + (dy)^2$$
and dividing through by $(dt)^2$ we get the convenient form, Equation (18.4-4),
$$\left(\frac{ds}{dt}\right)^2 = \left(\frac{dx}{dt}\right)^2 + \left(\frac{dy}{dt}\right)^2$$
for the arc length of a curve in parametric coordinates.

Example 19.12-1

Find the arc length of one arch of the cycloid. We have to find
$$\frac{dx}{dt} = a(1 - \cos t) \quad \text{and} \quad \frac{dy}{dt} = a \sin t$$

Square and add:
$$\left(\frac{ds}{dt}\right)^2 = a^2[1 - 2 \cos t + \cos^2 t + \sin^2 t] = a^2\left[4 \sin^2 \frac{t}{2}\right]$$

Hence the total length of the arch of the cycloid is
$$S = \int_0^{2\pi} 2a \sin \frac{t}{2} \, dt = 2a \, 2\left(-\cos \frac{t}{2}\right) \Bigg|_0^{2\pi} = 8a$$

which seems reasonable when you examine the drawing of the cycloid.

EXERCISES 19.12

1. Find the arc length of $y = t^2$, $x = t$ from $t = 0$ to $t = 1$.
2. Find the arc length of $y = t^3$, $x = t^2$, from $t = 0$ to $t = 1$.
3. Find the circumference of a circle when the curve is given in the form $x = a \cos \theta$, $y = a \sin \theta$.
4.*Develop the theory of the surface generated by the rotation of a curve about a line in parametric coordinate notation.

19.13 SUMMARY

Problems occur frequently in which there are several independent variables, and we have just touched a few of the possibilities. The *method* of going from the known one-dimensional case, to the two-dimensional case, to the three-dimensional case, and on up to the *n*-dimensional case should be clear. It is good mathematical practice to do so, even if you are only going to go to three dimensions, since the *n*-dimensional case gives a good check on the work, and sometimes will give a far easier approach than the one you first found in three dimensions. We found that n-dimensional space is very vast indeed; it has many non-intuitive properties that will cause you trouble when you try to optimize some design with many adjustable parameters.

The use of polar coordinates is very valuable in many situations, but the lack of unique labeling of the points means that you must be constantly on your guard for nonsense answers. The integration of the normal distribution as a definite integral is an example of the fact that often definite integrals can be found when the indefinite integral cannot be done in closed form.

We used the same "trick" twice, the integration of a suitable product of integrals of the same form, and as Polya [P] observes, a trick used twice is a method. It is something that you should remember as a method for getting some difficult definite integrals. The two applications we used are very standard examples of the method.

Cylindrical and spherical coordinates are very useful in the corresponding kinds of problems, especially in n dimensions.

Parametric coordinates are another useful coordinate system, but we have looked at it only slightly because the choice depends so much on the particular application.

PART IV

MISCELLANEOUS TOPICS

20

Infinite Series

A central concept in the calculus is the idea of a limit. To introduce this idea, we looked (in Section 3.6) at the representation of decimal numbers, including rational numbers. You have the feeling that as you take more and more digits of a number you are getting closer and closer to the limit of the (discrete) sequence of approximations, and you also feel there is a number that is the limit. In Section 3.5 we showed a simple method of approximating the zero of an equation by gradually building up more and more digits of the decimal representation of the number. The existence of a number that is the limit of a convergent sequence was explicitly stated in Assumption 4 of Section 4.5, where we said (but did not define clearly) that we were dealing with convergent sequences.

Later, in Section 7.2, we examined more closely the idea of a limit. In particular, given an infinite sum of terms, we examined the sequence of partial sums, and said that *the limit of the convergent sequence of the partial sums is the sum of the series*. We applied this to the geometric progression

$$S = a + ar + ar^2 + ar^3 + \cdots$$

When $|r| < 1$, we found that as $n \to \infty$ the sequence of partial sums

$$S_n = a + ar + ar^2 + \cdots + ar^{n-1}$$

$$= \frac{a}{1-r} - \frac{ar^n}{1-r}$$

approaches the limit

$$S = \frac{a}{1 - r}$$

Logically, the limit of a sequence of terms, such as the partial sums of a series, is simpler than the limit of a continuous variable we used in the definitions of the derivative and the integral. Nevertheless, we hastened on to the limit of a continuous variable, and must now return to a closer look at the idea of the limit of discrete sequences. We chose the psychological path, rather than the logical path.

The idea of a limit of a sequence is based on the idea of a *null sequence*, a sequence that approaches zero. The example of a null sequence we used in Section 7.2 was $\{1/n\}$ as n approached infinity. This sequence approached zero, although for no finite n does it take on the value zero.

From the null sequence, we reached the general case by noting that the difference between a sequence and its limit must be the null sequence. For example, in the case of a geometric progression, the sequence of partial sums is

$$S_n = \frac{a}{1 - r} - \frac{ar^n}{1 - r}$$

When $|r| < 1$, the sequence

$$r^n \to 0$$

is a null sequence as $n \to \infty$. It is still a null sequence when multiplied by the fixed number $a/(1 - r)$. Therefore, as $n \to \infty$ the sequence S_n will approach the limit $a/(1 - r)$. Since the limit of a series is defined to be the limit of its partial sums, we say that the geometric series *converges* to this limit

$$S = \frac{a}{1 - r}$$

This result was applied to a number of cases. In particular, we found that (as we expected) the decimal sequence 0.3333 . . . 3 (a finite number of 3's) approached $\frac{1}{3}$ as we took more and more digits. Slightly more surprising, we found that the corresponding sequence of 9's approached 1.0.

The idea of a limit requires the sequence to not only approach its limit, but to stay close once far enough along; the difference between the sequence and its limit is a null sequence. We had to make the explicit assumption in Section 4.5 that for any convergent sequence there is always a number that is its limit, because we are not able to prove, in an elementary fashion, that we would not be in the position of Pythagoras when he found that among the everywhere dense rationals there is no square root of 2 (there was no rational number awaiting him). Thus we assumed that, if a sequence is convergent, then there is a number that is its limit.

We further discussed the rules for combining limits and saw them effectively repeated in Section 7.3 for continuous variables (and in L'Hopital's rule, Section

15.5). In Section 15.4, we looked at Stirling's approximation to $n!$ and found that the approximation did *not* differ from the true value by a null sequence. Thus convergence is a condition that is useful and makes things easy to understand, but we can occasionally handle sequences that are not convergent.

20.2 MONOTONE SEQUENCES

We first examine sequences that are either monotone (constantly) increasing or are monotone decreasing. They are an important class of sequences and shed light on the general situation.

Suppose we have a monotone increasing sequence S_n that is bounded above by some number A. In mathematical symbols,

$$S_{n-1} \le S_n \le A$$

We see intuitively (Figure 20.2-1), that the sequence cannot grow indefinitely (since it is bounded) and cannot "wiggle" (because it is monotone), so surely it must converge. We prove this theorem by contradiction; we assume that it does not converge. To be a bit more careful, the failure to converge implies that there exists at least one number $\varepsilon > 0$ such that no matter how far out you go, and no matter which number S you choose as the limit, you can find a difference between the sequence S_n and its limit S that is bigger than the original number ε. Going beyond this point in the sequence, you can find another such increase, and beyond this new place in the sequence another such increase, and so on, thus as many such increases as you please.

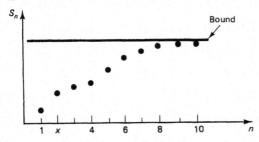

Figure 20.2-1 Monotone increasing sequence

But by picking enough of these, their sum (for a monotone increasing sequence) would add up to more than the distance ot the upper bound, a contradiction that the sequence was bounded. Thus you have the result:

Theorem 20.2-1.   A monotone increasing sequence that is bounded above must converge to a limit.

Obviously, the limit is not greater than the bound A and may be below it. A similar result for monotone decreasing sequences that are bounded below is easily found.

For a monotone increasing convergent sequence S_n of terms, the nth term must approach the limit. The sequence S_n comes from a series of positive (more carefully, nonnegative) terms, when we define

$$S_n = \sum_{k=1}^{n} u_k$$

since the sequence S_n is monotone increasing and $S_n - S_{n-1} = u_n \geq 0$. It follows that the nth term u_n of the series must approach zero. This is a necessary but not sufficient condition for convergence of a monotone series; $u_n \to 0$ must occur for convergence, but u_n may approach zero for a nonconvergent series.

Example 20.2-1

The series

$$\sum_{1}^{\infty} \frac{n^2}{n^2 + 7n + 13}$$

cannot converge since the limit, as n approaches infinity, of the value of the general term is 1.

Example 20.2-2

For $a > 1$, the sequence of partial sums

$$S_n = \sum_{k=0}^{n} \frac{1}{a^k + 1}$$

converges. We have only to find an upper bound to this monotone increasing sequence of terms S_n. It is easy to see that

$$u_n = \frac{1}{a^k + 1} < \frac{1}{a^k}$$

since

$$\frac{1}{u_n} = a^k + 1 > a^k$$

We have, therefore,

$$S_n < \sum_{k=0}^{n} \frac{1}{a^k} < \sum_{k=0}^{\infty} \frac{1}{a^k}$$

but this last is an infinite geometric progression with the first value 1 and the rate $|1/a| < 1$; hence this sum is

$$S_n < \frac{1}{1 - (1/a)} = \frac{a}{a - 1}$$

and provides an upper bound on the original series. We have shown that the given series converges, although we do not have a value for the sum, only an upper bound.

This is a special case of a fairly obvious result.

Theorem 20.2-2.   If each term of a series of nonnegative terms is bounded by the corresponding term of a second series, and if the second series converges, then the first series must converge.

Proof. In mathematical symbols, if (1) $\{a_n\}$ is a sequence of nonnegative numbers, (2) $a_n \leq b_n$, and (3) Σb_n converges, then Σa_n converges. The proof goes along the lines given in the above example. The Σb_n provides an upper bound to the monotone increasing sequence

$$S_n = \sum_{k=k_0}^{n} a_k$$

If the sum of a series of nonnegative terms does not converge, then necessarily it approaches infinity. Such a series is said to *diverge*.

Some of the results we have just obtained refer to more general series than monotone series. For example, the result that if a series converges then necessarily $u_n \to 0$ is such a result.

Theorem 20.2-3.   If a series of partial sums S_n converges, then $u_n \to 0$.

Proof. To prove this we first note that

$$S_n - S_{n-1} = u_n$$

Now we write (where S is the limit of the sequence S_n)

$$|u_n| = |S_n - S_{n-1}| = |(S_n - S) - (S_{n-1} - S)|$$
$$\leq |S_n - S| + |S_{n-1} - S|$$

But since the series is assumed to be convergent, both of these terms approach zero; therefore, u_n must approach zero.

Before going on, we need to note one more obvious theorem.

Theorem 20.2-4.   The convergence of any series is independent of any finite starting piece of the series, although of course the sum depends on it.

Proof. No matter how many terms you take of the beginning of the series, be it 100, 1000, 1 million, or any other finite number, the sum of them is also finite. If the series converges, then the *tail* of the series must also converge; and if the series diverges, then, since the sum of all the terms is infinite, the tail must total up to infinity *independent* of the sum of the first part you took separately.

This result is useful since it says that when determining the convergence or divergence of a series of terms, positive or otherwise, you need only study the values past *any* convenient point, and you can ignore any early part of the series you wish. If the series settles down to positive terms past some point, then it can be treated as a series of positive terms.

EXERCISES 20.2

1. Does $\Sigma\, 2/(3 + 1/n)$ converge?

2. Show that $\Sigma\, a/(1 + b^n)$ converges for $b > 1$.

3. Show that $\Sigma a/(b^n + (-1)^n)$ converges for $b > 1$.

4. If $u_n \to 0$ and $S_n = \Sigma\, (u_{n+1} - u_n)$, then S_n converges.

20.3 THE INTEGRAL TEST

The beginning student in infinite series soon begins to wonder why so much attention is paid to whether or not the series converges and so little attention to the number it converges to. The answer is simple; it is comparatively rare that the sum of a series can be expressed in a convenient closed form. It is much more likely that an integral can be done in closed form than can an infinite series be summed in closed form. The reason for this is that for integration we have a pair of powerful tools, integration by parts and change of variable; for infinite series we have only summation by parts (Section 20.4), and there is no comparable change of variable with its Jacobian. Thus, again, we see that, in a certain very real sense, discrete mathematics is harder than is the continuous calculus; we can less often expect to get a nice answer to the equivalent problem.

This being the case, it is natural to try to use integration as a tool for *testing* the convergence of series. (In Section 15.4 we used integration as a method of approximating finite and infinite sums.) We have the following result.

Theorem 20.3-1.   Given the monotone series of positive terms u_n

$$\sum_{k=1}^{\infty} u_n$$

then the integral

$$\int_a^{\infty} u(x)\, dx$$

converges or diverges with the sum, provided that for all x the function $u(x)$ is monotone decreasing [with $u_n = u(n)$ of course].

Proof. We have only to think of the fundamental result that the integral is the limit of the sum and draw the figure for the finite spacing of $\Delta x = 1$ (Figure 20.3-1). We see the upper and lower sums (over the monotone part, which is all that matters) bound the integral above and below. Their difference is not more than the projected total on the column to the left, not more than the size of the first term we are considering. We write

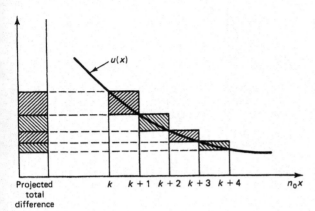

Projected
total
difference

Figure 20.3-1   Integral test

$$\text{upper sum} = \text{lower sum} + \text{difference}$$

We also have

$$\text{upper sum} \geq \text{integral} \geq \text{lower sum}$$

If the integral diverges, then the upper sum diverges and this differs from the lower sum by a fixed amount, so the lower sum must also diverge. Similarly, if the integral converges, then the lower sum must converge and so must the upper sum. In this simple case of only determing the convergence or divergence of the series, and not its actual sum, it is not necessary to use the more refined methods of estimating integrals by sums that we used in estimating Stirling's approximation to $n!$ (Section 15.4).

When examining this theorem, we see the condition that $u(x)$ must be monotone decreasing. Is it not enough that the original terms u_n be monotone decreasing? True, the proof we gave required $u(x)$ to be monotone, but perhaps another proof might not. A few moments' thought produces the example [$u(x)$ is monotone]

$$u_1(x) = u(x) + \sin^2 \pi x$$

which shows that although $u_1(x)$ takes on the values of the terms of the series u_n at the integers, nevertheless the total area under the curve is infinite since each loop of the $\sin^2 \pi x$ has the same finite area. Thus the monotone condition is necessary (at least some such constraint is necessary).

Example 20.3-1

The p-test. The series

$$\sum_{k=1}^{\infty} k^{-p}$$

occurs frequently. The corresponding integral (and we do not care about the lower limit since the convergence or divergence depends solely on the tail of the series), $p \neq 1$, is

$$\int_1^\infty x^{-p}\, dx = \lim\left[\frac{x^{1-p}}{1-p}\right]\Bigg|_1^{n\to\infty}$$

$$= \lim\frac{1-n^{1-p}}{p-1}, \qquad \text{as } n \to \infty$$

If $p > 1$, then the limit is $1/(p-1)$ since $n^{1-p} \to 0$. Next, if $p < 1$, the limit is infinite. Hence the series converges if $p > 1$ and diverges if $p < 1$. For the case $p = 1$, we can use the integral test and deduce the divergence from

$$\int \frac{dx}{x} = \ln x \to \infty$$

As an alternative approach we can look back to the approximation of Example 15.4-3:

$$h_n = \sum_1^n \frac{1}{k} - \ln n \to 0.57712 \ldots$$

Since $\ln n \to \infty$ as $n \to \infty$, it follows that the *harmonic series*

$$\sum_1^\infty \frac{1}{n}$$

diverges to infinity.

EXERCISES 20.3

Give the convergence or divergence of the following series.

1. $\Sigma\, 1/\sqrt{n}$
2. $\Sigma\, 1/n$
3. $\Sigma\, 1/n^{3/2}$
4. $\Sigma\, 1/n^2$
5. For $\varepsilon > 0$, $\Sigma\, 1/n^{1+\varepsilon}$ $\Sigma\, 1/n^{1-\varepsilon}$

20.4* SUMMATION BY PARTS

Summation has a discrete analog corresponding to integration by parts. Suppose we have

$$\sum_0^{n} u_k\, v_k$$

We write

$$V_n = \sum_0^n v_k \quad \text{hence} \quad V_n - V_{n-1} = v_n$$

The series may now be written in the equivalent forms (since $\Delta V_{k-1} = V_k - V_{k-1}$ and $\Delta V_{-1} = v_0$)

$$\sum_0^n u_k(V_k - V_{k-1}) = \sum_0^n u_k \, \Delta V_{k-1}$$

In this last form it resembles the usual integration by parts. If we write this out in detail, we have (remember that $V_0 = v_0$)

$$\sum = u_0 V_0$$

$$+ u_1(V_1 - V_0)$$
$$+ u_2(V_2 - V_1)$$
$$\cdots$$
$$+ u_n(V_n - V_{n-1})$$

Rearranging this in terms of the V's, we get

$$\sum = -V_0(u_1 - u_0)$$

$$-V_1(u_2 - u_1)$$
$$\cdots$$
$$-V_n(u_{n+1} - u_n) + V_n u_{n+1}$$

Hence we have, using $\Delta u_k = u_{k+1} - u_k$, and writing $v_{-1} = 0$,

$$\sum_0^n u_k \, \Delta V_{k-1} = u_{n+1} V_n - \sum_0^n V_k \, \Delta u_k \qquad (20.4\text{-}1)$$

which resembles integration by parts.

Just as we use integration by parts to eliminate powers of x, so too can we remove powers of n from a summation. The elimination of ln and arctan terms by integration by parts in integration does not work for series. The application of parts twice and finding the same summation again will work as it did in exponentials times sines and cosines.

Example 20.4-1

Consider the simple example

$$\sum_1^n ka^k$$

Using (20.4-1), we set $u_k = k$ and $\Delta V_{k-1} = a^k$. Then

$$V_n = \frac{1 - a^{n+1}}{1 - a}$$

and $\Delta u_k = 1$, and we have

$$\sum_0^n ka^k = (n + 1)\left\{\frac{1 - a^{n+1}}{1 - a}\right\} - \sum_0^n V_k$$

and the sum is now easily done by breaking it up into two terms, the second of which is a simple geometric progression (again). We leave the details to the student.

EXERCISES 20.4

1.* Show that $\Sigma (1/2^k) \sin ka = \{2 \sin a - 2^{-n}[2 \sin a(n + 1) - \sin a]\}/\{1 + 8 \sin^2 (a/2)\}$.

2. Sum $\Sigma k \cos kx$

20.5 CONDITIONALLY CONVERGENT SERIES

Many series of interest are not of constant sign, but have terms that may be positive or negative. Among these is the frequently occurring special class of *alternating series* whose terms alternate in sign.

Example 20.5-1

The *alternating harmonic series* is defined by

$$1 - \frac{1}{2} + \frac{1}{3} - \frac{1}{4} + \cdots = \sum_{n=1}^\infty \frac{(-1)^{n-1}}{n}$$

To sum this series, we recall, again, the result of Example 15.4-3:

$$h_n = (1 + \frac{1}{2} + \cdots + \frac{1}{n} - \ln n) \to \gamma$$

as n approaches infinity. Now consider

$$h_{2n} - h_n = \left(1 + \frac{1}{2} + \frac{1}{3} + \frac{1}{4} + \cdots + \frac{1}{2n - 1} + \frac{1}{2n}\right)$$
$$- 2\left(\frac{1}{2} + \frac{1}{4} + \frac{1}{6} + \cdots + \frac{1}{2n}\right) - \ln 2n + \ln n$$

where for the second part (from h_n) we have put a factor of 2 in front and compensated by dividing each term by 2. The result of adding these two lines changes the sign of all the terms with an even denominator; and this is the alternating harmonic series pattern for S_{2n}. But since both h_{2n} and h_n approach the same limit γ, their difference must approach zero. Hence

$$S_{2n} = 1 - \frac{1}{2} + \frac{1}{3} - \cdots + \frac{1}{2n} \to \ln \frac{2n}{n} = \ln 2$$

The S_{2n+1} approach the same limit as the S_{2n} since the difference $1/(2n + 1) \to 0$. Hence the alternating harmonic series converges, while the sum with all positive terms diverges. We say that such a series is *conditionally convergent*.

A useful theorem for dealing with alternating series whose terms are monotone decreasing in size is the following:

Theorem 20.5-1.   If the u_n are monotone decreasing to zero, then the series

$$\sum_0^\infty (-1)^n u_n$$

converges.

Proof. The proof is easy. (Note this time we begin with u_0 as the first term; hence there are small differences from the proof for the alternating harmonic series.) We write the even-indexed partial sums of the series out in detail:

$$S_{2n} = u_0 - u_1 + u_2 - u_3 + u_4 - \cdots + u_{2n}$$
$$= u_0 - (u_1 - u_2) - (u_3 - u_4) - \cdots - (u_{2n-1} - u_{2n})$$

From this we have that u_0 is an upper bound of S_{2n} (since each parentheses is a nonnegative number). Now writing the series in the form

$$S_{2n} = (u_0 - u_1) + (u_2 - u_3) + \cdots + (u_{2n-2} - u_{2n-1}) + u_{2n}$$

we see that these partial sums form a monotone increasing sequence that is bounded above by u_0, and hence must converge. Since

$$S_{2n+1} = S_{2n} + u_{2n+1}$$

and the $u_{2n+1} \to 0$, it follows that the partial sums for odd and even subscripts converge to the same limit. Therefore, the series converges.

If the series

$$\sum |u_n|$$

converges, we say that the series is *absolutely convergent*.
We need to prove the result:

Theorem 20.5-2.   If a series (alternating or not) is absolutely convergent, then it is also conditionally convergent.

Proof. It is not completely obvious that the positive and negative terms could not make the sum of the series oscillate indefinitely while still remaining finite in size, and thus the series would not converge. To show that this is impossible, we use a standard trick and define two series, one P_n consisting of the positive terms and one Q_n of the negative terms. To be able to name the terms easily, we supply zeros for the missing terms in each of the defined series. Thus we define

$$P_n = u_n \qquad \text{if } u_n \text{ is positive and zero otherwise}$$

$$Q_n = -u_n \qquad \text{if } u_n \text{ is negative and zero otherwise}$$

We have from the definition of P_n and Q_n for all finite sums

$$\sum u_n = \sum P_n - \sum Q_n$$

From the absolute convergence of the series, we see that

$$\sum P_n + \sum Q_n = \sum |u_n|$$

Hence each sum on the left (being positive) must converge separately, since each is separately bounded above (by the sum on the right) and is monotone increasing. Now Σu_n is the difference of two convergent series and hence also must converge.

Some of the results we proved for series with positive terms apply equally well for series with general terms. Thus for convergence it is necessary that $u_n \to 0$, as well as the fact that the convergence or divergence of a series depends only on the behavior of the tail of the series, and not on any finite number of beginning terms.

The main series for which we know the sum is a geometric progression when $|r| < 1$. Given any other series, we now ask how we can get the terms of the given series dominated by those of some suitable geometric series. The following *ratio test* is based on the convergence of a geometric series. Suppose we have a series u_n such that

$$\left| \frac{u_{n+1}}{u_n} \right| \to r < 1$$

as n approaches infinity. Then there is another number $r_1 < 1$ such that $r < r_1 < 1$. Next, from the limit we deduce that sufficiently far out

$$\left| \frac{u_{n+1}}{u_n} \right| < r_1$$

for all n larger than some starting value n_0. This means that for any $n \geq n_0$

$$|u_{n+1}| < r_1 |u_n|$$

$$|u_{n+2}| < r_1 |u_{n+1}| < r_1^2 |u_n|$$

$$\cdots$$

$$|u_{n+k}| < r_1 |u_{n+k-1}| < r_1^k |u_n|$$

Thus the original terms u_m $(m > n)$ are dominated by the corresponding terms of a geometric progression with rate $r_1 < 1$. This geometric progression is a convergent series (all, of course, sufficiently far out in the series). Hence we have the following ratio test.

Theorem 20.5-3.   If, for a series of positive terms,

$$\left| \frac{u_{n+1}}{u_n} \right| < r < 1$$

for all sufficiently large n, then the series converges; and if

$$\left| \frac{u_{n+1}}{u_n} \right| > r > 1$$

for all sufficiently large n, then the series diverges. This can be stated in the limiting form; if as n approaches infinity

$$\lim \left| \frac{u_{n+1}}{u_n} \right| \rightarrow r < 1 \quad \text{then the series converges}$$

$$\lim \left| \frac{u_{n+1}}{u_n} \right| \rightarrow r > 1 \quad \text{then the series diverges}$$

and if the limit is 1, no conclusion can be drawn.

The limiting form is slightly more restrictive than is the inequality form, but the difference is not great; u_{n+1}/u_n need not approach a limit and the result is true. The proof for the divergent part is very similar to that of the convergent part.

If the limit of the ratio is 1, then nothing definite can be concluded since we have both convergent and divergent series with the limit equal to 1 (as we will soon show, Example 20.6-1).

There is an important theorem concerning conditionally convergent series; if you permit a rearrangement of the order of the terms in the series, you can get any answer you wish! To prove this, we go back to the notation of P_n being either the corresponding positive term or else 0, while the Q_n refer to the negative terms. For a conditionally convergent series, both the series in P_n and Q_n must diverge to infinity.

Now we pick a number, say a positive number, that you want to be the limit of the rearranged series. First, we add terms from the P_n until we pass the given number. Next, we subtract terms from the Q_n until we pass under the number. Then we add terms from the P_n until we are again over; then subtract terms from the Q_n, and so on. Since the series was originally conditionally convergent, the terms of both the P_n and the Q_n must approach zero. Thus the alternate going above and below the chosen number will get smaller and smaller. On the other hand, since each diverged to infinity, we cannot run out of numbers to continue the process. The series, in the order we are taking the terms, will converge to the given number.

Notice that in the process we will displace numbers arbitrarily far from their original position in the series. If you restrict yourself to a maximum displacement of terms, then this result is not true; it depends on gradually displacing terms farther and farther from their original position.

This theorem explains why in defining the expectation of a random variable we included the absolute value of the random variable; we may not have a natural order, and if the series, or integral, were conditionally convergent, then depending on the order we chose we would get different results when we allow nonabsolute convergent series or integrals.

EXERCISES 20.5

1. Does $\Sigma \, (-1)^n/\sqrt{n}$ converge?

2. Give the interval of convergence of $\Sigma \, (-x)^n/n^2$.

3. Is $\Sigma \, (-1)^n/n^2$ absolutely convergent?

4. Is $\Sigma \, (-1)^n/\sqrt{n}$ absolutely convergent?

5. Write an essay on absolute convergent and conditional convergent integrals.

20.6 POWER SERIES

A particularly important class of infinite series is

$$\sum u_n x^n$$

or, what is almost the same thing,

$$\sum u_n (x - x_0)^n$$

The difference is merely a shift in the origin of the coordinate system. They are both called *power series*. In this section we treat only the first form, but see the exercises for the second form.

The convergence can be examined using the ratio test. For the first series, we have the condition that

$$\left| \left(\frac{u_{n+1}}{u_n} \right) x \right| < r < 1$$

for convergence, and similarly for divergence. The series with $x - x_0$ behaves similarly.

Example 20.6-1

For the geometric progression (note the change in notation from r to x)

$$a + ax + ax^2 + \cdots$$

the ratio test gives

$$|x| < 1$$

as we earlier found.
    For the series

$$\sum nx^n$$

the ratio test gives

$$\lim \left\{ \left| \frac{(n + 1)}{n} \right| x \right\} = |x|$$

Hence the series converges if $|x| < 1$ and diverges if $|x| > 1$. At $|x| = 1$, the series cannot converge since the nth term does not approach zero. The interval of convergence is $-1 < x < 1$.
    For the series

$$\sum \frac{x^n}{n}$$

we get

$$\lim \left\{ \left| \frac{nx}{(n + 1)} \right| \right\} = |x|$$

which converges for $|x| < 1$ and diverges for $|x| > 1$. At $x = +1$, the series is the harmonic series

$$\sum \frac{1}{n}$$

which diverges (Example 20.3-1), and at $x = -1$, you have the alternating harmonic series

$$\sum \frac{(-1)^n}{n}$$

which converges (Example 20.5-1). The interval of convergence is $-1 \leq x < 1$. Thus we have shown that at the ends of the interval of convergence the series can sometimes converge and sometimes diverge. It is easy to see (Example 20.3-1 with $p = 2$) that the series

$$\sum \frac{x^n}{n^2}$$

converges at both ends of the interval $-1 \leq x \leq 1$.

We need to prove the following important result.

Theorem 20.6-1. If a power series converges for some value x_0, then it converges for all x such that

$$|x| \leq |x_0|$$

Proof. The proof follows immediately from the observation that if the series converges for $x = x_0$ then the nth term must approach zero. Hence past some point we must have

$$|u_n x_0^n| < 1$$

We can therefore write, for $|x| < |x_0|$,

$$\sum |u_n x^n| = \sum |u_n x_0^n| \left|\frac{x}{x_o}\right|^n < \sum \left|\frac{x}{x_0}\right|^n = \frac{|x_0|}{|x_0 - x|}$$

and the series converges. Thus we always have, for power series, an *interval of convergence*, which can be (1) closed at both ends, (2) open at one end, or (3) open at both ends. The ends of the interval must be separately examined since, as we showed in Example 20.6-1, any particular situation can arise.

Can we integrate a power series term by term? If we stay inside (avoid the ends) the interval of convergence, then from the convergence assumption of the power series we can find some place far enough out so that the remaining terms total up to as small an amount as you wish. And in this closed interval there is a farthest out point that will apply for all the x values in the interval (see Section 4.5, assumption 3). When we integrate term by term the approximation of the finite series, the error will be multiplied by a fixed finite amount, the length of the path of integration. Thus the difference between the integral of the function and the integral of the finite series can again be as small as required. We conclude that we can indeed integrate a power series term by term and expect that the series will represent the corresponding integral, provided the length of the path of integration is finite and lies *inside* the interval of convergence.

Can we differentiate term by term? It is not so easy to answer because we see immediately that the differentiation process produces coefficients that are larger than the original series. But if the differentiated series converges, then, reversing the process and integrating it, we can apply the above result. Thus, if we differentiate a power series term by term and the resulting series converges, it is the derivative of the function.

EXERCISES 20.6

Find the interval of convergence.

1. $\sum x^n/(n^2 + 1)$
2. $\sum (x - x_0)^n$

3. $\Sigma (x - a)^n/n!$
4. $\Sigma (x - x_0)^n/n^2$

0.7 MACLAURIN AND TAYLOR SERIES

When dealing with power series, it is often convenient to change notation. Suppose we denote the power series by

$$f(x) = \sum_0^\infty \frac{a_n}{n!} x^n \qquad (20.7\text{-}1)$$

(we set $u_n = a_n/n!$). We see immediately that for $x = 0$ we have

$$f(0) = a_0$$

If we formally differentiate the series and then set $x = 0$, we get

$$f'(0) = a_1$$

A second formal differentiation gives

$$f''(0) = a_2$$

and in general we will get

$$f^{(n)}(0) = a_n$$

Therefore, the series can be written in the form

$$f(x) = \sum_0^\infty \frac{f^{(n)}(0)}{n!} x^n$$

This is the *Maclaurin (1698–1746) series.*

If, instead of evaluating the derivatives at the origin, we used the point $x = a$, we would have the corresponding *Taylor (1685–1731) series*

$$f(x) = \sum_0^\infty \frac{f^{(n)}(a)}{n!} (x - a)^n \qquad (20.7\text{-}2)$$

There has been the tacit assumption all along that both the differentiated series converged, and that the final series would represent the original function. Is there anything we can do to find the error term corresponding to a finite number of terms? It happens that there is. The method depends on integration by parts, and we have seen part of the formula before (Section 11.7 on the mean value theorem). We start with the obvious expression

$$f(x) = f(a) + \int_a^x f'(s) \, ds$$

where we have used the dummy variable of integrations to avoid confusion with the upper limit x. We pick the parts for the integration

$$U = f'(s), \qquad dV = ds$$
$$dU = f''(s) \, ds, \qquad V = -(x - s)$$

Notice the peculiar form we picked for the V. Certainly the derivative is correct, and by picking this form, then when $s = x$ the upper limit of the integration will drop out. We have

$$f(x) = f(a) - (x - s)f'(s)\big|_a^x + \int_a^x f''(s)(x - s) \, ds$$

$$= f(a) + (x - a)f'(a) + \int_a^x f''(s)(x - s) \, ds$$

We integrate by parts again, this time picking

$$U = f''(s), \qquad dV = (x - s) \, ds$$
$$dU = f^{(3)}(s)ds, \qquad V = \frac{-(x - s)^2}{2}$$

and we have, after putting in the limits of integration,

$$f(x) = f(a) + \frac{f'(a)}{1!}(x - a) + \frac{f''(a)}{2!}(x - a)^2$$
$$+ \int_a^x \frac{f^{(3)}(s)(x - s)^2}{2!} \, ds$$

Either by induction or the ellipsis method, after n stages we will have

$$f(x) = f(a) + f'(a)(x - a) + \cdots + \frac{f^{(n-1)}(a)(x - a)^{n-1}}{(n - 1)!} + R_n \quad (20.7\text{-}3)$$

where the remainder R_n is

$$R_n = \frac{1}{(n - 1)!} \int_a^x f^{(n)}(s)(x - s)^{n-1} \, ds \qquad (20.7\text{-}4)$$

If $R_n \to 0$ as $n \to \infty$, then the difference between the function and the sum of the first n terms of the power series will approach zero. Thus the power series will both converge and represent the function; the difference approaches zero.

The integral (20.7-4) can be reworked using the mean value theorem for integrals (Section 13.3). The factor $(x - s)^{n-1}$ is of constant sign in the interval $[a, x]$ of integration (even if $x < a$). As a result, we have

$$|R_n| = \left| f^{(n)}(\theta) \frac{1}{(n - 1)!} \int_a^x (x - s)^{n-1} \, ds \right|$$
$$= \frac{|f^{(n)}(\theta)| \, |(x - a)^n|}{n!} \qquad (20.7\text{-}5)$$

This is the derivative form of the remainder.

It often happens that

$$|f^{(n)}(s)| \le M_n$$

for some constants M_n. Then we have

$$|R_n| \le \frac{M_n|x - a|^n}{n!}$$

If this approaches zero as $n \to \infty$, then the expansion approaches the value of the function for the corresponding values of x. Unless M_n grows at least as fast as $n!$, then the factorial in the denominator will cause the whole term (for fixed x) to approach zero as $n \to \infty$. Therefore, the series expansion will converge and represent the function.

Thus we have the following important class of cases: if the derivatives are bounded by the corresponding terms of a geometric progression (which need not converge), then the series converges.

20.8 SOME COMMON POWER SERIES

The ability to find the general term of the Maclaurin and Taylor series evidently depends directly on the ability to find the nth derivative of a function. This we can do only in selected cases, which are fortunately the cases of main importance. For other functions we can get only as many terms as we are willing to compute derivatives, one after the other.

Example 20.8-1

For the function $y(x) = e^{ax}$, we have

$$y(x) = e^{ax}$$
$$y'(x) = ae^{ax}$$
$$y''(x) = a^2 e^{ax}$$

and, in general, the nth derivative acquires a coefficient a^n. If we expand this function $y(x) = \exp ax$ about the origin, we have for the derivative values

$$1, \quad a, \quad a^2, \quad \cdots \quad a^n, \quad \cdots$$

and hence the expansion

$$e^{ax} = 1 + ax + \frac{(ax)^2}{2!} + \frac{(ax)^3}{3!} + \frac{(ax)^4}{4!} + \cdots$$

The ratio test gives convergence for all x, since

$$\left| \frac{u_{n+1}}{u_n} \right| = \left| \frac{ax}{n} \right|$$

approaches 0 for all finite values a and x.

The special case of $x = 1/a$ gives

$$e = 1 + 1 + \frac{1}{2!} + \frac{1}{3!} + \frac{1}{4!} + \cdots$$

$$= 1 + 1 + \frac{1}{2} + \frac{1}{6} + \frac{1}{24} + \frac{1}{120} + \cdots$$

from which it is easy to compute a value for the constant e.

$$
\begin{array}{r}
e = 1.00000\ 00000 \cdots \\
1.00000\ 00000 \cdots \\
0.50000\ 00000 \cdots \\
0.16666\ 66667 \cdots \\
0.04166\ 66667 \cdots \\
0.00833\ 33333 \cdots \\
0.00138\ 88889 \cdots \\
0.00019\ 84127 \cdots \\
0.00002\ 48016 \cdots \\
0.00000\ 27557 \cdots \\
0.00000\ 02756 \cdots \\
0.00000\ 00251 \cdots \\
0.00000\ 00021 \cdots \\
\underline{0.00000\ 00002 \cdots} \\
2.71828\ 18286 \cdots
\end{array}
$$

We also need to check that R_n approaches zero as $n \to \infty$. The factorial in the denominator guarantees this.

Example 20.8-2

The other special case, $x = -1/a$, leads to series

$$e^{-1} = 1 - 1 + \frac{1}{2} - \frac{1}{3} + \frac{1}{4} - \cdots$$

whose corresponding sum is

$$e^{-1} = 0.3678794412 \cdots$$

This number arose in an earlier matching problem and was the limit of no match occurring when pairing off two sets of n corresponding items.

Example 20.8-3

Find the Maclaurin expansion of $y = \sin x$. We have

$$y = \sin x$$

$$y' = \cos x$$

$$y'' = -\sin x$$

$$y^3 = -\cos x$$

$$\text{etc.}$$

The corresponding values of the derivatives at $x = 0$ are

$$0, \quad 1, \quad 0, \quad -1, \quad \ldots$$

and these values repeat endlessly. Thus the expansion is

$$\sin x = x - \frac{x^3}{3!} + \frac{x^5}{5!} - \frac{x^7}{7!} + \cdots$$

$$= \sum_{0}^{\infty} (-1)^n \frac{x^{2n+1}}{(2n+1)!}$$

Again the ratio test shows that the series converges for all values, and the further test shows that $R_n \to 0$.

It should be noted that if we can find a convergent power series representation of a function then we have the values of the derivatives at the point of expansion.

Example 20.8-4

For $|x| < 1$, the function

$$\frac{1}{1 - x} = 1 + x + x^2 + x^3 + \cdots$$

Hence at $x = 0$ the nth derivative of $y = 1/(1 - x)$ is $n!$

EXERCISES 20.8

1. Find the Maclaurin expansion of $\ln (1 + x)$.
2. Find the Maclaurin expansion of $\ln (1 - x)$.
3. Find the Maclaurin expansion for $y = \ln [(1 + x)/(1 - x)]$.
4. Find the Maclaurin expansion of $\cos x$.
5. Find the first three terms of the expansion about $x = 0$ of $y = \tan x$.
6. Use the Maclaurin expansion of $\ln (1 + x)/(1 - x)$ from Exercise 3. For $x = 1/3$, this gives $\ln 2$. Evaluate it for the first three terms. What value of x gives $\ln 3$?
7. Using partial fractions and Example 20.8-4, find the values of the derivatives at the origin of the function $y = 1/(1 - x^2)$.

20.9 SUMMARY

The topic of infinite series is vast, and we have only touched on the matter. The convergence of an infinite series is made to depend on the convergence of the corresponding sequence of partial sums S_n. We studied, in turn, null sequences, monotone sequences, and then the general case. This is typical of doing new mathematics; the simpler cases are attacked first, and from them the general case is approached.

The power series is especially important, and we examined the expansion of a few of the common elementary transcendental functions, $\exp(x)$, $\sin x$, $\cos x$, and $\ln(1 + x)$; the last two are in the exercises. In the next chapter we will further examine this important topic of representing functions as infinite series.

21

Applications of Infinite Series

1.1 *THE FORMAL ALGEBRA OF POWER SERIES*

Power series play a prominent role in much of mathematics; hence we need to know how to handle them. It is evident from the linearity of both differentiation and summation that if

$$f(x) = \sum_{n=0}^{\infty} u_n x^n$$

$$g(x) = \sum_{n=0}^{\infty} v_n x^n$$

(21.1-1)

then

$$Af(x) + Bg(x) = \sum_{n=0}^{\infty} \{Au_n + Bv_n\}x^n$$

For multiplication of two power series (we ignore for the moment the problem of the convergence of the product), write the product

$$h(x) = f(x)g(x) = \sum_{n=0}^{\infty} w_n x^n$$

How can a term in x^n arise in the product? To avoid troubles, we first write the products with dummy variables of summation m and k:

$$\sum_{n=0}^{\infty} w_n x^n = \sum_{m=0}^{\infty} u_m x^m \sum_{k=0}^{\infty} v_k x^k$$

We see that the condition for the term x^n to arise is

$$m + k = n$$

Then, and only then, can we get a term x^n; the sum of the exponents of the individual terms in the product must add up to the value n. Rewriting this, we get

$$k = n - m$$

Thus each coefficient w_n of the product is given by corresponding finite sum of products

$$w_n = \sum_{m=0}^{n} u_m v_{n-m} \tag{21.1-2}$$

This is called the *convolution* of the two sequences of coefficients. Its structure is best seen in the Figure 21.1-1, where we have written on one strip the coefficients of the u_n series in order as you go down, and have written the coefficients of v_n in the opposite order. Each coefficient w_n is found from the sum of all the products when the two arrays are shifted an amount of exactly n. We have seen this before, for example in the sum of two random variables. Note that the convolution of the u_k with the v_m is the same as the convolution of the v_m with the u_k.

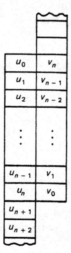

Figure 21.1-1   Convolution

Example 21.1-1

Consider the product $e^x e^y$. We get

$$\left(\sum_{m=0}^{\infty} \frac{x^m}{m!}\right)\left(\sum_{k=0}^{\infty} \frac{y^k}{k!}\right)$$

If w_n are the coefficients of the product, then the convolution gives

$$w_n = \sum_{m=0}^{n} \frac{x^m}{m!} \frac{y^{n-m}}{(n-m)!}$$

If (1) we write this as (multiply and divide by $n!$), (2) note the binomial coefficients appearing, and (3) apply the binomial expansion, we get

$$w_n = \sum_{m=0}^{n} \frac{n!}{m!(n-m)!} \frac{x^m y^{n-m}}{n!}$$

$$= \sum_{m=0}^{n} \frac{C(n,k)x^m y^{n-m}}{n!} = \frac{(x+y)^n}{n!}$$

We have, therefore,

$$\sum_{n=0}^{\infty} w_n = \sum_{n=0}^{\infty} \frac{(x+y)^n}{n!} = \exp(x+y)$$

This is, of course, what you expect since

$$e^x e^y = e^{x+y} \qquad\qquad (21.1\text{-}3)$$

We have *proved* that the function $\exp x$ is an exponential function.

Example 21.1-2

Division of power series is done much like the division of polynomials. Using

$$\tan x = \frac{\sin x}{\cos x}$$

$$= \frac{x - (x^3/3!) + (x^5/5!) - \cdots}{1 - (x^2/2!) + (x^4/4!) - \cdots}$$

we divide the two series:

$$
\begin{array}{r}
x + \dfrac{x^3}{3} + \left(\dfrac{2}{15}\right)x^5 + \cdots \\[2mm]
1 - \dfrac{x^2}{2!} + \dfrac{x^4}{4!} - \dfrac{x^6}{6!} \cdots \overline{\Big)\; x - \dfrac{x^3}{3!} + \dfrac{x^5}{5!} - \dfrac{x^7}{7!} \cdots} \\[2mm]
x - \dfrac{x^3}{2} + \dfrac{x^5}{4!} - \dfrac{x^7}{6!} \cdots \\[2mm]
\hline
\dfrac{x^3}{3} - \dfrac{x^5}{30} - \dfrac{x^7}{840} \cdots \\[2mm]
\dfrac{x^3}{3} - \dfrac{x^5}{6} + \dfrac{x^7}{72} \cdots \\[2mm]
\hline
\dfrac{2x^5}{15} - \dfrac{4x^7}{345} \cdots
\end{array}
$$

Thus, assuming that the quotient series converges, we have

$$\tan x = x + \frac{x^3}{3} + \frac{2x^5}{15} + \cdots \qquad (21.1\text{-}4)$$

We cannot expect convergence beyond the first zero of the denominator since there the quotient is infinite, and the power series, if it represented the quotient, would diverge there. We will simply state that this series does converge that far.

21.2 GENERATING FUNCTIONS

In Chapter 2 we introduced the idea of a generating function of the binomial coefficients. The expansion

$$(1 + t)^n = \sum_{k=0}^{n} C(n, k)t^k$$

is the generating function of the binomial coefficients $C(n, k)$. We also saw that by multiplying generating functions we could find various identities. For example the equation

$$(1 + t)^{n+m} = (1 + t)^n(1 + t)^m$$

regarded as generating functions, leads to various identities among the binomial coefficients when we equate the corresponding linearly independent terms on both sides (the corresponding powers of t). Because the expansion of a function in a power series about a point is unique, it follows that the corresponding coefficients of like powers are equal. We can differentiate and integrate such identities to get new identities.

We now regard the expansion

$$\sum_{k=0}^{\infty} u_n t^n$$

as a generating function of the numbers u_n [or of functions of x in case $u_n = u_n(x)$].

Example 21.2-1

We can regard the equation

$$\frac{1}{1 - t} = \sum_{k=0}^{\infty} t^k \qquad (21.2\text{-}1)$$

as the generating function of the numbers 1, 1, 1, ... If we differentiate this and then multiply through by t, we get

$$\frac{t}{(1 - t)^2} = \sum_{k=1}^{\infty} kt^k \qquad (21.2\text{-}2)$$

which is the generating function of the numbers 1, 2, 3, ...

Had we integrated the original expression (21.2-1) from 0 to t (and multiplied by -1), we would have found, for $|t| < 1$,

$$\ln(1 - t) = -\sum_{k=0}^{\infty} \frac{1}{k + 1} t^{k+1}$$

$$= -\sum_{k=1}^{\infty} \frac{1}{k} t^k \tag{21.2-3}$$

where in the second line we have shifted the dummy index of summation by 1. It is easy to verify that these two expansions are the same by simply writing out the first few terms and comparing the two forms.

Example 21.2-2

Given a probability distribution $p(n)$ for $n = 0, 1, 2, \ldots$, the corresponding generating function is

$$P(t) = \sum_{k=0}^{\infty} p(k)t^k \tag{21.2-4}$$

If we differentiate this expression, we get

$$P'(t) = \sum_{k=0}^{\infty} kp(k)t^{k-1} \tag{21.2-5}$$

The expected value of the distribution $p(k)$ (the mean or average) is

$$E\{K\} = P'(1) \tag{21.2-6}$$

If we multiply the derivative equation (21.2-5) by t and differentiate again, we get

$$\{tP'(t)\}' = tP''(t) + P'(t) = \sum_{k=0}^{\infty} k^2 p(k)t^{k-1}$$

Hence, putting $t = 1$, we have

$$E\{K^2\} = P''(1) + P'(1) \tag{21.2-7}$$

Recalling that the variance is

$$\mathrm{Var}\{K\} = E\{K^2\} - E^2\{K\}$$

we have that

$$\mathrm{Var}\{K\} = P''(1) + P'(1) - [P'(1)]^2 \tag{21.2-8}$$

Thus the generating function enables us to find the first and second moments of a distribution by the simple process of differentiation. Indeed, it is easy to see how to get the higher moments, if you want them, from the generating function of the distribution. The formula (21.2-8) for the variance is very useful.

A power series can, therefore, be regarded as a generating function of the coefficients. The student may notice that in this approach the convergence of the series is not important; it is the formal relationships that seem to matter. When we multiply two generating functions together, we get the corresponding generating function of the new set of numbers. We saw such a process when we multiplied two powers of $(1 + t)^k$ and compared the results. If the numbers are to be combined according to the corresponding convolution, then the corresponding coefficients of like powers are to be identified without regard to the convergence. The question of the convergence of generating functions is a delicate matter that has long been debated; it is, of course, nice if the series converges, but at times useful results are obtained by the formal manipulation of generating functions without regard to the convergence or divergence of any of the expansions.

The student may wonder how we get generating functions in cases where we do not know them already. The Fibonacci numbers in the next example will show how: (1) the assumption of the generating function of the Fibonacci numbers, (2) applying the known relationship that defines the Fibonacci numbers leads us to the generating function, and (3) the manipulation of the function to get it expanded in a formal power series, without the cost of trying to compute the general derivative, produces the coefficients.

Example 21.2-3

The Fibonacci numbers (Example 2.4-3) are defined by

$$f_{n+1} = f_n + f_{n-1}, \qquad f_0 = 0, \quad f_1 = 1$$

We assume (1) that we have the generating function of the Fibonacci numbers:

$$F(t) = \sum_{k=0}^{\infty} f_k t^k = \sum_{k=1}^{\infty} f_k t^k$$

If we multiply by t, we get

$$tF(t) = \sum_{0}^{\infty} f_k t^{k+1}$$

and shifting the dummy index of summation from $k + 1$ to k, we have

$$tF(t) = \sum_{1}^{\infty} f_{k-1} t^k$$

Evidently, multiplying the generating function by t lowers the index of summation by 1. This suggests that dividing by t would shift the index up by 1. We have

$$\frac{F(t)}{t} = \sum_{1}^{\infty} f_k t^{k-1} = \sum_{0}^{\infty} f_{k+1} t^k$$

We are now ready to do step (2). The fundamental defining equation for the Fibonacci numbers is

$$f_{n+1} - f_n - f_{n-1} = 0$$

We write out the three expansions in detail to see what is going on.

$$F(t) = f_0 + f_1 t + f_2 t^2 + f_3 t^3 + f_4 t^4 + f_5 t^5 + \cdots$$

$$tF(t) = \qquad f_0 t + f_1 t^2 + f_2 t^3 + f_3 t^4 + f_4 t^5 + \cdots$$

$$\frac{F(t)}{t} = f_1 + f_2 t + f_3 t^2 + f_4 t^3 + f_5 t^4 + f_6 t^5 + \cdots$$

We see that (remember $f_0 = 0$)

$$\frac{F(t)}{t} - F(t) - tF(t) = f_1 = 1$$

and the coefficients of all other powers are automatically zero because of the defining relation $f_{n+1} - f_n - f_{n-1} = 0$. Solve for $F(t)$ to get

$$F(t)\left\{\frac{1}{t} - 1 - t\right\} = 1$$

or

$$F(t) = \frac{t}{1 - t - t^2}$$

We have the generating function for the Fibonacci numbers. It is now a matter of getting it into a power series form (3). We could divide it out, but that appears to give a confusing result, so we first apply the method of partial fractions to the right-hand side.

$$\frac{t}{1 - t - t^2} = \frac{A}{t_1 - t} + \frac{B}{t_2 - t}$$

where t_1 and t_2 are the roots of the quadratic in the denominator

$$t^2 + t - 1 = 0$$

These roots are

$$t_1 = \frac{-1 - \sqrt{5}}{2} \quad \text{and} \quad t_2 = \frac{-1 + \sqrt{5}}{2}$$

After tedious algebra (and no one said that the method is easy to carry out, only that the ideas are simple), we get

$$A = \frac{t_1}{\sqrt{5}}, \qquad B = \frac{-t_2}{\sqrt{5}}$$

A little more algebra and we have

$$F(t) = \frac{1}{\sqrt{5}}\left\{\frac{t_1}{t_1 - t} - \frac{t_2}{t_2 - t}\right\}$$

$$= \frac{1}{\sqrt{5}}\left\{\frac{1}{1 - (t/t_1)} - \frac{1}{1 - (t/t_2)}\right\}$$

But recalling that for this quadratic equation the product of the roots $t_1 t_2 = -1$, we get

$$F(t) = \frac{1}{\sqrt{5}}\left\{\frac{1}{1 + t_2 t} - \frac{1}{1 + t_1 t}\right\}$$

We can now divide these out using the form

$$\frac{1}{1 + x} = 1 - x + x^2 - x^3 + x^4 - \cdots$$

to get finally for the coefficient of t^n

$$\frac{1}{\sqrt{5}}\{(-1)^n(t_1^n - t_2^n)\}$$

which upon further algebra yields the value given in Example 2.4-3:

$$f_n = \frac{1}{\sqrt{5}}\left[\left\{\frac{1 + \sqrt{5}}{2}\right\}^n - \left\{\frac{1 - \sqrt{5}}{2}\right\}^n\right]$$

Generalization 21.2-4

From the single example of the Fibonacci numbers, we see that with any given set of numbers defined by a difference equation with constant coefficients of the form

$$f_{n+1} = a_0 f_n + a_1 f_{n-1} + a_2 f_{n-2} + \cdots + a_k f_{n-k}$$

we can do a similar process and obtain the general term in a closed form, *provided* we can factor the denominator into linear factors. Indeed, if quadratic factors arose, we could, at some cost in effort, divide out the quadratic factors.

EXERCISES 21.2

1. The generating function of the sequence 1, 1, 1, ... is $1/(1 - t)$. Show that the square of it gives the numbers 1, 2, 3, ... What does the cube give?
2. If $p(n) = 1/2^{n+1}$ for $n = 0, 1, 2, \ldots$, find the generating function and check that $P(1) = 1$. Find the mean and variance of the distribution.
3. Using the recurrence relation $u_n = au_{n-1}$, $u_0 = 1$, find the generating function.
4.*Using the recurrence relation $u_{n+1} = au_n + bu_{n-1}$ with $u_0 = 0$, $u_1 = 1$, find the generating function.

21.3 THE BINOMIAL EXPANSION AGAIN

In Section 2.4 we found the binomial expansion for integer n:

$$(1 + x)^n = 1 + nx + \frac{n(n - 1)x^2}{2} + \cdots + x^n$$

$$= \sum_{k=0}^{n} C(n, k)x^k$$

where the binomial coefficients $C(n, k)$ are given by the formula

$$C(n, k) = \frac{n!}{k!(n - k)!} = \frac{n(n - 1)(n - 2) \ldots (n - k + 1)}{k!}$$

Looking at the expansion, we see that it is simply the Maclaurin expansion of the function $(1 + x)^n$. We can get it directly by

$$y(x) = (1 + x)^n \qquad\qquad\qquad y(0) = 1$$
$$y'(x) = n(1 + x)^{n-1} \qquad\qquad\quad y'(0) = n$$
$$y''(x) = n(n - 1)x^{n-2} \qquad\qquad y''(0) = n(n - 1)$$
$$\cdots \qquad\qquad\qquad\qquad\qquad \cdots$$
$$y^{(k)}(x) = n(n - 1) \cdots (n - k + 1)x^{n-k} \qquad y^{(k)}(0) = n(n - 1) \cdots (n - k + 1)$$
$$\cdots \qquad\qquad\qquad\qquad\qquad \cdots$$

To get the coefficient of x^k, the kth derivative must be divided by $k!$, and we see that we get exactly the same result as we earlier found for the binomial expansion.

Now that we have two different derivations of this important result, we are in a position to compare the assumptions behind the derivations and perhaps loosen the conditions. For the earlier proof we used mathematical induction and were pretty well confined to positive integer exponents n. In this new derivation we are not so limited. If we believe

$$\frac{dx^n}{dx} = nx^{n-1}$$

for the exponent n, *whether or not n is an integer*, then we have the corresponding binomial expansion. But we will now get an endless sequence of coefficients; the numerator factor will not produce a zero coefficient past some point *unless n* is an integer. It is natural to *extend* the definition of the binomial coefficients to noninteger values of n, and say

$$C(n, k) = \frac{n(n - 1)(n - 2) \cdots (n - k - 1)}{k!} \qquad (21.3\text{-}1)$$

for all real values of n and integer k. (Compare with Example 9.10-3.)

We will have, for noninteger exponents n, an infinite series

$$(1 + x)^n = \sum_{k=0}^{\infty} C(n, k)x^k \qquad (21.3\text{-}2)$$

What about the convergence? The ratio test gives

$$\left| \frac{u_{k+1}}{u_k} \right| = \left| \frac{(n - k)x}{k + 1} \right|$$

and as $k \to \infty$, we see that for $|x| < 1$ we have convergence. We will not examine the convergence at the ends of the interval.

The student of the foundations of what we know should note that the derivative

$$\frac{dx^n}{dx} = nx^{n-1}$$

was derived for rational numbers only, and by assumption 1 of Section 4.5 we extended it to the irrational exponents. However, *if* the student believes that for all real n

$$y = x^n$$

is equivalent to

$$\ln y = n \ln x$$

then differentiating this gives

$$\frac{1}{y}\frac{dy}{dx} = \frac{n}{x}$$

from which we get

$$\frac{dy}{dx} = \frac{ny}{x} = \frac{nx^n}{x} = nx^{n-1}$$

without use of assumption 1 of Section 4.5. Thus we have an alternative derivation for the derivative of a power of x, one that may seem to make fewer assumptions.

We need to examine the binomial coefficients for negative integers. We have the definition

$$C(n, k) = \frac{n(n - 1)(n - 2) \cdots (n - k + 1)}{k!}$$

If we replace n by $-n$, we get

$$C(-n, k) = \frac{(-1)^k n(n + 1)(n + 2) \cdots (n + k - 1)}{k!}$$

Thus we can write (for positive n) the negative binomial coefficients as

$$C(-n, k) = (-1)^k C(n + k - 1, k) \qquad (21.3\text{-}3)$$

This is the fundamental tool for converting binomial coefficients from negative exponents to positive exponents.

Example 21.3-1

Suppose we check this binomial expansion formula for positive exponents by observing that

$$(1 + x)^{1/2}(1 + x)^{1/2} = 1 + x$$

We have the Maclaurin expansion for the first few terms:

$$(1 + x)^{1/2} = 1 + \frac{1}{2}x + \frac{(1/2)(-1/2)}{2x^2} + \frac{(1/2)(-1/2)(-3/2)}{3!x^3}$$

$$\frac{(1/2)(-1/2)(-3/2)(-5/2)}{4!x^4} + \cdots$$

$$= 1 + \frac{x}{2} - \frac{x^2}{8} + \frac{x^3}{16} - \frac{5x^4}{128} + \cdots$$

We have now to multiply this series by itself (using detached coefficients); we have

$$
\begin{array}{cccccc}
1 & \dfrac{1}{2} & -\dfrac{1}{8} & \dfrac{1}{16} & -\dfrac{5}{128} & \cdots \\[2mm]
1 & \dfrac{1}{2} & -\dfrac{1}{8} & \dfrac{1}{16} & -\dfrac{5}{128} & \cdots \\[2mm]
\hline
1 & \dfrac{1}{2} & -\dfrac{1}{8} & \dfrac{1}{16} & -\dfrac{5}{128} & \cdots \\[2mm]
 & \dfrac{1}{2} & \dfrac{1}{4} & -\dfrac{1}{16} & \dfrac{1}{32} & \cdots \\[2mm]
 & & -\dfrac{1}{8} & -\dfrac{1}{16} & +\dfrac{1}{64} & \cdots \\[2mm]
 & & & \dfrac{1}{16} & \dfrac{1}{32} & \cdots \\[2mm]
 & & & & -\dfrac{5}{128} & \cdots \\[2mm]
\hline
1 & 1 & 0 & 0 & 0 & \cdots
\end{array}
$$

We can similarly check the binomial expansion for negative exponents by using

$$(1 + x)^{-1/2}(1 + x)^{-1/2} = \frac{1}{1 - x}$$

In this way we can find many identities among the binomial coefficients.

Example 21.3-2

First occurrence. We often want to know when something will happen for the first time. Consider the tossing of a coin, or other Bernoulli trial, with probability of success p and of failure $q = 1 - p$. For independent trials we have the probability of the first failure to occur on the kth trial equals: success, success, success, $\ldots$, success, failure. This probability is

$$ppp \cdots pq = p^{k-1}q$$

As a partial check of this, we examine the sum of all possible outcomes:

$$\sum_{k=1}^{\infty} p^{k-1}q = q \sum_{k=0}^{\infty} p^k = \frac{q}{1 - p} = 1$$

which is what it should be, since the first occurrence must occur sometime.

What is the expected value of this probability of first occurrence? We need the generating function of the probabilities. This generating function is (by definition)

$$G(t) = \sum_{k=1}^{\infty} (p^{k-1}q)t^k = qt \sum_{k=1}^{\infty} (pt)^{k-1}$$

$$= \frac{qt}{1 - pt}$$

To get the mean and variance of the distribution of first occurrence [see formula (21.2-8)], we need the first two derivatives of the generating function. We have

$$G(t) = \frac{qt}{1 - pt}$$

$$G'(t) = \frac{q}{1 - pt} + \frac{pqt}{(1 - pt)^2}$$

$$= \frac{q - pqt + pqt}{(1 - pt)^2}$$

$$= \frac{q}{(1 - pt)^2}$$

$$G''(t) = \frac{2pq}{(1 - pt)^3}$$

Evaluating these at $t = 1$, we get

$$E\{K\} = \frac{1}{q}$$

$$\text{Var}\{K\} = \frac{2pq}{q^3} + \frac{1}{q} - \left(\frac{1}{q}\right)^2 = \frac{2p + q - 1}{q^2} = \frac{p}{q^2}$$

Example 21.3-3

Suppose we now ask for the probability distribution of the time to the second failure (occurrence). Evidently, the first failure can occur at any time before the second, so considering all those times of first failure plus the second failure, which add up to the second failure at the kth trial, we see a convolution arising. We have a sum of all products of occurrence for the first and second failure such that the sum of their times is the given amount. The generating functions of each failure (first and second) are the same since the two are assumed to be independent. Hence, the generating function of the time to second failure is just the square of the generating function to first failure:

$$G^2(t) = \left(\frac{qt}{1 - pt}\right)^2$$

The individual terms involve the binomial coefficients of order -2. We have, using (21.3-3),

$$G^2(t) = (qt)^2 \sum_0^\infty (-1)^k C(-2, k)(pt)^k$$

$$= (qt)^2 \sum_0^\infty C(k+1, k)(pt)^k$$

Using $C(n, k) = C(n, n - k)$, we have

$$G^2(t) = (qt)^2 \sum_0^\infty C(k + 1, 1)(pt)^k$$

$$= (qt)^2 \sum_0^\infty (k + 1)(pt)^k$$

The probability distribution of the second failure is given by $(k - 1)p^{k-2}q^2$.

Case History 21.3-4

The Catalan numbers. The Catalan numbers occur in many different applications, but the background from which they come is, in each case, difficult to explain simply. The stages in the solution leading to them illustrate the use of many of the small ideas that we have just developed. The numbers u_n are defined by the convolution

$$u_{n+1} = u_0 u_n + u_1 u_{n-1} + \cdots + u_n u_0$$

or

$$u_{n+1} = \sum_0^n u_k u_{n-k}$$

with $u_0 = 1$. This convolution suggests immediately that we deal with generating functions. Let the generating function be

$$U(t) = \sum_{n=0}^\infty u_n t^n$$

We multiply the convolution equation by t^n and sum with respect to n. We see that

$$\frac{U(t) - 1}{t} = U^2(t)$$

This can be written as a quadratic equation

$$tU^2(t) - U(t) + 1 = 0$$

whose solution is

$$U(t) = \frac{1 \pm \sqrt{1 - 4t}}{2t}$$

There is no u_{-1} term; hence we must use the minus sign. We have, therefore,

$$U(t) = \frac{1 - \sqrt{1 - 4t}}{2t}$$

Our problem is now to get this expression as a power series in t. To do this, we begin with the binomial expansion of the radical:

$$\sqrt{1 - 4t} = \sum_0^\infty C\left(\frac{1}{2}, k\right)\left(-4t\right)^k$$

$$= 1 + \frac{1}{2}(-4t) + \frac{(1/2)(-1/2)}{2}(-4t)^2 + \cdots$$

We see that the general term

$$(-4t)^k C\left(\frac{1}{2}, k\right) = \frac{-(1)(1)(3)(5) \cdots (2k - 3)4^k t^k}{2^k k!}$$

$$= \frac{-(2k - 2)!}{2^{k-1}(k - 1)! k!} 2^k t^k$$

$$= -\left\{\frac{(2k - 2)!}{2^{k-1}(k-1)!}\right\} \frac{2^k t^k}{k!}$$

$$= -\left\{\frac{(2k - 2)!}{k!(k - 1)!}\right\} 2t^k$$

Hence

$$1 - \sqrt{(1 - 4t)} = \sum_1^\infty \left\{\frac{(2k - 2)!}{k!(k - 1)!}\right\} 2t^k$$

and our original expression becomes (in the second line shift the summation index)

$$\frac{1 - \sqrt{1 - 4t}}{2t} = \sum_1^\infty \frac{(2k - 2)! t^{k-1}}{k!(k - 1)!}$$

$$= \sum_0^\infty \frac{(2k)! t^k}{(k + 1)! k!}$$

$$= \sum_0^\infty \left\{\frac{C(2k, k)}{k + 1}\right\} t^k$$

The number u_n is (the coefficient of t^n)

$$u_n = \frac{C(2n, n)}{n + 1}$$

These are known as the *Catalan numbers*, and occur in numerous places in mathematics.

The values of the sequence of Catalan numbers can be either found one at a time from this formula, or else we can find a recurrence relation. To find a recursive formula, we write

$$u_{n+1} = u_{n+1}\left[\frac{u_n}{u_n}\right] = \left[\frac{u_{n+1}}{u_n}\right] u_n$$

$$= \left[\frac{C(2n + 2, n + 1)(n + 1)}{C(2n, n)(n + 2)}\right] u_n$$

$$= \left[\frac{(2n + 2)(2n + 1)(n + 1)}{(n + 1)(n + 1)(n + 2)}\right] u_n$$

$$= \left[\frac{2(2n + 1)}{n + 2}\right] u_n$$

The first few Catalan numbers are $u_o = 1$, $u_1 = 1$, $u_2 = 2$, $u_3 = 5$, $u_4 = 14$, $u_5 = 42$,

$u_6 = 132$, $u_7 = 429$, ... The size of the nth Catalan number can be found using Stirling's approximation for n!

EXERCISES 21.3

1. Write $C(-3, k)$ as a positive binomial coefficient.
2. Verify by expanding and multiplying out the first four powers of $(1 + x)(1 + x)^{-1} = 1$.
3. Verify by expanding and multiplying out the first four terms of $(1 + x)^{1/2}(1 + x)^{-1/2} = 1$.
4. Find the probability of the third occurrence at trial k.
5. Find the mean and variance of the probability distribution of the second occurrence.
6. Show that $(1 + x)^{1/3}(1 + x)^{2/3} = 1 + x$ through the fourth-degree terms.
7. Using the Stirling approximation, find an estimate of the nth Catalan number.

21.4 EXPONENTIAL GENERATING FUNCTIONS

The form of the generating function series we assumed is the standard one. However, at times it is worth using various alternative forms. The second most common form for a generating function of the numbers u_n is the *exponential generating function*, which has the form

$$F(t) = \sum_{0}^{\infty} \left(\frac{u_n}{n!}\right) t^n$$

The difference is that we now are dealing with the original numbers u_n divided by $n!$. Note that the formulas for the mean and variance do not apply with this generating function.

Example 21.4-1

Suppose we want the moments of the normal distribution centered about the origin (for convenience). The probability density function is

$$p(x) = \left(\frac{1}{\sigma\sqrt{2\pi}}\right) \exp\left(\frac{-x^2}{2\sigma^2}\right)$$

and the moments are

$$\mu_n = \int_{-\infty}^{\infty} x^n p(x)\, dx$$

We form the exponential generating function of the moments

$$M(t) = \sum_{n=0}^{\infty} \left(\frac{\mu_n}{n!}\right) t^n$$

We have carefully arranged that inside the integral there will be the sum

$$\sum_0^\infty \frac{(xt)^n}{n!} = \exp{(xt)}$$

Therefore, the moment generating function is of the form

$$M(t) = \int_{-\infty}^\infty e^{xt}\, p(x)\, dx$$

The exponent from the $p(x)$ will add to this to give us, in the exponent,

$$\frac{-x^2}{2\sigma^2} + xt = \frac{-1}{2\sigma^2}\{x^2 - 2\sigma^2 xt\}$$

This immediately suggests completing the square:

$$= \frac{-1}{2\sigma^2}\{x^2 - 2\sigma^2 xt + \sigma^4 t^2\} + \frac{\sigma^2 t^2}{2}$$

We now have

$$M(t) = \exp\left(\frac{\sigma^2 t^2}{2}\right) \frac{1}{\sigma\sqrt{2\pi}} \int_{-\infty}^\infty \exp\frac{(x - \sigma^2 t)^2}{2\sigma^2}\, dx$$

Change the variable of integration to $x - \sigma^2 t = y$. This gives

$$M(t) = \exp\frac{\sigma^2 t^2}{2} \int_{-\infty}^\infty p(y)\, dy = \exp\left(\frac{\sigma^2 t^2}{2}\right)$$

This is the generating function of the moments of the normal distribution. We now express this in powers of t. We have immediately, upon expanding the exponential in a Maclaurin expansion,

$$M(t) = 1 + \frac{(\sigma t)^2}{2} + \frac{(\sigma t)^4}{2^2 2} + \frac{(\sigma t)^6}{2^3 3!} + \frac{(\sigma t)^8}{2^4/4!} + \cdots$$

$$= \mu_0 + \mu_1 t + \frac{\mu_2 t^2}{2!} + \frac{\mu_3 t^3}{3!} + \frac{\mu_4 t^4}{4!} + \frac{\mu_5 t^5}{5!} + \frac{\mu_6 t^6}{6!} + \cdots$$

It remains to equate the like powers of t. Clearly, all the odd moments are zero (the integrand is odd over a symmetric range, so this clearly checks for all the odd powers of t). For the even powers, we get

$$\mu_0 = 1, \qquad \mu_2 = \sigma^2, \qquad \mu_4 = \sigma^4\left(\frac{4!}{8}\right) = 3\sigma^4, \qquad \mu_6 = 15\sigma^6, \cdots$$

The generating function approach gives us all the moments of the normal distribution at one time.

The method just used can be applied to many situations besides finding moments of a probability distribution. With variations you can find all the integrals of a family of the form

$$\int f^n(x)g(x)\, dx$$

provided you can do the corresponding integral that arises, depending on the kind of generating function you adopt. It is then a matter of expanding the generating function into powers of t and equating like terms.

EXERCISES 21.4

1. Using a generating function, find the moments of the distribution $p(x) = a \exp(-ax)$, $x \geq 0$.
2. Using a generating function, find the moments of the function $\sin \pi x$, $0 \leq x \leq 1$.

21.5 COMPLEX NUMBERS AGAIN

The expansion of the exponential function is

$$e^x = 1 + x + \frac{x^2}{2!} + \frac{x^3}{3!} + \frac{x^4}{4!} + \frac{x^5}{5!} + \cdots$$

Suppose we make the *formal* substitution

$$x = i\theta$$

where $i = \sqrt{-1}$. We will get

$$e^{i\theta} = 1 + (i\theta) + \frac{(i\theta)^2}{2!} + \frac{(i\theta)^3}{3!} + \frac{(i\theta)^4}{4!} + \frac{(i\theta)^5}{5!} + \cdots$$

and if we rearrange this in terms of real and imaginary numbers, (using $i^2 = -1$, $i^3 = -i$, $i^4 = 1$, $\cdots$), we get

$$e^{i\theta} = \left(1 - \frac{\theta^2}{2!} + \frac{\theta^4}{4!} - \cdots\right) + i\left(\theta - \frac{\theta^3}{3!} + \frac{\theta^4}{5!} - \cdots\right)$$

$$= \cos\theta + i\sin\theta \qquad (21.5\text{-}1)$$

This is the famous *Euler identity*. The formula is sometimes written as $\exp(i\theta) = \cos\theta$. The formula is certainly surprising at first, so we evaluate it for a number of values.

θ	Value
0	$\cos 0 + i\sin 0 = 1$
$\dfrac{\pi}{2}$	$\cos\dfrac{\pi}{2} - i\sin\dfrac{\pi}{2} = i$
π	$\cos\pi + i\sin\pi = -1$
$\dfrac{3\pi}{2}$	$\cos\dfrac{3\pi}{2} + i\sin\dfrac{3\pi}{2} = -i$
2π	$\cos 2\pi + i\sin 2\pi = 1$
$\cdots$	$\cdots$

We see that each additional increase of $\pi/2$ to the angle gives a multiplicative factor of i. Thus the results are consistent with the multiplicative property of the exponential function. The particular value

$$e^{\pi i} = -1$$

is very interesting as it shows that a close relationship exists between the two transcendental numbers, e and π, that dominate the calculus and its applications. (The number $\gamma = 0.577\ldots$ is a very different number in the sense that it is apparently unrelated to e and π.)

Let us explore the matter further and see if $\cos\theta + i\sin\theta$ is truly an exponential. We begin with

$$e^{i\theta} e^{i\phi} = e^{i(\theta + \phi)} \tag{21.5-2}$$

and write each term out in the sin and cos form.

$$(\cos\theta + i\sin\theta)(\cos\phi + i\sin\phi) = (\cos\theta\cos\phi - \sin\theta\sin\phi)$$
$$+ i(\sin\theta\cos\phi + \cos\theta\sin\phi)$$

Using the addition formulas of trigonometry, we have the result

$$(\cos\theta + i\sin\theta)(\cos\phi + i\sin\phi) = \cos(\theta + \phi) + i\sin(\theta + \phi)$$

This is the de Moivre (1667–1754) identity (of whom Newton once said, "Ask Monsieur de Moivre; he knows these things better than I do."). The identity (21.5-2) provides a very convenient way of remembering the addition formulas of trigonometry.

In this identity we can set $\theta = \phi$ and use the same methods as we used in Section 3.8 for studying the exponentials, to get the fact that Euler's and de Moivre's identities are true for all rational powers. Using the assumption 1 of Section 4.5, they are also true for all real numbers. To get to negative exponents, we can either replace i by $-i$ throughout, or a bit more reasonably note that

$$(\cos\theta + i\sin\theta)(\cos\theta - i\sin\theta) = \cos^2\theta + \sin^2\theta = 1$$

Hence $\cos\theta - i\sin\theta$ is the reciprocal of $\cos\theta + i\sin\theta$.

In some respects the de Moivre identity is the heart of trigonometry; it explains so much of what one sees. For example, it is what is behind the result (Section 2.3) where we showed that $\cos nx =$ polynomial of degree n in $\cos x$. It is simply the nth power of the $\cos x + i\sin x$, plus noting that the real parts have only $\sin^2 x$, which is easily converted to $\cos^2 x$ so the result is only in $\cos x$.

It is possible that you might have felt better about all the formalism if we had started with

$$\exp(ax)$$

and set $a = i$, but it would be formalism nevertheless.

We see that $|\exp(ix)| = 1$. If we plot $\cos\theta + i\sin\theta$ in the complex plane (Section 5.12), we find Figure 21.5-1. This figure should give you a bit more

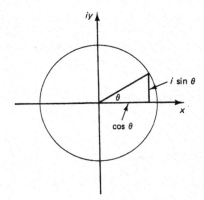

Figure 21.5-1 $e^{i\theta}$

confidence in the results obtained so far; you can see that the point described by

$$e^{i\theta} = \cos\theta + i\sin\theta$$

goes around the unit circle as θ goes from 0 to 2π, and the picture has the right periodicity for both sides of the equation. The geometric approach complements and reinforces the algebraic approach.

We write out the two Euler identities (21.5-1) (note the shift to the x variable),

$$e^{ix} = \cos x + i\sin x$$

$$e^{-ix} = \cos x - i\sin x$$

add, and divide by 2, to get

$$\cos x = \frac{e^{ix} + e^{-ix}}{2} \tag{21.5-3}$$

Similarly, by subtraction we get

$$\sin x = \frac{e^{ix} - e^{-ix}}{2i} \tag{21.5-4}$$

Thus we can convert from trigonometry to complex exponentials and back again as we please. This is useful, especially if you abbreviate e^{ix} by writing it as a new variable z:

$$e^{ix} = z \tag{21.5-5}$$

You have

$$\cos x = \frac{z + (1/z)}{2}$$

$$\sin x = \frac{z - (1/z)}{2i}$$

These substitutions reduce many trigonometric identities to problems in rational alge-bra where you do not need to recall the many trigonometric identities. In the sense of analytic geometry versus synthetic geometry, this is the analytic approach to trig-onometric identities. The analytic approach to trigonometry has the corresponding power. Note that

$$(e^{ix})^2 = z^2 = \cos 2x + i \sin 2x \qquad (21.5\text{-}6)$$

Example 21.5-1

Consider the expression (compare with Section 17.4)

$$\frac{1 - \cos x}{\sin x} = \left\{ 1 - \frac{z + (1/z)}{2} \right\} \Big/ \left\{ \frac{z - (1/z)}{2i} \right\}$$

$$= \frac{i\{2z - z^2 - 1\}}{z^2 - 1}$$

$$= \frac{i\{-(1 - z)^2\}}{(z - 1)(z + 1)}$$

$$= i\left\{ \frac{-(z - 1)}{z + 1} \right\}$$

$$= \left(\frac{1}{i} \right) \frac{z^{1/2} - z^{-1/2}}{z^{1/2} + z^{-1/2}}$$

$$= \frac{\sin (x/2)}{\cos (x/2)} = \tan \left(\frac{x}{2} \right)$$

In the above derivation, we have used mainly algebra and have used tri-gonometry only at the two ends. Of course, the derivation is much more rapid if you know all the various identities and when to use them; the above method is useful when you have no idea of what to do with the mess of trigonometric functions. The method is especially useful in handling "regular" identities.

Example 21.5-2

Consider the Euler identity

$$\frac{1}{2} + \cos x + \cos 2x + \cdots + \cos nx$$

Writing this in terms of the symbol z, we get

$$\frac{1}{2}\{z^{-n} + z^{-(n-1)} + \cdots + 1 + \cdots + z^{n-1} + z^n\}$$

This is clearly a geometric progression, and we get (arranging things in a symmetric form)

$$\frac{1}{2}\frac{z^{-n}\{1-z^{2n+1}\}}{1-z} = \frac{1}{2}\frac{z^{-(n+1/2)}-z^{(n+1/2)}}{z^{-1/2}-z^{1/2}}$$

$$= \frac{\sin\{(n+\frac{1}{2})x\}}{2\sin\left(\frac{x}{2}\right)}$$

which is the same as before, but a lot more directly obtained.

From the rules of exponents it follows that for complex exponents we have

$$\exp(a+ib) = (\exp a)(\cos b + i \sin b)$$

Example 21.5-3

In Example 16.5-7, we considered integrals of the form

$$\int e^{ax}\sin bx\, dx \quad \text{and} \quad \int e^{ax}\cos bx\, dx$$

and used integration by parts twice to get each answer. With complex exponentials we can get the two results much more easily. They are the imaginary and real parts of

$$\int e^{ax}e^{ibx}\, dx = \int \exp\{(a+ib)x\, dx$$

$$= \frac{\exp(a+ib)x}{a+ib}$$

$$= \exp(ax)\frac{\{\cos bx + i \sin bx\}(a-ib)}{a^2+b^2}$$

where we have ignored the constant of integration, and for the last equation have multiplied the numerator and the denominator by the complex conjugate $a - ib$ to make the denominator real. We have now to pick out the real and imaginary parts of the right-hand side. They are

$$\frac{a\cos bx + b\sin bx}{a^2+b^2}$$

$$\frac{a\sin bx - b\cos bx}{a^2+b^2}$$

which agrees with the earlier results (a here is $-a$ there).

What are we to think about all this formalism? Generally, most mathematicians are a lot happier when it is placed on a postulational basis of what appears to be more elementary assumptions. Engineers and scientists tend to feel that it *must be true* since the whole machinery works so well with the known parts of mathematics that they use daily. "Let the mathematicians make it more rigorous by the mathematician's standards if they wish," is their attitude, "we *know* it is true."

And this illustrates much of the progress of mathematics. What was at one time assumed to be obviously true by those using it, mathematicians have later analyzed more deeply and placed on a foundation of a postulational system that is logically more satisfying to them, and the results are generally more teachable. The postulational approach makes things more compact and makes them more certain in the sense that the assumptions are fewer, more visible, and easier to study for contradictions. But the engineer's and scientist's view is that whatever they believe is true is up to the mathematicians to account for; and if what the mathematicians report is not what the users want, then let the mathematicians try again. Of course, what was "obvious" to the engineers has more than once turned out to be false! The rigor of mathematics is the hygiene necessary to prevent these kinds of mistakes. This process describes much of the history of mathematics. The first formalizations are often not adequate and further work must be done. Thus mathematics is a cooperative effort (with much fighting between the two sides) of creating the useful, artistic structure we call mathematics. Mathematical rigor is seldom creative, but it is a necessary criticism of intuition if we are to avoid endless mistakes. We need to take advantage of the best of both approaches.

Of course, it is often the mathematician who first notices some interesting relationship and extends, generalizes, and abstracts it *before* it is needed by the users of mathematics. There is both a risk and an advantage of going forward without regard to the needs of the users; you are free from petty details, but you risk making an irrelevant generalization. Much useful mathematics has thus been found, and that which is ill generalized is soon forgotten.

EXERCISES 21.5

1. Find $\cos 6x$ in terms of powers of $\cos x$.

Using the z notation, complete the following identities:

2. $\cos^2 x - \sin^2 x$
3. $2 \sin x \cos x$
4. $\cos^3 x$ in terms of $\cos x$
5. $\tan 2x/(1 - \tan^2 x)$
6. $\sin x + \sin 2x + \sin 3x + \cdots + \sin nx$
7. $\cos x + 2 \cos 2x + 3 \cos 3x + \cdots + n \cos nx$

21.6 HYPERBOLIC FUNCTIONS

The Euler identities

$$\cos x = \frac{e^{ix} + e^{-ix}}{2}$$

$$\sin x = \frac{e^{ix} - e^{-ix}}{2i}$$

suggest looking at the corresponding real expressions. These are the *hyperbolic functions*

$$\cosh x = \frac{e^x + e^{-x}}{2}$$

$$\sinh x = \frac{e^x - e^{-x}}{2}$$

(21.6-1)

Cosh rhymes with "gosh," and sinh is pronounced "cinch." The other hyperbolic functions are defined similarly:

$$\tanh x = \frac{\sinh x}{\cosh x} = \frac{e^x - e^{-x}}{e^x + e^{-x}}$$

(21.6-2)

Their graphs are shown in Figure 21.6-1.

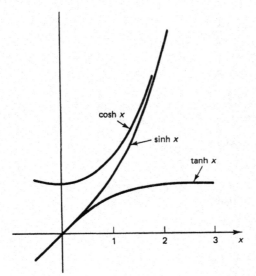

Figure 21.6-1   The hyperbolic function

Clearly, these functions are analogous to the trigonometric functions with minor differences. For example, we have

$$\cosh^2 x - \sinh^2 x = 1$$

$$1 - \tanh^2 x = \text{sech}^2 x$$

$$\text{ctnh}^2 x - 1 = \text{csch}^2 x$$

Similarly,

$$\frac{d \cosh x}{dx} = \sinh x$$

$$\frac{d \sinh x}{dx} = \cosh x$$

and there are no changes in sign as there are for the derivatives of the trigonometric functions.

You can show in a number of ways that the corresponding addition formulas are

$$\cosh(x + y) = \cosh x \cosh y + \sinh x \sinh y$$

$$\sinh(x + y) = \sinh x \cosh y + \cosh x \sinh y$$

For example, write out the right-hand sides as exponentials and then simplify.

The power series for the two main hyperbolic functions closely resemble those for the corresponding trigonometric functions (except that there are no changes in sign):

$$\cosh x = 1 + \frac{x^2}{2!} + \frac{x^4}{4!} + \frac{x^6}{6!} + \cdots$$

$$\sinh x = x + \frac{x^3}{3!} + \frac{x^5}{5!} + \frac{x^7}{7!} + \cdots$$

The inverse hyperbolic functions can be studied easily by simply interchanging the two variables and then solving for y. Thus we start with, for example,

$$\cosh x = y$$

and take the inverse function

$$x = \cosh y = \frac{e^y + e^{-y}}{2}$$

We have

$$e^{2y} - 2xe^y + 1 = 0$$

This is a quadratic exp y, which is easily solved for

$$e^y = x \pm \sqrt{x^2 - 1}$$

But since $y > 0$, we need to use the plus sign. Taking logs, we get

$$y = \ln(x + \sqrt{x^2 - 1})$$

This is suitable for the range $|x| > 1$.

EXERCISES 21.6

1. Write out all the definitions of the hyperbolic functions.
2. Differentiate all the hyperbolic functions.
3. Solve the quadratic that arises in finding the arctanh x. Be careful with the argument ranges.
4. Find the arcsinh x as a ln function.
5. Find by differentiation the power series of cosh x and sinh x.

1.7* HYPERBOLIC FUNCTIONS CONTINUED

Why are they called "hyperbolic functions?" They are related to the hyperbola

$$x^2 - y^2 = 1$$

much as the trigonometric (circular) functions are related to the circle

$$x^2 + y^2 = 1$$

In the first equation, $x = \cosh\theta$, and in the second, $x = \cos\theta$. Correspondingly, for the y we have $\sinh\theta$ and $\sin\theta$. Our problem is to identify the angle θ.

For the circular functions, we defined the angle θ as the ratio of the arc length cut out by the angle to the radius of the circle. For this analogy between the trigonometric and hyperbolic functions, we use the alternative definition of the angle as twice the area of the sector corresponding to the angle (see Figure 21.7-1). We need to find the shaded area to measure the "angle."

The proof is made easy if we refer the hyperbola to its asymptotes. Set

$$x - y = \sqrt{2}\,u, \qquad x + y = \sqrt{2}\,v$$

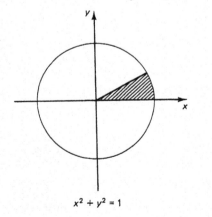

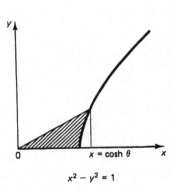

Figure 21.7-1   Hyperbolic angle

Then

$$x = \frac{u + v}{\sqrt{2}}, \qquad y = \frac{v - u}{\sqrt{2}}$$

The point A has coordinates

$$u = \frac{1}{\sqrt{2}}, \qquad v = \frac{1}{\sqrt{2}}$$

while the point P has coordinates

$$u = \frac{x - y}{\sqrt{2}}, \qquad v = \frac{x + y}{\sqrt{2}}$$

The angle θ, being the double of the area, is

$$\theta = 2 \int_{1/\sqrt{2}}^{(x+y)/\sqrt{2}} \frac{1}{2v} \, dv = \ln(x + y)$$

$$= \ln\{x \pm \sqrt{x^2 - 1}\} = \text{arccosh } x$$

where since $y > 0$ for positive angles the plus sign is used. This is the arccosh x as we asserted at the end of Section 21.6. From this it follows that

$$x = \cosh \theta$$

which shows that the angle plays the same role for both types of functions.

Thus the hyperbolic functions can be related to the hyperbola as the trigonometric (circular) functions are related to the circle. The analogy is often useful. However, it should be noted that when you represent some function in terms of the sinh x and cosh x for large positive x you are in trouble, since the two functions are both approximately

$$\frac{e^x}{2}$$

While, mathematically, the two functions cosh x and sinh x are linearly independent, in the presence of roundoff errors they are practically dependent (for $|x|$ large).

21.8 SUMMARY

In this chapter we have greatly extended the use of infinite series, especially power series. Since the power series is unique in any range, it follows that the powers of x are linearly independent for all powers, and we may equate the corresponding coefficients in two expressions that represent the same function in the same range. This is the basis for the generating function approach to many problems, especially combinatorial problems, where it is the main tool.

The binomial theorem was extended, by the usual approach, from integers to fractions, and then to all real exponents. Some tests of this new formula were given to both convince you that the formal expansions were consistent and to give you some practice in handling fractional exponents. Then a few of the numerous applications were indicated.

Next, we took on the tentative extension from the real numbers to the complex in the matter of power series. We simply asserted that the same formal power series we obtained for reals applied also for complex numbers. In particular, we used pure imaginary exponents to get the important Euler identities and found their relationship to the trigonometric functions. Thus we connected apparently different parts of mathematics together and found an inner unity. This phenomenon was discussed in Chapter 1 when we observed that the same kinds of things kept coming up in very different situations—the universality of mathematics. It follows that the inverse functions must similarly be related.

We then examined a close analogy to the trigonometric functions, called the *hyperbolic functions*. They again showed an inner structure that is not evident on the surface. This time we had to alter our original definition of angle to make the close analogy, something that should not surprise you. When you begin mathematics, there are often several ways of starting, several ways of defining things. As you progress and search for the inner unity, you sometimes have to go back and pick a different approach to make an analogy work. It is this inner unity, behind the surface diversity, that mathematicians find so beautiful about their subject. And this unity applies not only to pure mathematics, but often occurs in its various applications—the same mathematics is used in widely different fields. In a sense the mathematician is the last universalist; a sound knowledge of mathematics enables you to work in many different fields, and the contributions in your field are apt to reveal the unity behind the apparent diversity in many different fields of application.

22

Fourier Series

22.1 INTRODUCTION

In Chapter 20 we saw the expansion of a function in terms of powers of x. When we truncate the power series at some finite number of terms, we have a polynomial approximation to the function. Polynomials are widely used in much of mathematics, and perhaps are overused in statistics. The attraction seems to be that for paper and pencil computation they are easy to handle. With modern computers available, this advantage is not as large as past computation experience suggests. One fault with polynomials, among others, is that seldom does the polynomial approximation have much physical significance. Another fault is that as $|x| \to \infty$ polynomials rush off to $\pm\infty$, while most physical phenomena are bounded.

Instead of polynomials, the *Fourier* [J. B. J. Fourier, (1768–1830)] *functions*

$$
\begin{array}{llllll}
1 & \cos x & \cos 2x & \cos 3x & \ldots & \cos nx \ldots \\
& \sin x & \sin 2x & \sin 3x & \ldots & \sin nx \ldots
\end{array}
\tag{22.1-1}
$$

are much more useful in many situations. In this chapter we examine this topic more closely.

Not only are the Fourier functions often preferable to the use of polynomials, but history shows that many important results in mathematics came from the study of the representation of functions in terms of sines and cosines. Thus, besides their usefulness, the Fourier functions are of great theoretical interest in mathematics.

It is also worth noting, as you will soon see, that the Fourier series approach can handle discontinuous functions with fair ease, while polynomials have great trouble, and the theory of power series utterly fails you.

22.2 ORTHOGONALITY

The word *orthogonal* means *perpendicular* and represents a very great extension of the concept of two lines being perpendicular to each other.

Suppose we let U represent the point $(u_1, u_2, \ldots, u_n)$ in n-dimensional space. Similarly, let V represent the point $(v_1, v_2, \ldots, v_n)$. Consider the lines from the origin to these two points. Then the line between the two points is represented by $U - V$ (see Figure 22.1-1). Using the Pythagorean distance, these lines have the lengths

$$U^2 = \sum_{k=1}^{n} u_k^2, \qquad V^2 = \sum_{k=1}^{n} v_k^2$$

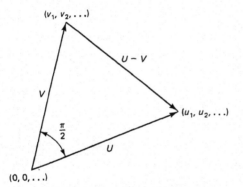

Figure 22.1-1 Perpendicular lines

The condition that U is perpendicular to V is

$$U^2 + V^2 = (U - V)^2$$

Expand these three sums and cancel the like terms. This leaves

$$\sum_{k=1}^{n} u_k v_k = 0$$

as the condition for perpendicularity (orthogonality) of the two vectors U and V in n-dimensional space. It is natural to *extend* this definition to integrals and to say that two functions $f(t)$ and $g(t)$ are *orthogonal* to each other, over the interval a to b, if

$$\int_a^b f(t)g(t)\, dt = 0$$

The Fourier functions (22.1-1) are *each* orthogonal to all the others. This means that we must prove (see Example 16.7-4)

$$\int_0^{2\pi} \cos mx \cos nx \, dx = \begin{cases} 0, & m \neq n \\ \pi, & m = n \neq 0 \\ 2\pi, & m = n = 0 \end{cases}$$

$$\int_0^{2\pi} \cos mx \sin nx \, dx = 0 \qquad\qquad (22.2\text{-}1)$$

$$\int_0^{2\pi} \sin mx \sin nx \, dx = \begin{cases} 0, & m \neq n \\ \pi, & m = n \neq 0 \\ 0, & m = n = 0 \end{cases}$$

We carry out the third integral this time (the first integral was done in Example 16.7-4); the second is 0, since a little thought shows that the integrand is positive as much as it is negative, so that the total must vanish). We need the trigonometric identities

$$\cos (a + b) = \cos a \cos b - \sin a \sin b$$

$$\cos (a - b) = \cos a \cos b + \sin a \sin b$$

from which, by subtraction, we get

$$\sin a \sin b = \frac{1}{2}[-\cos (a + b) + \cos (a - b)]$$

Using this, we get for the third integral

$$\int_0^{2\pi} \frac{1}{2}\{-\cos (m + n)x + \cos (m - n)x\}dx$$

$$= \frac{1}{2}\left\{ \frac{-\sin (m + n)x}{m + n} + \frac{\sin (m - n)x}{m - n} \right\} \Bigg|_0^{2\pi}$$

$$= 0, \qquad \text{if } m - n \neq 0$$

If $m = n \neq 0$, then the second term in the integrand is $\cos 0 = 1$ and hence integrates into 2π, and the factor $1/2$ in front reduces this to π, as it should. The last case for $m = n = 0$ means that the integrand is always 0, so the integral is also 0.

22.3 THE FORMAL EXPANSION

Suppose we try to represent a given function $f(x)$ in the form

$$f(x) = \frac{a_0}{2} + a_1 \cos x + a_2 \cos 2x + a_3 \cos 3x + \cdots$$

$$+ b_1 \sin x + b_2 \sin 2x + b_3 \sin 3x + \cdots \qquad (22.3\text{-}1)$$

The reason for the factor $1/2$ on the first term will soon become apparent. We multiply both sides by $\cos mx$ and integrate over the interval $[0, 2\pi]$. From the orthogonality relationships, we see that only one term will be other than 0, the term with coefficient a_m, and that this will be multiplied by π. Similarly, if we first multiply by sin mx and integrate, we get the only term, b_m. The result is

$$a_m = \frac{1}{\pi} \int_0^{2\pi} f(x) \cos mx\, dx$$

$$(22.3\text{-}2)$$

$$b_m = \frac{1}{\pi} \int_0^{2\pi} f(x) \sin mx\, dx$$

These formulas apply for $m = 1, 2, \ldots$. The case for $m = 0$ would give the integrated term as 2π, but the extra $1/2$ that we initially put in Equation (22.3-1) means that the coefficient a_0 is also computed by the same formula as for the other cosine terms. These coefficients, computed this way, are called the *Fourier coefficients* of the function.

It is important to note that due to the orthogonality of the Fourier functions the coefficients of the Fourier expansion are computed one at a time; there is no need to solve a set of simultaneous linear equations for the coefficients as we often have to do when we use the method of undetermined coefficients.

We have derived both the orthogonality and the expansion over the interval $[0, 2\pi]$, but it is often more convenient to use the equivalent interval $[-\pi, \pi]$, which is centered about the origin. The periodicity of the Fourier functions shows the equivalence.

Example 22.3-1

Let us formally expand the *step function* $(-\pi < x < \pi)$:

$$f(x) = -1 \text{ for } x < 0 \quad \text{and} \quad +1 \text{ for } x > 0$$

(see Figure 22.3-1). The function is an odd function; hence all the cosine (even) terms vanish. To verify this, we write (22.3-2)

$$a_m = \frac{1}{\pi}\left\{ \int_{-\pi}^0 + \int_0^\pi \right\}\{f(x) \cos mx\, dx\}$$

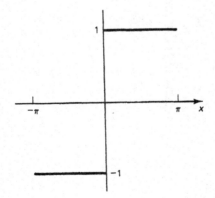

Figure 22.3-1 Step function

Putting in the values of $f(x)$, we get

$$a_m = \frac{-1}{\pi} \int_{-\pi}^{0} \cos mx \, dx + \frac{1}{\pi} \int_{0}^{\pi} \cos mx \, dx$$

Replace x by $-x$ in the first integral and we see that the two integrals exactly cancel.

We see from this one special case that for *any* odd function we need to compute only the sine terms whose coefficients are given by (22.3-2). Returning to our step function, we have

$$b_m = \frac{1}{\pi} \int_{-\pi}^{0} -\sin mx \, dx + \frac{1}{\pi} \int_{0}^{\pi} \sin mx \, dx$$

Again in the first integral we replace x by $-x$, and this time the two integrals combine:

$$b_m = \frac{2}{\pi} \int_{0}^{\pi} \sin mx \, dx$$

$$= \frac{2}{\pi} \frac{-\cos mx}{m} \Big|_{0}^{\pi}$$

$$= \frac{2}{\pi} \frac{1 - (-1)^m}{m}$$

Thus, all the even-indexed terms are zero, and the odd-indexed terms are

$$b_{2k-1} = \frac{4}{\pi(2k-1)}, \qquad k = 1, 2, 3, \ldots$$

We conclude that the function is, at least formally,

$$f(x) = \frac{4}{\pi} \left\{ \sin x + \frac{1}{3} \sin 3x + \frac{1}{5} \sin 5x + \cdots \right\}$$

The partial sums $f_n(x)$ are plotted in Figure 22.3-2, where we have plotted only the right

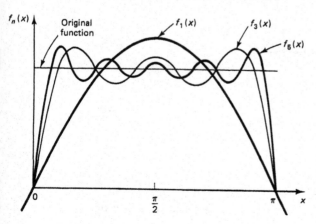

Figure 22.3-2 Fourier approximation

half of the graph since the negative part is the negative of what is shown. We see that the sequence of partial sums is "trying" to approximate the discontinuous function $f(x)$.

Example 22.3-2

Consider the function (Figure 22.3-3) that is 1 for $f(x)$ between $-\pi/2$ and $\pi/2$, and 0 otherwise. It is easy to see that the sine function, being odd, and the given function, being even, makes all the $b_n = 0$. Thus we have only to find the cosine term coefficients:

$$a_m = \frac{1}{\pi} \int_{-\pi}^{\pi} f(x) \cos mx \, dx$$

$$= \frac{2}{\pi} \int_{0}^{\pi} f(x) \cos mx \, dx$$

$$= \frac{2}{\pi} \int_{0}^{\pi/2} \cos mx \, dx$$

$$= \frac{2}{\pi} \frac{1}{m} \sin mx \Big|_{0}^{\pi/2}$$

Hence only alternate terms remain (with $a_0 = 1$), and the expansion is

$$f(x) = \frac{1}{2} + \frac{2}{\pi} \left[\frac{1}{1} \cos x - \frac{1}{3} \cos 3x + \frac{1}{5} \cos 5x \ldots \right]$$

The partial sums are shown in Figure 22.3-3.

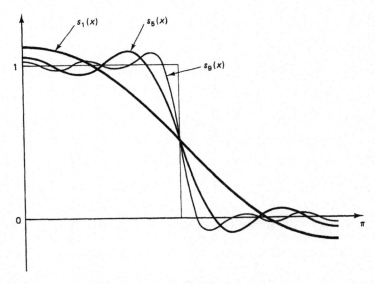

Figure 22.3-3   Fourier approximations

Example 22.3-3

Consider the function $y = x$ about the origin. It is an odd function, so only the sine terms occur. We have

$$b_m = \frac{1}{\pi} \int_{-\pi}^{\pi} x \sin mx \, dx$$

Using integration by parts (and symmetry), we have

$$b_m = \frac{2}{\pi} \left(\frac{-x \cos mx}{m} \bigg|_0^\pi + \int_0^\pi \frac{\cos mx}{m} dx \right)$$

$$= \frac{2}{\pi} \frac{\pi(-1)^m}{m} = \frac{2(-1)^m}{m}$$

Hence the expansion is (Figure 22.3-4)

$$x = 2 \left(\sin x - \frac{\sin 2x}{2} + \frac{\sin 3x}{3} - \cdots \right)$$

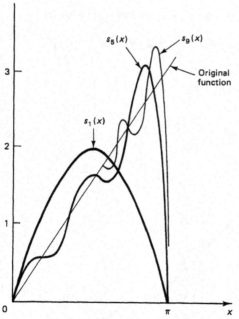

Figure 22.3-4 $y = x$

We see that it is easy to get the Fourier series coefficients of simple functions; it is a matter of doing the corresponding integrations. In applications (for example, in the important field of digital signal processing), it is often true that the function being expanded has only values -1, 0 or 1, or possibly $y = x$, and this makes the integration easy to carry out. Thus the above three examples are typical of such applications.

Given the formal expansion, it is convenient, for theoretical reasons, to eliminate the coefficients a_m and b_m. To do this, we need to change the dummy variable of integration from x to s; otherwise, we might confuse the different appearances of x. We have, from (22.3-1) and (22.3-2), formally,

$$f(x) = \frac{1}{\pi} \int_0^{2\pi} f(s) \left[\frac{1}{2} + \sum_{k=1}^{\infty} \{\cos ks \cos kx + \sin ks \sin kx\} \right] ds$$

$$= \frac{1}{\pi} \int_0^{2\pi} f(s) \left[\frac{1}{2} + \sum_{k=1}^{\infty} \cos k(s - x) \right] ds$$

Our rule for handling infinite series is to drop back to the sequence of partial sums, handle them, and then take the limit. Therefore, we now consider the partial sum (where we have moved the integration outside the finite summation):

$$f_n(x) = \frac{1}{\pi} \int_0^{2\pi} f(s) \left\{ \frac{1}{2} + \cos(s - x) + \cos 2(s - x) + \cdots + \cos n(s - x) \right\} ds$$

The quantity in braces was summed in Examples 2.6-3 and 21.5-2. The sum is

$$\frac{\sin\{(2n + 1)[(s - x)/2]\}}{2 \sin[(s - x)/2]}$$

Therefore, we have

$$f_n(x) = \frac{1}{\pi} \int_0^{2\pi} f(s) \frac{\sin\{(2n + 1)[(s - x)/2]\}}{2 \sin[(s - x)/2]} ds$$

We again assume that the function $f(s)$ is periodic. It is clear from the original formal representation as a sum of sines and cosines that we would get a periodic function, so this is not a real restriction. It does mean, however, that when thinking about Fourier series it is wise to think of them as drawn on a cylinder (see Figure 22.3-5). We make the change of variable from $s - x$ to s'. Set

$$s - x = s'$$

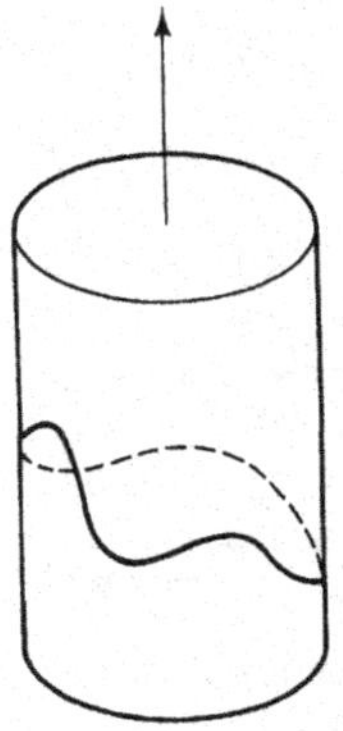

Figure 22.3-5 Function on a cylinder

The limits will shift, but due to the assumed periodicity they can still be taken over the same range; any interval of length 2π is equivalent to any other such interval. Dropping the prime on the s, we have

$$f_n(x) = \frac{1}{\pi} \int_0^{2\pi} f(s + x) \frac{\sin\{(2n + 1)(s/2)\}}{2\sin(s/2)} ds \qquad (22.3\text{-}3)$$

as a compact form for studying (1) the behavior of the partial sums of the Fourier series and (2) ultimately the question of the convergence of the Fourier series to the function.

EXERCISES 22.3

1. Expand the function $f(x) = 1$ for $|x| \le \pi/3$ and 0 otherwise.
2. Expand the function $f(x) = x$ for $|x| \le \pi/4$ and 0 elsewhere.
3. Expand the function $f(x) = 1$ for $|x| \ge \pi/4$ and 0 otherwise.

22.4 COMPLEX FOURIER SERIES

The student is always justly bothered by the formalism of complex numbers, but their advantage in many areas is such that it is necessary to become familiar with their use. In particular, a Fourier series is greatly simplified by their systematic use. We turn now to a deeper examination of complex numbers. We begin by repeating, using different words, much of what we have already said about complex numbers.

Let us review (repeating Section 4.8) how you handle such numbers as $\sqrt{2}$. In dealing mainly with rational numbers, when a $\sqrt{2}$ arises you segregate it. You then write any (every) number in the form

$$a + b\sqrt{2}$$

where a and b are nice rational numbers. You add and subtract such numbers, keeping the parts separated. When you multiply two such numbers, a $(\sqrt{2})(\sqrt{2})$ arises. You then recall that this product is by definition exactly 2. For example,

$$(a + b\sqrt{2})(c + d\sqrt{2}) = ac + bc\sqrt{2} + ad\sqrt{2} + bd\sqrt{2}\sqrt{2}$$

$$= (ac + 2bd) + (bc + ad)\sqrt{2}$$

In division, say finding the reciprocal of $a + b\sqrt{2}$, you proceed as follows:

$$\frac{1}{a + b\sqrt{2}} = \frac{1}{a + b\sqrt{2}}\left(\frac{a - b\sqrt{2}}{a - b\sqrt{2}}\right)$$

$$= \frac{a - b\sqrt{2}}{a^2 - 2b^2}$$

and you have the number in the canonical form again. The denominator cannot be zero since we have earlier proved that no rational number can equal $\sqrt{2}$.

When you face complex numbers, you have numbers of the form

$$a + ib$$

where $i = \sqrt{-1}$. It is convenient to represent complex numbers in the form of a picture called the *Argand diagram* (Section 5.12). We simply plot the numbers $x + iy$ with the real part along the x-axis and the imaginary part along the imaginary y-axis, as shown in Figure 22.4-1. The picture of the number seems to give a good deal of comfort to the beginner and is very useful for the expert as well. The simple fact is that the complex numbers are as "real" as are the real numbers; it is only your familiarity with the real numbers and the lack of experience with the complex numbers that gives you the uneasy feeling that they are somehow "imaginary numbers." The unfortunate name does not help; it would be better if they had another name, say *numbers of the second kind*, but custom is too strong to change it.

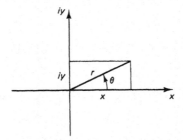

Figure 22.4-1 Argand diagram

To handle the arithmetic of complex numbers, you do as above, item for item, *except* that $i^2 = -1$ (instead of $\sqrt{2}^2 = 2$). Again using division as the example, we have as before (with the details deliberately suppressed to force you to think it through)

$$\frac{1}{a + ib} = \frac{1}{a + ib}\left(\frac{a - ib}{a - ib}\right) = \frac{a - ib}{a^2 + b^2}$$

The number $a - ib$ is said to be the *conjugate* of $a + ib$. Thus, to get the reciprocal of a complex number, you take the conjugate and divide by the quantity

$$a^2 + b^2$$

In Section 21.5 we found that formally (Euler's equations)

$$e^{ix} = \cos x + i \sin x$$

$$e^{-ix} = \cos x - i \sin x \qquad\qquad (22.4\text{-}1)$$

To solve for $\cos x$ and $\sin x$, we add and subtract these two equations to get

$$\cos x = \frac{e^{ix} + e^{-ix}}{2}$$

$$\qquad\qquad (22.4\text{-}2)$$

$$\sin x = \frac{e^{ix} - e^{-ix}}{2i}$$

Thus, sines and cosines may be written in the form of complex exponentials, and complex exponentials can be written in the form of sines and cosines; *there is a complete equivalence between the two forms*. But the amount of algebra in the two cases may be much different, as we saw in Section 21.5.

In the complex form the sin and cos are replaced by a pair of exponential functions, thus keeping the same amount of linear independence. But this brings up the matter that in the complex notation we have both positive and negative frequencies, since both positive and negative exponents occur.

In the rest of this section we simply repeat parts of Sections 22.2 and 22.3 with small changes in notation. The greater simplicity of the complex form should be apparent. Once again, complex numbers provide insight into what is going on, distasteful as they may be at first!

We have first the matter of the orthogonality over the interval $0 \le x \le 2\pi$ of the basis functions

$$e^{ikx} \tag{22.4-3}$$

for all integers k. In the complex form it is the product of the number by the conjugate that is orthogonal. Hence we must prove that the set of functions $\{\exp(ikx)\}$ $(-\infty < k < \infty)$

$$\int_0^{2\pi} e^{ikx} e^{-imx}\, dx = \begin{cases} 0, & \text{for } k \ne m \\ 2\pi, & \text{for } k = m \end{cases} \tag{22.4-4}$$

We have $(k \ne m)$

$$\int_0^{2\pi} e^{i(k-m)x}\, dx = \frac{e^{i(k-m)x}}{k-m}\Bigg|_0^{2\pi}$$

$$= \frac{1-1}{k-m} = 0, \qquad \text{for } k \ne m$$

(remember that $e^{ix} = \cos x + i \sin x$, and hence $\exp 2\pi i(k - m) = 1$ for all integers $k - m \ne 0$. For $k = m$, the integrand is 1, and the integral is 2π as required. We see that the orthogonality is much easier to prove and has a more compact representation than (22.2-1).

The formal expansion of a function into a complex Fourier series now has the form for both positive and negative frequencies:

$$f(x) = \sum_{k=-\infty}^{\infty} c_k e^{ikx} \tag{22.4-5}$$

where

$$c_k = \frac{1}{2\pi} \int_{-\pi}^{\pi} f(x) e^{-ikx}\, dx \tag{22.4-6}$$

The step of eliminating the appearance of the coefficients c_k (corresponding to the elimination of the earlier a_k and b_k) now takes the form

$$f(x) = \sum_{k=-\infty}^{\infty} \left[\frac{1}{2\pi} \int_{-\pi}^{\pi} f(s)e^{-iks}\, ds \right] e^{ikx}$$

The sequence of partial sums is then

$$f_n(x) = \frac{1}{2\pi} \sum_{k=-n}^{n} \left[\int_{-\pi}^{\pi} f(s)e^{-iks}\, ds \right] e^{ikx}$$

$$= \frac{1}{2\pi} \int_{-\pi}^{\pi} \left[f(s) \sum_{k=-n}^{n} e^{-ik(s-x)} \right] ds$$

But the sum is a geometric progression and has the sum

$$\frac{e^{i[n+(1/2)](s-x)} - e^{-i[n+(1/2)](s-x)}}{e^{i(s-x)/2} - e^{-i(s-x)/2}}$$

Divide both numerator and denominator by $2i$ to get

$$\frac{\sin\,[n + (1/2)](s - x)}{\sin\,(s - x)/2}$$

The partial sum is now

$$f_n(x) = \frac{1}{2\pi} \int_{-\pi}^{\pi} f(s)\, \frac{\sin[n + (1/2)](s - x)}{\sin\,(s - x)/2}\, ds$$

Again we use periodicity and shift the range of integration so that it centers about the point x by setting $s - x = s'$ (and drop the prime immediately):

$$f_n(x) = \frac{1}{2\pi} \int_{-\pi}^{\pi} f(s + x)\, \frac{\sin\,[n + (1/2)]s}{\sin\,(s/2)}\, ds$$

which corresponds to the result for the real case (22.3-3).

The student will note how much easier the complex derivation goes, how the messy trigonometry is avoided by the use of the summation of the geometric progression in place of the earlier identity we had to recall for the sum of the cos terms.

Example 22.4-1

Suppose we find the expansion in the complex form for the same function as in Example 22.3-1. We have

$$f(x) = -1 \text{ for } x < 0 \quad \text{and} \quad +1 \text{ for } x > 0$$

The c_m are given by

$$c_m = \frac{1}{2\pi} \int_{-\pi}^{\pi} f(s)e^{-ims}\, ds$$

$$= \frac{1}{2\pi} \int_{0}^{\pi} (e^{-ims} - e^{ims})\, ds$$

$$= \frac{i}{\pi} \int_0^\pi -\sin ms \; ds = \frac{i}{\pi m} \cos ms \; \Big|_0^\pi$$

$$= (1/(i\pi m))\{1 - (-1)^m\}$$

which gives alternate zero and nonzero coefficients. An examination of the two forms shows that both the real and complex expansions agree.

Note that for real functions the complex expansion has

$$c_k = \overline{c}_{-k}$$

and this makes the function real (since the two corresponding terms are conjugates of each other, and hence their sum is real).

EXERCISES 22.4

1. Compute $1/(3 - 4i)$.
2. Compute $(1 + i)^2$.
3. Compute $(1 - i)^3$.
4. Show that exp (ix) is a circle in the Argand diagram.
5. Differentiate the identity $\exp(ix) = \cos x + i \sin x$ and verify the result independently.
6. Differentiate the exponential representation of $\cos x$ and verify that the derivative is $\sin x$.
7. Expand $f(x) = 1$ for $|x| \le \pi/4$ and 0 elsewhere using the complex form.
8. Expand $f(x) = x$ for all x in the interval of definition using the complex form.

22.5 ORTHOGONALITY AND LEAST SQUARES

The purpose of this section is to show that the truncated Fourier series gives the best least squares fit. We begin (as we always do for infinite series) by considering the difference between the function $f(x)$ and a finite series of partial sums:

$$f_n(x) = \sum_{-n}^{n} C_k e^{ikt}$$

where we have written the coefficients as capital letters to distinguish them from the Fourier coefficients c_k. The least squares error is defined by (per usual)

$$M(C_k) = \frac{1}{2\pi} \int_{-\pi}^{\pi} |f(x) - f_n(x)|^2 \; dx \ge 0 \qquad (22.5\text{-}1)$$

The coefficients to be determined are the C_k. For a minimum, we cannot differentiate with respect to the C_k and set the derivatives equal to zero since we have absolute value signs, and these lead to conjugate functions and conjugate numbers; and we are unsure of how to handle them. Instead we proceed as follows (which gives another way of finding some maximum and minimum solutions). We (1) write the absolute squared term as a product of the quantity times its conjugate, (2) multiply out the terms from the sum, (3) substitute for the Fourier coefficients c_k, and (4) use the orthogonality of the terms to get from (22.5-1):

$$\frac{1}{2\pi} \int f(x)\bar{f}(x)\, dx + \sum_{-n}^{n} [-c_k\bar{C}_k - C_k\bar{c}_k + C_k\bar{C}_k] \geq 0$$

The terms in the sum suggest

$$|c_k - C_k||\bar{c}_k - \bar{C}_k| - c_k\bar{c}_k = |c_k - C_k|^2 - |c_k|^2$$

for each k. If we now return to the equation for the quality of the fit $M(C_k)$ (22.5-1), this gives:

$$M(C_k) = \frac{1}{2\pi} \int f(x)\bar{f}(x)\, dx - \sum_{-n}^{n} |c_k|^2 + \sum_{-n}^{n} |c_k - C_k|^2 \geq 0$$

and this will be a minimum *when and only when* all the $c_k = C_k$. Thus the integral of the square of the difference between the function and the best least squares approximation uses the Fourier coefficients and is given by (since $C_k = c_k$)

$$M(c_k) = \frac{1}{2\pi} \int_{-\pi}^{\pi} |f(x)|^2\, dx - \sum_{-n}^{n} |c_k|^2 \qquad (22.5\text{-}1)$$

This is the integral of the square of the failure to fit the function by the truncated series. In words, the integral of the absolute value of the function squared (divided by 2π), minus the sum of the squares of the absolute values of the coefficients, measures the least squares error of the fit. Thus in fitting a Fourier series you first find the integral of the square of the function (using the proper coefficient $1/2\pi$ in front) and then subtract the sum of the squares of the absolute values of the complex coefficients one by one as you find them. The difference measures the quality of the least squares fit at that stage of finding the coefficients and indicates whether or not to compute further coefficients. This inequality is known as *Bessel's inequality*. Using this inequality, you can sequentially find the coefficients of the Fourier expansion until you get the quality of fit you desire (as measured in the least squares sense).

The proof in the case of the real functions $\sin kx$ and $\cos kx$ is not very different. The coefficient in front of the integral is usually taken as $1/\pi$, and you use $a_0^2/2$ plus the sum of the squares a_k^2 and b_k^2. The details are similar otherwise.

An examination of the derivation of Bessel's inequality shows that it depends *only* on the orthogonality of the functions and not on the particular set of orthogonal functions. Thus you see the *generalization* that least squares and orthogonality are very closely associated.

From Bessel's inequality we see that the sequence of partial sums of the squares of the coefficients is bounded above by the integral of the square of the function. Since each coefficient squared is positive, the sequence of partial sums (of the squares of the coefficients) must be monotonically increasing, and hence the sequence must converge. From this we conclude:

Theorem. If $f^2(x)$ is integrable, then the individual Fourier coefficients must approach zero, that as $n \to \infty$

$$a_n = \frac{1}{\pi} \int_{-\pi}^{\pi} f(s) \cos ns \, ds \to 0$$

$$b_n = \frac{1}{\pi} \int_{-\pi}^{\pi} f(s) \sin ns \, ds \to 0$$

This does *not* prove that the series converges, since if the coefficients were $1/n$, the sum of the squares would converge, but the series with the coefficients themselves need not.

Rather than interrupt a later long derivation, we will here derive a related result:

$$\int_{-\pi}^{\pi} f(s) \sin (n + 1/2)s \, ds \to 0 \qquad (22.5\text{-}2)$$

We simply expand the $\sin(n + 1/2)s$ and note that in the two separate integrals we have a term, $\cos s/2$ or $\sin s/2$ that is of constant sign; hence the individual integrals, by the mean value theorem for integrals (Section 11.7) will have the form of $\cos \theta/2$ or $\sin \theta/2$ times one of the above integrals with this term removed.

EXERCISES 22.5

1. Find the Bessel inequality for the real functions using the divisor π instead of 2π.
2. Carry out the details of (22.5-2).

22.6 CONVERGENCE AT A POINT OF CONTINUITY

Just as with the Taylor series, the question soon arises, "Does the series converge, and if so, does it represent the function?" As with the Taylor series, if we can show the latter, then it will imply the former (since a_n and $b_n \to 0$). We therefore take up the problem of establishing under what conditions the series will represent the function. We, of course, are assuming that the integrals that define the Fourier coefficients exist, and this follows from the assumption that the function $f(x)$ is piecewise continuous with as many derivatives inside a piece as we need. In particular, to get a notation going, we assume that we have J pieces

$$x_j \leq x \leq x_{j+1}$$

with $x_0 = -\pi$ and $x_J = \pi$, and that inside each interval the function has as many derivatives as we use to prove the corresponding result. Technically, the intervals should be open intervals, because with the closed intervals we would effectively give the function two distinct values at the same point. However, rather than get lost in pedantic details, we will assume that the student can supply all the necessary circum-locutions to make it as rigorous as seems necessary. We will use the left- and right-hand limits of the function in the interval as the corresponding value of the function when operating on the jth interval (and of course the corresponding left- and right-hand derivatives); but when we are considering the behavior of the series at a point of discontinuity between two intervals, we will then have to be much more careful.

To show that at a point of continuity a series converges to the corresponding function, we need to find an expression for the difference between the function and the partial sum of n terms. In a previous section (22.4) we found the (real) representation

$$f_n(x) = \frac{1}{2\pi} \int_{-\pi}^{\pi} f(s + x) \frac{\sin\,[n + (1/2)]s}{\sin\,(s/2)} ds \qquad (22.6\text{-}1)$$

For the special function $f(x) = 1$ (for all x), we have

$$1 = \frac{1}{2\pi} \int_{-\pi}^{\pi} \frac{\sin\,(2n + 1)(s/2)}{\sin\,(s/2)} ds \qquad (22.6\text{-}2)$$

To be sure that this integral is correct, we need to look back at the way the sum arose (in the present notation). We had in the real notation

$$\frac{1}{2} + \cos s + \cos 2s + \cdots + \cos ns = \frac{\sin[n + (1/2)]s}{2 \sin\,(s/2)}$$

(which is a sum we obtained in Sections 2.2 and 2.5). Integrating this over the range of integration and dividing by π, we get 1 as the value of the integral. Since this (22.6-2) is an integration with respect to the dummy variable of integration s, we can multiply both sides by $f(x)$ and move the $f(x)$ inside the integral sign. We have, therefore

$$f(x) = \frac{1}{2\pi} \int_{-\pi}^{\pi} f(x) \frac{\sin\,[n + (1/2)]s}{\sin\,(s/2)} ds \qquad (22.6\text{-}3)$$

By subtracting the two equations (22.6-1) and (22.6-3), we now have the needed equation for the difference:

$$f_n(x) - f(x) = \frac{1}{2\pi} \int_{-\pi}^{\pi} \{f(s + x) - f(x)\} \frac{\sin\,(2n + 1)(s/2)}{\sin\,(s/2)} ds$$

If we write the function as

$$\frac{f(s + x) - f(x)}{\sin\,(s/2)} = \phi(s)$$

and suppose that $\phi^2(s)$ is integrable, then we can apply the last result of Section 22.4 to conclude that the difference approaches zero. Thus we have:

At a point of continuity the Fourier series converges to the function.

We have seen this in the examples in Section 22.3. In all cases at a point of continuity, the partial sums have apparently approached the function as $n \to \infty$.

22.7 CONVERGENCE AT A POINT OF DISCONTINUITY

At a point of discontinuity, the sequence of partial sums is presented with a difficulty; to what should it converge? We shall prove the important fact that the convergence at a point of discontinuity is the *average* of the two limits from the two intervals. That is, at a discontinuous point $x = x_i$,

$$f_n(x_i) \to \frac{f(x_i^+) + f(x_i^-)}{2}$$

where the superscripts $+$ and $-$ on the x_i refer to values slightly larger and slightly smaller than x_i (thus confining the function values for each term to their appropriate interval).

The proof goes much as the proof at a point of continuity except that we now need to break up the integral into two pieces, one for each interval. We will have

$$f_n(x) = \frac{1}{\pi} \int_{-\pi}^{\pi} \{f(x_i + s) + f(x_i - s)\} \frac{\sin (2n + 1)(s/2)}{\sin (s/2)} \left(\frac{ds}{2}\right)$$

When we construct the integral for $f(x)$, we naturally use

$$\frac{f(x_i^+) + f(x_i^-)}{2}$$

in place of the $f(x_i)$ we used before. Everything goes through much as before, although we have to modify the result we used to apply to the half-interval to prove the convergence. The details are not hard, only complicated, and are not likely to shed much light on how and why the Fourier series converges.

22.8 RATE OF CONVERGENCE

How fast does the Fourier series converge at a point of continuity? If it is rapid, then comparatively few terms will do for an approximation; but if slow, then many terms will be needed. It is necessary, therefore, to understand how we can decide, approximately, on the rate of convergence we can expect.

Consider the formula for the coefficient

$$a_n = \frac{1}{\pi} \int_{-\pi}^{\pi} f(x) \cos nx \, dx$$

If the function $f(x)$ consists of J intervals in each of which the function has all the derivatives we need, then we can write

$$a_n = \frac{1}{\pi} \sum_{j=1}^{J} \int_{x_{j-1}}^{x_j} f(x) \cos nx \, dx$$

where $x_0 = -\pi$, and $x_J = \pi$, and the other x_j mark the ends of the piecewise continuous intervals for the function and the derivatives. There is a similar formula for the b_n.

We integrate by parts for each interval:

$$a_n = \frac{1}{\pi} \sum_{j=1}^{J} \left\{ f(x) \frac{\sin nx}{n} \Big|_{x_{j-1}}^{y_j} - \frac{1}{\pi} \int_{x_{j-1}}^{y_j} f'(x) \frac{\sin nx}{n} dx \right\}$$

Now examine the terms that emerge from the integrated part. You see

$$f(x_i^-) \frac{\sin nx_i^-}{n} \quad \text{and} \quad f(x_i^+) \frac{\sin nx_i^+}{n}$$

and that the result from the upper limit of one interval will be exactly the same form, except for the algebraic sign, as that from the lower limit of the next interval (remember that the function is imagined to be on a cylinder and the two end values meet). If the function is continuous at that point x_i, the two terms will exactly cancel each other. From this we see that if the function $f(x)$ is continuous, *including* around the end, then the integrated part will drop out completely, and we will have only the integral part left. If the function is not continuous, then clearly the terms will not cancel completely for every n, and hence the Fourier coefficients of the expansion will decrease like $1/n$. If the function is continuous, then the next integration by parts will produce terms with $1/n^2$, and the series will converge absolutely.

Evidently, we can repeat the process on the derivative and conclude that if the derivative has no discontinuities, then the coefficients will decrease like $1/n^2$; if there are no discontinuities in the second derivative, then the coefficients of the Fourier series will decrease like $1/n^3$; and so on.

Thus we have the important fact that we can look at the function being expanded, on the real axis, note the lowest-level discontinuity, and deduce immediately the rate of decrease of the Fourier coefficients, the rate of convergence of the series. The convergence or nonconvergence of the Fourier series is a *local property* depending on whether or not the function $f(x)$ is continuous or discontinuous there. The rate of convergence is a *global property* and depends on the presence of discontinuities anywhere in the function (along the real line). This is in contrast with the convergence of a Taylor series, where we cannot tell from looking at the function along the real line what the radius of convergence will be. It takes a study of the function in the complex plane to understand the subject of convergence of Taylor series.

EXERCISES 22.8

1. Find the rate of convergence of the function $f(x) = 1 + x/\pi$ for $-\pi < x < 0$ and $f(x) = 1 - x/\pi$ for $0 < x < \pi$.

2. Find the rate of convergence of $f(x) = (1 - x^2/\pi^2)^2$.

3. Find the rate of convergence of $y = 1 - 3x^2 + 2x^4$ for the interval $-\pi < x < \pi$.

22.9 GIBBS PHENOMENON

The convergence of a Fourier series at a discontinuity must be a bit peculiar. Looking at the examples of discontinuous functions we expanded in Section 22.3, we see that near a discontinuity the partial sums have ripples. A closer examination shows that, as the number of terms in the partial sum increases, the ripples get compressed into a narrower and narrower region about the discontinuity; but it does not appear that the heights of the ripples decrease much.

The purpose of this section is to explore this matter closely and prove that the height of the ripples does not approach zero as $n \to \infty$. This phenomenon is known as *the Gibbs phenomenon* from the American scientist J. Willard Gibbs (1839–1903), who first explained it clearly (although it was published almost 50 years earlier by others!).

We take a function with a simple unit discontinuity at the origin as an illustration. Let $f(x)$ be $-1/2$ on the left and $1/2$ on the right of the origin. We have already found, in Section 22.3, the expansion for this $f(x)$

$$f(x) = \frac{2}{\pi} \sum_{k=0}^{\infty} \frac{\sin (2k + 1)x}{2k + 1}$$

But noting that each term of the sum can be written as

$$\int_0^x \cos (2k + 1)s \, ds$$

we have for the partial sums (we can interchange the finite summation and integration operations)

$$f_n(x) = \frac{2}{\pi} \int_0^x \sum_0^n \cos (2k + 1)s \, ds$$

The sum of cosines inside the integral can be condensed as follows:

$$s_n = \cos s + \cos 3s + \cos 5s + \cdots + \cos (2n + 1)s$$

To sum this series, multiply through by the sine of half the spacing, $\sin s$:

$$(\sin s)s_n = \frac{1}{2}\{\sin 2s + \sin 4s - \sin 2s + \sin 6s - \sin 4s \cdots$$

$$+ \sin 2(n + 1)s - \sin 2ns\}$$

$$= \frac{1}{2} \sin 2(n + 1)s$$

Hence

$$s_n = \frac{\sin 2(n + 1)s}{2 \sin s}$$

The integral becomes

$$f_n(x) = \frac{1}{\pi} \int_0^x \frac{\sin 2(n + 1)s}{\sin s} ds$$

To find the maxima and minima of this function $f_n(x)$, you follow the rule: differentiate, set equal to zero, and solve the resulting equation. But the derivative of this integral, by the fundamental theorem of the calculus, is just the integrand

$$\frac{1}{\pi} \frac{\sin 2(n + 1)x}{\sin x}$$

This means that at the extremes

$$2(n + 1)x = k\pi$$

for some integer k. The first maximum is when $k = 1$, $x = \pi/2(n + 1)$. At this point the function has the value

$$f_n\left(\frac{\pi}{2(n + 1)}\right) = \frac{1}{\pi} \int_0^{\pi/2(n+1)} \frac{\sin 2(n + 1)s}{\sin s} ds$$

We change variables to make things neater; set $2(n + 1)s = u$. We get (putting in a u in the numerator and denominator)

$$f_n\left(\frac{\pi}{2(n + 1)}\right) = \frac{1}{\pi} \int_0^\pi \left[\frac{\sin u}{u}\right]\left[\frac{u/2(n + 1)}{\sin\left(\dfrac{u}{2(n + 1)}\right)}\right] du$$

As $n \to \infty$, the second square bracket $\to 1$; hence the integral approaches

$$\frac{1}{\pi} \int_0^\pi \frac{\sin u}{u} du$$

This is easily found from tables (or else a numerical integration) to be 1.17898 . . . . Converting this to the percentage size of the overshoot, we get

$$\frac{1.17898 \ldots}{2} = \frac{1}{2} + 0.08949 \ldots$$

and similarly the undershoot of the next extreme value is

$$0.04858 \ldots$$

The limit of the ripple of the finite Fourier approximation thus has about a 9% overshoot and a 5% undershoot. This phenomenon can be seen on many oscilloscopes when a voltage is suddenly applied; the oscilloscope does not pass all of the high frequencies, but necessarily truncates them, and therefore is apt to show this spike in the trace on the face of the tube. The physics and the mathematics agree in this case.

Fourier apparently believed that, as the Fourier series converged, then at a point of discontinuity the limit function included the vertical line (see Figure 22.9-1). (He did not know the Gibbs phenomenon.) The usual definition of a function, which requires it to be single valued, does not allow this to be the limit function, and the limit of the Fourier series is generally drawn with the vertical line omitted.

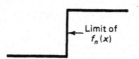

Limit of
$f_n(x)$

Figure 22.9-1   Limiting Fourier series at a discontinuity

EXERCISES 22.9

1. Sum the cosine series using complex exponentials.

22.10 THE FINITE FOURIER SERIES

It is a remarkable fact that the Fourier functions of sines and cosines are orthogonal over a set of equally spaced points as well as orthogonal over the continuous interval. Given the interval $0 \leq x \leq L$, the set of 2N points

$$x_k = \frac{Lk}{2N}, \qquad k = 0, 1, \ldots, 2N - 1$$

is equally spaced. We have for the complex form of the Fourier series the conditions

$$\sum_{k=0}^{2N-1} \exp\left(\frac{2\pi i p x_k}{L}\right) \exp\left(\frac{-2\pi i q x_k}{L}\right) = \begin{cases} 0, & p \neq q \\ 2N, & \text{if } p = q \end{cases}$$

To prove this is a matter of simply summing the geometric progression

$$1, r, r^2, \ldots, r^{2N - 1}$$

where

$$r = \exp\left[\frac{2\pi i (p - q)}{2N}\right]$$

If $r \neq 1$, then the sum is

$$\frac{1 - r^{2N}}{1 - r}$$

But

$$r^{2N} = \exp\left[2\pi i(p - q)\right] = 1$$

for $p \neq q$. Hence the sum is zero. If $p = q$, then each term is 1 and therefore the sum is $2N$.

To get this into the "real form," we need to write the sines and cosines in terms of the exponentials and see what results. We have

$$\sum_{p=0}^{2N-1} \cos \frac{\pi k p}{N} \cos \frac{\pi m p}{N}$$

$$= \frac{1}{4} \sum \left[\exp\left\{ \frac{i\pi(k + m)p}{N} \right\} + \exp\left\{ \frac{i\pi(k - m)p}{N} \right\} + \exp\left\{ \frac{i\pi(-k + m)p}{N} \right\} \right.$$

$$\left. + \exp\left\{ \frac{i\pi(-k - m)p}{N} \right\} \right]$$

$$= \begin{cases} 0, & |k \pm m| \neq 0, 2N, 4N, \ldots \\ N, & |k - m| = 0, 2N, 4N, \ldots \\ 2N, & |k \pm m| \text{ both} = 0, 2N, 4N, \ldots \end{cases}$$

Similarly, for sines,

$$\sum_0^{2N-1} \sin \frac{\pi k p}{N} \sin \frac{\pi m p}{N} = \begin{cases} 0, & |k \pm m| \neq 0, 2N, 4N, \ldots \\ N, & |k - m| = 0, 2N, 4N, \ldots \\ -N, & |k + m| = 0, 2N, 4N, \ldots \\ 0, & |k \pm m| \text{ both} = 0, 2N, \ldots \end{cases}$$

The sine times cosine terms are easily shown to be zero.

As a result, for the restricted set of $2N$ functions

1. $\cos \dfrac{2\pi x}{L}, \quad \cos \dfrac{4\pi x}{L}, \ldots, \quad \cos \dfrac{2\pi(N - 1)x}{L}, \quad \cos \dfrac{2\pi N x}{L}$

$\sin \dfrac{2\pi x}{L} \quad \sin \dfrac{4\pi x}{L}, \ldots, \quad \sin \dfrac{2\pi(N - 1)x}{L}$

only one of the above nonzero sum conditions will arise, and we have the orthogonality of the $2N$ functions over the set of $2N$ equally spaced points in the original interval of length L.

The formal expansion in the Fourier coefficients, now labeled A_k, B_k, has only $2N$ terms,

$$f(x) = \frac{A_0}{2} + \sum_{k=1}^{N-1} \left[A_k \cos \frac{2\pi k x}{L} + B_k \sin \frac{2\pi k x}{L} \right] + \frac{A_n}{2} \cos \frac{2\pi N x}{L}$$

where *both* end terms have a coefficient $1/2$ so that the same formulas

$$A_k = \frac{1}{N} \sum f(x_p) \cos \frac{2\pi k x_p}{L}$$

$$B_k = \frac{1}{N} \sum f(x_p) \sin \frac{2\pi k x_p}{L}$$

apply for the proper values k.

In the complex notation, the corresponding equations are

$$f(x) = \sum_{k=-N+1}^{N} c_k \exp\left(\frac{2\pi i k x}{L}\right)$$

$$c_k = \frac{1}{2N} \sum_{p=0}^{2N-1} f(x_p) \exp\left(\frac{-2\pi i k x_p}{L}\right)$$

Note the great similarity of the two equations in the complex form.

The famous *fast Fourier transform* is simply a very efficient way of computing the finite Fourier series coefficients. It does not compute the coefficients of the continuous function.

The $2N$ coefficients are enough to exactly reproduce the function data. If fewer terms are used, then we have a least squares fit (in analogy with the continuous case including the Bessel inequality).

The relationship between the finite discrete Fourier series and the continuous expansion is of great interest. Suppose we have the continuous expansion with lowercase letters a_k and b_k,

$$f(x) = \frac{a_0}{2} + \sum_{k=1}^{\infty} a_k \cos\left(\frac{2\pi k x}{L}\right) + b_k \sin \frac{2\pi k x}{L}$$

and we compute the sums that lead to the discrete coefficients A_k and B_k. We get

$$\sum_{p=0}^{2N-1} f(x_p) \cos\left(\frac{2\pi k x_p}{L}\right) = NA_k = N(a_k + a_{2N-k} + a_{2N+k} \ldots)$$

The terms on the right arise from the *aliasing* of the frequencies as indicated by the orthogonality conditions. As a result, we have

$$A_k = a_k + \sum_{m=1}^{\infty} (a_{2Nm-k} + a_{2Nm+k})$$

This shows clearly the *aliasing* when you go from the continuous model to the discrete model, (Figure 22.10-1). Imagine the frequencies along a ribbon, and then fold the ribbon as shown in the figure. At any frequency in the "folded interval," all points lying above a given frequency will appear as if they were the base frequency. Hence we speak of the "folding frequency," also called the *Nyquist* (1889–1976) *frequency*. This aliasing is familiar to those who have watched "Westerns" in the movies and seen the stagecoach wheels appear to go backwards, as well as to those who have used a

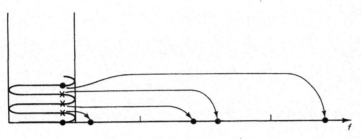

Figure 22.10-1 Aliasing

stroboscope that flashes its light at equally spaced intervals and makes the rotating equipment appear to be almost standing still.

If the convergence of the original Fourier series is rapid, then the difference between the discrete coefficients and the continuous coefficients is not much; but if the convergence is slow (there are discontinuities in the function), then the difference may be large. One cannot assert that by chance the sum will not add up to zero or a small number, but for this to happen for *all* the coefficients seems very unlikely! The corresponding formula for the sine coefficients is

$$B_k = b_k + \sum_1^\infty (-b_{2Nm-k} + b_{2Nm+k})$$

The formula for the A_0 term is just the same as computing the area under the curve using the trapezoid rule and therefore is worth special attention:

$$A_0 = a_0 + 2 \sum_{m=1}^\infty a_{2Nm}$$

a_0, is, of course, the area under the curve $y = f(x)$; hence the error $A_0 - a_0$ in the trapezoid rule is the sum of aliased terms.

In general, in numerical work on a computer you deal with equally spaced samples and not the continuous function. Thus the finite Fourier series is of relevance in these situations, and the aliasing gives a clear picture of what the relationship is between the model you want and the model you compute.

EXERCISES 20.10

1. Expand the function $y = x$, $-\pi < x < \pi$ in a finite Fourier series.
2. Expand $y = 1$ for positive values and -1 for negative values of x in the interval $-L/2 < x < L/2$.
3. Fill in the details for the derivation of Bessel's inequality.
4. Show that for $2N$ points the corresponding finite Fourier series exactly reproduces the data.

22.11 SUMMARY

In this chapter we have introduced the important concepts of orthogonality and Fourier series. Many of the results depend only on the orthogonality and not on the specific Fourier functions. The Fourier series gives the best least squares fit in the interval for the given set of functions, and this result depends only on the orthogonality of the functions. Often the details of the representation suggest physical effects in the phenomenon being studied; hence, frequently, the Fourier series is preferable to the polynomial approach.

The convergence of the Fourier series at points of continuity and discontinuity was investigated (using methods developed earlier) and produced the Gibbs phenomenon (also true for other sets of orthogonal functions, although the amount of overshoot may not be the same).

It is important to remember that the underlying functions, the sines and cosines, are periodic, and this implies that the function being expanded is also periodic and should be thought of as being on a cylinder. The rate of convergence can be found by an inspection of the kinds of discontinuities in the function and derivatives, and this represents a very useful property in practice.

The finite Fourier series is what we measure and compute with, and its relationship to the continuous Fourier series, which is what we tend to think about, is of fundamental importance.

23

Differential Equations

A differential equation is a relationship among $x, y, y', y'', \ldots y^{(n)}$, where the primes are derivatives of the function $y = y(x)$. For example,

$$\left(\frac{dy}{dx}\right)^2 = x^2 + y^2$$

$$\frac{d^2y}{dx^2} = \left(\frac{dy}{dx}\right)^4 y - x \sin y$$

are differential equations. These are *ordinary differential equations* because they involve only ordinary derivatives. If there were partial derivatives, then the equations would be called *partial differential equations*, but we will not study these.

The *order* of a differential equation is the order of the highest derivative appearing. Thus the first example is of first order and the second is of second order. The *degree* of the equation is the power of the highest derivative appearing. The first is of second degree and the second is of first degree.

We have already studied the special case of

$$\frac{dy}{dx} = f(x)$$

and found that the solution (the indefinite integral) involves an arbitrary additive constant. In the more general case of a first-order first-degree equation

$$\frac{dy}{dx} = f(x, y)$$

the solution will also involve a single arbitrary constant, but it need no longer be an additive constant C. In this chapter we will look mainly at first-order differential equations, but we will look briefly at two special cases of second-order equations of the form

$$\frac{d^2y}{dx^2} = f(x, y, y')$$

whose solution will have two arbitrary constants in the general solution. The topic of differential equations is so vast that we can take only a very limited look at it in this book.

23.2 WHAT IS A SOLUTION?

There are three different kinds of answers we can give to the question, "What is meant by the solution of a differential equation?"

First, there is what might be called the *algebraic answer*, which is simply "A solution is a function that when substituted, along with its derivatives, into the equation exactly fits the equation." As an example, we have

$$\frac{dy}{dx} = -xy \quad \text{which has the solution} \quad y = C \exp\left(\frac{-x^2}{2}\right)$$

where C is an arbitrary constant. Direct differentiation and substitution into the differential equation gives an *identity in x*. Another example is

$$y'' + y = 0 \quad \text{has the solution} \quad y = C_1 \cos x + C_2 \sin x$$

where both C_1 and C_2 are arbitrary constants. Again it is easy to check that this is a solution. In both cases it is a matter of simply differentiating the solution and then substituting into the differential equation to verify that you have a solution. In this respect it resembles verifying that you have found the right integral; you simply differentiate and note that you have the integrand.

Second, there is the *geometric answer*, "The solution of a first-order differential equation is a curve along which the slope at any point is exactly the derivative at that point as given by the differential equation." (Similar results apply for higher-order equations.) This introduces the idea of the *direction field* of a first-order differential equation, which is a picture at each point of which the slope element (as sketched in the Figure 23.2-1) has slope given by the differential equation. An example is

$$\frac{dy}{dx} = x^2 - y^2$$

We could take a close mesh of points, compute the slope at each point, and then draw the corresponding curves through the points such that at each point the curve has the indicated slope. Roughly, we would get a curve with the appropriate slope through each point of the picture (Figure 23.2-1).

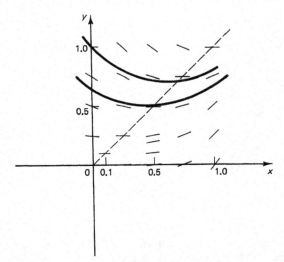

Figure 23.2-1   Direction field,
$y' = x^2 - y^2$

In this approach it is much easier, generally, to locate the curves along which the slope is a constant, the *isoclines* (iso-, same; clines, slope). For the above differential equation, the isocline curves are the hyperbolas

$$x^2 - y^2 = c$$

The slope along each of these curves is c. Figure 23.2-2 shows the direction field, and some possible solutions to the equation are sketched in.

In general, points on the isocline along which $y' = 0$ represent local extremes of the solutions. In the above example, these curves of maxima and minima are the two straight lines

$$y = \pm x$$

If you differentiate the differential equation to get the second derivative and then set this equal to zero (and eliminate the first derivative, if it occurs, by using the original differential equation), you get the curves along which the inflection points of the solution must lie. In the example this gives

$$y'' = 2x - 2yy' = 0$$

Divide out the 2 and eliminate the y' to get the curve

$$x - y(x^2 - y^2) = 0$$

This curve is indicated by the dotted line in Figure 23.2-2.

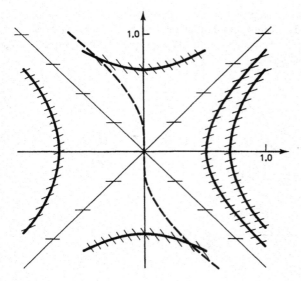

Figure 23.2-2   Isoclines

There is a third answer to the question, "What is a solution?" The differential equation describes the *local* conditions that the function must obey at every point; the solution is the *global* answer. Thus solving a differential equation means going from the local description to the global description.

We have said that through each point in the plane there is a single solution, but this can be misleading. Consider the differential equation

$$y' = \sqrt{1 - x^2 - y^2}$$

Using isoclines it is easy to make a crude sketch of the direction field (Figure 23.2-3). We see immediately that outside the circle of radius 1 there is no solution since the slope is imaginary. The solution to this differential equation is confined to a limited area of the plane.

EXERCISES 23.2

Sketch the direction fields and find the curve of max–min.

1. $y' = 1 - x^2 - y^2$
2. $y = xy' + (y')^2$
3. $y' = x/(x + y + 1)$
4. Find the curve of inflection points of $y' = x^2 + y^2$.

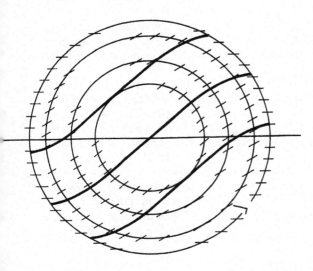

Figure 23.2-3 $y' = \sqrt{1 - x^2 - y^2}$

23.3 WHY STUDY DIFFERENTIAL EQUATIONS?

A differential equation is a local description of how a system changes. Given that a system is in some state now and subject to a given set of "forces," what is the immediate response of the system? When we solve this, we get the global description of the corresponding phenomenon.

As we currently understand the world, many of the things we are interested in fall into the general pattern of being expressed by a differential equation. In astronomy the planets are subject to various forces, chiefly gravitational. The planets have a present position and present velocity, and the gravitational interactions, via Newton's laws of motion, describe the main part of the evolution of the solar system. Newton's law

$$f = ma = m\,\frac{d^2s}{dt^2}$$

(where f = force, m = mass, and s = position) is a differential equation. Similarly, in physics and chemistry the basic laws (as we currently understand them) connect the present state of the system with the immediate future state. Even in biology we have such simple examples as the following: with an adequate food supply the growth of a bacterial colony is proportional to its size, or

$$\frac{dS}{dt} = kS$$

where $S = S(t)$ is the size of the colony at time t, and k is the constant of proportionality. In the social sciences we may have a situation that is more or less stable; then at a given date a new law becomes effective. As a result, the social (or economic) system reacts to the new conditions by making local changes. Thus the corresponding model of the social system is apt to be a set of differential equations connecting the variables and their rates of change.

Each field of application requires a knowledge of the details of the field, the rules (laws) of what will arise in the immediate future from the given state and given forces. Mathematics will seldom supply much help (beyond dimensional analysis) with the basic formulation of the model you are studying; but once the model is formulated, mathematics can tell you how the model would evolve, and hence (insofar as the modeling you have done is relevant and accurate) you will be able to anticipate the consequences that would occur if the proposed model is followed. In particular, in the social sciences it is necessary to model systems rather than experiment on real live models that include humans themselves. Modeling, or as it is often called, *simulation*, is the answer to the perennial question, "What if . . . ?" Often differential equations will aid in answering such questions (provided *you* have made a relevant model).

23.4 THE METHOD OF VARIABLES SEPARABLE

The easiest differential equations to solve, beyond mere integration, are those in which the variables can be separated as in the following examples.

Example 23.4-1

Solve the differential equation

$$y' = -xy$$

We write y' as dy/dx and then separate the variables (using differentials) to obtain two integrals. The detailed steps are shown below.

$$\frac{dy}{dx} = -xy$$

$$\frac{dy}{y} = -x \, dx$$

$$\int \frac{dy}{y} = -\int x \, dx + C$$

$$\ln y = \frac{-x^2}{2} + C$$

$$y = \exp\left(\frac{-x^2}{2} + C\right) = \exp\left(\frac{-x^2}{2}\right) e^C$$

$$= C_1 \exp\left(\frac{-x^2}{2}\right)$$

where we have written the constant $e^C = C_1$.

The constant of integration for differential equations no longer appears as an additive constant; it may appear almost any place and in any form. The disguise of a constant can cause the student trouble until it is realized that both results for the differential equation

$$y' = 2(x + a)$$

whose solutions are

$$y = (x + a)^2 + C \quad \text{and} \quad y = x^2 + 2ax + C$$

are the same; they differ by the meaning of the C. The C in the second solution is merely

$$C + a^2$$

of the first solution. Both are constants, so for suitable choice of C in either case you can get the same solution.

Example 23.4-2

Integrate the differential equation

$$y' = y^2 + a^2$$

We have

$$dy = (y^2 + a^2) \, dx$$

$$\frac{dy}{y^2 + a^2} = dx$$

Integrating both sides, we get

$$\frac{1}{a} \arctan \frac{y}{a} = x + C$$

$$\frac{y}{a} = \tan a(x + C)$$

$$y = a \tan a(x + C)$$

This is a *one-parameter family of solutions* of the differential equation. Often only one particular solution is wanted. We are in a position similar to that for the method of undetermined coefficients; we must impose the proper number of conditions to determine the unique solution. In this case, *suppose* we want the solution through the origin (0, 0). We then have

$$0 = a \tan a(C)$$

It is natural to pick $C = 0$, although other values such as $C = k\pi/a$ will give the same result. The particular solution is, therefore,

$$y = a \tan ax$$

Direct substitution will verify that this fits both the equation and the assumed boundary (or initial) condition.

Example 23.4-3

Integrate

$$xy' + (x^2 - 1)(y + 1) = 0$$

Taking slightly larger steps, we get in sequence

$$\frac{dy}{y + 1} + \frac{x^2 - 1}{x}\, dx = 0$$

$$\frac{dy}{y + 1} + \left(x - \frac{1}{x}\right) dx = 0$$

$$\ln(y + 1) + \frac{x^2}{2} - \ln x = C$$

$$\ln \frac{y + 1}{x} = C - \frac{x^2}{2}$$

$$\frac{y + 1}{x} = e^C \exp\left(\frac{-x^2}{2}\right) = C \exp\left(\frac{-x^2}{2}\right)$$

$$y = Cx \exp\left(\frac{-x^2}{2}\right) - 1$$

where we have calmly written e^C as a constant also labeled C. We could give one of the constants a subscript to distinguish the multiple use of the same symbol, but it is rarely done in practice. You simply replace any convenient form of a constant by C (being slightly careful not to confuse yourself in case there are several different Cs).

   If we tried to find a solution to this differential equation that passes through the origin, we would have

$$0 = -1$$

and the constant C does not appear. This, plus the simple fact that the equation is a contradiction, shows that there is no solution that passes through the origin. If we picked the point $x = 1$, $y = 0$, we would have

$$0 = Ce^{-1/2} - 1$$

and hence $C = e^{1/2}$ and the particular solution is

$$y(x) = x \exp\left\{-\left(\frac{x^2}{2} - \frac{1}{2}\right)\right\} - 1$$

Example 23.4-4

Integrate

$$y' = e^{-(x+y)}$$

We separate the variables:

$$e^y\, dy = e^{-x}\, dx$$

$$e^y = C - e^{-x}$$

$$y = \ln(C - e^{-x})$$

in the general solution of the differential equation. Direct substitution to check the solution gives

$$y' = \frac{e^{-x}}{C - e^{-x}} = e^{-x} \exp\{-\ln (C - e^{-x})\}$$

$$= \frac{e^{-x}}{C - e^{-x}}$$

If we reduce the problem to integrals, then even if the integrals cannot be done in closed form we regard the solution of the differential equation as having been found. In this method of separation of variables, it is sufficient, in theory, to write out the integrals. In practice, of course, the integrals must be done, but that is now a technique that you have presumably mastered and to give the details at this point would obscure the ideas of how to solve differential equations. However, in the exercises you are expected to work out the integrals to give you further experience in this art. But do not confuse the basic simplicity of the method of solution with the complexity of some of its steps.

Example 23.4-5

The astronomer Sir John F. W. Herschel (1792–1871) gave the following derivation of the normal distribution. Consider dropping a dart at a target (the origin of the coordinate system) on a horizontal plane. We suppose that (1) the errors do not depend on the coordinate system used, (2) the errors in perpendicular directions are independent, and (3) small errors are more likely than large errors. These all seem to be reasonable assumptions of what happens when we aim a dart at a point.

Let the probability of the dart falling in the strip $x, x + \Delta x$ be $p(x)$, and of falling in the strip $y, y + \Delta y$ be $p(y)$ (see Figure 23.4-1). Hence the probability of falling in the

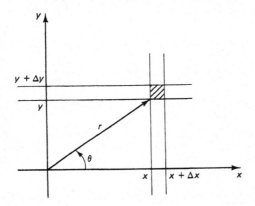

Figure 23.4-1

shaded rectangle is

$$p(x)p(y) \, \Delta x \, \Delta y = g(r) \, \Delta x \, \Delta y$$

where r is the distance from the target to the place where the dart falls. From this we conclude

$$g(r) = p(x)p(y)$$

Now $g(r)$ does not depend on θ; hence we have

$$\frac{\partial g(r)}{\partial \theta} = 0 = p(x) \frac{\partial p(y)}{\partial \theta} + p(y) \frac{\partial p(x)}{\partial \theta}$$

But we know that

$$x = r \cos \theta$$

$$y = r \sin \theta$$

and we can write

$$\frac{\partial p(x)}{\partial \theta} = \frac{\partial p(x)}{\partial x} \frac{\partial x}{\partial \theta} = p'(x)(-y)$$

$$\frac{\partial p(y)}{\partial \theta} = \frac{\partial p(y)}{\partial y} \frac{\partial y}{\partial \theta} = p'(y)(x)$$

This gives

$$p(x)p'(y)(x) + p(y)p'(x)(-y) = 0$$

Separating variables in this differential equation, we get

$$\frac{p'(x)}{xp(x)} = \frac{p'(y)}{yp(y)} = K$$

Both terms must be equal to a *constant K* since x and y are independent variables. This is a crucial step, so think this through! Changing either variable cannot change the other term; hence each term must be a constant. The equations are essentially the same, so we need only treat one of them. We have

$$\frac{p'(x)}{p(x)} = Kx$$

$$\ln p(x) = \frac{Kx^2}{2} + C$$

$$p(x) = A \exp\left(\frac{Kx^2}{2}\right)$$

But since small errors are more likely than large ones, K must be negative, say

$$K = -1/\sigma^2$$

(thus we get the formula in the standard form). This means that

$$p(x) = A \exp\left(\frac{-x^2}{2\sigma^2}\right)$$

The condition that the integral of a probability distribution over all possible values must total to 1 means that we can determine the constant of integration A to be

$$p(x) = \left\{\frac{1}{\sqrt{2\pi}\,\sigma}\right\} \exp\left(\frac{-x^2}{2\sigma^2}\right)$$

which is the normal distribution with mean zero and variance σ^2.

This derivation is seductively innocent, but in the final analysis the normal distribution rests on arbitrary assumptions. The name, *normal distribution*, implies that somehow you should "normally" expect to see it, but often random variables come from other than normal distributions. Furthermore, experience shows that, while the main part of the distribution often looks like a normal curve, the tails of the distribution may be either too small or too large. There are other sources of the normal distribution. For example, the *central-limit theorem*, which we do not prove here, states loosely that the sum of a large number of small independent errors will closely approximate the normal distribution. We saw something like this in Example 12.6-1.

EXERCISES 23.4

Integrate the following using the variables separable method.

1. $y' = c$
2. $yy' = x$
3. $xy' = y$
4. $y' = xy$
5. $xy' = (1 + x)(1 + y)$
6. $y' = xy - x - y + 1$
7. $y' = y \ln x$
8. $y' = x^m y^n$
9. Show that the sum of 12 independent random variables, from a uniform distribution $0 \leq x \leq 1$, minus 6 has mean zero and variance 1.

23.5 HOMOGENEOUS EQUATIONS

There is a large class of differential equations that can be solved by reducing them to variables separable. They are of the form of *homogeneous equations*, where by "homogeneous equations" we mean that all the terms are of the same degree in the variables. Thus the terms

$$xy, \quad x^2, \quad y^2, \quad \frac{y^3}{x}, \quad x^2 + y^2, \quad \frac{x^3 + xy^2}{x + y}$$

are all homogeneous of degree 2. The test is that, on substituting xt for x and yt for y, the variable t factors out of all the terms and emerges, in this case as t^2, multiplying the original expression.

If the equation is homogeneous, then the change of variable, $y = vx$, to the new variable v in place of y will produce an equation whose variables are separable, and we are reduced to the previous case.

Example 23.5-1

Integrate

$$xy' = \frac{x^2 + y^2}{y}$$

This is a homogeneous equation of degree 1. Set $y = vx$ and get

$$x(xv' + v) = \frac{x^2 + v^2x^2}{vx}$$

$$xv' + v = \frac{1 + v^2}{v} = \frac{1}{v} + v$$

$$xv' = \frac{1}{v}$$

$$v\,dv = \frac{1}{x}\,dx$$

$$\frac{v^2}{2} = \ln x + C$$

$$x = A \exp\left\{\frac{1}{2}\left(\frac{y}{x}\right)^2\right\}$$

An alternative expression, and one that you might prefer, is to solve for y. We have

$$v^2 = \frac{y^2}{x^2} = 2 \ln x + C$$

$$y = \pm x[2 \ln x + C]^{1/2}$$

Example 23.5-2

Integrate

$$y' = \frac{y}{x} + f\left(\frac{y}{x}\right)$$

for a continuous function $f(y/x)$. We make the usual change of variables

$$y = vx$$

to get

$$xv' + v = v + f(v)$$

and separate variables

$$\frac{dv}{f(v)} = \frac{dx}{x}$$

and we have reduced the problem to quadratures.

$$\int \frac{dv}{f(v)} = \ln x + C$$

$$x = A \exp \left\{ \int \frac{dv}{f(v)} \right\}$$

Unless we are given a particular function $f(y/x)$, we can go no further.

It is worth noting that the use of polar coordinates

$$x = r \cos \theta, \qquad y = r \sin \theta$$

will also lead to a variables separable equation, and at times the use of polar coordinates may shed more light on the situation than the use of the more compact change $y = vx$.

Example 23.5-3

Using the equation of Example 23.5-1, we get

$$x = r \cos \theta, \qquad y = r \sin \theta$$

and the equation becomes

$$(r \cos \theta) \frac{r' \sin \theta + r \cos \theta}{r' \cos \theta - r \sin \theta} = \frac{r^2}{r \sin \theta}$$

$$\frac{r' \sin \theta + r \cos \theta}{r' \cos \theta - r \sin \theta} = \frac{1}{\sin \theta \cos \theta}$$

$$r' \sin^2 \theta \cos \theta + r \cos^2 \theta \sin \theta = r' \cos \theta - r \sin \theta$$

$$r' \cos \theta [\sin^2 \theta - 1] + r \sin \theta [1 + \cos^2 \theta] = 0$$

$$r' \cos^3 \theta = r \sin \theta (1 + \cos^2 \theta)$$

$$\frac{r'}{r} = \frac{\sin \theta}{\cos^3 \theta} + \frac{\sin \theta}{\cos \theta}$$

$$\ln r = \frac{1}{2} \sec^2 \theta - \ln \cos \theta + C$$

$$\ln (r \cos \theta) = \frac{1}{2} \tan^2 \theta + C'$$

$$\ln x = \frac{1}{2} \left(\frac{y}{x} \right)^2 + C'$$

$$x = A \exp \left\{ \frac{1}{2} \left(\frac{y}{x} \right)^2 \right\}$$

as before. In this case the amount of algebra is distinctly larger than before.

EXERCISES 23.5

Solve the following equations.

1. $y' = (x + y)/x$
2. $y' = xy/(x^2 + y^2)$
3. $y' = (x^2 + y^2)/2x^2$
4. $(x + y)y' = x - y$
5. $(2x^2 - y^2)y' + 3xy = 0$

23.6 INTEGRATING FACTORS

It is evident that we can write a first-order, first-degree, differential equation in the form

$$M(x, y)\, dy + N(x, y)\, dx = 0$$

Sometimes this can be put in the form of the derivative of a function of the variables x and y. For example, if we had

$$x\, dy + 2y\, dx = 0$$

then multiplying through by x gives

$$x^2\, dy + 2xy\, dx = 0$$

and this is the derivative of

$$x^2 y = C$$

When we have the situation where the equation is the derivative of a function, we call it an *exact differential* of the function, and we can solve the problem.

Example 23.6-1

Solve

$$2(x + 1)y\, dy + y^2\, dx = 0$$

This is the derivative of

$$(x + 1)y^2 = C$$

and we have the solution. (It could also have been solved by the variables separable method.)

Example 23.6-2

Solve

$$y' + y = e^{-x}$$

This can be written in the form

$$e^x y' + e^x y = 1$$
$$(e^x y)' = 1$$
$$e^x y = x + C$$
$$y = (x + C)e^{-x}$$

If you want the particular solution passing through the point $x = 0, y = 1$, then you must have

$$1 = C$$

Therefore, the particular solution through $(0, 1)$ is

$$y = (x + 1)e^{-x}$$

Often the equation as it stands is not an exact derivative of some function, but by multiplying by a suitable function it can be made exact. Such a function is called an *integrating factor*.

Example 23.6-3

Solve $xy' + ny = f(x)$. A little study suggests the product form; hence we need to multiply through by x^{n-1} to get

$$x^n y' + nx^{n-1} y = x^{n-1} f(x)$$

We now have

$$(x^n y)' = x^{n-1} f(x)$$

and integrating both sides gives us

$$y = x^{-n} \int x^{n-1} f(x) \, dx + Cx^{-n}$$

as the solution.

Example 23.6-4

Integrate

$$y + (\tan x)y' = \sin x$$

Rewriting this in the form (multiply by $\cos x$)

$$(\cos x)y + (\sin x)y' = \sin x \cos x$$

leads to

$$(y \sin x)' = (\sin x \cos x)$$

and integrating this we get

$$y \sin x = \frac{1}{2} \sin^2 x + C$$

or finally

$$y = \frac{\sin x}{2} + \frac{C}{\sin x}$$

Example 23.6-5

Integrate

$$xy' - y = x^2 e^{-x}$$

The difference in signs between the two terms on the left suggests a quotient, so we divide by x^2 to get

$$\frac{xy' - y}{x^2} = \left(\frac{y}{x}\right)' = e^{-x}$$

from which we get

$$\frac{y}{x} = C - e^{-x}$$

$$y = Cx - xe^{-x}$$

Finding an integrating factor is evidently an art requiring experience with derivatives so that you can imagine how to construct a function which upon differentiation will lead to your expression. There is not a unique integrating factor, and different ones can lead to different forms of the answer.

The condition that there exists an integrating factor depends on the fact that the exact differential equation must come from the solution

$$f(x, y) = C$$

by differentiation. Hence, if the integrating factor is called $r(x, y)$, then

$$\frac{\partial f(x, y)}{\partial y} \frac{dy}{dx} + \frac{\partial f(x, y)}{\partial x} = 0$$

means that

$$\frac{\partial f(x, y)}{\partial y} = r(x, y) M(x, y) \quad \text{and} \quad \frac{\partial f(x, y)}{\partial x} = r(x, y) N(x, y)$$

This is only slightly useful in searching for an integrating factor.

When we have the differential equation in the exact form, it remains to find the solution. Since we know

$$\frac{\partial^2 f(x, y)}{\partial x\, \partial y} = \frac{\partial^2 f(x, y)}{\partial y\, \partial x}$$

if the derivatives exist and are continuous, then

$$\frac{\partial(rM)}{\partial y} = \frac{\partial(rN)}{\partial x}$$

This provides us with a method for solving the equation when it is an exact differential. Integrate

$$\frac{\partial u}{\partial x} = rM$$

inserting a function of y for the arbitrary constant. Substitute this into

$$\frac{\partial u}{\partial y} = rN$$

and determine the arbitrary function of y. The solutions is then

$$u = C$$

By symmetry we could, of course, reverse the order and integrate with respect to x first.

Example 23.6-6

Integrate

$$(3x^2y + 8xy^2)\, dx + (x^3 + 8x^2y + 12y^2)\, dy = 0$$

We have

$$\frac{\partial(3x^2y + 8xy^2)}{\partial y} = 3x^2 + 16xy = \frac{\partial(x^3 + 8x^2y + 12y^2)}{\partial x}$$

which shows that the equation is exact. From

$$\frac{\partial u}{\partial x} = 3x^2y + 8xy^2$$

we have, upon integration with respect to x,

$$u = x^3y + 4x^2y^2 + K(y)$$

Then we have

$$\frac{\partial u}{\partial y} = x^3 + 8x^2y + 12y^2 = x^3 + 8x^2y + \frac{dK(y)}{dy}$$

$$\frac{dK(y)}{dy} = 12y^2$$

$$K(y) = 4y^3$$

The solution is, therefore,

$$x^3y + 4x^2y^2 + 4y^3 = C$$

EXERCISES 23.6

Integrate the following:

1. $xy' + 4y = 3$
2. $xy' - 4y = 3$
3. $y + \operatorname{ctn} xy' = \cos x$
4. $(x + y^2)\, dy + (y - x^2)\, dx = 0$
5. $(xy + 1)/y\, dx + (2y - x)/y^2\, dy = 0$
6. $(4y^2 - 2x^2)/(4xy^2 - x^2)\, dx + (8y^2 - x^2)/(4y^3 - x^2y)\, dy = 0$

23.7 FIRST-ORDER LINEAR DIFFERENTIAL EQUATIONS

An important class of differential equations has the form

$$\frac{dy}{dx} + P(x)y = Q(x)$$

This equation is called a *linear first-order differential equation*. It is linear in y and its derivatives and is of first order. We look for an integrating factor, and after a few moments we see that if we multiply through by

$$r(x) = \exp\left\{ \int P(x)\, dx \right\}$$

we will have

$$\exp\left\{ \int P(x)\, dx \right\}\frac{dy}{dx} + \left[P(x)\exp\left\{ \int P(x)\, dx \right\} \right]y = Q(x)\exp\left\{ \int P(x)\, dx \right\}$$

This can be written as

$$\left\{ \exp\left(\int P(x)\, dx \right)y(x) \right\}' = Q(x)\exp\left\{ \int P(x)\, dx \right\}$$

Upon integration, this becomes

$$y(x) = \exp\left\{ -\int P(x)\, dx \right\}\left[\int Q(x)\exp\left\{ \int^x P(s)\, ds \right\} dx + C \right]$$

where we have had to change to the dummy variable s in the integral on the right-hand side to avoid confusion between what are the variables of integration and of the problem. Of course, as indicated, the variable x is to be put in the integral in the exponent when it is done.

We have reduced the problem of solving first-order linear differential equations to a matter of doing two *quadratures* (ordinary integrations), as they are called. We conventionally regard the problem as solved as a problem in differential equations. Of course, in practice a great deal more work may be required to get the solution in a useful form. The habit of regarding the problem in differential equations as solved when it is reduced to quadratures is an example of "divide and conquer"—attack one phase of the problem at a time. The general solution looks messy, but when we can solve $\int [P(x)\ dx]$ in closed form, it often makes things look simple. We do not try to remember the form of the result and substitute into it, but rather remember the method of derivation and repeatedly use it. It is the methods of doing mathematics, rather than the results, that are important.

Example 23.7-1

Integrate

$$2x\frac{dy}{dx} + y = x^2$$

We first put it into the standard form

$$y' + \frac{1}{2x}y = \frac{x}{2}$$

We have

$$P(x) = \frac{1}{2x}$$

and

$$\int P(x)\ dx = \frac{1}{2}\ln x$$

Hence

$$\exp\left\{\frac{1}{2}\ln x\right\} = \sqrt{x}$$

Multiply the standard form through by this integrating factor to get

$$\sqrt{x}\,y' + \frac{1}{2\sqrt{x}}y = \frac{x^{3/2}}{2}$$

Integrate both sides to get

$$\sqrt{x}\,y = \frac{x^{5/2}}{5} + C$$

or

$$y = \frac{x^2}{5} + \frac{C}{\sqrt{x}}$$

At this point we check that we have no errors by differentiating and manipulating the terms that emerge.

$$2xy' + y = 2x\left(\frac{2x}{5} - \frac{C}{2x^{3/2}}\right) + \frac{x^2}{5} + \frac{C}{\sqrt{x}}$$

$$= \frac{4x^2}{5} - \frac{C}{\sqrt{x}} + \frac{x^2}{5} + \frac{C}{\sqrt{x}} = x^2$$

as it should.

Example 23.7-2

Integrate

$$y' + 2xy = e^x$$

The integrating factor, upon integrating $P(x) = 2x$, and taking its exponential, is

$$\exp x^2$$

and we have

$$\{(\exp x^2)y\}' = e^x \exp x^2 = \exp (x^2 + x)$$

and we can now do the indicated integration on the left to get

$$y = \exp (-x^2)\left\{ \int^x \exp (s^2 + s)\, ds + C\right\}$$

The integral on the right cannot be done in closed form.

An alternative approach to first-order linear differential equations, and one which has the great advantage that it generalizes to nth order equations, is the following. Given the equation

$$y' + P(x)y = Q(x)$$

we first study the *homogeneous* equation (the right-hand side set equal to zero)

$$y' + P(x)y = 0$$

This is easily solved as follows:

$$\frac{dy}{y} + P(x)\, dx = 0$$

$$\ln y + \int P(x)\, dx = C$$

$$y = C \exp \left\{ -\int P(x)\, dx \right\}$$

Suppose, next, that we can find one solution $Y(x)$ of the original *complete* equation; that is, we can find a particular $Y(x)$ such that

$$Y' + P(x)Y = Q(x)$$

Then the general solution of the complete equation is

$$y(x) = Y(x) + C \exp \left\{ -\int P(x)\, dx \right\}$$

We check this by substitution into the original equation:

$$Y' + C$$

$$\times \left[-P(x) \exp \left\{ -\int P(x) \, dx \right\} \right] + P(x) \left[Y + C \exp \left\{ -\int P(x) \, dx \right\} \right] = Q(x)$$

The terms with the constant C term all cancel out, and the other terms in Y also cancel since they are a solution of the complete equation. We see that one single solution of the complete equation causes the right-hand side to cancel, leaving the constant of integration term to satisfy the homogeneous equation.

How do you find the supposed solution $Y(x)$? By guessing! Or any other method you care to use (for example, the method of variation of parameters to be studied in Section 24.4).

Example 23.7-3

Consider the equation

$$y' + 3y = x^2$$

It is reasonable to guess the form for a solution of the complete equation as

$$Y(x) = ax^2 + bx + c$$

This gives, on putting it into the given differential equation,

$$(2ax + b) + 3(ax^2 + bx + c) = x^2$$

Equating coefficients of like powers of x, we have

$$3a = 1$$

$$2a + 3b = 0$$

$$b + 3c = 0$$

Their solution is easily found to be

$$a = \frac{1}{3}, \qquad b = \frac{-2}{9}, \qquad c = \frac{2}{27}$$

hence the particular solution is

$$Y(x) = \frac{1}{27} \{9x^2 - 6x + 2\}$$

The solution of the homogeneous equation

$$y' + 3y = 0$$

is

$$\frac{dy}{y} + 3 \, dx = 0$$

$$\ln y + 3x = C$$

$$y = C \exp \{-3x\}$$

The general solution of the complete equation is therefore

$$y(x) = \frac{1}{27}\{9x^2 - 6x + 2\} + C \exp\{-3x\}$$

Example 23.7-4

Consider the equation

$$y' + y = \sin x$$

We naturally try for the particular solution $Y(x)$:

$$Y(x) = A \sin x + B \cos x$$

Hence we get

$$A \cos x - B \sin x + A \sin x + B \cos x = \sin x$$

and equate like terms in cos and sin (they are linearly independent functions). We get the equations

$$A + B = 0$$

$$-B + A = 1$$

The solution is $A = \frac{1}{2}$, $B = -\frac{1}{2}$, so we have

$$Y(x) = \frac{\sin x - \cos x}{2}$$

We have now to solve the homogeneous equation

$$y' + y = 0$$

It is easy to get the solution

$$y = Ce^{-x}$$

So the complete solution is

$$y(x) = \frac{\sin x - \cos x}{2} + Ce^{-x}$$

This is easily checked by direct substitution.

Example 23.7-5

Integrate

$$y' + y = e^{-x}$$

We try for the particular solution

$$Y(x) = Ae^{-x}$$

This leads to

$$A(-e^{-x}) + Ae^{-x} = e^{-x}$$

which is a contradiction, $0 = e^{-x}$. Next we try, after some thought,

$$Y(x) = Axe^{-x}$$

$$Ax(-e^{-x}) + Ae^{-x} + Axe^{-x} = e^{-x}$$

Hence we get $A = 1$, and the particular solution is

$$Y(x) = xe^{-x}$$

The homogeneous equation

$$y' + y = 0$$

has the solution

$$y = Ce^{-x}$$

So the general solution of the complete equation is

$$y(x) = xe^{-x} + Ce^{-x} = e^{-x}(x + C)$$

This shows indirectly why the original trial for the particular solution failed; the trial solution was a solution of the homogeneous equation. When this occurs, putting an extra factor of x in front will lead to a solution.

Differential equations usually arise from the nature of the original problem, but at times they are "manufactured" by the mathematician, as the following example shows.

Example 23.7-6

Suppose you have the integral, as a function of the parameter a,

$$F(a) = \int_0^\infty \exp\left[-\left(\frac{a^2}{x^2} + x^2\right)\right] dx$$

After other approaches fail, we try to find a differential equation for the function $F(a)$ (assume for the moment that $a > 0$).

$$\frac{dF(a)}{da} = -2a \int_0^\infty \frac{1}{x^2} \exp\left[-\left(\frac{a^2}{x^2} + x^2\right)\right] dx$$

Now for $a \neq 0$ set $x = a/t$. The integral becomes

$$\frac{dF(a)}{da} = -2a(-1) \int_0^\infty \frac{t^2}{a^2} \exp\left[-\left(t^2 + \frac{a^2}{t^2}\right)\right]\left(-\frac{a}{t^2}\right) dt$$

Hence we have the differential equation

$$F' + 2F = 0$$

$$\frac{dF}{F} + 2\,da = 0$$

$$\ln F = -2a + C$$

$$F = Ce^{-2a}$$

for $a > 0$. But we see immediately from the original problem that this form also applies for $-a$ (we need to avoid $a = 0$). Hence the solution is of the form

$$F(a) = Ce^{-2|a|}$$

To find C, we use $a = 0$, which is an integral we already know.

$$F(0) = \int_0^\infty \exp\left(-x^2\right) dx = \frac{\sqrt{\pi}}{2}$$

We have, therefore, the solution for the integral:

$$F(a) = \frac{\sqrt{\pi}}{2} \exp\left(-2|a|\right)$$

We seem to have the answer, but we made the transformation for $a \neq 0$, and we used the value $a = 0$ to determine the constant of integration. So we need to examine the process a bit more closely.

Did the integral for the derivative $F'(a)$ converge? Well, as $x \to 0$ the exponent for $a \neq 0$ beats out the factor $1/x^2$. But let us look at the integrand more closely. The main effect is the exponent, so we study the function (the exponent only)

$$g(x) = \frac{a^2}{x^2} + x^2$$

The maximum occurs when the derivative

$$g'(x) = \frac{-2a^2}{x^3} + 2x = 0$$

This occurs at $x = \sqrt{a}$, $g(\sqrt{a}) = 2a$. At this point the original integrand has the value

$$e^{-2a}$$

and while this is not the maximum of the integrand [only of the exponential factor $g(x)$ that was used to locate the position], it gives us a clue as to the shape of the integrand. (We could find the actual maximum if we wished, but it is not worth the trouble as we want to understand the shape of the integrand, not its detailed behavior.) As $a \to 0$, this tends to peak up, as does the true maximum of the integrand. Thus the peak of the integrand narrows as $a \to 0$, and the integral approaches the limiting area $\sqrt{\pi}/2$. We are uneasy to say the least. Looking at the integrand at $x = 0$, we see that

for $a \neq 0$,   integrand value $= 0$

for $a = 0$,   integrand value $= 1$

The integrand value at $x = 0$ has a discontinuity as $a \to 0$. However, the width of the peak approaches 0, and we think that we are safe in assuming that the integral itself is continuous at $a = 0$. We see from this one example the need for greater rigor, even if it happens to lie beyond the range of this book.

We can apply some reasonableness checks. We have the result

$$\int_0^\infty \exp\left[-\left(\frac{a^2}{x^2} + x^2\right)\right] dx = \frac{\sqrt{\pi}}{2} C e^{-2|a|}$$

How does it check against what we can see? Examining the integrand, we see the symmetry implied by the $|a|$ in the answer. As a gets large, we see that the result decays exponentially as we feel that it should from examining the integrand. It has the right value (we made it that way) at $a = 0$. Thus the answer is plausible, but we wish that we had more rigor to support the conclusion; it is the discontinuity of the integrand, as a function of a, at $x = 0$, that has us worried a bit. But the jump is of unit size and is confined to a small interval that decreases as $a \to 0$, so it seems safe. The need for further study of mathematics is apparent.

EXERCISES 23.7

Solve the following:

1. $y' + 1/xy = x$
2. $y' + xy/2 = x$
3. $y' + y = x^2 - 1$
4. $y' - 2y/x = x^4$
5. $y' - 3y = \exp\{3x\} + \exp\{-3x\}$
6. $y' + ay = \cos t$

23.8 CHANGE OF VARIABLES

Just as with integration, a suitable change of variable can often reduce an apparently hopeless case to a tractable one. There are no simple rules in either case; only a careful examination of the equation can suggest a suitable transformation. We are therefore reduced to giving you a selection of possible substitutions to try.

Example 23.8-1

Integrate

$$y' = f(ax + by + c)$$

Set $ax + by + c = u$; then the equation becomes

$$\frac{1}{b}(u' - a) = f(u)$$

$$u' = bf(u) + a$$

and we are reduced to quadratures.

Example 23.8-2

Integrate the Bernoulli equation

$$y' = A(x)y + B(x)y^n$$

After some trial and error, we are led to try the substitution

$$y = z^m$$

and we search for the m that will make the equation tractable. We get

$$mz' = A(x)z + B(x)z^{mn-m+1}$$

This suggests that if we set

$$mn - m + 1 = 0$$

or

$$m = \frac{1}{1 - n}$$

the equation is then

$$\frac{1}{(1 - n)} z' = A(x)z + B(x)$$

and we have a linear first-order differential equation, which we know how to solve.

EXERCISES 23.8

Solve the following:

1. $y' = y + y^2$
2. $y' + y^2 = 1$

23.9 SPECIAL SECOND-ORDER DIFFERENTIAL EQUATIONS

In this section we examine two special cases of second-order differential equations. They are of considerable importance, both theoretically and practically. The general second-order equation may be imagined to be solved for the second derivative (they usually arise in practice in this form)

$$y'' = f(x, y, y')$$

The general case is too general for a first course, so we limit ourselves to two special cases.

The first case is

$$y'' = f(x, y')$$

with no y term present. We merely set $y' = z$. The equation is now a first-order equation

$$z' = f(x, z)$$

When we find the solution of this equation (if we can), then it is a matter of one further integration with its corresponding additive constant to get the final answer.

Example 23.9-1

Solve

$$y'' = \frac{y'}{x} + 1$$

We set $y' = z$ and have

$$z' = \frac{z}{x} + 1$$

$$xz' - z = x$$

Divide through by x^2 and you have the exact equation:

$$\left(\frac{z}{x}\right)' = \frac{1}{x}$$

$$\frac{z}{x} = \ln x + C_1$$

$$z = x \ln x + C_1 x$$

$$z = y' = x \ln x + C_1 x$$

$$y = \frac{x^2}{2} \ln x - \frac{x^2}{4} + C_1 \frac{x^2}{2} + C_2 = \frac{x^2}{2} \ln x + Cx^2 + C_2$$

The second form of a second order differential equation we can handle is

$$y'' = f(y, y')$$

with no x term present. We set $y' = v$ (velocity if you wish for a mnemonic). We have

$$\frac{d^2y}{dx^2} = \frac{d\{dy/dx\}}{dx} = \frac{dv}{dx} = \frac{dv}{dy}\frac{dy}{dx} = v\frac{dv}{dy}$$

Hence the equation becomes

$$v\frac{dv}{dy} = f(y, v)$$

which is a first-order equation. We solve this (if we can) and then substitute $v = dy/dx$ to get another first-order equation to solve.

Example 23.9-2

Consider the important equation

$$y'' + k^2 y = 0$$

If we use this method (for this equation another method is preferable but it illustrates the method), we are led to

$$v\frac{dv}{dy} + k^2 y = 0$$

The variables are separable, and we get finally

$$v^2 + k^2 y^2 = C^2$$

where it is convenient to make the constant positive. Solving for v, we get (choosing the plus sign for the square root)

$$v = \frac{dy}{dx} = \sqrt{C^2 - k^2 y^2}$$

Again the variables are separable, and we have

$$dy/\sqrt{C^2 - k^2 y^2} = dx$$

Integrating, we get

$$\frac{1}{k} \arcsin \frac{ky}{C} = x + C'$$

Take the sine of both sides and solve for y:

$$y = \frac{C}{k} \sin (kx + C'')$$

But C/k is a constant, so we may write the solution in the forms

$$y = \frac{C}{k} \sin (kx + C'') = C_1 \cos kx + C_2 \sin kx$$

where C_1 and C_2 are suitable constants. Had we chosen the minus sign in the square root, things would not have come out differently.

In both of the cases we studied we have reduced the second-order differential equation to a sequence of two first-order differential equations. The importance of these special cases is that they occur frequently in practice.

EXERCISES 23.9

Solve the following:

1. $y'' + ay' = f(x)$
2. $y'' + ay' + by = c$

23.10 DIFFERENCE EQUATIONS

Difference equations occur frequently in discrete mathematics, and it is fortunate that the method for solving them closely parallels that for differential equations. We will consider here only the simplest examples.

Consider the first-order linear difference equation

$$y_{n+1} + ay_n = f(n)$$

We do the same as we did for differential equations *except* that in place of e^{mx} we try r^n as a trial solution (think of the constant $r \leftrightarrow e^m$). The first step is to examine the corresponding homogeneous equation, which is

$$y_{n+1} + ay_n = 0$$

We try a solution of the form $y = r^n$. The result is

$$r^{n+1} + ar^n = 0$$

so $r = -a$. The general solution of the homogeneous equation is

$$y_n = C(-a)^n$$

The second step is to find one solution of the complete equation. One method is to guess at the solution of the complete equation, and when we find one, any one at all, then the general solution of the complete equation is this particular solution of the complete equation plus the general solution of the homogeneous equation.

Example 23.10-1

Solve

$$y_{n+1} = 2y_n + n, \qquad \text{with } y_0 = 1$$

We naturally try, for the particular solution of the complete equation, an expression of the form

$$Y_n = an + b$$

We get, upon direct substitution,

$$a(n + 1) + b = 2(an + b) + n$$

from which we get, when we equate the linearly independent terms,

$$a = 2a + 1, \qquad a + b = 2b$$

whose solutions are

$$a = -1, \qquad b = -1$$

Hence the particular solution is

$$Y_n = -(n + 1)$$

The solution of the homogeneous part is easily found. Assume

$$y_n = r^n$$

You get

$$r^{n+1} = 2r^n$$

from which $r = 2$. The general solution of the complete equation is therefore

$$y_n = C2^n - (n + 1)$$

To fit the intitial condition $y_0 = 1$, we have

$$y_0 = 1 = C - 1$$

from which $C = 2$, and the final solution is

$$y_n = 2^{n+1} - n - 1$$

Example 23.10-2

Suppose that there are a men and b women and that n chips are passed out to them at random. What is the probability that the total number of chips given to the men is an even number?

We regard this as a problem depending on n. We start with $n = 0$. The probability that the men have an even number of chips when no chips have yet been offered is

$$P(0) = 1$$

We next ask how can the stage n arise? It comes from the stage of $n - 1$ plus the process of passing out one more chip at random. The $n - 1$ state is either that the men have an even number of chips, $P(n - 1)$, or that they have an odd number, $1 - P(n - 1)$. From these we build up the state n by taking these two probabilities and multiplying them by the corresponding probabilities of the nth chip going to a man or a woman. Thus we have the difference equation

$$P(n) = \frac{b}{a + b}P(n - 1) + \frac{a}{a + b}[1 - P(n - 1)]$$

We write this difference equation in the canonical form

$$P(n) - \frac{b - a}{b + a}P(n - 1) = \frac{a}{a + b}$$

For the moment call $(b - a)/(b + a) = A$. The solution of the homogeneous equation is clearly

$$CA^n$$

for some constant C. We next need a particular solution of the complete equation. With the right-hand side a constant, we try a constant, say B. We have

$$B - AB = \frac{a}{a + b}$$

or

$$B = \frac{a}{a + b}\left[\frac{1}{1 - A}\right] = \frac{a}{a + b}\left[\frac{a + b}{(a + b) - (b - a)}\right]$$

$$= \frac{a}{2a} = \frac{1}{2}$$

Thus the general solution of the difference equation is

$$P(n) = \frac{1}{2} + C\left(\frac{b - a}{b + a}\right)^n$$

It remains to fit the initial condition $P(0) = 1$. This gives immediately $C = \frac{1}{2}$, and the final solution is

$$P(n) = \frac{1}{2} + \frac{\{(b - a)/(b + a)\}^n}{2}$$

This can be written in a more symmetric form:

$$\frac{1}{2}\left[\frac{(b + a)^n}{(b + a)^n} + \frac{(b - a)^n}{(b + a)^n}\right]$$

Let us examine this solution for reasonableness. If $a = 0$, no men, then the solution is

$$P(n) = \frac{1}{2} + \frac{1}{2} = 1$$

and certainly the number of chips given at random to the men is an even number (0 is an even number). If $a = b$, then we have

$$P(n) = \frac{1}{2}$$

and this by symmetry is clearly correct. Finally, if there were no women, $b = 0$, then

$$P(n) = \frac{1}{2} + \frac{(-1)^n}{2}$$

which alternates between 0 and 1 and again is correct. Thus we have considerable faith in the result found.

EXERCISES 23.10

1. Suppose that a system can be in one of two states, A or B, and that if it is in state A then at the next time interval it has probability p_A of going to state B; while if it is in state B it has probability p_B of going to state A. Find the probability of being in state A at time n. Check your answer by the three cases you can easily solve: (a) $p_A = 0$, (b) $p_A = p_B$, (c) $p_A = 1$, $p_B = 0$.
2. Using complex notation, find the integrals $\int_0^\infty [\exp(-ax) \sin x \, dx]$ and the corresponding one with $\cos x$ by means of differentiating.

23.11 SUMMARY

Differential equations are of fundamental importance in many fields of application of mathematics. Generally, a good deal of knowledge of the field of application is needed to set up the initial differential equations that represent the phenomenon you are interested in. Once set up, it becomes, to a great extent, a matter of solving them using the mathematical tools available.

The method of direction fields tends to be forgotten as trivial and uninteresting, but on several occasions the author has disposed of an important physical problem over lunch by sketching the direction field on the back of a place mat; the sketch showed the nature of the solution, and that was all that was needed to understand the phenomenon.

It is surprising how many practical problems yield to the elementary methods of (1) variables separable and (2) homogeneous equations leading to variables separable, or (3) have an obvious integrating factor. The method of change of variables depends, as it does in integration, on a sudden insight into what might make the problem solvable, and there are few general rules.

Linear differential equations are very important, and their study will be continued in the next chapter.

Finally, we examined linear first-order difference equations. The theory is almost exactly parallel to the theory of linear differential equations. We have not pursued the corresponding analogies between the other methods of solving differential equations and the corresponding difference equations. Generally, integration is replaced by summation, and there are only a few other changes to be considered. You can make an abstraction based on the two special cases, linear differential equations and linear difference equations, to get at the general linear problem, but this lies properly in a course in linear algebra, not in a first course such as this one.

24

Linear Differential Equations

24.1 INTRODUCTION

The second-order linear differential equation

$$p_0 \frac{d^2y}{dx^2} + p_1 \frac{dy}{dx} + p_2 y = f(x) \tag{24.1-1}$$

with constant coefficients (lowercase letters) plays a large role in both mathematics and applications. The generalization to nth-order equations

$$p_0 \frac{d^n y}{dx^n} + p_1 \frac{d^{n-1}y}{dx^{n-1}} + \cdots + p_n y = f(x) \tag{24.1-2}$$

is fairly immediate and brings in no significantly new ideas. Furthermore, many of the ideas apply to equations with variable coefficients:

$$P_0(x) \frac{d^n y}{dx^n} + P_1(x) \frac{d^{n-1}y}{dx^{n-1}} + \cdots + P_n(x)y = f(x) \tag{24.1-3}$$

(uppercase letters for the coefficients).

When you begin mathematics, each problem is more or less self-contained. But with experience you gradually learn to keep, in one corner of your mind, some possible generalizations of the problem you are working on. Often in the middle of a problem a generalization will illuminate the situation and enable you to understand more clearly what is going on in the particular case. We will, therefore, work on all three forms (24.1-1, 24.1-2, 24.1-3) more or less in parallel. At the center of the stage will be the simple second-order case with constant coefficients, but the extensions, or generalizations if you prefer, are close by.

We naturally need an existence theorem to assure us that what we are looking for actually exists. It is beyond the range of this book to prove it (most beginning texts on differential equations have the proof), but we will state the theorem, which will probably seem at least plausible to you from the simple idea of a direction field, if not from other experiences in the previous chapter. The fact is that mathematicians for many years solved differential equations before they had the appropriate existence theorems. Often, only after the importance is understood, does the existence theorem get studied.

Theorem.   In any closed interval $a \leq x \leq b$ in which the coefficients $P_0(x) \neq 0$, $P_1(x)$, . . . , $P_n(x)$, and $f(x)$ are continuous there exists one and only one solution $y = y(x)$ of Equation (24.1-3), which is continuous together with its first n derivatives in the interval, and such that the function and its first $n - 1$ derivatives take on prescribed values at a point x_0 in the interval.

The theorem applies to the general case of an nth-order differential equation with variable coefficients; hence it covers the other two cases, (24.1-1) and (24.1-2), as well. Notice that the solution and/or its derivatives can fail to be continuous only if one or more of the coefficients fail to be continuous. Also note, if one or more of the coefficients has a discontinuity, it is reasonable to expect that we can splice pieces of solutions together; the ending conditions of the solution in one interval will supply the starting conditions in the next interval. See Example 24.3-1.

The theorem is intuitively appealing because we start with all the lower-order derivatives (including the zeroth derivative), and the differential equation gives us the value of the higest derivative. These derivatives tell us how to step forward a very small step (as we imagined in the direction field approach), and we then get new values of the lower derivatives to use for the next step.

The theory has a large role for the *homogeneous* equation, which is simply the original equation (24.1-3) with the right-hand side replaced by 0.

$$P_0(x) \frac{d^n y}{dx^n} + P_1(x) \frac{d^{n-1} y}{dx^{n-1}} + \cdots + P_n(x)y = 0$$

The term $f(x)$ on the right-hand side of (24.1-3) is often called the *forcing term*; it is what drives the system. Without this term, the differential equation clearly has the trivial solution

$$y(x) = 0$$

meeting the initial conditions of all zeros. In many cases you start with this solution, and at time $t = t_0$ the forcing term enters into the system, forcing the solution to respond first through the nonzero values of the highest derivative and gradually more strongly through all the lower-order derivatives changing from zero.

From the existence theorem there are n different *fundamental* solutions of the homogeneous equation, one for each set of initial conditions of the form: one initial value of $y_0, y_0', \ldots, y_0^{(n-1)}$ equal to 1 and all the others equal to 0. Any other set of initial conditions is simply a linear combination of these fundamental solutions. Thus there are n linearly independent solutions of the homogeneous equation. Given these n linearly independent solutions, Section 4.7 assures us that the equations to determine the particular solution that fits any given equation (to determine the particular solution that fits the given data) will be solvable (for linear differential equations).

4.2 SECOND-ORDER EQUATIONS WITH CONSTANT COEFFICIENTS

We begin with the case of the homogeneous equation with constant coefficients

$$p_0 y'' + p_1 y' + p_2 y = 0 \qquad (24.2\text{-}1)$$

Recall that the derivative of e^x is again an exponential. This strongly suggests that you try looking for a solution of the form

$$y = Ce^{mx} \qquad (24.2\text{-}2)$$

since then each term will be the same exponential but with different coefficients. You see, upon putting this function into the equation, that you get the condition

$$C\{p_0 m^2 + p_1 m + p_2\}e^{mx} = 0$$

Obviously, the solution fits or not regardless of the coefficient $C \neq 0$. The exponential cannot be zero for any finite value x; hence the quadratic factor

$$p_0 m^2 + p_1 m + p_2 = 0 \qquad (24.2\text{-}3)$$

must supply the needed zero values. This is called the *characteristic equation*. From the standard quadratic equation formula for the roots, we get

$$m = \frac{\{-p_1 \pm \sqrt{p_1^2 - 4p_0 p_2}\}}{2p_0}$$

for the two roots. There are three cases to be considered: (1) two real distinct roots, (2) two real equal roots, and (3) two complex conjugate roots. We take them up in turn.

Example 24.2-1

Solve the homogeneous equation

$$y'' + 4y' + 3y = 0$$

We try (neglecting the coefficient C) $y = e^{mx}$ as a possible solution and get the characteristic equation

$$m^2 + 4m + 3 = 0$$

$$(m + 3)(m + 1) = 0$$

Hence the roots are $m = -1$ and $m = -3$. We have the two linearly independent solutions

$$e^{-3x} \quad \text{and} \quad e^{-x}$$

each of which fits the equation. Indeed, we see that *any* linear combination

$$y(x) = C_1 e^{-3x} + C_2 e^{-x}$$

is a solution, since we get, on putting these solutions into the differential equation and arranging the terms by the coefficients C_1 and C_2,

$$C_1\{(-3)^2 + 4(-3) + 3\}e^{-3x} + C_2\{(-1)^2 + 4(-1) + 3\}e^{-x} = 0$$

since each brace is the characteristic equation evaluated at a root.

If you think of the two fundamental solutions, $y_1(0) = 1$, $y_1'(0) = 0$ and $y_2(0) = 0$, $y_2'(0) = 1$, you see that any solution of the equation must be of this form, and this is why it is called the *general solution*.

When we assign the initial conditions of the existence theorem, we get, supposing that the initial point is $x = 0$,

$$C_1 + C_2 = y(0)$$

$$-3C_1 - C_2 = y'(0)$$

from which we can get the particular values of $C_1 = -\{y(0) + y'(0)\}/2$ and $C_2 = \{3y(0) + y'(0)\}/2$ to meet the given conditions. Again, this resembles the method of undetermined coefficients.

Example 24.2-2

Solve the homogeneous equation

$$y'' + 4y' + 4y = 0$$

The characteristic equation is

$$m^2 + 4m + 4 = 0$$

whose roots are $m = -2$ and $m = -2$. We do not have two linearly independent solutions to form the general solution. Some examination of the situation will gradually suggest that maybe a term like

$$y = xe^{-2x}$$

might work. We try

$$y' = -2xe^{-2x} + e^{-2x}$$

$$y'' = 4xe^{-2x} - 2e^{-2x} - 2e^{-2x}$$

We have

$$x\{4 + 4(-2) + 4\}e^{-2x} + \{-4 + 4\}e^{-2x} = 0$$

and xe^{-2x} is a solution. Thus we have the two linearly independent solutions

$$e^{-2x} \quad \text{and} \quad xe^{-2x}$$

and the general solution is

$$y(x) = C_1 e^{-2x} + C_2 x e^{-2x}$$

A study of what happened shows that in the general case of a double root trying x times the solution will also fit the equation and will lead us to the linearly independent second solution.

Example 24.2-3

Solve the homogeneous equation

$$y'' + 4y' + 13y = 0$$

The characteristic equation is

$$m^2 + 4m + 13 = 0$$

The two roots are

$$m_k = -2 \pm \sqrt{4 - 13} = -2 \pm 3i \qquad (k = 1, 2)$$

The general solution will be

$$y(x) = C_1 \exp(m_1 x) + C_2 \exp(m_2 x) \tag{24.2-4}$$

If C_1 is the conjugate of C_2, then the sum of the terms will be real, and you have a real solution.

You may prefer to work with a real solution at all stages; in that case recall the Euler identities

$$e^{ix} = \cos x + i \sin x$$

$$e^{-ix} = \cos x - i \sin x$$

Hence

$$e^{(-2 \pm 3i)x} = e^{-2x} e^{\pm 3ix} = e^{-2x} \{\cos 3x \pm i \sin 3x\}$$

This suggests that the two "real" linearly independent solutions are

$$e^{-2x} \cos 3x \quad \text{and} \quad e^{-2x} \sin 3x$$

and the general solution is

$$y(x) = C_1 e^{-2x} \cos 3x + C_2 e^{-2x} \sin 3x \tag{24.2-5}$$

Direct substitution verifies that this is a solution. For the cosine term we get, ignoring the front constant C_1,

$$e^{-2x} [4 \cos 3x + 6 \sin 3x - 9 \cos 3x - 8 \cos 3x - 6 \sin 3x + 13 \cos 3x] = 0$$

And similarly for the sine term.

Thus either the complex (24.2-4) or the real (24.2-5) form of the solution may be used; the first gives us the difficulty of thinking about complex numbers while the second involves more trouble in differentiation and subsequent algebra.

Once we have an existence theorem, then we need not worry too much about how we find the solution, *provided* we are prepared to verify by direct substitution that it is the unique solution satisfying the initial conditions. This is a powerful fact with regard to solving differential equations; the ultimate test is, "Does it fit?" If it does, fine; if not, too bad! The mere existence theorem is at times very valuable in practice.

We have seen how to solve a second-order linear homogeneous differential equation with constant coefficients. We next need to describe what kinds of solutions we find. If the real parts of the characteristic roots are negative (for real or complex roots), then the solution decays exponentially to zero. If the roots are real, then the solution is fairly smooth. But if the roots are complex, then the size of the imaginary part gives the frequency of wiggling. If the real part of the roots is positive, then the solution grows like an exponential as $x \to \infty$. If the roots are equal, then you have terms

$$e^{mx} \quad \text{and} \quad xe^{mx}$$

and they are, depending on the sign of m, growing in size for positive m, and ultimately decaying for negative m.

The initial conditions (stated in the existence theorem) determine the specific behavior of the solution.

For the sake of completeness, the special case of a root equal to 0 is of interest (see Section 23.9). In this case the corresponding solution is

$$e^{mx} = e^{0x} = 1$$

and in the general solution this appears as the coefficient of the constant C_1 (or C_2) $= (1)^n = 1$; and if the other root is negative, then the solution exponentially approaches this constant (which the initial conditions determine). However, notice that if the characteristic root is zero then there is no y term, and the differential equation is really a differential equation in y' of one order lower. Of course, the solution of this lower-order equation will require a final integration to get the $y(x)$.

In summary, given a second-order linear differential equation with constant coefficients

$$p_0 y'' + p_1 y' + p_2 y = 0$$

we try solutions of the form e^{mx} [Equation (24.2-2)] and are led immediately to the characteristic equation (24.2-3),

$$p_0 m^2 + p_1 m + p_2 = 0$$

which is a quadratic, and has roots m_1 and m_2.

1. If the roots are distinct, then the general solution is

$$y(x) = C_1 \exp{(m_1 x)} + C_2 \exp{(m_2 x)}$$

2. If the roots are multiple, then the general solution is

$$y(x) = C_1 \exp(mx) + C_2 x \exp(mx)$$

3. If the roots are complex, $a + ib$, then the general solution can be written either in the form of case 1 or else, if you prefer in one of the two following forms:

$$y(x) = C_1 \exp(ax) \cos bx + C_2 \exp(ax) \sin bx$$

$$= e^{ax}\{C_1 \cos bx + C_2 \sin bx\}$$

The solutions are linearly independent, so we can fit any two conditions we please. Typically, they are given in the form of the starting value plus the starting derivative of $y(x)$.

Example 24.2-4

Given the equation $y'' + y = 0$, find the solution that satisfies $y(0) = 1$, $y'(0) = 0$. We have the characteristic equation

$$m^2 + 1 = 0$$

whose solutions are i and $-i$. Hence the general solution is

$$y(x) = C_1 \cos x + C_2 \sin x$$

We now impose the two initial conditions

$$y(0) = 1 = C_1$$

$$y'(0) = 0 = C_2$$

These are trivial to solve and the desired solution is

$$y(x) = \cos x$$

EXERCISES 24.2

Find the general solution of the following:

1. $y'' + 3y' + 2y = 0$
2. $2y'' + 5y' + 2y = 0$
3. $y'' - 4y' + 4y = 0$
4. $y'' + 3y' - 4y = 0$
5. $y'' + k^2 y = 0$
6. Fit the solution of Exercise 1 to the conditions $y(0) = 1$, $y'(0) = 0$.
7. Fit the solution of Exercise 3 to $y(0) = 0$, $y'(0) = 1$.
8. Discuss the equation $y'' = 0$ from the point of view of this general method.

24.3 THE NONHOMOGENEOUS EQUATION

The nonhomogeneous equation with constant coefficients

$$p_0 y'' + p_1 y' + p_2 y = f(x) \tag{24.3-1}$$

is next on the list of equations to study. We first note that, if we have one *particular solution* $Y(x)$ to this equation, then the general solution (involving two arbitrary constants) is

$$y(x) = C_1 e^{m_1 x} + C_2 e^{m_2 x} + Y(x) \tag{24.3-2}$$

By direct substitution we have

$$C_1\{p_0 m_1^2 + p_1 m_1 + p_2\} \exp(m_1 x) + C_2\{p_0 m_2^2 + p_1 m_2 + p_2\} \exp(m_2 x)$$
$$+ \{p_0 Y''(x) + p_1 Y'(x) + p_2 Y(x)\} = f(x)$$

The first two braces are zero because the exponentials are solutions of the homogeneous equation, and the third brace exactly equals $f(x)$ because we assumed that $Y(x)$ was a solution of the nonhomogeneous equation (24.3-1).

The problem remains, therefore, "How do we find a particular solution of the complete (nonhomogeneous) equation?" The answer is that we have several methods. In the first method it is a matter of luck to a great extent. We simply guess and try. In Section 24.4 we will give a second, analytical method of finding the solution. Remember we need only one particular solution of the complete equation, and no matter how we get it, if it fits, then it is a solution of its equation.

A little thought shows that, if the right-hand side is a polynomial of degree m, then trying a polynomial of degree m with undetermined coefficients will (probably) do the trick. Occasionally, we will have to raise the degree slightly. We see upon inspection that the highest-degree term in the guessed solution will come *only* from the y term. The next will come from the next highest term of the guess from the y term, and the highest degree term when differentiated in the y' term will also occur. Thus the system of equations to be solved is triangular and always solvable (except in degenerate cases). We are safe this time; we can solve the system when the right-hand side is a polynomial. Also, we can, with more labor, solve the equation when the forcing term $f(x)$ is piecewise a polynomial. We can find the solution for a given set of initial conditions, but the general case gets into messy notation.

Example 24.3-1

Suppose we have the equation

$$y'' + 3y' + 2y = f(x)$$

with $f(x) = 0$ except in the first interval $0 \le x \le 1$ where $f(x) = 1$ (Figure 24.3-1). This forcing function may be thought of as a unit pulse in electricity or the sudden application of a new economic or legal law, together with the later cancellation of it. We also suppose that we have the initial condition $y = 0$, $y' = 0$. This means that nothing was happening before the pulse.

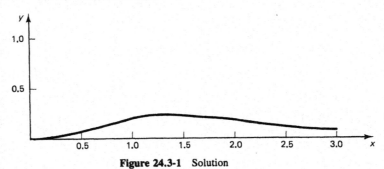

Figure 24.3-1 Solution

A particular solution for the first interval ($0 \leq x \leq 1$) can be found as follows. Try $y = C$ as the particular solution of the complete equation. We get $C = 1/2$. This gives us the general solution for this first interval [you get the characteristic equation $m^2 + 3m + 2 = (m + 1)(m + 2) = 0$]:

$$y(x) = C_1 e^{-x} + C_2 e^{-2x} + \frac{1}{2}$$

When the initial conditions are applied, we get

$$0 = C_1 + C_2 + \frac{1}{2}$$

$$0 = -C_1 - 2C_2$$

from which (by addition) $C_2 = 1/2$ and therefore $C_1 = -1$. The solution for the first interval is, therefore,

$$y(x) = \frac{1 + e^{-2x}}{2} - e^{-x}$$

$$= \frac{1 - 2e^{-x} + e^{-2x}}{2} = \frac{1}{2}(1 - e^{-x})^2$$

Before going on, we check that it fits the initial conditions and the differential equation.

Next, we find the conditions at the end of the interval. At $x = 1$ we have for this solution

$$y(1) = \frac{1}{2}\left(1 - \frac{1}{e}\right)^2$$

$$y'(1) = e^{-1} - e^{-2} = \frac{1 - (1/e)}{e}$$

These are the initial conditions for the next interval ($x > 1$) in which we have only the homogeneous equation to solve, since $f(x) = 0$ in this interval.

$$y(x) = C_1 e^{-x} + C_2 e^{-2x}$$

$$y'(x) = -C_1 e^{-x} - 2C_2 e^{-2x}$$

At $x = 1$, the solution must fit the initial conditions; that is,

$$\frac{1}{2}\left(1 - \frac{1}{e}\right)^2 = C_1 \frac{1}{e} + C_2 \frac{1}{e^2}$$

$$\frac{1}{e} - \frac{1}{e^2} = -C_1 \frac{1}{e} - 2C_2 \frac{1}{e^2}$$

Multiply both equations through by e^2 to get

$$eC_1 + C_2 = \frac{1}{2}(e - 1)^2$$

$$eC_1 + 2C_2 = -e + 1$$

from which the solution is easy to find $(x \geq 1)$:

$$y(x) = [e - 1]e^{-x} + \frac{[1 - e^2]e^{-2x}}{2}$$

See Figure 24.3-1.

In this example you have seen that you can handle discontinuous forcing functions. Generalizing from this single example, you see that, given any forcing function, *if* you can approximate it in various intervals by polynomials, then the above method, slightly extended, will get you the solution. You can mechanize the arithmetic on a computer since it is essentially the same for each interval. A discontinuous coefficient in the differential equation can be handled similarly and need not be discussed further.

Example 24.3-2

Consider the differential equation

$$y'' + 4y' + 3y = 1 - x^2$$

The homogeneous equation

$$y'' + 4y' + 3y = 0$$

has the characteristic equation

$$m^2 + 4m + 3 = (m + 1)(m + 3) = 0$$

Hence the corresponding solution is

$$y = C_1 e^{-x} + C_2 e^{-3x}$$

For the complete equation with the quadratic on the right-hand side, we would naturally guess at a quadratic as a solution. We try the method of undetermined coefficients

$$Y = ax^2 + bx + c$$

When this is put into the differential equation, we get

$$2a + 4(2ax + b) + 3(ax^2 + bx + c) = 1 - x^2$$

Equating like powers of x, we have

$$3a = -1$$

$$8a + 3b = 0$$

$$2a + 4b + 3c = 1$$

from which we get in turn $a = -1/3$, $b = 8/9$, and $c = -17/27$. Thus the final solution is

$$y = \frac{-17 + 24x - 9x^2}{27} + C_1 e^{-x} + C_2 e^{-3x}$$

The guess method applies to forcing functions having terms in polynomials, exponentials, sines and cosines, and combinations of them. These solutions of the nonhomogeneous equation are comparatively easy to guess, although the algebra may be tedious.

If a term in the forcing function is a solution of the homogeneous equation, then it will cancel out and cannot be equated to the right-hand side term. It is necessary, therefore, to raise the degree by a factor of x (or x^2 in case of a double root). See Example 24.4-2 in the next section.

Example 24.3-3

Consider the differential equation

$$y'' + 4y' + 3y = \sin x$$

We naturally try the guessed solution of the form

$$Y(x) = A \sin x + B \cos x$$

Differentiation and substitution lead to

$$(-A \sin x - B \cos x) + 4(A \cos x - B \sin x) + 3(A \sin x + B \cos x) = \sin x$$

Since $\sin x$ and $\cos x$ are linearly independent, we equate like terms to get the equations

$$-A - 4B + 3A = 1$$

$$-B + 4A + 3B = 0$$

These equations reduce to

$$2A - 4B = 1$$

$$4A + 2B = 0$$

whose solution is $A = \frac{1}{10}$, $B = -\frac{2}{10}$. The general solution of the complete equation is therefore

$$y(x) = \frac{\sin x - 2 \cos x}{10} + C_1 e^{-x} + C_2 e^{-3x}$$

EXERCISES 24.3

Solve the following:

1. $y'' + 4y' + 4y = x$, $y(0) = 1$, $y'(0) = 0$
2. $y'' + 2y' + y = \exp(-x)$, $y(0) = 0$, $y'(0) = 0$. *Hint:* Try Ax^2e^{-x}.
3. $y'' + y' + y = 1$, $y(0) = 1$, $y'(0) = -1$
4. $y'' + ay' + by = x^3$, $y(0) = 0$, $y'(0) = 1$
5. $y'' + 3y' + 2y = \exp(-x) + \exp(-2x)$, $y(0) = 1$, $y'(0) = 0$

24.4 VARIATION OF PARAMETERS METHOD

The guess method is of limited effectiveness in finding a particular solution of the complete equation since you may at times have little or no idea of what form to assume. The method of *variation of parameters* reduces the question to a matter of integration only. The equation we start with is of the form

$$p_0 y'' + p_1 y' + p_2 y = f(x) \tag{24.4-1}$$

The plan is to first solve the homogeneous system and then consider the general solution, *not* with unknown constant coefficients but rather with unknown *functions of x*. Thus we write the solution as

$$Y(x) = C_1(x) \exp(m_1 x) + C_2(x) \exp(m_2 x) \tag{24.4-2}$$

where, of course, the m_i are the roots of the characteristic equation (which we are assuming are distinct for purposes of presentation). We have introduced two arbitrary *functions* of x, $C_1(x)$ and $C_2(x)$ and can therefore impose two conditions (in x).

We first differentiate our trial solution (24.4-2):

$$Y'(x) = C_1(x)m_1 \exp(m_1 x) + C_2(x)m_2 \exp(m_2 x)$$
$$+ C_1'(x) \exp(m_1 x) + C_2'(x) \exp(m_2 x)$$

At this point we impose one of the two conditions we have available. We will require that the second part of the derivative expression be identically zero:

$$C_1'(x) \exp(m_1 x) + C_2'(x) \exp(m_2 x) = 0 \tag{24.4-3}$$

This prevents any second derivatives of the $C_i(x)$ from arising. We now differentiate what is left of the guessed solution (the top line of the derivative):

$$Y''(x) = C_1(x)m_1^2 \exp(m_1 x) + C_2(x) m_2^2 \exp(m_2 x)$$
$$+ C_1'(x)m_1 \exp(m_1 x) + C_2'(x)m_2 \exp(m_2 x)$$

When we multiply the $Y''(x)$ by p_0, the $Y'(x)$ by p_1 and $Y(x)$ by p_2, and add, we find that the terms in C_1 and C_2 all cancel *because* the m_i satisfy the characteristic equation. This leaves us with the second part of the $Y''(x)$ expression involving the $C_i'(x)$ equated to $f(x)$. This equation, together with the earlier condition we imposed (24.4-3), gives us two equations:

$$C_1'(x)m_1 \exp{(m_1x)} + C_2'(x)m_2 \exp{(m_2x)} = f(x)$$

$$C_1'(x) \exp{(m_1x)} + C_2'(x) \exp{(m_2x)} = 0$$

We can solve these equations for both $C_1'(x)$ and $C_2'(x)$ as follows. Multiply the bottom equation by $-m_2$ and add to get

$$C_1'(m_1 - m_2) \exp{(m_1x)} = f(x)$$

or
$$C_1' = \frac{f(x) \exp{(-m_1x)}}{m_1 - m_2} \qquad (24.4\text{-}4)$$

Next multiply the bottom equation by $-m_1$ and add to get

$$C_2'(m_2 - m_1) \exp{(m_2x)} = f(x)$$

or
$$C_2' = \frac{-f(x) \exp{(-m_2x)}}{m_1 - m_2} \qquad (24.4\text{-}5)$$

If the two integrals that arise from (24.4-4) and (24.4-5) can be done, then we have the solution of the complete equation. If they cannot be done, we have still reduced the problem to one of doing two quadratures (integrations).

The first time you meet the method it usually appears mysterious (and you wonder how you could have invented it), so let us review the process. You take the standard solution of the homogeneous linear differential equation and change the two constants C_1 and C_2 into variables $C_1(x)$ and $C_2(x)$. Then you differentiate this expression to get the needed derivatives for the differential equation. Along the way, having introduced two arbitrary functions, $C_1(x)$ and $C_2(x)$, you impose convenient conditions that at the first stage of differentiation the terms in the first derivative cancel out. This choice means that you will not have to deal with the second derivatives of the undetermined functions $C_1(x)$ and $C_2(x)$. Then you find the second derivative and put the derivatives into the original differential equation. This gives the second constraint on the two functions $C_1(x)$ and $C_2(x)$. Both equations involve only the first derivatives of the introduced functions, $C_1(x)$ and $C_2(x)$, and, solving these two linear equations, you are led to two integrals to determine the unknown functions. You then have the solution. The method is not completely arbitrary, but rather has an underlying motivation that is the secret of the success of the method.

Example 24.4-1

Given the equation

$$y'' + 2y' + y = f(x)$$

the homogeneous equation immediately leads you (via the characteristic equation) to the solution

$$y = C_1e^{-x} + C_2xe^{-x}$$

You now regard $C_1 = C_1(x)$ and similarly for $C_2 = C_2(x)$. Thus you try the solution for the complete equation:

$$Y(x) = C_1(x)e^{-x} + C_2(x)xe^{-x}$$

$$Y'(x) = C_1(x)(-1)e^{-x} + C_2(x)(-xe^{-x} + e^{-x}) + [C_1'(x)e^{-x} + C_2'(x)xe^{-x}]$$

You set the term in the brackets equal to zero (to avoid differentiating the coefficients again). As a result you have, on differentiating again,

$$Y''(x) = C_1 e^{-x} + C_2(x)(xe^{-x} - 2e^{-x}) + C_1'(x)(-e^{-x}) + C_2'(x)(-x + 1)e^{-x}$$

When you form $Y'' + 2Y' + Y = f(x)$, you get

$$C_1(x)\left[e^{-x} - 2e^{-x} + e^{-x}\right] + C_2(x)\left[xe^{-x} + 2(-x + 1)e^{-x} + (x - 2)e^{-x}\right]$$

$$+ C_1'(x)(-e^{-x}) + C_2'(x)(-x + 2)e^{-x} = f(x)$$

The two brackets cancel out exactly (as they should) and you have the pair of equations (when you multiply through by e^x)

$$-C_1'(x) + (-x + 2)C_2'(x) = f(x)e^x$$

$$C_1'(x) + xC_2'(x) = 0$$

These are easy to solve. By addition, you get

$$2C_2'(x) = f(x)e^x$$

Using this in the second equation, you get

$$C_1'(x) = \frac{-xf(x)e^x}{2}$$

Hence you have

$$C_1(x) = -\int \frac{xf(x)e^x}{2}\, dx$$

$$C_2(x) = \int \frac{f(x)e^x}{2}\, dx$$

as a pair of integrals that determine the coefficients for the particular solution of the complete equation. To this particular solution you naturally add the general solution of the homogeneous equation to get the general solution of the original problem. Notice that you do not use the additive constant that arises in the integration, since such terms will be covered when you add the solution of the homogenous equation. You cannot carry the solution further without having a specific $f(x)$, but these steps are now well within your experience, since you learned to integrate some chapters ago.

Example 24.4-2

Solve using variation of parameters

$$y'' + y = \sin x$$

The homogeneous equation has the solution

$$y(x) = C_1 \cos x + C_2 \sin x$$

You therefore assume the form for the solution of the complete equation

$$y(x) = C_1(x) \cos x + C_2(x) \sin x$$

The first differentiation gives

$$y' = -C_1(x) \sin x + C_2(x) \cos x + [C_1'(x) \cos x + C_2'(x) \sin x]$$

and you set the bracket involving the derivatives of the functions $C_1(x)$ and $C_2(x)$ equal to zero. Another differentiation gives

$$y'' = -C_1(x) \cos x - C_2 \sin x + [-C_1'(x) \sin x + C_2'(x) \cos x]$$

Putting these in the original differential equation, you get only the terms from the bracket of y'' as one equation plus the bracket from the first differentiation as the second equation:

$$-C_1'(x) \sin x + C_2'(x) \cos x = \sin x$$

$$C_1'(x) \cos x + C_2'(x) \sin x = 0$$

To eliminate $C_2'(x)$, multiply the top equation by $-\sin x$ and the bottom by $\cos x$ and add to get

$$C_1'(x) = -\sin^2 x$$

If you multiply the top equation by $\cos x$ and the lower by $\sin x$ and add, you get

$$C_2'(x) = \sin x \cos x$$

Hence you have

$$C_1(x) = \int (-\sin^2 x)\, dx = \frac{-x}{2} + \frac{\sin 2x}{4}$$

$$C_2(x) = \int \sin x \cos x\, dx = \frac{\sin^2 x}{2}$$

and now that you know $C_1(x)$ and $C_2(x)$ you have your solution:

$$C_1 \cos x + C_2 \sin x - \frac{x \cos x}{2} + (2 \sin x \cos x)\frac{\cos x}{4} + \frac{\sin^3 x}{2}$$

which can be simplified further, if you wish:

$$C_1 \cos x + C_2 \sin x - \frac{x}{2} \cos x + \frac{1}{2} \sin x (\cos^2 x + \sin^2 x)$$

Now changing the meaning of the constant C_2, you have

$$C_1 \cos x + C_2 \sin x - \frac{x}{2} \cos x$$

as a more compact form for the solution.

The method of variation of parameters is *not* an isolated trick. The next example shows a small variant that is useful.

Example 24.4-3

Given the equation

$$P_0(x)y'' + P_1(x)y' + P_2(x)y = 0 \qquad (24.4\text{-}6)$$

with variable (or constant) coefficients, suppose $u(x)$ is a known solution. Then, to get the general solution, we write (in the spirit of variation of parameters but in different notation)

$$y(x) = u(x)v(x)$$

$$y' = u'v + v'u$$

$$y'' = u''v + 2u'v' + v''u$$

Substitute into (24.4-6) to get

$$(uP_0)v'' + (2P_0u' + P_1u)v' + (P_0u'' + P_1u' + P_2u)v = 0$$

The last term is zero since u is a solution of (24.4-6). If we write

$$v' = z \quad \text{then} \quad v'' = z'$$

and we have the first-order linear equation

$$uP_0z' + (2P_0u' + P_1u)z = 0$$

Divide by uP_0 to get it in the standard form:

$$z' + \left(\frac{2u'}{u} + \frac{P_1}{P_0}\right)z = 0$$

The integrating factor is

$$\exp\left\{\int\left[\frac{2u'}{u} + \frac{P_1}{P_0}\right]dx\right\} = u^2 \exp\left\{\int\frac{P_1}{P_0}dx\right\}$$

Hence we have

$$\left(u^2 \exp\int\frac{P_1}{P_0}dx\, z\right)' = 0$$

from which we get

$$z = \frac{dv}{dx} = \frac{C_1}{u^2}\exp\left(-\int\frac{P_1}{P_0}dx\right)$$

and finally

$$y = uv = C_1u\int\frac{1}{u^2}\exp\left(-\int\frac{P_1}{P_0}dx\right)dx + C_2u$$

as the general solution.

Thus, knowing one solution of a second-order linear homogeneous differential equation, you can get the general solution as the result of two quadratures.

EXERCISES 24.4

Solve the following:

1. $y'' + 5y' + 4y = f(x)$
2. $y'' + 6y' + 9y = f(x)$
3. $y'' + y' + y = f(x)$
4. If $\exp(-x)$ is a solution of $y'' + xy' + (x - 1)y = 0$, then find the second solution.
5. Show that, given any two functions $u_1(x)$ and $u_2(x)$ having second derivatives, the determinant

$$\begin{vmatrix} y & y' & y'' \\ u_1 & u_1' & u_1'' \\ u_2 & u_2' & u_2'' \end{vmatrix} = 0$$

gives a linear second-order differential equation having $u_1(x)$ and $u_2(x)$ as solutions. Generalize.

24.5 nth-ORDER LINEAR EQUATIONS

The treatment of nth-order linear differential equations with constant coefficients is a simple generalization of the first- and second-order equations. The main difference is in the handling of the homogeneous equation. The assumption of a solution in the form [see (24.2-2]

$$e^{mx}$$

leads to the corresponding nth-order characteristic equation [see (24.2-3)]

$$p_0 m^n + p_1 m^{n-1} + \cdots + p_n = 0$$

It is a fact that real polynomials are composed of real linear and real quadratic factors. This is true because complex roots must occur in conjugate pairs, since no one can tell $-i$ from $+i$. It is an arbitrary convention when we plot points in the complex plane and put one of them as positive. Hence the product of the corresponding complex factors

$$[m - (a + ib)][m - (a - ib)] = [(m - a) - ib][(m - a) + ib]$$

$$= (m - a)^2 + b^2$$

$$= m^2 - 2am + (a^2 + b^2)$$

is a real quadratic function. The rules for finding the corresponding n linearly independent solutions are the same (almost) as before. For simple real roots, you try

$$e^{mx}$$

For multiple roots of multiplicity k, you try the k functions

$$e^{mx}, \ xe^{mx}, \ x^2 e^{mx}, \ \cdots, \ x^{k-1} e^{mx}$$

to get the corresponding k solutions. For complex roots $a \pm ib$, you try (if you prefer this form to the complex form)

$$e^{ax} \cos bx \quad \text{and} \quad e^{ax} \sin bx$$

and if there are multiplicities of the roots, then you include the proper powers of x in front. In this way, for each root, real or complex, you get one corresponding linearly independent solution of the homogeneous equation.

The solution of the complete equation, which we have labeled $Y(x)$, can be found by the guess method (Section 24.3) or by the method of variation of parameters (Section 24.4). The system of linear equations that arises from the equating of the terms in $C_i'(x)$ to zero at *each* stage except the last differentiation, together with the result from the differential equation, can always be solved (we do not explicitly prove this fact here) for the $C_i'(x)$; and hence by their corresponding n integrations, the solution for the nth-order linear differential equation with constant coefficients can be found.

Example 24.5-1

Find the solution of

$$y''' + 3y'' + 3y' + y = e^{-x}$$

The characteristic equation is

$$m^3 + 3m^2 + 3m + 1 = (m + 1)^3 = 0$$

and the characteristic roots are $m = -1, -1, -1$. Thus the solution of the homogeneous equation is

$$y = (C_1 + C_2 x + C_3 x^2) e^{-x}$$

When we try to find a particular solution of the complete equation, we see that the right-hand side is a solution of the homogeneous equation, and we are driven to try, finally,

$$Y(x) = Ax^3 e^{-x}$$

We have in turn

$$Y'(x) = A(-x^3 e^{-x} + 3x^2 e^{-x})$$
$$Y''(x) = A(x^3 e^{-x} - 6x^2 e^{-x} + 6x e^{-x})$$
$$Y'''(x) = A(-x^3 e^{-x} + 9x^2 e^{-x} - 18x e^{-x} + 6 e^{-x})$$

When we multiply these equations for the derivatives by 1, 3, 3, 1 and add, we get all the higher terms canceling (which checks the differentiation) and have left

$$6A e^{-x} = e^{-x}$$

Hence $A = \frac{1}{6}$. The general solution is, therefore,

$$y(x) = \left(C_1 + C_2 x + C_3 x^2 + \frac{x^3}{6} \right) e^{-x}$$

Example 24.5-2

Solve

$$y''' + y = x^2$$

The characteristic equation is

$$m^3 + 1 = (m + 1)(m^2 - m + 1) = 0$$

and the roots are

$$m_1 = -1, \qquad m_2 = \frac{1 + i\sqrt{3}}{2}, \qquad m_3 = \frac{1 - i\sqrt{3}}{2}$$

The solution of the homogeneous equation is

$$y(x) = C_1 e^{-x} + e^{x/2} \left\{ C_2 \cos\left(\frac{\sqrt{3}}{2} x\right) + C_3 \sin\left(\frac{\sqrt{3}}{2} x\right) \right\}$$

For the complete equation, we guess the form

$$Y(x) = ax^2 + bx + c$$

and find immediately that we have

$$ax^2 + bx + c = x^2$$

from which $a = 1$, $b = 0 = c$. Thus the complete solution is

$$y(x) = C_1 e^{-x} + e^{x/2} \left\{ C_2 \cos\left(\frac{\sqrt{3}}{2} x\right) + C_3 \sin\left(\frac{\sqrt{3}}{2} x\right) \right\} + x^2$$

Example 24.5-3

Suppose we solve, by the method of variation of parameters,

$$y''' + 6y'' + 11y' + 6y = x$$

The characteristic equation is

$$m^3 + 6m^2 + 11m + 6 = (m + 1)(m + 2)(m + 3) = 0$$

and the solution of the homogeneous equation is

$$y = C_1 e^{-x} + C_2 e^{-2x} + C_3 e^{-3x}$$

For the solution of the nonhomogeneous equation, we try

$$Y(x) = C_1(x)e^{-x} + C_2(x)e^{-2x} + C_3(x)e^{-3x}$$

and we get after the first differentiation

$$Y' = -C_1 e^{-x} - 2C_2 e^{-2x} - 3C_3 e^{-3x}$$

when we remember to set

$$C_1' e^{-x} + C_2' e^{-2x} + C_3' e^{-3x} = 0$$

The next differentiation gives

$$Y'' = C_1 e^{-x} + 4C_2 e^{-2x} + 9C_3 e^{-3x}$$

when we set

$$-C_1' e^{-x} - 2C_2' e^{-2x} - 3C_3' e^{-3x} = 0$$

The third differentiation gives

$$Y''' = -C_1 e^{-x} - 8C_2 e^{-2x} - 27C_3 e^{-3x} + C_1' e^{-x} + 4C_2' e^{-2x} + 9C_3' e^{-3x}$$

When we multiply these equations by the coefficients of the differential equation, the terms in the coefficients cancel, while the terms in the derivatives of the coefficients give (bringing the equations all together)

$$C_1' e^{-x} + C_2' e^{-2x} + C_3' e^{-3x} = 0$$

$$-C_1' e^{-x} - 2C_2' e^{-2x} - 3C_3' e^{-3x} = 0$$

$$C_1' e^{-x} + 4C_2' e^{-2x} + 9C_3' e^{-3x} = x$$

These equations have the solution

$$C_1' = \frac{xe^x}{2}, \qquad C_2' = -xe^{2x}, \qquad C_3' = \frac{xe^{3x}}{2}$$

whose integrals are

$$C_1 = \frac{\{x - 1\}e^x}{2}$$

$$C_2 = -\left\{\frac{x}{2} - \frac{1}{4}\right\}e^{2x}$$

$$C_3 = \frac{\{x/3 - 1/9\}e^{3x}}{2}$$

Hence the solution is

$$y(x) = C_1 e^{-x} + C_2 e^{-2x} + C_3 e^{-3x} + \frac{x}{6} - \frac{11}{36}$$

The solution could, of course, have been found easier by the guess method, but we were illustrating the method of variation of parameters for nth-order linear equations and wanted an easily checked solution.

EXERCISES 24.5

Solve the following:

1. $y^{iv} + y = x^3$
2. $y^{iv} - y = \sin x$
3. $y''' + y'' + y' + y = f(x)$
4. $y''' - 2y'' + y' - 2y = x$

4.6 EQUATIONS WITH VARIABLE COEFFICIENTS

Much of the theory for equations with constant coefficients goes over fairly directly
to equations with variable coefficients:

$$P_0(x)y^{(n)}(x) + P_1(x)y^{(n-1)}(x) + \cdots + P_n(x)y(x) = f(x)$$

where we use capital letters for the variable coefficients. By direct substitution, you
see that, if you have one solution $Y(x)$ of the complete equation, then you are again
reduced to solving the homogeneous equation to get the n linearly independent terms
that form the general solution; the terms from the $Y(x)$ combine with the $f(x)$ to reduce
the equation to the homogeneous form.

The next step, finding the characteristic equation, *fails* because the solutions are
no longer simple exponentials. However, *if* you can find n linearly independent
solutions of the homogeneous equation, then you can take the corresponding combination with constant coefficients along with the $Y(x)$ to form the general solution, since
the equations are solvable by Section 4.7.

A careful examination of the method of variation of parameters will show that
nowhere did the form of the solutions matter; if you can solve the homogeneous
equation for the n linearly independent solutions, then the method of variation of
parameters will lead you to the corresponding n integrals for the particular solution of
the complete equation.

Unfortunately, there is no general method for finding the n linearly independent
solutions of the homogeneous equation. As in Example 24.4-3, each known solution
of the homogeneous equation can be used to reduce the order of the equation by one.
There are methods for getting the power series expansions about the origin, or any
other point, but they lie beyond this course.

There is one particular case that occurs frequently and can be reduced to the case
of linear differential equations with constant coefficients; it is the *Euler* equation

$$p_0 x^n y^{(n)}(x) + p_1 x^{n-1} y^{(n-1)}(x) + \cdots + p_n y(x) = f(x)$$

where the kth derivative is multiplied by x^k. The change of independent variable

$$x = e^t \quad \text{or} \quad t = \ln x$$

changes the derivatives as follows:

$$xy' = \frac{x\,dy}{dx} = x\left(\frac{dy}{dt}\right)\left(\frac{dt}{dx}\right) = \frac{x}{x}\frac{dy}{dt} = \frac{dy}{dt}$$

$$x^2 y'' = x^2\left(\frac{d}{dx}\frac{dy}{dx}\right) = x^2\left(\frac{d}{dx}\right)\left(\frac{1}{x}\frac{dy}{dt}\right)$$

$$= \frac{x^2\,d}{dx}\frac{(-1/x^2)\,dy}{dt} + \frac{(1/x^2)\,d^2y}{dt^2}$$

$$= \frac{d^2y}{dt^2} - \frac{dy}{dt} = D(D-1)y$$

where $D = d/dt$ is regarded as an operator.

Repeating this process we get

$$x^3 y^{(3)} = D(D - 1)(D - 2)$$
$$x^4 y^{(4)} = D(D - 1)(D - 2)(D - 3)$$

and so on. Using these substitutions, you see that an equation with constant coefficients emerges.

In a sense, the prevalence of Euler-type equations is an indication of the importance of making the natural choice of the independent variable. The change of variables reduces the case to the constant coefficient case studied earlier.

EXERCISES 24.6

1. Solve $x^2 y'' + 4xy' + 2y = x$.
2. Solve $x^2 y'' + y = 0$.

24.7 SYSTEMS OF EQUATIONS

A single differential equation of order y can be written as a system of n first-order equations by a simple change of notation. Let the nth-order differential equation be

$$y^{(n)}(x) = f(x, y, y', \ldots, y^{(n-1)})$$

We set

$$z = y'$$
$$u = z' = y''$$
$$v = u' = z'' = y'''$$

etc.

As a result, we have the above $n - 1$ of these equations in the new variables plus the the original equation in the new variables.

Example 24.7-1

Reduce the following equation to a system of first-order equations.

$$y''' + xy'' - (\sin x)y' + (e^x)y = f(x)$$

We get

$$y' = z$$
$$z' = u$$
$$u' = -xu + (\sin x)z - (e^x)y + f(x)$$

as the corresponding system of first-order equations.

For linear systems, we can do the reverse process (provided the needed derivatives exist) and convert the system to a single nth-order equation. It is a matter of eliminating the variables after suitable differentiations.

Example 24.7-2

Given the system

$$y' = ay + bz + f(x)$$
$$z' = cy + dz + g(x)$$

find an equivalent second-order equation in y. Since we will need a y'', we differentiate the top equation to get

$$y'' = ay' + bz' + f'(x)$$

Now eliminate the z' using the second equation

$$y'' = ay' + b\{cy + dz + g(x)\} + f'(x)$$
$$= ay' + bcy + bdz + bg(x) + f'(x)$$

We now eliminate the z in this equation by using the top equation of the original pair.

$$y'' = ay' + bcy + d\{y' - ay - f(x)\} + bg(x) + f'(x)$$
$$= (a + d)y' + (bc - ad)y - df(x) + bg(x) + f'(x)$$

and we have the required equation. The method may be systematized using determinants in the general case.

EXERCISES 24.7

1. Reduce to a system of three equations $y''' + xy'' - y' + xy = \sin x$.
2. Convert Example 24.7-2 to a single equation in z.
3. Reduce the equation $P_0(x)y'' + P_1(x)y' + P_2(x)y = f(x)$ to a system of equations.

24.8 DIFFERENCE EQUATIONS

The theory of linear difference equations with constant coefficients

$$a_0 y_n + a_1 y_{n-1} + a_2 y_{n-2} + \ldots + a_k y_{n-k} = f(n)$$

is very much like the theory of linear differential equations. The main difference is that, by custom, in place of the trial solution

$$e^{mx}$$

you use

$$r^n$$

which is the same thing once you write $e^m = r$ and note that $x \to n$. There are, of course, other differences, such as integrations are replaced by summations. The differences are best illustrated by examples.

Example 24.8-1

Fibonacci equation. The defining conditions for the Fibonacci numbers are

$$f_{n+1} = f_n + f_{n-1}, \qquad f_0 = 0 \qquad f_1 = 1$$

This is a homogeneous linear difference equation as it stands, so we start with the trial solution

$$f_n = r^n$$

and, after dividing out the r^{n-1} factor, we have the characteristic equation

$$r^2 - r - 1 = 0$$

whose solutions are

$$r_1 = \frac{1 + \sqrt{5}}{2} \quad \text{and} \quad r_2 = \frac{1 - \sqrt{5}}{2}$$

Hence the general solution of the homogeneous equation is

$$C_1 \left[\frac{1 + \sqrt{5}}{2} \right]^n + C_2 \left[\frac{1 - \sqrt{5}}{2} \right]^n$$

We now impose the two initial conditions

$$f_0 = 0 = C_1 + C_2$$

$$f_1 = 1 = \frac{C_1(1 + \sqrt{5})}{2} + \frac{C_2(1 - \sqrt{5})}{2}$$

Subtract the top equation from twice the bottom equation to get

$$2 = \sqrt{5}(C_1 - C_2)$$

from which we get

$$C_1 = \frac{1}{\sqrt{5}} = -C_2$$

and the solution is

$$f_n = \frac{1}{\sqrt{5}} \left\{ \left[\frac{1 + \sqrt{5}}{2} \right]^n - \left[\frac{1 - \sqrt{5}}{2} \right]^n \right\}$$

which is what we had before (Example 2.4-3).

Case History 24.8-2

Evaluate the integral

$$I(k) = \int_0^\pi \frac{\cos k\theta - \cos k\phi}{\cos \theta - \cos \phi} \, d\theta$$

We do not see how to do it immediately, so we have to think a bit. Finally, we remember that the cosine function satisfies a three-term recurrence (difference) relation (in n for all x),

$$\cos (n + 1)x + \cos (n - 1)x = 2 \cos x \cos nx$$

Maybe we can get a three-term difference equation for $I(k)$. Looking at this carefully, we start with the sum

$$I(k + 1) + I(k - 1)$$

and examine the terms in the numerator, which for the θ terms gives

$$\cos (k + 1)\theta + \cos (k - 1)\theta = 2 \cos \theta \cos k\theta$$

There are, of course, the corresponding terms in ϕ.

We now have for the numerator of the difference of the terms

$$2 \cos \theta \cos k\theta - 2 \cos \phi \cos k\phi$$

After some trial and error (considering the denominator carefully), we think to write it as (subtract and add $\cos \phi \cos k\theta$)

$$2[\cos \theta \cos k\theta - \cos \phi \cos k\theta + \cos \phi \cos k\theta - \cos \phi \cos k\phi]$$

This can now be written as

$$2[\{\cos \theta - \cos \phi\} \cos k\theta + \cos \phi\{\cos k\theta - \cos k\phi\}]$$

Remembering the denominator, we now have

$$I(k + 1) + I(k - 1) = 2 \int_0^\pi \cos k\theta \, d\theta + 2 \cos \phi \, I(k)$$

The integral is zero, so we have the desired second-order difference equation in the form

$$I(k + 1) - 2 \cos \phi \, I(k) + I(k - 1) = 0$$

We have found a second-order difference equation for the integral based on our observation that the cosine function satisfies a second-order difference equation.

The characteristic equation of the difference equation is

$$r^2 - 2 \cos \phi \, r + 1 = 0$$

whose roots are

$$r_m = \cos \phi \pm \sqrt{(\cos^2 \phi - 1)} = \cos \phi \pm i \sin \phi = \exp \{\pm i\phi\}$$

Hence the general solution is

$$I(k) = C_1 \exp \{i\phi k\} + C_2 \exp \{-i\phi k\}$$

To find the two constants C_1 and C_2, we need two values of the integral. In the case of $k = 0$, the numerator is identically zero, but there is a single point in the range of integration where the denominator is zero. If we believe that this isolated point should, from the use of limits, be zero (the numerator is identically zero in the whole interval), then

$$I(0) = 0$$

For the case $k = 1$, the integral is clearly π.

Using these two values, we get

$$I(0) = C_1 + C_2 = 0$$

$$I(1) = \pi = C_1 \exp(i\phi) + C_2 \exp(-i\phi)$$

The solution of these equations is

$$C_1 = \frac{2i\pi}{\sin \phi} = -C_2$$

Hence

$$I(k) = \frac{\pi(\sin k\phi)}{\sin \phi}$$

Since you might have a slight doubt about the integral in the case $k = 0$, we check the solution for $k = 2$. We have

$$I(2) = \int_0^\pi \frac{\cos 2\theta - \cos 2\phi}{\cos \theta - \cos \phi} \, d\theta$$

But

$$\cos 2\theta - \cos 2\phi = 2(\cos^2 \theta - \cos^2 \phi)$$

Hence the integrand becomes

$$I(2) = 2 \int_0^\pi (\cos \theta + \cos \phi) \, d\theta$$

$$= 2\pi \cos \phi$$

We now compare this with the general solution for $k = 2$:

$$I(2) = \frac{\pi(\sin 2\phi)}{\sin \phi} = \frac{\pi(2 \sin \phi \cos \phi)}{\sin \phi}$$

and we see that they are the same.

Example 24.8-3

Solve the linear difference equation

$$y_{n+1} - 4y_n + 4y_{n-1} = n$$

The homogeneous equation is

$$y_{n+1} - 4y_n + 4y_{n-1} = 0$$

whose characteristic equation is

$$r^2 - 4r + 4 = (r - 2)^2 = 0$$

from which the corresponding solution is

$$y_n = C_1(2)^n + C_2 n(2)^n$$

We now need one solution of the nonhomogeneous equation. Looking at the right-hand side, we guess at

$$Y_n = An + B$$

We get

$$A(n + 1) + B - 4\{An + B\} + 4\{A(n - 1) + B\} = n$$

Equating like powers of n, we get

$$A - 4A + 4A = 1$$

$$A + B - 4B - 4A + 4B = 0$$

From the top equation, $A = 1$. From the lower, we get $B = 3$. Hence the complete solution is

$$y_n = C_1 2^n + C_2 n 2^n + n + 3$$

This is easily checked by direct substitution into the original equation.

EXERCISES 24.8

1. Analyze the generalized Fibonacci equation $f_{n+1} = af_n + bf_{n-1}, f_0 = 0, f_1 = 1$.
2. Solve $y_{n+1} - y_{n-1} = n^2$.
3. Solve $y_n = y_{n-1} + \sin nx, y_0 = 0$.

24.9 SUMMARY

Linear differential equations occur frequently; most are of first and second order, but occasionally you meet a third- or fourth-order equation. The reason they occur is that our current theories about the world assume a great deal of linearity and involve relations between the function and various rates of change. But the argument is partly circular; the reason the theories are linear is that we can solve linear equations with comparative ease, and we have great trouble with nonlinear equations, algebraic or differential.

The very common case of constant coefficients (things do not vary with the independent variable of either time or space) is solved by reducing the problem to the characteristic equation via the trial solution $\exp(mx)$. The one difficult step is finding the zeros of this characteristic equation; the rest is routine algebra and calculus. Using the method of variation of parameters, you can reduce the problem to integrations of functions that you may or may not be able to do in closed form, but that at worst can be tabulated easily.

You have seen that the roles of the homogeneous and nonhomogeneous equations are the same for constant and variable coefficients, and you may have noticed some similarity with linear algebraic equations. Linear systems have the property that various solutions of the homogeneous equations can be added, and that it suffices to have one solution to the nonhomogeneous equation. That is the key to success in the whole field. Indeed, the additivity of the solutions is what makes linear problems, wherever you meet them, comparatively easy.

We have not gone deeply into the relationship to the idea of linear independence, although obviously it is in the background when we asserted that the n solutions of the homogeneous equation formed the general solution to the equation; we were asserting that they were linearly independent and that no other solution was linearly independent. The proof as indicated in Section 24.1 (and from the earlier solution of linear algebraic equations) seems to be reasonable. A more rigorous treatment is beyond an introduction to differential equations. But this should point out the importance of linear algebra as an abstract field worth studying both for its intrinsic interest and for the wealth of its applications.

25

Numerical Methods

25.1 ROUNDOFF AND TRUNCATION ERRORS

A computing machine is finite, both in size and speed; and, as you have seen, mathematics regularly uses the infinite, both (1) in the representation of numbers and (2) in processes such as differentiation, integration, and infinite series. When we try to carry out our mathematical operations on a machine, the finite size of the machine gives rise to *roundoff*, and the finite processes used by the finite-speed machine give rise to *truncation errors*.

As an example of roundoff, consider the representation of the number one-third as a decimal in an eight-decimal-digit machine:

$$\frac{1}{3} = 0.33333333$$

There is roundoff due to the dropped digits past the last retained digit. Thus, most numbers that arise during a computation will be rounded off. The effect of roundoff appears most seriously when two numbers of about the same size are subtracted; then, because the leading zeros are shifted off (in scientific work), the rounded digits move over into the more significant digit positions of the number.

Example 25.1-1

Due to roundoff effects, you can sometimes get spectacular results. Suppose that ϵ is so small that ϵ^2 is less than 10^{-8}. Then it follows that

$$(1 + \epsilon)(1 - \epsilon) - 1 = -\epsilon^2 = 0$$

on an eight-decimal-place computer. Also, for a small number ϵ,

$$(1 + \epsilon) - (1 - \epsilon) = 2\epsilon$$

At best the leading digits of this result are the first digits of ϵ, and there is a significant loss of relative accuracy.

Example 25.1-2

Consider

$$f(x) = \sqrt{x + 1} - \sqrt{x}$$

for large x. There will be a great deal of cancellation, but if you remember what you did when using the delta process for the function $y(x) = \sqrt{x}$ you rearrange this in the form

$$f(x) = (\sqrt{x + 1} - \sqrt{x})\left(\frac{\sqrt{x + 1} + \sqrt{x}}{\sqrt{x + 1} + \sqrt{x}}\right) = \frac{1}{\sqrt{x + 1} + \sqrt{x}}$$

which is easily evaluated without serious loss of accuracy.

The methods of rearranging a computation before evaluation to avoid serious loss of accuracy resemble, many times, the methods we used to rearrange an expression before taking the limit in the four-step Δ process.

Another tool that is sometimes useful in avoiding roundoff is the mean value theorem,

$$f(x + a) - f(x) = f'(\theta)a$$

where θ is in the range x to $x + a$.

Example 25.1-3

Evaluate

$$f(x) = \ln (x + 1) - \ln x$$

for large x. By the mean value theorem, we have that their difference is, for some θ in the interval $0 < \theta < 1$,

$$x + \frac{1}{\theta}$$

A reasonable choice of θ might be $\theta = 1/2$. Hence we have, approximately,

$$f(x) = \ln (x + 1) - \ln x = \frac{1}{x + (1/2)}$$

Example 25.1-4

Evaluate, for large n (see Figure 25.1-1)

$$\int_n^{n+1} \frac{dt}{1 + t^2} = \arctan (n + 1) - \arctan n$$

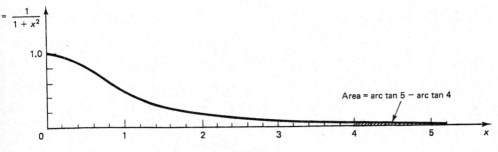

Figure 25.1-1

For large n, there will be a great deal of cancellation of the leading digits of the two numbers since both are near $\pi/2$. If we apply some trigonometry, we find that from the identity (16.1-16)

$$\tan (\theta - \phi) = \frac{\tan \theta - \tan \phi}{1 + \tan \theta \tan \phi}$$

Taking the arctan of both sides gives

$$\theta - \phi = \arctan \left(\frac{\tan \theta - \tan \phi}{1 + \tan \theta \tan \phi} \right)$$

We now write $\theta = \arctan a$ and $\phi = \arctan b$ to get the identity we need:

$$\arctan a - \arctan b = \arctan \frac{a - b}{1 + ab}$$

Hence we have

$$\int_n^{n+1} \frac{dt}{1 + t^2} = \arctan \left(\frac{1}{1 + n(n + 1)} \right)$$

which can be evaluated accurately, and furthermore involves only one arctan evaluation.

Rule. It is better to anticipate roundoff errors and rearrange the formulas than it is to try to force your way through stupid forms of writing and then evaluating the formulas. The claim that "double precision arithmetic is the answer to all roundoff problems on a computer" is only partly true. Often the *methods* in the calculus are necessary to handle the roundoff errors that arise.

A final tool from the calculus is the use of the power series representation of functions. The words "truncation error" come from the approximation of infinite series using only the leading terms, *truncating* the series. This is exactly what we did in the theory of infinite series. We studied the sequence of partial sums. In computing we are forced to settle for some finite sum, and we cannot take the limit. We will deal with this matter only slightly, as the serious theory is inappropriate for an elementary course.

Example 25.1-5

We use Example 25.1-3 again.

$$\ln(x + 1) - \ln x = \ln\frac{x + 1}{x}$$

$$= \ln\left[\frac{[x + (1/2) + (1/2)]}{[x + (1/2)] - (1/2)}\right]$$

Set $x + 1/2 = u$. We have the more symmetric form

$$\ln\left[\frac{u + (1/2)}{u - (1/2)}\right] = \ln\left[\frac{1 + (1/2u)}{1 - (1/2u)}\right]$$

But we have (set $1/2u = t$)

$$\ln\left[\frac{1 + t}{1 - t}\right] = \ln(1 + t) - \ln(1 - t)$$

and, using the power series expansions of the two logs, we get

$$\ln\left[\frac{1 + t}{1 - t}\right] = t - \frac{t^2}{2} + \frac{t^3}{3} - \frac{t^4}{4} + \cdots$$

$$+ \left(t + \frac{t^2}{2} + \frac{t^3}{3} + \frac{t^4}{4} + \cdots\right)$$

$$= 2t + \frac{2t^3}{3} + \frac{2t^5}{5} + \cdots$$

Using this expression, with $t = 1/2u$ and $u = x + 1/2$, you get

$$\ln(1 + x) - \ln x = \frac{2}{2u} + \frac{2(1/2u)^3}{3} + \cdots$$

$$= \frac{1}{x + (1/2)} + \frac{1/12}{[x + (1/2)]^3} + \cdots$$

and from the first term we have the same approximation as before (Exercise 25.1-3), but we now have some indication of the higher-degree correction terms in the approximation.

EXERCISES 25.1

1. Evaluate for large x, $1/\sqrt{x} - 1/\sqrt{x + 1}$.
2. Evaluate for small x, $(1 - \cos x)/\sin x$.
3. Evaluate for large x, $1/\sqrt{x - 1} - 1/\sqrt{x + 1}$.
4. Evaluate for large x, $(1 + x)^{1/3} - x^{1/3}$.
5. For x near 4, evaluate $(1/\sqrt{x} - 1/2)/(x - 4)$.
6. For x near zero, evaluate $\{(e^{2x} - 1)/(e^x - 1)\}^{1/2}$.

7. For small x, evaluate $(x - \sin x)/(x - \tan x)$.

8. Show that for small x, $x\{e^{(a+b)x} - 1\}/\{(e^{ax} - 1)(e^{bx} - 1)\}$ is approximately $(a + b)/ab$.

5.2 ANALYTIC SUBSTITUTION

Let us review how we handled past encounters with numerical methods. In Newton's method, Section 9.7, we replace the curve locally with the tangent line that passes through the point and has the same slope as the curve whose zero we want. We replace an intractable function with one that we can handle (a straight line). We then simply find the zero of the straight line and use that as the next approximation to the zero of the function.

When we faced numerical integration (Section 11.11), for both the trapezoid and midpoint rules, we analytically substituted a local straight line for the given curve, the secant line in the trapezoid rule, and the tangent line at the midpoint of the interval in the midpoint rule. For Simpson's rule (Section 11.12), we approximated the function over a double interval by a parabola and then integrated the parabola over the double interval as if it were the function (locally).

We immediately *generalize* and observe that we can make an *analytic substitution* of any function (that we can handle) for the given function (that we cannot handle), and then act as if it were the original function. The *quality* of the approximation is a matter that we will later look into briefly. Polynomials are widely used because they are generally easy to handle, but the Fourier functions, $\sin kx$ and $\cos kx$, are increasingly being used. As you will see, the details of the derivation depend on the particular set of approximating functions you pick, but *the form of the formula you derive is the same*. Consequently, the amount of computing you will do when using the formula will be about the same (the formula will *not* directly involve the particular functions you use in the approximation process). The major tool will be the familiar method of undetermined coefficients.

We have used the method of undetermined coefficients several times to find local polynomials passing exactly through the data, and we have at times used polynomials to get a least squares fit. Thus, in setting out to create a formula to use in numerical methods you must decide:

1. What information (sample points, function, and/or derivative values) to use.
2. What class of approximating functions to use (polynomials, sines and cosines, or exponentials).
3. What criterion to use to select the specific function from the class you adopted (exact match at the given points, least squares, etc.).
4. Where you want to apply the criterion.

This last point is difficult to explain at the moment, but it may become slightly clearer as we go on.

25.3 POLYNOMIAL APPROXIMATION

We begin with the polynomial approximation of a function when we are given some data (samples) about the function. The samples may be at any place and involve the function as well as various values of derivatives. Although you tend to think that the derivative is usually messier than is the original function, when you ask about the computing time you find that, once you have the special functions (sin, cos, $\sqrt{\ }$, ln and exp), which arise in the original function, you also have most of the special functions that arise in the derivative; and it is the special functions that occupy most of the machine time on a computer even though each seems to be but one instruction. When you differentiate a function, no new awkward pieces arise, but a sine can give rise to a cosine, and conversely; generally, once you have paid the cost of evaluating the parts of a function, the cost of a derivative is cheap.

Example 25.3-1

Suppose we want a polynomial that takes on the (general) values $y(0)$, $y(1)$, $y'(0)$, and $y'(1)$. With four conditions, we need four undetermined coefficients, a cubic:

$$y(x) = ax^3 + bx^2 + cx + d \qquad (25.3\text{-}1)$$

The four conditions give rise to the four equations

$$y(0) = d$$
$$y(1) = a + b + c + d$$
$$y'(0) = c$$
$$y'(1) = 3a + 2b + c$$

Eliminating $c = y'(0)$ and $d = y(0)$, we get

$$a + b = y(1) - y(0) - y'(0)$$
$$3a + 2b = y'(1) - y'(0)$$

These two equations are easily solved for a and b:

$$a = y'(1) + y'(0) - 2\{y(1) - y(0)\}$$
$$b = 3\{y(1) - y(0)\} - 2y'(0) - y'(1)$$

and we have the cubic through the given data.

A specific, useful case occurs when $y(0) = y'(0) = y'(1) = 0$ and $y(1) = 1$ (see Figure 25.3-1). The coefficients are $a = -2$, $b = 3$, $c = 0$, and $d = 0$. The polynomial through the data is

$$y(x) = -2x^3 + 3x^2 = x^2(3 - 2x)$$

It is easy to verify that this equation satisfies the given data.

Example 25.3-2

Suppose we use the general cubic polynomial (25.3-1) to analytically approximate a curve in the interval, and from this get an approximation to the integral (another integration formula)

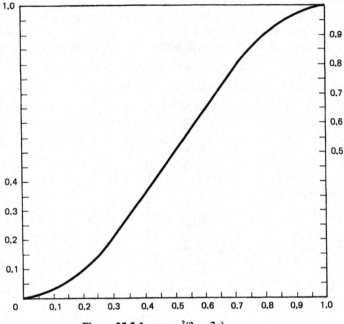

Figure 25.3-1 $y = x^2(3 - 2x)$

$$J = \int_0^1 y(x)\, dx$$

We integrate the approximating cubic of Example 25.3-1 to get

$$J = \int_0^1 \{ax^3 + bx^2 + cx + d\}\, dx$$

$$= \frac{a}{4} + \frac{b}{3} + \frac{c}{2} + d$$

Direct substitution of the values of the coefficients produces the formula

$$J = \frac{1}{2}\{y(0) + y(1)\} + \frac{1}{12}\{y'(0) - y'(1)\}$$

This formula makes a very interesting *composite* integration formula for functions with continuous first derivatives, since it leads to

$$\text{area} = A = \text{trapezoid rule} + \frac{1}{12}\{y'(0) - y'(n)\}$$

The two end derivatives are the only correction terms needed to gain the local accuracy of a cubic.

Example 25.3-3

Suppose all you know about the function is $y(0)$, $y'(0)$, $y''(0)$, and $y(1)$, and you want the integral over the interval 0 to 1.

First you find the interpolating polynomial using undetermined coefficients:

$$y(x) = Ax^3 + Bx^2 + Cx + D$$

To get the formula, you impose the four conditions:

$$y(0) = D$$

$$y'(0) = C$$

$$y''(0) = 2B$$

$$y(1) = A + B + C + D$$

from which you get

$$A = y(1) - y(0) - y'(0) - \frac{y''(0)}{2}$$

and you have the interpolating polynomial.

To get the approximating integral, you integrate this polynomial:

$$J = \frac{A}{4} + \frac{B}{3} + \frac{C}{2} + D$$

$$= \frac{1}{4}\left\{ y(1) - y(0) - y'(0) - \frac{y''(0)}{2} \right\} + \frac{1}{3}\frac{y''(0)}{2} + \frac{1}{2}y'(0) + y(0)$$

$$= \frac{1}{24}[18y(0) + 6y(1) + 6y'(0) + y''(0)]$$

This is the approximation formula for the integral in terms of the given data.

EXERCISES 25.3

1. Find the quadratic through $(0, 1)$, $(1, 2)$, $(2, 3)$.
2. Find the quadratic through $(-1, a)$, $(0, b)$, $(1, c)$.
3. Find the cubic through $(0, 1)$, $(1, 0)$, $(2, 1)$, and $(3, 0)$.
4. Using $f(1/3)$ and $f(2/3)$, find an integration formula of the form $\int_0^1 f(x)\, dx = af(1/3) + bf(2/3)$.

25.4 THE DIRECT METHOD

In all the preceding examples, the final formula was linear in the given data. If you think about the process, you see that this is inevitable; it is impossible that products or powers of the data will arise using the methods we have used (think about dimensional analysis). You know, therefore, that the *form* of the final answer is a linear expression in the given data. This suggests that by using the method of undetermined coefficients you can go *directly* to the final answer.

How can this be done? Since the interpolation formula is true for any data of the given form, the integration formula is also true for that data. In particular, the integration formula must be true for the particular functions $y(x) = 1, x, x^2, \ldots,$ x^{m-1}, where $m - 1$ is the highest power used in the interpolation formula. Conversely, since integration is a linear process, if the integration formula is true for the individual powers of x, then it must be true for any linear combination of the powers, for any polynomial of degree $m - 1$.

The plan is to set up the form with undetermined coefficients and then impose the conditions that the formula be exact for as many consecutive powers of x as there are coefficients, beginning with $y(x) = 1$, and going on to $y(x) = x$, $y(x) = x^2, \ldots,$ as far as you can go.

Example 25.4-1

Derive by the direct method the integral in Example 25.3-2.

Given $y(0)$, $y(1)$, $y'(0)$, and $y'(1)$, the form to choose is

$$J = \int_0^1 y(x)\, dx = Ay(0) + By(1) + Cy'(0) + Dy'(1)$$

Next write out the equations for the conditions that the equation be exactly true for each power of x as indicated on the left. These are the *defining equations*:

$$
\begin{array}{lll}
1: & 1 = A + B \\[2mm]
x: & \dfrac{1}{2} = & B + C + D \\[2mm]
x^2: & \dfrac{1}{3} = & B \quad + 2D \\[2mm]
x^3: & \dfrac{1}{4} = & B \quad + 3D
\end{array}
$$

In the first line $y = 1$, so $J = 1$ and $y(0) = 1$, $y(1) = 1$, $y'(0) = 0$, and $y'(1) = 0$. Hence the A and B terms are present and the C and D terms are not. In the second line the integral of $y(x) = x$ over the interval 0 to 1 is $1/2$; $y(0) = 0$ for $y = x$, $y(1) = 1$, $y'(0) = 1$, and $y'(1) = 1$. Hence the B, C, and D are present, but the A is missing. Similarly, for the third line you use $y(x) = x^2$ and for the fourth line you use $y(x) = x^3$.

To solve these four equations, simply subtract the third from the fourth to get

$$D = \frac{1}{4} - \frac{1}{3} = \frac{-1}{12}$$

Use that D in either of the bottom two equations, say the last one, and you have

$$B = \frac{1}{4} - 3\left(\frac{-1}{12}\right) = \frac{1}{4} + \frac{1}{4} = \frac{1}{2}$$

From the top equation, you get immediately

$$A = 1 - B = 1 - \frac{1}{2} = \frac{1}{2}$$

Finally, from the second equation (the only one involving C)

$$C = \frac{1}{2} - B - D = \frac{1}{2} - \frac{1}{2} - \left(\frac{-1}{12}\right) = \frac{1}{12}$$

and we have the same formula

$$J = \frac{1}{2}\{y(0) + y(1)\} + \frac{1}{12}\{y'(0) - y'(1)\}$$

as before, but with a great deal less labor!

Example 25.4-2

Given data for $y(x)$ at $x = -1$, 0, and 1, find an approximation for the integral

$$J = \int_{-1}^{1} \frac{y(x)}{\sqrt{1 - x^2}}\, dx$$

The form we use is

$$J = ay(-1) + by(0) + cy(1)$$

and we impose the three conditions that the formula be true for $y = 1$, for $y = x$ and for $y = x^2$. We need the integrals

$$J_n = \int_{-1}^{1} \frac{x^n}{\sqrt{1 - x^2}}\, dx \qquad (n = 0, 1, 2)$$

For $n = 0$, it is an arcsine, and the value is π. For $n = 1$, the integrand is odd, so the integral is 0. For $n = 2$, the integral, when you use the substitution $x = \sin\theta$ to get rid of the radical, becomes a Wallis integral with the value $\pi/2$ (remember both limits). The defining equations are, therefore,

$$1: \qquad \pi = \quad a + b + c$$

$$x: \qquad 0 = -a \qquad + c$$

$$x^2: \qquad \frac{\pi}{2} = \quad a \qquad + c$$

From the second equation, $a = c$, from the bottom equation, $a = c = \pi/4$, and from the top, $b = \pi/2$. Hence the formula is

$$J = \int_{-1}^{1} \frac{y(x)}{\sqrt{1 - x^2}}\, dx = \frac{\pi}{4}[y(-1) + 2y(0) + y(1)]$$

This differs from Simpson's formula by both (1) the front coefficient (now $\pi/4$ instead of 1/3), and (2) the middle coefficient is twice the end coefficients rather than four times as large. It shows the effect on the end coefficients due to the factor

$$\frac{1}{\sqrt{1 - x^2}}$$

in the integrand. Notice that we did not approximate the whole integrand by a polynomial, only the unknown function $y(x)$.

Generalization 25.4-3

We see that given almost any data and any linear operation (integration, interpolation, and differentiation are examples of such operations), we can set up the defining equations that make the formula exactly true for $f(x) = 1, x, x^2, \ldots$, as far as there are arbitrary coefficients to be determined. We need to compute the *moments* of the linear operator (typically, we have used integration with or without a weight factor and have at times used derivatives as well as function values). If the moments cannot be found analytically (in closed form), then you can find them numerically, once and for all, using some suitable formula such as Simpson's, at a suitably small spacing. Once found, the moments are then the left-hand sides of the defining equations. The right-hand sides depend only on the given data locations; one side depends only on the form assumed and the other only on the formula being approximated. The solution of the linear equations determines the formula.

 We see a further generalization: we did not have to use the powers of x as the criterion for exactness; we could use *any* set of linearly independent functions. In particular, we might use the first few terms of a Fourier expansion if that seemed suitable. The arithmetic to find the formula might be more difficult, but the arithmetic in the use of the formula would be the same; only the numerical values of the unknown coefficients would be different.

 You are not restricted to integration formulas; the derivation of any linear formulas would go just about the same: find the "moments" of the operation on the left-hand side by some method, write down the terms on the right, solve the system of equations, and you have the formula!

This method will work even when there is no interpolating function through the data! The formula will still be exactly true for the functions that you imposed. The method is very general, and the generality is necessary if we are to cope with the vast number of possible formulas that can exist. You are now in a position to derive any formula you think fits the situation you face, rather than select a formula that happens to have been used in the past.

Example 25.4-4

Using polynomial approximation, find an estimate of the derivative at the origin of a function given the data $y(-1), y(0), y(1), y'(-1),$ and $y'(1)$. We set up the formula with undetermined coefficients:

$$y'(0) = ay(-1) + by(0) + cy(1) + dy'(-1) + ey'(1)$$

and impose the conditions that the formula be exact if the function $y(x) = 1, x, x^2, x^3,$ x^4. The defining equations (whose solution gives the formula) are, therefore,

$$1: \quad 0 = \quad a + b + c$$

$$x: \quad 1 = -a \quad + c \quad + d + e$$

$$x^2: \quad 0 = \quad a \quad + c - 2d + 2e$$

$$x^3: \quad 0 = -a \quad + c \quad + 3d + 3e$$

$$x^4: \quad 0 = \quad a \quad + c \quad - 4d + 4e$$

Inspecting these equations suggests (after a little thought) that we should exploit the obvious symmetry in the problem. If we set $a = -c$ and $d = e$, then we have imposed two conditions but have eliminated the third and fifth equations. The top equation then gives $b = 0$. We have left to solve the second and fourth equations:

$$1 = -2a + 2d$$

$$0 = -2a + 6d$$

The solution of these two equations (it was clearly worth the thought to find the symmetry!) is $d = -1/4$ and $a = -3/4$. The formula is, therefore,

$$y'(0) = \frac{3}{4}[y(1) - y(-1)] - \frac{1}{4}[y'(1) + y'(-1)]$$

It is easy to verify that this formula is correct by checking that it does give the right values for the first five powers of x, beginning with the zeroth.

EXERCISES 25.4

1. Find Simpson's half-formula, $\int_{-1}^{0} f(x)\, dx = af(-1) + bf(0) + cf(1)$.
2. Find the formula for $-\int_{0}^{1} f(x) \ln x\, dx = af(0) + bf(1/2) + cf(1)$.
3. Find $\int_{0}^{\infty} e^{-x}\, dx = af(0) + bf(2)$.
4. Explain why the formula in Example 25.4-4 has no $y(0)$ term.
5. Discuss finding the solution of the simultaneous equations independent of the operation you are approximating. Apply this to the given data $y(-1)$, $y(0)$, and $y(1)$. Check your answer by applying it to find Simpson's formula.

25.5 LEAST SQUARES

We have been finding the approximating curve to use in analytic substitution by making the curve *exactly* fit the given data. If you have a large amount of data and want to use an approximating curve that has comparatively few parameters, you are naturally attracted to the method of least squares. If the given data are "noisy," then it is plainly foolish to have the approximating curve go exactly (within roundoff) through the data. The resulting curve will generally have many wiggles, and if you were trying to find the derivative of the data, the resulting derivative would hardly look reasonable. Integration, being a "smoothing" operation, is less vulnerable than is differentiation to wiggles in the approximating curve (but still violent wiggles are something to avoid).

Given any function values, the fitting of least squares polynomials has been discussed in Sections 10.5 and 10.8, as well as noting in Section 21.5 that the partial sums of the Fourier series approximation give the corresponding least squares fit. When you find the least squares fit and operate on it as if that were the original curve, you will get the answer.

Example 25.5-1

From the data $(-2, 0)$, $(-1, 0)$, $(0, 1)$, $(1, 1)$, and $(2, 2)$, find the area under the curve from -2 to 2 using a least squares approximating straight line.

To find the line, we first make the table:

x	y	x^2	xy	
-2	0	4	0	
-1	0	1	0	
0	1	0	0	
1	1	1	1	
2	2	4	4	
0	4	10	5	Sums

The normal equations for the straight line

$$y = a + bx$$

are

$$a \sum 1 + b \sum x = \sum y$$

$$a \sum x + b \sum x^2 = \sum xy$$

which are

$$5a + 0b = 4$$

$$0a + 10b = 5$$

From this we get $a = 4/5$, $b = 1/2$. The line looks about right, (see Figure 25.5-1), so we go ahead. Next, the integral

$$\int_{-2}^{2} y(x)\, dx = \int_{-2}^{2} (a + bx)\, dx = 4a = 16/5 = 3.2$$

and we did not need the coefficient b, nor part of the table we computed. It pays to look ahead before you plunge into the routine details.

Example 25.5-2

Fit the given data $(1, 5)$, $(2, 3)$, $(3, 2)$, $(4, 1)$, and $(5, 1)$ by the curve

$$y(x) = ax + b + \frac{c}{x}$$

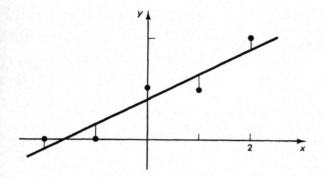

Figure 25.5-1 Least squares line

and then estimate the derivative at $x = 3$, the midpoint.

This time (having learned from the previous example) we set up the normal equations first. We need to recall from Section 16.8 that we first write the least squares condition

$$m(a, b, c) = \sum_1^5 \left\{ y - \left(ax + b + \frac{c}{x} \right) \right\}^2$$

to differentiate with respect to each parameter, and rearrange. We get for the normal equations

$$a \sum x^2 + b \sum x + c \sum 1 = \sum xy$$

$$a \sum x + b \sum 1 + c \sum \frac{1}{x} = \sum y$$

$$a \sum 1 + b \sum \frac{1}{x} + c \sum \frac{1}{x^2} = \sum \frac{y}{x}$$

We now make the appropriate table:

x	y	x^2	xy	$\dfrac{1}{x}$	$\dfrac{1}{x^2}$	$\dfrac{y}{x}$
1	5	1	5	1.0000	1.0000	5.0000
2	3	4	6	0.5000	0.2500	1.5000
3	2	9	6	0.3333	0.1111	0.6667
4	1	16	4	0.2500	0.0625	0.2500
5	1	25	5	0.2000	0.0400	0.2000
15	12	55	26	2.2833	1.4636	7.6167

The normal equations are, therefore,

$$55a + 15b + 5c = 26$$

$$15a + 5b + 2.2833c = 12$$

$$5a + 2.2833b + 1.4636c = 7.6167$$

Before solving these, let us ask what coefficients are needed? We have

$$y(x) = ax + b + \frac{c}{x}$$

Hence

$$y'(3) = a - \frac{c}{9}$$

We therefore eliminate b in the first stage of the solution. We get from the first two equations, and the second two equations,

$$10a - 1.8500c = -10$$

$$1.8499a + 4.2092c = 2.1368$$

From these $c = 0.4096$ and $a = -0.9242$. The derivative is, therefore,

$$y'(3) = -0.9697$$

It is worth finding $b = 4.9855$ to plot the data and curve to check the reasonableness of the derivative estimate (see Figure 25.5-2).

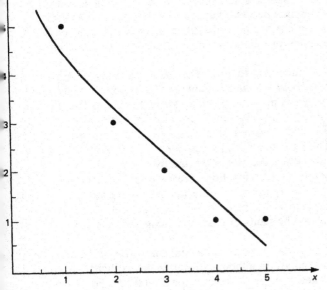

Figure 25.5-2   Least squares fit

EXERCISES 25.5

1. Generalize Example 25.5-1 to $2N + 1$ equally spaced points.
2. Find the integral under the curve in Example 25.5-2 $[1, 5]$.
3. Given the data $(0, 1)$, $(1, 3)$, $(2, 2)$, $(3, 4)$, and $(4, 4)$, find the integral under the least squares straight line from 0 to 4.
4. Using the data of Exercise 3, fit a quadratic and estimate the derivative at the origin.

25.6 ON FINDING FORMULAS

Which formula to use? That is a typical question in numerical methods *when* you stop and think about what you are doing. The direct method allows you to derive many different formulas to do the same job; for example, a low-order polynomial over a short interval or a higher-order polynomial over a longer interval (trapezoid over one interval and Simpson over a double interval are examples). The four questions raised in Section 25.2 must all be answered *either* consciously by choice or implicitly by picking a formula without thinking. We are better prepared to discuss the four choices now that a few numerical methods have been given.

1. What information to use? Generally, the closer the information is to where you are getting the answer, the better. It is almost true that a value of a derivative at a point is as good as another function value at some other place. As we noted, derivatives are apt to be much cheaper to compute than you at first think because of the large amount of machine time necessary to evaluate the special functions (including square root). As you can see from the direct method, each piece of information effectively determines one unknown parameter in the assumed form (Section 25.4).

2. What class of approximating functions to use? The great tendency of most people is to use polynomials *without any careful thinking*. But the direct method shows that the resulting formula is no more complex for other sets of linearly independent functions than it is for the polynomial formula. The derivation may be more difficult, but that is hardly a matter of great importance *if* the formula is to be used extensively. The right functions to use, generally, come from the nature of the problem and cannot be determined abstractly.

3. What criterion to use to select the specific function from the class? We have seen mainly exact matching and a little of least squares. There are other criteria such as minimax—the maximum error in the interval shall be minimum. This is an increasingly popular criterion as the faults of exact matching and least squares become better understood.

4. Where to apply the criterion? As you go on in numerical analysis, you will find that this question arises again and again, although it is often not explicitly stated but rather quietly assumed when you use the conventional methods. For example, when finding the zeros of a polynomial, is it the values of the zeros that you

want to be accurate, the vanishing of the polynomial at the zeros to be accurate, or do you want to be able to reconstruct the polynomial from the computed zeros? Due to the inaccuracies of computing, these three things are not the same! Again, when finding a minimum, a broad curve leaves the position of the minimum ill determined, and a sharp minimum determines the location of the minimum quite well (see Figure 25.6-1). But when judged by the performance of the system, the broad minimum means that the performance is not degraded much for small changes, but the sharp minimum means that the system is sensitive to small changes. Which one do you want? Do not confuse, in numerical work with its inevitable small errors, accuracy in the solution values with accuracy in the property wanted.

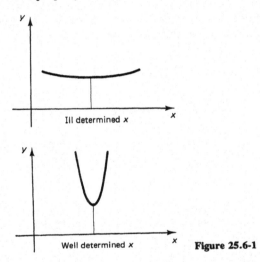

Ill determined x

Well determined x   **Figure 25.6-1**

The author's opinion is that with the direct method available we have reached the point in numerical methods where you should fit the formula to the problem, not the problem to the formula. In practice, the gathering of the data is usually so expensive in comparison to the data processing that care should be taken to get as much as you can from the given data.

EXERCISES 25.6

1. Analyze the solution of the quadratic $x^2 + 80x + 1 = 0$ for accuracy in the three senses given above, item 4.
2. Derive the equivalent of Simpson's formula making it exact for 1, $\cos \pi x/2$, and $\sin \pi x/2$ for the double interval $0 \leq x \leq 1$.
3. Derive the polynomial extension of Simpson's formula when you use both the function and derivative values at each point. Discuss the corresponding composite formula.

25.7 INTEGRATION OF ORDINARY DIFFERENTIAL EQUATIONS

Once we can handle numerical integration, it is natural to turn to the numerical solution of ordinary differential equations. Frequently, the equations that arise cannot be solved in closed form, and numerical methods are necessary since the importance of the original problem is such that we must take some action. The numerical solution of ordinary differential equations is a vast subject, and we can give only the simpler methods and ideas involved.

The simplest method for integrating differential equations comes directly from the direction field (Section 23.2). You recall that after marking the slopes of the direction field you draw a smooth curve through them as best you can. This process can be mechanized easily. We will take a definite differential equation with a known solution (for check purposes) to illustrate the method before giving the general case.

Example 25.7-1

Given the differential equation

$$y' = 1 + y^2 \quad \text{and} \quad y(0) = 0$$

find the solution. The slope at the point $(0, 0)$ is, from the differential equation, 1. Suppose we go along this direction for a short distance, say $\Delta x = 0.2$. We will then be at the point

$$(0.2, 0.2)$$

(see Figure 25.7-1). At this point the differential equation says that the slope is 1.04. We go along this direction for another step of $\Delta x = 0.2$ and get the point

$$(0.4, 0.408)$$

We repeat this process again and again. It is time to make a table of the numbers so we will not get lost, and so we can compare our solution with the known solution $y = \tan x$.

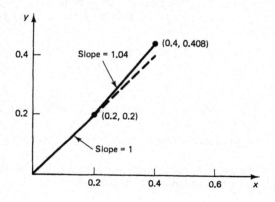

Figure 25.7-1 Numerical solution

TABLE 25.7·1

x	y	y'	$y' \Delta x$	$\tan x$
0	0.000	1.00	0.200	0.000
0.2	0.200	1.04	0.208	0.203
0.4	0.408	1.17	0.234	0.423
0.6	0.624	1.41	0.282	0.684
0.8	0.724	1.52	0.304	1.030
1.0	1.028			1.557

The answer is surprisingly good considering the step size. Obviously, decreasing the step size by a factor of 2 would decrease the error per step by about a factor of 4; but there would be twice as many steps, so the result would be about twice as accurate. Thus trying to get crude answers by this method is reasonable, but it is not a method for getting accurate answers without using a lot of computer time and incurring a lot of roundoff error in the process.

The fault of the method is immediately apparent: you are always using the slope that was and not the slope that is going to be. The crude formula we used cannot do what you did by eye in anticipating the new slope at the end of the interval as you drew the solution curve for the direction field solution. We need a better method that uses this simple observation.

Euler's method does exactly this. It is a method that looks ahead, and then from the sample of the slope at the final end of the interval makes a new guess. In particular, with this new information Euler's method tries again and does a better job at moving forward the step Δx by averaging the two slopes at the ends of the interval.

Let the equation and initial conditions be

$$y' = f(x, y) \quad \text{and } y_0 = y(x_0)$$

where we are starting at the point (x_0, y_0). We begin by assuming that the method is already going, and later we will discuss how to get started. We assume that you have the two most recent points on the curve (Figure 25.7-2). We will number the points

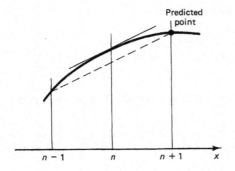

Figure 25.7-2 Euler's predictor

by subscripts, which no longer are the arguments of the function. The formula

$$p_{n+1} = (y_{n+1}) = y_{n-1} + 2y_n'\Delta x$$

reaches from the previous point y_{n-1}, uses the current slope y_n', and *predicts* (hence the letter p) the next point (see Figure 25.7-2). At this point we make an estimate of the slope (from the derivative given by the differential equation):

$$p_{n+1}' = f(x_{n+1}, p_{n+1})$$

Now using this slope together with the slope at x_n, we can get an *average* slope for the interval $[x_n, x_{n+1}]$, and hence the formula

$$c_{n+1} = y_n + \frac{p_{n+1}' + y_n'}{2}\Delta x$$

gives a *corrected* value for the new point $n + 1$ (a more reliable point, probably, than the predicted value).

It remains to say how we can get started. One way is to realize that, when you differentiate the given differential equation, you get a formula for the second derivative at the initial point. Differentiation again gives the third derivative, and so on, so you can build up a Taylor series about the initial point and thus reach to the first point, which you need to get started.

Example 25.7-2

For the particular equation we are using

$$y' = 1 + y^2$$

we have the successive derivatives

$$y'' = 2yy'$$

$$y''' = 2yy'' + 2(y')^2$$

and at the point (0, 0) we have

$$y' = 1, \quad y'' = 0, \quad y''' = 2$$

Hence for the expansion about the origin we have

$$y(x) = 0 + 1x + \frac{0x^2}{2} + \frac{2x^3}{3!} + \cdots$$

So at the point $x = 0.2$, we have, approximately,

$$y(0.2) = 0.2 + \frac{0.008}{3} + \cdots = 0.203 \cdots$$

as the first point from which Euler's method takes off. We now make the beginning of the Table 25.7-2.

However, we note that the predicted value has an error of approximately $\Delta x^3 y'''(\theta)/3$ and the corrector has an error of about $-(\Delta x)^3 y'''(\theta)/12$, so the difference is approximately

TABLE 25.7-2

x	y	y'	p	p'	c	$\dfrac{p-c}{5}$
0	1	1				
0.2	0.203	1.0412				
0.4	0.422	1.178	0.41648	1.173	0.424	−0.002
0.6	0.683	1.466	0.674	1.455	0.685	−0.002
0.8	1.027	2.047	1.008	2.016	1.031	−0.004
1.0	1.546		1.502	3.256	1.557	−0.011

$$p - c = \frac{5(\Delta x)^3 y'''(\theta)}{12}$$

of which $\frac{1}{5}$ is due to the corrector (never mind that the θ values are not the same in the two formulas; we are only estimating an error). Thus we add this amount to the corrected value to get the final value

$$y_n = c_n + \frac{p_n - c_n}{5}$$

Correct value $= 1.557$; error $= 0.011$

This answer is significantly better than the answer from the crude method. This "mopping up" of the residual error in the corrector by adding $(p - c)/5$ is usually worth the effort as it is easy to do, and the computation of $p - c$ gives you some control over the *accuracy per step* of the integration. When $p - c$ gets too large, you need to halve the interval, using the formula

$$y_{n-(1/2)} = \tfrac{1}{2}(y_n + y_{n-1}) - \tfrac{1}{8}(y'_n - y'_{n-1})$$

and, of course, using the differential equation to get the corresponding derivative. Alternatively, this can be done by simply restarting at the point where too much accuracy was being lost. If $p - c$ gets too small, you can double the interval size and save computing effort. Thus, by changing step size as appropriate, you put the effort where it is needed and not where it is not.

In the days of computing by hand, the numerical solution of a differential equation was not undertaken lightly; but with hand calculators, especially programmable hand calculators, let alone small personal computers, the numerical solution of an ordinary differential equation is merely a matter of selecting a prewritten program, and programming the subroutines that describe the particular functions involved in the differential equation, along with the starting values and the desired accuracy.

We next turn briefly to the problem of integrating a system of first-order ordinary differential equations (Section 24.7). The method is exactly the same as for a single equation *except* that each equation has its values predicted in parallel, each derivative is then evaluated, each corrected value is computed (in parallel), and the final results for each variable are obtained before going on to the next interval. It is merely a great deal more work, since usually the right-hand-side functions for the derivatives involve a great deal more computing.

EXERCISES 25.7

1. Integrate $y' = y^2 + x^2$ from 0 to 1 using spacing $h = 0.2$, starting with $y(0) = 1$.
2. Integrate $y' = -y$ from 0 to 2 using $h = 0.2$ starting with $y(0) = 1$.
3. Integrate $y' = -xy$ starting with $y(0) = 1$, using spacing $h = 0.1$ and going to $x = 2$.
4. Derive the midpoint formula used in this section.
5. Integrate the system $y' = z$, $z' = -y$, $y(0) = 0$, $y'(0) = 1$, using $h = 0.1$ out to $x = 3.2$.

25.8 FOURIER SERIES AND POWER SERIES

A power series is a local approximation at a point; the partial sums fit the curve to higher- and higher-order derivatives. As a result, the quality of the approximation of the partial sum to the function is very good near the point about which the expansion is made, but gets poorer and poorer as you go away from the point. It is a local fit. On the other hand, the Fourier series is a least squares fit in the whole interval. No one point is made to be exceptionally good; the fit is designed to be uniformly good in the whole interval in the least squares sense. The fit is global (for the interval). Thus the two types of expansions do rather different things.

There is an interesting relationship between the two types of expansions, however. Suppose we have a power series partial sum

$$f_n(x) = a_0 + a_1 x + a_2 x^2 + \cdots + a_{n-1} x^{n-1} + a_n x^n$$

First we note that the way anyone who thinks about the matter evaluates a polynomial, as the above truncated power series is, by the *chain* method:

$$f_n(x) = a_0 + [\ldots x\{a_{n-2} + x(a_{n-1} + x a_n)\} \ldots]$$

starting at the innermost multiplication and working outward.

Suppose, next, we make the change of variable

$$x = \cos t$$

The expansion is now

$$f_n(\cos t) = a_0 + a_1 \cos t + \cdots + a_{n-1} \cos^{n-1} t + a_n \cos^n t$$

This is a series in the *powers* of cos t; and for a Fourier series we want an expansion in *multiples* of the angle. There is a simple way of converting from the power series in cos t to the form of a Fourier series by using the identity

$$\cos t \cos kt = \frac{1}{2} \cos (k + 1)t + \frac{1}{2} \cos (k - 1)t$$

In words, the cosine of a given frequency when multiplied by cos t becomes one-half of the amount at the next higher frequency, plus one-half at the next lower frequency.

We use an inductive proof. The starting inner terms of the chain form are

$$a_{n-1} + a_n \cos t$$

This is a Fourier series! Multiply this by cos t and apply the above trigonometric identity (see Figure 25.8-1). The a_{n-1} term becomes a cos term, but the coefficient of the cos term is divided by 2 and sent to both the one higher and the one lower frequency. We also add in the term a_{n-2}. We again have a Fourier series:

$$\left(a_{n-2} + \frac{1}{2}a_n\right) + a_{n-1} \cos t + \frac{1}{2}a_n \cos 2t$$

It is easy to see the general inductive step. Thus after the $n - 1$ stages in Figure 25.8-1, you have the Fourier series coefficients of the function $f_n(\cos t)$.

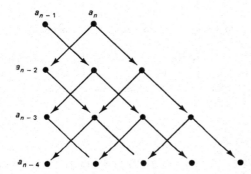

Figure 25.8-1 Conversion to Fourier series

There are several things to note. First, the process is clearly reversible. Second, in going from the power series form to the Fourier series form, the highest coefficient gets divided by 2^{n-1}, and, in general, whatever the rate of decrease of the a_k coefficients of the power series may be (as k increases), the coefficients of the Fourier series form decrease much more rapidly. But note that the sum of the coefficients is the same for both expansions; there is no loss in the sum of the coefficients.

If we now dropped the highest-frequency term of the Fourier series form, the change in the function would be not more than the size of the coefficient, since $|\cos t| \leq 1$ for all t. Indeed, we can drop as many terms from the Fourier series as we like as long as the sum of the absolute values of the coefficients does not exceed the error we are willing to commit.

This is the basis for a method called *economization*. We want to end up with an approximate least squares fit to a function whose power series we know. The plan is to take (1) more than enough terms of the power series, (2) convert to the Fourier series form, (3) drop the high-order terms, committing an additional error not more than the sum of the absolute values of the dropped terms, (4) convert back to the polynomial form, and the result is that we have a fit that is much like the Fourier series least squares approximation. The final transformation to the variable x merely stretches the x-axis and does not affect the quality of the fit very much (of course it affects the weights used in the fitting, but this is not large).

If we think in terms of the equal ripple property of the cosines, then what is going on is a bit clearer. We get the Fourier series form of the function and, since it is generally very rapidly converging, the error after dropping several terms tends to look like the last term dropped. The error committed is approximately equal ripple. Thus this equal ripple property is preserved when you go back from the t variable to the x variable. Economization tends to produce, from the power series approximation, an equal-ripple approximation good in the whole interval. The degree of the equal-ripple approximation can be significantly lower than the degree of the power series polynomial approximation of the same accuracy.

EXERCISES 25.8

1. Beginning with five terms of the exponential in the interval $(0, 1)$, economize the series to three terms. Compare with three terms of the power series.
2. Find a series in x for the sine function in the interval $(0, \pi/2)$ using a cubic final approximation.
3. Describe two ways of getting back from the Fourier series to the original power series.

25.9 SUMMARY

We have taken a very brief look at the vast and continually growing field of numerical analysis. Its main purpose is to find numerical solutions to problems that we cannot solve analytically, although at times we do evaluate functions that have been found in a suitable formula but that are too messy for us to grasp their meaning.

One of the main ideas of numerical analysis is analytic substitution; for the function we cannot handle, we substitute one we can and use some method of approximation so that the result we obtain is closely related to the answer we want. Polynomials are the classical functions to use, along with exact matching of the data, but more modern methods also use the Fourier series and other functions, as a basis; and we use exact matching of the data as well as least squares and other criteria for selecting the particular function from the class.

We have given you a powerful method for finding any of a very large class of formulas, polynomial or otherwise, and its great possibilities in adapting the formula to fit the problem should be understood.

26

Epilogue

In the prologue (Chapter 1) we discussed the explosive growth of knowledge, especially in mathematics, and the rapidity of change. This leads to the hopelessness of trying to equip the student with all the specific results that may be needed. Since mathematics is the language of science, it is equally true that the explosive growth of the applications of specific results of mathematics to science presents serious problems. Instead of presenting mainly results, we chose to emphasize the *methods* of mathematics. The examples used were chosen either because they illustrated the point or else were both illustrative and useful (probably!). Methods that apply in science generally were specifically mentioned when appropriate.

We adopted a deliberate policy of including as many examples as we could of the general methods used in mathematics and also commonly occurring in applications. Many points of importance in the examples were not mentioned explicitly in the text. For example, we regularly estimated the area or volume of an integral by independent means. The hidden purpose was to inculcate in your mind the idea that mathematics is not just the manipulation of abstract symbols but is also often connected with measurable things in the real world. Sometimes we used Monte Carlo methods of estimating the integrals to further emphasize the usefulness of random processes; randomness is something to be used constructively rather than always avoided! The teaching of science is slowly coming to this realization.

We also tried to teach you to *read* mathematical equations by frequently putting the results in words. The purpose was to give you practice in "listening to what equations have to say." Hence many of the results were stated in words as well as in equations. We also frequently stated things in both algebraic and geometric language to give alternative views of the same thing. This, too, is a useful habit to acquire.

Again, dimensional analysis was not explicitly discussed, but was used implicitly many times. For example, in the discussion of the mean value theorem, in the discussion of van der Waals' equation, and in many other places we used the methods of dimensional analysis.

26.2 METHODS OF MATHEMATICS

As earlier stated, the few widely agreed upon methods of mathematics are:

Extension

Generalization

Abstraction

These are all somewhat the same thing. Their importance is that *only* by such an approach can you hope to master the large mass of specific details that arise in the use of mathematics (as well as in mathematics itself!). There have been many examples and exercises with the label "extend," "generalize," or "abstract." As we advanced in the book, we deliberately made things more abstract to develop your abilities in this direction. Making everything as easy as possible would have defeated this goal of learning to do mathematics.

But we also, by example and by specific statement, gave many rules such as: instead of proving $a = b$ it is often better to prove $a - b = 0$. The way you go about doing mathematics is important. If you expect to cover a lot of ground, an efficient approach is essential! Another word for the same thing is "style." We have thought about style throughout the book, a style that will be useful to you not only in mathematics but in science and engineering generally. It is hoped that you have both consciously and unconsciously absorbed a good style. It is one thing to take over another's style, but is more important for you to develop your own.

We also gave examples of how to find and prove theorems. Theorems do not arise in a vacuum and then require a proof. Experience shows that often the hypotheses of the theorem come from the proof you finally find, hence the name *proof-driven theorems* [L]. It is only after you have found some kind of a proof that you know what you have to assume for the hypotheses of the theorem. Of course, you then generalize and extend the theorem you have just found to try to find the limits that it can be extended to, and which are the conditions you first assumed but you now see can be removed. You rarely find the final theorem the first time. It is important to realize this; otherwise, you are apt to become discouraged when you look at the final polished results that are published in books and journals. The published material is the result of hours and hours of rethinking, finding alternative proofs, and final polishing. Unfortunately, many mathematicians have the subconscious standard of beauty that correlates mathematical elegance with surprise. They arrange the final presentation in as surprising a form as they can, and you have little chance of understanding how they originally found the results. They also seem to prefer to give you the results rather than the methods.

Another powerful tool for doing mathematics is reasoning by analogy and similarity, and we have exploited it repeatedly. For example, area and probability are closely related. Symmetry is another tool of mathematics that we have repeatedly illustrated.

You have seen that *in the calculus* the infinite is handled by dropping back to the finite case and ultimately taking the limit. You saw this potential infinity in the following:

1. Numbers: the limit of the partial sums
2. Missing values: limit of function
3. Tangent line: limit of secant line
4. Integral: limit of a sum
5. Indeterminate forms: limit of function values
6. Improper integrals: limit of finite case
7. Infinite series: limit of partial sums
8. Fourier series: limit of partial sums

It is the same simple idea in each case, in spite of the surface diversity. It could be claimed that the calculus is the systematic use of the limit method for dealing with infinity.

Similarly, you saw the method of undetermined coefficients, beginning with mathematical induction and going through the whole book; if you can find the form of the answer, then by imposing the conditions of the problem you can determine the coefficients of the form. It is a slow but sure method of doing many problems. There is often a clever method of solving a problem, but the method of undetermined coefficients saves learning or devising a lot of isolated tricks. Yes, it is often inelegant, but it is a tool having broad applications.

The book also included a lot of material on linear independence, because it is a fundamental tool of mathematics. We also gave examples of proving the impossibility of doing something. It is important to be able to prove that something is impossible to do, rather than that you are merely stupid.

You have seen abstraction, extension, and generalization many times. It is an attitude you need to develop to the point where it is almost automatic; when a specific problem is solved, how general is the method and what other kinds of similar problems can you now do? It is for this reason we have repeatedly advocated that you do a few problems but study them carefully. A general approach will solve more specific problems than you can possibly do one at a time.

If you now review the book, you should see how comparatively few ideas are involved, how things that once seemed confusing are now obvious and clear. The deliberate use of the spiral of learning, which tends to encourage "chunking" of ideas, means that you can finally see the simplicity behind the face of complexity. The book was written mainly by thinking and only occasionally was a "copy from some other source" used. Thus you see that the author "knows" comparatively little; he mainly constructed the details from general principles he had learned over a lifetime of practice. Many items were recalled from having pondered what the mathematics was that had proved to be successful in some applications and how far the same methods could be extended to other situations. It is necessary, apparently, for individuals to make the abstractions for themselves *if* the abstraction is to become useful later on. What you learn for yourself is often what turns out to be most useful in later life. Thus many of the abstractions have not been specifically labeled, and you have been encouraged to find them yourself so they will be your own.

26.3 APPLICATIONS

We have shown how the things you have learned are applicable in many places, within mathematics itself as well as in other fields. The main fields of application we used for the calculus, probability and statistics, were chosen for their importance in modern science (as compared with the classical application of the calculus to mechanics), as well as the range of applications generally in life. In neither case was a complete course attempted; only those parts were used that illustrated the use of the mathematics that you had just learned.

The case histories were included to give you a slight understanding of the actual use of mathematics in practice. They were, of course, only small parts of much larger units of thought, but reality is too large to encompass completely in a single book.

The rich variety of applications we have given should prove the point that mathematics is indeed the language of science and engineering, as well as increasingly of other fields as they become more precise in their statements of results. A few of the examples were chosen for mere teaching purposes, but an effort was made to select useful results even when we did not explain their use at that time. The attempt to teach so many different ideas naturally gave you trouble in mastering things. Since you will likely take later courses in both probability and statistics, the concepts will get further reinforcement. The exposure here will hopefully facilitate their later complete mastery. We took the risk of including too much variety for the beginner to digest.

Many of the illustrations are useful in other fields. For example, many of the specific integrals were chosen because they arise in many places. One such case is the integral

$$\int_0^x \frac{dz}{\sqrt{x - x^2}}\, dx = 2 \arcsin \sqrt{x}$$

which arises in a key result in probability theory. Similarly, as you go on in mathematics, probability, and statistics, you can expect to meet other specific examples we have given.

26.4 PHILOSOPHY

Unlike almost all mathematics courses, we have deliberately engaged your attention on the philosophical foundations of mathematics. We have *not* tried to give you the answers to many of the points raised, but rather it was intended that they be sufficiently disturbing to you that you would think through your own position as to the bases of mathematics and its applications. If you have skipped such items, it is your loss. It is hoped that you have seen through any glib answers that you may have been given and have been willing to think for yourself. Thinking for yourself is perhaps the single most important habit you could have learned from this book.

To encourage thinking for yourself, we have gone out of the way to produce examples of the use of mathematics that seem to contradict "common sense"; surfaces of revolution whose volume was finite but whose surface was infinite, lines with peculiar lengths, optimization problems with no shortest path, and the like. The mathematics you have learned must be tempered with "common sense" when it is applied and action taken based on the results. There is a definite gap between the theories of mathematics and the various fields of applications. In a sense, society has already exploited most of the applications where the mathematical model is close to our observations of the real world; most (but not all!) new applications will have larger gaps between the mathematical model and what is observed.

Mathematics is both useful and intellectually interesting in its own right, and we have often engaged in what might be called "metamathematics," the study of why mathematics is the way it is. It is an interesting topic of endless fascination. To name but one example, why are the sin and cos functions of trigonometry not exact duals of each other? Why do you usually find more cosines than you find sines as you scan material involving them? We showed how trigonometric identities can be reduced to rational functions with complex coefficients, another example of the partial equivalence of different fields.

Pure mathematicians are people who find mathematics is their main interest, mathematicians have an interest in both mathematics and in its applications, while applied mathematicians care mainly about applications. There is room for a wide variety of interests in the coming world in which you can use what you have learned and extend known results to get new results. Much of mathematics and its applications still remains to be discovered (created if you prefer).

It is hoped that you have seen how mathematics is used and have a somewhat accurate impression of its role in science and engineering as well as in other fields. It is also hoped that you have some idea of your future needs for mathematics in your coming career.

This book is a serious attempt to prepare you to push forward the frontiers of knowledge, rather than merely fill you full of past results. The author believes in the approach based on regeneration rather than retrieval of knowledge. It is hoped that the earlier parts of the book, which displayed the methods of mathemetics, prepared you for the later more rapid discussion of many of the applications. It is the methods of the mathematics that produce the results, and, because of the vastness of current knowledge, learning the methods of mathematics appears to be the only hope for preparing you to play your part in future developments.

Appendix A

Table of Integrals

NOTES ON USING THE TABLE

Integration is a difficult art. Given a differentiable function, you can differentiate it and thus get the corresponding integral. Therefore, no table can hope to be complete, and the design of a table is the problem of including those integrals that are likely to occur and excluding those that are not so that the table is compact and easy to use.

There is a large, random component in integration in closed form, but there is also a systematic component. Without some system, you would have to look through, on the average, one-half of the table to find the particular integral you want; with a systematic structure, you are guided directly to the region where you can expect to find it. Unfortunately, a change of variable can greatly change the form of the integrand, and thus you will have to look at some other place for the changed form.

The table is arranged as a number of short subtables for your convenience. They are ordered, as in the text, roughly, from rational algebraic, to quadratic irrationalities, ln, exp, trig, and inverse trig, followed by a short table of definite integrals. Each new function goes through, roughly, the order of the earlier tables in the complexity of the integrand, so a particular combination of functions in an integrand is found under the most complex function, and within that under the next most complex function, and so on. This structure has not been imposed rigidly as it would produce many (probably) useless integrals in the table. Instead, we have given a number of short subtables running from simple to complex.

If you do not find the desired integral (or a substitution to make), then you need to think of a change of variable. You can also think of the form that the answer might take and then try the method of undetermined coefficients.

The constant of integration has been left off the results, and the limitations of the validity have not always been indicated. A division by zero should alert you to the alternative ln expression. Reduction formulas end with the number of the integral you need to finish the recursion. A few integrals that cannot be done in closed form have been given as an infinite series.

FOUR GENERAL FORMULAS

1. Integration by parts

$$\int U \, dV = UV - \int V \, dU$$

is used: (a) to eliminate terms like ln x, arcsin x, arctan x, and x^n (and other functions), by choosing these as U; (b) for exp x, sin x and cos x (and other functions) one or more applications of (1) can sometimes (with additional manipulations) produce the original integrand again.

2. Change of variable

$$\int f(x) \, dx = \int f(g(t))g'(x) \, dx$$

is used in both directions.

3. Fundamental theorem of calculus

$$\frac{d}{dx} \int^x f(t) \, dt = f(x)$$

is the basis for the guess method:
(a) study the table and guess the form,
(b) apply (3),
(c) equate the linearly independent terms,
(d) if you can solve these equations and produce an identity then you have the integral.

4. Basic power formula

$$\int x^n \, dx = \begin{cases} \dfrac{x^{n+1}}{n+1} \\ \ln|x| \end{cases} \quad (n \neq -1)$$

This peculiar effect of two different forms for the answer, depending on the presence of a division by 0, makes for apparent irregularity throughout the table.

BASIC ALGEBRAIC FORMULAS

5. $\int \dfrac{dx}{a^2 + x^2} = \dfrac{1}{a} \arctan \dfrac{x}{a}$

6. $\int \dfrac{dx}{a^2 - x^2} = \dfrac{1}{2a} \ln \dfrac{a + x}{a - x}$ $(a^2 > x^2)$

7. $\int \dfrac{dx}{x^2 - a^2} = \dfrac{1}{2a} \ln \dfrac{x - a}{x + a}$ $(x^2 > a^2)$

These three possibilities produce a corresponding choice in the table.

Any rational function in x can be handled by the method of partial fractions (Sections 17.2 and 17.3). You will end up with form (16) below, or one of the following, where we have written

$$X = a + bx + cx^2 \quad \text{and} \quad 4ac - b^2 = q > 0$$

8. $\int \dfrac{dx}{a + bx + cx^2} = \dfrac{2}{\sqrt{q}} \arctan \dfrac{2cx + b}{\sqrt{q}}$

9. $\int \dfrac{dx}{(a + bx + cx^2)^2} = \dfrac{2cx + b}{qX} + \dfrac{2c}{q} \int \dfrac{dx}{X}$ (8)

10. $\int \dfrac{dx}{(a + bx + cx^2)^n} = \dfrac{2cx + b}{q(n - 1)X^{n-1}} + \dfrac{2(2n - 3)c}{q(n - 1)} \int \dfrac{dx}{X^{n-1}}$ (8)

Note especially the Hermite–Ostrogradskii method (Problem 17.3-8*) for finding the algebraic part of the answer. This leaves a fraction in the denominator with only simple zeros to be factored.

Terms in $(a + bx)^n$ $(m, n \geq 1)$

11. $\int (a + bx)^n \, dx = \dfrac{(a + bx)^{n+1}}{b(n + 1)}$

12. $\int x(a + bx)^n \, dx = \dfrac{x(a + bx)^{n+1}}{b(n + 1)} - \dfrac{(a + bx)^{n+2}}{b^2(n + 1)(n + 2)}$

13. $\int x^m (a + bx)^n \, dx$

$$= \dfrac{x^m(a + bx)^{n+1}}{b(n + 1)} - \dfrac{m}{b(n + 1)} \int x^{m-1}(a + bx)^{n+1} \, dx \quad (11)$$

14. $\int \dfrac{(a + bx)^n}{x} \, dx = \sum\limits_{k=1}^{n} C(n, k) a^{n-k} b^k \dfrac{x^k}{k} + a^n \ln|x|$

15. $\int \dfrac{(a + bx)^n}{x^m} \, dx = -\dfrac{(a + bx)^{n+1}}{a(m - 1)x^{m-1}} + \dfrac{b(n - m + 2)}{a(m - 1)} \int \dfrac{(a + bx)^n}{x^{m-1}} \, dx$ (14)

16. $\int \dfrac{dx}{(a + bx)^n} = \begin{array}{l} \dfrac{-1}{b(n - 1)(a + bx)^{n-1}} \quad (n > 1) \\[3mm] \dfrac{1}{b} \ln |a + bx| \quad (n = 1) \end{array}$

17. $\int \dfrac{x \, dx}{(a + bx)^n}$

$= \begin{array}{l} \dfrac{-x}{b(n - 1)(a + bx)^{n-1}} + \dfrac{1}{b(n - 1)} \int \dfrac{dx}{(a + bx)^{n-1}} \quad (n > 1) \qquad (16) \\[4mm] \dfrac{x}{b} - \dfrac{a}{b^2} \ln |a + bx| \quad (n = 1) \end{array}$

18. $\int \dfrac{x^m \, dx}{(a + bx)^n}$

$= \begin{array}{l} \dfrac{-x^m}{b(n - 1)(a + bx)^{n-1}} + \dfrac{m}{b(n - 1)} \int \dfrac{x^{m-1} \, dx}{(a + bx)^{n-1}} \quad (n > 1) \qquad (16) \\[4mm] \dfrac{x^m}{bm} - \dfrac{a}{b} \int \dfrac{x^{m-1}}{a + bx} \, dx \quad (n = 1) \end{array}$

19. $\int \dfrac{dx}{x(a + bx)^n}$

$= \begin{array}{l} \dfrac{1}{a(n - 1)(a + bx)^{n-1}} + \dfrac{1}{a} \int \dfrac{dx}{x(a + bx)^{n-1}} \quad (n > 1) \qquad (19) \\[4mm] \dfrac{1}{a} \ln \left| \dfrac{x}{a + bx} \right| \quad (n = 1) \end{array}$

20. $\int \dfrac{dx}{x^m(a + bx)^n}$

$= \dfrac{-1}{(m - 1)x^{m-1}(a + bx)^n} - \dfrac{nb}{m - 1} \int \dfrac{dx}{x^{m-1}(a + bx)^{n+1}} \quad (m \neq 1) \qquad (19)$

The substitution $a + bx = u$, $x = (u - a)/b$, $dx = \dfrac{du}{b}$ effectively interchanges the two terms.

Terms in $\sqrt{a + bx}$ $(n > 0)$

21. $\int \sqrt{a + bx} \, dx = \dfrac{2}{3b}(a + bx)^{3/2}$

22. $\int x\sqrt{a + bx}\; dx = \frac{2x}{3b}(a + bx)^{3/2} - \frac{4}{15b^2}(a + bx)^{5/2}$

23. $\int x^n\sqrt{a + bx}\; dx$

$$= \frac{2x^n(a + bx)^{3/2}}{(2n + 3)b} - \frac{2an}{(2n + 3)b}\int x^{n-1}\sqrt{a + bx}\; dx \qquad (21)$$

24. $\int \frac{\sqrt{a + bx}}{x}\; dx = 2\sqrt{a + bx} + a\int \frac{dx}{x\sqrt{a + bx}} \qquad (29)$

25. $\int \frac{\sqrt{a + bx}}{x^n}\; dx = \frac{(a + bx)^{3/2}}{a(n - 1)x^{n-1}} - \frac{b(5 - 2n)}{2a(n - 1)}\int \frac{\sqrt{a + bx}}{x^{n-1}}\; dx \qquad (24)$

26. $\int \frac{dx}{\sqrt{a + bx}} = \frac{2}{b}\sqrt{a + bx}$

27. $\int \frac{x\; dx}{\sqrt{a + bx}} = \frac{-2(2a - bx)}{3b^2}\sqrt{a + bx}$

28. $\int \frac{x^n\; dx}{\sqrt{a + bx}} = \frac{2x^n}{b}\sqrt{a + bx} - \frac{2n}{b}\int x^{n-1}\sqrt{a + bx}\; dx \qquad (23)$

29. $\int \frac{dx}{x\sqrt{a + bx}} = \begin{cases} \dfrac{1}{\sqrt{a}} \ln \left| \dfrac{\sqrt{a + bx} - \sqrt{a}}{\sqrt{a + bx} + \sqrt{a}} \right| & (a > 0) \\[4mm] \dfrac{2}{\sqrt{-a}} \arctan \left(\dfrac{\sqrt{a + bx}}{\sqrt{-a}} \right) & (a < 0) \end{cases}$

30. $\int \frac{dx}{x^n\sqrt{a + bx}} = -\frac{\sqrt{a + bx}}{a(n - 1)x^{n-1}} - \frac{(2n - 3)b}{2(n - 1)a}\int \frac{dx}{x^{n-1}\sqrt{a + bx}} \qquad (29)$

The substitution $a + bx = u^2$, $x = \dfrac{u^2 - a}{b}$, $dx = \dfrac{2u\; du}{b}$ removes the radical.

Terms in $\sqrt{a^2 - x^2}$ $\quad (a^2 \geq x^2)$

31. $\int \sqrt{a^2 - x^2}\; dx = \frac{1}{2}\left[x\sqrt{a^2 - x^2} + a^2 \arcsin \left(\frac{x}{a}\right) \right]$

32. $\int x\sqrt{a^2 - x^2}\; dx = -\frac{1}{3}(a^2 - x^2)^{3/2}$

33. $\int x^n\sqrt{a^2 - x^2}\; dx$

$$= \frac{-x^{n-1}(a^2 - x^2)^{3/2}}{n + 2} + \frac{n - 1}{n + 2}a^2 \int x^{n-2}\sqrt{a^2 - x^2}\; dx \qquad (31), (32)$$

34. $\displaystyle\int \frac{\sqrt{a^2 - x^2}\,dx}{x} = \sqrt{a^2 - x^2} - a \ln \left| \frac{a + \sqrt{a^2 - x^2}}{x} \right|$

35. $\displaystyle\int \frac{\sqrt{a^2 - x^2}\,dx}{x^n}$

$$= -\frac{(a^2 - x^2)^{3/2}}{a^2(n - 1)x^{n-1}} + \frac{n - 4}{a^2(n - 1)} \int \frac{\sqrt{a^2 - x^2}}{x^{n-2}}\,dx \qquad (34),\ (31)$$

36. $\displaystyle\int \frac{dx}{\sqrt{a^2 - x^2}} = \arcsin\left(\frac{x}{a}\right)$

37. $\displaystyle\int \frac{x\,dx}{\sqrt{a^2 - x^2}} = -\sqrt{a^2 - x^2}$

38. $\displaystyle\int \frac{x^n\,dx}{\sqrt{a^2 - x^2}} = \frac{-x^{n-1}\sqrt{a^2 - x^2}}{n} + \frac{a^2(n - 1)}{n} \int \frac{x^{n-2}\,dx}{\sqrt{a^2 - x^2}} \qquad (36),\ (37)$

39. $\displaystyle\int \frac{dx}{x\sqrt{a^2 - x^2}} = \frac{1}{a} \ln \left| \frac{x}{a + \sqrt{a^2 - x^2}} \right|$

40. $\displaystyle\int \frac{dx}{x^n\sqrt{a^2 - x^2}} = -\frac{\sqrt{a^2 - x^2}}{a^2(n - 1)x^{n-1}} + \frac{n - 2}{a^2(n - 1)} \int \frac{dx}{x^{n-2}\sqrt{a^2 - x^2}} \qquad (39),$
(36)

The substitution $x = a \sin\theta$, $\sqrt{a^2 - x^2} = a \cos\theta$, $dx = a \cos\theta\,d\theta$ removes the radical.

Terms in $\sqrt{a^2 + x^2}$

41. $\displaystyle\int \sqrt{a^2 + x^2}\,dx = \frac{1}{2}[x\sqrt{a^2 + x^2} + a^2 \ln |x + \sqrt{a^2 + x^2}|]$

42. $\displaystyle\int x\sqrt{a^2 + x^2}\,dx = \frac{1}{3}(a^2 + x^2)^{3/2}$

43. $\displaystyle\int x^n\sqrt{a^2 + x^2}\,dx$

$$= \frac{x^{n-1}(a^2 + x^2)^{3/2}}{n + 2} - \frac{(n - 1)a^2}{n + 2} \int x^{n-2}\sqrt{a^2 + x^2}\,dx \qquad (42),\ (41)$$

44. $\displaystyle\int \frac{\sqrt{a^2 + x^2}\,dx}{x} = \sqrt{a^2 + x^2} - a \ln \left| \frac{a + \sqrt{a^2 + x^2}}{x} \right|$

45. $\displaystyle\int \frac{\sqrt{a^2 + x^2}\,dx}{x^n}$

$$= -\frac{(a^2 + x^2)^{3/2}}{a^2(n - 1)x^{n-1}} - \frac{n - 4}{a^2(n - 1)} \int \frac{\sqrt{a^2 + x^2}\,dx}{x^{n-2}} \qquad (44),\ (41)$$

46. $\int \frac{dx}{\sqrt{a^2 + x^2}} = \ln |x + \sqrt{a^2 + x^2}|$

47. $\int \frac{x \, dx}{\sqrt{a^2 + x^2}} = \sqrt{a^2 + x^2}$

48. $\int \frac{x^n \, dx}{\sqrt{a^2 + x^2}} = \frac{x^{n-1}\sqrt{a^2 + x^2}}{n} - \left(\frac{n-1}{n}\right)a^2 \int \frac{x^{n-2} \, dx}{\sqrt{a^2 + x^2}}$ (47), (46)

49. $\int \frac{dx}{x\sqrt{a^2 + x^2}} = -\frac{1}{a} \ln \left| \frac{a + \sqrt{a^2 + x^2}}{x} \right|$

50. $\int \frac{dx}{x^n\sqrt{a^2 + x^2}}$

$$= -\frac{\sqrt{a^2 + x^2}}{a^2(n-1)x^{n-1}} - \frac{n-2}{a^2(n-1)} \int \frac{dx}{x^{n-2}\sqrt{a^2 + x^2}}$$ (49), (46)

The substitution $x = a \tan \theta$, $\sqrt{a^2 + x^2} = \sec \theta$, $dx = a \sec^2 \theta \, d\theta$ removes the radical.

Terms in $\sqrt{x^2 - a^2}$ $(x^2 \geq a^2)$

51. $\int \sqrt{x^2 - a^2} \, dx = \frac{1}{2}[x\sqrt{x^2 - a^2} - a^2 \ln |x + \sqrt{x^2 - a^2}|]$

52. $\int x\sqrt{x^2 - a^2} \, dx = \frac{(x^2 - a^2)^{3/2}}{3}$

53. $\int x^n\sqrt{x^2 - a^2} \, dx$

$$= \frac{x^{n-1}(x^2 - a^2)^{3/2}}{n + 2} + \frac{a^2(n-1)}{n+2} \int x^{n-2}\sqrt{x^2 - a^2} \, dx$$ (52), (51)

54. $\int \frac{\sqrt{x^2 - a^2} \, dx}{x} = \sqrt{x^2 - a^2} - a \arcsec \left(\frac{x}{a}\right)$

55. $\int \frac{\sqrt{x^2 - a^2}}{x^n} \, dx$

$$= \frac{(x^2 - a^2)^{3/2}}{a^2(n-1)x^{n-1}} + \frac{n-4}{a^2(n-1)} \int \frac{\sqrt{x^2 - a^2}}{x^{n-2}} \, dx$$ (54), (51)

56. $\int \frac{dx}{\sqrt{x^2 - a^2}} = \ln |x + \sqrt{x^2 - a^2}|$

57. $\int \frac{x \, dx}{\sqrt{x^2 - a^2}} = \sqrt{x^2 - a^2}$

58. $\int \dfrac{x^n \, dx}{\sqrt{x^2 - a^2}} = \dfrac{x^{n-1}\sqrt{x^2 - a^2}}{n} + \dfrac{a^2(n - 1)}{n} \int \dfrac{x^{n-2} \, dx}{\sqrt{x^2 - a^2}}$ (57), (56)

59. $\int \dfrac{dx}{x\sqrt{x^2 - a^2}} = \dfrac{1}{a} \operatorname{arcsec} \left(\dfrac{x}{a} \right)$

60. $\int \dfrac{dx}{x^n\sqrt{x^2 - a^2}}$

$$= \dfrac{\sqrt{x^2 - a^2}}{a^2(n - 1)x^{n-1}} + \dfrac{n - 2}{a^2(n - 1)} \int \dfrac{dx}{x^{n-2}\sqrt{x^2 - a^2}}$$ (59), (56)

The substitution $x = a \sec\theta$, $\sqrt{x^2 - a^2} = a \tan\theta$, $dx = a \sec\theta \tan\theta \, d\theta$ removes the radical.

Terms in $\sqrt{2ax - x^2}$ $(a > 0)$

61. $\int \sqrt{2ax - x^2} \, dx = \dfrac{1}{2}\left[(x - a)\sqrt{2ax - x^2} + a^2 \arcsin\left(\dfrac{x - a}{a} \right) \right]$

62. $\int x\sqrt{2ax - x^2} \, dx = -\dfrac{(2ax - x^2)^{3/2}}{3} + a \int \sqrt{2ax - x^2} \, dx$ (61)

63. $\int x^n\sqrt{2ax - x^2} \, dx$

$$= -\dfrac{x^{n-1}(2ax - x^2)^{3/2}}{n + 2} + \dfrac{a(2n + 1)}{n + 2} \int x^{n-1}\sqrt{2ax - x^2} \, dx$$ (62)

64. $\int \dfrac{\sqrt{2ax - x^2}}{x} \, dx = \dfrac{(2ax - x^2)^{3/2}}{ax} + \dfrac{2}{a} \int \sqrt{2ax - x^2} \, dx$ (61)

65. $\int \dfrac{\sqrt{2ax - x^2}}{x^n} \, dx$

$$= \dfrac{(2ax - x^2)^{3/2}}{(3 - 2n)ax^n} + \dfrac{n - 3}{a(2n - 3)} \int \dfrac{\sqrt{2ax - x^2}}{x^{n-1}} \, dx$$ (64), (61)

66. $\int \dfrac{dx}{\sqrt{2ax - x^2}} = 2 \arcsin\sqrt{\dfrac{x}{2a}} = \arccos\left(1 - \dfrac{x}{a} \right)$

67. $\int \dfrac{x \, dx}{\sqrt{2ax - x^2}} = -\sqrt{2ax - x^2} + a \int \dfrac{dx}{\sqrt{2ax - x^2}}$ (66)

68. $\int \dfrac{x^n \, dx}{\sqrt{2ax - x^2}} = -\dfrac{x^{n-1}\sqrt{2ax - x^2}}{n} + \dfrac{(2n - 1)a}{n} \int \dfrac{x^{n-1} \, dx}{\sqrt{2ax - x^2}}$ (67)

69. $\int \dfrac{dx}{x\sqrt{2ax - x^2}} = -\dfrac{\sqrt{2ax - x^2}}{ax}$

70. $\displaystyle\int \frac{dx}{x^n\sqrt{2ax - x^2}} = \frac{-\sqrt{2ax - x^2}}{a(2n - 1)x^n} + \frac{n - 1}{a(2n - 1)}\int \frac{dx}{x^{n-1}\sqrt{2ax - x^2}}$ (69)

The substitution $x = 2a \sin^2 \theta$, $\sqrt{2ax - x^2} = 2a \sin \theta \cos \theta$, $dx = 4a \sin \theta \cos \theta\, d\theta$ removes the radical.

Terms in ln x

71. $\displaystyle\int \ln x\, dx = x \ln x - x$

72. $\displaystyle\int x \ln x\, dx = \frac{x^2}{2} \ln x - \frac{x^2}{4}$

73. $\displaystyle\int x^n \ln x\, dx = \frac{x^{n+1}}{n + 1} \ln x - \frac{x^{n+1}}{(n + 1)^2}$

74. $\displaystyle\int \frac{\ln x}{x}\, dx = \frac{(\ln x)^2}{2}$

75. $\displaystyle\int \frac{\ln x}{x^n}\, dx = \frac{-\ln x}{(n - 1)x^{n-1}} - \frac{1}{(n - 1)^2 x^{n-1}}$

76. $\displaystyle\int \frac{dx}{\ln x} = \ln|\ln x| + \ln x + \frac{(\ln x)^2}{2 \cdot 2!} + \frac{(\ln x)^3}{3 \cdot 3!} + \cdots$

77. $\displaystyle\int \frac{x\, dx}{\ln x} = \int \frac{e^{-y}}{y}\, dy$ for $y = -2 \ln x$ (84)

78. $\displaystyle\int \frac{x^n\, dx}{\ln x} = \int \frac{e^{-y}}{y}\, dy$ for $y = -(n + 1) \ln x$ (84)

79. $\displaystyle\int \frac{dx}{x \ln x} = \ln|\ln x|$

80. $\displaystyle\int \frac{dx}{x^n \ln x} = \int \frac{e^{-y}}{y}\, dy$ for $y = (n - 1) \ln x$ (84)

Recall that $\displaystyle\int f(x) \ln g(x)\, dx = F(x) \ln g(x) - \int \frac{F(x)}{g(x)} g'(x)\, dx$, where $F(x) = \displaystyle\int_0^t f(t)\, dt$

Terms in e^x

81. $\displaystyle\int e^{ax}\, dx = \frac{e^{ax}}{a}$

82. $\displaystyle\int xe^{ax}\, dx = \frac{e^{ax}}{a^2}(ax - 1)$

83. $\displaystyle\int x^n e^{ax}\, dx = e^{ax} \sum_{k=0}^{n} \frac{(-1)^k n!}{(n-k)!} \frac{x^{n-k}}{a^{n+1}}$

84. $\displaystyle\int \frac{e^{ax}}{x}\, dx = \ln x + \frac{ax}{1!} + \frac{(ax)^2}{2\cdot 2!} + \frac{(ax)^3}{3\cdot 3!} + \cdots$

85. $\displaystyle\int \frac{e^{ax}}{x^n}\, dx = \frac{-e^{ax}}{(n-1)x^{n-1}} + \frac{a}{n-1} \int \frac{e^{ax}}{x^{n-1}}\, dx$ (84)

86. $\displaystyle\int \frac{dx}{1+e^x} = \ln\left(\frac{e^x}{1+e^x}\right)$

87. $\displaystyle\int \frac{dx}{a+be^{mx}} = \frac{1}{am}[mx - \ln(a+be^{mx})]$

88. $\displaystyle\int \frac{dx}{ae^{mx}+be^{-mx}} = \frac{1}{m\sqrt{ab}}\arctan\left(e^{mx}\sqrt{\frac{a}{b}}\right)$ $(ab>0)$

89. $\displaystyle\int \frac{dx}{ae^{mx}-be^{-mx}} = \frac{1}{2m\sqrt{ab}}\ln\left|\frac{\sqrt{a}e^{mx}-\sqrt{b}}{\sqrt{a}e^{mx}+\sqrt{b}}\right|$

90. $\displaystyle\int e^{-a^2x^2}\, dx = x - \frac{a^2x^3}{3} + \frac{a^4x^5}{5\cdot 2!} - \frac{a^6x^7}{7\cdot 3!} + \cdots$

91. $\displaystyle\int x e^{-a^2x^2}\, dx = -\frac{e^{-a^2x^2}}{2a^2}$

92. $\displaystyle\int x^{2n} e^{-a^2x^2}\, dx = \frac{-x^{2n-1}e^{-a^2x^2}}{2a^2} + \frac{2n-1}{2a^2} \int x^{2(n-1)} e^{-a^2x^2}\, dx$ (90)

93. $\displaystyle\int x^{2n+1} e^{-a^2x^2}\, dx = \frac{-x^{2n}e^{-a^2x^2}}{2a^2} + \frac{n}{a^2} \int x^{2n-1} e^{-a^2x^2}\, dx$ (91)

RATIONAL TRIGONOMETRIC FUNCTIONS

94. $\displaystyle\int \sin ax\, dx = -\frac{\cos ax}{a}$

95. $\displaystyle\int \cos ax\, dx = \frac{\sin ax}{a}$

96. $\displaystyle\int \tan ax\, dx = -\frac{1}{a}\ln|\cos ax|$

97. $\displaystyle\int \cot ax\, dx = \frac{1}{a}\ln|\sin ax|$

98. $\displaystyle\int \sec ax\, dx = \frac{1}{a}\ln|\sec ax + \tan ax|$

99. $\int \csc ax \, dx = -\dfrac{1}{a} \ln |\csc ax + \ctn ax|$

100. $\int \sec^2 ax \, dx = \dfrac{1}{a} \tan ax$

101. $\int \csc^2 ax \, dx = -\dfrac{1}{a} \ctn ax$

102. $\int \sin^2 ax \, dx = \dfrac{1}{2a}[ax - \sin ax \cos ax]$

103. $\int \cos^2 ax \, dx = \dfrac{1}{2a}[ax + \sin ax \cos ax]$

104. $\int \sin^3 x \, dx = -\dfrac{\cos x}{3}[2 + \sin^2 x]$

105. $\int \cos^3 x \, dx = \dfrac{\sin x}{3}[2 + \cos^2 x]$

106. $\int \sin^n x \, dx = \dfrac{-\sin^{n-1} x \cos x}{n} + \dfrac{n-1}{n} \int \sin^{n-2} x \, dx$ (94), (102)

107. $\int \cos^n x \, dx = \dfrac{\cos^{n-1} x \sin x}{n} + \dfrac{n-1}{n} \int \cos^{n-2} x \, dx$ (95), (103)

108. $\int \tan^n x \, dx = \dfrac{\tan^{n-1} x}{n-1} - \int \tan^{n-2} x \, dx$ (96)

109. $\int \cos^m x \sin^n x \, dx = \begin{cases} \dfrac{\cos^{m-1} x \sin^{n+1} x}{m+n} + \dfrac{m-1}{m+n} \int \cos^{m-2} x \sin^n x \, dx \\[3mm] -\dfrac{\sin^{n-1} x \cos^{m+1} x}{m+n} + \dfrac{n-1}{m+n} \int \cos^m x \sin^{n-2} x \, dx \end{cases}$

110. $\int \dfrac{\sin^n x}{\cos^m x} \, dx = \begin{cases} \dfrac{-\sin^{n-1} x}{(n-m)\cos^{m-1} x} + \dfrac{n-1}{n-m} \int \dfrac{\sin^{n-2} x}{\cos^m x} \, dx \\[3mm] \dfrac{\sin^{n+1} x}{(m-1)\cos^{m-1} x} + \dfrac{m-n-2}{m-1} \int \dfrac{\sin^n x}{\cos^{m-2} x} \, dx \end{cases}$

111. $\int \dfrac{\cos^m x}{\sin^n x} \, dx = \begin{cases} \dfrac{-\cos^{m+1} x}{(n-1)\sin^{n-1} x} - \dfrac{m-n+2}{n-1} \int \dfrac{\cos^m x}{\sin^{n-2} x} \, dx \\[3mm] \dfrac{\cos^{m-1} x}{(m-n)\sin^{n-1} x} + \dfrac{m-1}{m-n} \int \dfrac{\cos^{m-2} x}{\sin^n x} \, dx \end{cases}$

112. $\displaystyle\int \frac{dx}{\cos^m x \sin^n x} = \begin{cases} \dfrac{-1}{(n-1)\sin^{n-1} x\,\cos^{m-1} x} + \dfrac{m+n-2}{n-1}\displaystyle\int \dfrac{dx}{\cos^m x\,\sin^{n-2} x} \\[4mm] \dfrac{1}{(m-1)\sin^{n-1} x\,\cos^{m-1} x} + \dfrac{m+n-2}{m-1}\displaystyle\int \dfrac{dx}{\cos^{m-2} x\,\sin^n x} \end{cases}$

For a rational function in $\sin x$ and $\cos x$, the "tangent half-angle" substitution

$$\sin x = \frac{2t}{1+t^2}, \qquad \cos x = \frac{1-t^2}{1+t^2}, \qquad dx = \frac{2\,dt}{1+t^2}$$

$$t = \tan \frac{x}{2} = \frac{1-\cos x}{\sin x} = \frac{\sin x}{1+\cos x}$$

produces an integrand that is rational in t.

113. $\displaystyle\int \frac{dx}{a + b\cos x}$

$$= \begin{cases} \dfrac{2}{\sqrt{a^2 - b^2}}\arctan\left[\dfrac{\sqrt{a^2+b^2}\,\tan\dfrac{x}{2}}{a+b}\right] & (a^2 > b^2) \\[6mm] \dfrac{1}{\sqrt{b^2-a^2}}\ln\left|\dfrac{a+b+\sqrt{a^2+b^2}\,\tan\dfrac{x}{2}}{a+b}\right| & (b^2 > a^2) \end{cases}$$

114. $\displaystyle\int \frac{dx}{a + b\sin x}$

$$= \begin{cases} \dfrac{2}{\sqrt{a^2-b^2}}\arctan\dfrac{a\tan(x/2)+b}{\sqrt{a^2-b^2}} & (a^2 > b^2) \\[6mm] \dfrac{1}{\sqrt{b^2-a^2}}\ln\left|\dfrac{a\tan(x/2)+b-\sqrt{b^2-a^2}}{a\tan(x/2)+b+\sqrt{b^2-a^2}}\right| & (b^2 > a^2) \end{cases}$$

115. $\displaystyle\int \frac{dx}{a^2\cos^2 x + b^2\sin^2 x} = \frac{1}{ab}\arctan\frac{b\tan x}{a}$

116. $\displaystyle\int \sqrt{1-\cos x}\,dx = -2\sqrt{2}\cos\frac{x}{2}$

117. $\displaystyle\int \sqrt{1+\cos x}\,dx = -2\sqrt{2}\sin\frac{x}{2}$

118. $\displaystyle\int \sqrt{1-\sin x}\,dx = -2\left(\sin\frac{x}{2}+\cos\frac{x}{2}\right)$

119. $\displaystyle\int \sqrt{1+\sin x}\,dx = -2\left(\sin\frac{x}{2}-\cos\frac{x}{2}\right)$

120. $\displaystyle\int \frac{dx}{\sqrt{1-\cos x}} = -\sqrt{2}\ln\tan\frac{x}{4}$

121. $\displaystyle\int \frac{dx}{\sqrt{1 + \cos x}} = -\sqrt{2}\ \ln \tan\left(\frac{x + \pi}{4}\right)$

122. $\displaystyle\int \frac{dx}{\sqrt{1 - \sin x}} = -\sqrt{2}\ \ln \tan\left(\frac{x}{4} - \frac{\pi}{8}\right)$

123. $\displaystyle\int \frac{dx}{\sqrt{1 + \sin x}} = -\sqrt{2}\ \ln \tan\left(\frac{x}{4} + \frac{\pi}{8}\right)$

124. $\displaystyle\int \sin mx \sin nx\ dx = \frac{\sin (m - n)x}{2(m - n)} - \frac{\sin (m + n)x}{2(m + n)}$

125. $\displaystyle\int \sin mx \cos nx\ dx = -\frac{\cos (m - n)x}{2(m - n)} - \frac{\cos (m + n)x}{2(m + n)}$

126. $\displaystyle\int \cos mx \cos nx\ dx = \frac{\sin (m - n)x}{2(m - n)} + \frac{\sin (m + n)x}{2(m + n)}$

127. $\displaystyle\int x \sin x\ dx = \sin x - x \cos x$

128. $\displaystyle\int x \cos x\ dx = \cos x + x \sin x$

129. $\displaystyle\int x^n \sin x\ dx = -x^n \cos x + n \int x^{n-1} \cos x\ dx \qquad (130),\ (128)$

130. $\displaystyle\int x^n \cos x\ dx = x^n \sin x - n \int x^{n-1} \sin x\ dx \qquad (129),\ (127)$

131. $\displaystyle\int \frac{\sin x}{x}\ dx = x - \frac{x^3}{3 \cdot 3!} + \frac{x^5}{5 \cdot 5!} - \frac{x^7}{7 \cdot 7!} + \cdots$

132. $\displaystyle\int \frac{\cos x}{x}\ dx = \ln x - \frac{x^2}{2 \cdot 2!} + \frac{x^4}{4 \cdot 4!} - \frac{x^6}{6 \cdot 6!} + \cdots$

133. $\displaystyle\int \frac{\sin x}{x^n}\ dx = -\frac{\sin x}{(n - 1)x^{n-1}} + \frac{1}{n - 1}\int \frac{\cos x}{x^{n-1}}\ dx \qquad (134),\ (132)$

134. $\displaystyle\int \frac{\cos x}{x^n}\ dx = -\frac{\cos x}{(n - 1)x^{n-1}} - \frac{1}{n - 1}\int \frac{\sin x}{x^{n-1}}\ dx \qquad (133),\ (131)$

135. $\displaystyle\int \frac{x\ dx}{\sin x} = x + \frac{x^3}{3 \cdot 3!} + \frac{7x^5}{3 \cdot 5 \cdot 5!} + \frac{31x^7}{3 \cdot 7 \cdot 7!} + \frac{127x^9}{3 \cdot 9 \cdot 9!} + \cdots$

136. $\displaystyle\int \frac{x\ dx}{\cos x} = \frac{x^2}{2} + \frac{x^4}{4 \cdot 2!} + \frac{5x^6}{6 \cdot 4!} + \frac{61x^8}{8 \cdot 6!} + \frac{1385x^{10}}{10 \cdot 8!} + \cdots$

137. $\displaystyle\int e^{ax} \sin bx\ dx = e^{ax}\left[\frac{a \sin bx - b \cos bx}{a^2 + b^2}\right]$

138. $\displaystyle\int e^{ax} \cos bx \, dx = e^{ax}\left[\frac{a \cos bx + b \sin bx}{a^2 + b^2}\right]$

INVERSE TRIGONOMETRIC FUNCTIONS

139. $\displaystyle\int \arcsin x \, dx = x \arcsin x + \sqrt{1 - x^2}$

140. $\displaystyle\int \arccos x \, dx = x \arccos x - \sqrt{1 - x^2}$

141. $\displaystyle\int \arctan x \, dx = x \arctan x - \frac{1}{2} \ln (1 + x^2)$

142. $\displaystyle\int x \arcsin x \, dx = \frac{1}{4}[(2x^2 - 1) \arcsin x + x\sqrt{1 - x^2}]$

143. $\displaystyle\int x^n \arcsin x \, dx = \frac{x^{n+1} \arcsin x}{n + 1} - \frac{1}{n + 1}\int \frac{x^{n+1} \, dx}{\sqrt{1 - x^2}}$ (38)

144. $\displaystyle\int x \arctan x \, dx = \frac{1}{2}(1 + x^2) \arctan x - \frac{x}{2}$

145. $\displaystyle\int x^n \arctan x \, dx = \frac{x^{n+1} \arctan x}{n + 1} - \frac{1}{n + 1}\int \frac{x^{n+1}}{1 + x^2} \, dx$ (5), (4)

DEFINITE INTEGRALS

146. $\displaystyle\int_0^\infty x^{n-1} e^{-x} \, dx = (n - 1)! = \Gamma(n)$

147. $\displaystyle\Gamma(x) \, \Gamma(1 - x) = \frac{\pi}{\sin \pi x}$

148. $\displaystyle\int_0^\infty e^{-x^2} \, dx = \frac{\sqrt{\pi}}{2} = \Gamma\left(\frac{1}{2}\right) = 1.7724538509 \ldots$

149. $\displaystyle\int_0^1 x^{m-1}(1 - x)^{n-1} \, dx = \beta(m, n) = \frac{\Gamma(m) \, \Gamma(n)}{\Gamma(m + n)}$

150. $\displaystyle\int_0^\infty \frac{x^{p-1}}{1 + x} \, dx = \frac{\pi}{\sin p\pi}$ $(0 < p < 1)$

151. $\displaystyle\int_0^\infty \frac{dx}{(1 + x)\sqrt{x}} = \pi$

152. $\displaystyle\int_0^\infty \frac{dx}{a^2 + x^2} = \frac{\pi}{2a}$ $(a > 0)$

153. $\displaystyle\int_0^{\pi/2} \sin^n x \, dx$ $\displaystyle= \frac{(n-1)(n-3)\cdots(1)}{n(n-2)\cdots(2)} \frac{\pi}{2},$ $\qquad n = 2, 4, 6, \ldots$

154. $\displaystyle\int_0^{\pi/2} \cos^n x \, dx$ $\displaystyle\frac{(n-1)(n-3)\cdots 2}{n(n-2)\cdots 3},$ $\qquad n = 1, 3, 5, \ldots$

155. $\displaystyle\int_0^\infty \frac{\sin mx}{x} \, dx =$ $\begin{cases} \dfrac{\pi}{2}, & m > 0 \\[2mm] 0, & m = 0 \\[2mm] -\dfrac{\pi}{2}, & m < 0 \end{cases}$

156. $\displaystyle\int_0^\infty \frac{\sin^2 x}{x^2} \, dx = \frac{\pi}{2}$

157. $\displaystyle\int_0^\infty \frac{\sin^3 x}{x^3} \, dx = \frac{3\pi}{8}$

158. $\displaystyle\int_0^\infty \frac{\sin^4 x}{x^4} \, dx = \frac{\pi}{3}$

159. $\displaystyle\int_0^\infty \frac{\cos mx}{1 + x^2} \, dx = \frac{\pi}{2} e^{-|m|}$

160. $\displaystyle\int_0^\infty \frac{\sin x}{\sqrt{x}} \, dx$

161. $\displaystyle\int_0^\infty \frac{\cos x}{\sqrt{x}} \, dx$ $\Biggr\} = \sqrt{\dfrac{\pi}{2}}$

162. $\displaystyle\int_0^\infty \frac{\cos ax - \cos bx}{x} = \ln\left|\frac{b}{a}\right|$

163. $\displaystyle\int_0^\infty \frac{\arctan ax - \arctan bx}{x} \, dx = \frac{\pi}{2} \ln\left|\frac{a}{b}\right| \qquad (a, b > 0)$

164. $\displaystyle\int_0^\infty \frac{dx}{a^2 \sin^2 x + b^2 \cos^2 x} = \frac{\pi}{2ab}$

165. $\displaystyle\int_0^\infty \frac{e^{-ax} - e^{-bx}}{x} \, dx = \ln\left|\frac{b}{a}\right| \qquad (a, b > 0)$

166. $\displaystyle\int_0^\infty e^{-a^2 x^2} \, dx = \frac{\sqrt{\pi}}{2a}$

167. $\displaystyle\int_0^\infty x e^{-x^2} \, dx = \frac{1}{2}$

168. $\displaystyle\int_0^\infty x^2 e^{-x^2} \, dx = \frac{\sqrt{\pi}}{4}$

169. $\int_0^\infty e^{-[x^2+(a^2/x^2)]} \, dx = \dfrac{\sqrt{\pi}\,e^{-2\,|a|}}{2}$

170. $\int_0^\infty e^{-ax} \sin bx \, dx = \dfrac{b}{a^2 + b^2}$ $(a > 0)$

171. $\int_0^\infty e^{-ax} \cos bx \, dx = \dfrac{a}{a^2 + b^2}$ $(a > 0)$

172. $\int_0^\infty e^{-a^2x^2} \cos bx \, dx = \dfrac{\sqrt{\pi}}{2a} e^{-b^2/4a^2}$

173. $\int_0^1 (\ln x)^n \, dx = (-1)^n n!$

174. $\int_0^1 \sqrt{\ln \dfrac{1}{x}} \, dx = \dfrac{\sqrt{\pi}}{2}$

175. $\int_0^1 \dfrac{\ln x}{1 + x} \, dx = \dfrac{-\pi^2}{12}$

176. $\int_0^1 \dfrac{\ln x}{1 - x} \, dx = \dfrac{-\pi^2}{6}$

177. $\int_0^1 \ln \left(\dfrac{1 + x}{1 - x} \right) \dfrac{dx}{x} = \dfrac{\pi^2}{4}$

178. $\int_0^1 \dfrac{\ln x}{\sqrt{1 - x^2}} \, dx = \dfrac{-\pi \ln 2}{2}$

179. $\int_0^\infty \ln \dfrac{e^x + 1}{e^x - 1} \, dx = \dfrac{\pi^2}{4}$

180. $\int_0^{\pi/2} \ln \sin x \, dx$
  $= \dfrac{-\pi \ln 2}{2}$

181. $\int_0^{\pi/2} \ln \cos x \, dx$

182. $\int_0^\infty e^{-x} \ln x \, dx = -\gamma = -0.577215 \ldots$

Appendix B

Some Geometric Formulas

1. Triangle $A = \dfrac{ab}{2}$

2. Parallelogram $A = ab$

3. Circle Circumference $C = 2\pi r$
Area $A \qquad = \pi r^2$

4. Sphere Surface $= 4\pi r^2$
Volume $= \dfrac{4\pi}{3} r^3$

5. Right circular cylinder Side surface $= 2\pi rh$
Total surface $= 2\pi r(r + h)$
Volume $\qquad = \pi r^2 h$

6. Right circular cone

Slant height $= \sqrt{a^2 + r^2}$

Side surface $= \pi r \sqrt{a^2 + r^2}$

Total surface $= \pi r(r + \sqrt{a^2 + r^2})$

Volume $= \dfrac{\pi r^2 h}{3}$

7. Pyramid

Volume $= \dfrac{(\text{base area})\, a}{3}$

8. Rectangular box

Body diagonal $= \sqrt{a^2 + b^2 + c^2}$

Surface $= 2(ab + bc + ca)$

Volume $= abc$

Appendix C

The Greek Alphabet

Greek Alphabet		Names	Pronunciation
A	α	Alpha	Al'-fa
B	β	Beta	Bay'-ta
Γ	γ	Gamma	Gam'-ma
Δ	$\delta\ \partial$	Delta	Del'-ta
E	ϵ	Epsilon	Ep'-sa-lon
Z	ζ	Zeta	Zay'-ta
H	η	Eta	Ay'-ta
Θ	θ	Theta	Thay'-ta
I	ι	Iota	Eye'-owe-ta
K	κ	Kappa	Kap'-pa
Λ	λ	Lambda	Lam'-da
M	μ	Mu	Mew
N	ν	Nu	Noo
Ξ	ξ	Xi	Zī*
O	o	Omicron	Om'-i-kron
Π	π	Pi	Pī
P	ρ	Rho	Row
Σ	σ	Sigma	Sig'-ma
T	τ	Tau	Taw
Υ	υ	Upsilon	Up'-sa-lon
Φ	ϕ	Phi	Fī
X	χ	Chi	Kī
Ψ	ψ	Psi	Sī
Ω	ω	Omega	Owe'-may-ga

*Long i as in "eye."

Answers to Some of the Exercises

The answers to all the examples and exercises in this book are known, hence their purpose cannot be to "get the answer." Their purpose is to show you, and to give you practice in, a mathematical style of thinking. Thus the answers are only a partial check; you may "get the answer" and yet use a poor style. Since the inevitable error rate (in a first edition) is probably between 1% and 10%, you can expect at times to be right and the book to be wrong.

Exercises 2.2

3. Fails at $n = -40$ **5.** Some people consider 1 as neither a prime nor a composite number.

Exercises 2.3

1. 5050   **2.** 338,350   **3.** m^3 **4.** $3m + 1$ **5.** $1/m$ **9.** 2^{n-1}

13. Verify some value, say $n = 0$. Then do an induction up through the positive integers and then an induction through the negative integers.

Exercises 2.4

1. $a^5 + 5a^4b + 10a^3b^2 + 10a^2b^3 + 5ab^4 + b^5$

2. $a^5 - 5a^4b + 10a^3b^2 - 10a^2b^3 + 5ab^4 - b^5$

3. 1, 9, 36, 84, 126, 126, 84, 36, 9, 1

4. $1 + 10x + 45x^2 + 120x^3 + 210x^4 + 252x^5 + 210x^6 + 120x^7 + 45x^8 + 10x^9 + x^{10}$

5. $a^7 - 7a^6b + 21a^5b^2 - 35a^4b^3 + 35a^3b^4 - 21a^2b^5 + 7ab^6 - b^7$

6. $p^n + C(n, 1)p^{n-1}q + C(n, 2)p^{n-2}q^2 + \cdots + q^n$

7. $p^n - C(n, 1)p^{n-1}q + C(n, 2)p^{n-2}q^2 + \cdots + (-1)^n q^n$

8. $2^5 = 1 + 5 + 10 + 10 + 5 + 1$ **9.** $0 = 1 - 5 + 10 - 10 + 5 - 1$

10. $2^4 = 16$

11. $\dfrac{(p + q)^n + (p - q)^n}{2} = p^n + C(n, 2)p^{n-2}q^2 + C(n, 4)p^{n-4}q^4 + \cdots$

12. $2^{n-1} = 1 + C(n, 2) + C(n, 4) + C(n, 6) + \cdots$

13. 2^n

15. Add $n!$ to both sides and cancel the $(n + 1)!$ on both sides, thus reducing the problem to one lower n. You end the reduction at $1(1!) = 2! - 1$.

Exercises 2.5

1. $(5050)^2$ **2.** 15,050   **3.** 2,348,350

5. $n(n + 1)(2n + 1)(3n^2 + 3n - 1)/30$

6. $k(k - 1)(k - 2)(k - 3)(k - 4)/6! = e$, other coefficients as before

9.* You can find the coefficients and check, or else show the equality for at least eight different values of n.

Exercises 2.6

1. $2^{n+1} - 1$ **2.** $3(2^n - 1)$ **3.** $2 - 1/2^n$ **4.** 15,251   **5.** 1300

8. $n \geq [\log \{(r - 1)C + a\} - \log a]/\log r$

9. A factor more than 1 million.

Exercises 3.2

1. 4   **2.** 1   **3.** 7   **4.** 1

5. No. There would be one extra stage.

6. *I* would ignore the signs, find the answer, and then attach the proper signs.

Exercises 3.3

1. The step size would be negative.

2. $\frac{1}{4} - \frac{1}{160}, \frac{1}{4} - \frac{2}{160}, \ldots, \frac{1}{4} - \frac{7}{160}$

3. Denominators have only factors of 2 and 5.

4. Denominators are powers of 2.

5. $a + \left(\dfrac{b - a}{N + 1}\right)k, k = 1, 2, \ldots, N$

Exercises 3.4

1. Not the ratio of two integers

5. At the evenness step

7.* Set $x = p/q$, multiply through by q^n. Argument then resembles that for Exercise 6 (applied twice).

Exercises 3.5

1. 1.732   **2.** 2.236   **3.** 1.2599   **4.** 1.7099
5. Between 0.195 and 0.196   **6.** Between 0.6640 and 0.6641
8. Between 1.532 and 1.533, and between 0.3472 and 0.3473

Exercises 3.6

1. 0.272727 . . .   **2.** 78.8333 . . .   **3.** $\frac{4}{33}$ **4.** $3 + \frac{145}{999}$ **5.** $\frac{50}{33}$ **6.** $\frac{1}{900}$
8. Keep doubling the number and each time a 1 passes the decimal point you get a 1 in the corresponding binary digit.
10. $h(1 + k)/(1 - k)$

Exercises 3.7

1. All numbers are positive, or at most one negative number.
2. No *simple* rule exists.   **4.** Start with $(x - y)^2 \geq 0$.

Exercises 3.8

1. (a) x^{13}, (b) x, (c) x^{42} **2.** (a) x^2, (b) x^{-7}, (c) $x^{-7/6}$

Exercises 4.3

1. Same "staircase" pattern, always rising as you go to the right
3. Monotone decreasing from 0° to 180°, monotone increasing from 180° to 360°, repeated endlessly.
4. No   **5.** $(x - 1)/(x + 1)$

Exercises 4.4

2. For $y(|x|)$ you take the absolute value (a positive or 0 number) and use it to compute the function y; for $|y(x)|$ you compute the function and then take the absolute value. If $y(x) = x - 1$, you get $y(|x|) = |x| - 1$ and $|y(x)| = |x - 1|$.
3. For $x \leq 0$, $y = 0$. For $x \geq 0$, $y = x$.
4. For $x \leq 0$, $y = 0$. For $0 \leq x \leq 1$, straight line slope = 1. For $x \geq 1$, $y = 1$.
5.* $\frac{1}{2}(|x| - 2|x - 1| + |x - 2|)$ **6.** No

Exercises 4.6

1. A linear combination of $1, x, x^2, \ldots, x^4$ with constant coefficients

2. $P + Q = x^3 + x^2 + x$; $P - Q = x^3 - x^2 + x + 2$; $PQ = x^5 + x^2 - x - 1$

Exercises 4.7

1. No linear combination (with proper coefficients) can be identically zero unless all the coefficients are zero.
2. $(x - 2)(x - 3)(x - 4)$ **3.** $(x + 1)(x - 1)(x - 2)(x - 3)$
4. $(5x^2 + 3x + 2)/2$ **5.** x^3

Exercises 4.8

1. $7 + i$ **2.** $24 + 7i$ **3.** 25 **4.** $(-1 + 18i)/25$ **5.** $9/5$
6. $a^2 - b^2 + 2abi$ **7.** $a^3 - 3ab^2 + i(3a^2b - b^3)$
8. $(a^2 + b^2)^4 = a^8 + 4a^6b^2 + 6a^4b^4 + 4a^2b^6 + b^8$
13.* $c = \dfrac{\pm \sqrt{a + \sqrt{a^2 + b^2}}}{2}$; $d = b/2c$ if you eliminate d first; if you eliminate e, the formula appears slightly different, but is the same.

Exercises 5.1

2. y coordinate $= 0$; x coordinate $= 0$ **3.** $x = y$ and $x = -y$
4. Square on its corner   **5.** Square with side length $= 2$

Exercises 5.2

1. (a) 5, (b) 5, (c) 0, (d) 3   **3.** $d(p_1, p_2) = \sqrt{5}$, $d(p_1, p_3) = \sqrt{5}$, $d(p_2, p_3) = \sqrt{2}$
4. The points are all on one line.   **5.** $d(p_1, p_3) = d(p_2, p_3) = \sqrt{5}$
6.* Yes; $D(x, y)$ satisfies the distance conditions.

Exercises 5.3

1. The coordinates satisfy the equation of the line.
2. The coordinates satisfy both equations.
5. $x = \frac{3}{2}$, $y = -\frac{1}{2}$ **6.** $x = \frac{21}{25}$, $y = \frac{-28}{25}$ **7.** y-axis; x-axis
8. $45°$ line through the origin   **9.** Point $(0, 0)$ satisfies the equation.

Exercises 5.4

2. $7x + 3y = 4$ **3.** $y = 3x/2$ **4.** $y = 2x + 1$
5. $y = 3x$, $y = 3x/2$, $y = 2x + 1$ **7.** $B = 0$, $A = 0$, $C = 0$
8. Horizontal line, vertical line, line through the origin

Exercises 5.5

1. 3   2. -3 3. $\frac{1}{2}$ 4. $-\frac{3}{2}$ 5. $-p/q$
6. $\tan \phi = -3$ 7. $\tan \phi = 13$ 10. 3

Exercises 5.6

1. $x + y + 1 = 0$ 2. $y = x$ 3. $y = 1$ 4. $y = -x$ 5. $y = -3x$
6. $x + y - 2 = 0$ 7. $x + y - 2 = 0$ 8. $x + y + a = 0$, slope $= -1$
9. Slope $= \frac{3}{4}$, x intercept $= \frac{19}{3}$ 10. Slope $= -1$, y intercept $= -1$
11. $a = -2$, $b = -\frac{4}{3}$ 12. $c = -\frac{5}{7}$, $b = \frac{-5}{9}$ 13. $(0, 1)$
14. $\left(\dfrac{p + r}{2}, \dfrac{q + s}{2}\right)$ 15. $\left(\dfrac{2p + r}{3}, \dfrac{2q + s}{3}\right)$, $\left(\dfrac{p + 2r}{3}, \dfrac{q + 2s}{3}\right)$
16.* $\left(\dfrac{(n - k)p + kr}{n}, \dfrac{(n - k)q + ks}{n}\right)$, $k = 1, 2, \ldots, n - 1$

Exercises 5.8

1. $-\dfrac{5}{13}x - \dfrac{12}{13}y - \dfrac{36}{13} = 0$ 2. $\dfrac{4}{\sqrt{17}}x - \dfrac{1}{\sqrt{17}}y - \dfrac{5}{\sqrt{17}} = 0$
3. $\dfrac{4}{5}x + \dfrac{3}{5}y - \dfrac{12}{5} = 0$ 4. -1 5. 0   6. $-\dfrac{1}{\sqrt{2}}$
7. $\tan \theta = +1$, $p = \dfrac{17}{2\sqrt{2}}$
8. Divide each equation by its $\pm\sqrt{A^2 + B^2}$ and equate to $+$ or $-$ the distance of the other.
10. $\dfrac{Ax + By}{\pm\sqrt{A^2 + B^2}} = 0$, either sign

Exercises 5.9

1. $x' - y' = 0$, $x' + y' = 0$ 2. $Ax' + By' + C + Ak + Bk = 0$
3. $3x' + 4y' = 0$, $x' - y' = 0$ 4. Set $C_1 = C_2 = 0$ and put primes on the variables.

Exercises 5.10

1. $-\frac{1}{2}$ 2. 0   3. $(a - b)c$ 4. $\frac{1}{2}[ab + bc + ca - a^2 - b^2 - c^2]$

Exercises 5.12

2. 5   5. $\sqrt{2}/2$ 7. 1, 1, 1, 1

Exercises 6.2

1. $x^2 + y^2 + 2x - 2y + 1 = 0$ 2. $x^2 + y^2 + 6x + 8y = 0$
3. $x^2 + y^2 - 2x - 2y - 2 = 0$ 4. $x^2 + y^2 = 2$
5. No circle; points lie on a line
6. $x^2 + y^2 - 5x - 12y = 0$ 7. $x^2 + y^2 - 12.8x - 10y = 0$
8. No circle possible; radius too small
9. Same as Exercise 8

Exercises 6.3

1. Add 9   2. Add $\frac{27}{4}$ 3. Add $\frac{9}{20}$ 4. Outside
5. $x^2 + y^2 + 4x + 6y = 36$ 6. $x^2 + y^2 + 10x - 24y = 0$ 7. $F = 13$
8. $x^2 + y^2 - \frac{20}{3}x + \frac{2}{3}y + 4 = 0$, $(\frac{10}{3}, \frac{-1}{3})$, $r = \dfrac{\sqrt{65}}{3}$

Exercises 6.4

1. $\dfrac{(x + 2)^2}{4^2} + \dfrac{(y - 3)^2}{4^2} = 1$ 2. $-\dfrac{[x + (1/3)^2]}{23/9} + \dfrac{(y + 1)^2}{23/12} = 1$

3. $\dfrac{(x - 2)^2}{3^2} - \dfrac{(y + 1)^2}{(3/2)^2} = 1$ 4. $\dfrac{[x + (1/2)^2]}{2} + \dfrac{[y - (1/2)^2]}{2} = 1$

5. $\dfrac{(x + 2)^2}{3} - \dfrac{(y - 2)^2}{2} = 1$ 6. $5x^2 + 3y^2 - 14x - 6y = 0$

Exercises 6.5

4. Center at $(1, -1)$ 5. $\dfrac{(x + 2)^2}{4} + \dfrac{(y - 1)^2}{9} = 1$ 6. $x^2 + 9_y^2 - 18y = 0$
7. As $k \to 0$, ellipse approaches the point $(0, 0)$; $k < 0$ imaginary.

Exercises 6.6

4. $(x + 2)^2 + (y - 3)^2 = -2$
5. $(x + 2)^2 - y^2 = 0$, pair of lines $y = \pm(x + 2)$
6. Hyperbola goes to two lines as $k \to 0^+$ and up the other axis for $k < 0$.
7. Ellipse shape

Exercises 6.8

4. Imaginary pair of lines   **5.** Pair of parallel lines, $y = \pm 2$
6. Pair of coincident lines, $x = -2, x = -2$

Exercises 6.9

1. Angle 45°; equation, $4(x')^2 = 2$ **2.** $(x')^2 - (y')^2 = 2$

Exercises 6.11

3. (a) About y-axis; (b) about x-axis.   **4.** About 45° line   **5.** About origin

Exercises 7.3

1. a **2.** a/c **3.** ad **4.** a/c **5.** a/b **6.** 0

Exercises 7.4

1. 0   **2.** 2   **3.** 2   **4.** 3   **5.** n **6.** na^{n-1} **7.** $\dfrac{1}{2\sqrt{a}}$

Exercises 7.5

1. $4x^3$ **2.** $2ax + b$ **3.** $-1/x^2$ **4.** $-1/(x + 1)^2$
5. $\dfrac{1}{2\sqrt{x}}$ **6.** $20x^3$ **7.** $\dfrac{a}{2\sqrt{ax + b}}$ **8.** $\dfrac{-2x}{(x^2 + 1)^2}$

Exercises 7.6

1. $0, 1, 2x, 3x^2, 4x^3, 19x^{18}, 1, nx^{n-1}$ **2.** $12x, 12x^2, 12x^3, 12x^5, 12x^{11}$
3. $a, 3cx^2, nbx^{n-1}, 0, 12ax^3$ **4.** $4x^3$ **5.** $21x^2$ **6.** $34x$

Exercises 7.7

1. $2x$ **2.** $3x^2 - 6x$ **3.** $3x^2 + 2x + 1$ **4.** $8x - 2$ **5.** $x^2 + x + 1$
6. $2ax + b$ **7.** $nax^{n-1} + mbx^{m-1}$ **8.** $n(a + 1)x^{n-1}$ **9.** nx^{n-1}
10. $18x(x^2 + 1) + 16x(x^2 - 1)$ **11.** $2nx(a^2 + x^2)^{n-1} - 2mx(a^2 - x^2)^{m-1}$
12. $3(x + 1)^2 + 6(x + 1) = 3(x + 1)(x + 3)$

Exercises 7.8

1. $2x$ **2.** $4x^3$ **3.** $4x(x^2 + 1)$ **4.** 1   **5.** $3x^2 + 12x + 11$
6. $2x[2x^2 + a^2 + b^2]$

7. $2x[(x^2 - b^2)(x^2 - c^2) + (x^2 - a^2)(x^2 - c^2) + (x^2 - a^2)(x^2 - b^2)]$

8. $2/(x + 1)^2$ 9. $-a/(x - a)^2$ 10. $(b - a)/(x + b)^2$

11. $x(x^3 + 3a^2x - 2a^3)/(x^2 + a^2)^2$ 12. $-4a^2b^4x/(a^2x^2 - b^4)^2$

Exercises 7.9

1. $-1/x^2, -2/x^3, -3x^{-4}, -25x^{-6}$ 2. $-1/(a + x)^2$ 3. $-2x/(a^2 + x^2)^2$

4. $-4x/(a^2 + x^2)^3$ 5. $4x/(a^2 - x^2)^3$ 6. $1/\sqrt{2x}, 1/\sqrt{x}$

7. $1/\sqrt{x}, \sqrt{2}/(2\sqrt{x})$ 8. $2x^{-1/3}$ 9. $\frac{1}{2}\sqrt{x}$ 10. $-x^{-4/3}$

11. $\frac{1}{2}(a + x)^{-1/2}$ 12. $\frac{1}{2}(b - x)^{-3/2}$ 13. $-x/\sqrt{a^2 - x^2}$ 14. $x/(a^2 - x^2)^{3/2}$

15. $\dfrac{a^2 + x^2}{(a^2 - x^2)^2}$ 16. $\dfrac{1}{2\sqrt{x}} - \dfrac{1}{2x^{3/2}}$ 17. $-a^2/(x^2 - a^2)^{3/2}$ 18. $\dfrac{1}{4\sqrt{x + \sqrt{x}}}$

19. $\frac{4}{3}x^{1/3} - \frac{4}{3}x^{-7/3}$ 20. $\dfrac{1}{\sqrt{x}(\sqrt{x} + 1)^2}$ 21. $\dfrac{1}{2\sqrt{x}\sqrt{\sqrt{x} - 1}(\sqrt{x} + 1)^{3/2}}$

Exercises 7.10

1. $\frac{1}{3}x^{-2/3} + \frac{4}{3}x^{-1/3} + 4x^{1/3}$ 2. $\dfrac{-[bx^2 + 2(a + c)x + b]}{(x^2 - 1)^2}$

3. $(m - np)p^{m-1}(1 - p)^{n-m-1}$ 4. $4x^3 - 18x^2 + 22x - 6$

5. $-(2x + 1)/x^2(x + 1)^2$ 6. $\dfrac{-1}{2x^{3/2}}$ 7. $-\dfrac{4}{3x^{7/3}}$ 8. $\frac{1}{5}x^{-6/5}$

9. $\dfrac{x^2 - 2cx + ac + bc - ab}{(x - c)^2}$ 10. $\dfrac{1}{2\sqrt{x}}$

Miscellaneous Exercises

1. $4x^3$ 2. $\dfrac{1}{(x + 1)^2}$ 3. $\dfrac{1 + x^2}{(1 - x^2)^2}$ 4. $3x^2 - 2x + 1$ 5. $\dfrac{-1}{\sqrt{1 - s}(1 + s)^{3/2}}$

6. $\dfrac{1}{2\sqrt{x}}$ 7. $4x + 1$ 8. $8uvx$ 9. $\dfrac{-4a^2x}{(x^2 - a^2)^2}$

10. $1 + \dfrac{1}{2\sqrt{x + 1}}$ 11. $\dfrac{2a^2t}{(a^2 - t^2)^2}$ 12. $\dfrac{-3(1 - \sqrt{x})^2}{2\sqrt{x}}$ 13. $\dfrac{-(a^{2/3} - x^{2/3})^{1/3}}{x^{1/3}}$

14. $16x(1 + x^2)\{1 + (1 + x^2)^2\}[1 + \{1 + (1 + x^2)^2\}^2]$ 15. $\dfrac{3\sqrt{x}}{2}$

16. $\dfrac{1}{8\sqrt{x}\sqrt{1 + \sqrt{x}}\sqrt{1 + \sqrt{1 + \sqrt{x}}}}$ 17. $\dfrac{-8x}{(x^2 - 1)^2}$ 18. $-\dfrac{3}{2}x^{-5/2}$

19. $2x + 3a$ 20. $\dfrac{1}{4\sqrt{u + 1}\sqrt{x + 1}}$ 21. $\dfrac{-a}{t\sqrt{2at - t^2}}$ 22. $3x\sqrt{x^2 + a^2}$

23. $(2 - 5t)t(1 - t)^2$ **24.** $\dfrac{-t}{2(1 - t^2)^{3/2}}$ **25.** $\dfrac{-2(2y + a)}{[y(y + a)]^3}$

Exercises 8.1

1. $y = 3x - 2, x + 3y = 4$ **2.** $y = 4x - 3, x + 4y = 5$ **3.** $x + y = 2, y = x$
4. $y = x + 1, x + y - 1 = 0$ **5.** $x - 2y + 1 = 0, 2x + y = 3$
6. $x + 2y = 3, y = 2x - 1$ **7.** $y = nx - n + 1, x + ny = n + 1$

8. $y = (2ax_1 + b)x - ax_1^2 + c, y = y_1 - \dfrac{x - x_1}{2ax_1 + b}$.

Exercises 8.2

1. $\dfrac{x}{\sqrt{x^2 + 1}}, \dfrac{1}{(x^2 + 1)^{3/2}}$ **2.** $nx^{n-1}, n(n - 1)x^{n-2}$ **3.** $\dfrac{x}{(1 - x^2)^{3/2}}, \dfrac{1 + 2x^2}{(1 - x^2)^{5/2}}$

4. $-\dfrac{1}{x^2}, \dfrac{2}{x^3}$ **5.** $3! \ n(n - 1)(n - 2)x^{n-3}, \dfrac{-3!}{x^4}, \dfrac{3}{8}x^{-5/2}$ **7.** $\dfrac{(-1)^n n!}{x^{n+1}}$

8. $\dfrac{(-1)^{n-1}(2n - 2)!}{2^{2n-1}(n - 1)!}$ **9.** $\dfrac{m!}{(m - n)!}(a + bx)^{m-n}b^m$ for $(m \geq n)$, 0 for $m < n$

Exercises 8.3

1. $-b/a$ **2.** -1 **3.** -1 **4.** -1 **5.** -1 **6.** Not defined
7. $x + y = 2a, y = x$ **8.** $x + y = 2, y = x$
9. $ay + bx = \sqrt{2}ab, by - ax = \dfrac{b^2 - a^2}{\sqrt{2}}$ **10.** Try $x^n + y^n = 2a^n$

Exercises 8.4

1. 0   **2.** 2   **3.** $\dfrac{1}{a}$ **4.** Infinite   **5.** $\dfrac{6}{13\sqrt{13}a}$

Exercises 8.5

1. Base $= \sqrt{2}$ side   **2.** Extra tin on rim for one thing
3. Depth $= (a + b - \sqrt{a^2 + b^2 - ab}/6$ **4.** $r = h$, half as high   **5.** $r/h = \sqrt{2}/2$
6. $r = 2R/3, h = H/3$
7. Width $= \dfrac{P(6 + \sqrt{3})}{33}$; height of side $= P\left(\dfrac{5 - \sqrt{3}}{22}\right)$; total height $=$ width
8. $x = a/\sqrt{2}, y = b/\sqrt{2}$ **9.** From sphere of radius $a = ca^{3/2}(a^{3/2} + b^{3/2})^{-1}$
10. $d = \sqrt{2}h$ **11.** From $A, CA^{3/2}/(A^{3/2} + B^{3/2})$ **12.** Area $= 16$

13. $V = \left(\dfrac{5}{5\pi}\right)\dfrac{5}{3}$ **14.** Width $= \dfrac{2P_1}{4 + \pi}$, side height $= \dfrac{P_1}{4 + \pi}$ **15.** 90° angle

16. (a) No; (b) yes; (c) yes; (d) no; (e) depends on how area is counted; the shape that has maximum area for a given perimeter

Exercises 8.6

1. $2\sqrt{3}/3$ **2.** $\mp\sqrt{3}/2$
3. Zeros, $-1, -1, 0, 1, 1$, max, min $\pm 1/\sqrt{5}, \pm 1$, inflection points $0, \pm\sqrt{3/5}$

Exercises 9.1

1. $\dfrac{1}{6}\dfrac{dy}{dx}, \dfrac{1}{18}, \dfrac{d^2y}{dx^2}$ **2.** $\dfrac{e + f(dy_1/dx_1)}{b + c(dy_1/dx_1)}$

Exercises 9.4

1. $v = 2t - 1, a = 2$ **2.** 45°   **3.** 0   **4.** 40

Exercises 9.5

2. $6\pi r h, 3\pi r^2$ **3.** $-3/40$ m/s   **4.** Rate of decrease $= 5/[\pi h(2\pi r - h)]$
6. $\sin\theta_1 = C_1/C_2$, if $C_1 < C_2$

Exercises 9.6

1. $\dfrac{dI}{dx} = 2d\left\{\dfrac{I_2(c - x)}{[h_2^2 + (c - x)^2]} - \dfrac{I_1 x}{[h_1^2 + x^2]^2}\right\}$

2. $\dfrac{dz}{dt} = (43{,}600 - 1200\sqrt{2})\dfrac{t}{z}, z^2 = 100[436 - 120\sqrt{2}]t^2$

3. $10\left[\dfrac{H_2}{(c - x)^3} - \dfrac{H_1}{x^3}\right]$ **4.** 120/7   **5.** $\dfrac{dC}{dt} = \dfrac{2\pi R}{C}$ **6.** $vR/2$

Exercises 9.7

1. 1.2599   **2.** $x_{n+1} = \dfrac{2x_n + (N/x_n^2)}{3}$

4. $x_{n+1} = \dfrac{(n - 1)x_n + (N/x^{*\,n-1})}{n}$ **5.** 0.755

Exercises 9.8

1. 1, 1, 1, 2   **2.** 1, 1, 1, 2, 3

Exercises 9.10

1. $n(n + 1)2^{n-2}$ **2.** $\dfrac{x(x + 1)}{(1 - x)^3}$ **3.** 2, 6   **5.** 3^n **6.** $(2n)3^{n-1}$ **7.** 0

Exercises 9.11

1. Coefficient of t^n in $(1 - t^2)^n$, in particular 0 if n is an odd number.

Exercises 9.13

1. $dy = 12x^3 dx$ **2.** $dy = \dfrac{-x \, dx}{\sqrt{a^2 - x^2}}$ **3.** $dy = \dfrac{-2 \, dx}{(1 + x)^2}$ **4.** $dy = \dfrac{dx}{2\sqrt{x}}$

5. $2x \, dx + 2y \, dy = 0$ **6.** $3y^2 dx + 6xy \, dy + 6xy \, dx + 3x^2 dy = 0$

7. $\dfrac{dx}{2\sqrt{x}} + \dfrac{dy}{2\sqrt{y}} = 0$ **8.** $dy = 2x(x^2 - 6)(3x^2 + 2) \, dx$

9. $dy = 8x(1 + x^2)^3 dx$

Exercises 10.1

1. $2x - 3y + 4z - 5 = 0$ **2.** $x + y + z - 1 = 0$
3. $6x + 3y + 2z = 6$ **5.** $D = 0$ **6.** $C = 0$ **7.** Each parallel to a given line
8. $A_1/A_2 = B_1/B_2 = C_1/C_2$ **12.** $x = 1, 2y + z + 1 = 0$
13. $x = 2, y = 1, z = 3$. One generalization is to make the right-hand sides a, b, c; another, increase the number of variables and equations with the right-hand sides consecutive integers.

Exercises 10.2

1. $x^2 + y^2 + z^2 = 16$ **2.** $x^2 + y^2 + z^2 + 6x - 8y = 0$
3. Center $(1, 2, -3)$, $r = 5$ **4.** Center $(\frac{1}{2}, -\frac{1}{2}, -\frac{3}{2})$, $r = \sqrt{\pi}/2$
8. $x^2 + y^2 + z^2 - x - y - z = 0$

Exercises 10.3

1. $\dfrac{\partial z}{\partial x} = 2x, \dfrac{\partial z}{\partial y} = 2y, \dfrac{\partial^2 z}{\partial x \, \partial y} = 0$

2. $\dfrac{\partial z}{\partial x} = \dfrac{x}{\sqrt{x^2 + y^2 - 1}}, \dfrac{\partial z}{\partial y} = \dfrac{y}{\sqrt{x^2 + y^2 - 1}}, \dfrac{\partial^2 z}{\partial x \, \partial y} = \dfrac{-xy}{(x^2 + y^2 - 1)^{3/2}}$

3. $\dfrac{\partial z}{\partial x} = \dfrac{1}{y}, \dfrac{\partial z}{\partial y} = \dfrac{-x}{y^2}, \dfrac{\partial^2 z}{\partial x \, \partial y} = \dfrac{-1}{y^2}$

4. $\dfrac{\partial z}{\partial x} = 2xy + y^2, \dfrac{\partial z}{\partial y} = x^2 + 2xy, \dfrac{\partial^2 z}{\partial x \, \partial y} = 2(x + y)$

5. $\dfrac{\partial z}{\partial x} = \dfrac{2y}{(x + y)^2}, \dfrac{\partial z}{\partial y} = \dfrac{-2x}{(x + y)^2}, \dfrac{\partial^2 z}{\partial x \, \partial y} = \dfrac{2(x - y)}{(x + y)^3}$

6. $x + y + z = \sqrt{3}$

Exercises 10.4

1. 5.1   **2.** -3.66 **3.** $-5\frac{30}{7}$

Exercises 10.5

2. $y = \dfrac{3}{10}(6 - x)$

3. $y = -x$

4. $y = \dfrac{x}{10} \sum x_n y_n + \dfrac{1}{5} \sum y_n$

5. The summation would go to $n - 1$ instead of n.

Exercises 10.6

1. $y = \dfrac{113}{1025}x^2$ **2.** $\sum_{j=0}^{5} S_{j+k}c_j = \sum_{i=1}^{n} (x_i)^k y_i$, for $k = 0, 1, \ldots, 5$

Exercises 10.8

1. $a \sum u_1^2 + b \sum u_1 u_2 + c \sum u_1 u_3 = \sum y u_1$

$a \sum u_1 u_2 + b \sum u_2^2 + c \sum u_2 u_3 = \sum y u_2$

$a \sum u_1 u_3 + b \sum u_2 u_3 + c \sum u_3^2 = \sum y u_3$

Exercises 11.6

1. $\dfrac{x^4}{4} + C$ **2.** $\dfrac{x^2}{2} + \dfrac{x^3}{3} - \dfrac{x^4}{4} + C$ **3.** $ax + \dfrac{bx^2}{2} + \dfrac{cx^3}{3} + C$ **4.** $\dfrac{2x^{3/2}}{3} + C$

5. $3x + 2\sqrt{x} + C$ **6.** $cx - \dfrac{2x^{3/2}}{3} + C$ **7.** $\dfrac{2}{5}x^{5/2} - \dfrac{2}{3}x^{3/2} + \dfrac{1}{2}x^{1/2} + C$

8. $\dfrac{(a^2 + x^2)^5}{10} + C$ **9.** $\dfrac{1}{3}(x^2 - a^2)^{3/2} + C$ **10.** $\dfrac{(x^3 - x)^4}{4} + C$

11. $\dfrac{2(x - 1)^{3/2}}{3} + C$ **12.** $2\sqrt{x + a} + C$ **13.** $\dfrac{2x^{3/2}}{3} - 2x^{1/2} + C$

14. $\dfrac{(1 + x^{3/2})^4}{6} + C$ **15.** $\dfrac{3}{8}(a^2 + x^2)^4 + C$

Exercises 11.9

1. $\frac{1}{4}$ **2.** $\frac{2}{15}$ **3.** 0 **4.** $\frac{1}{3}$ **5.** 2 **6.** $\frac{58}{105}$ **7.** a **8.** $a^3/3$
9. $(10\sqrt{2} - 8)/3$ **10.** $8\pi/105$ **11.** $8\pi/315$ **12.** $\pi/6$ **13.** $7\pi/20$
14. $2\pi a^5/15$ **15.** $4\pi/15$ **16.** $\pi/6$ **17.** $28\pi/3$ **18.** $\pi/5$
19. $1/(n + 1),\ \pi/(2n + 1),\ 2\pi/(n + 2)$ **20.** $1/(n + 1)$ **21.** $\dfrac{2^{n+2} - (n + 2)}{(n + 1)(n + 2)}$

Exercises 11.10

1. $\dfrac{(1 + x^2)^4}{8} + C$ **2.** $\dfrac{(x^{2/3} + 1)^3}{2} + c$ **3.** $-1/6$ **4.** $1/3$ **5.** $\dfrac{1}{2(k + 1)}$
6. 3 **7.** $\dfrac{t^2}{2} + \dfrac{2t^{3/2}}{3} + C$

Exercises 12.4

1. $P(10, 7) = \dfrac{10!}{3!}$ **2.** $2 \times 4 \times 7 = 56$ **3.** $9 \times 3 \times 4 = 108$ **4.** $\frac{1}{6}$ **5.** $\frac{1}{6}$
6. $P(2) = \frac{1}{36},\ P(3) = \frac{2}{36},\ P(4) = \frac{3}{36},\ P(5) = \frac{4}{36},\ P(6) = \frac{5}{36}\ P(7) = \frac{6}{36},\ P(8) = \frac{5}{36},$
$P(9) = \frac{4}{36},\ P(10) = \frac{3}{36},\ P(11) = \frac{2}{36},\ P(12) = \frac{1}{36}$
7. $P(6) = 0.414,\ P(7) = 0.531$ **8.** $\dfrac{10}{10} \cdot \dfrac{9}{10} \cdot \dfrac{8}{10} \cdots \dfrac{10 - k + 1}{10}$
9.
AB	BA	CA	DA	EA
AC	BC	CB	DB	EB
AD	BD	CD	DC	EC
AE	BE	CE	DE	ED
10. $P(10)\{\text{no duplicate}\} = 0.5031$ **11.** 5×21^2
12. Problem of no duplicate
$$P(k) = \frac{1000}{1000} \cdot \frac{999}{1000} \cdots \frac{1000 - k + 1}{1000}$$
$$P(50) = 0.2877 \ldots$$

Exercises 12.5

1. $\dfrac{1}{52}$ **2.** $\dfrac{52!}{(13!)^4}$ **3.** $\dfrac{P(n, k)}{C(n, k)} = k!$ **4.** $C(n, k)$ **5.** $C(52, 7)$ **6.** $C(20, 5)$
7. $C(12, 5)/2^{12}$ **9.** $1 - 1(1 - \frac{1}{6})(1 - \frac{2}{6}) \ldots (1 - \frac{k}{6})$

Exercises 12.6

3. Professor, $\dfrac{3s}{s} = 5$, student $= s + \dfrac{2k^2}{3s}$ **4.** 6
5. (a) $p^2 + q^2$, (b) $2pq(p^2 + q^2)$, (c) $(2pq)^2(p^2 + q^2) \cdots$, expected value $1/(p^2 + q^2)$

Exercises 12.10

1. $m_3 = np(1 - p)(2 - p) + 3n^2p^2(1 - p) + n^3p^3$; central moment $= np(1 - p)(2 - p)$

2. $m_1 = \dfrac{p}{q}$, $m_2 = \dfrac{p}{q} + \dfrac{2p^2}{q^2}$, $m_3 = \dfrac{1}{q} + 6\left(\dfrac{p}{q}\right)^2 + 6\left(\dfrac{p}{q}\right)^3$

3. Mean $= \dfrac{p}{q}$, variance $= \dfrac{p}{q} + \dfrac{p^2}{q^2}$

Exercises 13.2

1. $m_1 = 1/2$, $\sigma^2 = 3/10$ 2. $m_1 = 0$, $\sigma^2 = 1/6$ 3. $m_1 = 0$, $\sigma^2 = 4/3$

4. $m_1 = 0$, $\sigma^2 = 1/5$ 7. $\dfrac{16\pi}{100}$

Exercises 13.3

1. $\theta = 1/(k + 1)$

Exercises 13.8

1. $7/16$ 2. $\dfrac{2\sqrt{3} - 1}{3}$ 3. $2/3$

4. $13/24$ 5. $7/15$ (*Hint:* Use symmetry twice)

Exercises 14.2

2. (*a*) Reduce to degree $N(x) <$ degree $D(x)$.
   (*b*) In the derivative, denominator degree $\geq 2 +$ numerator degree; hence $\neq 1/x$.

Exercises 14.3

1. $2 \ln 3$ 2. $2 \ln 2 + \ln 3$ 3. $3 \ln 3$

4. $\frac{1}{2} \ln 2$ 5. $\frac{1}{4} (\ln 2)^2$ 6. $3 \ln 2 + \ln 3$

7. $\frac{3}{2} \ln 3$ 8. $3 \ln 3 - 3 \ln 2$ 9. $-8 \ln 2$

10. 1 11. 2 12. $3/2$ 13. $\frac{1}{4}(\ln 2)^2$

14. 2 15. $1/2$

Exercises 14.4

4. $n = \dfrac{\ln 2}{\ln (1 + r)}$

Exercises 14.5

1. $\dfrac{1}{2x\sqrt{\ln x}}$ 2. $x + 2x \ln x$ 3. $\dfrac{3(x^2 - 1)}{x^3 - 3x}$

4. $nx^{n-1} \ln x$ 5. $\ln|1 + x| + C$ 6. $\dfrac{1}{b}\ln|a + bx| + C$

7. $-\dfrac{1}{b(a + bx)} + C$ 8. $\dfrac{1}{2}\ln(1 + x^2) + C$

9. $\frac{1}{2}\ln(a^2 + x^2) + C$ 10. $-\dfrac{1}{2(a^2 + x^2)} + C$

12. $\frac{1}{4}(\ln x)^4 + C$ 13. $\dfrac{\ln^{a+1}x}{a + 1} + C$ 14. $\frac{1}{2}(5 - 4e + e^2) + C$

15. $\ln|x| + \dfrac{\ln^2 x}{2} + C$ 16. $\ln|1 + \ln x| + C$

Exercises 14.7

1. $(x^4/4) \ln x - (x^4/16) + C$
3. $x \ln^n x - nx \ln^{n-1} x + n(n - 1) \ln^{n-2} x + \cdots + (-1)^n n! \, x + C$
6. Requires only integer exponent on $\ln x$
7. Too rare a form to remember

Exercises 14.8

6. $\text{var} = \dfrac{b^2 - 1}{2 \ln b} - \dfrac{(b - 1)^2}{\ln^2 b}$

Exercises 14.9

1. 1,2   2. 2,3   3. $1/a,\ 1/a$
4. 1/2   5. 1

Exercises 14.10

1. $\displaystyle\sum_{k=0}^{N} Q_N(x) \ln^k x + C$, where N = degree in $P(x, \ln x) = 0$

2. $Q_2(x) \ln^2 x + Q_1^{(x)} \ln x + Q_0(x)$

$Q_2(x) = \displaystyle\int_0^x P_2(x)\, dx$

$Q_1(x) = \displaystyle\int_0^x \left(P_1 - \dfrac{Q_2}{x}\right) dx$

$Q_0(x) = \displaystyle\int_0^x \left(P_0 - \dfrac{Q_1}{x}\right) dx$

Exercises 15.2

1. $\exp\left(x - \frac{1}{x}\right)\left(1 + \frac{1}{x^2}\right)$

2. $nx^{n-1} \exp(x^n)$

3. $-2(x^3 - x) \exp(-x^2)$

4. $(-x^n + nx^{n-1}) \exp(-x)$

5. $\dfrac{e^{-x}}{(1 + e^{-x})^2}$ 6. $\dfrac{e^{ay}[a + (a + b)e^{bx}]}{(1 - e^{bx})^2}$

7. $(-x^{n+1} + nx^{n-1})e^{-x^2/2}$

8. $\max = n^n e^{-n}$ at $x = n$

9. No maximum, upper bound $= 1$

10. At $x = 1$, $y = 1/\sqrt{e}$ 11. $x = \pm 1$, $y = 1/\sqrt{e}$

12. $x = n \pm \sqrt{n}$, $y = (n \pm \sqrt{n})^n e^{-(n \pm \sqrt{n})^2/2}$

13. $\frac{1}{3} \exp(x^3) + C$ 14. $e^x(x - 1) + C$

15. $e^x(x^2 - 2x + 2) + C$ 16. $-(x^2 + 2) \exp(-x^2/2) + C$

Exercises 15.3

1. $\pi/4$ 2. Mean $= n + 1$, $\sigma^2 = n + 1$

3. $2\pi[2 - (5/e)]$

Exercises 15.5

1. 0   2. 0   3. 2   4. 1/4!   5. 1/4!

Exercises 15.6

1. $2/a^3$ 2. $9/a^4$

Exercises 15.7

1. a/k

Exercises 15.8

1. $\mu = \dfrac{(n + 1)}{a}$, $\sigma^2 = \dfrac{(n + 1)}{a^2}$

Exercises 15.11

1. $\dfrac{(2\sigma^2)^{k-1/2}}{\sigma\sqrt{2\pi}} \Gamma(k + \frac{1}{2})$, k even; 0, k odd

2. $\Gamma(1.462) = 0.886$ approx.

Exercises 15.12

1. $-(x^3 + 3x^2 + 6x + 6)e^{-x}$
2. $-(x^4 + 4x^3 + 12x^2 + 24x + 24)e^{-x}$
3. $-\left(\dfrac{x^2}{a} + \dfrac{2x}{a^2} + \dfrac{2}{a^3}\right)e^{-ax}$

Exercises 16.5

1. $-2 \sin 2x$, $-2x \sin x^2$, $-2 \sin x \cos x = -\sin 2x$
2. $2 \tan x \sec^2 x$ 3. $2 \sec^2 x \tan x$
13. $1/(1 + \cos x)$ 14. $-1/(1 - \cos x)$
15. $2(1 + \sin x) \cos x$ 16. $\sec x(\tan^2 x + \sec^2 x)$
17. $-\csc x(\operatorname{ctn}^2 x + \csc^2 x)$

18. $\dfrac{-2 \cos x}{(1 + \sin x)^2}$ 19. $\dfrac{2 \sin x}{(1 + \cos x)^2}$ 20. 0

Exercises 16.6

1. $\dfrac{-\cos (ax + b)}{a} + C$ 2. $\dfrac{x}{2} + \dfrac{\sin x \cos x}{2} + C$
3. $\dfrac{\cos^3 x}{3} - \cos x + C$ 4. $-\operatorname{ctn} x - x + C$
5. $-x^2 \cos x + 2x \sin x + 2 \cos x + C$
6. $\frac{2}{3}(\tan x)^{3/2} + C$ 7. $\ln |1 + \tan x| + C$
8. $e^{ax}\left(\dfrac{a \cos bx - b \sin bx}{a^2 + b^2}\right) + C$

9. $\dfrac{-\cos x}{2 \sin^2 x} - \frac{1}{2} \ln |\csc x + \cot x| + C$
16. $1/4$

Exercises 16.7

1. $n!$ 2. 0 for $m \neq n$, π for $m = n \neq 0$, 2π for $m = n = 0$
3. 0   5. $2a^{n+1}W_{n+1}$ 6. $2a^n W_n$ 7. $2a^{n+2}(W_n - W_{n+2})$

Exercises 16.8

1. $x \operatorname{Arctan} x - \frac{1}{2} \ln(1 + x^2) + C$
2. $x \operatorname{Arcsin} x + \sqrt{1 - x^2} + C$

Exercises 16.10

6. $\dfrac{-5\sqrt{5-2x^2}}{2} + C$ **7.** $\dfrac{-5}{4}\ln|5-2x^2| + C$

8. $\dfrac{x\sin 3x}{3} + \dfrac{\cos 3x}{9} + C$ **9.** $\ln|x+1| + C$ **10.** $\dfrac{-1}{x-2} + C$

11. $\dfrac{e^{2x}}{2} - 2x - \dfrac{e^{-2x}}{2} + C$ **12.** Arctan $e^x + C$

13. $(x-1)\ln(1+\sqrt{x}) - \dfrac{x}{2} + \sqrt{x} + C$

14. $\dfrac{x^2}{2} - \tfrac{1}{2}\ln(x^2+1) + C$ **15.** $\dfrac{(x+1)^{3/2}}{3} - \sqrt{x^2+1} + C$

16. $\dfrac{e^{2x}[2\sin 3x - 3\cos 3x]}{13} + C$ **17.** $\dfrac{3y}{8} - \dfrac{\sin y}{2} + \dfrac{\sin y \cos y}{8} + C$

18. $\tfrac{2}{3}(\text{Arcsin } x)^{3/2} + C$ **19.** $-2t + 2\ln|\sec t| + 3\tan t + \dfrac{\tan^2 t}{2} + C$

20. $\dfrac{-1}{2(1-\cos\theta)^2} + C$ **21.** $e^{-t}\dfrac{(-\cos at + a\sin at)}{1+a^2} + C$

22. $\tfrac{1}{2}[\dfrac{\sin 5\theta}{5} - \sin\theta] + C$ **23.** $\arcsin\dfrac{x}{a} + C$ **24.** $-\sqrt{a^2-x^2} + C$

25. $\dfrac{a^2}{2}\arcsin\dfrac{x}{a} - \dfrac{x\sqrt{a^2-x^2}}{2} + C$ **26.** $\dfrac{x^2-1}{2}\ln(1-x) - \dfrac{x^2}{4} - \dfrac{x}{2} + C$

27. $\tfrac{1}{2}[(2x-1)\arcsin\sqrt{x} + \sqrt{x-x^2}] + C$

28. $-e^{-x}(x^2+2x+2) + C$ **29.** $\tfrac{1}{6}\sin 3t + \tfrac{1}{2}\sin t + C$

30. $\dfrac{x}{2}\sqrt{x^2+a^2} - \dfrac{a^2}{2}\ln|x+\sqrt{x^2+a^2}| + C$ **31.** $\dfrac{x^2}{2} - 2x + \ln|x| + C$

32. $-\cos x + \dfrac{\cos^3 x}{3} + C$ **33.** $\dfrac{3\pi}{16}$ **34.** $\dfrac{\pi}{2a}$ **35.** 1

36. 0   **37.** ∞ **38.** 8/3   **39.** 1/4

Exercises 17.2

1. $\dfrac{4}{x+3} - \dfrac{3}{x+2}$ **2.** $\dfrac{-3}{x+2} - \dfrac{2}{(x+2)^2} + \dfrac{3}{x+1} - \dfrac{1}{(x+1)^2}$

3. $\dfrac{1}{2}\left[\dfrac{1}{x+1} - \dfrac{8}{x+2} + \dfrac{9}{x+3}\right]$ **4.** $\dfrac{1}{x+a} - \dfrac{a}{(x+a)^2}$

5. $\dfrac{1}{a^3}\left[\dfrac{1}{x} - \dfrac{1}{x+a} - \dfrac{a}{(x+a)^2} - \dfrac{a^2}{(x+a)^3}\right]$

6. $\dfrac{1}{b^3}\left[-\dfrac{2}{x} + \dfrac{b}{x^2} + \dfrac{2}{x+b} + \dfrac{b}{(x+b)^2}\right]$

Exercises 17.3

1. $\dfrac{x}{x^2 + 1} - \dfrac{x}{(x^2 + 1)^2}$ 2. $x - \dfrac{x}{x^2 + 1}$ 3. $1 - \dfrac{a^2}{x^2 + a^2}$

4. $\dfrac{1}{x^2 + a^2} - \dfrac{a^2}{(x^2 + a^2)^2}$ 5. $\dfrac{x}{(x^2 + a^2)^2}$

6. $\dfrac{1}{2}\left[\dfrac{x}{x^2 + 1} - \dfrac{2x}{x^2 + 2} + \dfrac{x}{x^2 + 3}\right]$

7. $\dfrac{1}{a^6}\left[\dfrac{1}{x} - \dfrac{x}{x^2 + a^2} - \dfrac{xa^2}{(x^2 + a^2)^2} - \dfrac{xa^4}{(x^2 + a^2)^3}\right]$ 8. $\dfrac{x}{(x + 1)^2} = \dfrac{1}{x + 1} - \dfrac{1}{(x + 1)^2}$

Exercises 17.4

1. $\dfrac{1}{a} \ln \tan \dfrac{x}{2} + C$ 2. $x - \dfrac{a}{\sqrt{1 - a^2}} \ln \left(\dfrac{\sqrt{\dfrac{1 + a}{1 - a}} + \tan \dfrac{x}{2}}{\sqrt{\dfrac{1 + a}{1 - a}} - \tan \dfrac{x}{2}} + C\right)$

3. $x + \dfrac{2}{\sqrt{a^2 - 1}} \arctan \left(\sqrt{\dfrac{a - 1}{a + 1}} \tan \dfrac{x}{2}\right) + C$

4. $-\dfrac{1}{b} \ln (a + b \cos x) + C$

Exercises 17.5

2. $\dfrac{x}{2} - \dfrac{\sin 2x}{4} + C$ 3. $\dfrac{\sin^{a + 1} x}{a + 1} - \dfrac{\sin^{a + 3} x}{a + 3} + C$

4. $\dfrac{3x}{8} + \dfrac{\sin 2x}{4} + \dfrac{\sin 4x}{32} + C$

5. $\dfrac{x}{16} - \dfrac{\sin 4x}{64} + \dfrac{\sin^3 2x}{48} + C$

6. $\dfrac{5x}{16} + \dfrac{\sin 2x}{4} - \dfrac{\sin^3 2x}{48} + \dfrac{3 \sin 4x}{64} + C$

7. $\dfrac{x}{8} - \dfrac{\sin 4x}{32} + C$ 8. $\dfrac{1}{128}\left[3x - \dfrac{\sin 4x}{2} + \dfrac{\sin 8x}{8}\right] + C$

9. $\dfrac{1}{\cos x} + C$

Exercises 17.6

1. $I_n = \dfrac{-\operatorname{ctn}^{n-1} x}{n - 1} - I_{n-2}$, $I_1 = \ln |\sin x| + C$, $I_0 = x + C$

2. $I_n = \dfrac{\sin x}{(n - 1) \cos^{n-1} x} + \dfrac{n - 2}{n - 1} I_{n-2}$, $I_1 = \ln |\sec x + \tan x| + C$ $I_0 = x + C$

3. $I_n = I_{n-1} - \dfrac{1}{(n-1)x^{n-1}}$, $I_0 = -|\ln|1 - x| + C$

4. $I_n = I_{n-1} - \dfrac{x^n}{n}$, $I_0 = -\ln|1 - x| + C$

5. $I_n = -\dfrac{n}{2}I_{n-1} + \dfrac{x^2}{2}\ln^n x$, $I_0 = (x^2/2) + C$

6. $f(x)\ln x - \displaystyle\int \dfrac{f(x)}{x}\, dx$,   pick $f(0) = 0$

Exercises 17.7

1. π

2. π

3. $\left[\dfrac{1}{n} - \dfrac{1}{n - 1} + \dfrac{1}{n - 2} + \cdots + (-1)^n \ln 2\right]$

4. $2t^2 - 4t + 4\ln|t + 1| + C$

5. $-2\ln|\cos \sqrt{x}| + C$

6. Set $x = e^t$, $\int f(t)e^t\, dt$

7. Set $x = \sin\theta$, $\dfrac{\theta \sin^3\theta}{3} + \dfrac{1}{3}\cos\theta - \cos^3\theta + C$

8. Set $x = \tan\theta$, $\theta \tan\theta + \ln|\cos\theta| + C$

9. Set $x = e^t$, $\dfrac{te^{2t}}{2} - \dfrac{e^{2t}}{4} + C$

10. Set $x = a\sin^2\theta$, $\dfrac{a^2}{4}\left(\theta - \dfrac{\sin 4\theta}{4}\right) + C$

11. Set $x = t^{12}$.

Exercises 17.8

2. $2W_{n+1}$

3. $I_n = \dfrac{-x^{n-1}\sqrt{1 - x^2}}{n} + \dfrac{n - 1}{n}I_{n-2}$, $I_1 = -\sqrt{1 - x^2}$, $I_0 = \arcsin x$

4. $a^{n-1}\int \tan^n\theta\, d\theta$, etc.

5. $-\dfrac{1}{2(x^2 + a^2)} + C$

6. $\dfrac{32\pi a^3}{105}$

Exercises 18.3

1. $y = nx - (n - 1)$

2. $y = \pm x/\sqrt{3}$

3. Max − min $\tan bx = -a/b$, infl. $\tan bx = \dfrac{b^2 - a^2}{2ab}$

4. $\tan x = \pi - 2x$

5. $\dfrac{1}{4}\dfrac{b^2}{a(a^2 + b^2)}$

6. $\dfrac{\pi(2n)!}{2^{2n+1}(n!)^2}$

8. Falls at a rate $16/\pi$

9. $y = \dfrac{8(\pi - 3)}{\pi^2} + \dfrac{(24 - 7\pi)8}{\pi^3}x$

Exercises 18.4

2. $\dfrac{2}{27}[(1 + 9x)^{3/2} - 1]$

3. $\dfrac{8}{243}\left[\left(1 + \dfrac{81}{4}y\right)^{3/2} - 1\right]$

4. $\ln|\sec x + \tan x|$

Exercises 18.6

3. $\pi\left\{\sqrt{2} + \ln(\sqrt{2} + 1) - \dfrac{\sqrt{e^2 - 1}}{e^2} - \ln\left(1 + \dfrac{\sqrt{e^2 - 1}}{e}\right)\right\}$

4. $\dfrac{12\pi}{5}a^2$

Exercises 18.7

1. πa^3 **2.** $\dfrac{4a^3}{3}$ **4.** $\dfrac{2\pi a^2 c}{3}$

5. $\dfrac{16}{3}ab^2$

Exercises 18.9

1. $\frac{1}{4}$ distance from base   **2.** $\frac{3}{8}c$ **3.** $\dfrac{r^2}{2}$ **4.** $\dfrac{3r^2}{10}$

5. Total force $= \dfrac{16}{3}W$

Exercises 18.10

1. $\left[\dfrac{2\pi a^3}{3}h + \dfrac{\pi a^4}{4}\right]W$ **2.** $\dfrac{5}{12}\pi a^4 W$ **3.** $\dfrac{W}{2}$

Exercises 18.11

1. $\dfrac{2\pi ac}{(a^2 + c^2)^{3/2}}$ 2. $2\pi\left[1 - \dfrac{c}{\sqrt{c^2 + R^2}}\right]$ 3. $\dfrac{1}{a} - \dfrac{1}{b} \to \dfrac{1}{a}$

Exercises 19.1

1. $2/(15\sqrt{3})$ 2. $\dfrac{\pi}{2}$ 3. $\dfrac{31}{60}$ 4. $\dfrac{1}{54}$

Exercises 19.2

1. π 2. $\dfrac{\pi a^4}{2}$ 3. π 4. $\dfrac{4}{3}$ 6. $\dfrac{a^3}{24}$

Exercises 19.4

1. $2\pi a,\ \pi a^2$ 2. a 3. $2a/n$

Exercises 19.5

1. $P = \dfrac{b \ln b}{(b - 1)^2} - \dfrac{1}{b - 1}$

2. $h(z) = \dfrac{1}{z^2}\displaystyle\int_{1/b}^{z} x\,f(x)\,g(x/z)\,dx + \dfrac{1}{bz^2}\displaystyle\int_{z}^{1} x\,f(x)\,g(x/bz)\,dx$

4. Yes

Exercises 19.6

3. $4(u^2 - v^2)$

Exercises 19.7

2. $a^3/210$

Exercises 19.11

1. $3\pi a^2$ 2. $5\pi^2 a^3$

Exercises 19.12

1. $\dfrac{1}{4}\left[2\sqrt{5} + \ln(2 + \sqrt{5})\right]$ 2. $\dfrac{2}{27}\left[13^{3/2} - 8\right]$

Exercises 20.2

1. No. $u_n \to 1$

2. $\sum \dfrac{a}{1 + b^n} < a \sum \dfrac{1}{b^n} = \dfrac{a}{1 - (1/b)} = \dfrac{ab}{b - 1}$ Yes

3. $\sum \dfrac{a}{b^n + (-1)^n} \leq a \sum \dfrac{1}{(1/2)b^n} = 2a \sum \dfrac{1}{b^n} = \dfrac{2ab}{b - 1}$ Yes

4. $S_n = u_{n+1} \to 0$, hence convergent

Exercises 20.3

1. D 2. D 3. C 4. C 5. C, D

Exercises 20.5

1. Yes   2. $|x| \leq 1$ 3. Yes   4. No

Exercises 20.6

1. $|x| \leq 1$ 2. $|x - x_0| < 1$ 3. All x 4. $|x - x_0| \leq 1$

Exercises 20.8

1. $\ln(1 + x) = x - \dfrac{x^2}{2} + \dfrac{x^3}{3} - \dfrac{x^4}{4} + \cdots$

2. $\ln(1 - x) = -x - \dfrac{x^2}{2} - \dfrac{x^3}{3} - \dfrac{x^4}{4} - \cdots$

3. $\ln\left[\dfrac{1 + x}{1 - x}\right] = +2\left[x + \dfrac{x^3}{3} + \dfrac{x^5}{5} + \cdots\right]$

4. $\cos x = 1 - \dfrac{x^2}{2!} + \dfrac{x^4}{4!} - \dfrac{x^6}{6!} + \cdots$

5. $\tan x = x + \dfrac{x^3}{3} + 2\dfrac{x^5}{15} + \cdots$

6. $\dfrac{842}{1215} = 0.693 \ldots, x = \frac{1}{2}$

7. $y(0) = 1,\ y'(0) = 0,\ y''(0) = 1,\ y'''(0) = 0, \ldots$

Exercises 21.2

1. $\dfrac{2}{2}, \dfrac{3 \cdot 2}{2}, \dfrac{4 \cdot 3}{2}, \dfrac{5 \cdot 4}{2}, \dfrac{6 \cdot 5}{2}, \cdots, C(k + 1, 2)$

2. $\dfrac{2}{2 - t}, \mu = 1, \sigma^2 = 2$

3. $\dfrac{1}{1 - at}$ 4. $U(t) = \dfrac{t}{1 - at - bt^2}$

Exercises 21.3

1. $(-1)^k C(k + 2, k)$

4. $(qt)^3 \sum_{0}^{\infty} \dfrac{(k + 1)(k + 2)}{2} (pt)^k$

5. Mean $= \dfrac{d}{dt} G^3(t) \bigg|_{t=1} = \dfrac{3}{q}$

7. $\dfrac{2^{2n}}{(n + 1)\sqrt{\pi m}}$

Exercises 21.4

1. $m_k = \dfrac{k!}{a^n}$

2. $e^{xt} \left[\dfrac{t \sin \pi x - \pi \cos \pi x}{\pi^2 + t^2} \right]$

Exercises 21.5

1. $32 \cos^6 x - 48 \cos^4 x + 18 \cos^2 x - 1$

Exercises 21.6

1. $\sinh x = \dfrac{e^x - e^{-x}}{2} = \dfrac{1}{\operatorname{csch} x}$

    $\cosh x = \dfrac{e^x + e^{-x}}{2} = \dfrac{1}{\operatorname{sech} x}$

    $\tanh x = \dfrac{e^x - e^{-x}}{e^x + e^{-x}} = \dfrac{1}{\tanh x}$

2. $\dfrac{d}{dx} \sinh x = \cosh x$

    $\dfrac{d}{dx} \cosh x = \sinh x$

    $\dfrac{d}{dx} \tanh x = \operatorname{sech}^2 x$

    $\dfrac{d}{dx} \operatorname{ctnh} x = -\operatorname{csch}^2 x$

    $\dfrac{d}{dx} \operatorname{sech} x = -\operatorname{sech} x \tanh x$

    $\dfrac{d}{dx} \operatorname{csch} x = -\operatorname{csch} x \operatorname{ctnh} x$

3. $y = \dfrac{1}{2} \ln \left(\dfrac{1 + x}{1 - x} \right)$

4. $y = \ln (x + \sqrt{x^2 + 1})$

5. $\cosh x = 1 + \dfrac{x^2}{2} + \dfrac{x^4}{4!} + \cdots$

    $\sinh x = x + \dfrac{x^3}{3!} + \dfrac{x^5}{5!} + \cdots$

Exercises 22.3

1. $f(x) = \dfrac{1}{3} + \dfrac{2}{\pi} \displaystyle\sum_{k=1}^{\infty} \left[\dfrac{\sin (\pi k/3)}{k} \right] \cos kx$

2. $f(x) = \dfrac{1}{4} + \dfrac{2}{\pi} \displaystyle\sum_{k=1}^{\infty} \left[\dfrac{\sin (\pi k/4)}{k} \right] \cos kx$

3. $f(x) = \dfrac{3}{4} - \dfrac{2}{\pi} \displaystyle\sum_{k=1}^{\infty} \left[\dfrac{\sin (\pi k/4)}{k} \right] \cos kx$

Exercises 22.4

1. $(3 + 4i)/5$ 2. $2i$ 3. $-2 - 2i$

7. $\dfrac{2}{\pi} \displaystyle\sum_{k=-\infty}^{\infty} \left[\dfrac{\sin (\pi k/4)}{k} \right] e^{ikx}$

8. $\dfrac{1}{i} \left[\displaystyle\sum_{1}^{\infty} \dfrac{(-1)^k e^{ikx}}{k} - \sum_{1}^{\infty} \dfrac{(-1)^k e^{-ikx}}{k} \right]$

Exercises 22.5

1. $\dfrac{1}{\pi} \displaystyle\int_{-\pi}^{\pi} f^2(x) \, dx \geq \dfrac{a_0^2}{2} + \sum_{k=1}^{N} (a_k^2 + b_k^2)$

Exercises 22.8

1. $f(x)$ is continuous but $f'(x)$ is not; hence rate like $1/k^2$.
2. $f(x), f'(x), f''(x)$ all continuous, but $f'''(x)$ not continuous; hence rate like $1/k^4$.
3. Rate $= 1/k^2$

Exercises 23.2

4. $x + y(x^2 + y^2) = 0$

Exercises 23.4

1. $y = cx + C$
2. $y^2 = x^2 + C$
3. $y = Cx$
4. $y = Ce^{x^2/2}$

5. $y = Cxe^x - 1$

6. $y = C \exp (x^2/2 - x) + 1$

7. $y = Cx^x e^{-x}$

8. $\left.\begin{array}{c} \dfrac{-1}{(n-1)y^{n-1}} \\ \ln y \end{array}\right\} = \begin{cases} \dfrac{x^{m+1}}{m+1} + C \\ \ln x + C \end{cases}$

Exercises 23.5

1. $y = x \ln x + Cx$

2. $y = C \exp \left(\dfrac{x^2}{2y^2}\right)$

3. $\dfrac{2x}{x-y} = \ln x + C$

4. $x^2 - 2xy - y^2 = C$

5. $y^4(5x^2 - x^2)^3 = C$

Exercises 23.6

1. $y = \frac{3}{4} + cx^{-4}$

2. $y = -\frac{3}{4} + cx^4$

3. $y = (\cos x) \ln |\sec x| + C \cos x$

4. $y^3 + 3xy - x^3 = C$

5. $\dfrac{x^2}{2} + \dfrac{x}{y} + 2 \ln y = C$

Exercises 23.7

1. $y = \dfrac{x^2}{3} + C/x$

2. $y = 2 + Ce^{-x^2/4}$

3. $y = (x - 1)^2 + Ce^{-x}$

4. $y = \dfrac{x^5}{3} + Cx^2$

5. $y = xe^{3x} - \dfrac{1}{6}e^{-3x} + Ce^{3x}$

6. $y = \dfrac{\sin t + a \cos t}{1 + a^2} + Ce^{-at}$

Exercises 23.8

1. $Ce^{-x} - 1$

2. $y = \dfrac{Ce^{2x} - 1}{Ce^{2x} + 1}$

Exercises 23.9

1. $y = \int e^{-ax} \left\{ \int^x e^{at} f(t)\, dt \right\} dx + C_i e^{-ax} + C_2$

2. $m_i = \dfrac{-a \pm \sqrt{a^2 - 4b}}{2}$

   $y = C_1 e^{m_1 x} + C_2 e^{m_2 t} + \dfrac{c}{b}$

Exercises 23.10

1. $P_A(n) = C(1 - p_A - p_B)^n + \dfrac{p_B}{p_A + p_B}$

Exercises 24.2

1. $y = C_1 e^{-x} + C_2 e^{-2x}$
2. $y = C_1 e^{-x/2} + C_2 e^{-2x}$
3. $y = e^{2x}(C_1 + C_2 x)$
4. $y = C_1 e^x + C_2 e^{-4x}$
5. $y = C_1 \cos kx + C_2 \sin kx$
6. $y = 2e^x - e^{-2x}$
7. $y = xe^{2x}$

Exercises 24.3

1. $y = \frac{1}{4}\left\{ (1 + x)e^{-2x} + x - 1 \right\}$

2. $y = \dfrac{x^2}{2} e^{-x}$

3. $y = \dfrac{-2}{\sqrt{3}} e^{-x/2} \sin \dfrac{\sqrt{3}}{2} x + 1$

4. $y = C_1 e^{m_1 x} + C_2 e^{m_2 x} + \dfrac{x^3}{b} - \dfrac{3a}{b^2} x^2 + \left(\dfrac{6a^2}{b^3} - \dfrac{6}{b^2} \right) x - \dfrac{6a^3}{b^4} + \dfrac{12a}{b^3}$,

   provided $m_1 \ne m_2$ and neither $= 0$
5. $y = (2 + x)e^{-x} - (1 + x)e^{-2x}$

Exercises 24.4

1. $y = \dfrac{e^{-x}}{3} \int^x e^t f(t)\, dt + \dfrac{e^{-4x}}{3} \int^x e^{4t} f(t)\, dt + C_1 e^{-x} + C_2 e^{-4x}$

2. $y = -e^{-3x} \int^x te^{3t} f(t)\, dt + xe^{-3x} \int^x e^{3t} f(t)\, dt + C_1 e^{-3x} + C_2 xe^{-3x}$

3. $y = \dfrac{e^{m_1 x}}{m_1 - m_2} \int^x e^{-m_1 t} f(t)\, dt + \dfrac{e^{m_2 x}}{m_2 - m_1} \int^x e^{-m_2 t} f(t)\, dt$
   $+ C_1 e^{m_1 x} + C_2 e^{m_2 x}$, where m_1, m_2 are roots of $m^2 + m + 1 = 0$

4. $y = e^{-x} \int^x e^{(2t - t^2/2)} \, dt$

Exercises 24.5

1. $y = \sum_{k=1}^{4} C_k e^{m_k x} + x^3 \qquad m_k = \dfrac{\pm 1 \pm i}{\sqrt{2}}$

2. $y = C_1 e^x + C_2 e^{-x} + C_3 \cos x + C_4 \sin x - \dfrac{x \cos x}{4}$

3. $y = C_1 e^{-x} + C_2 \cos x + C_3 \sin x + \dfrac{e^{-x}}{2} \int^x f(t) e^t \, dt + \dfrac{\cos x}{2} \int^x f(t)(\sin t - \cos t) \, dt$

$\qquad + \dfrac{\sin x}{2} \int^x f(t)(-\sin t - \cos t) \, dt$

4. $y = C_1 e^{-2x} + C_2 \cos x + C_3 \sin x \dfrac{-x}{2} - \dfrac{1}{4}$

Exercises 24.6

1. $y = \dfrac{C_1}{x^2} + \dfrac{C_2}{x} + \dfrac{x}{6}$

2. $y = C_1 \cos(\ln x) + C_2 \sin(\ln x)$

Exercises 24.7

1. $y' = u$
$u' = v$
$v' = -xv + u - xy + \sin x$

2. Interchange letters $a, c;\ b, d;$ and f, g.

3. $y' = z$
$z' = \dfrac{-P_1}{P_0} z - \dfrac{P_2}{P_0} y - \dfrac{f}{P_0}$

Exercises 24.8

2. $y_n = C_1 + C_2(-1)^n + \dfrac{n^3 - n}{6}$

3. $y_n = \dfrac{\sin x}{\cos x - 1 - \sin^2 x}(\cos nx - 1) + \dfrac{\cos x - 1}{\cos x - 1 - \sin^2 x}$

Exercises 25.1

1. $\dfrac{1}{\sqrt{x}\sqrt{x+1}(\sqrt{x+1} + \sqrt{x})}$

2. $\tan \dfrac{x}{2}$

3. $\dfrac{2}{\sqrt{x+1}\sqrt{x-1}(\sqrt{x+1} + \sqrt{x-1})}$

4. $\dfrac{1}{(1+x)^{2/3} + (1+x)^{1/3}x^{1/3} + x^{2/3}}$

5. $\dfrac{-1}{2\sqrt{x}(\sqrt{x}+2)}$

6. $(e^x + 1)(e^x - 1)^{1/2}$

7. $-\dfrac{1 - (x^2/20) + (x^4/840) + \cdots}{2 + (4/5)x^2 + (34/105)x^4 + \cdots}$

Exercises 25.3

1. $y = x + 1$

2. $y = b + \dfrac{c-a}{2} + \dfrac{a - 2b + c}{2}x^2$

3. $y = \frac{1}{3}[3 - 10x + 9x^2 - 2x^3]$

4. $\displaystyle\int_0^1 f(x)\,dx = \dfrac{f(1/3) + f(2/3)}{2}$

Exercises 25.4

1. $\dfrac{5f(-1) + 8f(0) - f(1)}{12}$

2. $\dfrac{17f(0) + 20f(1/2) - f(1)}{36}$

3. $\dfrac{f(0) + f(2)}{2}$

Exercises 25.5

3. $y = \dfrac{14 + 7x}{10}, \qquad A = 11.2$

5. $81/20$

Exercises 25.6

2. $\dfrac{4(1 - \sqrt{2}) + \sqrt{2}\,\pi}{2\pi}\left[f(0) + f(1) + \dfrac{(4 - \pi)(\sqrt{2} - 1)}{\pi}f\!\left(\dfrac{1}{2}\right)\right]$

3. $\frac{1}{15}[7f(-1) + 16f(0) + 7f(1)] + \frac{1}{15}[f'(-1) - f'(1)]$

Exercises 25.7

2. $y = e^{-x}$

3. $y = \exp(-x^2/2)$

5. $y = \sin x$

Index